Quantum Magnetic Resonance Imaging Diagnostics of Human Brain Disorders

Quantum Magnetic Resonance Imaging Diagnostics of Human Brain Disorders

Madan Kaila and Rakhi Kaila

ELSEVIER

AMSTERDAM • BOSTON • HEIDELBERG • LONDON • NEW YORK • OXFORD
PARIS • SAN DIEGO • SAN FRANCISCO • SINGAPORE • SYDNEY • TOKYO

Elsevier
32 Jamestown Road London NW1 7BY
30 Corporate Drive, Suite 400, Burlington, MA 01803, USA

First edition 2010

British Library Cataloguing-in-Publication Data
A catalogue record for this book is available from the British Library

Library of Congress Cataloging-in-Publication Data
A catalog record for this book is available from the Library of Congress

ISBN: 978-0-323-16507-5

This book has been manufactured using Print On Demand technology. Each copy is produced to order and is limited to black ink. The online version of this book will show color figures where appropriate.

Transferred to Digital Printing in 2016

Table of Contents

Preface

This book has been written so that readers can use any of the chapters independently. The subject matter of one chapter is not sequentially related to the other two, although the theme and the aim underlying each chapter are the same. That aim is to bring to the attention of medical professionals the recent developments that have taken place in the field of magnetic resonance imaging (MRI), and the application of MRI to the diagnosis of human brain disorders.

Over the past decade or so, the concept of multi-quantum coherences among the atoms and molecules in the brain, has come to the fore. It has been demonstrated that applications of MRI, using these multi-quantum coherences, in brain imaging can be very rewarding. With these new techniques, it has become possible to image quantum chemical and physical events in space and time, and to use this information in the interpretation of the resulting images to make detailed, molecular-level diagnoses of human brain disorders. In fact, MRI should also enable one to identify, and even predict, tumor development at very early stages.

This book presents three chapters. Chapter 1 discusses recent theories, experiments, and other developments suggesting that the human brain is a quantum computer that follows normal principles of quantum science. The brain handles processing of problems in a parallel manner. It receives information from many sensory organs of the body simultaneously, and plans action, spontaneously, in a coherent, parallel manner. This is different from the sequential handling of data that we have become used to in the electronic computer. The simplified mathematics of quantum science is presented in this chapter, but mainly with reference to practical illustrations, tables, figures, and the like. This chapter also presents an investigation of the close relationship between the brain, as a quantum computer, and intermolecular quantum coherence imaging.

Chapter 1 may not interest some medical professionals (e.g., the clinical doctors) who are directly involved in the diagnostics field, but will certainly interest medical scientists who are involved in research and development in the area of brain science. Thus, clinicians can skip this chapter if they wish to, or make a partial, introductory study of the quantum science applied to imaging. This will not jeopardize their understanding of the information in the following two chapters. Chapters 2 and 3 are directed toward MRI and its application to imaging and the diagnosis of human brain disorders. Chapter 2 is mainly devoted to conventional MRI of the brain. All the basic techniques of imaging are covered. This has been done using practical illustrations, without much delving into the basic physics of such imaging, although some fundamental aspects and concepts, such as the meaning of the basic dipole magnetic moment and its implication in imaging, are included in Chapter 2. Chapter 3 takes a different approach to imaging, in which quantum imaging and the development of more advanced knowledge of quantum science play a prominent role. First, the quantum mechanics of the matrices representing atomic spins is briefly explained. Their relationship to quantum magnetic resonance imaging (QMRI) is then highlighted.

This is not done in the manner of a normal textbook. Instead, it is done as an information-based initiative. There are two reasons for this presentation. First, the authors want the knowledge presented to remain fairly easily understandable to most of the audience, and not become too complicated, particularly for medical professionals (e.g., clinical doctors). Nevertheless, medical scientists with a good background in physics, chemistry, and mathematics (PCM) will find it very stimulating, and are challenged to make it a base for further learning. Second, the area of QMRI is fairly new, and textbook-level formal PCM treatment is not yet easy to prepare, particularly in an easily understandable format, for the wider scientific community.

Specific Appreciation

The authors express specific gratitude to the producers of the intellectual material used in the preparation of this book. The producers referred to are the authors of the research articles, and the publishers who made available in wide circulation the material used. In particular, appreciation for the material used is expressed to: John Wiley & Sons, Elsevier, American Institute of Physics (AIP), American Physical Society (APS), NATURE, SCIENCE, American Association of Physicists in Medicine (AAN), Radiological Society of North America (RSNA), Proceedings of the National Academy of Science, USA (PNAS), The Institute of Physics, UK (IOP), Wm. C. Brown, Taylor and Francis, Springer Verlag, SCIENCE, and others whose material has been used. Advancement of science for the coming generations can be possible only with free semination and dissemination of the knowledge produced.

Acknowledgments

It would not have been possible for us to write this book without the facilities and courtesies extended by the School of Physics, and the Faculty of Science, at the University of New South Wales. In addition to use of information technology facilities, the award of the honorary position of Vesting Fellow, at the School of Physics, to Dr Madan Kaila greatly assisted in enabling this work. The wonderful e-journal access provided by the University of New South Wales (UNSW) library is highly appreciated. The support of the Head of the School of Physics, Prof. Richard Newbury, and of the Dean of the faculty of science is particularly appreciated.

Appreciation of those whose efforts contributed to the creation of this work would be incomplete without thanking the late Dr Mulkh Raj Kaila (father of Dr Madan Kaila) and the late Dr Krishan Lal Kaila (eldest brother of Dr Madan Kaila). Their contributions to the educational career of Madan Kaila, in the field of science, led to the creation of this work. We wish to reiterate our sincerest tributes, in memory of Dr Krishan Lal Kaila and Dr Mulkh Raj Kaila, by including their portraits here. But for their sacrifice, this work would never have been possible.

Left: Brother of Madan Kaila: Dr Krishan Lal Kaila, 1932 (Lahore) to 2003 (Delhi); Emeritus Scientist, National Geophysical Research Institute, Hyderabad, India. *Right*: Father of Madan Kaila: Dr Mulkh Raj Kaila, 1910 (Lahore) to 1957 (Delhi).

Special Note to the Reader

This book is written with a wide audience in mind. Chapter 1 is devoted to the physical aspects of quantum information science (QIS). The later parts of that chapter use bio-chemo-medical studies as a base for exploring new directions in the field of magnetic resonance imaging (MRI). Chapter 2 is intended to lay a foundation for visualizing quantum principles as the base for MRI. It includes illustrations of the techniques currently in use in the field of imaging. Chapter 3 develops the relationship of MRI to noninvasive diagnostics of human brain disorders.

The three chapters can be easily read individually. They stand alone and contain references to relevant sources for further reading. Nevertheless, reading the chapters in the order in which they appear in the book gives one some education about MRI science, with the diverse and overlapping information required for learning and relearning. The illustrations, figures/tables, and other features are numbered sequentially for the chapters, although they include references to their sources and are integrated with the text. This allows readers to find more detailed information quickly, in a logical manner.

Our purpose in writing this book was to induce in the community an awareness of the relationship between QIS and the medical science. The book can be a rich source of inspiration for educators to further simplify QIS so as to bring it to the level of those who have little background in physics, chemistry, and mathematics (PCM).

1 Biomedical Quantum Computer

Table of Contents

1.1 Introduction

1.1.1 Similarities and Differences Between Classical and Quantum Mechanical Information Processing

This chapter is designed to benefit the overall scientific community involved in the research and development of quantum mechanical information processing (QMIP). This community itself has evolved, using widely different theoretical and experimental approaches over the years. Unfortunately, the research at present is not realizing its full potential, in part because groups tend to work in isolation from each other and because the overall scientific community shares few common goals. This chapter intends to reduce this communication gap. It also intends to create grounds for workers to familiarize themselves about the various useful technologies that are currently being pursued.

The basics of QMIP were founded along with the birth of quantum mechanics (QM) itself. This happened during the 1920s and 1930s. The enormous success of QM in explaining properties of matter, at solid-state, atomic, subatomic, and other levels, has proven unparalleled in the history of the development of physics. Today, society enjoys the technologies that resulted from these scientific breakthroughs, albeit with little recognition of their origins. Solid-state (SS) quantum physics has resulted in applications as varied and widely available as electronic computers (ECs), mobile communications, space travel, and many others.

Until recently, quantum mechanical principles have not been the subject of intensive development in connection with QMIP. However, the quantum computer (QC) has suddenly become a popular research area, and this field has had to revisit quantum mechanical principles right from the basics. The issues that now confront the scientific community are at the grassroots level. For intensive research and

Quantum Magnetic Resonance Imaging Diagnostics of Human Brain Disorders. DOI: 10.1016/B978-0-12-384711-9.00001-4

development in the areas of quantum cryptography, quantum teleportation, and the QC, one has to rely heavily on in-depth analysis of the quantum-level physics. A rapid development of suitable technologies that can put QMIP into practice would then become feasible. The motivation behind QC and QMIP research is not only the high speed and large amount of information processing that is possible, but also the development of unbreakable methods of transporting secret information (encryption). In QMIP, this will be enabled using normal communication channels.

QMIP is similar to classical information processing (CIP) in that it uses binary logic; present-day ECs, such as desktop computers, use 0 and 1 digit information coding. This is achieved through low and high voltage applied to a silicon gate (switch). High voltage closes the gate (code 1) and lets the electrical signal go through to the next switch; the low voltage (code 0) does not allow conduction. The "not" (0) and "one" (1) digitally coded electrical signals are the tools for coding information through large (8 bits, 16 bits, etc.) binary numbers. In CIP, the signal progresses in a sequential order using unit sets of an eight-digit bit or a sixteen-digit bit basic component. In QMIP, the coding is still binary but the creation of the bits is quite different. QMIP also drastically differs in manipulation and transport of the bit information. The QM energy states used in computation may be, for example, the position and momentum, spin up or spin down, etc., of an elementary particle such as an electron. These could further be divided into substates. The electron spin may exist anywhere between 0 and 2π directions. Control of spin can be achieved by external application of an appropriate direction and magnitude of a magnetic field. One may even use the internal magnetic field of the nucleus around which the electrons precess in an atom.

These substates add further dimension to the storage of information. The states are then manipulated through their evolution in time for the desired information transport. The elementary particles used in QMIP can be atoms, electrons, photons, and so forth. Theoretical schemes and algorithms have used particle states to design a QC. Unfortunately, there is an oversupply of theoretical schemes in the literature—significant implementation of these schemes in practice is a different matter. It is not easy to code on electron spin orientation, direction of photon polarization, or the like. The required quantum states of the particles are difficult to maintain and process. Some success has been achieved, though, by using the technique of liquid-state nuclear magnetic resonance (LSNMR). In QMIP, 0 and 1 codes are replaced by a wave function $|\psi\rangle$.

The eigen (energy values) states of the wave functions—say, for instance, $|0\rangle$ the ground state and $|1\rangle$ the first excited state—can be coded as 0 and 1, respectively. One can also have linear superpositions of two (or more) energy-state wave functions $|0\rangle$ and $|1\rangle$, for example, or use the degeneracy (more than one state per energy) of a single function, the hyperfine splitting. The qbits in LSNMR are among different phase states (e.g., the superposed states of the nuclear spins). These superposed states can be created by rotation of the spin states. This is done by the application of radio frequency (RF) field pulses in an orthogonal plane. The RF field is superposed onto a direct-current (DC) magnetic field applied in a certain direction.

The interaction between the nuclear spin magnetic moments of the neighboring nuclei controls the evolution of the qbits. One measures the state of the qbits by using NMR spectroscopy (absorption or emission). The conjoining of all these events is handled through a quantum gate (QG). The structure of a QG is quite different from that of a classical gate (CG). It is necessary to test the qbit at every step for any error and then to control the state for further progression. The configuration of a QG is also different in different systems. The rotations of spins in LSNMR may be replaced by a totally different technique in another system, such as a polarizer/analyzer combination for a photon qbit in an optical computer.

The measured result in a QG is the *expectation*, i.e., the most probable value of the wave function in that state. The expectation value of the states is a finite probability between 0 and 1. By suitable implements, it can be made very close to 1. Though there is an analogy in bit coding in the sense of CIP, the technology required to implement a QG would be quite different. It is the multidimensional ascendancy—i.e., the involvement of the multi-bits for transporting the information in polynomial time—that makes the QMIP such a powerful tool, as well as such a difficult and maybe impossible one. LSNMR is the only technique that can boast of some degree of achievement in the field so far. The qbits can be designed using atoms, molecules, ions, and so forth, configured in a coherent enclosed environment. This coherent environment may be devised through an electrodynamic cavity where laser beams control interactions between energy states of the particles. Another technique is to confine electrons in a small group in the neighboring quantum dots (QDs) in the solid state. The measurement and transport of the electrons between the bits may be controlled by the application of an external terahertz (THz) electric field. The entanglement of discrete atomic energy levels (Rydberg atom) provides many high-purity states for storage and transport of information. Nevertheless, all these high-flying ambitions do not come without a price. As the number of qbits increases, so does the decoherence between the states. The chance of accumulating errors in time and space in the operation of the whole system is quite high. Still, despite the myriad problems facing QMIP, appreciable progress has been made in some areas. Several theoretical and experimental techniques have been developed, using various areas of physics (e.g., optics, magnetism) to put this complex and difficult scheme into practice [1–55].

1.1.2 LSNMR and QMIP

LSNMR has been successful in implementing a multi-qbit QC [1–9, 17]. One may note that in LSNMR the huge size of the equipment involved would prohibit this technology from providing a desktop computer. Still, its ability to work at room temperature and the ready availability of well-developed NMR technology are certainly advantages not to be ignored. Furthermore, the materials involved are in liquid state and can be easily prepared in the laboratory. These may in fact be simple, naturally occurring molecules (e.g., DNA). LSNMR provides an easy-to-manipulate technology that would at least allow one to better understand the basics of QM and thus lead to quicker realization of QMIP. One should remember that progression in quantum computing through understanding means less time and thus less money spent in the hunt for a QC.

1.1.3 Incomplete Description of QM States

The basic principles of QM inhibit simultaneous measurement of the physical properties of the discrete quantum states (e.g., the position and momentum, spin up and spin down, of an electron; various polarization states of a photon with respect to some specified axis). If you make a measurement on one state, the other one gets disturbed, thus both cannot be measured simultaneously. Violation of this principle of QM can, however, be attained by following strict constraints involving *hidden variables* (HVs). This chapter specifically includes a section on mathematical deductions and practical examples to illustrate the HVs. The Einstein, Podolsky, and Rosen (EPR) entanglement of quantum states [30], Bell's inequalities (BIs) [52], Shor's factoring algorithm (SFA) [36], and others are covered in this chapter, with illustrations of applications in various technologies. Quantum processes are at work all the time in real life, whether it is within a single atom or an ensemble of atoms—liquid state, solid state, etc. Einstein, Podolsky, and Rosen were driven by the principle of common sense and of physical reality in real life. A good example of the QMIP system at work is in the human body: the DNA QMIP. With this mechanism, nature cleverly carries out the computation process by passing on genetic information from one generation to the next. Over the past decade or so, there have been efforts to prepare quantum states using DNA in laboratories [51] and to explore their secrets for use in QMIP.

1.1.4 Solid-State Technology and QMIP

Let us now examine the status of QMIP in the solid state. There is a prevalent active research component, the area of nano-silicon technology, but the outcome so far is not encouraging. In the solid state, preparation of the qbits requires a very high degree of precision. There have been efforts to prepare electron spin and/or charge qbits in quantum wells, QDs, etc., using nano-fabrication technology. This too has yielded no worthwhile results so far. There is also a need to have a cryogenic infrastructure to provide mK order of temperatures for the bits to function. This only adds to the complexity of the processes involved. So far, no experimental device has yet been fabricated that demonstrates that a charge-/spin-based solid-state technique in QMIP actually proceeds in computation to a significant level. Progress in experimentally solving the prevalent decoherence problem (multiscaling of qbits, etc.) in the solid state has been very slow [10–16]. One reason for disappointment in this area may be the fact that there is not a good enough understanding of the basics of the electron charge and/or spin state in a localized environment such as a QD.

Over the years, a common perception has developed that one should stick to applied research only. However, one should remember that applied research is pure research and pure research is applied research. It is like the chicken-and-egg paradox: which came first, nobody knows. One thing is sure though, one does not come without the other. All the technologies that society enjoys today are the fruits of the curiosity research done by physicists, chemists, mathematicians, and other "pure" scientists over more than a century. Why, then, has this culture of pure research, evolved over a century, been declared unnecessary today? In fact, basically destroyed? It seems that the research on QMIP has rekindled the flame of basic research. It is a well-known fact that in order for a few geniuses (provided there are some) to make breakthrough progress, the toil of many others who may not be geniuses is required. The background forms a necessary part of the process. Despite reaping the fruits of both pure and applied research, over decades, society seems to have cut itself off at the knees by ignoring and undervaluing the basic research culture. Had it not, the QMIP field would otherwise have been much closer to success by now.

In short, those in the solid-state technology area lack understanding of the basics. That is why this area has not been able to progress at the rate it should. Slow progress in solid-state quantum-computing stands in stark contrast to the fast developments made in the field of nano-fabrication. Maybe there is no answer to the crippling problem of decoherence. It may be wise to take caution from the case of the amorphous silicon solar cell, which has not replaced single-crystal technology despite the efforts made and noise propagated over the past few decades.

1.1.5 QMIP Coded on Light

Quantum information can also be coded on light [18–24]. A *photon* is a particle that does not have charge. It moves at the velocity of light and is less susceptible to decoherence than other particles. In light, encoding can be done using a photon's different polarization states, each of which can be used as a carrier of information. A laser beam of photons can be split into coherently intimate orthogonal states, in accordance with the basic requirements of the principles of QM. The orthogonal states are unitary transformations of the same entity, and thus a good preparation for a basic pure qbit. Coherent manipulation of semiconductor quantum bits with terahertz radiation can provide a high speed in QMIP [25, 26]. Terahertz electromagnetic radiation is a dark invisible light. A THz probe in QMIP would be extremely useful in imaging, remote sensing, surveillance, and the like. Although trapped-ion techniques have seen considerable progress over the years [27–29], given the lack of basic understanding of the QM processes involved, any significant degree of further progress seems unlikely. The only way out is basic research that goes ahead of the applied research.

The following commentary about the basic limitations on the description of QM, produced in 1935 by Einstein et al. [30] and thus called the EPR paradox, still requires many explanations today. It is worth quoting the outlines of the paradox here: "two or more physical quantities can be regarded as simultaneous elements of reality, only when they can be simultaneously measured or predicted ...since either one or the other, but not both simultaneously, can be predicted, they are not simultaneously real ... the wave function thus does not provide a complete description of the physical reality ... whether or not such a description exists (sometimes referred to as the phenomena of wave-function collapse)... " The EPR paradox is under test more than ever before, because of QMIP research. One can find a myriad of theoretical and experimental techniques presently being pursued to scrutinize the paradox. Though the references listed in this chapter constitute only a brief and partial listing, they provide the reader with vast latitude and longitude of information in the field. One needs to familiarize oneself with the present status of the field both in theory and experiment.

A successful experiment would require an efficient algorithm to find a technique that can implement a viable solution to the problems involved. There is no dearth of theoretical algorithms in the literature, but their practical implementation to achieve the desired goal is far from realization. Efforts are in place to use the available theoretical quantum algorithms. Designs for practical gates capable of manipulating qbits have been detailed in the literature (some recent ones can be found in Refs. [31–56]).

1.1.6 General Concept: Controlled-NOT Gate and 2-Bit Operation [57]

This section is included to help those who do not have an adequate knowledge of CGs based on the Boolean algebraic operations. A notable aspect of current QMIP research is the quest to realize a scalable quantum-computing architecture. In the most common paradigm, the two required elements are a single-qbit arbitrary rotation gate and a 2-bit maximally entangling gate, that is, a controlled-NOT (CNOT) gate.

There are many two-level quantum systems suitable for encoding qbits and realizing single-qbit rotations, such as spin-½ systems, simple harmonic (SHM) oscillators, phonons, superconducting systems (charge, phase, flux), optical systems, and so on. Common to these architectures is the need for accurate characterization of the systems that generate the universal gate set. A 2-qbit gate will be critical for any realization of a QC. A *QG* is a unitary operation that by definition maps pure states to pure states. In the standard model, a 2-qbit QG transforms states of the 2 qbits to the states of 2 qbits.

In principle, a 2-qbit QG could output superposition states and entangled states, and could also disentangle states, depending on the nature of the gate. The task of determining whether a prototype gate is working is not a simple matter. The prototype may decohere the qbits and thus not generate the correct amount of entanglement or introduce phase errors. Myriads of potential experimental faults might occur. The exclusive-OR (XOR) gate is the classical 2-bit CNOT gate where a control bit flips the state of a target bit. As shown in Table 1.1, there are only four possible inputs and therefore output states.

General concept: *CNOT gate and 2-bit operation*. The simplest characterization of 2-bit gate is a straightforward matter of entering each possible input and measuring the output state. If the gate conforms to Table 1.1, it is at least an XOR gate. A more complete characterization, which allows measurement of the error probabilities, is arrived at by measuring the probability of each of the four possible output states for each of the four inputs. This yields a truth table, as shown in Table 1.2.

Table 1.1 Two-Input–Two-Output Table for an Ideal Classical CNOT (XOR) Gate

(CNOT)in	(CNOT)out
00	00
01	01
10	11
11	10

Table 1.2 Four-Input–Four-Output Truth Table for an Ideal Classical CNOT (XOR) Gate

	(00)out	(01)out	(10)out	(11)out
(00)in	1	0	0	0
(01)in	0	1	0	0
(10)in	0	0	0	1
(11)in	0	0	1	0

The numbers indicate the probability of achieving the selected output state for a given input state.

In the quantum case, the inputs are qbits, which can exist in an arbitrary complex superposition of a classical bit, e.g., $|\psi\rangle$ in $= \alpha|0\rangle + \beta|1\rangle$, where $|\alpha|^2 + |\beta|^2 = 1$. The gate has infinitely many possible inputs and characterization is not a simple matter of exhausting all possible inputs. Table 1.3 shows the output states of a CNOT gate. They may be entangled, i.e., the states may have correlations that may not necessarily be replicated by the classical modes of physics. Input–output table for a quantum CNOT gate depicts $|\text{CNOT}\rangle$ in, out. With the logical inputs, the table is as for a classical CNOT. In contrast, with control-superposition inputs, the outputs vary between the separable and the maximally entangled, $|\alpha|^2 + |\beta|^2 = 1$, $|\gamma|^2 + |\delta|^2 = 1$. Because the CNOT is reversible, the input and the outputs can be swapped, and Table 1.3 is still correct.

Table 1.3 A Signature of the QG Operation is the Generation of the Entangled Output States

Input Label	$\|\text{CNOT}\rangle$in	$\|\text{CNOT}\rangle$out
Logical	$\|00\rangle\|01\rangle$ $\|10\rangle\|11\rangle$	$\|00\rangle\|01\rangle$ $\|11\rangle\|10\rangle$
Control superposition	$(\alpha\|0\rangle + \beta\|1\rangle)\ (\alpha\|0\rangle + \beta\|1\rangle)$	$(\alpha\|00\rangle + \beta\|11\rangle)\ (\alpha\|01\rangle + \beta\|10\rangle)$
Control and target superposition	$(\alpha\|0\rangle + \beta\|1\rangle)\ (\gamma\|0\rangle + \delta\|1\rangle)$	$(\alpha\gamma\|00\rangle + \alpha\delta\|01\rangle)$ $(\beta\delta\|10\rangle + \beta\gamma\|11\rangle)$

Thus, it is necessary to identify and preferably quantify these outputs. A beginning point is to measure a series of correlations or coincidence probabilities. This is done between the control and target arms with the aim of identifying uniquely the quantum correlations.

1.1.7 Nondeterministic Polynomial Time Complete and Satisfiability [17]

A QC can operate in parallel on all its possible inputs at the same time, but the amount of information that can be extracted from the result is limited by the phenomenon of wave-function collapse. One can, however, use a computational model whereby one measures the expectation value rather than a random eigenvalue of the observable. Such an expectation value of a QC can solve a nondeterministic polynomial time complete (NP-C) problem. Results have been realized, to a large extent, by using NMR spectroscopy. The qbits are prepared in this system through macroscopic ensembles of the molecules and the nuclear spins. In this technique one identifies a manifold of statistical spin states, called *pseudo-pure state*. Its mathematical description is isomorphic to that of an isolated spin system.

The LSNMR QC very much resembles a DNA computer (DNA QC). Currently, in a DNA QC, one increases the amount of the sample to increase the capacity. LSNMR QCs and DNA QCs provide the reader with useful practical illustrations of the QM operators in work in a QC. One can easily comprehend the intricacies of the QM processes involved through these simpler illustrations. This may then become the necessary and desired groundwork for research investigations in diverse directions. NP-C problems are a class of computationally intractable problems of particular interest, both because they are "polynomially equivalent to one another" and because they are encountered in many important applications.

In an LSNMR QC, spins in each molecule of a liquid-state ensemble are largely isolated from the spins in all other molecules. Each molecule is thus effectively an independent QC. What one observes is the sum of certain observables over all molecules in the sample. This is proportional to the ensemble average expectation value of the observable. This machine performs a calculation using quantum parallelism at the molecular level and then amplifies its results to the microscopic level via a form of classical parallelism. A *mixed state* is a statistical mixture of independent quantum systems that are not in the same quantum state. The vectors (or wave functions) in the Hilbert space of a system of spins are called *spinors*. The standard basis here consists of the joint eigenvectors, of the total and the z-component, of the spin angular momentum. The encoding used in QC maps each integer k, in the range $[0, 2^n - 1]$ to the kth basis element versus a particular ordering of the basis. For a single-spin-½ particle, the basis consists of the spins down (antiparallel to the z-axis) state, represented by the vector [1, 0] and is denoted by the "bra" $\langle 0|$. The spin "up" state is represented by the vector [0, 1] and is denoted as $\langle 1|$. The corresponding column vectors are denoted by the kets $|0\rangle$ and $|1\rangle$, respectively. The basis vectors of an n-spins system are formed by taking the tensor or Kronecker products (KP) of the basis vectors of its constituent spins. This is done in some arbitrary but fixed order. The general definition of the KP of an $M \times N$ matrix $\mathbf{A}$ with an $M' \times N'$ matrix $\mathbf{B}$ is the matrix $MM' \times NN'$ of $\mathbf{A} \otimes \mathbf{B}$. The basis vectors of the 2-spin system are $|00\rangle$, $|01\rangle$, $|10\rangle$, and $|11\rangle$. In a QC, any logical operation on a register in the basis state $|k\rangle$ must transform the register to another basis state $|1\rangle$. Energy dissipation rapidly destroys microscopic order, so the ability to store information in a QC must essentially be completely isolated from the surroundings.

Isolated quantum systems naturally evolve by unitary transformations that are necessarily reversible. This prevents one from implementing conventional logic gates such as the AND gate. The quantum

$$\mathbf{A} \otimes \mathbf{B} = \begin{bmatrix} a_{11}\mathbf{B} & \cdots & a_{1N}\mathbf{B} \\ \vdots & \ddots & \vdots \\ a_{M1}\mathbf{B} & \cdots & a_{MN}\mathbf{B} \end{bmatrix}$$

$$\text{where } \mathbf{A} = \begin{bmatrix} a_{11} & \cdots & a_{1N} \\ \vdots & \ddots & \vdots \\ a_{M1} & \cdots & a_{MN} \end{bmatrix}$$

XOR gate is represented as follows:

$$\begin{matrix} [1\ 0\ 0\ 0] & \equiv & \langle 00| & \rightarrow & \langle 00| & \equiv & [1\ 0\ 0\ 0] \\ [0\ 1\ 0\ 0] & \equiv & \langle 01| & \rightarrow & \langle 01| & \equiv & [0\ 1\ 0\ 0] \\ [0\ 0\ 1\ 0] & \equiv & \langle 10| & \rightarrow & \langle 11| & \equiv & [0\ 0\ 0\ 1] \\ [0\ 0\ 0\ 1] & \equiv & \langle 11| & \rightarrow & \langle 10| & \equiv & [0\ 0\ 1\ 0] \end{matrix} \Leftrightarrow \begin{bmatrix} 1 & 0 & 0 & 0 \\ 0 & 1 & 0 & 0 \\ 0 & 0 & 0 & 1 \\ 0 & 0 & 1 & 0 \end{bmatrix}$$

Any gate can be made reversible by copying some of the input bits to the output bits. For example, the quantum XOR gate, which copies the first input bit and overwrites the second bit with its output, has the truth table shown. A nontrivial complex linear combination of the basis vectors is called a *coherent superposition*. Unitary transformations are linear. The result of operating on a superposition is the same superposition of the results of the transformation, applied to the individual basis states. In this sense a QC can operate in parallel on exponentially many basis states at once. There are, however, serious problems with this approach. The first is that it promises to be very difficult to build a QC of any significant size. Such a construction requires the precise assembly, isolation, control, and measurement of atomic-scale systems.

A more fundamental problem lies in the fact that the amount of information that can be extracted from even perfectly precise measurements of a quantum system (QS) is extremely limited. The act of observation irreversibly alters the system. One can place a 2-spin system in the superposition $C_0|00\rangle + C_1|01\rangle + C_2|10\rangle + C_3|11\rangle = [C_0{}^* C_1{}^* C_2{}^* C_3{}^*]\dagger$. Here * denotes the complex conjugate and † the Hermitian transpose and measures the z-component of the spins. One will observe one of the four basis states $k = 0{-}3$, each with probability $|k^2| = CkCk^*$. The system is forced into the observed state by the act of observation. All subsequent measurements of the same observable will therefore yield the same result.

This is the phenomenon of wave-function collapse. It is not possible to completely determine the state of the system (i.e., the coefficients C_k) from a finite number of measurements on identically prepared copies of a given quantum system. If, instead of random eigenvalue, the result of measurement of an observable is the expectation value of that observable, we have what is called an *expectation value QC* (EVQC).

One assumes in this model that the expectation value can be measured to an arbitrarily high degree of precision and in an amount of time proportional to the number of digits in the result. One can show that an EVQC can solve the NP-C problem, namely, the satisfiability (SAT) problem in liner time. The mixed states of spin systems are described by a generalization of the spinors, known as the *density matrix.* This density matrix enables one to compute the statistical properties of the ensemble. This matrix ψ of the pure state is obtained from the corresponding spinor ψ as $\psi = |\psi\rangle\langle\psi|$. Such a density matrix is a projection operator that has only one nonzero eigenvalue.

1.1.8 The Density Matrix Concept [17]

As an example, in the case of a 2-spin system in the state $|00\rangle$, we have $\psi = |00\rangle\langle 00| = \mathrm{Diag}(1, 0, 0, 0)$. The density matrix of a mixed state is the average over a representative ensemble of pure states; for instance, $\psi = \int_{(\psi)} p(\psi)|\psi\rangle\langle\psi|\,\mathrm{d}\psi$, where $p(\psi)$ is the probability density of the pure state described by the spinor ψ, and (ψ) denotes the set of all unit norm spinors. Thus, a density matrix can be an arbitrary positive semi-definite Hermitian matrix. The diagonal elements are the relative populations of the various energy levels; the off-diagonal elements represent coherent correlations in the phases of processing spins in pairs of energy levels across the sample. One can work with a reduced density matrix of size 2^n, where n is the number of spins in a single molecule. This is instead of the density matrix of size 2^N. Here $N = nM$ (M is the number of molecules in the ensemble), the total number of spins in the sample. The (reduced) density matrix evolves in time according to the Louville–von Neumann equation, $\mathrm{d}(\psi)/\mathrm{d}t = \mathrm{i}[\psi, H] = \mathrm{i}(\psi H - H\psi)$. H is a matrix representation of the spin Hamiltonian of the molecule and $[\psi, H]$ denotes the matrix commutator. This has the general solution $\psi(t) = U(t)\psi(0)U^{\dagger}(t)$, where $U(t)$ is time (t) dependent unitary matrix. The ensemble average of the expectation value of any observable K is obtained as the trace product $\mathrm{Tr}(K\psi)$ of the corresponding matrix K with the density matrix. One defines a pseudo-pure state to be one that has a density matrix that can be shifted by adding a mixture of the unit matrix to it. This is to obtain a scalar multiple of the density matrix of a pure state. The molecules of a sample in a pseudo-pure state are in a statistical mixture of quantum states. Nevertheless, a preponderance of one particular state is present. This manifests itself when one adds up the magnetization of all the molecules. The result is that we can "emulate" a QC by NMR spectroscopy on macroscopic liquid samples in open test tubes. The number of operations required is identical to the number of operations executed by the QC. The only difference is that one can determine the state of the system in terms of its expectation values, without wave-function collapse. This shows that the requirement for an exponential-state space is logically and physically distinct from the probabilistic aspects of quantum computing. A standard technique in NMR spectroscopy is to use pulses of RF radiation to transform the state of the spins by unitary transformations. Because the inequivalent spins in a molecule generally have distinct resonance frequencies, the frequency range of these pulses can be made selective for single spins.

This selective pulse imparts sufficient energy to rotate the net magnetization of the *k*th spin by $\pi/2$ and is in phase with the imaginary component of the carrier corresponding to the unitary matrix

$$1 \otimes \cdots \otimes 1 \otimes U_{\pi/2} \otimes 1 \otimes \cdots \otimes 1 = U_{\pi/2} \equiv (1/\sqrt{2})\begin{pmatrix} 1 & 1 \\ -1 & 1 \end{pmatrix}$$

Here the matrix $U_{\pi/2}$ occurs as the *k*th factor of the KP. For a 2-spin system, one has the energy levels as shown in Figure 1.1. The density matrix can be expressed as follows: $\psi = [(1-\alpha)1 + \langle 2\alpha|\psi\rangle\langle\psi|)]/[(1-\alpha)2^n + 2\alpha]\ (-1 \leq \alpha \leq 1)$. Here $|\psi\rangle$ is a unit spinor and $\alpha = -\Sigma_i hw_i/(kT)$.

Here w_i represents the frequencies corresponding to various energy levels, h the Planck's constant, k the Boltzmann constant, and T the temperature of the ensemble. The four dashed double-headed arrows are the transitions by the rules of angular momentum and they connect pairs of states that differ by only a single-spin flip (Figure 1.1). The spins are generally also coupled to one another, either by space dipole–dipole interactions or by a bond effect called the *scalar coupling*. This coupling causes the energy differences associated with various transitions to be generically distinct. Tipping both spins in the xy plane with a nonselective $\pi/2$ pulse $U_{\pi/2} \otimes U_{\pi/2}$ causes them to precess in phase, generating detectable (macroscopic) rotating magnetic moment. The real part of the Fourier transform (FT) of the resulting signal gives an NMR spectrum, with two pairs of peaks shown below. The intensity of each pair is proportional to the total population difference between states in which the corresponding spin is up and those in which it is down.

The quantum XOR gate flips one of these spins, given that the other spin is up. This in turn corresponds to the transitions 1–3 and 2–3 in an output of the first and second spins, respectively (Figure 1.2).

One can implement the quantum XOR gate by a single RF pulse, the frequency range of which spans only the peak of one of these transitions. It imparts sufficient energy to invert the populations of the corresponding pair of states. Thus, the

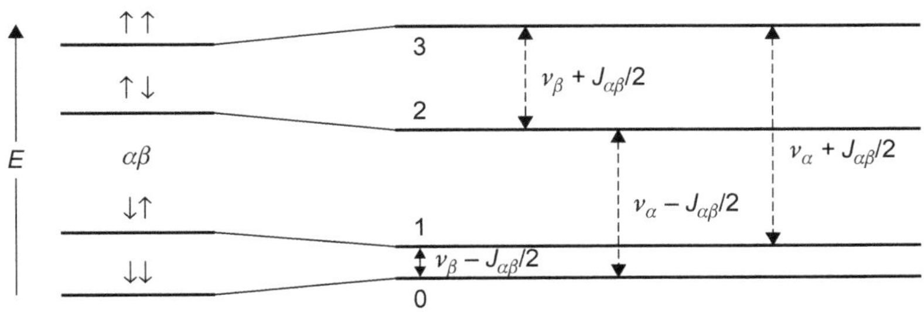

Figure 1.1 The four energy levels associated with 2 spins α and β have resonance frequencies ν_α and ν_β. The levels when there is no coupling between the spins are shown on the left and those with a coupling of $J_{\alpha\beta}$ on the right. The allowed transitions between the energy levels are indicated with dashed double-headed arrows.

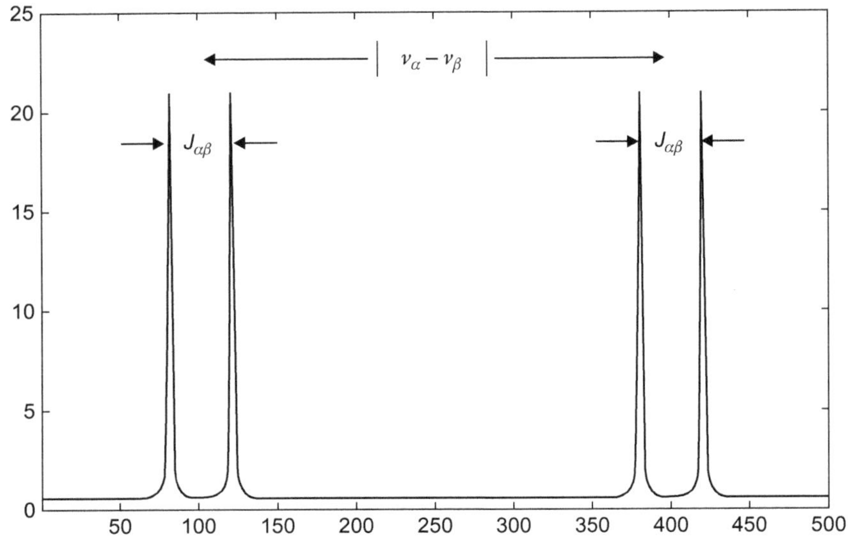

Figure 1.2 Simulated NMR spectrum of a 2-spin system with $\nu_\alpha = 100\,\text{Hz}$, $\nu_\beta = 400\,\text{Hz}$, and $J_{\alpha\beta} = 40\,\text{Hz}$. From left to right, the four peaks correspond to the transitions $0 \leftrightarrow 1$, $2 \leftrightarrow 3$, $0 \leftrightarrow 2$, $1 \leftrightarrow 3$. This spectrum would be obtained by applying a nonselective $\pi/2$ pulse to the equilibrium state and Fourier transforming the resulting signal.

unitary transformation U_f needed to compute the Boolean function in SAT can be performed by an appropriate pulse sequence. The expectation value of the operator S in the SAT algorithm corresponds to the sum of the intensities of the 2^n peaks that are obtained by flipping the $(n+1)$th spin. For a 2-spin system, the matrix $S = S_2$ can be written in terms of the operators $P_{01} = \text{Diag}(1, -1, 0, 0)$ and $P_{23} = \text{Diag}(0, 0, 1, -1)$, the expectation values of which give the population differences between the subscript states (and hence the intensities of the corresponding peaks) as

$$S_2 = (1 \otimes 1 - P_{01} - P_{23})/2 = (1 \otimes 1 - I_z)/2$$

where $I_z = \text{Diag}(1, -1)$ is the matrix of the operator for the z-component of the spin. Finally, a superposition in which all the basis states are equiprobable is easily obtained. This is done by applying a $\pi/2$ pulse with a frequency range that spans all the peaks due to the flips of the input spins but which misses those peaks due to the output spin. In solids, the dominant mechanism of spin–spin relaxation is the dipole–dipole interaction, which typically results in decoherence times of a small fraction of a millisecond. In contrast, in fluids the dipole–dipole interaction is averaged to zero by the rapid rotational diffusion of nearby molecules, leading to a decoherence time that can be on the order of many seconds. The intramolecular scalar couplings between the spins, however, are not averaged to 0. The number of selective pulses that can be used in one experiment can be over a thousand. It is possible that error-correcting schemes like those proposed to control decoherence and other errors in QC can be developed for NMR QCs to extend the time available for the computations.

1.1.9 DNA QC: QGs [51]

The following is a QG illustration for the case of a DNA computer [51]. Consider a Boolean formula $F=(x \vee y) \wedge (\bar{x} \vee \bar{y})$. The variables x and y are Boolean; i.e., their values can be 0 or 1. Usually one thinks of 0 as "false" and 1 as "true." The $\vee$ is the logical OR function, i.e., $x \vee y = 0$ only if $x = y = 0$ (otherwise 1).

The symbol $\wedge$ represents a logical AND function, i.e., $x \wedge y = 1$, if $x = y = 1$ (otherwise 0). $\bar{x}$ denotes the negation of x, i.e., $\bar{x}$ is 0 if $x = 1$ and 1 if $x = 0$. The same holds for $\bar{y}$. The SAT problem is to find the Boolean values for x and y that make the formula F true. In this example, $x = 0$ and $y = 1$ works, as does $x = 1$ and $y = 0$, whereas $x = y = 0$ does not. The formula F consists of two clauses. The first is $(x \vee y)$ and the second is $(\bar{x} \vee \bar{y})$. A *clause* is a formula that is of the form $\ell_1 \vee \cdots \vee \ell_k$, where each ℓ_k is a variable or its negation. In general, a SAT problem consists of a Boolean formula of the form $C_1 \wedge \cdots \wedge C_m$, where each C_i is a clause. The question is then to find values for the variables so that the whole formula has the value 1. The reason for the name of this problem is that making all of the clauses true is often called "satisfying" the clauses. The current best method essentially tries all the 2^n choices for n variables. A DNA QC model in a first-order approximation can be expressed as follows.

The strands of DNA consist of sequences $\alpha_1 \cdots \alpha_k$ over the alphabet {A, C, G, T}. Double strands of DNA consist of two DNA sequences, $\alpha_1 \cdots \alpha_k$ and $\beta_1 \cdots \beta_k$, that satisfy the Watson–Creek (W–C) complementary condition. For each $i = 1 \cdots k$, α_i and β_i must be complements, i.e., A$\leftrightarrow$T or C$\leftrightarrow$G. Complementary sequences of the strands anneal in an antiparallel fashion, as shown below:

$$5'-\alpha_1-\alpha_2-\alpha_3\cdots-3'-\updownarrow\updownarrow\updownarrow 3'-\beta_1\beta_2\beta_3\cdots-5'$$

Here $5'$ and $3'$ refer to the chemically distinct ends of the DNA strands. A number of simple operations can be performed on test tube samples (ensemble) that contain DNA strands. (1) Synthesize a large number of copies of any short single strand (at least 20 nucleotides). (2) Create a double strand of DNA from the complementary single strands by allowing them to anneal. (3) Extract those sequences that contain some consecutive pattern of length l, assuming that the pattern is $\delta_1 ... \delta_l$, where each δ_i is in {A, C, G, T}. (4) Detect a DNA strand $\alpha_1 ... \alpha_k$ that will be removed only if for some i, $\delta_1 = \alpha_{ij}$, $\delta_2 = \alpha_{i+1}, ..., \delta_k = \alpha_{i+k-1}$. The last operation simply determines whether or not there are any DNA strands at all in the test tube. (5) Amplify the operation to replace all of the DNA strands in the test tube. In this computation model one starts with one fixed test tube. The test tube is the same for all computations. The set of DNA in the test tube corresponds to the simple graph G_n (Hamiltonian path) shown in Figure 1.3. It encodes 2-bit numbers.

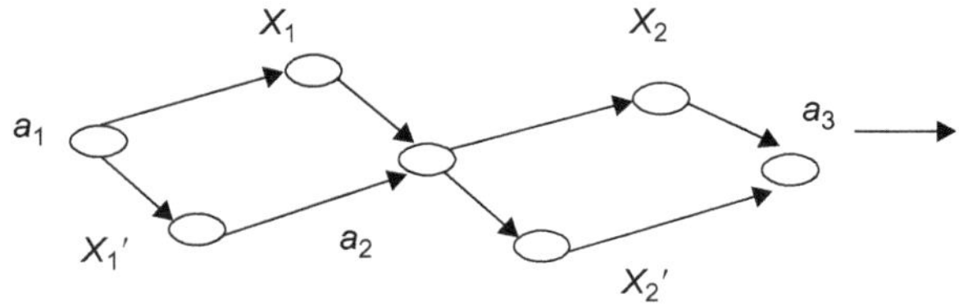

Figure 1.3 Following $F = (x \vee y) \wedge (\bar{x} \vee \bar{y})$, the graph is encoded into a test tube of DNA as follows. Each vertex of the graph is assigned a random pattern of length l (20 here), from {A, C, G, T}. This name of the vertex has two parts: the first half is denoted by p_i and the second half by q_i. Thus, $p_i q_i$ is the name associated with the ith vertex. The test tube was filled with the following kinds of strands. (1) For each vertex, put many copies of a $5' \rightarrow 3'$ DNA sequence, of the form $p_i q_i$, into the test tube. (2) For each edge from $i \rightarrow j$, place many copies of a $3'-5'$ DNA sequence of the form $^{-}q_i {}^{-}p_i$ (denotes the sequence, i.e., the Watson–Creek complement of x). (3) Add a $3' \rightarrow 5'$ sequence of length $l/2$, complementary to the first half of the initial vertex, to the test tube. (4) Similarly, add a $3' \rightarrow 5'$ sequence complementary to the last half of the final vertex to the test tube (i.e., add $^{-}p_i$ and $^{-}q_i$). The key is that every legal path in G_n corresponds to a correctly matched sequence of vertices and edges. If you consider any path in the graph, it naturally consists of a sequence that alternates vertex, edge, vertex, edge.

The graph G_n has nodes a_1, x_1, x_1', a_2, x_2, x_2', a_3,..., a_{n+1}, with edges from a_k to both x_k and x_k' and from both x_k and x_k' to a_{k+1}. The paths of length $n+1$ that start at a_1 and end at a_{n+1} are assumed to be in the initial test tube.

The graph is constructed so that all the paths that start at a_1 and end at a_{n+1} encode an n-bit binary number. At each stage a path has exactly two choices. If it takes the vertex with an unprimed label, it will encode a, 1. If it takes the vertex with a primed label, it will encode a, 0. Therefore, the path $a_1 x_1' a_2 x_2' a_3$ encodes the binary number 0 (a rough schematic is shown in Figure 1.3).

1.1.10 QMIP Differences from Analogue-to-Digital Conversion

To fully comprehend the difference in capability between CIP and QMIP, when applied to the area of imaging, e.g., one must comprehend the basic differences in the operation of the two systems. It is advisable to start by reading a text on the EC. For the sake of completion, the following summary may be helpful.

CIP (like QMIP) works on the 0 and 1 (binary) digit encoding system. In an EC, this is achieved by using a sequential progression of an electrical signal through digital gates (switches) fabricated in the silicon wafer. The signals propagated are the high (conducting) and the low (nonconducting) voltage states of the gates. At present, the size of the basic blocks of information transport (i.e., 8 bits, 16 bits, or 32 bits [the highest so far achieved]) limits the data-processing capacity of the system. Miniaturization to micro- or nanometer space level through compaction of the components in a single silicon chip has delivered MHz–GHz rates of computation. QMIP also works on the existing binary computation concept, but builds on a completely different footing. There is no limit on the scale of the quantized phase (time)-state space available in QMIP. It works in a polynomial time system. As an example, the time-space in its smallest unit for a photon can be in units of a millionth of a fraction of the time of rotation of polarization direction, over the angle 2π. One can thus process as quantum states the discrete states of polarization of a photon. These millions of tiny cells in time (angle)-space are the qbits in QMIP. One can use the frequency domain technologies, such as optical or terahertz spectroscopies, for manipulation of the qbits. QMIP thus offers a much faster and greater information storage and processing capacity. QMIP is intrinsically digital (quantum). Quantum image processing would use the polynomial time evolution of the quantum energy states in atoms, molecules, liquid-state materials, solid-state materials, and so on, as microprocessor blocks. In analogue-to-digital conversion (ATDC), an optical scan is used to scrutinize the contrasting parts of the image pixel by pixel, in relation to the adjacent regions. This is converted to electrical signals proportionally point by point in the object. The result is an electrical digital format of the picture. The electrical output is then remotely transmitted to another destination. There it is then reprocessed using a receiver working in a reverse order.

ATDC has an upper bound. One needs to have ever-smaller pixel size to get better images that provide ever-finer details. However, this amounts to having to deal with a much larger amount of information obtained over smaller areas processed; then and only then can one reproduce the original image in an exact manner. QMIP provides an infinite capacity in this context, as it operates in polynomial time-space. The analogy drawn in the literature in the use of binary digits 0 and 1 used in CIP and QMIP is an oversimplified picture. Although the QMIP concept may be put into practice by yet unknown technology, the magnitude of the difficulties involved in building a QC is enormous. Processing of information in the up and down electron spin states is not as easy as translating the information in low (0) and high (1) voltage states in CIP. While exploring the application of some prominent quantum algorithms with reference to mainly existing technologies, this chapter highlights the underlying immensity of the problems involved.

1.2 Theory: Theoretical Schemes and Algorithms in QMIP

1.2.1 The EPR Entangled States [30]

There are abundant examples of correlated states in QM systems. One of these is the quantum charge and/or spin-up and spin-down state of an electron in a localized solid-state environment such as a QD. The others are nuclear spins of molecules in a liquid-state ensemble (NMR), and discrete energy levels in single or correlated atoms in vacuums, gaseous state, etc. According to QM, no experiment exists that can measure spin up of one electron without affecting the correlated spin down of another electron. However, according to the principle of reality of the physical world to which QM so well applies, it is believed that there must be some hidden constraints (variables) under which this may be possible.

Using mathematical logical deductions from the EPR paradox, one discovers some constraints called as *BIs* [52]. The BIs provide one with some guidelines for exploitation of the discrete quantum states in QMIP applications. The next section presents a summary of the logical steps that lead to these constraints, called the *violation of Bell's inequalities (VBIs)*, and their relationship to experiments. This, we feel, is necessary for the sake of meeting the aim of this chapter. The following section is particularly devoted to the underlying mathematical logic involved. It provides the appraisal required to understand the relationship between the VBIs and an actual operational system.

1.2.2 Bell's Inequalities [52]

1.2.2.1 Specific Locality Condition

Let Aa be the vector representing the spin of an electron in the direction a and Bb be the vector representing the spin of a second electron in direction b. a and b are the unit vectors in the orthogonal space in which the vectors A and B are projected. The absolute magnitude of the spin of an electron (Aa or Bb) is $1/2(h/2\pi)$, where h is Planck's constant. For convenience in mathematical development, it is taken as 1. In QM language, one can say with certainty that the scalar product $Aa \cdot Bb = \pm 1$ (in case the spins are in the same direction or in the opposite direction). This product is represented in QM by a self-adjoint operation in Hilbert space.

The expectation (most probable) value E of the observable is

$$[E(a, b)]_\psi = \langle \psi | (\sigma_1 a) \cdot (\sigma_2 b) | \psi \rangle = -a \cdot b \tag{1.1}$$

Here σ_1 and σ_2 are the familiar Pauli spin matrices, with expectation values 1 and -1. When the spin analyzers are parallel, we get

$$[E(a \cdot a)]_\psi = -1 \quad \text{for all } a \quad \text{(Partial QM Agreement)} \tag{1.2}$$

Thus, one can predict with certainty the result B, by previously obtaining the result on A. Because ψ (the wave function describing a particular spin state) does not determine the result of an individual measurement, there may be a more complete specification of the state in which this determination is manifest. Call this state λ. This new state λ may have many dimensions, discrete or continuous. The different parts of this state may interact with either part, 1 or 2 (Figure 1.4 [52]), of the apparatus used to perform the experiment.

Let Λ be the space of the states λ with some dimension. One requires a set of subsets of Λ to be defined so that the probability measures can be defined upon it. If the distribution function for the state λ on the space Λ is ρ, then

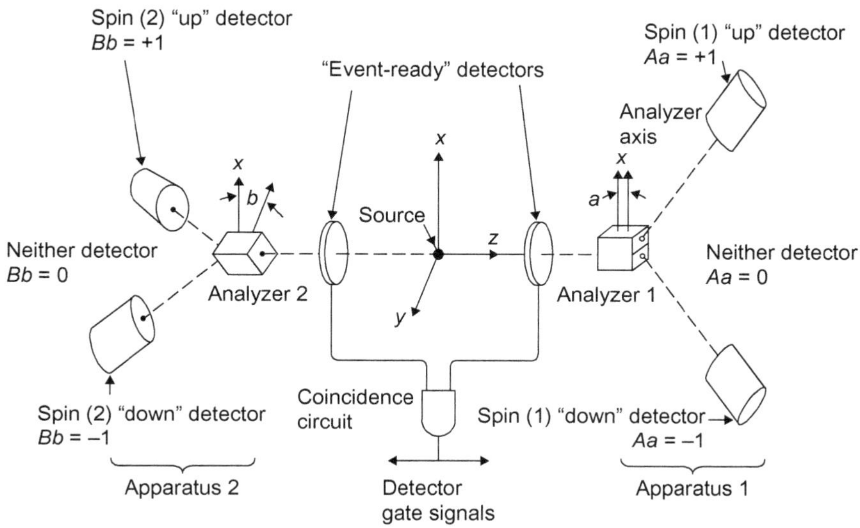

Figure 1.4 "Event-ready" detectors signal both arms that a pair of particles has been emitted for a given gate signal. The result on either arm is assigned the value $+1$ if the corresponding spin-up detector responds; -1 if the spin-down detector responds; and 0 if neither detector responds.

$$\int_{\Lambda} \mathrm{d}\rho = 1 \tag{1.3}$$

In a deterministic theory, $Aa \cdot Bb$ has definite value $(Aa \cdot Bb)_{\lambda}$ for the state λ. The theory is valid for all a and b and for all λ Ξ (space) Λ so that

$$(Aa \cdot Bb)\ (\lambda) = \ Aa(\lambda)Bb(\lambda) \quad \text{(Locality Condition)} \tag{1.4}$$

That is, once the state λ is specified and the particles have been separated, the measurements of Aa can depend only upon λ and a, and not b. Likewise for B, in this perspective. A realistic and deterministic theory that denies the existence of "action at distance" is local in this sense. The expectation value of $(Aa \cdot Bb)$ under this theory then would be

$$E(a,\ b) = \int_{\Lambda} Aa(\lambda)Bb(\lambda)\,\mathrm{d}\rho \tag{1.5}$$

If the locality conditions (1.4) and (1.2), for partial agreement with QM, are both satisfied, then the expectation values satisfy a simple inequality. This inequality is then an alternative prediction to that of QM, for the expectation value of

$(Aa \cdot Bb)$. The peculiarities made by this inequality are quantitatively different from those of equation (1.1). Equation (1.2) can hold if and only if

$$Aa(\lambda) = -Bb(\lambda) \tag{1.6}$$

This holds for all $\lambda \;\Xi$ (space) Λ. Using equation (1.6), one can calculate the following function, which involves three different possible orientations of the analyzers. $E(a,\ b) - E(a,\ c) = -\int_\Lambda [Aa(\lambda)Ab(\lambda) - Aa(\lambda)Ac(\lambda)]\mathrm{d}\rho = -\int_\Lambda Aa(\lambda)Ab(\lambda)[1 - Ab(\lambda)Ac(\lambda)]\mathrm{d}\rho$. Because $A,\ B = \pm 1$, the last expression can be written as $|E(a,\ b) - E(a,\ c)| \leq \int_\Lambda [1 - Ab(\lambda)Ac(\lambda)]\mathrm{d}\rho$. Using equations (1.3), (1.5), and (1.6), one gets

$$|E(a,\ b) - E(a,\ c)| \leq 1 + E(b,\ c) \tag{1.7}$$

Inequality (1.7) is one of the families of the inequalities. A simple instance of the disagreement between the predictions (1.1) and (1.7) is provided by taking a, b, etc., to be coplanar with c, making an angle of $2\pi/3$ with a and b making an angle of $\pi/3$ with both a and c. Then $a \cdot b = b \cdot c = 1/2$, $(a \cdot c) = -1/2$. For these directions

$$|[E(a,\ b)]_\psi - [E(a,\ c)]_\psi| = 1, \quad \text{while } 1 + [E(b,\ c)]_\psi = 1/2 \tag{1.8}$$

These values do not satisfy the inequality (1.7). Hence, the QM prediction and inequality (1.7) are incompatible, at least for some pairs of analyzer orientations. Thus, no deterministic hidden-variables theory satisfying equation (1.2) and the locality equation (1.3) can agree with all of the QM predictions concerning spins of a pair of spin-½ particles in the singlet state. The preceding argument leads to formulations that provide direct experimental predictions for systems that can actually be performed in a laboratory. By itself, though, the preceding analysis is insufficient to do this, because of its reliance upon the existence of a pair of analyzer orientations for which there is a perfect condition. Unfortunately, equation (1.2) (the locality condition) cannot hold in an actual experiment. Any real detector will have an efficiency less than 100%, and any real analyzer will have some attenuation as well as some leakages into its orthogonal channel. It is quite possible that imperfections are inherently correlated with measurement and detection processes in a way that depends upon λ.

In the idealized systems, whenever a particle is observed at one apparatus, an associated particle is always observed at the other apparatus (Figure 1.4 [52]). The selection of the subensemble of observed particles from among all those emitted by the source depends only upon the collimator and source geometry, and can have no dependence upon the parameters a and b. Hence, ρ was defined for the observed particle, and one can then be confident that it is independent of a and b.

1.2.2.2 General Locality Condition

In this case the system under examination is not restricted to spin-½ particles alone. The system may be any discrete-state, quantum mechanically correlated system,

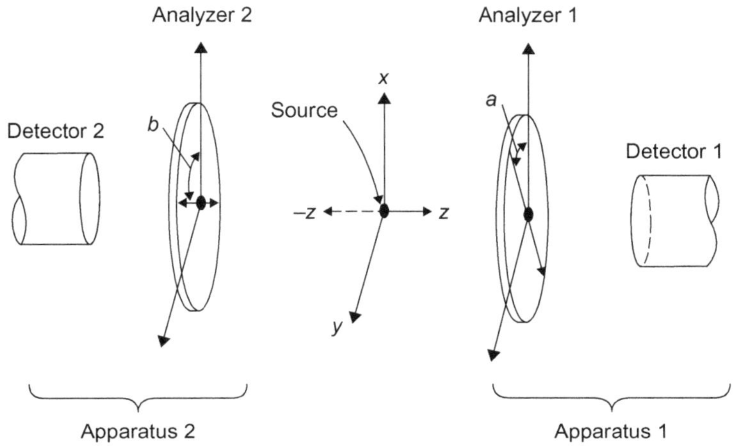

Figure 1.5 Apparatus configuration used. A source emitting particle pairs is viewed by two apparatuses. Each apparatus consists of an analyzer and an associated detector. The analyzers have parameters a and b, which are externally adjustable. In this example, a and b represent the angles between the analyzer axes and a fixed reference axis, as shown in Figure 1.6.

such as particle emissions. However, not all QM-correlated systems are predicted to violate the resulting inequalities. Now let a and b be any apparatus parameter under control by the experimenter. Aa and Bb are now the outcomes at apparatus 1 and 2 (Figure 1.4 [52]). The explanatory diagram in Figure 1.5 is reproduced by the courtesy of Ref. [52] as a practical illustration. The locality says that the outcome (or the probability of the outcomes) of a measurement performed on one part of a composite system is independent of what aspects of the other component of the experiment one chooses to measure. It by no means excludes the possibility of obtaining knowledge concerning system 2 from an examination of system 1: the state λ contains information common to both systems. A measurement on one of these presumably reveals some of this. Nor does it prevent a measurement on one component of a composite system locally disturbing that component.

The two systems are well separated. The values of A and B, $Aa(\lambda)$ and $Bb(\lambda)$, are now bounded by 1 and not $= 1$, i.e.,

$$|Aa(\lambda)| \leq 1, \quad |Bb(\lambda)| \leq 1 \tag{1.9}$$

$$E(a, b) = \int_{\Lambda} Aa(\lambda)Bb(\lambda)\, \mathrm{d}\rho \tag{1.10}$$

One includes in this ensemble only those particles that have previously triggered the event-ready detectors. Thus, one is assured that the distribution ρ and the range of Λ are independent of a and b.

Taking a', b' to be alternative settings for analyzers 1 and 2, and following the logical steps as in the previous section, one gets the inequality

$$-2 \leq E(a, b) - E(a, b') + E(a', b) + E(a', b') \leq 2 \tag{1.11}$$

The QM prediction $[E(a, b)]_{QM}$ can be written as

$$[E(a, b)]_{QM} = Ca \cdot b \tag{1.12}$$

Here the coefficient C is bounded by 1 for actual systems and $= \pm 1$ only in the ideal case. Suppose we take a, a', b, b' to be coplanar vectors with angle $\Phi = \pi/4$; then

$$[E(a, b) - E(a, b') + E(a', b) + E(a', b')]_{QM} \leq 2\sqrt{2}C \tag{1.13}$$

There is a wide range of values for C for which the prediction by inequality (1.11) disagrees with that by equation (1.12). The relevance of VBIs to experiments becomes more evident when one proceeds as follows. Suppose that during a period of time while the adjustable parameters have the values a and b, the source emits N particles of the two-component systems of interest. For this period, denote by $N_1(a)$ and $N_2(b)$ the number of counts at detectors 1 and 2, respectively. Denote by $N_{12}(a, b)$ the number of simultaneous counts from the two detectors, i.e., coincidence outputs (Figure 1.5). When N is sufficiently large, the probabilities for these results corresponding to the whole ensemble (with due allowance for random errors) are given as below:

$$p_1(1) = N_1(a)/N, \quad p_{12}(1) = N_{12}(a)/N, \ \ldots \tag{1.14}$$

The different p's are the corresponding different probabilities.

In terms of the probabilities, the BI can be written as

$$[p_{12}(a, b) - p_{12}(a, b') + p_{12}(a', b) + p_{12}(a', b')]/[p_1(a') + p_2(b)] \leq 1 \tag{1.15}$$

For the experiment, one uses $R(a, b)$ as the rate of coincidence counts. Taking $r_1(a)$ and $r_2(b)$ as the rate of single particle detections by either apparatus, the above BI can be written in terms of the measurable count rates as

$$[R(a, b) - R(a, b') + R(a', b) + R(a', b')]/[r_1(a') + r_2(b)] \leq 1 \tag{1.16}$$

The inequalities (1.15) and (1.16) are a general prediction for any local realistic natural phenomena. Experiments based on VBIs involve pairs of particles. The parameters a and b, considered abstractly, are taken to be the orientation angles relative to some reference axis in a fixed plane. In most of these experiments, the method of preparing the pairs of polarized particles attempts to achieve cylindrical symmetry about a normal to fixed plane and reflection symmetry with respect to planes through the normal.

This symmetry is exhibited in the QM predictions for detection rates and correlations. It is as follows. (Group 1): $[p_1(a)]_{QM}$ and $[r_1(a)]_{QM}$ are independent of a; $[p_2(b)]_{QM}$ and $[r_2(b)]_{QM}$ are independent of b; $[p_{12}(b)]_{QM}$, $[R(a,\ b)]_{QM}$, and $[E(a,\ b)]_{QM}$ are functions only of $|a-b|$. (Group 2): The corresponding predictions for the local realistic theories exhibit the symmetries $p_1(a) \equiv p_1$ and $r_1(a) \equiv r_1$, independent of a; $p_2(a) \equiv p_2$ and $r_2(a) \equiv r_2$, independent of b; $p_{12}(a,\ b) \equiv p_{12}(|a-b|)$, $R(a,\ b) \equiv R(|a-b|)$, and $E(a,\ b) \equiv E(|a-b|)$. One must emphasize two points concerning the two groups of symmetry relations. They do not simply follow from the corresponding QM symmetry relations (group 1) or from the symmetry of the experimental arrangement. This is because one does not know what symmetry-breaking factors may lurk at the level of the HVs. No harm is done by assuming the second group of symmetry conditions, as they are susceptible to experimental verification. Taking, a, a', b, b', so that $|a-b| = |a'-b| = |a'-b'| = (1/3)|a-b'| = \Phi$, one gets

$$3E(\Phi) - E(3\Phi)| \leq 2 \quad \text{(from equation (1.11))} \tag{1.17}$$

$$\begin{aligned} S(\Phi) &= [3p_{12}(\Phi) - p_{12}(3\Phi)]/[(p_1 + p_2)] \\ &= [3R(\Phi) - R(3\Phi)]/[(r_1 + r_2)] \leq 1 \text{ (from equation (1.15)) (BI)} \end{aligned} \tag{1.18}$$

$S(\Phi)$ is the correlation function of the two particles; simultaneous events [52].

1.2.2.3 Evaluation of an Experimental Scheme

For an experimental arrangement with a configuration like that shown in Figure 1.5 [52], the QM predictions can be written as $p_{12}(\Phi) = (1/4)\eta_1\eta_2 f_1 g[\varepsilon^1_+ \varepsilon^2_+ + \varepsilon^1_- \varepsilon^2_- F\cos(n\Phi)]$, $[p_1]_{QM}$

$$= \frac{1}{2}\eta_1 f_1 \varepsilon^1_+;\ [p_2]_{QM} = \frac{1}{2}\eta_2 f_2 \varepsilon^2_+;\ \varepsilon^i_+ = \varepsilon^i_M + \varepsilon^i_m;\ \varepsilon^i_+ = \varepsilon^i_M - \varepsilon^i_m \tag{1.19}$$

Here η's are the respective effective quantum efficiencies of the detectors, and ε_M and ε_m are the maximum and minimum transmissions of the analyzers relative to the pertinent orthogonal basis. The functions f's are the collimator efficiencies, i.e., the probability that an appropriate emission enters apparatus 1 or 2. The function g is the conditional probability that, given that emission 1 enters apparatus 1, emission 2 will enter apparatus 2. The function F is a measure of the initial state purity and the inherent QM correlation of the two emissions. The values of n are 1 or 2, depending upon whether the experiment is performed with fermions or bosons. Applying equation (1.19) to equation (1.18), one finds, for the case of the apparatus efficiencies 1, 2, chosen as $= 1$, the QM prediction can be written as

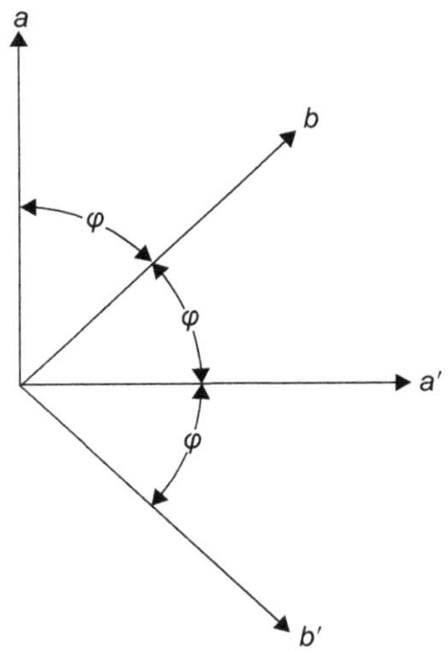

Figure 1.6 Optimal orientations for a, a' and b, b'. If the correlation is of the form $C_1 + C_2 \cos n\varphi$, then the maximum violation of the inequalities occurs at $n\varphi = \pi/4$.

$$S_{QM} = \frac{1}{4}\eta g\{2\varepsilon_+ + [3\cos(n\Phi) - \cos(3n\Phi)]F(\varepsilon_-^2/\varepsilon_+)\} \tag{1.20}$$

Selecting the optimum value of $\Phi = \pi(4n)$, the condition for the VBI becomes

$$\eta g\varepsilon + [\sqrt{2}(\varepsilon_-/\varepsilon_+)^2 F + 1] > 2(\text{QM}) \tag{1.21}$$

Thus, in a correlation experiment with the values in the domain specified by the VBI equation (1.21), one is capable of distinguishing the prediction from the inequality, $S(\Phi) \leq 1$, i.e., equation (1.18) and that of the QM, i.e., equation (1.20) (Figure 1.6).

1.2.3 VBIs Deduced from the Output States of a Nondegenerate Parametric Amplifier [54]

The nondegenerate parametric amplifier (NOPA) is an alternative, finite-dimensional, spin-based approach that presents a local realist model for position and momentum measurements on the original EPR state. The Wannir function (WF) generally has all the properties of a probability measure, but the drawback is that it can be negative. A proper probability measure must always be positive. In the particular case illustrated here, the WF is positive. If one uses the measurement of the parity, one can interpret the WF as a correlation function for the parity measurements, and then the regularized EPR states are nonlocal. One can create pseudo-spin operators based on parity and from them derive a violation of the

Clauser–Horne–Shimony (CHSH) inequality (basically the same as VBI). This is done using the $|NOPA\rangle$ state [46]. The following is a summary of the mathematical logical steps involved. This approach is associated with a number operator. The pseudo-spin operators in this scheme can be represented as $s_z = \Sigma_{n=0}{}^{\infty}\ (-1)^n|n\rangle\langle n|$, $s_+ = (s_-)\dagger = \Sigma_{n=0}{}^{\infty}(-1)^n|2n\rangle\langle 2n+1|$, $s_x = s_+ + s_-$, $s_y = -i(s_+ + s_-)$; these satisfy the usual commutation relations, $[s_z, s_\pm] = \pm 2s_\pm$, $[s_+, s_-] = s_z$. Now the pseudo-spin operator $s = (s_x, s_y, s_z)$ in a sense corresponds to the spin operator σ used for a normal spin-½ system. In the general case, a measurement of the pseudo-spin can be made along a certain direction a. It can be done by using the operator $a \cdot s$. For the present purpose, a planar variation is sufficient.

This variation is $s_\theta = \cos(\theta)s_z + \sin(\theta)s_x$. The two parts of the bipartite system considered here can be subjected to individual measurement of the above type. The NOPA entangled state can then be represented as $|NOPA\rangle = (1/\cosh r)\ \Sigma_{n=0}{}^{\infty}\tanh^n r|n\rangle \otimes |n\rangle$. The parameter r is referred to as the *squeezing parameter*. It is a measure of the amount of squeezing in the system. This state is sometime referred to as a *regularized EPR state*, and at infinite squeezing, $(r \to \infty)$, the state approaches the idealized EPR state [30]. One can then write $\langle NOPA|S_\alpha S_\beta'|NOPA\rangle = [\cos\alpha\cos\beta + K\sin\alpha\sin\beta](S_\alpha S'_\beta \equiv S_\alpha \otimes S_\beta)$. $\otimes$ refers to the multiplication of the two operators. K is a (strictly increasing) function of r. $K = 0$ when $r = 0$ and $K \to 1$, when $r \to \infty$. Thus, K can be used as a measure of the amount of squeezing in this state.

The results of the individual measurements, $S_\alpha \otimes I$ and $I \otimes S_\beta$ (I is an identity operator), are denoted by S_α and S_β', respectively. These are the classical values ± 1, registered from the measurement.

The question now is whether these results can be described under the assumption of a local realism. This is tested as follows. *Realism*: There is a classical probabilistic model. The results depend on an HV, λ, such that $S_\alpha = S_\alpha(\lambda)$, $S_\beta' = S_\beta'(\lambda)$. *Locality*: The model is local such that the measurement settings at one subsystem do not affect the other subsystem; i.e., $S_\alpha(\lambda)$ is independent of β and $S_\beta'(\lambda)$ is independent of α. *Restriction*: The measurement results are restricted in size, $S_\alpha(\lambda) \leq 1$, $S_\beta'(\lambda) \leq 1$.

When this is the case, we have the CHSH inequality $|E(S_\alpha S'_\gamma) + E(S_\alpha S'_\delta)| + E(S_\beta S'_\gamma) - E(S_\beta S'_\delta)| \leq 2$. However, using the $|NOPA\rangle$ state, one gets $|\langle S_0 S'_\gamma\rangle + \langle S_0 S'_{-\gamma}\rangle| + |\langle S_{\pi/2}S_{+\gamma}\rangle\langle S_{\pi/2}S'_{-\gamma}\rangle| = 2(|\cos\gamma| + K\xi|\sin\gamma|)$, with $\gamma = \arctan(K)$. The maximum of the RHS is obtained at $2\sqrt{(1+K^2)} > 2$. The conclusion is that 1, or 2 (above), or both must fail since 3 always holds in this setting. In the ideal case, there is violation at any nonzero squeezing. In a noisy setting, the violation will be lowered by the noise.

So, a certain lowest amount of squeezing will be required. Note that when $K = 1$, the violation will be as large as that generated by the singlet state in the original setting. This corresponds to an infinite squeezing parameter, r, i.e., the original EPR state [30]. Unfortunately, infinite squeezing cannot be achieved in practice. The angle γ at which there is maximum violation depends on the squeezing. The detector efficiency problem is less of an issue than in the usual BI, because lost photons in an optical implementation, for example, will not imply that

Table 1.4 The Truth Table

B	*B′*	*S*	*S′*	*SS′*	*f*(*B*, *B′*)
0	0	+ 1	+ 1	+ 1	0
0	1	+ 1	− 1	− 1	1
1	0	− 1	+ 1	− 1	1
1	1	− 1	− 1	+ 1	0

The function *f* proves to be exclusive-OR (XOR), since the multiplication of spin values is conveniently interpreted here as a test whether the spin values are equal or not.

experimental runs are dropped from the statistics. Instead, such losses will change the measured parity, introducing noise into the statistics and leading to a situation similar to that of the ion-trap experiment, in which "dark" events lead to increased noise in the experimental data.

One need not use auxiliary assumptions such as the nonenhancement assumption. The idea is to derive violation of local realism from continuous-variable systems using only the assumptions 1−3 above. One can group the number of states two-by-two in even parity and in odd parity and extend the groups. The generalized BIs here are $-|E(S_{\alpha,d}S'_{\gamma,d}) + (S_{\alpha,d}S'_{\delta,d})| + |E(S_{\beta,d}S'_{\gamma,d}) - E(S_{\beta,d}S'_{\delta,d})| \leq 2$ and $|\langle S_{0,}\ S'_{\gamma,d}\rangle + \langle S_{0,d}S'_{-\gamma,d}\rangle| + |\langle S_{\pi/2,d}S'_{\gamma,d}\rangle - \langle S_{\pi/2,d}S'_{-\gamma,d}\rangle| = 2(\cos\gamma + Kd|\sin\gamma|)$. One can construct an infinite computing hierarchy of pseudo-spin systems, each violating a BI. The S_z, 2^k operator of each pseudo-spin system corresponds to one of the bits of the number operator $n(n_z)$, and each bit is a qbit. Measurement technique can be transferred from the language of spin ($\pm 1/2$) to the language of bits (± 1). One performs measurement of the multiplication of spin-measurement results (SS') in the BIs in the function $f(B, B')$. This corresponds in the bit language to the truth table shown in Table 1.4.

1.2.4 EPR Entangled States

1.2.4.1 Ion-Trap Computer [46, 53]

Certain VBIs involving correlations between binary digits of particular observables can be used to detect interesting nonclassical correlations in a particular entangled state. The state considered here is of two separated sets of N ions [46]. The system associated with the entangled state is comprised of a NOPA and two linear ion traps. Each trap lies within an optical cavity and contains N identical ions. The schematics of this system are shown in Figures 1.7 and 1.8. The NOPA operates below some threshold, and its two external output fields first pass through Faraday isolators. Each of these then feeds into a linearly damped optical cavity via a lossy mirror. Each cavity is aligned such that its axes coincide with the x-axis. It is closed at one end by a perfectly reflecting mirror. In addition, each cavity supports a cavity mode of frequency ω_c. Within both cavities lie N identical two excited level ions of mass M, charge Z, and internal transition frequency ω_a. These ions are

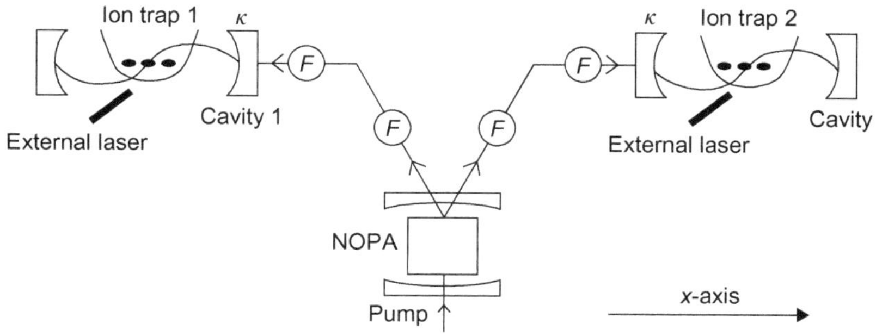

Figure 1.7 Schematic diagram for the system associated with the $2N$-way entangled state. The system consists of, first, a subthreshold optical-mode generated parametric amplifier without output modes passing through Faraday isolators (represented by an F cavity enclosed in a circle) and then feeding into linearly damped optical cavities. These cavities are aligned along the x-axis and each has one ideal mirror and one lossy mirror (with damping constant K). Inside each cavity is a harmonic ion trap that confuses N identical ions (each represented by a black circle) in a linear chain parallel to the x-axis. Experimental lasers of frequency ω_L are incident on the first ions in both traps from a direction perpendicular to the x-axis.

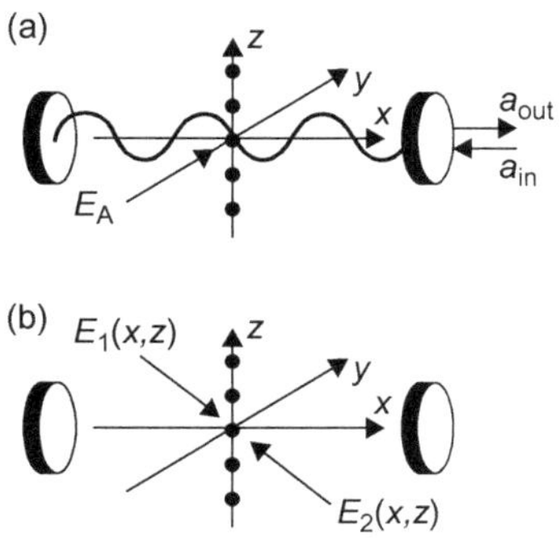

Figure 1.8 Schematic for state transfer between light and collective motional modes of a string of trapped ions. (a) First, a single-ion interaction with the cavity field mode is used to "receive" an incoming quantum state, after which (b) auxiliary lasers couple the single-ion and collective vibrational modes.

trapped in a linear configuration parallel to the x-axis by harmonic potential (a linear trap) and hence are tightly confined in the y and z directions. The traps are aligned such that the jth trap is centered on a node of the cavity field. Finally, the external lasers of frequency ω_L, whose beams are perpendicular to the x-axis, are incident on the first ions of both traps.

1.2.5 N-State Rydberg Atom Data Register Information Processing [44]

Consider that one prepares an N-state quantum register composed of $\log_2 N$ bits. Perform an operation with one state phase-shifted (flipped) from all the others. The number of operations then required to find the flipped state is order $\sqrt{N}(O\sqrt{N})$ steps. This is obtained following Grover's algorithm [32]; a classical algorithm would require at least $N/2$ steps. The algorithm performs an inversion about the average defined by the unitary transformation D on the state vector $D_{ij} = 2/N$ if $i \neq j$, and $D_{ii} = -1 + 2/N$. This operation amplifies the flipped state and attenuates the others. One can refer to the storage and retrieval of the information in the quantum phase of a coherent superposition state of energy levels through a highly excited atom. The algorithm works by considering a quantum system composed of multiple subsystems, each subsystem thus having an N-dimensional state space. As an example, such a subsystem consists of N Rydberg states in a single cesium (Cs) atom. This is provided by the quantum system, which is a collection of these Cs atoms in an effusive atomic beam. If the two energy states are designated as $|0\rangle$ and $|1\rangle$, then the superpositions $[|0\rangle + \exp(i\Phi)|1\rangle]/\sqrt{2}$ are also possible states of the qbit. Here the real number Φ is the quantum-phase difference between the probability amplitudes of the two levels in the superposition.

In the scheme considered here for the Cs atoms, three optical laser pulses intersect the Cs atoms beam. The first one is a 10-ns laser beam with a line width of approximately $0.5\,\text{cm}^{-1}$. This is tuned to $1.08\,\mu\text{m}$. The beam excites the Cs atom from the 6s ground state to the 7s state via two-photon absorption. (One neglects the hyperfine interaction, which is unresolved and has no role in the experiment.) The 7s state provides a reservoir of probability amplitude that is coherently transferred to the quantum register. One uses an optically driven unitary transformation that couples the subspace of N Rydberg levels to produce and search the contents of the register. A number of high-flying Rydberg np states are accessible from the 7s state. This is done by absorption of a photon from an ultra-fast Ti:sapphire chirped-pulse-amplified laser pulse. This contains approximately $100\,\text{cm}^{-1}$ of coherent bandwidth centered at a wavelength of 785 nm ($0.785\,\mu\text{m}$).

This is the coherent bandwidth contained in a 150-fs pulse, and it can access Rydberg np states in the range $n = 29-39$. One uses this laser twice in the experiment. The first pulse programs the quantum register with some information. The second exhausts the quantum search for the information. The optical radiation is programmed with a computer-controlled pulse shape, which generates a spectral mask such that N eigenstates were excited. The pulse program also placed information in the quantum register, by setting the phase of each quantum state relative to the 7s state. Because the $\langle \text{np}|z|7\text{s}\rangle$ matrix elements are relatively real, setting the phase is a simple matter of adjusting the relative phase of the optical radiation at each resonant frequency $\omega_{\text{np-7s}}$. For example, if $N = 5$ with principal quantum numbers ni, $i = 1-5$, one can store a binary number 00100 by exciting states, $i = 1, 2, 3, 4, 5$. This is done with a radiation field of the form $E\cos(\omega_{\text{np-7s}})t$, where state $i = 3$ is excited with radiation $E\cos(\omega_{\text{n3p-7s}})t$. There is a complication: different n-states evolve with a different time dependence $\exp(-i\omega_n t)$, so that any information stored as quantum phase in the N-state register r will also evolve. The effect is wave packet motion; one adjusts for it by programming the phase of each state to compensate for the phase evolution so that the phases are easy to interpret at the time they are read out. In the limit of lowest-order perturbation theory, the unitary transformation that transfers population from the 7s tail state to the N-state

Rydberg quantum data register can be represented by a matrix P, whose elements are $P_{ii} = 1$, $P_{0i} = P_{i0}{}^* = a_i \exp(i\Phi)$, $a \ll 1$ and $i \geq 1$, and $P_{ij} = 0$, or all other i, and $j \geq 1$. The state ($i = 0$) retains most of the probability amplitude. To match the database search problem, one makes two simplifications. The Rydberg levels are scaled to equal amplitudes and they are prepared in such a way that at some target time they are relatively real. Information is stored as flipped phase, i.e., $\Phi = \pi$ for some of the states. The preparation of the states is thus represented by $P \rightarrow A$. In this example, the populations of each of the states in the N-state register are $= |\varepsilon|^2$, but one of the states is flipped. At the target time, a second programmed laser pulse with the same amplitude as in A performs a quantum unitary operation. This is represented by matrix B on these levels by redistributing population through the 7s reservoir state. This pulse amplifies the flipped state or states and suppresses the rest of the states in the wave packet. Although amplitudes are the same as those in pulse A, the phases are all set to π, i.e., to the phase used for the flipped states. This pulse is efficient because it performs its amplification in a single step. It is universal because it will decode any binary number stored as quantum phase. After pulse B, all that remains is to collect the information from the storage medium. The only states that retain population are those whose phases were originally flipped when information was stored.

Each Cs atom is an identical subsystem of the total atomic beam. One uses state-selective ramped-field ionization to identify the states that were amplified in the atoms. A time-varying uniform electric field F is ramped from zero to several kilovolts per centimeter in 1 μs. The ionized electrons are measured using standard particle multipliers with detection efficiency near 1. Thus, the value of the electric field at the instant of ionization constitutes a measurement of N for the atom. According to the rules of QM, the possibility of finding the atom in the N state is proportional to the square of the amplitude for that component of the coherent state. The Rydberg states survive for several microseconds, which makes ramped-field ionization possible. However, the readout pulse B must arrive within a much shorter time.

The radiative lifetime of the reservoir state is of the order of several microseconds. The coherence time is also limited by the velocity of the distribution of the Cs atoms in the laser beam by black-body radiation and by Rydberg–Rydberg interactions. The total number of different values that can be stored as a phase in an N-state quantum register is M^{N-1}. Here M is the number of different phases that are used. In the simplest case of $M = 2$, one maps 2^{N-1} different numbers into the phase space of N Rydberg states. Retrieval of the information can still be accomplished with a single query of the databases; i.e., with a single universal unitary operator B. One can test this by loading a computer-generated random number between 1 and 2^{N-1} into pulse A and performing the retrieval experiment a number of times. One need not confine information storage to $M = 2$ quantum phases per state. Other algorithms can be implemented by other unitary transformations, such as the application of ultra-fast shaped terahertz pulses. Entanglement of additional degrees of freedom, such as spin and orbital angular momentum, will also extend the reach of the system (Figure 1.9).

(a)

$$A\Psi = \begin{pmatrix} 1 & \varepsilon & -\varepsilon & \cdot & -\varepsilon \\ -\varepsilon & 1 & 0 & \cdot & 0 \\ \varepsilon & 0 & 1 & \cdot & 0 \\ \cdot & \cdot & \cdot & \cdot & \cdot \\ \varepsilon & 0 & 0 & \cdot & 1 \end{pmatrix} \begin{pmatrix} 1 \\ 0 \\ 0 \\ \cdot \\ 0 \end{pmatrix} = \begin{pmatrix} 1 \\ -\varepsilon \\ \varepsilon \\ \cdot \\ \varepsilon \end{pmatrix} \begin{matrix} 7\mathrm{s} \\ n\mathrm{p} \\ (n+1)\mathrm{p} \\ \cdot \\ (n+N-1)\mathrm{p} \end{matrix}$$

(b)

$$B\Psi = \begin{pmatrix} 1 & \varepsilon & \varepsilon & \cdot & \varepsilon \\ -\varepsilon & 1 & 0 & \cdot & 0 \\ -\varepsilon & 0 & 1 & \cdot & 0 \\ \cdot & \cdot & \cdot & \cdot & \cdot \\ -\varepsilon & 0 & 0 & \cdot & 1 \end{pmatrix} \begin{pmatrix} 1 \\ -\varepsilon \\ \varepsilon \\ \cdot \\ \varepsilon \end{pmatrix} = \begin{pmatrix} 1 \\ -2\varepsilon \\ 0 \\ \cdot \\ 0 \end{pmatrix}$$

Figure 1.9 The unitary transformations (a) create the Rydberg data register and (b) retrieve the locations of the flipped bits.

1.2.6 The Scaling of the qbits: SFA [36]

SFA can factor large numbers in polynomial time. Despite significant progress, error effects from decoherence and imperfect QGs are still a major problem. Quantum error correction (QEC) and fault-tolerant quantum computation (FTQC) allow error-free calculations in the presence of imperfect quantum components. However, the implementation of such schemes often requires large numbers of qbits and complicated quantum circuits to create simple fault-tolerant gates. For large-scale quantum algorithms, the delicate interference between the computational states required for quantum computation is generally thought to require components with precision $1n_p$. Here $n_p = KQ$ represents the number of locations where an error can occur during an operation utilizing K qbits and Q elementary steps.

Current circuits designed using SFA assume that any arbitrary pair of qbits can be coupled as nonlinear nearest neighbor (NLNN). However, several architectures, most notably in solid-state models, are restricted to a single line of qbits with nearest neighbor (NN) interaction only. For the theory of the NLNN case, the reader is referred to Ref. [55]. Here we include only an illustration for the NN case [36] (Figures 1.10 and 1.11).

1.2.7 Deutsch–Jozsa Algorithm Ion-Trap Quantum Computer [9]

In a Deutsch–Joza (D–J) algorithm ion-trap quantum computer (ITQC), one uses four functions (fns) for mapping. There is one input bit ($a = 0, 1$) mapped onto one output bit ($f(a) = 0, 1$). The four f_n's can be divided into two constant ones, $f_1(a) = 0$, $f_2(a) = 1$, and two balanced ones, $f_3(a) = 1$, $f_4(a) = \mathrm{NOT}\ a$. An unknown f_n is characterized as constant or balanced by evaluating $f(0) \oplus f(1)$, which yields 0 or 1 for a constant (or balanced) f_n; $\oplus$ denotes addition modulo 2. This evaluation classically requires two f_n calls, whereas the D–J quantum algorithm allows one to obtain the desired information with a single evaluation of the unknown f. The truth table is shown in Table 1.5.

The third line is the effect of the logic function U_{fn} on the qbit w. CNOT is a CNOT operation. Z-CNOT is a zero CNOT. The control bit in cases 3 and 4 is the

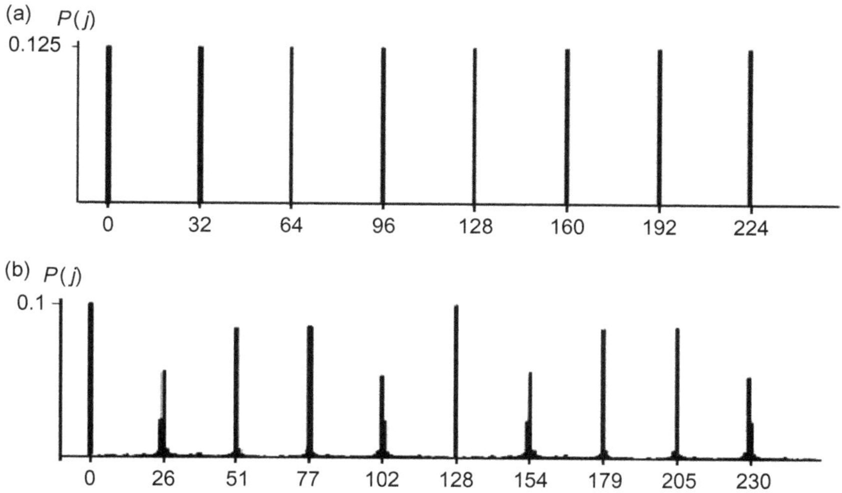

Figure 1.10 Probability of different measurements j at the end of quantum period finding with total number of states $2^{2L} = 256$: (a) period $r = 8$ and (b) period $r = 10$.

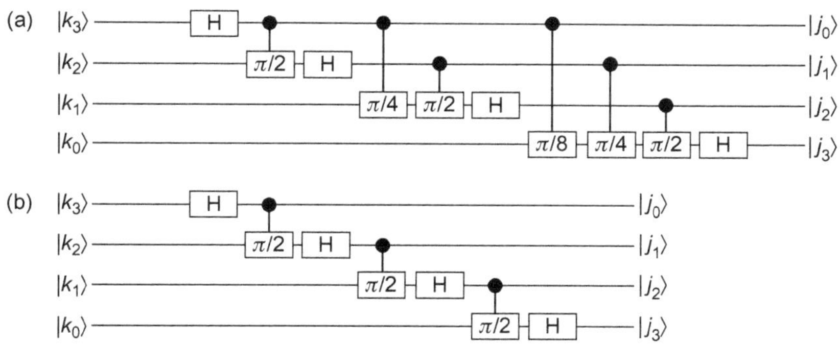

Figure 1.11 Circuit for a 4-bit (a) QFT. A controlled $\pi/2^d$ rotation gate requires a number of qbits and a number of fault-tolerant gates that grow polynomially with d. (b) Approximate FT with $d_{\max} = 1$.

input bit a. The 2 qbits required for the D–J algorithm are encoded in the electronic state, and in the phonon (vibrational) quantum number of the axial vibration mode of the single trapped ion (Figure 1.13 [9]). Qbit operations are realized by applying laser pulses on the "carrier" or "blue sideband" of the electronic quadrupole transition. The 2 qbits required for the D–J algorithm are encoded as follows. Electronic state: $S_{1/2}(m = -1/2) \equiv |0\rangle \equiv |S\rangle$, $D_{5/2}(m = -1/2) \equiv |1\rangle \equiv |D\rangle$. Phonon quantum number state (the axial vibration mode), single trapped ion: $n_Z = 0_z \equiv |1\rangle$ and $n_Z = 1_z \equiv |0\rangle$. The wR_{y-}, R_y on $|w\rangle$ manipulates for implementing U_{fn} into an

Table 1.5 The Truth Table

Functions	**Constant Function Case 1** f_1	**Constant Function Case 2** f_2	**Balanced Function Case 3** f_3	**Balanced Function Case 4** f_4
$F(0)$	0	1	0	1
$F(1)$	0	1	1	0
$W \oplus f(a)$	ID	NOT	CNOT	Z-CNOT

The third line is the effect of the logic function U_{fn} on the qbit w: ID denotes the identity, CNOT is a controlled-NOT operation, Z-CNOT is a zero controlled-NOT, and the control bit in cases 3 and 4 is the input bit a.

optimized pulse sequence, $R_{y-w}U_{fn}R_{yw}$. As the operations act also on the motional state, one implements them with pulses on the carrier and the blue axial sideband. However, sideband pulses operate on both qbits simultaneously. Thus, for operations on $|w\rangle$ alone, one first swaps the information from $|w\rangle$ to $|a\rangle$ with a sequence of three blue sideband pulses; one then rotates $|a\rangle$ as desired and swaps back. For a swap operation, one might be tempted to use a single π-pulse on the blue sideband. However, applying this to the state $|00\rangle = |SI_Z\rangle$ leads to a population of states, with two phonons outside the computational subspace. Therefore, one uses a composite pulse sequence consisting of three pulses, whose lengths are chosen such that starting from $|S, I_Z\rangle$ the ion is rotated by π, 2π, and π, respectively. As a result, the ion is rotated by 4π back to $|S, I_Z\rangle$ independent of the pulses' relative phase. In addition, using the blue sideband ensures that $|11\rangle|D, 0_Z\rangle$ also stays unchanged, as required for the swap operation. The desired swap operation $|S, 0_Z\rangle \leftrightarrow |D, 1_Z\rangle$ is possible because, compared to the $|S, 1_Z\rangle \leftrightarrow |D, 2_Z\rangle$ transition, the Rabi frequency for the $|S, 0_Z\rangle \leftrightarrow |D, 1_Z\rangle$ transition is smaller by $1/\sqrt{2}$. So, in this manifold the three pulses' lengths correspond to the rotation angles of $\pi/\sqrt{2}$, $2\pi/\sqrt{2}$, and $\pi/\sqrt{2}$. It can be shown that by choosing the laser-atom phase of the second pulse to be across $(\cot^2(\pi/\sqrt{2}) = \pi(0.3033)$ relative to the first and third pulses, the populations of $|10\rangle = |D, 1_Z\rangle$ and $|01\rangle = |\mathbf{S}, \mathbf{0_Z}\rangle$ are exchanged (Figures 1.12 and 1.13).

1.2.8 Cirac Zoller CNOT QG [3]

In a Cirac Zoller (C–Z) CNOT QG, two Ca^+ ions are held in a linear Paul trap and are individually addressed. This is achieved using focused laser beams. The pure qbits are prepared by superposition of two long-lived electronic (excitation) states of each ion. C–Z proposed a string of ions in a linear trap to serve as a quantum memory where the qbit information is carried by two internal states of each ion. Computational operations are carried out by addressing each ion individually with a laser beam. Single-qbit rotations are performed using coherent excitations by a single laser pulse driving transitions between the qbit states. For a 2-qbit CNOT operation, C–Z proposed to use the common vibration of an ion string to convey the information for a conditional operation (the vibrational mode is called the "bus-mode"). This can be achieved with a sequence of three steps after the ion string has

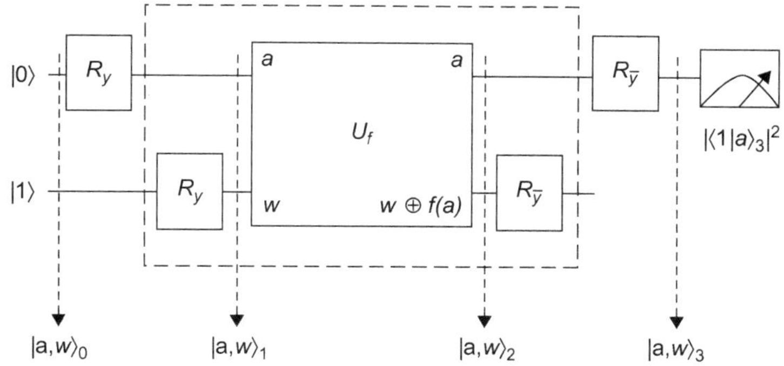

Figure 1.12 Quantum circuit for implementing the D–J algorithm with basic quantum operations. The upper line shows the input qbit $|a\rangle$ ("which side of the coin" information); the lower line shows an auxiliary working qbit $|w\rangle$ (corresponding to the channel on which the answer is provided). The rotations R_y create superpositions $|a\rangle_1 = (|0\rangle + |1\rangle)/\sqrt{2}$ and $|w\rangle_1 = (|0\rangle - |1\rangle)/\sqrt{2}$ from the inputs $|a\rangle 0 = |0\rangle$ and $|w\rangle 0 = |1\rangle$. The box U_f represents a unitary operation specific to each of the functions f_n, which applies f_n to a and adds the result to w modulo 2. Table 1.5 lists the logic operations required for transforming $|w\rangle$ into $W \oplus f(a)$. The output of the box is $|a, w\rangle_2 = (|0, w_{in} \oplus f_n(0)\rangle + |1, w_{in} \oplus f_n(1)\rangle)/\sqrt{2}$. Up to an overall sign $|w\rangle$ is left unchanged, but the positive superposition $(|0\rangle + |1\rangle)/\sqrt{2}$ on $|a\rangle$ is transformed into a negative superposition $|a\rangle_2 = (|0\rangle - |1\rangle)/\sqrt{2}$ if f is balanced; otherwise it is unchanged. After the final rotations R_y, a measurement on $|a\rangle$ is performed with result $|a\rangle_3 =$ either $|0\rangle$ or $|1\rangle$. Because of the sign change in $|a\rangle_2$ if f is balanced, $|\langle 1|a\rangle_3|^2 = f_n(0) \oplus f_n(1)$, i.e., $|a\rangle_3$, yields the desired information whether the function f_n is balanced or constant. The working qbit w resumes its initial value $|w\rangle_3 = |w\rangle 0 = |1\rangle$ logical value assigned to the respective states. Solid lines show carrier transitions; dashed lines show blue sideband transitions.

been prepared in the ground state ($n = 0$) of the bus-mode. First, the quantum information of the control ion is mapped onto the vibrational mode. The entire string of ions is moving and thus the target ion participates in the common motion.

Then, conditional upon the motional state, the target ion's qbit is inverted. Finally, the state of the bus-mode is mapped back onto the control ion. Mathematically, this amounts to performing the operation $|\varepsilon_1\rangle|\varepsilon_2\rangle \rightarrow |\varepsilon_1 \oplus \varepsilon_2\rangle$. This describes the logical state of the 2 qbits in question and $\oplus$ denotes addition modulo 2. Quantum information is encoded in trapped Ca^+ ions employing the electronic $|S_{1/2}, m_j = -1/2\rangle$ and $|D_{5/2}, m_j = -1/2\rangle$ levels of a narrow quadrupole transition near 729 nm.

The ions are prepared in the ground state of the trap's harmonic oscillator potential by laser cooling. Manipulation of the internal qbit state is performed by exciting the ions on the $|S_{1/2}\rangle - |D_{5/2}\rangle$ resonance ("carrier") transition, while the vibrations' degree of freedom is manipulated on the "blue" detuned sideband that is on a transition which changes both the electronic and motional degrees of freedom.

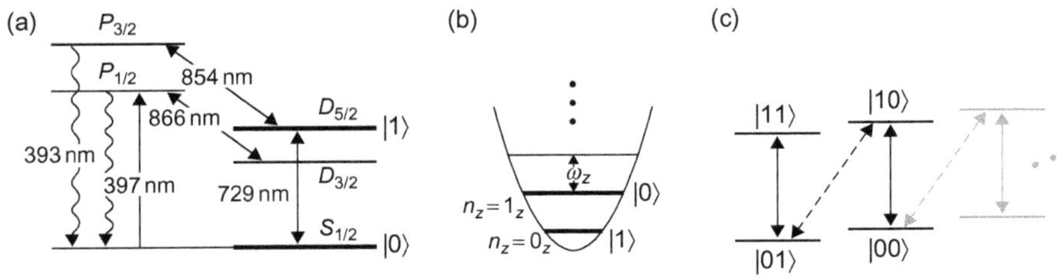

Figure 1.13 Quantum mechanical energy levels relevant for the ITQC. (a) Ca^+ level scheme. The upper and lower electronic states $S_{1/2}$ $(m = -1/2)$ and $D_{5/2}$ $(m = -1/2)$ of the narrow quadrupole transition ($\tau D \sim 1$ s) at 729 nm serve to implement one of the qbits, $|a\rangle$. Coherent radiation of a titanium–sapphire laser at 729 nm drives the qbit transition. Lasers at 397, 866, and 854 nm are used for the excitation of resonance fluorescence, for Doppler cooling, and for optical pumping. (b) The lowest two number states, $n_z = 0_z$, 1_z, of the axial vibration motion in the trap form the other qbit, $|w\rangle$. (c) The combination of electronic states and energy eigenstates of the harmonic oscillator potential spans the computational subspace. Numbers in the ket notation denote the quantum logical values assigned to the respective states. Solid lines show carrier transitions; dashed lines show blue sideband transitions.

To switch between carrier (R) and sideband (R^+) rotations, one shifts the laser frequency with an acousto-optic modulator. The phase of the light field is controlled via the phase of the RF driving the acousto-optic modulator. Additional phase shift due to light shifts arises because we have to drive sideband transitions (which couple much more weakly than carrier transition) with high laser intensity. One cancels the unwanted light shift with an additional off-resonant laser field, including a light shift of equal strength but opposite sign. Phase gate and CNOT gate operations that change the electronic statue of a single ion, conditional upon the state of motion, have been implemented previously using composite pulse techniques. For an implementation of the single-ion CNOT gate operation, one uses a combination of a pair of Ramsey pulses and a phase gate operation. The computational subspace for this operation consists of the states $|S, 0\rangle$, $|D, 0\rangle$, $|S, 1\rangle$, $|D, 1\rangle$, i.e., the electronic states of the ion, $|S\rangle$, $|D\rangle$, and the phonon state $|n = 0\rangle$, $|n = 1\rangle$. The description of the phase gate operation is briefly tabulated in Table 1.6 and Figure 1.14.

1.2.9 LSNMR Quantum Information Processing [1]

The 7 qbits are provided by the coupled spin-½ nuclei of transcrotonic acid (Figure 1.15 [1]). The scheme involves a coherent manipulation of 7 qbits starting from the cat state of 3 qbits. The *cat state* consists of an equal superposition of two states, one with all spins up and the other with all spins down. A 3-qbit manipulation is first carried out. The deviation density matrices are used for describing states of the nuclei. The states are described by the traceless part of the density

Table 1.6 The Truth Table

	$\|S, n=1\rangle$	$\|D, n=1\rangle$	$\|S, n=0\rangle$	$\|D, n=0\rangle$
$\|S, n=1\rangle$	−1	0	0	0
$\|D, n=1\rangle$	0	−1	−1	0
$\|S, n=0\rangle$	0	0	−1	0
$\|D, n=0\rangle$	0	0	0	1

With a 2π-rotation on the blue sideband, the $|S, 0\rangle$ and $|D, 1\rangle$ state amplitudes acquire a phase factor -1. The state $|D, 0\rangle$ is not affected by this operation and its phase does not change. However, transition starting from the $|S, 1\rangle$ state would yield a rotation of $2\pi/\sqrt{2}$, because sideband Rabi frequency depends on the phonon number. This shortcoming can be circumvented by using a combination of four pulses on the target ion possessing rotational angles (i.e., lengths of Rabi pulses) about rotation axes (i.e., phases of the exciting radiation). Together with two carrier $\pi/2$ pulses (Ramsey) before and after, this sequence completes the single-ion CNOT gate operation.

matrix up to an overall scale. The thermal equilibrium state of the molecule with one proton (H) and two ^{13}C nuclei (C_1 and C_2) at high field in a liquid is given by $\mu_H\sigma_z^{(H)} + \mu_C\sigma_z^{(C_1)} + \mu_C\sigma_z^{(C_2)}$, where μ_H and μ_C are the nuclear magnetic moments. The standard Pauli matrices are used as an operator basis and superscripts on operators refer to the nucleus the operator acts on. The cat-state benchmark for this system begins by eliminating signal from the carbon nuclei to obtain the initial states $\sigma^{(H)}$. Next, a sequence of QGs is used to achieve the state $\sigma_z^{(H)}\sigma_z^{(C_1)}\sigma_z^{(C_2)}$ (Figure 1.16 [1]). This is a sum of several coherent states.

In particular, it contains the states $|000\rangle\langle 111| + |111\rangle\langle 000|$. This is the deviation density matrix for the cat state $(|000\rangle + |111\rangle)/\sqrt{2}$. Here 0 and 1 represent the "down" and "up" spins, respectively. If each qbit is rotated by a phase Φ around the z-axis, the three-coherent state rotates by 3Φ and all other coherences rotate by 0, Φ, or 2Φ. This can be used to label the three states and eliminate all other states. After labeling the three states, coherence is decoded to the state $\sigma_z{}^{(H)}|00\rangle\langle 00|$ (Figure 1.16 [1]) and observed on the proton. The resulting spectrum is compared to a reference spectrum obtained after applying a 90° RF pulse to the proton in the initial thermal state. The reference spectrum has four peaks corresponding to the states $|00\rangle$, $|01\rangle$, $|10\rangle$, and $|11\rangle$ of the carbon nuclei. The spectrum obtained after decoding the cat state should have a single peak, ideally with the same intensity as that of the corresponding peak in the reference system.

1.2.10 Linear Optical Quantum Computation [22]

In linear optical quantum computation (LOQC), the simplest optical elements one can use are phase shifters and beam splitters (BSs). These elements generate the evolutions that can be implemented by passive linear optics. The evolutions preserve the total photon creation operator. The operator is defined by $a^{(l)}\dagger|n\rangle = \sqrt{(n+1)}|n+1\rangle$. Let U be the unitary operator applied to a state by such an evolution. Using $U|0\rangle = |0\rangle$ gives $Ua^{(l)}\dagger|0\rangle = Ua^{(l)}\dagger U\dagger U|0\rangle = Ua^{(l)}\dagger U\dagger|0\rangle = \sum_k u_{kl}a^{(k)}\dagger|0\rangle$.

The coefficients u_{kl} introduced by these equations define a matrix u, which must be unitary. Conversely, for every unitary u, there is a sequence of phase shifters and BSs that implements the corresponding operation up to a global phase. For an optical element X, let $u(X)$ be the unitary matrix associated with X according to the

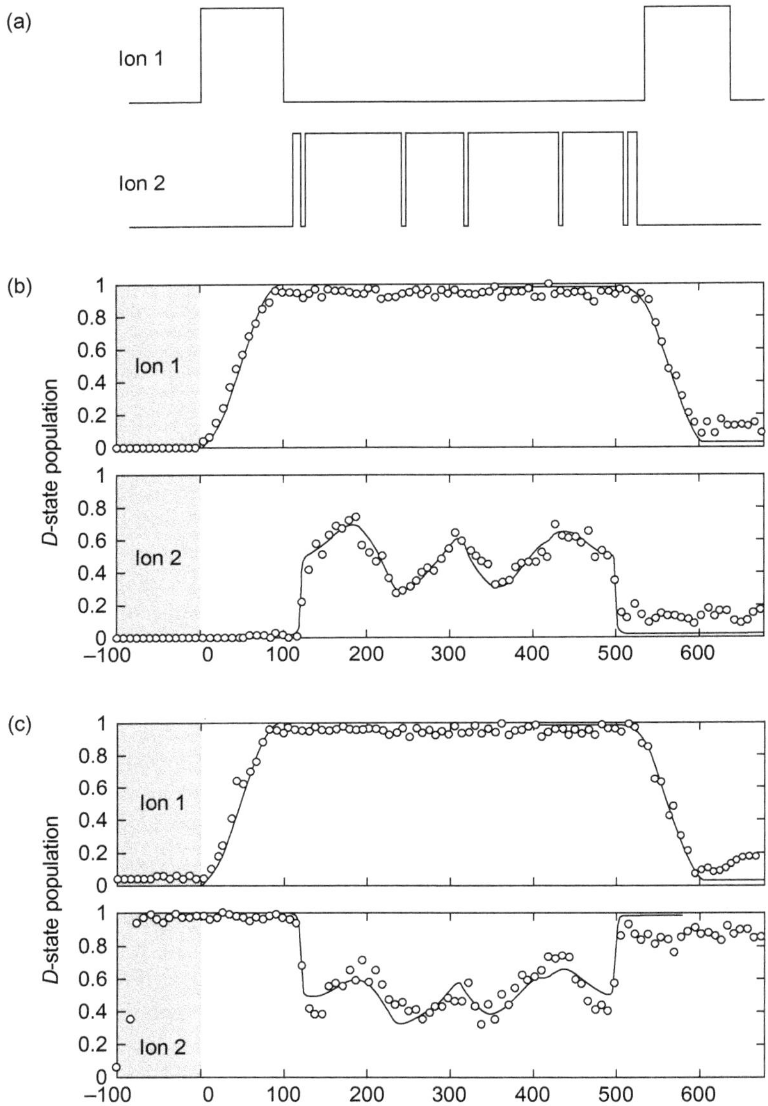

Figure 1.14 State evolution of both qbits under the CNOT operation. For this, the pulse sequence (a) is truncated as a function of time and the $D_{5/2}$ state probability is measured. The solid lines indicate the theoretically expected behavior. They do not represent a fit. Input parameters for the calculations are the independently measured Rabi frequencies on the carrier and sideband transitions and the addressing error. The initial state preparation is indicated by the shaded area and drawn in all figures for negative time values. The actual Cirac–Zoller CNOT gate pulse sequence starts at $t = 0$. After mapping the first ion's state (control qbit) with a p-pulse of length 95 ms to the bus-mode, the single-ion CNOT sequence (consisting of six concatenated pulses) is applied to the second ion (target qbit) for a total time of 380 ms. Finally, the control qbit is reset to its original value with the concluding p-pulse applied to the first ion for 95 ms.

	M	H_1	H_2	C_1	C_2	C_3	C_4
M	–969.4						
H_1	6.9	–3560.3					
H_2	–1.7	15.5	–2938.2				
C_1	127.5	3.8	6.2	–2327.0			
C_2	–7.1	156.0	–0.7	41.6	–18599.2		
C_3	6.6	–1.8	162.9	1.6	69.7	–15412.8	
C_4	–0.9	6.5	3.3	7.1	1.4	72.4	–21685.1

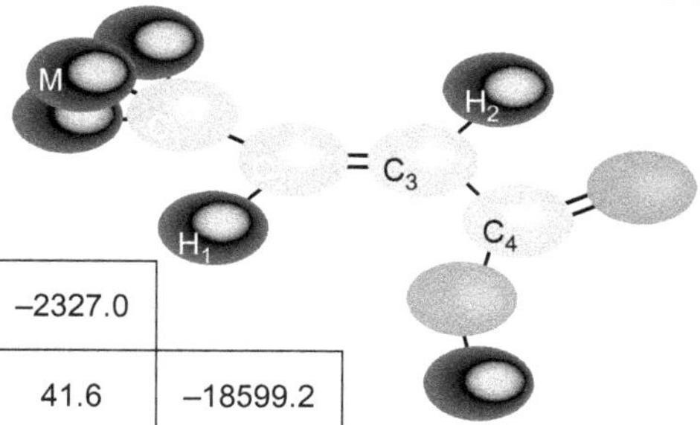

Figure 1.15 Characteristics of crotonic acid. Molecular structure of transcrotonic acid together with a table of the chemical shifts (on the diagonal) and *J*-coupling constants (below the diagonal). The chemical shifts are given with respect to reference frequencies of 500.13 MHz (protons) and 125.76 MHz (carbons) on the 500 MHz spectrometer used. The decoherence times (T_2*) were greater than 2 s.

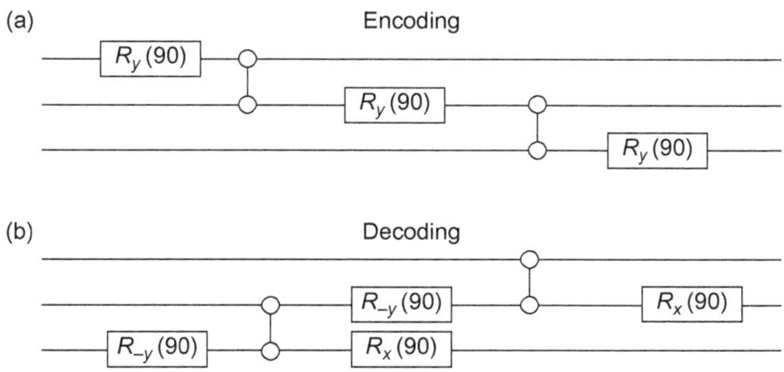

Figure 1.16 Quantum networks for the cat-state benchmark. (a) Encoding of the deviation matrix$\sigma_{z//}$ into $\sigma_y\sigma_y\sigma_x$ by using a cascade of 1-qbit rotation ($R_y(90)$) and 2-qbit operations (vertical bars) given by *J*-coupling gates. A *J*-coupling gate is given by the unitary operator $\exp(-\sigma_x\sigma_x\pi/4)$. A three coherence $|000\rangle\langle111| + |111\rangle\langle000|$ is contained in the output, which can be labeled using a magnetic gradient or phase cycling. (b) Decoding the coherence to a pseudo-pure state is accomplished by a similar inverse cascade. The output state is $\sigma_x|00\rangle\langle00|$. Both networks generalize by extending the cascade to more qbits.

above rules. The unitary matrix associated with phase shifter P_θ is $u(P_\theta) = e^{i\Phi}$. The unitary matrix associated with BS $B_{\theta,\Phi}$ is

$$u(B_{\theta,\Phi}) = \begin{pmatrix} \cos(\theta) & -e^{i\Phi}\sin(\theta) \\ -e^{i\Phi}\sin(\theta) & \cos(\theta) \end{pmatrix}$$

The phase shifters and BSs applied to a Bosonic qbit's modes preserve the qbit state space. Their effect can therefore be expressed in the qbit basis using the standard Pauli operators σ_x, σ_y, σ_z. Thus, all 1-qbit rotations can be implemented with linear optics.

Readout is accomplished by measuring a mode with a photodetector that destructively determines whether one or more photons are present in a mode. One assumes that photodetectors can be applied at any time and that the measurement results can be used to control other optical elements. One also needs a photon counter that destructively counts the number of photons in a mode. An approximate photon counter that suffices for this purpose can be designed by using BSs and multiple photodetectors. To measure a mode, one can use BSs to obtain the photon evenly over N modes and then use a photodetector on each. The desired count is the number of detectors that "see" photons. LOQC is based on a series of nondeterministic operations with increasing probability of success.

Box 1.1 QGs and Networks

Quantum information processing (QIP) is accomplished by applying QGs and measurements to prepare qbits. The gates evolve the state according to the laws of QM. The power of QIP depends on the ability to implement enough evolutions using the available gates. If all unitary evolutions can be approximated up to a global phase, the set of gates is called *universal*. Standard quantum computation relies on universal gate sets where each gate acts on 1 or 2 qbits. One such gate set consists of the 1-qbit rotations $U_\Phi = \exp(-i\sigma_u\varphi/2)$, $U = X, Y$, or Z, where φ can be restricted to $\varphi = 45^\circ$ and either the conditional sign flip (see text) or one of the 90° rotations $(UV)^{(12)}_{90^\circ} = \exp(-i\pi\sigma^{(1)}_{(u)}\sigma^{(2)}_{(v)}/4)$ with $U, V = X, Y$, or Z.

A sequence of state preparations, QGs, and measurements is called a *quantum network*. Quantum networks can be depicted by time-space diagrams, with time lines of qbits given by lines running from left to right, and gates by elements that intercept the lines. Our conventions for depicting 1-qbit gates are:

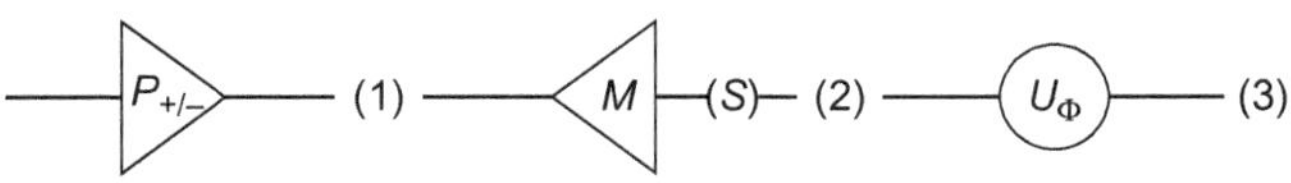

(1) is a preparation gate, with $P = X, Y$, or Z corresponding to preparations of σ_x, σ_y, or σ_z eigenstates. For example, if $P_z = Z_+$, the $|0\rangle$ state is prepared. (2) is a measurement gate, where $M = X, Y$, or Z corresponds to measurements

in the eigenbasis of σ_x, σ_y, or σ_z. The symbol S denotes the measurement outcome, which can be $+1$ or -1. (3) is a 1-qbit rotation around $U = X$, Y, or Z by angle φ (in degrees by default). Two-qbit gates are denoted by:

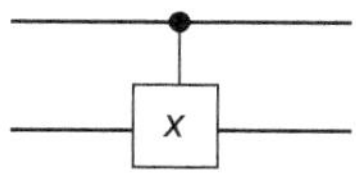

(4)

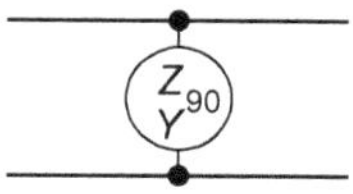

(5)

(4) is a conditional sign change by phase x and applies x only to the state $|11\rangle$. (5) is a $(ZY)^{(12)}_{90°}$ rotation.

Many of the gates are equivalent up to 1-qbit rotations. Here are some equivalences used in the text:

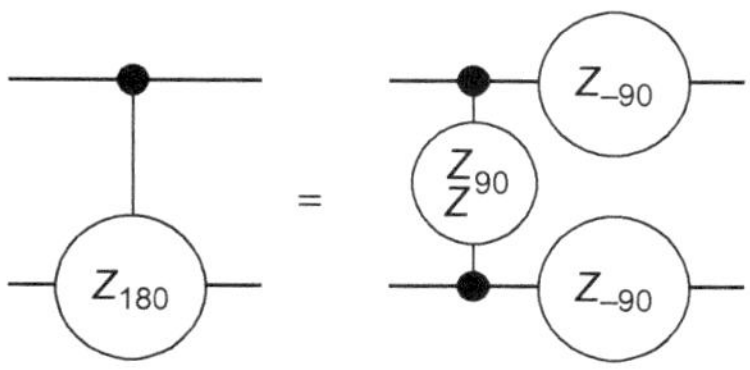

(6)

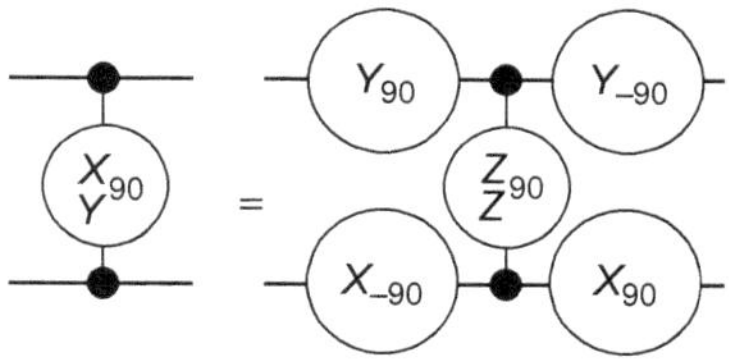

(7)

(7) expresses one gate by conjugating another by $Y^{(1)}_{-90°}$ and $X^{(2)}_{90°}$.

Optical networks are similar to quantum networks except that the basic systems are optical modes. The basic elements of an optical network drawing are:

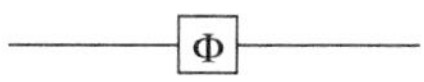

(8)

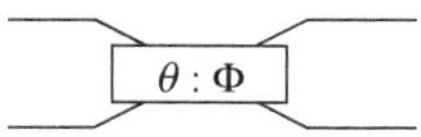

(9)

(8) shows a phase shifter $\mathbf{P}_\Phi$ and (9) a BS $\mathbf{B}^{(12)}_{\varphi,\Phi}$, where mode 1 is the top mode. If $\varphi = 0$, only angle θ may be given in a diagram.

State preparation is like (1), with P_Φ replaced by 0 or 1, for the number of photons inserted into the mode. Measurement is like (2), with M replaced by $\mathbf{n}$ and S by R, for the number of photons detected.

1.2.11 *Tomography and Spectroscopy [2]*

In a QC, it is important to be able to determine the state of a quantum system and to measure the properties of its evolution. State determination can be achieved by using tomography. Spectroscopy can be used to probe the energy spectrum associated with the state's evolution. Tomography and spectroscopy (TAS) are both aspects of the same QC that can be represented by a "scattering" experiment. Versions of this circuit also play a crucial role in many of the quantum algorithms that represent marked improvements over the best classical counterparts. Typically, when analyzing a quantum system, one is interested in the evolution of an initial state over a given time period. This is implemented on a QC by decomposing the evolution operator for the time period into a sequence of elementary operations called *QGs* (a combination of elementary gates forms a *quantum circuit*). The circuit representing an evolution can be readily modified to account for experimental conditions under one's control or to include interactions with additional degrees of freedom used as probes. In a typical QC, a family of quantum algorithms uses many independent instances of the experiment required to implement the scattering circuit. The measurements yield the expectation values $\langle\sigma_z\rangle$ and $\langle\sigma_y\rangle$ of the Pauli spin operators for the ancilla qbit used as a probe particle. The expectation value obtained by use of the scattering circuit has the property $\langle\sigma_z\rangle = \mathrm{Re}[\mathrm{Tr}(U\rho)]$, $\langle\sigma_z\rangle = \mathrm{Im}[\mathrm{Tr}(U\rho)]$. Here ρ represents the initial state (density matrix) of the system. The state ρ and the operator U appear symmetrically here. As a result, the scattering circuit can be used to measure properties of U or ρ, by using known instances of ρ or U, respectively. One first prepares the system in a completely mixed initial state $\rho = I/N$, where $N = 2^n$ is the dimensionality of the qbit state space of the system and I is the identity operator. The final measurement yields $\langle\sigma_z\rangle = \mathrm{Re}[\mathrm{Tr}(U)/N]$, which is proportional to the sum of eigenvalues of U and provides global information about the spectral density of U. The value of the spectral density near specific energies can be obtained as shown in Figure 1.17 [2]. This is like the scattering circuit except that the controlled operation U consists of a pair of FTs in an extra n_2 qbit and a controlled evolution of $U^t = \exp(-iht)$ in between. The number of qbits n_2 and the timescale δ determine the resolution and the range with which the spectral density is obtained. The circuit of Figure 1.18 [2] can be shown to yield

$$f(E) = \mathrm{Tr}(U^t|E\rangle\langle E| \otimes \rho) = \left(\frac{1}{N_2}\right)\sum_{t=0}^{N_2-1} \exp(i4\pi E_t/N_2)\mathrm{Tr}(U^t\rho), \quad N_2 = 2^{(N_2)}$$

The function $f(E)$ is related to the spectral density at energy $= 2E/(N_2\delta)$ by "smoothing" on scales of order $2/(N_2\delta)$ and by identifying energies that differ by multiples of $1/\delta$ where one has used frequency units for energy normalization such that $\Sigma_{E=0}^{N_2-0} f(E) = 1$. This means that one can think of $f(E)$ as representing the probability that the energy content of ρ is in a region of about $\pm 1/(N_2\delta t)$. To be able to distinguish $f(E)$ from 0 without excessive measurement accuracy requires the energy content of ρ to be concentrated in the region considered.

1.2.12 QDs: Terahertz Electrodynamics [26]

In this scheme the information is stored in the two lowest electronic states of doped QDs. Three disks of a semiconductor (e.g., GaAs) are embedded within another semiconductor with a larger band gap (e.g., $Al_xGa_{1-x}As$) (Figure 1.19 [26]). The central disk (GaAs) is taller than the outer two. The barriers between the disks are sufficiently low to allow an electron to rapidly tunnel between them. The structure

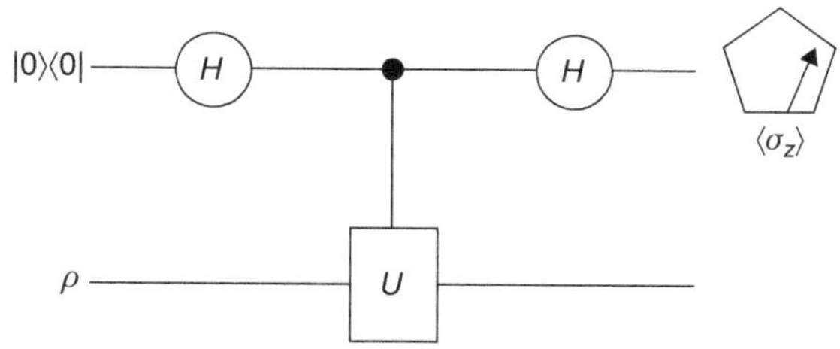

Figures 1.17 The scattering circuit. The circuit represents a sequence of instructions for applying operations to quantum systems. The horizontal lines represent the time lines of the quantum systems of interest. The operations are applied in left-to-right order. In this case, there are two systems. The bottom system models the physical system of interest, and is initially in the state ρ (a density matrix). The top system is an ancilla qbit acting as a probe particle. It is initialized in the state $|0\rangle$, the density matrix of which is $|0\rangle\langle 0|$.The qbit can be thought of as a spin-½ particle with 0 and 1 representing the "down" and "up" states, respectively. The circuit consists of the following steps: (1) Apply a Hadamard transform H to the ancilla qbit. Because $H|0\rangle = (|10\rangle + |11\rangle/\sqrt{2}$ and $H|1\rangle = (10\rangle - |11\rangle)/\sqrt{2}$, the new state of the qbit is $(|0\rangle + |1\rangle/\sqrt{2}$. (2) Apply a "controlled-U" operator, which does nothing if the state of the ancilla is $|0\rangle$ and applies U to the system if the ancilla is in state $|1\rangle$. (3) Apply another HG to the ancilla and perform measurements of its spin polarizations along the z- and y-axes. The polarization along the z-axis $\langle\sigma_z\rangle$ yields Re[Tr(σU)] for any unitary operator U that can be controlled.

of the QD that participates in the quantum computation must have one and only one electron. The lowest two electronic-energy levels $|0\rangle$ (ground) and $|1\rangle$ (excited) within the QD potential well will form the qbits that store the quantum information. The third level, $|02\rangle$, will serve as an auxiliary state to perform conditional rotations of the state vector of the qbit (Figure 1.20).

1.2.13 Quantum Sidebands Optical Field: EPR Entanglement [21]

In this scheme, the spectral sideband correlations of a single-mode squeezed light are transferred to entanglement between two spatial modes. The entanglement thus produced would be of a quite different character from that produced by two-mode squeezing. It would involve a pairwise correlation between only a single sideband on each of the beams. The system used is essentially a Mach–Zehnder interferometer (MZI). Here the path length for one of the interferometer arms is much greater than for the other. This introduces a time delay between the two arms of τ of the unbalanced (U) MZI (Figure 1.21).

1.3 Experimental Techniques

1.3.1 Shor's Factoring Algorithm [56]

To find prime factors of an L-digit integer N, the numbers of steps required in a classical computer (CC) increase exponentially with L. QCs, in contrast, can factor

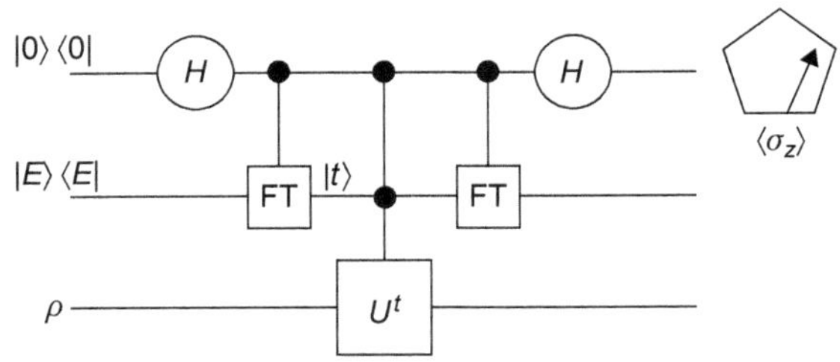

Figure 1.18 Circuit for evaluating the spectral density of a Hamiltonian H modulated by the energy populations of the state ρ. It requires the ability to conditionally realize $U^t = \exp(-iht)$. Time t is expressed in units of an appropriately chosen scale δ, as indicated by the coefficient of the Hamiltonian in the exponent. The second ancillary register, formed by n_2 qbits, is prepared in the initial state $|E\rangle$, where E represents the energy at which we wish to evaluate the spectral density. After the first controlled FT, the logical basis states of the second register represent time. The middle controlled operation maps the computational states $|1\rangle \oplus |t\rangle_2 \oplus |n\rangle_3$ into $|1\rangle \oplus |t\rangle_2 \oplus U^t|n\rangle_3$; with no effect if the first qbit is in the state $|0\rangle$. The second FT completes the circuit, to enable deterministic evaluation of the spectral density. Note that use of the FT, and not its inverse, is crucial for transferring the desired signal to the output. Without the ancilla qbit and if the second register is prepared in $|0\rangle$ and measured at the end, this is a version of the phase-estimation algorithm. For sufficiently large n_2, the phase-estimation algorithm can yield a randomly sampled eigenvalue of H. The circuit shown here has the advantage that only 1 qbit need be measured. The desired spectral information is given by the qbit's polarization—a much simpler measurement that can be performed even if we only have access to ensembles of these systems without the ability to measure individual members, such as in NMR quantum computation.

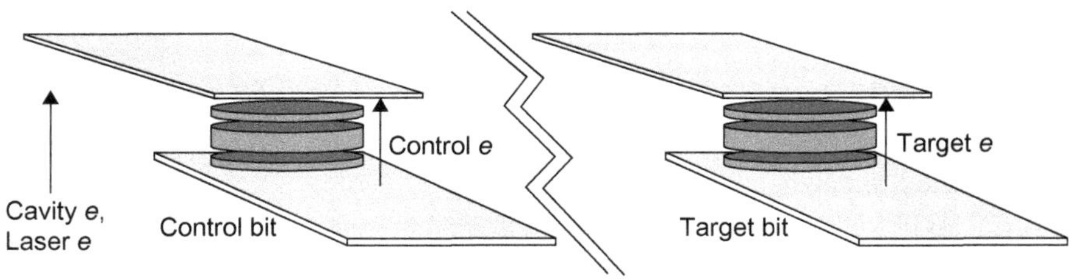

Figure 1.19 Fundamental elements of the proposed QC. Each set of QDs contains one electron, and is individually addressable by a pair of gate electrodes. One QD is chosen to be a control bit, and the other a target bit, for a CNOT operation. Many fundamental elements are embedded in a single-mode cavity.

integers in polynomial time using Shor's algorithm. In this experimental technique, NMR spectroscopy of perfluorobutadienyl iron complex with inner two carbons labeled ^{13}C was used. Factorization of $N = 15$ (whose prime factors are 3 and 5) was implemented. Out of 7 qbits, 5 of fluorine (^{19}F) and 2 of carbon (^{13}C) qbits were entangled.

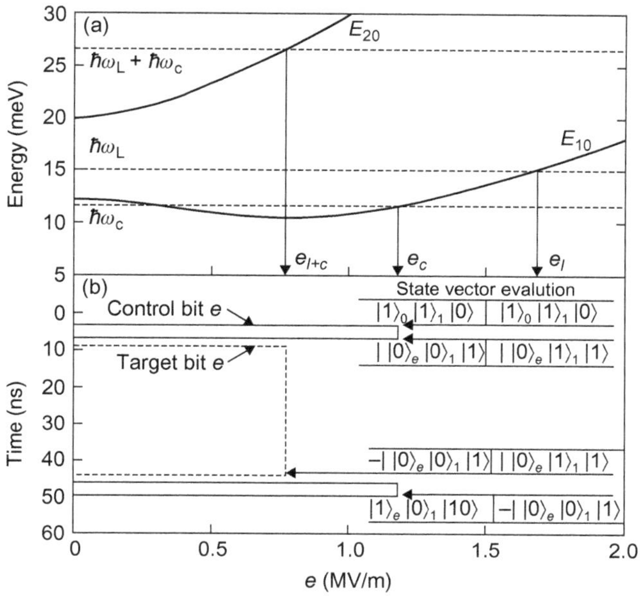

Figures 1.20 (a) Transition energies between states $|0\rangle$, $|1\rangle$ and (E_{10}) and $|0\rangle$ and $|2\rangle$ (E_{20}) versus applied electric field, and photon energies of a cavity mode ($\hbar\omega_c$), a laser ($\hbar\omega_L$), and the sum ($\hbar\omega_L$) + ($\hbar\omega_c$). The E_{10} transition resonates with ($\hbar\omega_c$) and ($\hbar\omega_L$), at electric fields e_c and e_l, respectively. The E_{20} transition resonates with the two-photon transition with energy ($\hbar\omega_L$) + ($\hbar\omega_c$) at electric field $e_l + e_c$. (b) A sequence of electric field pulses to a control and a target bit that are used in a CNOT gate. First, a "π" pulse is applied to the control bit, transferring a photon to the cavity and multiplying the state vector by i (the state vectors i, j), if and only if the control bit is 1. Then a "2π" pulse is applied to the target bit, multiplying the state vector by -1 if and only if there is a photon in the cavity and the target bit is in its ground state. Finally, a second "π" pulse is applied to the control bit, removing the photon from the cavity, returning the control bit to the excited state, and again multiplying the state vector by i. The state vectors in which the control bit is 0 are unaffected by the sequence of electric field pulses, and thus are not shown. One-bit rotations can be effected by applying an appropriately timed pulse with amplitude e_l. As shown by Cirac and Zoller, the gate shown here, together with 1-bit rotations on the target bit, results in a CNOT operation.

SFA works by using a QC to quickly determine the period of the function $f(x) = a^x \bmod N$ (the remainder of a^x divided by N). Here a is a randomly chosen small number with no factors in common with N. From this period, one can find factor N with high probability. The two main components of this algorithm are the modular exponentiation (computation of $a^x \bmod N$) and the inverse quantum Fourier transform (QFT). For $N = 15$, it requires coherent control over 7 qbits. Here a may be 2, 4, 7, 8, 11, 13, or 14. If one picks up $a = 2$, 7, 8, or 13, one finds that $a^4 \bmod 15 = 1$ and therefore all $a^{2k} \bmod N = 1$ for $k \leq 2$. In this case $f(x)$ simplifies

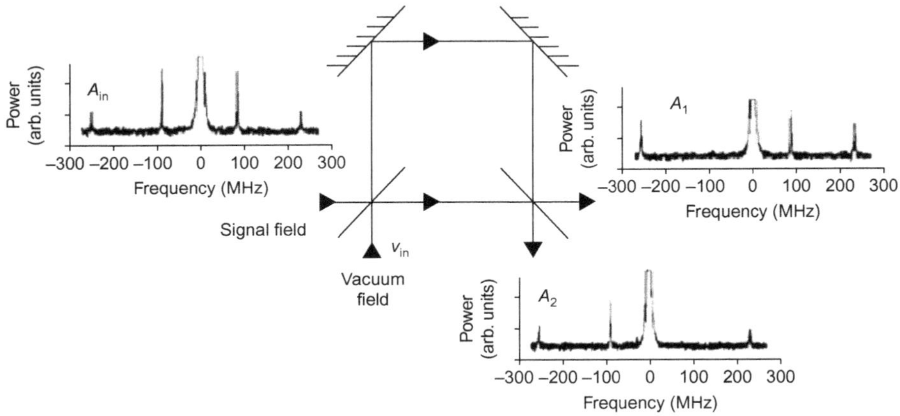

Figure 1.21 A schematic diagram of the system along with spectra illustrating successful operation at the classical level. All the spectral measurements were made with a scanning confocal Fabry−Pérot cavity, with a free spectral range of 500 MHz and a line width of approximately 2 MHz. The truncated carrier is shown at zero frequency, the PM sidebands are at ±90.5 MHz, and residual mode-mismatch peaks (with less than 1% power) are at ±250 MHz.

to multiplications controlled by just 2 qbits; $x = k - 1$ ($x = 0 \leq x \leq k - 1$, x is the bit number) $= x_0$ and x_1. If $a = 4$, 11, or 14, then $a^2 \bmod 15 = 1$ ($k = 1$), so only x_0 is relevant. Thus, the first register can be as small as 2 qbits ($k = 2$). However, 3 qbits ($k = 3$) allow for the possibility of detecting more periods. Together with $m = [\log_2 15] = 4$ qbits to hold $f(x)$, one needs 7 qbits total. The experimental investigation in this technique was based on $a = 11$ and $a = 7$. In a static magnetic field, each spin I has two discrete energy eigenstates, $|0\rangle$ (spin up) and $|1\rangle$ (spin down), described by the Hamiltonian $H_0 = -\sum_i (h/2\pi)\omega_i I_{zi}$, where $\omega_i/2\pi$ is the transition frequency between $|0\rangle$ and $|1\rangle$ and I_z is the z-component of the spin angular momentum operator. All 7 spins in this molecule are well separated in frequency $\omega_i/2\pi$ and interact pairwise via the j-coupling described by $H_j = \Sigma_{i<j}(h/2\pi)I_{ij}I_{zi}I_{zj}$. The initial state of the 7 qbits is $|\psi\rangle = |0000001\rangle$ (Figure 1.22). Experimentally one starts from thermal equilibrium. The density matrix is then given by $\rho_{th} = \exp\{-[(H_0/k_BT)/2^7]\}$ with $k_BT \geq (h/2\pi)$ at room temperature. Thus, each spin is in a statistical mixture of $|0\rangle$ and $|1\rangle$. One converts ρ_{th} into a 7-spin effective pure state ρ_1 via temperature averaging. ρ_1 constitutes a suitable initial state for SFA because it generates the same signal as $|\psi\rangle$ up to a proportional constant. ρ_1 is highly mixed and remains separable under unitary transforms (Figure 1.23).

1.3.2 Deutsch's Algorithm LSNMR QC [8]

It is not easy to efficiently simulate the behavior of a QM system. The system is not just confined to its eigenstates; instead, it can exist in any superposition of

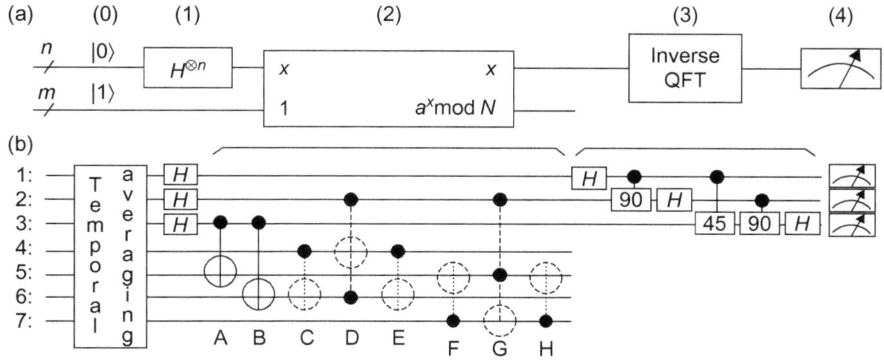

Figure 1.22 Quantum circuit for SFA. (a) Outline of the quantum circuit. Wires represent qbits and boxes represent operations. Time goes from left to right. (0) Initialize a first register of $n = 2[\log_2 N]$ qbits to $|0\rangle \oplus \cdots \oplus |0\rangle$ (for short $|0\rangle$) and a second register of $m = [\log_2 N]$ qbits to $|0\rangle \oplus \cdots \oplus |0\rangle \oplus |1\rangle$. (1) Apply a Hadamard transform H to the first n qbits, so the first register reaches $\Sigma x = 0$ to $(2^n - 1)|x\rangle/\sqrt{(2^n)}$. (2) Multiply the second register by $f(x) = a^x \text{mod} N$ (for some random $a < N$ that has no common factors with N) to get $|\psi_z\rangle \Sigma x = 0$ to $(2^n - 1)|x\rangle|1\rangle\langle a^x \text{mod} N\rangle/\sqrt{(2^n)}$. As the first register is in a superposition of 2^n terms $|x\rangle$, the modular exponentiation is computed for 2^n values of x in parallel. (3) Perform the inverse QFT on the first register, giving $|\psi_3\rangle = \Sigma y = 0$ to $(2n - 1)\Sigma x = 0$ to $(2n - 1)$exp $(2\pi/xy/2^n|y\rangle|a^x \text{mod} N\rangle)/2^n$ where interference causes only terms $|y\rangle$ with $y = c(2^n/r)$ (for integer c) to have a substantial amplitude, with r the period of $f(x)$. (4) Measure the qbits in the first register. On an ideal single-QC, the measurement outcome is $c(2^n)/r$ for some c with high probability, and r can be quickly deduced from $c(2^n/r)$ on a CC via continued fractions. (b) Detailed quantum circuit for the case $N = 15$ and $a = 7$. Control qbits are marked by filled circles; $\oplus$ represents a NOT operation and 90° and 45° represent z rotations over these angles. The gates shown in dotted lines can be removed by optimization, and the gates shown in dashed lines can be replaced by simpler gates.

them. The space needed to describe the system is a very large one. As an example, a system containing N spin-½ particles inhabits a Hilbert space of dimension 2^N. This evolves under a series of transformations described by the matrices, containing 4^N elements. In QMIP, one needs a set of gates identical to that in CIP, such as the AND, OR, and NOT gates. QGs differ, however, in that they must be reversible. This is because the evolution of any system can be described by a series of unitary transformations which are themselves reversible. It must be possible to reconstruct the input bits knowing only the design of the gate and the output bits. Thus, every input bit must be in some sense in the outputs. One trivial consequence of this is that the gate must have exactly as many outputs as inputs. For this reason, the gates of CIP are not reversible. It is, however, possible to construct reversible equivalents of AND and OR gates in which the inputs are preserved. In the case of CIP, the NAND gates are universal. It can be shown that certain gates or combination of gates is universal for quantum computing. In particular, it can be shown that the combination of a general qbit rotation with the 2-bit CNOT gate is universal.

i	$\omega_i/2\pi$	$T_{1,i}$	$T_{2,i}$	J_{7i}	J_{6i}	J_{5i}	J_{4i}	J_{3i}	J_{2i}
1	−22052.0	5.0	1.3	−221.0	37.7	6.6	−114.3	14.5	25.16
2	489.5	13.7	1.8	18.6	−3.9	2.5	79.9	3.9	
3	25088.3	3.0	2.5	1.0	−13.5	41.6	12.9		
4	−4918.7	10.0	1.7	54.1	−5.7	2.1			
5	15186.6	2.8	1.8	19.4	59.5				
6	−4519.1	45.4	2.0	68.9					
7	4244.3	31.6	2.0						

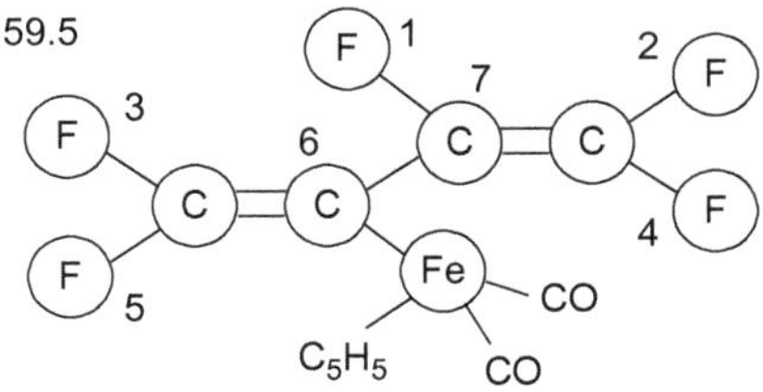

Figure 1.23 Structure and properties of the QC molecule, a perfluorobutadienyl iron complex with the inner two carbons ^{13}C-labeled. Based on the measured J_{13C19F} values, one concludes that the placement of the iron is as shown, different from that derived from infrared spectroscopy. The table gives the $\omega i/2\pi$ (in Hz) at 11.7 T, relative to a reference frequency of approximately 470 and 125 MHz for ^{19}F and ^{13}C, respectively; the longitudinal (T_1, inversion recovery) and transverse (T_2, estimated from a single-spin-echo sequence) relaxation time constants (in s); and the J-couplings (in Hz). Ethyl (2-^{13}C) bromoacetate was converted to ethyl, 2-fluoroacetate by heating with AgF, followed by hydrolysis to sodium fluoroacetate using NaOH in MeOH. This salt was converted to 1,1,1,2-tetrafluoroethane using MoF_6, and was subsequently treated with two equivalents of n-butyl lithium followed by I_2 to provide trifluoroiodoethene. Half of the ethylene was converted to the zinc salt, which was recombined with the remaining ethylene and coupled using $Pd(Ph_3P)_4$ to give (2,3-^{13}C) hexafluorobutadiene. The end product was obtained by reacting this butadiene with the anion obtained from treating $[(\pi\text{-}C_5H_5)Fe(CO)_2]_2$ with sodium amalgam. The product was purified with column chromatography, giving a total yield of about 5%. The sample at 0.88 ± 0.04 mol% in perdeuterated diethyl ether was dried using 3-A molecular sieves, filtered through a 0.45-μm syringe filter, and flame-sealed in the NMR sample tube using three freeze-thaw vacuum degassing cycles. All experiments were performed at 30°C.

Single-qbit rotations are easily implemented in LSNMR QC, as they correspond to rotations within the subspace corresponding to a single spin.

Such rotations can be achieved using RF fields. One particularly important single-bit gate is the Hadamard gate (HG) that performs the rotational transformations. The HG is self-inversive. It can be used to convert superposition of states back into eigenstates for later analysis. The 2-bit gates correspond to rotations within subspaces, corresponding to 2 spins. They require some kind of spin−spin interaction for their implementation. In LSNMR, the scalar spin−spin coupling (J-coupling) has the correct form for the construction of CNOT. This gate inverts the value of 1 qbit when another qbit (the control bit) has some specific value, such as $|1\rangle$. The truth table is shown in Table 1.7. It is also necessary to have some way of reading out information about the final quantum state of the system and thus obtaining the result of the calculation. In most implementations of a QC, this amounts to finding which of the eigenstates a two-level system is in. It is possible to obtain equivalent information by exciting the spin system and observing the resulting NMR spectrum. Different qbits correspond to different spins and thus give rise to signals at different resonance frequencies. The eigenstate of a spin before the excitation can be determined from the relative phase (absorption or emission) of the NMR signal. The Deutsch's algorithm (DA) is shown in Table 1.8.

QCs of necessity use reversible logic, and so it is not possible to implement the binary function, f, directly. It is, however, possible to design a propagator U_f that captures f within a reversible transformation by using a system with 2 input qbits and 2 output qbits, as shown in Figure 1.24. One uses quantum circuits that may be drawn by analogy to classical electronic circuits (CEC).

The lines are used to represent "wires" down which qbits "flow." It is possible to begin with a superposition of states. The calculation can begin with the second bit in the superposition $(|0\rangle - |1\rangle)/\sqrt{2}$.

It can also begin with the first qbit also in superposition of states $(|0\rangle + |1\rangle)/\sqrt{2}$. In this instance the first bit then ends up in the superposition $(|0\rangle \pm 1\rangle)/\sqrt{2}$. An LSNMR computer is an ensemble of spins in a statistical mixture of states. Such a system is most conveniently treated using a density matrix that can describe either a mixture or a pure state. The density matrix can be decomposed in the product

Table 1.7 Truth Table for the CNOT Gate

Input		**Output**	
0	0	0	0
0	1	0	1
1	0	1	1
1	1	1	0

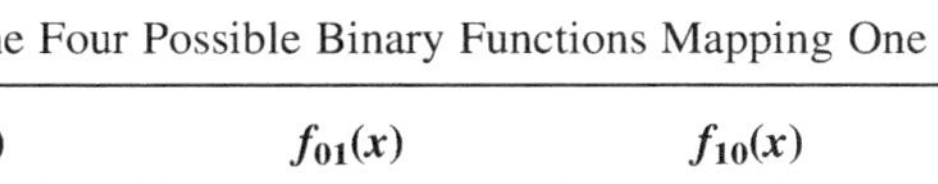

The first qbit (the control qbit) is unchanged by the gate, whereas the second qbit is flipped if the control qbit is in state 1, effectively implementing an XOR gate.

Table 1.8 The Four Possible Binary Functions Mapping One Bit to Another

X	$f_{00}(x)$	$f_{01}(x)$	$f_{10}(x)$	$f_{11}(x)$
0	0	0	1	1
1	0	1	0	1

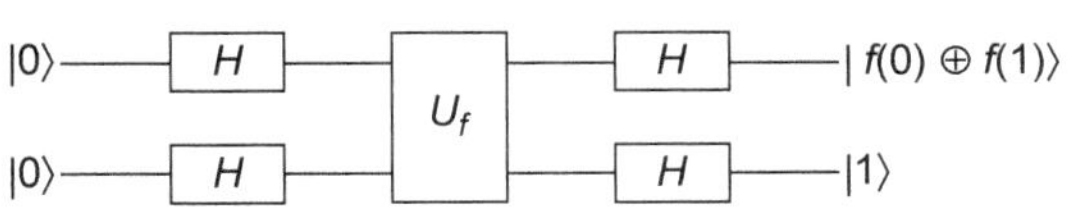

Figure 1.24 A quantum circuit for solving Deutsch's problem.

operation with the basis $\rho_{01} = (I_Z - S_Z - 2I_ZS_Z + 1/2E)/2$ (E is the Hamiltonian). One ignores the multiples of the unit matrix. This gives rise to no observable effects in any NMR experiment. The density matrix can be reached from the thermal equilibrium, e.g., the density matrix $(I_Z + S_Z)$, by series of RF and field gradient pulses.

The unitary transformation matrix corresponding to the Hadamard operator on a single spin can be written as

$$H = (1/\sqrt{2})\begin{pmatrix} 1 & 1 \\ 1 & -1 \end{pmatrix}$$

This corresponds to a 180° rotation around an axis tilted at 45°, between the z- and x-axes. Such a rotation can be achieved directly, using an off-resonance pulse or using a three-pulse sandwich such as 45°, −180°, −45°. The unitary transformation corresponds to the four possible propagators U_f. Each propagator corresponds to flipping the state of the second qbit under certain conditions as shown in Figure 1.25. In the technique described here, one uses a 2-spin system.

This constitutes a 50-mM solution of the pyrimidine base cytosine in D_2O. There is a rapid exchange of the two amine protons and the single amide proton with the deuterated solvent. It leaves two remaining proteins forming an isolated 2-spin system. The system is operated at a basic frequency of 500 MHz. It can be used both for the implementation of a classical algorithm to analyze $f(0)$ and $f(1)$ and for the implementation of DA. The pulse sequences differ only in the placement of the 90°_y pulses. An NMR QC is capable of implementing the classical algorithm, as it is simple to determine $f(0)$. The other value, $f(1)$, can be determined in a very similar way.

1.3.3 *Optical Magnetic Electron Spin Quantum State [16]*

A nitrogen vacancy defect center in diamond (NVDC-D) offers great promise for measuring quantum states of a single electron. The defect consists of a bound state of a

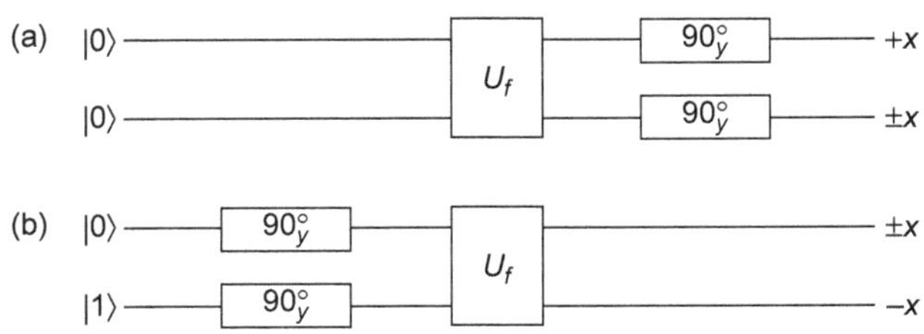

Figure 1.25 Modified quantum circuits for the analysis of binary functions on an NMR QC. (a) A circuit for the classical analysis of $f(0)$; the normal circuit (Table 1.7) is followed by 90° pulses to excite the NMR spectrum. Clearly $f(1)$ can be obtained in a similar way. (b) A circuit for the implementation of DA, with Hadamard operations replaced by 90° $\pm \nu$ pulses. The final 90° excitation pulses cancel out the 90°_y pulses, and four pulses can be omitted.

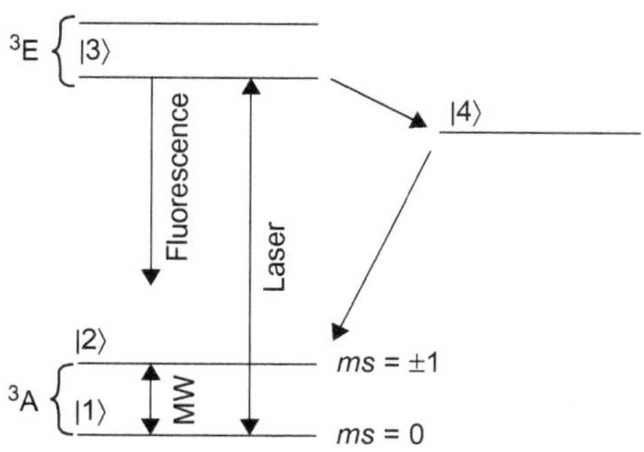

Figure 1.26 Energy-level scheme of the NVDC-D. The excited spin state manifold can be approximated by a single-spin sublevel, because nonresonant optical excitation was used in the experiment. The grayed-out lines correspond to the $ms = \pm 1$ sublevels.

substitutional ^{14}N atom and a vacancy in the adjacent site. This is known to have a strongly dipole-allowed optical transition, which is between the electron spin triplet ground state ^{3}A and a first excited spin triplet state ^{3}E (Figure 1.26 [16]). The fluorescence emission strongly depends on the electron spin quantum state. It leads to optically detected magnetic resonance of a single defect center. Electron spin relaxation time T_1 of the defect center is of the order of 1 ms at room temperature. The spin coherence is influenced by the optical excitation and coupling to the additional degrees of freedom like single ^{14}N nuclear spin of the NV center. The NV center in diamond has an electron paramagnetic ground state. This spin state can be read out directly via optical excitation and subsequent fluorescence relaxation. A confocal microscope experimental technique allows coupling of the microwave pulsed magnetic fields. The NV defect is subjected to continuous laser irradiation ($\lambda = 514$ nm).

1.3.4 Optical Continuous-Variable EPR Entanglement [37]

In this experimental technique, the Kerr nonlinearity of an optical fiber is utilized to produce two amplitude-squeezed beams with nonlinear interaction of each beam, uncoupled to the other. The two amplitude-squeezed beams are made to interfere at a 50/50 BS. In the vein sum, the squeezing is obtained for the amplitude quadratures. The difference squeezing is produced for the phase quadratures. A passively mode-locked Cr^{4+}:YAG (yttrium–aluminum–garnet) laser is used to produce optical pulses at a center wavelength of 1505 nm with full width at half-maximum (FWHM) bandwidth of 130 fs and repetition frequency of 163 MHz. These pulses are injected into an asymmetric fiber sagnac interferometer. The sagnac loop consists of an 8-m-long polarization-maintaining fiber and a BS. The fiber has a birefringence characterized by a beat length of 1.95 mm for 1505 nm light. This supports the s- and p-polarization states with negligible cross talk. The BS of the interferometer has 91% (90%) reflectivity and 9% (10%) transmittivity for the s(p) polarization states. This provides a strong and a weak counter-propagating pulse within the sagnac loop. There is one pair for each polarization. Because of the Kerr nonlinearity, the strong pulses acquire an intensity-dependent phase shift during propagation (Figure 1.27).

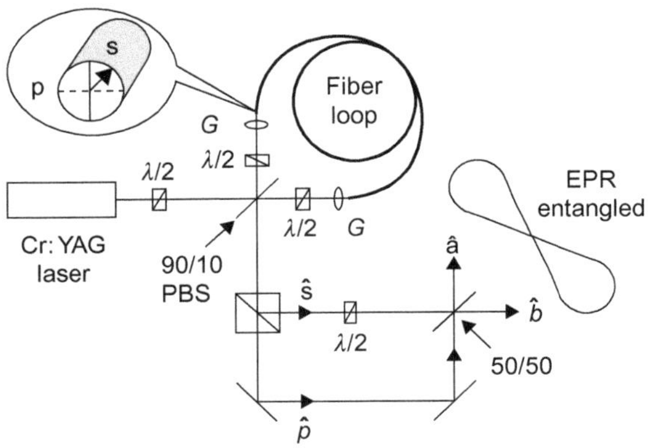

Figure 1.27 Schematic of the experimental setup. $\lambda/2$: half-wave plate; *G*: gradient index lens; PBS: polarizing beam splitter; 50/50: BS with 50% reflectivity; and 90/10: BS with 90% reflectivity. s and p are the two squeezed beams from the respective polarization states. *a* and *b* are the EPR entangled output beams. Inset shows the polarization directions of the input beam to the fiber.

1.3.5 Terahertz Laser Solid-State Quantum Bits [25]

An electron bound to a shallow donor atom (e.g., a Si or S atom) in *n*-GaAs has a hydrogenic character with a large Bohr radius of approximately 10 nm. The hydrogenic 1s-2p$^+$ ($m = 1$) transition lies in the terahertz (THz) frequency range and may be tuned from 1 to over 5 THz by the application of a magnetic field of just a few tesla. Here *m* is the quantum number associated with the projection of the orbital angular momentum onto the magnetic-field axis. The choice of the magnetic field is made such that the 2p$^+$ state is resonant with the ionized state. The three processes characterizing the photoconductive response of the system (Figure 1.28) create conditions such that most of the electrons excited to the 2p$^+$ state do not return to the ground state. Instead, they are ionized to the conduction band, where they contribute to the macroscopic conductivity of the sample. The 1s-2p$^+$ state of the donors is mapped onto up and down states of fictitious spins in a magnetic field. Intense terahertz pulses are used to coherently rotate the spins through angles $\gg 2\pi$. Such arbitrary 1-bit rotations are a requirement for quantum information processes. The sample is driven by intense terahertz radiation provided by a free electron laser (FEL). A fixed frequency of 2.52 THz is chosen for this technique. The 1s-2p$^+$ transition is tuned with a magnetic field of 3.5 T. To generate the short terahertz pulses, the light-activated semiconductor switches were used to "slice" short segments from the 2-μs-long pulses from a FEL. One determines the temporal profile of the photocurrent transient.

The exponential decay of the photocurrent is determined by the electron recapture rate. This is a constant sample parameter. Excited electrons are known to ionize within the first nanosecond after the end of the terahertz pulse. The integrated photocurrent is thus proportional to the total fraction of electrons excited out of the 1s ground state at the end of the pulse. The Rabi oscillation frequency is seen to increase with THz fields. However, the rate at which the oscillations are damped also increases, resulting in a number of cycles of oscillation that are roughly independent of the THz field.

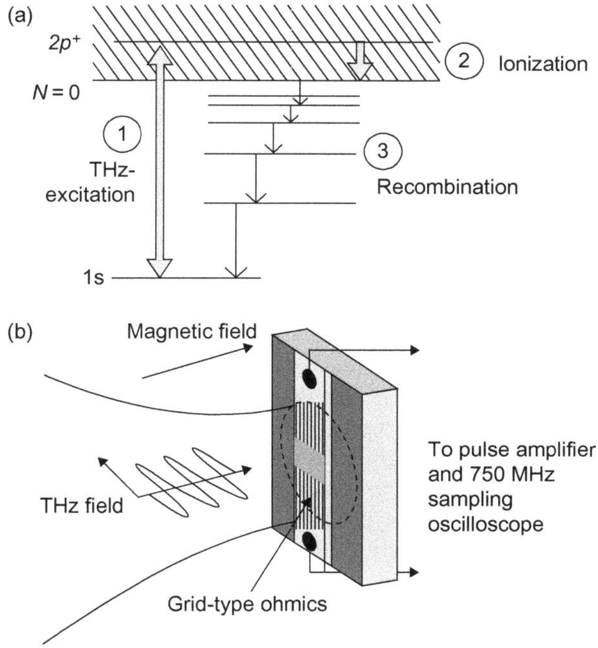

Figure 1.28 Schematic drawing of terahertz photoconducting processes and sample. (a) Processes: (1) After the "switch-on" of the terahertz field, E_{THz} (the bound electron population) oscillates between the ground (1s) and excited ($2p^+$) states. (2) Electrons left in the $2p^+$ state at the end of the terahertz pulse are ionized into the conduction-band continuum of conduction-band states associated with the lowest ($N = 0$) Landau level. Thus, the transient increase in conductivity is proportional to the number of electrons excited out of the 1s state by the terahertz pulse. (3) The free conduction-band electrons recombine with donor ions. (b) Sample geometry: The sample was fabricated from a 15-μm-thick GaAs layer grown by molecular-beam epitaxy on a semi-insulating GaAs substrate. The electron density was $2.8 \times 10^{14}\,cm^{-3}$ with mobility 140,000 $V^{-1}\,s^{-1}\,cm^2$ at 77 K. Quasi-CW photoconductivity spectra indicate that a single donor species (sulfur) dominates. The sample had a 20-nm Si-doped ($\sim 10^{18}\,cm^{-3}$) cap layer to facilitate the formation of ohmic contacts (AuNiGe), which was etched off the remaining sample area after alloying the contacts. The active region of the sample (100 μm × 100 μm) was defined by wet etching and the contacts. This active region was much smaller than the size of the terahertz focus (500-μm FWHM), ensuring uniform illumination. To minimize distortions of the terahertz fields by the ohmic contacts, the contacts were made of 10 equally spaced 10-μm AuNiGe strips perpendicular to E_{THz}. The back of the sample had an impedance-matched antireflection coating to eliminate internal reflections. The sample was placed in the bore of a superconducting solenoid, and immersed in liquid helium at 2 K. The sample was connected directly to a 40 Ω coaxial cable that fed through a "bias-tee" into a broadband pulse amplifier and a 750 MHz digital oscilloscope. The sample was biased with an electric field of 1 kV/m, below the threshold for impact ionization.

One models the dynamics of the system using the density matrix formalism for a two-level system, including relaxation into the conduction band.

The Rabi frequency is given as $\Omega_R = (eE_{THz}x_{12}/(h/2\pi))$. Here e is the electronic charge, E_{THz} the amplitude of the terahertz electric field, and x_{12} the dipole matrix element of the 1s-2p$^+$ transition. The mechanisms that damp the observed Rabi oscillations are extrinsic to the qbits and much faster than the predicted intrinsic decoherence.

1.3.6 Optical EPR Entanglement: Continuous Variables [20]

The quadrature amplitude squeezing of the output vacuum mode can be formed from the superposition of the original signal and the idler modes (named superposed mode). This can be experimentally measured. The phase squeezing of the superposed bright-output modes and the quantum correlation between the quadrature-phase amplitudes of the output bright signal and idler modes are directly inferred. One is not able to measure the bright-phase squeezing with self-homodyne detection. The technique demonstrated here uses the frequency-degenerate twin beams from a NOPA, the EPR beams. They can produce the quantum entanglement required for teleportation of the quantum (Figure 1.29).

1.3.7 The Biomedical QC

1.3.7.1 Nondeterministic Polynomial Time Complete SAT Problem [17]

An expectation value quantum computer (EVQC) can be realized from a molecular ensemble (e.g., an NMR QC). It can solve the NP-C problem in a much easier

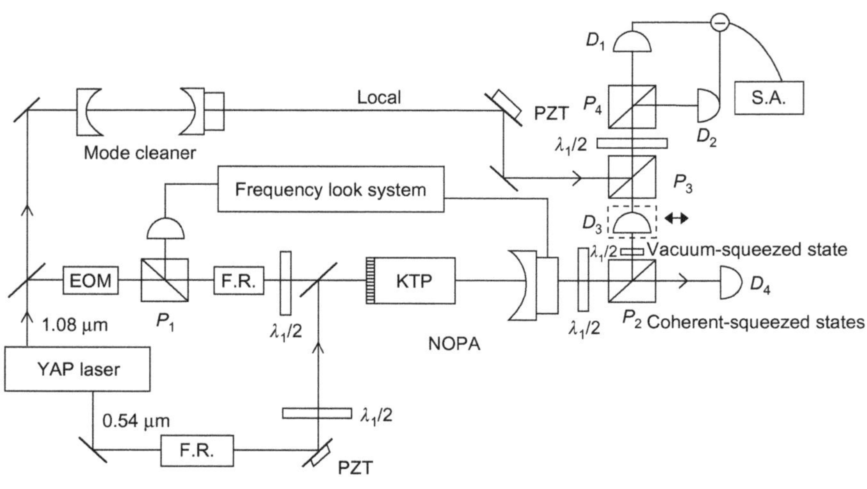

Figure 1.29 The experimental setup.

way. The NP-C problem is also referred to as the satisfiability or SAT. In this problem, one is given a Boolean function $f\{0, 1\}$ in basis 2 and dimension 2^{n+1}. This can be expressed as a logical computation of clauses where each clause is a disjunction of Boolean variables or their negations. The variables are assigned values from the set $\{0, 1\}$ and such an assignment is said to satisfy f, if the value of f on the assignment is 1. The problem then is to determine if there exists a satisfying assignment for the function f. The state w of the system is given by $|w\rangle$ versus a fixed basis.

It represents some observables K. Represent each assignment, $x \Xi \{0, 1\}^n$, together with the state of the $(n + 1)$th output bit, $y \Xi \{0, 1\}^n$, by an elementary basis vector $|x, y\rangle$. This is done in a Hilbert space of dimension 2^{n+1}. It is possible to convert the Boolean operation f into a sequence of unitary transformations U_f. When applied to any single-basis state $|x, 0\rangle$, it yields $U_f\ |x, 0\rangle = |x, f(x)\rangle$. If one prepares the input in the superposition of all of its possible states given by $\Sigma x \Xi \{0, 1\}^n C_{(x,0)}|x, 0\rangle$, where $|x, 0\rangle = 2^{-n/2}$ and $C(x, 1) = 0$, for all x, then the result of performing computation on this superposition is $U_f\ \Sigma x \Xi \{0, 1\}^n C_{(x,0)}|x, 0\rangle = \Sigma x \Xi \{0, 1\}^n C_{(x,0)}|x, f(x) \equiv \Sigma x \Xi \{0, 1\}^n d(x,f(x)|x, f(x)\rangle \equiv Z_f$. There exists an observable S with the Hermitian matrix S such that the expectation value of S is $\langle Z_f|S|Z_f\rangle = 2^{-n}|\{x \Xi \{0, 1\}^n|f(x) = 1\}$, i.e., 2^{-n} times the number of satisfying assignments. If the basis states $|x, y\rangle$ inherit the order of the integer for each bit string (x, y) in the binary representation, then S is the diagonal matrix $S = 1 \otimes 1 \otimes \cdots \otimes 1 \otimes |1><1| = \mathrm{Diag}(0, 1, \ldots, 0, 1)$, where 1 is the 2×2 identity matrix and Diag is the diagonal matrix of its arguments. Here $\langle Z_f|S|Z_f\rangle = \Sigma x \Xi \{0, 1\}^n\ d(x, 1)|^2$ and each nonzero output value of S is the function of all assignments that satisfy the given function. The time required to program U_f to prepare the superposition and to measure the expectation value of S to the required precision of 2^{-n} are all $O(n)$. This shows that an EVQC can solve SAT in linear time. All NP-Cs are polynomially reducible to SAT. It proves that an EVQC can solve NP-C problem in polynomial time.

1.3.7.2 A Biomolecular DNA QC: Aldeman–Lipton Model [51, 58, 59]

Biomolecular computers work at the molecular level. Biological and mathematical operations have some similarities. DNA, the genetic material that encodes for living organisms, is stable and predictable in its reactions. Therefore, DNA can be used to encode information for mathematical systems. After all, it is the major information storage molecule in living cells. Billions of years of evolution have tested and refined this wonderful informational molecule. The special enzymes in which DNA resides can either duplicate the information in DNA molecules or transmit this information to other DNA molecules.

A DNA is a polymer strung together from monomers called *deoxyribonucleotides*. Distinct nucleotides are detected only with their bases. Those bases are respectively abbreviated as adenine (A), guanine (G), cytosine (C), and thymine (T). The two strands of a DNA can, under appropriate conditions, form a double strand if the respective bases are the Watson–Crick complements of each other. This means A matches T and C matches G. Also, 3′-end matches 5′-end.

The single strands 5′-ACCTGGATGTAA-3′ and 3′-TGGACCTACATT-5′ can form a double strand. One can call the strand 3′-TGGACCTACATT-5′ as the complementary strand of 5′-ACCTGGATGTAA-3′ and simply denote 3′-TGGACCTA CATT-5′ by ACCTGGATGTAA. The length of a single-stranded DNA is the number of nucleotides comprising the single strand. If a single-stranded DNA includes 20 nucleotides, it is called a 20 mer. The length of a double-stranded DNA (where each nucleotide is base paired) is counted by the number of the pairs. If one makes a double-stranded DNA from a single-stranded 20 mer, the length of the double-stranded DNA is 20 base pairs; this is written as 20 bp.

The Aldeman–Lipton (A–L) model is as follows. A (test) tube is a set of molecules of DNA. It is a multi-set of finite strings over the alphabet {A, C, G, T}. In a typical DNA computer, the input strands are separate from the operator strands. The massive parallelism that DNA computing potentially offers may make it competitive with electronic computing. It is likely, however, that situations will arise in which one would like the ability to carry out computations with DNA for which competitiveness with ECs is not an issue. For example, the implementation of logic functions via DNA may aid in the assembly of complex structures from DNA. Strands of DNA capable of making logical decisions could conceivably also serve as components of smart drug delivery systems or gene-manipulation systems. In the model illustrated here, one uses a way of carrying out Boolean logical functions through hybridization and ligation. These are the two basic DNA chemical operations. Operation of the computation machinery consists of the operator strands separated from the input strands. This would be essential for applications in self-assembly and smart drug delivery. This environment provides the inputs on which decisions are to be based (Figure 1.30).

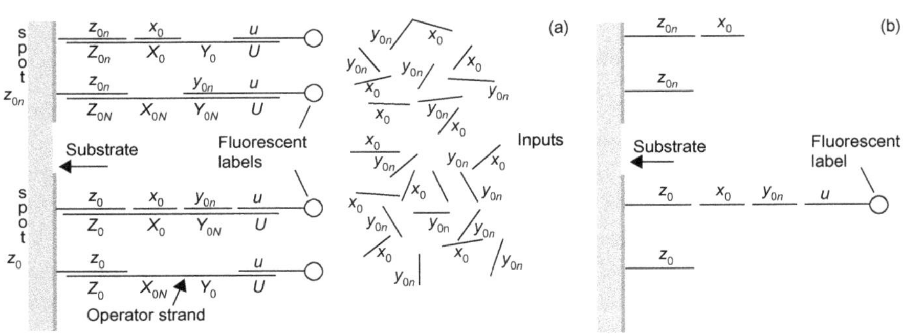

Figure 1.30 Illustration of how the mathematical (logical) operations can be implemented in DNA: (a) illustrates the operator strands, input strands, and fluorescent dye-labeled strands assembled through hybridization; (b) illustrates the DNA that remains attached to the hybridization plate after ligation and washing. For the case illustrated, spot z_0 will fluoresce, whereas z_{0n} will not.

1.4 Summary and Conclusion

In the sections that follow, some Appendix Figures, do not have their full text description. The descriptive material below each figure, provides preliminary visual impression of the matter. The reader should see the reference quoted, for details.

1.4.1 NMR and DNA QC (Appendices A1.1–A1.4)

1.4.1.1 LSNMR QC

1.4.1.1.1 Homonuclear Example: The Basic CNOT Gate Operation [5, 61–63]

The basic CNOT gate operation is emphasized in this conclusion because it is simple to follow. It reinforces the reader's ability to apply the basics of QM to any type of QC, including a solid-state (SS) QC. It provides a summary of the basic concepts involved. This should help to enhance the reader's comprehension of the basics of a QC, especially for a beginner in the field. At present, the field of quantum computing is in a state of disarray, such that any idea about QM principles, whether incorporating existing or new knowledge, could be a savior in many respects. In the completion of any QC, one requires preparation, manipulation, coherent evolution, and measurement of the pure quantum states. In an ensemble of molecules prepared for an LSNMR QC, there is a statistical mixture of pure states. These are the states of the nuclear spins in thermal equilibrium at room temperature. Destructive interference between the states (which is very likely) would eliminate the coherence needed for the quantum algorithms (QAs) employed in the computer to progress. The LSNMR is the simplest experiment where one can actually observe the progress of a quantum mechanical system, starting from the lowest dimension of quantum computing. In the example chosen here, the ensemble of the system is a homonuclear 2-spin (same kind of nuclei in a molecule) molecular system (Figure 1.31 [5]).

The molecule involved is (2, 3)-dibromothiophene, as shown in Figure 1.31. In a strong DC magnetic field ($B_z \sim 5\,\mathrm{T}$), the precession frequencies of the two hydrogen nuclei in the molecule are both approximately 200 MHz. These precession frequencies of the two hydrogen nuclei are vastly different from those due to the chemical shift ($\sim$kHz). The chemical shifts arise from the unique chemical environments of the nuclei. Single-spin rotations may be accomplished by applying RF pulses. The RF pulses are resonant at one of the proton (hydrogen) frequencies. They cause the addressed spin to rotate around the transverse axis (B_x, B_y) while still processing around B_z. The spin on one molecule may also interact with the spin of the other molecule through the dipolar or electron-mediated interactions. These nonlinearities are used to accomplish the required logic operations. The ensemble of these molecules at room temperature can be represented by its density matrix $\rho = I(1/2^N) + \rho_\Delta$. The first term describes the equilibrium (thermal) part of the ensemble.

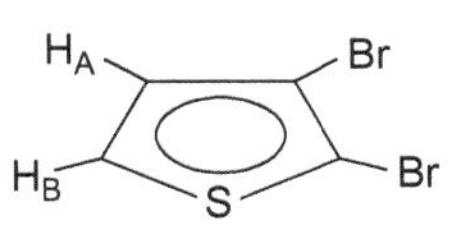

Figure 1.31 (2, 3)-Dibromothiophene.

It is proportional to the identity matrix I. The second part is the deviation (from equilibrium) density matrix. Under the action of a unitary transformation, e.g., the free evolution of the imposed pulse sequences, the identity part of the density matrix will not change. The dynamics of the ensemble of the molecules, if they are noninteracting, can be approximated by analyzing the deviation density matrix. This applies for the macroscopic deviation from the identity. The response of the system can be quantified through the quantum evolution equations. The macroscopic signal observed reflects quantum dynamics within a decoherence timescale. This timescale allows the use of thousands of RF pulses. In this particular example, chosen for the sake of illustration, the deviation density matrix can be represented as

$$\rho_\Delta = \alpha \times \begin{pmatrix} 3 & 0 & 0 & 0 \\ 0 & -1 & 0 & 0 \\ 0 & 0 & -1 & 0 \\ 0 & 0 & 0 & -1 \end{pmatrix}$$

Here $\alpha = (h/2\pi)\omega/(2k_BT)$, where h is Planck's constant, ω the frequency of the RF pulse (corresponding to a particular energy transition, between 2-spin states), and T the temperature of the ensemble.

This representation means that of the 10^{23} (Avogadro number) molecules, $(1/4 + 3\alpha)10^{23}$ are in the $|\downarrow\downarrow\rangle$ (spin-down) state and $(1/4 - \alpha)10^{23}$ are in each of the $|\downarrow\uparrow\rangle$, $|\uparrow\downarrow\rangle$, and $|\uparrow\uparrow\rangle$ states. The system is a statistical mixture of the four possible states. The excess (or deficient) single-state spin population among a uniformly populated background of levels behaves like a pure state. Suppose now that the chemical shifts of the N spins are small compared to their average frequency ω. This means that $|\omega_i - \omega_j| \ll \omega$. In that case, for $N = 2$ (spins) one gets

$$\rho_\Delta = (2\alpha/4) \times \begin{pmatrix} 1 & 0 & 0 & 0 \\ 0 & 0 & 0 & 0 \\ 0 & 0 & 0 & 0 \\ 0 & 0 & 0 & -1 \end{pmatrix}$$

(see Ref. [5] for details). One can take the second spin system as a label. It tells us, by comparison with the first one, which of the two pure states $|\uparrow\rangle$ or $|\downarrow\rangle$ the first spin system was prepared in. Mathematically, this is represented by the block-diagonal structure of the deviation density matrix shown above. One takes the status of the first spin as a pure state, conditioned on the result of measurement of the second spin. The experimental arrangement now acts so as to have unitary transforms (RF absorption and emissions) occur for the first spin alone and nothing to happen to the second spin. Now represent U as the unitary transform effected by a series of pulse sequences to the spin system. It may be a 1-bit operation carried out by a QC. The meaning of the result will be that the output is either $U|\downarrow\rangle\langle\downarrow|U$ or $U|\uparrow\rangle\langle\uparrow|U$. These two states are distinguished by the state of the second spin. Experimentally, this difference is determined in the final readout of the state.

Thus, one distills $k = 1$ qbits from the $N = 2$ thermal spins. The concept is easily generalized to $N > 2$ spins. In a nutshell, in this technique, one first identifies equally populated states that naturally exist in a thermal ensemble. Then one performs the unitary transforms to group together those states to form a uniform background. Against this background, one can then extract a differently populated single pure state.

One can label this group using some other group of spins. In this manner one creates a state in which a few of the spins are in pure states, conditioned on the state of the others. Thus, one can distill a 2-qbit pure state from 4 spins, a 3-qbit spin state from 6 spins, and so on. The next step is to accomplish a logical state readout. A general quantum computation produces as output a pure state $|\psi\rangle = U_{QC}|\psi_0\rangle$. The result is then found from measurement of the probabilities $|\langle\psi|\psi\rangle|^2 d\psi$. In an LSNMR QC, only ensemble averages are accessible for measurement, and only certain spin states contribute to an observable.

In a conventional NMR experiment, what is experimentally detected is the net transverse magnetization, approximately $n(\mu_x + i\mu_y) \sim n\gamma(h/2\pi)\mathrm{Tr}[\rho_\Delta(I_x + I_y)]$. Here n = the total number of spins contributing μ_x and μ_y, the x- and y-components of the nuclear magnetic moments; γ = the gyromagnetic ratio; h = Planck's constant; and Tr is the trace of the deviation density matrix having components I_x and I_y. The signal in the pick-up coil gives the coefficients of the I_x and I_y terms. An arbitrary computation can be designed to move the result to these terms. Measurement of the free induction decay can identify the coherence terms that contain the result. The net measurement of the readout is done by measuring the precession frequency (its amplitude and phase). Given the ability to perform arbitrary single-bit operations, the next element required for quantum logic gates is the nonlinear interaction between spins such as the CNOT gate operation.

This gate conditionally flips 1 spin based on the value of the other. The interaction terms in the Hamiltonian, $H \sim -\mu$ (magnetic moment), $B \sim -\gamma(h/2\pi)(B_z I_Z)$ provide the desired nonlinearity.

For the 2-spin case with scalar coupling, a CNOT can be implemented. This is done through a controlled phase shift preceded and followed by rotations given by the sequence $C_{AB} = R_{YA}(-90)R_{ZB}(-90)\ R_{ZA}(270 = -90)R_{ZB}(180)\ R_{YA}(90)$, as shown below. One is faced with the problem that the usual output from an LSNMR QC is nondeterministic, as it averages out in an ensemble. This problem can be addressed with Shor's quantum factoring algorithm, Appendix A1.3. One can optimize memory size and speed of the computation by implementing the SFA. The core of the SFA is the modular exponentiation function (MEF) $= x^a (\mathrm{mod}\, N)$, where N = the total number of spins participating in the system and a = the number of optimum qbits required for the QC. This is a probabilistic algorithm that enables a QC to find a nontrivial factor of a large composite number N. This is done in a time-bounded frame from above by a polynomial in $\log(N)$. A substantial squeezing of the needed memory space can be achieved without sacrificing much in speed. A QC capable of storing $5K + 1$ qbits can run the SFA to factor a K-bit number N, in a timescale of the order $O(K^3)$. The SFA fails if N is even or a prime power $N = p^\alpha$, p prime. The smallest complete integer N that can be successfully factored using the SFA method is $N = 15$ (Figure 1.32 [5]).

For any positive integer x, with $x < 15$ and the greatest common divider (gcd) of $(x, 15) = 1$, i.e., for $x = 1, 2, 3, 4, 7, 8, 11, 13, 14$, one has $x^4\ (\mathrm{mod}\, 15) = 1$.

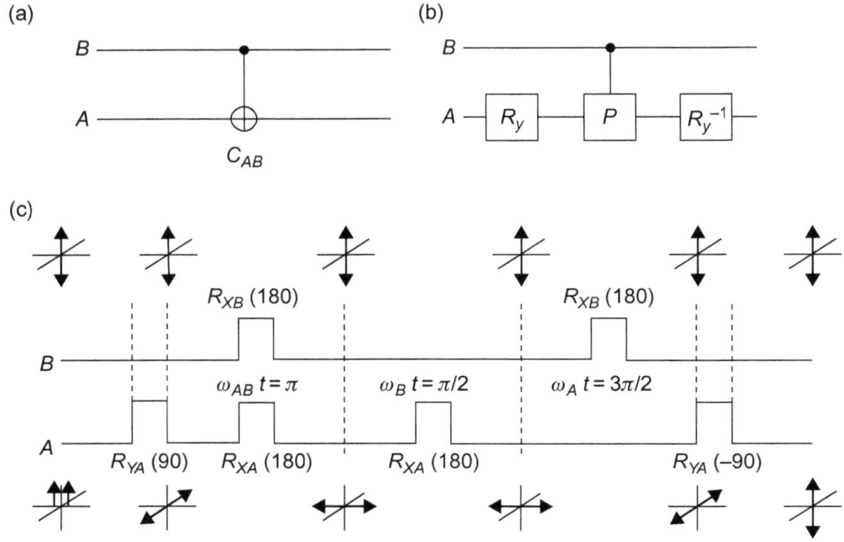

Figure 1.32 (a) A CNOT gate acting on 2 qbits; (b) the CNOT gates implemented by a controlled phase-shift gate (specified by a unitary matrix with diagonal elements $\{1, 1, 1, -1\}$), preceded and followed by $\pi/2$ rotations; (c) the pulse sequence corresponding to the components in (b).

Therefore, considering $x^a = x_1^{2a} x_0^{a}$, only the first 2 bits of a are relevant in the computation of x^a. The state of a QC can be expressed by the QFT, $(\text{MEF})_{\text{QFT}} = (1/2^{L/2})\Sigma_a |a\rangle_\text{i} |x^a \pmod N\rangle_\text{o}\ |g(a)\rangle$. The i and o represent the input and the output states, L the length of the qbit number, and $|g(a)\rangle$ the garbage stored in the scratch-pad space. If one were to perform the QFT on $|a\rangle I$, one would be probing the periodicity properties of the MEF$\rightarrow x^a (\text{mod}\, N) \otimes g(a)$. This results from the unitary transform performed on the input state i, i.e., U: $|a\rangle_\text{i} |0\rangle_\text{o} \rightarrow |a\rangle_\text{i} |x^a \pmod N\rangle_\text{o}$. This would be quite different from the periodicity properties of $x^a(\text{mod}\, N)$ in which one is really interested. The garbage in the scratch pad must be erased.

This erasure has to be a reversible operation, i.e., a unitary operation, so as not to destroy the coherence of the computation. One can run the calculation to completion by producing the MEF. Copy the result from the output register to another ancillary register, and then run the computation backward to erase both the output register and the scratch pad. For $N = 15$, the SFA is the best way that actually computes x^a on a QC.

One can do even better if one is willing to allow the CC to perform the calculation of x^a. Obviously, this strategy will fail dismally for large values of K: the classical calculation will require exponential time. Still, if one's goal is merely to construct the entangled state, $(1/2^L)\Sigma_a|a\rangle_\text{i}|x^a(\text{mod}\, N)\rangle_\text{o}$, while using QC resources as sparingly as possible, then the classical computation of x^a is the most efficient procedure for small K. So, imagine that $x < 15$, and the gcd of $(x, 15) = 1$ is randomly chosen.

The CC then generates a lookup table by computing the 4-bit number $x^a(\mathrm{mod}\,15)$, for $a = 0, 1, 2, 3$. The CC then instructs the QC to execute a sequence of operations that prepares the above state. These operations require no scratch space at all, so only $L + K = 6$ qbits of storage are needed to prepare the entangled state. The worst case (most complex lookup table) is for $x = 7$ or 13.

1.4.1.1.2 Heteronuclear Example: SFA [56]

This illustration is included as a summary to the chapter, to apprise one as to how the complexity of the system builds up as one increases the number of nuclei (including nuclei of different kinds) participating in the system. In this example, an ensemble of bromoacetate molecules forms an LSNMR QC. The QC has five ^{19}F and two ^{13}C nuclear spins participating in the computations. The thermal equilibrium is represented by the density matrix (see Section 1.3.1) $\rho_{\mathrm{th}} = \exp(-H_0/k_BT/2^7)$. It is converted into a 7-spin effective pure state via temporal averaging. This constitutes a suitable state for the SFA. The simplest meaningful instance of the SFA is the factorizing of $N = 15$. A standard classical circuit technique used in an EC is based on the following mathematical identity:

$$a^x = a^{\{x_k\}} = 2^{\{(n-1)x_{n-1}\}} \ldots 2^{\{4x_2\}} 2^{\{2x_1\}} 2^{\{x_0\}}$$

Here x_k is the binary digit of x, where $k = n - 1, n - 2, \ldots 0$ and n is the qbit number. One uses the MEF consisting of several multiplications of a^x where $x = 2^k \mathrm{mod}\, N$ for all k, $0 = \leq k \leq (n - 1)$. The powers of a (i.e., 2^k) can be efficiently precomputed on an electronic CC by repeated squaring of a. Among the numbers $a = 2, 7, 8$, or 13, one finds for $a = 7$, $15/7^4 = 15/2401 = 1/160.06667 = 1/(160 + 1/15)$. So here, $x = 2^{\{(n-1)x_{n-1}\}} = 4 = 2^2$ and $n = 2$ and $x_k = (n - 1)x_{n-1} = x_1, x_0$. Thus, the first register can be as small as 2 qbits. However, 3 bits would allow more periods to be detected. Also, with $N = 15$, $\log_2[15]$, i.e., 4 qbits to hold $f(x)$, one needs 7 bits total (Figure 1.23 [56]). In contrast, among $a = 4$, 11, or 14, $a^2 \mathrm{mod}\, 15 = 1$ for $a = 11$ demonstrates $15/121 = 1/(8 + 1/15)$. Then $n = 1$ and $x_k = (n - 1)x_{n-1} = x_0$. So, only x_0 is valid. The quantum circuit (Figure 1.22 [56]) for the case of LSNMR was realized, for the case of $a = 7$, with spin-selective RF pulses separated by time intervals of free evolution. The pulse sequence is designed such that the resulting transformation of the spin states corresponds to the computational states in the algorithm. Upon completion of this sequence, one estimates the state of the first 3 qbits $\rho \sim \Sigma_c W_c |c2^3/r\rangle\langle c2^3/r|$ via NMR spectroscopy. The periodicity P in the amplitude of the output was thus obtained. Then r can be worked out from $r = 2^n/P$. In the cases $a = 7$ and 11, the experiments confirmed RF pulse spectra corresponding to the factorization $15 = 5 \times 3$ [56].

1.4.1.2 The DNA QC (Appendix A1.4) [58–60]

It has been demonstrated that the A–L model can be used successfully to determine the NP-C (SAT 0) problem in a DNA QC. Recently, lots of bio-algorithms and their experimental applications for a biology-based solution to the problem have appeared in the scientific literature. The customary approach in DNA quantum computing is to analyze the problem in reference to a graph, $G = (V, E)$. The graph is about finding a subset C of the vertices that maximizes the number of edges that have one vertex in C, and one not in C (Appendix A1.4). If C is a solution to the maximum cut problem, then so is V/C. The graph G defines such a problem. It is easy to see that $C = \{A_6, A_3\}$. Equivalently, $C = \{A5, A4, A2, A1\}$ is a solution to the NP-C problem. DNA, the genetic material that encodes for living organisms, is stable and its reactions are predictable, so it can be used to encode information for mathematical systems. A DNA is a polymer strung together from monomers called deoxyribonucleotides (A, G, C, T). The two strands of DNA can form a double strand if the respective bases are the Watson–Crick (W–C) complements of each other. One can call 3′-TGGACCTACATT-5′ the complementary strand of 5′-ACCTGGATGTAA-3′. Then, mathematically, 3′-TGGACCTACATT-5′ $\equiv \underline{\text{ACCTGGATGTAA}}$. In the A–L model, a test tube contains a set of molecules of DNA. It is a multi-set of finite strings over the alphabet {A, G, C, T}. Given such a tube, one can perform a set of operations. This set can be implemented with a constant number of biological steps for the DNA strands. The complexity of each manipulation is O(1) steps.

It has been determined that the solutions of the maximum cut problem can be figured out in $O(n^2)$ steps using DNA molecules. The solution of the NP-C problem needs polynomial time in an EC. In a DNA QC, it is done in a polynomial number of steps in space. However, a DNA QC has two major advantages: parallelism and complementarity. The A–L model has been extended to solve problems such as the set-cover problem and the subset-sum problem. Investigation of the application of DNA computing to data analysis and machine-learning tasks has recently also gained some momentum. In a DNA QC, the technique is to code information in the so-called *DNA records*. The presentation of analysis is in the form of tables using attributes, and data have been developed therefrom. The studies carried out demonstrate five fundamental relational algebra operations, i.e., the selection, projection, union, set difference, and Cartesian product, on such a database. These five basic operations perform most of the data-retrieval operations on current databases.

Other operations exist; e.g., join, intersection, and division operations. These can be expressed in terms of the above five basic operations. The algorithms developed in the field basically are generic solutions to the five basic operations. One can express the specification of the five fundamental relational algebra operations on the DNA strands. However, the assumptions made in these studies are not very realistic. It is expected that all generic manipulations will run error free. From a mere programming perspective, the proposed algorithms present only ad hoc solutions. It seems, though, that it may be possible to solve the error problems. The programming structure of some of the algorithms is not efficient. It can be improved by the use of recursive function calls. There are also fundamental practical problems. For example, some of the algorithms perform exhaustive searches. From a genetic engineering point of view, the problem currently has no practical solution. Performance and scalability, including DNA databases with many labels, DNA records, and attributes, are other problems of serious concern.

1.4.1.3 Human Brain Information Processing (Appendices A1.7 and A1.8) [65, 66]

1.4.1.3.1 Quantum Model of the Brain

A quantum model of the brain associates consciousness with coherent superposition of the quantum states in the brain. One conjectures that the collective quantum phenomena produce coherent states in the brain. Nerve cells are responsible for much of the communication within the human body. The structure of nerve cells consists of an array of dendrites, which gather input from other neurons; a cell body; and an axon, possibly branched, along which nerve impulses are transmitted to other cells. The axon is filled with microtubules (MTs) and neurofilaments, which are generally known as *cytoskeleton*. These filament proteins are arranged parallel to the axon. Each neuronal MT is typically about 100-μm (10^{-6}-m) long and spans more than 10^5 tubulin subunits. The network of cytoskeleton tubes is interconnected by high-molecular-weight proteins known as MT-associated proteins (MAPs). Information processing and energy transport have been proposed as secondary MT functions. It is known that tubulins are assembled from guanosine triphosphate (GTP)-rich tubulin dimers and that this GTP is hydrolyzed rapidly after addition of the tubulin subunit. Some of the energy released in the process may be stored in the lattice through a conformational change of the protein dimer. It is supposed that it propagates along MT through a sequence of dipole flips as the lattice reorients to accommodate the additional energy. These conformational changes or flips are believed to be the result of a mobile electron. The electron may be localized at one of the two binding states in the tubulin molecule. Movement of the electron from one binding state to the other causes the tubulin dimer and its electric dipole to orient. These two states may be identified by the location of the free electron.

This may be an oversimplified picture of a more intricate, and as yet unknown, information-processing capability of neuronal MTs. This model of the MT cellular automata is based on the belief that cytoskeleton behaves as a cellular nervous system. The automatas' behavior within MTs could explain their capacity for intelligent organization.

1.4.1.3.2 Dipole Electric Field: Brain MTs (Appendices A1.9 and A1.10) [67]

MTs are cylinders whose exterior surface (diameter ~ nm) has 13 arrays of protein dimers (tubulins). The interior of the cylinders contain ordered water molecules. The MTs have an electric dipole moment and a resultant electric field. They represent a dipole due to individual dipolar charges of each tubulin monomer. The MT polarity is closely connected with its functional behavior, which can be regulated by phosphorylation and dephosphorylation of MAP. GTP molecules are bound to tubulins in the heterodymer. After polymerization, when the heterodymer is attached to the MT, the GTP bound to the β-tubulin is hydrolyzed to guanosine diphosphate (GDP).

In contrast, the GTP molecule of the α-tubulin is not hydrolyzed. The MTs present a calm dynamic instability. They are the communication channel of information and exhibit a spin-glass (S-G) phase. The cylinders have randomly oriented small surface regions with a resultant electric field. The electric field of each MT influences the environment around it and the fields of other MTs. This influence extends over distances of about a micron (10^{-6} m) and includes electric field pulses in the frequency range of THz.

This information propagation happens most efficiently around the normal human body temperature of 300 K. In this molecular model of human brain information processing, it is found that the maximum information storage is obtained in the S-G phase. In this phase, there are domains with many energy levels, which can store information. The interaction among the domains is due to the electric field generated by the oscillating dipoles. The electric field is emitted to the neighboring area, producing many channels among the domains in the MT. Information is stored in the MT walls and propagates along the axial direction of the MT.

1.4.2 Solid-State QC

1.4.2.1 Single Electron Charge Spin QC/Quantum Dot Cellular Automata (Appendices A1.5 and A1.6) [15]

The developments in semiconductor technology that have taken place recently (e.g., nano-fabrication and progress in the field of nano-electronics) are worthy of commendation, but these advances are no comfort to those involved in the development of a single electron charge spin quantum computer (SEC/S QC). The readout of a scalable single-electron-based QC relies on the ability to secure the state of a single electron. The electron charge qbit development based on quantum dot cellular automata (QDCA) looks, from a distance, like an attractive proposition, but whether it will deliver the goods remains to be seen. The concept of using charge or spin as a basis for the development of a QC has so far proven to be only an exercise in theoretical fantasy. Manipulation of electron spin or charge in the solid state to create even a single bit has proved to be a formidable, if not an impossible, task.

Therefore, there has been a shift in this quarter. It is now felt that it may be more fruitful to use nuclear spins in the solid state, instead of electron spin or charge. This is because a nucleus lives deep inside a single atom, relatively free from the broader noisy environment of the lattice. The sample QDCA should illustrate the problems facing an solid-state QC in designing an all-electronic QC. QDCA cells can be constructed using GaAs/AlGaAs heterostructures. One can also use a QDCA architecture, employing buried dopants in semiconductors.

The structure provides a very strong confined potential for the electrons. The energy levels are well characterized. To utilize the well-developed Si technology, there are proposals to use a charge-based qbit for the QC using P donors in Si. Conventionally, QDCA relies on incoherent evolution, governed by T_1, the relaxation time explained below, to mediate transitions between the logical states. For P donors in Si, this relaxation time is on the order of approximately milliseconds. In contrast, an alternative ultra-fast (picosecond) switching mechanism (coherent evolution between defined system eigenstates) has also been found feasible, although so far only theoretically [15]. The simplest QDCA is a cell composed of four QDs containing two mobile electrons that can move between the dots via tunnel junctions. The electrons tend to occupy diagonally opposite sites to minimize the energy due to coulomb interactions. These two ground computational states are labeled 0 and 1 (Appendices A1.5 and A1.6). The "e" indicates the position of the electrons. The next highest energetic states are the noncomputational states and ideally are only transiently populated during correct operation. If the two dots are placed next to each other, one cell influences the state of the other cell via capacitative coupling. In an array of cells, when the first cell is switched from one computational state to the other, the rest of the chain relaxes to minimize the energy of the total system. The result of this relaxation is to transfer the state information of the initial cell along the chain without net electron flow and with minimum energy dissipation. The speed at which this switching occurs is governed by the incoherent tunneling state of the junctions. The inverse of this is referred to as the T_1 relaxation time. One can use buried P donors in Si and call this system BDCA. The basic idea is to construct an array of four ionized donors that contain two free electrons, therefore mimicking the layout of a conventional QDCA. This also represents a limit in terms of miniaturization for this form of nano-computing, as each potential well is created by only one donor atom. To provide a convenient formalism, one constructs an effective Hamiltonian using the pseudo-spin approach to describe the BDCA system. By defining each pair of P donors and their shared electrons as a single pseudo-spin object, one can specify two states—top (T) and bottom (B)—which localize the position of the electron. Each BDCA cell then consists of a pair of these objects, where the computation states are $|TB\rangle = |0\rangle$ and $|BT\rangle = |1\rangle$.

1.4.2.2 QDs: Terahertz Electrodynamic Cavity QC [26]

In the solid state, the qbits must be well isolated from the decoherence influence of the environment. They must also be manipulated individually to initialize the computer. Subsequently the qbits have to perform quantum logic operations and measure the result of the computation. In semiconductors, present technology allows only the nearest-neighbor qbits to be coupled. A detailed knowledge of the neighborhood decoherence is thus required in multiscaling the qbits. It is felt that a solid-state computer, if it can be realized, will be the only one to produce a QC with a large number—approximately 10^3—of qbits operating in parallel. The technological challenge in designing QGs capable of manipulating an arbitrary number of qbits is immense. In the GaAs/AlGaAs/GaAs/AlGaAs/GaAs scheme, the lowest electronic states of the QDs are coupled by THz cavity photons. The cavity is a metal enclosure that houses the qbits of the QDs. Recent developments in the area of THz spectroscopy provide the impetus for progress in this direction [64]. The photons in the cavity act as a data bus which can couple an arbitrary pair of QDs. A sequence of adiabatic voltage pulses is applied to the individual QDs. This can effect a CNOT operation involving any 2 qbits in the computer. However, there is a need to develop new types of QDs gated and loaded with single electrons. Limited-mode THz cavities with extremely high-quality factor (Q) have yet to be developed. Also, detectors for single THz photons have yet to be fabricated and tested [64]. In one proposed scheme ([25]), one uses intense pulses of THz radiation to induce coherent damped Rabi oscillations in an *n*-GaAs semiconductor matrix. The quantum-confined electron (donor atom) is coherently manipulated as in an atom-electron even while sharing space with a large number of atoms ($\sim 10^5$ atoms). The two low-lying states of the donor impurities are used as the qbits. The spins of the electron are manipulated by the application of a magnetic field (the angular momentum projection upon the field) and the strong electric field of the THz radiation. The THz field is acquired from a FEL. The readout of the qbits is performed while the electrons are in the ionization band of the solid. This is done through measurement of the photocurrent produced within $a \sim ps$. The strong electric fields of the THz radiation enable coherent rotation of fictitious spins through angles $>2\pi$. Such arbitrary 1-bit rotations are the basic requirements for QMIP. The magnetic field is set corresponding to the online resonance condition, which is tuned to the highest THz field. One to two cycles of Rabi oscillations (corresponding to the dipole strength of 1s-$2p^+$ transition) can be generated in this technique.

1.4.2.3 Optoelectronic Spin Quantum State [16]

One can generate coherent oscillations among spin sublevels (magnetic) of the ground state in diamond nano-crystals at room temperature. The coherence time is long enough ($\sim\mu s$) to allow coherent manipulation of a single electron spin. Hyperfine interactions of the electron spin with the ^{14}N nuclear spin in NVDC-D present abundant degrees of freedom for the design of an optoelectronic spin quantum state (OSS) QC. This scheme offers easier laser-microwave field manipulations and a promising processing technology. The basics of the incomplete description of the QM states, as emphasized by the EPR paradox, has evolved into optical fields being used as an important direction of approach. A typical example is the generation of continuous-variable EPR (CV-EPR) entanglement using an optical fiber interferometer.

1.4.3 Optical Quantum Computer

Quantum effects are particularly easy to observe in optical systems. Efficient quantum computation is possible using only BSs, phase shifters, single photon sources, photodetectors, and so on. One exploits the feedback from the photodetectors, which are robust against errors from photon loss and detector efficiency. The basic elements are accessible to an experimental investigation with current technology. Optical (Bosonic) qbits are defined by the states of the optical modes. An *optical mode* is a physical system whose state space consists of a superposition of the number states $|n\rangle$, where $n = 0, 1, 2, \ldots$ gives the number of photons in the matrix. In addition to the instances of an ideal quantum system, a complete implementation of a QC requires a means for state preparation. It also requires the ability to apply sufficiently powerful QGs and readout methods. To process information, these elements are combined in quantum networks. The initial state is the vacuum state $|0\rangle$ in which there are no photons in any of the modes. The basic element that adds photons to the initial state may be a single photon source.

It can be used to make the state of any given mode; e.g., the one-photon state $|1\rangle$. It is sufficient to be able to prepare this state nondeterministically. This means that the state preparation has a nonzero probability of success, and one can determine whether or not it succeeded. Efforts are also in progress in other areas of solid-state quantum-computing possibilities, such as the electron charge, spin, atom/ion trap, photon, and so on.

1.4.3.1 Kerr Cell Nonlinearity EPR Entanglement of States [37]

In this technique, the Kerr cell nonlinearity in optical fiber is exploited for the generation of two independent, optically squeezed beams. The interference is created at a BS and EPR entanglement is obtained between the output beams. The quantum correlations in the two noncommuting observables, amplitude and phase, are characterized by the sum and difference squeezing of the entangled beams. The corresponding variances of the sum and difference photocurrents are the measured quantities.

1.4.3.2 Hydrogen Impurity Atom-Like Polarization States of an Electron in n-GaAs [25]

It is known that decoherence of the qbits in semiconductors is a serious problem, but it must be overcome to enable the design of an solid-state QC. Even a single-qbit operation of an electron is proving to be extremely difficult. Hydrogen atom-like excitation states of electrons bound to donor impurities in a solid-state environment in a semiconductor can serve as model qbits. A good technique would be to use intense pulses of THz radiation to induce coherent damped oscillations in the population of the two low-lying states. These states are the states of the donor impurities in GaAs. In this scheme, quantum-confined extrinsic electrons in GaAs can be coherently manipulated. In this solid-state environment, quantum information can be coded on light by using two orthogonal optical modes. The polarization states can be used to encode two discrete states as well as the continuous-variable states. A key feature of the polarization is its ability to analyze in any basis by using a series of half and quarter plates (or a rotatable Babinet compensator), a polarizing beam splitter (PBS), and intensity measurements. Analyzing a complete set of basis—say, linear, diagonal, and circular—allows the amplitude and the relative phases of the polarization modes to be determined.

1.4.3.3 Entangled Single-Mode Quantum Optical Sidebands [18, 21]

In this scheme, a pair of optical sideband modes are separated from an average or carrier frequencies by a radio or microwave frequency. Much like polarization analysis, full characterization in this encoding would reveal the amplitude and relative phases of the sidebands. However, unless there is equal power in the two frequency modes, the homodyne detection is insufficient to fully characterize the system. The RF analyzer is directly analogous to an polarization system. It comprises a rotatable Babinet compensator constructed using an Nd:YAG laser operating at 1064 nm. Phase-modulation (PM) sidebands were imposed on the laser beam at 90.5 MHz with an electro-optic modulator. A relatively strong depth of the original laser power was transferred to each of the PM sidebands. Spectral measurements were performed using a scanning confocal Fabry–Perot cavity with a free spectral range of 500 MHz and a line width of approximately 2 MHz.

The optical spectrum analyzer (OSA) allowed unambiguous identification of the carrier frequency-shifted carrier and the sidebands of interest. Homodyne detection allows measurement of quadrature amplitude and phase fluctuations, respectively, in Fourier space. EPR entanglement has been recognized as a basic resource of any continuous-variable quantum information protocol. One can separate the quantum sidebands of a single spatial mode of a squeezed optical beam into two separate spatial beams. When entangled, these two beams would involve a pairwise correlation between only a single sideband on each of the two beams. The path-length difference between the two separated beams is such that the quantum sidebands at a particular RF are decomposed into separate spatial beams. This path-length difference is determined by the frequency at which sideband separation is required. One can thus separate the positive and negative sidebands of a phase-modulated optical field. Applying this to a single-mode squeezed light provides spatially separated entangled beams.

1.4.3.4 Linear Optics Quantum Computer [22]

It has been accepted that there have to be nonlinear couplings between optic modes of photons to achieve quantum correlations. Having such couplings at sufficient strength is possible, but is technically difficult. However, QMIP with linear optics is also possible. A qbit can be realized by one photon in two optical modes (such as horizontal or vertical polarization). A nondeterministic QC can be designed using linear optics, based on a nonlinear sign shift between 2 qbits. That system uses two additional photons and post selection. The sign shift has a certain probability of success, and one can determine whether or not it succeeded. The linear optics model has features not available to classical deterministic or probabilistic computation. The probability of success of the QGs can be increased arbitrarily close to 1.

The result is based on using entangled states prepared nondeterministically and quantum teleportation. Thus, quantum computation is possible in principle with linear optics, though the resources required are not easy to obtain. It is further seen that with quantum coding, the resources for obtaining accurate encoded qbits are very efficient. One can hope that the goal of LOQC can be achieved in the future if there are rapid achievements in the design of suitable optical components.

1.4.3.5 Quadrature-Phase Squeezing and EPR Correlation of Bright Light Field [20]

Interest in EPR beams is very keen because of the successful teleportation experiments in continuous quantum variables. In this technique, the entangled EPR beams are generated by combining two independent squeezed vacuum fields. These are produced using a subthreshold-degenerate optical parametric oscillator (DOPO) at a 50/50 BS. These instruments play a key role in transforming the quantum information. The bright EPR beams cannot be produced using a degenerate optical parametric amplifier (DOPA), because the degenerate signal and idler modes cannot be separated. Due to technical problems and inadequacies, the bright quadrature-phase squeezed light and bright EPR beams have been difficult to realize experimentally. The principal difficulty in these experiments is to control the frequency degeneration of the signal and idler modes with perpendicular polarization. Another problem is achieving double resonance of the two nondegenerate subharmonic modes in the cavity. One solves this problem by injecting frequency-degenerate seed waves into a nondegenerate parametric oscillator (NOPO) below its oscillation threshold; hence the name narrow-band NOPA. Further, one locks the cavity on the frequency of seed waves, thereby obtaining the brighter twin beams with rigorously degenerate frequency and orthogonal polarization. To ensure stable phase squeezing of the NOPA output, one actively locks the injected subharmonic signal and harmonic pump field in phase, to achieve maximum parametric amplification.

1.4.3.6 Quantum Interaction-Free Measurement

1.4.3.6.1 Optical Quantum Zero Scheme (Appendices A1.11 and A1.12) [68]

This illustration of an optical quantum zero scheme (QZS) for quantum interaction-free measurement (QIFM) exemplifies the application of two amplitude-enhanced states of a single photon to QMIP. These are the horizontal and vertical polarizations of a photon. The polarization states are divided into two different spatial paths *a* and *b* (Figure A1.11, Appendix A1.11). One can perform measurements on the amplitude-enhanced states at the detectors D_5(aH), D_2(aV), D_3(bH), and D_4(bV). The first (in position shown) 50/50 BS performs $\pi/2$ rotation. A single photon is split into 2-qbit, switched states, i.e., On and Off. The two On–Off states are the superposed states $|\text{Off}\rangle \sim |\text{Off}\rangle + |\text{On}\rangle$ and $|\text{On}\rangle \sim -|\text{Off}\rangle + |\text{On}\rangle$, respectively. The four quantum states of the photon—$|00\rangle$, $|01\rangle$, $|10\rangle$, and $|11\rangle$—can be superposed by the transform rotation R at the 50/50 BS. The probability of detection of aH in D_5 can be made close to 1. This is done by choosing the path lengths Off and On, such that there is a destructive interference at D_1 for aH. This results in certain detection at D_5 of aH. In fact, because the optical circuits are not composed of 100% perfect components, optical detection at D_1 (the quantum interrogation detector) is about 1/4 (1 out of the 4 possible states 00, 01, 10, 11), instead of being 0. Therefore, the probability of detection at D_5 is slightly less than 1. The probability of detecting aV-D_2, bH-D_3, and bV-D_4 would be close to 1/2. One can improve the probability of detection at D_5 by choosing 5/95 BSs instead of 50/50 BSs.

A further improvement becomes possible by using the following technique, called the *QZS*, which can function much more effectively. The On/Off switch on entrance in the optical circuit for the photon (Figure A1.11, Appendix A1.11) is successively rotated $R(\theta)$ in time by small steps, $\Delta\theta\pi/(2N)$ ($N \gg 1$). This results in states $|\text{Off}\rangle \to \cos\theta|\text{Off}\rangle + \sin\theta|\text{On}\rangle$ and $|\text{On}\rangle \to -\sin\theta|\text{Off}\rangle + \cos\theta|\text{On}\rangle$. The output registers are monitored at each step. If the state is 00, the measurements on the output registers result in detecting 00 without affecting the evolution due to rotations. This leaves the system in $|\text{On}\rangle|00\rangle$ after N rotations. But if the state is not 00, the system evolves as $|\text{Off}\rangle|00\rangle(R)\rightarrow(\cos\theta|\text{Off}\rangle + \sin\theta|\text{On}\rangle)|00\rangle(\text{GSA})\rightarrow$ $(\cos\theta|\text{Off}\rangle|00\rangle + \sin\theta|\text{On}\rangle)|xy\rangle|(xy = 01,\ 10,\ 11)$ (Detect)$\rightarrow \sim\cos\theta|\text{Off}\rangle|00\rangle$. In this equation, GSA = Grover's search algorithm.

The measurement on the output register results in the state $|00\rangle$ keeping the computer in the $|\text{Off}\rangle|$ state with the probability $\cos^2(\pi/2N)$. After total of N cycles, it leads to 1, as $N\rightarrow\infty$. Thus, if the GSA-amplified state is not 00, the final state is $|\text{Off}\rangle\ |00\rangle$ without the computer running. If the GSA state is $|00\rangle$, the final state is $|\text{On}\rangle|00\rangle$, i.e., this time the computer is running. In this approach one can at best exclude a single value of the GSA state. To circumvent this problem, one uses the chained quantum zero scheme (CQZS, Figure A1.12(a), Appendix A1.12). This allows one to counterfactually use the Off switch state—the computer not running. This determines the actual GSA state with efficiency of approximately 1. One can also use another strategy of placing one QZS inside another one; call this the third additional switch-Off′. This will avoid the computer running if the GSA state is not 00. This strategy leads to defining an additional rotation R', $\Delta\theta'\pi/2N'$. One then needs to couple $|\text{Off}''\rangle$ to $|\text{Off}\rangle$. A single photon starts in cavity Off′. Using active optical elements (pocket cells), a small amount of amplitude is exchanged between Off′ and Off via a BS. A possible implementation of the scheme is shown in Figure A1.12, Appendix A1.12.

If the state is not 00, the small-amplitude component effectively stays in cavity Off. But if it is 00, then first all the small-amplitude component is transferred to cavity On and then, via the PBS in cavity On, is actively absorbed in A_1. Now the entire procedure, starting with the amplitude exchange between Off′ an Off, is repeated N' times. At the end of $N' \times N$ total cycles, if the GSA state is 00 ($\neq$00), then the photon will be measured in cavity Off′ (Off) with probability approaching 1 (Figure A1.12(e), Appendix A1.12) as $N'\rightarrow\infty$ ($N'/N\rightarrow\infty$). In neither of these cases does the computer run. One can then reinterrogate for the other elements one by one. One thus identifies the GSA state counterfactually by changing the connections to the algorithm (Figure A1.12(b), Appendix A1.12).

1.4.3.6.2 Quantum Mechanical Interaction-Free Imaging Devices (Appendix A1.13) [69]

This illustration of QMIP is a typical example of how fundamental research can lead to unimaginable unique benefits. Interaction-free measurement (IFM) can develop into a very useful in situ imaging system in the field of biology and medicine (investigation of cells, etc.). Here interaction of radiation with an object can be at a minimum level. In an efficient scheme referred to as the Elitzur–Vaidman (EV) scheme, an arbitrary object—classical or quantum—can affect the interference of a simple quantum particle with itself (Appendix A1.13). The noninterference of the particle allows the presence of the object to be inferred without the particle and object ever directly interacting. In the EV IFM scheme, the measurement is interaction free at most half of the time. One needs, to use high-efficiency IFM systems to achieve high-efficiency IFI. This normally means using high-quality components in the optical circuits (normally an interferometer). It leads to the probability of achieving IFMs arbitrarily close to unity. This means that the probability of absorption of the particle by the object can be made equal to zero. The possibility of detecting the presence of an object without ever interfering with it led to the suggestion of interaction-free imaging (IFI), the optical imaging of photosensitive objects with much less than the classically expected light being absorbed or scattered by the object. One of the current limitations in imaging of biological systems is the fact that the incident radiation causes damage to the object. Now there is a possibility of evading this limitation via IFI beams. Figure A1.13 (Appendix A1.13) shows a sketch of a Mach–Zehnder polarizing interferometer. It allows effective tuning of the beam-splitter reflectances. The first half-wave plate ($\lambda/2$) is set so that the light input to the interferometer is linearly polarized at from the vertical axis. The first PBS splits the light into horizontal ($T_1 = \sin^2\theta$) and vertical ($T_1 = \cos^2\theta$) components (e.g., $\theta = 45°$ gives $R_1 = 0.5$). If no object is present, the second PBS recombines the beams to the original polarization, which then is rotated back to the vertical by the second $\lambda/2$ plate so that the light is always detected at D_1. If an object is present, however, the interference is modified or distorted. In the latter case, only the horizontal component is transmitted by the interferometer, the vertical component being absorbed by the object. In QM terms, only the probability amplitude of the horizontal polarization path contributes to the final probabilities. The horizontally polarized output is rotated toward the vertical axis by the second $\lambda/2$ plate, so that some counts occur at D_{IFM} ($T_2 = \cos^2\theta$). These counts are the IFMs.

In experimental practice, it was found necessary to lock the interferometer so that one port, the IFM port, was at a null. This was done with an additional laser (a He–Ne laser at 632 nm) and a simple fringe slope locking system.

1.4.4 Ion-Trap Computer [3, 9]

In this technique, two $^{40}Ca^+$ ions are in a common mode of vibration (the *bus-mode*) in a linear trap. In addition, each ion can be excited to its individual energy state (*carrier mode*). The qbit is encoded in a superposition of the $S_{1/2}(|0\rangle)$ ground state and the metastable $D_{5/2}$ state ($|1\rangle$). The Ca ions are manipulated on the $S_{1/2} \rightarrow D_{5/2}$ quadrupole transition near 729 nm by a laser beam. An electro-optical beam deflector switches the beam between the ions. The cycle states are created by the Doppler cooling of the individual ion followed by that of the bus-mode. The qbit manipulations required for CNOT operations are realized by applying laser pulses with well-defined phases on the carrier or the blue side (bus) mode of the electronic quadrupole transition. First, a π-pulse applied to the blue sideband of the first ion (i.e., the control bit) maps its internal state to a corresponding state of the bus-mode. The phonon number n of the bus-mode also forms a qbit, where $|n = 0\rangle(|n = 1\rangle)$ represents the logical state $|1\rangle(|0\rangle)$. With the quantum information of the control bit in the vibrational mode, one addresses the second ion (i.e., the target bit) and prepares a single-ion CNOT gate operation between the ion and the bus-mode. The second ion's internal state is flipped if no vibration is present in the bus-mode, i.e., if the bus qbit is $|1\rangle$. This operation consists of a pair of Ramsay pulses enclosing a composite phase gate. Finally, a π-pulse on the blue sideband is applied to the first ion. It restores the controlling qbit and the bus-mode to their original states.

1.4.5 TAS State Evolution [2]

In an experimental setting, tomography is used to determine the state of a quantum system, whereas spectroscopy is required for obtaining information about the energy spectrum of the system. One uses tomography to directly measure Wigner function (WGF). WGF is a basic tool to represent the state system in phase space. The discrete WGF (DWGF) can be defined as $A(q, p) = \underline{U}^q R \underline{V}^{-p} \exp(i2\pi pq)/(2N)$, where U is the shift operator in the computational basis $\underline{U}|q\rangle = |q+1\rangle$. $\underline{V}$ is the shift in the basis related to the computational basis via the discrete Fourier transform (DFT). R is the reflection operator $(R|n\rangle = N - n\rangle)$ and N is the dimension of the Hilbert space. The DFT for the state is evaluated at the point (q, p) in phase space by using the scattering circuit with $U = A(q, p)$ to obtain $W(q, p) = \mathrm{Tr}[A(q, p)\rho]/(2N) = \sigma_Z/(2N)$. The function is defined in the grid of $2N \times 2N$ points. It can be negative and relates traces of operator points to phase-space average, $\mathrm{Tr}(\rho_1, \rho_2) = N\Sigma_{q,\,p} W_1(q, p) W_2(q, p)$ and provides positive probabilities when projected on any skew line in phase space. The quantum circuit that implements $A(q, p)$ can be decomposed into a sequence of elementary steps. The controlled $\underline{U}$, $\underline{V}$, R operations can be implemented via efficient circuits, which require elementary gates that depend polynomially on $\log N$. $W(q, p)$ can be measured for a practical computer, for example, by using the molecule of trichloroethylene dissolved in chloroform. The proton ^1H and the two strongly coupled ^{13}C nuclei (C_1 and C_2) store the 3 qbits. The C_1 can be used as a probe particle and the pair $H{-}C_2$ can store that state, whose WGF is measured. The coupling constants in this example are $J_{HC1} = 208$ Hz, $J_{HC2} = 9.1$ Hz, and $J_{C1C2} = 103.1$ Hz; and C_1 and C_2 with the chemical shift $\delta_{C1C2} = 908.9$ Hz. One determines the value of $W(q, p)$ for each of the independent 16 phase-space points. Each of these circuits corresponds to a different sequence of RF pulses and delays. The number of pulses in each sequence depends on q, p and varies from 5 to 17. One uses temporal averaging to obtain, from the part of the density matrix ρ that deviates from the identity, the four pseudo-pure initial states, whose WGF can be determined [2].

1.4.6 The Outlook: The Physical Quantum Computer (Appendices A1.1–A1.6, A1.11–A1.13) [61–64, 68, 69]

One should realize that an understanding of the basics of QM is very tightly linked to any success in realizing a QC in any system. This held true for the tremendous number of applications developed in physics—quantum or otherwise—in the past century, and there is nothing to suggest that it will not hold true in the twenty-first century. A wealth of knowledge is available in the field of QMIP. Enough modeling studies have been developed, but no practical implementation of the QC, in any recognizable form, is anywhere in sight. One must remember, too, that the QC has to compete with the EC. The technologies that can be implemented have to be convenient to operate.

An invention that operates at room temperature has a better chance of being ultimately accepted. QMIP, because of its intrinsic nature of parallel processing, offers enormous power in quantity and speed. To realize a QC in practice, though, one must first be able to manipulate the functionality of the materials involved, in a nano-scale dimension in a physical device. Second, a terahertz speed in operation time is desired; in fact, it is paramount to realizing the dream of putting QMIP into practice. It is important that there be an output in the shape of a tangible device, and the device operate at 300 K. This demands integration of the required functions of electron, photon, spin, etc., quanta in some form of a compact scaffolding of extraordinary precision and performance. Do the solid-state, optical, and other components have the capacity to deliver what it takes to make a QC? In a more practical approach, one asks the following questions. Are there quantum mechanical processes happening in the human brain? There is evidence to suggest that quantum computing may be operative in human brain functioning at 300 K. Does a mixed, solid-state—liquid-state phase, such as that in the human brain, hold the answer? Silicon solid-state technology has proven futile so far. An electronic charge-cellular automata model (Appendices A1.5 and A1.6) is presumed to fit the data information-processing technique through neural activities in the human brain. However, the resources required to put the same model into operation in the solid state either have not been worked out or are beyond the reach of modern-day technology. An all-optical approach to the QC suffers from the exponentially increasing number of the components required.

This chapter has provided the reader with a bank of information to examine and compare regarding what suitable technologies are available at the moment or may be developed in the future. There is room to improve the existing technologies and discover new ones. One may have to take a direction in which CIP is used as the initial point and build the QMIP from there on. Shor's quantum factoring algorithm is based on factoring large numbers into prime numbers. This would be immensely useful in CIP-to-QMIP transition. The presently available, well-developed technologies (e.g., nano-solid state, optical-squeezed light) offer a good start in the nonbiomedical quantum endeavor. However, it has been found in practice that electron charge and/or spin-based solid-state schemes are not easy to multiscale in qbit numbers. The development of milli-Kelvin (10^{-3} K) operative temperatures to eliminate temperature-dependent decoherence has not paid the dividends expected.

The charge-based automata model (Appendices A1.5 and A1.6) could be used to put into practice QMIP in the solid state, but the decoherence problems may stall any headway in this direction. Some still believe that solid-state technology, as in the past, could provide the required high-density integration of the qbits on a miniaturized scale, but the advances in this field are nowhere near substantiating that belief. Going too far with a bad choice can be very harmful in the long run. Amorphous silicon solar cell technology is a good reminder in this respect: it never delivered what it initially promised.

Much wider dissemination of QM knowledge in the field would be very beneficial. This is the basic aim of this chapter. Grassroots knowledge of QM has to be spread over disciplines as far-ranging as engineering, chemistry, biology, and even medicine. This basically means restarting the study and development of "pure" QM science as if we were in the beginning of the last century. Over the years, the infrastructure of research in basic sciences such as physics, chemistry, and mathematics has been reduced to minimum. Unfortunately, this is the course adopted in nations where the available literate person power and the infrastructure of research can be made easily available. The basic philosophy of scientific research and advancement, so prevalent in the past, must be brought "back to the future." Human development and advancement have been abandoned over the years.

It seems that basic research and pure science have been buried by the thirst for materialistic advancement. The pundits (the managers and accountants) have gone astray in their assumption that human development is not material advancement. The practice of keeping research dollars confined to applied research, marketing, and so on is both selfish and very harmful. In the past, an independent, free-thinking mind was allowed to evolve. This freedom resulted an enormous wealth of new technologies, scientific infrastructures (for further advancement), and a well-trained and well-occupied workforce (person power). The vast number of technologies developed from the basic research, which society unmindfully enjoys today, is evidence of the importance of that basic research. However, the present system has closed its doors to the basic researchers, who have the ability to start from scratch and make new inventions.

Now more than ever, there is a need to widely communicate and educate people about the science of QM. This should be done right from the secondary education level. Engaging young minds in QM science and its applications can be very rewarding for the community at large and for advancement of both pure and applied science. An intellectually absorbed younger generation is much better than one that is not. More and more young people in the society today are occupied in the use of bran rather than brain. Universities produce more and more degrees in the field of finance and business services and move society away from basic intellectual development. Recent hysteria motivated by an increased occurrence of mental health problems may stimulate in more research in health sciences, and thus may result in some changes. Young minds have to be educated to realize that quantum science is functioning in each human body; there is increasing evidence that the human brain is a quantum brain. The quantum mechanical processes exploited in producing better contrast in brain magnetic resonance imaging (MRI) systems is a good advancement and an evidence of the fruits that QMIP research can deliver.

1.4.7 The Outlook: Bio-Chemo-Medical QMIP (Appendices A1.4–A1.10, A1.14, A1.15, and A1.25) [70–86]

1.4.7.1 The Quantum Model of the Human Brain: Simple Theory

Biomedical applications of gradient-field nuclear magnetic resonance imaging (NMRI) techniques have shed new light on the way one looks at the quantum science operative in the human body [72–77, 79, 80, 85, 86]. The dipolar molecular interactions creating intermolecular multiple-quantum coherences (iMQCs) are leading to a high-contrast imaging technique for the human brain. In this technology one uses the presence of a gradient in the magnetic field in the x, y, and z directions. This allows quantum correlations among distant molecules in a planar environment to be observed and analyzed. Two magnetic gradient-field pulses separated by a delay τ generate not only the conventional spin echo after delay τ, but also the additional echoes at delays 2τ, 3τ. The two basic transverse RF pulse sequences used in NMR are correlation spectroscopy (COSY) and correlation spectroscopy revamped by asymmetric Z-gradient echo detection (CRAZED).

The additional echoes are explained as due to a distant dipolar demagnetizing field B_d. It depends explicitly on the local magnetization produced by the distant field. Molecular diffusion in the liquid state only averages dipolar couplings between spins, which are separated by much less than the distance between molecules (diffuse on an NMR timescale). The distant spins effectively generate an additional magnetic field. This leads to an additional term $\mathrm{d}M/\mathrm{d}t = \gamma M \times B_\mathrm{d}$ that must be added into the conventional NMR Bloch equations. Solution of the new Bloch equations predicts multiple echoes, in agreement with the experiment. The conventional nonquantized Bloch differential equations are linear as to magnetization; so is their density matrix treatment counterpart [83]. The new 2D Bloch equations include the demagnetizing field and are also nonlinear in M. Doubling M doubles B_d and changes the time of evolution of the RF pulses applied. The nonlinearity implies that the molecular dynamics in the ensemble are drastically altered.

The analysis, either through the Bloch equations or conventional density matrix theory, is thus not valid. In the case of solid-state NMR, line-broadening effects of the dipolar interactions have been recognized for quite a while. Quadrupolar NMR analysis in solids has paved a new way toward solution of these earlier problems [78], but in liquids the scenario is quite different. To keep treatment of the problem tractable, one may assume, to a first-order approximation, that there is only 1 spin per molecule or that the spins are magnetically fully equivalent. The Hamiltonian for N spins interacting via the dipole–dipole interactions, in the presence of the magnetic-field gradient G, with s_i representing the unit direction vectors for the spins, can be written in QM formalism as [85] $H/(\hbar = h/2\pi) = \Sigma^N i = 1[(\Delta\omega_0 + \gamma G s_i)]I_{zi} + \Sigma^N i = 1\Sigma^N j = 1$, $j \neq 1$ $D_{ij}(3I_{zi}I_{zj} - I_iI_j)D_{ij} = (\mu_0/4\pi)(\gamma^2\hbar/4)[(1-3\cos^2\theta_{ij})/r_{ij}{}^3] = 188.7[(1-3\cos^2\theta_{ij})/r_{ij}{}^3]$ rad/s (for ^{1}H; r_{ij} in nm).

The first term in $H/\hbar$ is the Zeeman term, reflecting the interaction with the applied magnetic field B_{0z}; $\omega_0 = \gamma B_{0z}$ is the Larmor frequency and γ the gyromagnetic ratio. In the rotating frame, ω_0 will be replaced with the resonance offset $\Delta\omega$. The second term is the scalar part of the dipole–dipole interaction, r_{ij} is the separation between spins i and j, θ_{ij} is the angle between internuclear vectors, and μ_0 is the vacuum permeability. It is seen that for nearest protons, D_{ij} is on the order of 10^5 rad/s and the intrinsic relaxation time T_2 in the absence of these couplings is on the order of 100 ms. Thus, even a single pair of isolated spins (e.g., isolated water molecule in deuterated ice) can coherently interact for a time T such that $D_{ij}T \gg 1$, and the bilinear operators in the dipolar Hamiltonian thus permit extensive coherence transfer.

In a real solid, the slow ($1/r^3$) falloff of the dipolar interaction with distance then implies that the number of significant and different coherence transfer pathways grows rapidly in each successive derivative. The dipolar interactions are usually not obvious in liquids. This is because of the angular dependence for a spherical sample filled with an isotropic liquid, $\Sigma D_{ij} = 0$. Thus, adding up the effects of the dipolar couplings in all directions makes the dipolar effects vanish, if the magnetization is not modulated. However, if the sample is not spherical, or if the magnetization is modulated, angular averaging does not destroy the dipolar effects. The critical difference between liquids and solids is that in a liquid, every pair of spins interacts coherently and for a time t such that $D_{ij}t \ll 1$. One can see this by taking the time necessary for diffusion to cover the spins by the internuclear separation as a rough estimate [72] of the coherent interaction time t, making $D_{ij} \sim [(\mu_0/4\pi)(\gamma^2\hbar/4)[(1 - 3\cos^2\theta_{ij})/r_{ij}^3][(r_{ij}^2/2D)] \leq \mu_0\gamma^2\hbar/(16\pi D r_{ij}) = 8.2 \times 10^{-8}/r_{ij}$ (water; r_{ij} in nm). This is actually an overestimate for very distant spins ($r \gg 150$ nm for water), because in that case t is limited by relaxation, not by diffusion effects. As no individual coupling makes an important contribution to the time evolution, the mean field approximation will be valid. What matters are the sums over coupling elements. The number of spins in a spherical shell of radius r is proportional to r^2, so in fact the near-neighbor interactions that dominate solid line shapes are unimportant in liquids.

Strictly speaking, the spin and spatial degrees of freedom make the dipoles' couplings time dependent. One can ignore the nearby spins in evaluating the sums and integrals and use a cutoff distance to produce a time-dependent effective Hamiltonian that does not depend on this distance. It is possible to demonstrate that the restriction $D_{ij}t \ll 1$ does not eliminate the effects of dipolar couplings. The number of coupled spins is very large. Hence, various summations over the dipole couplings will in fact be substantial. However, the restriction makes calculation of the free induction decay tractable (as opposed to the case of solids) so as to obtain a variety of exact results.

1.4.7.2 Quantum Model of the Dipole–Dipole Interactions of Spins in Solutions in the Human Brain: The Experiment

1.4.7.2.1 Experimental Modeling [72–76]

The decoherence problems in NMR and DNA QMIP are not as severe as in the SS, optical, or other systems. In fact, one should expect otherwise. The biomedical areas may be used as a resource for developing the building blocks of the basic knowledge required to develop a solid-state QC. It is believed that the human brain performs like a QC. Thus, it may be better to start at this level to understand how to build a laboratory-scale physical QC. Recent developments in MRI of the brain demonstrate that quantum coherence of proton spins, separated by distances from 1 micron to 1 mm, are sustained for times longer than milliseconds (ms). Although these unentangled quantum couplings are not the type of quantum processes that are likely to prove useful in brain function, they nevertheless demonstrate that mesoscopic quantum coherence can indeed survive, including in the brain's milieu. Quantum modes in peptides, DNA, and proteins are not stable.

Entangled superpositions are required for quantum computation. In this regard, biology can take advantage of quantum modes in clever ways. Recently, a model known as the orchestrated objective reduction (OOR) has been put forward to simplify understanding of, and reduce the complexity in, exposing the systems in operation in the human brain. Be aware that the OOR model is very speculative in nature; nevertheless, it provides a modeling platitude. It conjectures that due to some forces, the origin of which is not yet clear, there is an environment of isolation that permits quantum computation to proceed within the neuronal system of the human brain (Appendices A1.7–A1.10, A1.14, and A1.15). Earlier models proposed CIP among the tubulin dimers. The tubulins make MTs behave as molecular-level automata, regulating real-time cellular behavior. MTs are poised to mediate between a tubulin-based quantum computation and the classical function of the neurons. The neuronal cytoplasm in which the MTs are embedded alternates between the phases of isolated quantum superposition/computation. The solid or gelatinous "gel" attains polymerization, and the classical states of the input/output communicate with the liquid solution or "sol." The input to and the output from each OOR event evolve as classical MT cellular automata. This occurs in the form of patterns of tubulin conformational states to regulate synaptic function membrane activities, and attachment sites for MAP. Input from synaptic activities may be provided by metabotropic receptors that interfere between membrane synaptic functions and the internal cytoskeleton/MTs. MAPs also interconnect MTs in bundles or networks, and the OOR model suggests that MAPs regulate, or orchestrate, MT quantum states by their particular attachment sites on a microtubular lattice. The question is how environmental decoherence can be avoided, and quantum superposition be sustained, long enough for the system to reach the threshold for the OOR. It is well known that the technological quantum devices (e.g., the solid state) often require extremely low temperatures to avoid decoherence through environmental interaction.

Nevertheless, it is suggested that the delicate quantum coherence survives in the warm, wet, and noisy milieu of the brain—at least long enough for quantum computation to play a neurophysiological role. One may say it seems unlikely. It is felt that the orthodox neuronal network mechanism of decoherence would destroy a superposed MT-associated quantum state on a timescale of the order of 10^{-13} s. In OOR, the superpositions of the conformational state of a tubulin driver occur at the level of each of the proteins' atomic nuclei. The separated states are coupled to delocalizable electrons, resulting in the creation of a hydrophobic pocket of the tubulin dimer protein. It points to a process of conformational change in the dimer-controlled quantum van der Waals (London) forces. Superposition occurs not only at the level of a mass distribution from itself, but concomitantly at the level of the underlying space-time geometry. According to the OOR (Penrose) model, the gravitational collapse criterion for the objective reduction superpositions, involving different space-time geometries, is considered inherently unstable. The rate of collapse is determined by a measure of the difference in geometries. The surrounding single ionic forces tend to cancel the coherence over the distances of importance in the brain. Therefore, the forces meddling between the tubulin and its environment should instead be characterized by the dipolar interactions. As in the coulomb case of interactive charges, the force resulting from the dipole potential constitutes only a phase factor in the evolution of the (reduced) density matrix traced over the environmental degrees of freedom [70].

There are tidal (electromagnetic) effects (due to surrounding ions) that determine the leading contribution to the rate of decoherence. It is well known that lasers maintain quantum coherence against thermal disruption at room temperature. The dynamical timescale for lasers is determined by the rate at which the system is pumped by an incoherent source of energy. Similar events may be happening in the appropriate dynamical timescales in the MT case. It might be determined by the characteristic processes that might counteract decoherence by scattering. This may be due to actin gelatin in sol-gel cycles and the ordering of the water. In the OOR model, actin gelatin encases MTs during the quantum computation phase. Afterward, the gel liquefies to an aqueous form suitable for communication between the MT sites and the external environment. Such alternating phases can explain how input from and output to the environment can occur without disturbing quantum isolation.

Recent modeling studies [65–67] have found that electromagnetic fields, whether due to single charges or the dipoles, make decoherence time a direct function of temperature. If the human brain is able to perform quantum computation at ambient temperatures, then attempts to achieve the same in the solid state, through milli-Kelvin temperatures, are certainly not without flaws. The studies suggest that the human physiological temperature of 310 K is just right for isolation of the OOR quantum state.

Does this quantum state make available just the right time—of nearly a tenth of a millisecond—for the quantum computation in living cells? However, it is presumed that the milli-Kelvin temperature is the only recourse for quantum computation in the solid state. Wouldn't it be better to find out what actually happens in the context of the human brain first? The human brain does follow some kind of information theory as it keeps the day-to-day body functions running. One should remember that OOR is only an idea, yet to be tested by experimentation. The model might then be further developed, based on the observed results. In the brain, one expects longer-lived (10^{-4}–10^{-5} s) quantum coherent states at 300 K. This is contrary to the principles of thermodynamics, statistical mechanics, and kinetic theory. These principles dictate that increased thermal agitation should have a disruptive effect on the formulation and preservation of quantum coherence.

Any cause for optimism is thus very dim, but the OOR model may yet be helpful. Some of the enigmatic features of the cognitive processes occurring in the brain (e.g., consciousness) might be understood via a quantum approach. One can endeavor to close the gap of knowledge between the techniques being tried for development of a solid-state QC and the information processes happening in the human brain. Are there some unknown electromagnetic and/or gravitational interactions operative in the information processes that occur in the human biological system? This is the question researchers should be asking themselves. We have included in Appendices A1.7–A1.10, A1.14, and A1.15, illustrative pictures of an axon enclosing MTs, configurational arrangement of tubulins in MTs, nearest-neighbor dipole–dipole interaction forces, and the like. An interdisciplinary researcher may find it a good illustration from the point of view of the learning process about information theory and the progressing experimental research.

1.4.7.2.2 Practical Applications: iMQC Imaging in the Human Brain (Appendices A1.14–A1.16) [66, 67, 71–76, 80–82, 88]

The basics of conventional NMRI are well known. In the most basic form, it directly observes single-quantum (SQ), single-spin coherences resulting from the magnetization of the ensemble of molecules. The newly developed 2D NMR (gradient along x, y, and z directions) allows one to exploit observation of coherences between states in a multispin system. These coherences are made to evolve silently. This happens during a successfully incremented time interval and is detected after transformation into magnetization. In order for the transformation to take place, a net coupling between the spins must exist. In the case of spins, one can distinguish two types of couplings. The scalar coupling acts through chemical bonds. On the other hand the dipolar couplings, which are much stronger, act through space. Scalar couplings are commonly used to effect transformation of multispin coherences into magnetization. The 2D NMR can provide information about the connectivity between spins in a molecule. However, dipole couplings to a first degree of approximation are normally not observable in liquids. The dipolar coupling strength between 2 spins, scales is approximately $3\cos^2\theta - 1$. Here θ is the angle between the interspin vector and the main magnetic field. This coupling averages to zero when integrated over all directions. In the case of short-range dipolar interactions, the interspin vector samples all directions on an NMR timescale through molecular diffusion. This is not true for long-range dipolar interactions.

There θ is almost constant in time. In that case the distribution of spins is quasi-continuous. The dipolar interactions of spins average to zero in space as long as the liquid is magnetically isotropic. Magnetic-field gradient pulses applied during the experiment can break this isotropy. Thus, long-range dipolar couplings can be reintroduced by the experimenter in a controlled manner. In this way, the structure of a sample can be probed on the distance scale on which the dipole couplings act. Most importantly, this distance may be tuned through the choice of experimental parameters. The gradient pulse causing the reappearance of the dipolar coupling due to the correlation gradient can be thought to wind up a helix of magnetization along its axis. The dipolar interactions that are introduced in this fashion act over a distance scale of approximately one-half pitch of the helix. Hence, the larger the area under the gradient pulse, the shorter the correlation distance. In practice, the correlation distance ranges from tens to hundreds of micrometers. This is far above the microscopic range provided by scalar couplings, yet below the size of an imaging voxel. Functional magnetic resonance imaging (fMRI) can measure brain activity noninvasively, with spatial resolution of millimeters and a temporal resolution of seconds.

The technique is based on the blood oxygen level dependent (BOLD) effect. The BOLD effect is thought to arise from localized changes in the concentration of the strongly paramagnetic deoxyhemoglobin molecules in the brain. Blood flow increases within seconds near the site of activation and overcompensates for the increased metabolic demand, resulting in decreased deoxy-hemoglobin and increased (diamagnetic) oxyhemoglobin contents. The ensuing changes in susceptibility gradients across capillaries and venous blood vessels result in an increase of the apparent transverse relaxation time (TRT) of the spins. Thus, an image the intensity of which is weighted by the TRT will show neuronal activations, through the secondary effects of blood oxygenation, as an increase in signal intensity. The sensitivity of this technique can be improved by using instead intermolecular multiple-quantum coherence imaging (iMQCI). Here, multispin coherences have a higher sensitivity to susceptibility gradients than single-spin coherences. An intermolecular zero-quantum coherence (iZQC) evolves at the difference of the SQ frequencies of the 2 spins involved; hence, the zero-quantum (ZQ) signal intensity is a function of the distribution of the susceptibility gradients. The BOLD signal is a function of the average strength of these gradients within a voxel [73], although it is still a function of blood oxygenation.

It has been demonstrated that it is possible to perform medical imaging based on multiple-quantum coherences (MQCs). This may be due to flipping of two or more separated spins simultaneously.

One can image using five orders (2, 1, 0, −1, −2) of quantum coherence. In the study (Appendices A1.20–A1.22 [75]), it was achieved using a multiple-correlation spectroscopy revamped by asymmetric *Z*-gradient echo detection (multiCRAZED) pulse sequence. Each echo was independent of the others, and acquired the same amplitude as would be found in a normal sequence that selected only one of these orders. In this way, the simultaneous acquisition of double-quantum (DQ), ZQ, and SQ images permits evaluation of the differences in signal intensity. Another illustration on quantum coherence among distant spins can be found in an important medical application [76] (Figure A1.24, Appendix A1.24) that uses a double-quantum filter (DQF). The application concerns localized detection of glutathione (GSH), in vivo, in the human brain.

It is operated in combination with point-resolved spectroscopic sequence (PRESS). In this case, frequency-selective pulses instead of nonselective pulses are used. This is done both for generating double-quantum coherences (DQCs) and for converting them back into single-quantum coherences (SQCs). This localized DQF scheme is illustrated in Appendix A1.23. A chemical shift-selective sequence (CHESS) consists of three Gaussian pulses, each followed by a gradient crusher. It was used for water suppression. A 90° excitation pulse and two 180° pulses, which form the PRESS, were used in achievable spatial localization. The duration of the first echo time t_1 was fixed at 5 ms. A frequency-selective 90° Gaussian pulse was applied symmetrically at each side of the last slice-selective 180° pulse. Each of these two 90° pulses was 10 ms in duration (bandwidth 90 Hz at half height). They had a frequency set to that of CH proton of cysteine moiety of GSH at 4.56 ppm. During the first TE/2 period, the CH_2 protons of the cysteine moiety evolved into antiphase terms via its *J*-coupling with CH protons, which was subsequently converted into MQCs by the first selective 90° Gaussian pulse. Because only the CH of cysteinyl groups is excited by the Gaussian pulse, the antiphase terms from the CH_2 are fully converted into DQCs regardless of the phase accumulation of the slice-selective pulses. The phase-calibration scans for determining the optimal phase for the DQ excitation pulses are avoided. Further, the signal was stated from the CH proton of the cysteinyl group. Appendices A1.4–A1.10 and A1.14–A1.34 provide the reader with illustrations and links to follow to become part of the research and development in the biomedical aspects of QMIP. The field of NMR has evolved over more than half a century and has now become a very good example of a direct connection between the basic principles of QM and their practical applications. However, there is still a need to comprehend the technology, the associated technical terms that have evolved, and the quantum mechanical interpretation of the results.

This is a necessity if we are to make good progress in the science of QMIP in general. Appendix A1.25 presents an illustration, for a starting point, in a generic 2D NMR system. The pulse sequence is divided into four parts. An initial set of pulses and delays applied during the preparation period transforms the initial density matrix $\rho \sim 1 - (h\omega_0/kT)I_Z$ into a more complicated form. The preparation period can be viewed as generating a unitary transformation U on the initial state of the system. The system is then allowed to evolve without pulses for an evolution period of length t_1. After the evolution period, a (possibly different) set of pulses and delays during the mixing period provides another unitary transformation V, which transfers populations and coherences into different matrix elements.

The observable signal I_x or I_y is then acquired at many different times t_2 in the detection period [77]. The MQCs correspond to superposition of energy levels with $\Delta M \neq 1$. They are not observed in the simplest NMR spectra, such as those produced by Fourier transforming the free induction decay, after a single pulse. In that case the preparation sequence is a single pulse and the mixing pulse is omitted. The only observables in NMR are the components M_x and M_y of the transverse magnetization. These operators are proportional to the sums of the corresponding single-spin angular momentum operators, e.g., $M_x \sim \sum^N i = 1(I_{xi}) = \sum^N i = 1(I_i^+ + I_i^-)$. They involve only a single spin at a time, as opposed to more complex forms such as $(I_i^+ I_i^-)$, and only connect states with $\Delta M \pm 1$. The second reason for the absence of MQCs in a normal spectrum has to do with the form of the equilibrium density matrix. It contains only single-spin operators, and the rotation by a single short pulse will produce only single-spin operators. A simple sequence that produces a multiple-quantum spectrum (MQS) is illustrated in Appendix A1.25. The preparation period acts to convert ρ (the density matrix) from its equilibrium form into operators such as I_1^+ and I_2^+. The first pulse converts $I_z \equiv \sum^N i = 1 I_{zi}$ into the single-quantum, single-spin operator I_x. The total Hamiltonian H of the system would also include the term H_{int}. This is the secular portion of all additional internal parameters such as the spin couplings $\sum i > jJ_{ij}I_iI_j$ or chemical shift differences. The couplings in H_{int} can then act during long decay τ to produce one-quantum, multiple-spin operators in ρ. The second pulse then converts these operators into multiple-quantum, multiple-spin operators. These operators evolve during t_1 and are partially transferred back into observable (single-quantum, single-spin) operators by the third pulse and delay t_2.

The signal $S(\tau, t_1, t_2)$ is acquired for many different times of t_1. One often requires many values of τ or t_2 to be saved as well. Fourier transformation with respect to t_1 can then yield frequencies that correspond to transitions between states with $\Delta M = N \gg 1$.

1.4.8 The Future of Research and Development in QMIP [73–88]

The development of technologies based on QMIP will depend upon continuous research to improve the theory as well as the experiments. Both are equally important; one cannot progress without the other. Becoming involved in experimental work is not easy, as such work requires a fully established laboratory. These are not available just anywhere. Even if there is an appropriate laboratory nearby, there may be no easy access to it. As an example, almost all hospitals have NMRI facilities for clinical purposes; however, any given hospital may devote very little time on the machine to basic research. Applied theoretical modeling is a useful route to making a worthwhile contribution, and this is in the interest of medical practitioners.

We wish to close this chapter by including, in this section, references to the introductory basics, for scientists who have the ability or the potential to acquire introductory knowledge of QM. In fact, development of this kind of "person power" is very timely. Such scientists can bridge the gap between a medical doctor and technical staff. Appendices A1.26–A1.32 have been added to help such enthusiasts to trace back the starting concepts, mathematical formalism, technology, and the like, so as to be able to apply QM concepts and principles to real-life situations. For example, in the gradient magnetic-field approach to NMRI, one must include the dipolar–dipolar molecular interactions in the full Hamiltonian (energy operator of the system). This will be particularly important with reference to high-contrast brain imaging. There will be energy transitions due to the molecular chemical shifts $\Sigma_k \sigma_k I_{zk}$ and the intermolecular scalar couplings $\Sigma_{kl} J_{kl} I_k I_l$. The numerical values of σ_k and J_{kl} depend on the electronic state and the molecular orientations. The orientations would be very important in solids. In liquids, nuclear spin transitions are induced without changing electronic or vibrational state. QM predicts that the correct time evolution of any system is found by solving the time-dependent Schrödinger equation $i(h/2\pi)(\partial\psi/\partial t) = H\psi$.

The linearity of the equation implies that the wave function at time t is related to the wave function at any other time (say $t = 0$) such that $\psi(t) = U\psi(0)$. U depends only on the Hamiltonian H (the total energy of the system written in QM formalism) and not on the wave function. The H is independent of the wave function. Equivalently, it is possible to expand the wave function in some convenient basis set $\psi(t) = \Sigma_i c_i(t)\phi_i$, and then to define a density matrix P with elements $P_{ij} = c_i(t)c_j(t)$. The time evolution of P is then given exactly by $i(h/2\pi)(\partial P/\partial t) = [H, P]$; $P(t) = UP(0)U^+$. Here U^+ is the complex conjugate of U.

The basic mathematical formalism, established using the density matrix approach for specific cases of homonuclear and heteronuclear molecules, has led to the development of complex mathematical equations for deriving signal intensity.

In the liquid state, one involves a 2D interacting quantum system through the application of an external gradient magnetic field. It thus becomes easier to excite and measure distant dipolar molecular interactions created by the shielding demagnetizing field. Their involvement with the environment (e.g., the brain tissues) can be projected onto a visual system, as either a soft- or a hard-copy image. It is also possible that a real-time imaging system using THz spectroscopy will be developed, although this area of spectroscopy is itself still under development [87]. In the solid state, progress in the NMRI field is far from close to being established, as is in the liquid state. Some illustrations are provided here for readers, so that they can better visualize the complex skills/background required and the benefits expected (Appendices A1.30 and A1.31 [88]).

This section presents a glimpse of the diversities involved in the QM field. One needs to develop a good comprehension of the discipline of QM to be able to gainfully participate in the field of QMIP. The reader who is not familiar with QM is referred to a standard text to start with. The field of medical science needs a lot of intellectual input from physics, chemistry, and mathematics (PCM). Without the help of PCM experts, the mysteries of the human body will remain unsolvable. Diseases like epilepsy, Alzheimer's, and many others lack proper diagnostics and treatment. This will improve in real time as their PCM basis becomes well understood. Quantum NMRI of the human brain provides a technology operative at the human physiological temperature of 310 K. This will provide a convenient and an unparallel source of knowledge for understanding and using the mysteries of quantum science to benefit humanity.

The data generated from quantum MRI measurements of human brain activity will be helpful in understanding the principles of quantum science operative in the human body. This data will also extend our understanding of the artificial systems, such as solid state, optical. This wealth of knowledge generated by medical quantum science would be of immense value to researchers working in the development of physical and biological systems of QMIP. In the field of MRI, the ultimate objective is to develop clinically useful imaging protocols. iZQCI may be capable of detecting tumors that are too small to be seen with conventional imaging. The idea is to produce optimal pulse sequences that can provide acceptable signal levels at the most popular imaging-field strength of approximately 1.5 T. This should also discriminate between malignant and benign tissue better than conventional MRI (Appendices A1.17–A1.19 [74]).

One should realize that the money and time spent on MRI research are worthless unless the knowledge accumulated is translated into easily digestible educational material for the benefit of future generations. One method that has been adapted is the quantum matrix mechanics approach, through which we can model and understand the complex evolution behavior of magnetic-field excited spins. This technique lacks a pictorial visualization of the behavior of spins behind the scenes. The second approach is product operator formalism, whereby successive steps of algebra can be written in reference to a step-by-step pictorial representation of the states of the spins. A physical picture depicting intermediate steps is highly educational and illuminating for those not conversant with quantum algebra. Appendices

A1.26–A1.28 show an illustration of efforts in this direction. They show that one can work out a detailed picture of the evolution of multiple spins (homonuclear as well as heteronuclear) starting from a simple 2-spin (IS) model to the more complex triple- and multi-quantum coherences. In the past, high-resolution NMR has used double- and triple-quantum transitions, by the direct, continuous-wave, slow-passage technique. This mode excludes ZQ transitions. It can be regarded as overriding formally forbidden transitions by the use of a strong RF field. With the new concepts being put forward today, it is necessary to postulate the MQC. This must be detected indirectly in a 2D experiment. The basic pulse for exciting DQC, in a homonuclear spin system, is $90°(X) - \tau - 180°(X) - \tau - 90°(X) - t_1 - 90°(X)$.

One looks for the pathways that eventually lead to observation of S-spin signals. The behavior of the I spins is related by symmetry considerations. One looks for an answer to the question of how to account for the special sensitivity of MQC to the transmitter effect, RF phase shifts, and magnetic-field gradients, and why the MQS remains unaffected by the active spin–spin couplings. One can take the usual shortcut of omitting chemical shift evolution whenever it is refocused by a 180° pulse. The preparation period extends to a point just before the second $90°(X)$. There are some unusual properties of DQC.

For example, the chemical shifts δ_I and δ_S are measured with respect to the transmitter frequency. A transmitter shift of Δ Hz affects both δ_I and δ_S, and results in a displacement of the DQ frequency by 2Δ Hz. In analogous fashion, a phase shift Φ of an RF pulse translates into a phase shift 2Φ of the DQC. It is of central importance for phase-cycling techniques designed to separate DQC from SQC. Finally, the application of a magnetic-field gradient, during DQ evolution, spreads out both the frequencies δ_I and δ_S, and disperses the DQC, twice as fast as an SQC. Consequently, after reconversion into an SQC, refocusing requires a gradient pulse of twice the intensity or twice the duration. The evolution of spin period is peculiar in the sense that the required correlation motion takes place between individual spins within a molecule; this feature cannot be directly related to the motion of a macroscopic vector associated with a single chemical site. Nevertheless, by momentarily focusing attention on the behavior of spins at the molecular level, it is possible to reconcile the microscopic and macroscopic models. Such a pictorial representation of MQ phenomena seems to provide a more transparent alternative to the product operator algebra. There are two contributions to the observed signals. Although the homonuclear experiment monitors the DQC, ZQC is also involved throughout the evolution and only disappears, through cancellation, with an equal and opposite polarization transfer component. Because this does not occur in the heteronuclear case, both ZQC and DQC frequencies are observed. This would also be true of the homonuclear case if site-selective excitation were used [81].

Brain gamma-aminobutyric acid (GABA) has long been of interest to neurologists and psychiatrists, owing to its integral role in such disorders as anxiety/panic, substance abuse/addiction, and epilepsy. GABA is the prime inhibitory neurotransmitter in the mammalian brain, exerting its effects via GABAergic synapses. The concentration of GABA in the brain is relatively low, ranging from 0.5 to 1.4 mmol/cm^3, but GABA is heterogeneously distributed throughout the brain. Past studies with animals and humans have consistently revealed a nearly twofold elevation of GABA levels in cortical gray matter (GM) compared with cerebral white matter (WM), a tissue-dependent difference not seen with more commonly measured metabolites such as choline (Cho), creatine (Cr), and *N*-acetylaspartate (NAA), which differ between tissues to a lesser extent than GABA. With the advent of magnetic resonance spectroscopy (MRS), scientists and clinicians have gained the ability to noninvasively detect and measure brain GABA levels in human subjects. However, with its low abundance and complicated multiresonance NMR structure, accurate and precise measurement of brain GABA in vivo is very difficult.

Several techniques using proton MRS have been devised to optimize GABA detection sensitivity, and also to localize GABA measurements to specific brain regions. The 2D *J*-resolved MRS imaging (JMRSI) acquisition sequence used a slice-selective spin-echo MRSI scheme modified to incrementally acquire spectra at each phase-encode step with increasing echo time to sample the *J*-coupling of the metabolites.

Once a global manual shim was completed, a set of T_1-weighted sagittal images was acquired, clearly displaying the central anatomy of the brain for slab positioning. The 3-cm-thick MRSI slab was positioned using the sagittal image that most clearly displayed the anterior commissure–posterior commissure (AC–PC) line on the defining slice; see Appendix A1.30, Figure A1.30(a) and (b). The T_1-weighted oblique image sets were separated into three binary tissue maps: GM, WM, and cerebrospinal fluid (CSF); see Appendix A1.31, Figure A1.31. This study provides strong evidence that in vivo, brain GABA concentration is significantly higher in the cortical GM than WM. Consistent elevations in cortical GM GABA levels (0.96–0.24 mm) versus WM (0.44–0.16 mm) have been demonstrated across a population of healthy human subjects. This work demonstrates the potential utility of this 2D, *J*-resolved GABA MRSI technique in clinical MRS studies of neurological and neuropsychiatric disorders in which pathology is specific to tissue type; see Appendix A1.31, Figure A1.31 [88]. Multiple-spin echoes (MSEs) and iMQCs in highly polarized systems have generated tremendous interest, as well as controversy, in the NMR community over the past few years.

These phenomena have been described using either classical theory for the demagnetizing field or quantum mechanical density matrix treatments. To date, both treatments have led to fully quantitative predictions of the signals for simple sequences, such as CRAZED experiments. One can determine the connection between the demagnetizing field and intermolecular dipolar coupling. The residual dipolar couplings between distant spins are responsible for the dipolar demagnetizing field, and give rise to the iMQCs. From the classical viewpoint, these phenomena are due to the demagnetizing field produced by the spatial modulation of the nuclear magnetization arising in the sample following the second pulse in the CRAZED sequence. Though some theoretical issues still remain to be addressed, intermolecular dipolar interaction effects have lost much of their mystical character and are becoming useful tools in NMR. Recently, there has been great interest in the potential of the MQC or MSE contrast mechanisms for MRI, because these contrast mechanisms may provide improved detection of tumors and eliminate the need for contrast agent injection. A CRAZED-like sequence was incorporated into a spin-echo imaging sequence for acquisition of DQC images; see Appendix A1.33 (Figure A1.33 [80]). As is applied in the standard CRAZED pulse sequence for spectroscopic experiments, gradients with a 1:n ratio of total gradient areas, before and after the $\beta(\pi/3)$ RF pulse, were used to select the n-quantum coherence order. For the DQC, $n = \pm 2$.

The DQCs are excited by the first $\pi/2$ pulse, and the β pulse transforms the DQCs back to SQCs. Because there is no homonuclear J-coupling in water protons, the detected signals result exclusively from the residual dipolar interaction. The first π pulse refocuses chemical shifts and magnetic-field inhomogeneities while retaining long-range dipolar couplings. The second π pulse removes the effects of inhomogeneity and chemical shift. When the second $\pi/2$ pulse in the standard CRAZED experiments is replaced with a $\pi/3$ (for $n = -2$ quantum transition) or a $2\pi/3$ (for $n = 5 + 2$ quantum transition) pulse, the maximum signal from the DQCs is increased by a factor $3\sqrt{3}/4$. It is simpler to design a smaller flip angle for uniform excitation, so a $\pi/3$ pulse ($n = 2$) was used for the β RF pulse. The phase-encoding and -dephasing pulses for the readout gradient were placed immediately before acquisition to avoid any interactions with coherence selections.

In Figure A1.34 [80], Appendix A1.34, DQC and conventional SQC T_2-weighted images from two slices of a brain are shown. Almost complete elimination of signals at the magic angle is evident. DQC images in a brain slice with different echo time (TE) values are displayed in Figure A1.35 [80], Appendix A1.35(c). There are apparent changes in contrast with TE among different types of tissue. Also, the signal intensities in the CSF space increase with TE. With region of interest (ROI) analysis, it was found that signal intensities in the a GM and WM decrease with TE but at rates different from that of conventional SQC imaging; there are deviations from exponential decays. The apparent paradox of the CSF signal "growth" with increases in TE is seen in accordance with the predictions of signal changes for long T_2 spins. There are two practical limitations on the existing DQF approaches for GSH in vivo editing. The first is that extra scans are required to achieve maximal editing yield. Because slice-selective pulses in the sequence induce phase increments, the phase of the DQ excitation pulse has to be set correctly to eliminate this phase accumulation and maximize the signal-to-noise ratio (SNR) by additional phase-calibration scans. The second is the influence of water presaturation pulses on the GSH yield, which has not been fully addressed in some studies.

With the proximity of the chemical shifts between the CH proton of GSH cysteinyl group (4.56 ppm) and water (4.70 ppm), the CH proton is generally assumed to be completely destroyed by water saturation pulses. The suppression of the CH proton of the GSH cysteinyl group seems to eliminate its influence on the detection of the signal arising from the CH_2 protons. However, if the CH protons are not completely suppressed by the water saturation pulses, the residual CH proton signal will be converted into CH_2 signal at 2.95 ppm by J-coupling. Because this residual CH proton signal depends on water saturation pulses that may vary from experiment to experiment, using water suppression may lead to variations in detected GSH signal. Recently, a new DQF has been introduced and combined with the

PRESS localization sequence. The distinguishing feature of the new filter is that frequency-selective pulses, instead of nonselective pulses, are used for both generating DQCs and converting them back into SQCs. With this modification, both problems described above are eliminated. The new sequence is successfully demonstrated on a phantom and in five healthy volunteers.

The specific localized DQF sequence is shown in Appendix A1.23 (Figure A1.23) [76]. A CHESS consisting of three Gaussian pulses, each followed by a gradient crusher, was used for water suppression. A 90° excitation pulse and two 180° pulses, which form the PRESS, were used to achieve spatial localization. The duration of the first echo time, t_1, was fixed (5 ms) and kept as short as possible, as this period was merely used for spatial localization. A frequency-selective 90° Gaussian pulse was applied symmetrically at each side of the last slice-selective 180° pulse.

Each of these two 90° pulses was 10 ms in duration (bandwidth 90 Hz at half height) and had a frequency set to that of the CH proton of cysteine moiety of GSH at 4.56 ppm. During the first TE/2 period, the CH_2 protons of the cysteine moiety evolved into antiphase terms via *J*-coupling with the CH proton, which was subsequently converted into MQCs by the first selective 90° Gaussian pulse. Because only the CH of the cysteinyl group is excited by the Gaussian pulse, the antiphase terms from the CH_2 are fully converted into DQCs regardless of the phase accumulation of the slice-selective pulses. Thus, phase-calibration scans for determining optimal phase for the DQ excitation pulses are avoided. Further, the signal produced by the CH proton of the cysteinyl group, which had been severely reduced by the water saturation pulses, was left in antiphase state by the first frequency-selective 90° Gaussian pulse and not converted into observable CH_2 signals around 2.95 ppm by *J*-coupling. This minimized potential contamination of the CH proton signal by the GSH signal if the CH proton was incompletely suppressed by the water saturation pulses. The signal acquisition started before the full TE, with exact timing. It led to in-phase peaks for the GSH, an NAA peak around 2.75 ppm, and an enhanced SNR for GSH. Due to the strong couplings between the two CH_2 protons, signal detection starting from the full echo does not ensure a pure absorption GSH lineshape. Thus, it is not possible to choose TE to achieve a pure absorption GSH peak and maximize the GSH yield at the same time. In this work, TE was chosen to be 70 ms. G_1 and G_2 were applied to crush magnetizations that did not originate from the DQC. At the onset of G2, the magnetization of interest had been converted into SQC by the second 90° Gaussian pulse. The ^1H spectra in Appendix A1.24 were acquired from the left parietal lobe of the five volunteers. The spectra from their right parietal lobes were similar to those shown here. A GSH signal at 2.95 ppm is clearly detected in all cases. A fraction of NAA resonance (2.4–2.8 ppm) coedits with GSH and becomes the dominant peak in the edited spectrum due to its high concentration. Due to their higher SNRs, the zeroth-order phasing for in vivo spectra was carried out based on the coedited NAA peaks around 2.4–2.8 ppm without extra scans. The peak between 1.30 and 1.40 ppm in the in vivo spectra was attributed to macromolecules.

References

[1] E. Knill, R. Laffamee, R. Martinez, C.H. Tseng, Nature 404 (2000) 368–370.
[2] C. Miquel, J.P. Paz, M. Saraceno, E. Knill, R. Laffamme, C. Negrevergne, Nature 418 (2002) 59–62.
[3] F. Schmidt-Kaler, H. Häfner, M. Riebe, S. Gulde, G.P.T. Lancaster, T. Desuschle, et al., Nature 422 (2003) 408–411.
[4] M.A. Nielson, E. Knill, R. Laflamme, Nature 396 (1998) 52–55.
[5] N.A. Greshenfeld, I.L. Chuang, Science 275 (1997) 350–356.
[6] J.A. Jones, M. Mosca, R.H. Hansen, Nature 393 (1998) 344–346.
[7] I.L. Chuang, L.M.K. Vandersypen, X. Zhou, D.W. Leung, S. Lloyd, Nature 393 (1998) 143–146.
[8] J.A. Jones, M. Mosca, J. Chem. Phys. 109 (1998) 1648–1653.
[9] S. Gulde, M. Riebe, G.P.T. Lancaster, C. Becher, J. Eschner, H. Hafner, et al., Nature 421 (2003) 48–50.
[10] A.D. Greentree, A.R. Hamilton, L.C.L. Hollenberg, R.G. Clark, Phys. Rev. B 71 (2005) 113310-1-3
[11] C.D. Hill, L.C. Hollenberg, A.G. Fowler, C.J. Wellard, A.D. Greentree, H.S. Goan, Phys. Rev. B 72 (2005) 045350-1-9
[12] M.A. Wilde, M. Rhode, Ch. Heyn, D. Heitmann, D. Grundler, U. Zeiner, et al., Phys. Rev. B 72 (2005) 165429-1-10
[13] R. Brenner, T.M. Bueher, D.J. Reilly, J. Appl. Phys. 97 (2005) 034501-1-5
[14] G.A. Prinz, Science 282 (1998) 1660–1663.
[15] J.H. Cole, A.D. Greentree, C.J. Wellard, L.C.L. Hollenberg, S. Prawer, Phys. Rev. B 71 (2005) 115302-1-10
[16] F. Jelezko, T. Gaebel, I. Popa, A. Gruber, J. Wrachtrup, Phys. Rev. Lett. 92 (2004) 076401-1-4
[17] D.G. Cory, A.F. Fahmy, T.F. Havel, et al., Proc. Nat. Acad. Sci. USA (94) (1997) 1634–1639.
[18] E.H. Huntington, G.N. Milford, C. Robilliard, T.C. Ralph, Opt. Lett. 30 (2005) 2481–2483.
[19] E.H. Huntington, T.C. Ralph, J. Opt. 4 (2002) 123–128.
[20] Y. Zhang, H. Wang, X.L.J. Jing, C. Xie, K. Peng, Phys. Rev. A (62) (2000) 023813-1-4
[21] E.H. Huntington, G.N. Milford, C. Robilliard, T.C. Ralph, O. Glöckl, U.L. Andersen, et al., Phys. Rev. A (71) (2005) 041802 (R)-1-4
[22] E. Knill, R. Laffamme, G.J. Milburn, Nature 409 (2001) 46–52.
[23] N. Korolkova, G. Leuchs, R. Loudon, T.C. Ralph, C. Silberhorn, Phys. Rev. A (65) (2002) 052306-1-12
[24] D. Bouwmeester, J.-W. Pan, K. Mattle, M. Eibel, H. Weinfurter, A. Zeilinger, Nature 390 (1997) 575–579.
[25] B.E. Cole, J.B. Williams, B.T. King, M.S. Sherwin, C.R. Stanley, Nature 410 (2001) 60–63.
[26] M.S. Sherwin, A. Imamoglu, T. Montroy, Phys. Rev. A 60 (1999) 3508–3514.
[27] M. Riebe, H. Haffner, C.F. Roos, W. Hansel, J. Benhelm, G.P.T. Lancster, et al., Nature 429 (2004) 734–737.
[28] J.P. Barjaktarevic, G.J. Milburn, R.H. McKenzie, Phys. Rev. A 71 (2005) 012335-1-5
[29] Q.A. Turchette, C.S. Wood, B.E. King, C.J. Myatt, D. Leibfried, W.M. Itano, et al., Phys. Rev. Lett. 81 (1998) 3631–3634.

[30] A. Einstein, B. Podolsky, N. Rosen, Phys. Rev. 47 (1935) 777–780.
[31] E. Hagley, X Maitre, G. Nogues, C. Wunderich, M. Brune, J.M. Raimond, et al., Phys. Rev. Lett. 79 (1997) 1–5.
[32] L.K. Grover, Phys. Rev. Lett. 79 (1997) 325–328.
[33] P. Walther, K.J. Resch, T. Rudolph, E. Schenck, H. Weinfurter, V. Vedral, et al., Nature 434 (2005) 169–176.
[34] K.J. Resch, S.H. Myrskog, J.S. Lundeen, A.M. Steinberg, Phys. Rev. A 64 (2001) 056101-1-3
[35] D.F.V. James, P.G. Kwiat, W.J. Munro, A.G. White, Phys. Rev. A (2001) 052312-1-15
[36] A.G. Fowler, L.C.L. Hollenberg, Phys. Rev. A 70 (2004) 032329-1-7
[37] Ch. Silberhorn, P.K. Lam, O. Weiβ, F. König, N. Korolkova, G. Leuchs, Phys. Rev. Lett. 86 (2001) 4267–4270.
[38] J.L. O'Brien, G.L. Pryde, A. Gilchrist, D.F.V. James, N.K. Langford, T.C. Ralph, et al., Phys. Rev. Lett. 93 (2004) 080502-1-4
[39] G. Leuchs, T.C. Ralph, C. Silberhorn, N. Korolkova, J. Modern Optics 46 (1999) 1927–1939.
[40] L.K. Grover, Science 280 (1998) 228–232.
[41] G. Brassard, Science 275 (1997) 627–628.
[42] B. Koiller, X. Hu, H.D. Drew, S. Ds Sarma, Phys. Rev. Lett. 90 (2003) 067401-1-4
[43] E. Knill, R. Laflamme, W.H. Zurek, Science 279 (1998) 342–345.
[44] J. Ahn, T.C. Weinacht, P.H. Bucksbaum, Science 287 (2000) 463–465.
[45] C.H. Bennett, D.P. DiVincenzo, Nature 404 (2000) 247–255.
[46] D.T. Pope, G.J. Milburn, Phys. Rev. A 69 (2004) 052102-1-11
[47] B. Schumacher, Phys. Rev. A 51 (1995) 2738–2747.
[48] P.G. Kwiat, E. Waks, A.G. White, I. Appelbaum, P.H. Eberhard, Phys. Rev. A 60 (1999) R773–R776.
[49] A.G. White, D.F.V. James, P.H. Eberhard, P.G. Kwiat, Phys. Rev. Lett. 83 (1999) 3103–3107.
[50] R.A. Kaindl, M.A. Carnahan, D. Hagele, R. Lovenich, D.S. Chemla, Nature 423 (2003) 734–738.
[51] R.J. Lipton, Science 268 (1995) 542–545.
[52] J.F. Clauser, A. Shimony, Rep. Prog. Phys. 41 (1978) 1881–1927.
[53] S. Parkins, E. Larsabal, Phys. Rev. 63 (2000) 012304-1-17
[54] J.-A. Larson, Phys. Rev. A (67) (2003) 022108-1-8
[55] S.J. Devitt, A.G. Fowler, L.C.L. Hollenberg, arXiv:quant-ph/0408081 v3 28 Jan 2005, 1–11.
[56] L.M.K. Vandersypen, M. Steffen, G. Breyta, C.S. Yannoni, M.H. Sherwood, I.L. Chuang, Nature 414 (2001) 883–887.
[57] A.G. White, A. Gilchrist, G.J. Pryde, J.L. Obrien, M.J. Bremner, N.K. Langford, arXiv: quant-ph/0308115V1'22 August 2003 (www.quantinfo.org).
[58] B. Yurkee, A.P. Mills Jr., S.L. Cheng, Biosystems 52 (1999) 165–174.
[59] D. Xiao, W. Li, Z. Zhang, L. He, Biosystems 82 (2005) 203–207.
[60] A. Schuster, Biosystems 81 (2005) 234–246.
[61] J. Jones, Prog. NMR Spectrosc. 38 (2001) 325–360.
[62] V. Vedral, A. Barenco, A. Ekert, Phys. Rev. A 54 (1996) 147–153.
[63] D. Beckman, A.N. Chari, S. Devabhaktuni, J. Preskill, Phys. Rev. A 54 (1996) 1034–1063.
[64] M.M. Kaila, J. Supercond. 18 (2005) 1–5.
[65] L.P. Rosa, J. Faber, Phys. Rev. E 70 (2004) 031902-1-6

[66] J.A. Brown, T.A Tuszynski, Phys. Rev. E 56 (1997) 5834–5840.
[67] J. Faber, R. Portugal, L.P. Rosa, Biosystems 83 (2006) 1–9.
[68] O. Hosten, M.T. Rakher, J.T. Barreiro, N.A. Peters, P.G. Kwiat, Nature 439 (2006) 949–952.
[69] A.G. White, J.R. Mitchell, O. Nairz, P.G. Kwiat, Phys. Rev. A 58 (1998) 605–613.
[70] S. Hagan, S.R. Hamerroff, J.A. Tuszynski, Phys. Rev. E 65 (2002) 061901-1-11
[71] A. Tuszynaski, S. Hameroff, M.V. Satarics, B. Trpisova, M.L.A. Nip, J. Theor. Biol. 174 (1995) 371–380.
[72] S. Lee, W. Richter, S. Vathyam, W.S. Warren, J. Chem. Phys. 105 (1996) 874–900.
[73] W. Richter, M. Richter, W.S. Warren, H. Merkle, P. Andersen, G. Adriany, et al., Magn. Reson. Imag. 18 (2000) 489–494.
[74] R.R. Rizi, S. Ahn, D.C. Alsop, S. Garrett-Rose, M. Mescher, W. Richter, et al., Magn. Reson. Med. 43 (2000) 627–632.
[75] X. Tang, H. Ong, K. Shannon, W.S. Warren, Magn. Reson. Imag. 21 (2003) 1141–1149.
[76] T. Zhao, K. Heberlein, C. Joans, D.P. Jones, X. Hu, Magn. Reson. Med. 55 (2006) 676–680.
[77] Q. He, W. Richter, S. Vathyam, W.S. Warren, J. Chem. Phys. 98 (1993) 6779–6800.
[78] S.E. Ashbrook, M.J. Duper, Concepts Magn. Reson. 28A (2006) 183–248.
[79] Y. Assaf, Y. Cohen, Magn. Reson. Med. 43 (2000) 191–199.
[80] J. Zhong, Z. Chen, E. Kwok, Magn. Reson. Med. 43 (2000) 335–341.
[81] R. Freeman, Concepts Magn. Reson. 10 (1998) 63–84.
[82] S. Ahn, W.S. Warren, L. Sanghyuk, J. Magn. Reson. 128 (1997) 114–129.
[83] F. Bloch, J. Magn. Reson. 102 (1956) 104–135.
[84] J. Battiste, R.A. Newmark, Progr. NMR Spectrosc. 48 (2006) 1–23.
[85] S. Vathyam, S. Lee, W.S. Warren, Science 272 (1996) 92–96.
[86] W.S. Warren, S. Ahn, M. Mescher, M. Garwood, K. Ugurbil, W. Richter, et al., Science 281 (1998) 247–251.
[87] M.M. Kaila, Hot electron non-equilibrium high temperature superconductor THz radiation sensing and the integrated electron cooling, Chapter, Book: New Topics in Superconductivity Research, ISBN: 1-59459-985-0, Nova Science Publishers, Hauppauge, NY, USA, 2006.
[88] J. Eric Jensen, B.deB. Frederick, P.F. Renshaw, NMR Biomed. 18 (2005) 570–576.

Appendix A1.1 [61]

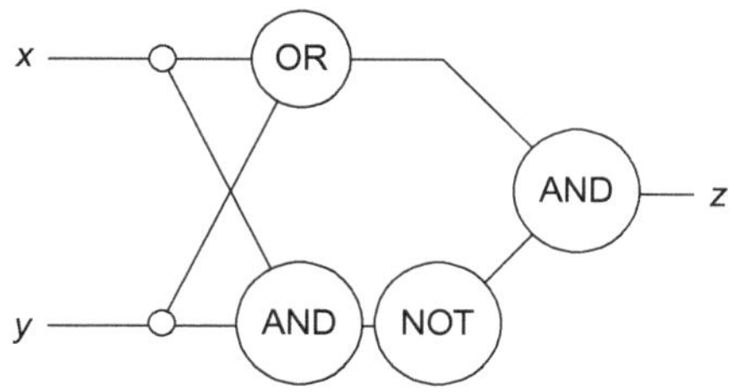

Figure A1.1 A circuit to compute the XOR function, $z = x$ XOR y, using AND, OR, NOT gates. Note that this circuit uses three implicit gates: two *C* gates (shown as small circles) where wires split into two (to copy the input) and one swap gate (where the wires cross over). These implicit gates are fairly easy to implement in traditional ECs, but can cause problems in other designs and therefore cannot simply be ignored. Some authors even consider the wires that interconnect gates as nontrivial gates in their own right.

Appendix A1.2 [61]

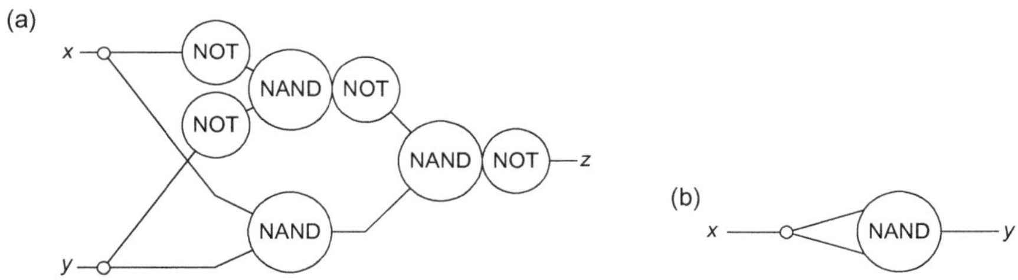

Figure A1.2 (a) A circuit to compute the XOR (XOR) function, $z =$ XOR y, using only NAND and NOT gates. This is not the best such circuit. Simpler circuits are known. This one preserves the basic structure seen in Figure A1.1. (b) A circuit to implement a NOT gate, $y =$ NOT x, using a NAND gate. By combining circuits (a) and (b), it is possible to implement XOR using only NAND gates. Any other function may be computed in a similar fashion, and so NAND is a universal gate. Note, however, that both circuits use implicit clone gates, and (a) also uses one implicit swap gate.

Appendix A1.3 [62]

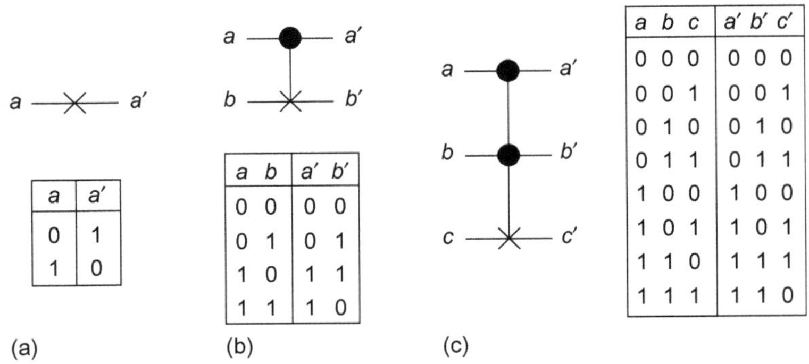

a	a′
0	1
1	0

a b	a′ b′
0 0	0 0
0 1	0 1
1 0	1 1
1 1	1 0

a b c	a′ b′ c′
0 0 0	0 0 0
0 0 1	0 0 1
0 1 0	0 1 0
0 1 1	0 1 1
1 0 0	1 0 0
1 0 1	1 0 1
1 1 0	1 1 1
1 1 1	1 1 0

Figure A1.3 Truth tables and graphical representations of the elementary QGs used for the construction of more complicated quantum networks. The control qbits are graphically represented by a dot, the target qbits by a cross. (a) NOT operation. (b) control-NOT. This gate can be seen as a "copy operation" in the sense that, after the action of the gate, a target qbit (b) initially in the state 0 will be in the same state as the control qbit. (c) Toffoli gate. This gate can also be seen as control–control-NOT: the target bit (c) undergoes a NOT operation only when the two controls (a) and (b) are in state 1.

Appendix A1.4 [59]

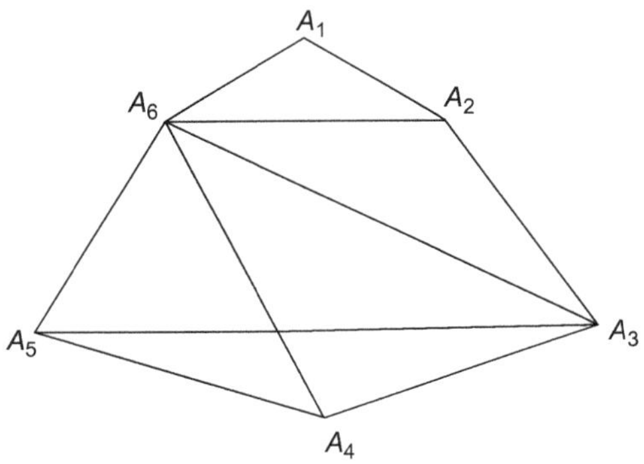

Figure A1.4 Graph *G*.

Appendix A1.5 [15]

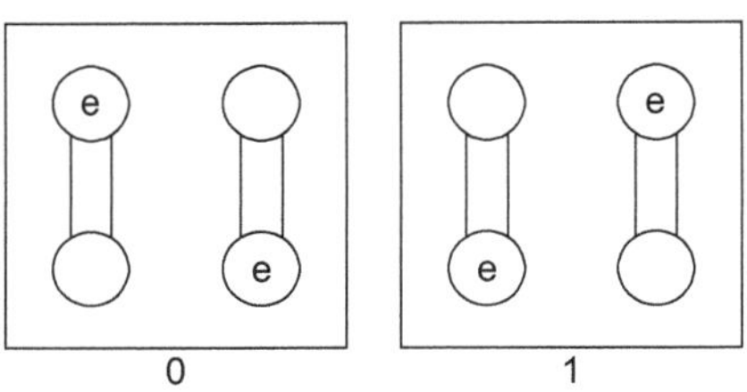

Figure A1.5 Two possible states for a basic QDCA cell, where the 0 and 1 states constitute the ground or "computational" states and "e" labels the position of electrons.

Appendix A1.6 [15]

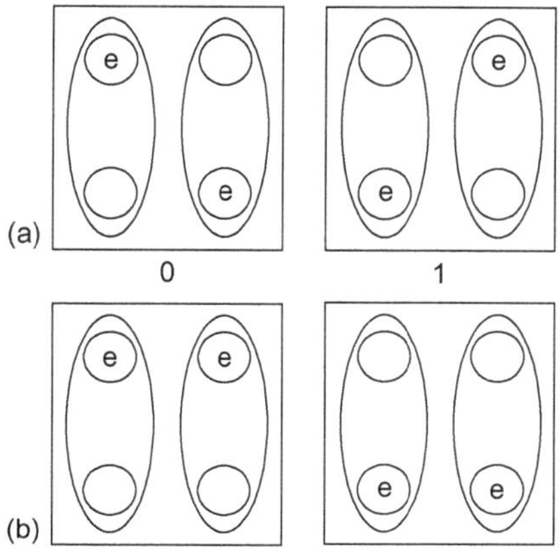

Figure A1.6 (a) The ground states of the buried donor cellular automata cell where the positions of the electrons "e" are designated by top (a) and bottom (b). These computational states are referred to as |TB⟩ and |BT⟩, and are assigned the logical values of 0 and 1, respectively. (b) The excited or noncomputational states are labeled |TT⟩ and |BB⟩, respectively, and correspond to the first excited state of the system.

Appendix A1.7 [65]

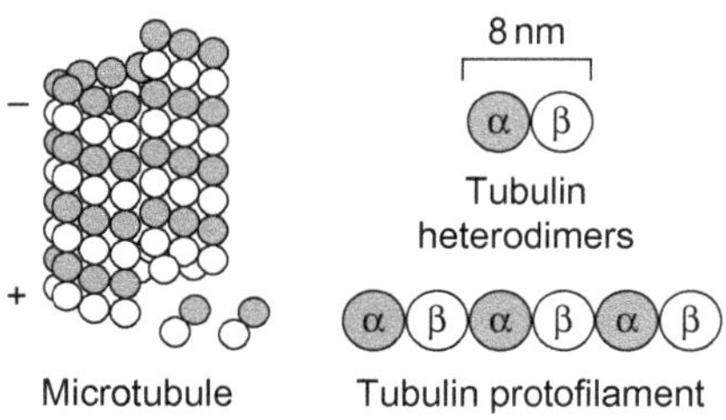

Figure A1.7 Illustration of MT, tubulin, and protofilament.

Appendix A1.8 [66]

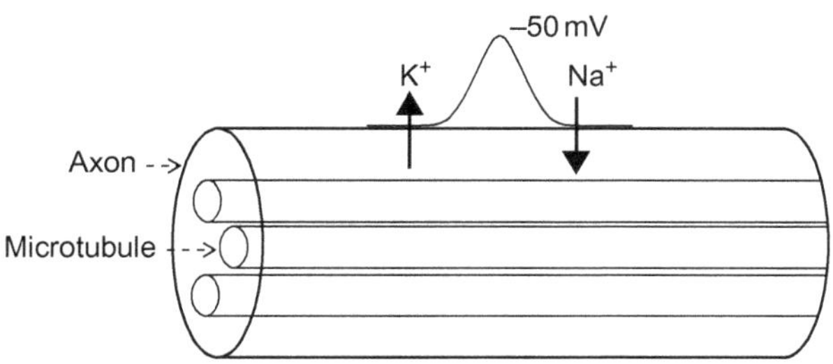

Figure A1.8 An action potential moving an axon containing three MTs. An electric field is caused by the potential difference once the ions are displaced, and a magnetic field is caused by the moving ions. The potential difference is typically 50 mV and its magnitude is represented by the curve above the axon. Given the membrane thickness, the transient electric field may be as high as 10^5 V/m within the membrane.

Appendix A1.9 [67]

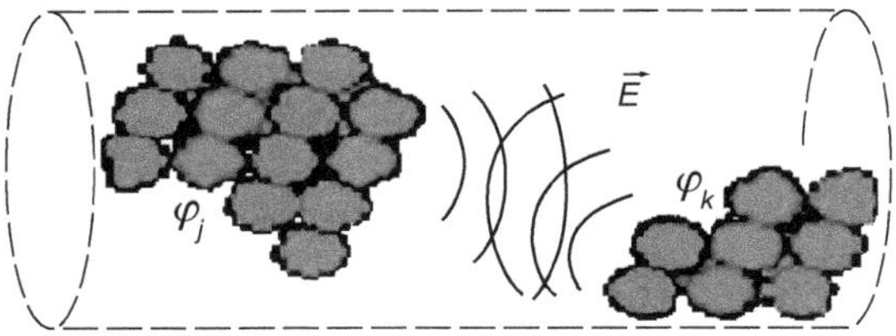

Figure A1.9 A representation of the communication between domains accomplished by the electromagnetic field on the MTs.

Appendix A1.10 [67]

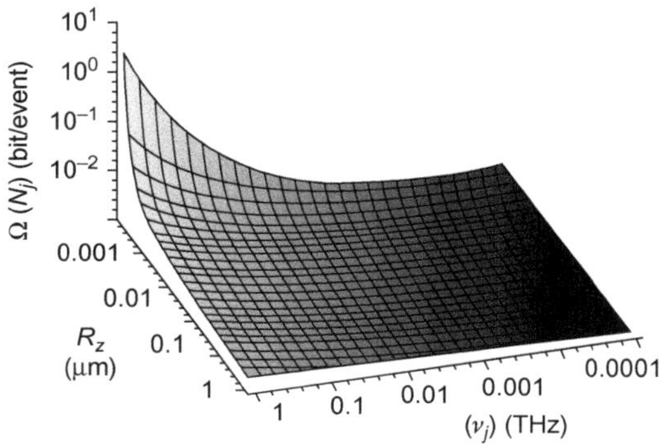

Figure A1.10 Communication capacity: frequency ν_j, distance R_z, when $T \sim 300$ K.

Appendix A1.11 [68]

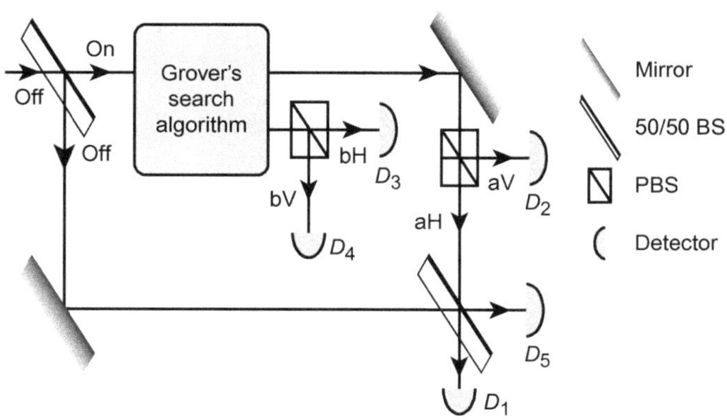

Figure A1.11 An optical realization of counterfactual computation. By means of a 50/50 BS (which serves as a $\pi/2$ rotation), an H-polarized single photon is in superposition of passing and not passing through the algorithm, encoding the "operating switch" in different spatial modes, "On" and "Off." Then, on a second 50/50 BS, the two histories are interfered only if the photon after the algorithm is in the mode $|aH\rangle$. The modes $|aH\rangle$ and $|aV\rangle$ are distinguished via a PBS that transmits H and reflects V.

Appendix A1.12 [68]

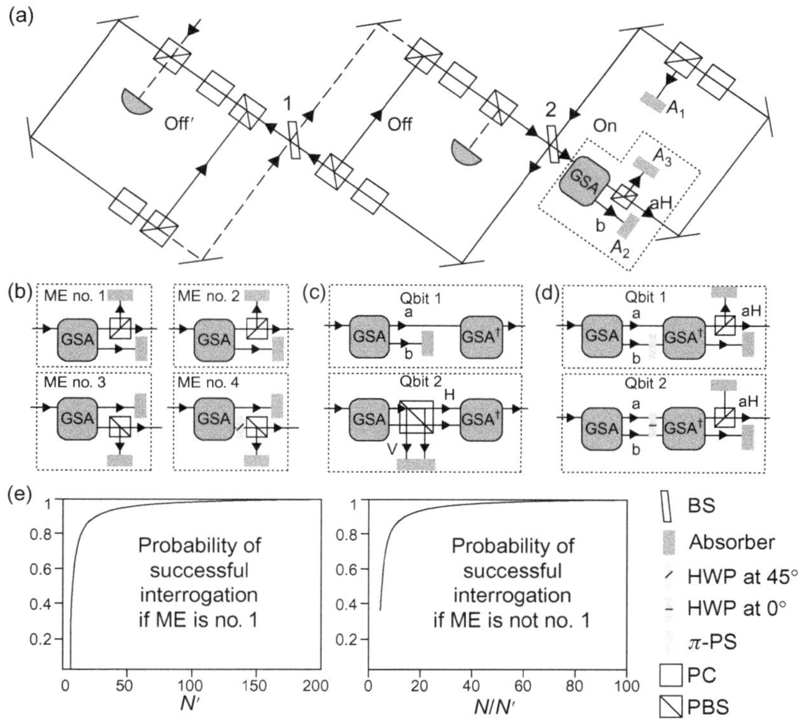

Figure A1.12 Proposed setup for the CQZS effect. Three cavities correspond to three states of the switch $|\text{Off}'\rangle$, $|\text{Off}\rangle$, and $|\text{On}\rangle$, separated by BSs—the rotation operators. Pocket cells (PCs) rotate the polarization by 90° on demand; half-wave plates (HWPs) at 45° rotate the polarization by 90°. (a) Interrogation for element number 1. A_1–A_3 are absorbers.
(b) Settings for the interrogation of different elements. (c) Setup configurations for qbit-by-qbit interrogation. $\text{GSA}^\dagger$ undoes the action of GSA. (d) Configurations for error suppression. π-PS induces a π-phase shift on path b; HWP at 0° induces a π-phase shift on polarization V. (e) Probability of successful interrogation for the setup in (a), as function of the cycling parameters N' and N (numerically evaluated).

Appendix A1.13 [69]

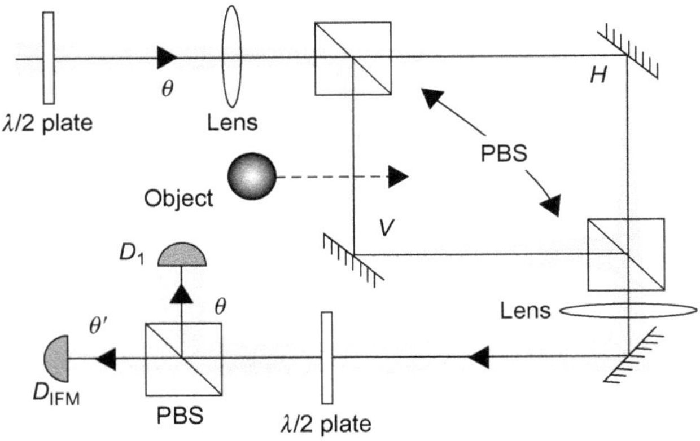

Figure A1.13 Polarizing MZI. PBS denotes the polarizing beam splitter and λ/2 the half-wave plate at 670 nm. The locking laser (not shown) entered from the top port of the first PBS and exited from the side port of the second.

Appendix A1.14 [71]

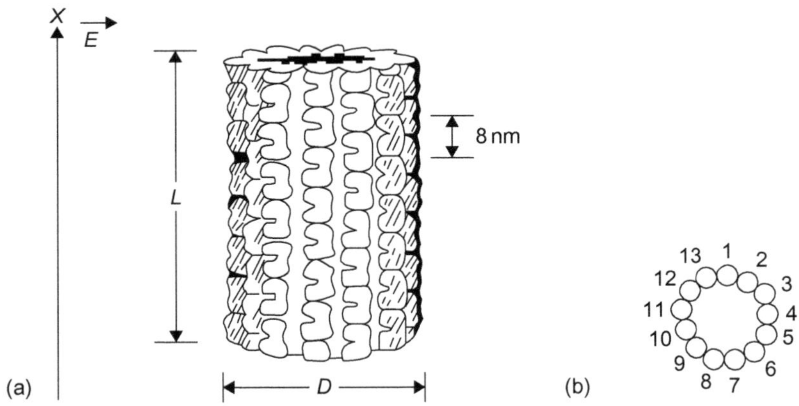

Figure A1.14 An illustration of a microtubular arrangement.

Appendix A1.15 [71]

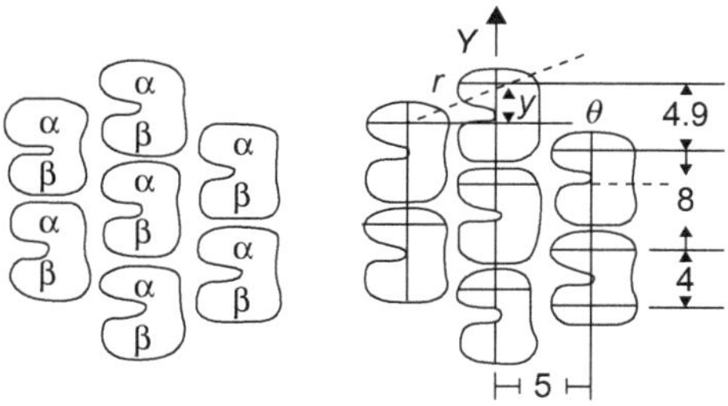

Figure A1.15 A tubulin dimer and its NNs.

Appendix A1.16 [72]

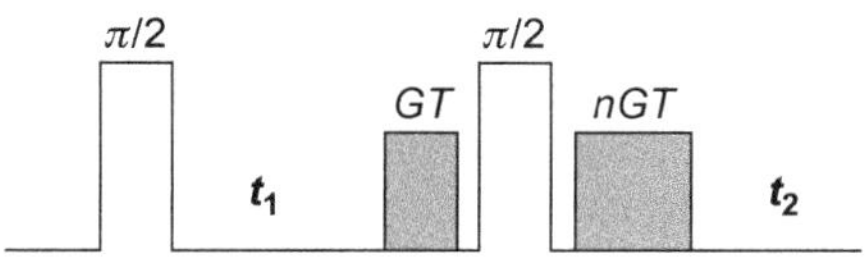

Figure A1.16 The CRAZED pulse sequence. This is a two-pulse COSY sequence modified with an n-quantum filter at the second gradient pulse. If $n \neq 1$, no signal is predicted in conventional NMR theory, but experimentally this sequence produces large signals that are explained by the quantum mechanical and classical treatments presented.

Appendix A1.17 [74]

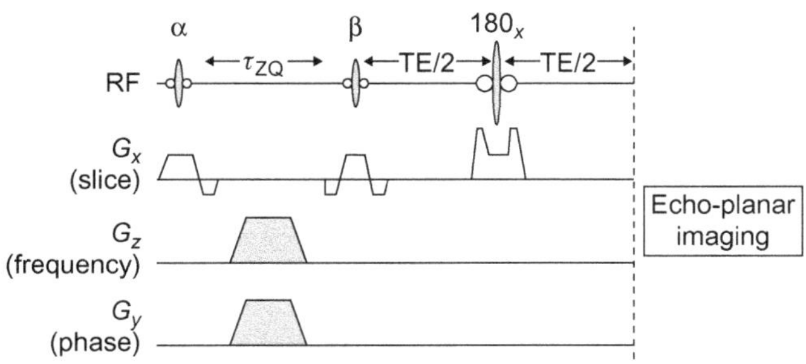

Figure A1.17 Schematic diagram of an iZQC imaging pulse sequence. A standard-spin echo-planar imaging sequence was modified to include a slice-selective preparation pulse and filter gradient between the normal excitation and refocusing pulses.

Appendix A1.18 [74]

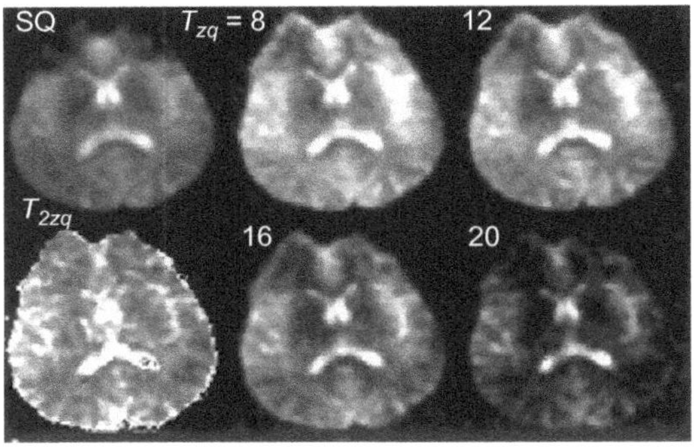

Figure A1.18 Conventional spin-echo-planar image (top left) is compared to four different iZQC images using spin-echo-planar imaging (right). The iZQC intensity and contrast are different from any conventional image, particularly in regions with large susceptibility variations. For short values of T_{zq}, the iZQC has a magnetization-squared weighting. The figure at the bottom left shows a "map" of iZQC relaxation times by pixel (white, 60 ms; black, 10 ms). Other imaging parameters include field of view (FOV) 24 cm × 24 cm, matrix size 128 × 64, slice thickness of 10 mm, TE = 60 ms, NEX = 16, and TR = 4 ms.

Appendix A1.19 [74]

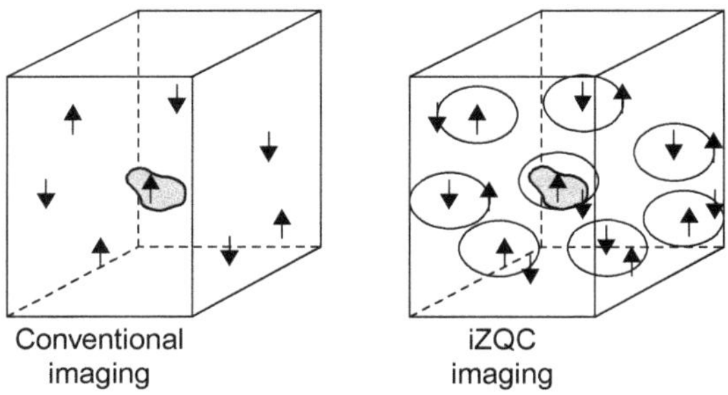

Figure A1.19 iZQCs are expected to reveal resonance-frequency differences between pairs of spins separated by approximately $d = (\pi/gGT)$, where G and T are the strength and length of the correlation gradient, respectively. Typically $d \sim 100$ mm, which is much smaller than voxel dimensions. Thus, iZQC images can be more sensitive than conventional images to subvoxel structure.

Appendix A1.20 [75]

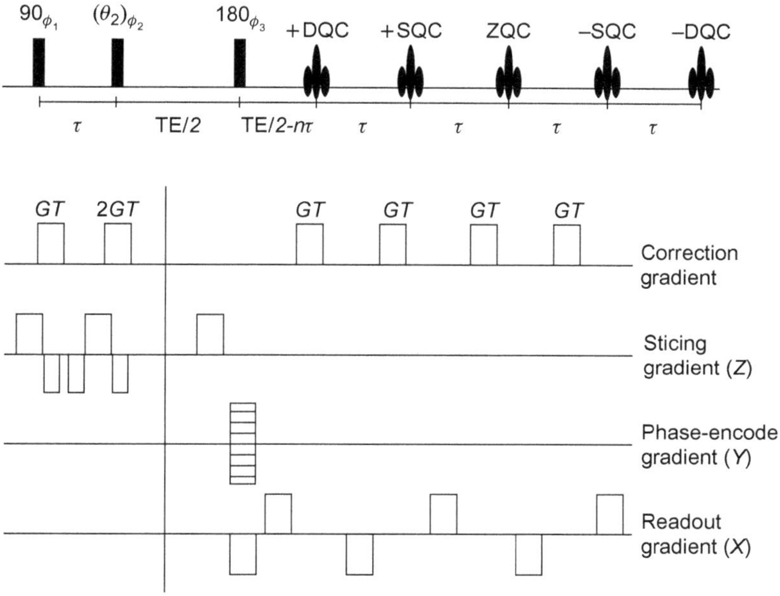

Figure A1.20 MultiCRAZED pulse sequence, adapted as an imaging sequence. The case illustrated here does single-line imaging for each pulse-sequence repetition; all five echoes sample the same line in *k*-space. The sequence can be adapted in the usual way for other *k*-space sampling, e.g., as an echo-planar imaging sequence.

Appendix A1.21 [73]

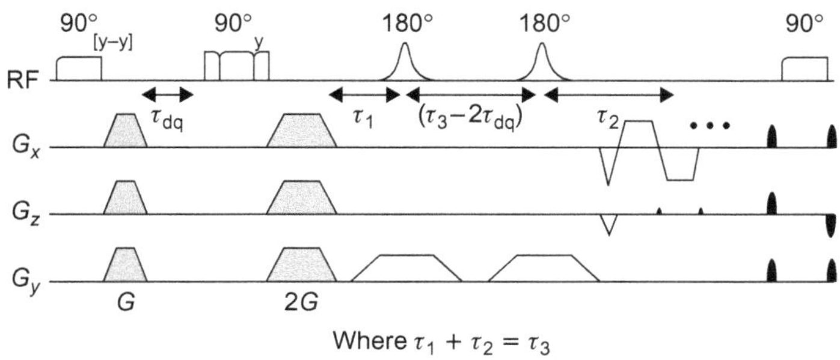

Figure A1.21 Intermolecular double-quantum coherence (iDQC) pulse sequence. The iDQC preparation period (including the pair of correlation gradients) is followed by a double-spin echo and a segmented echo-planar image readout sequence. Note that the timing is modified from the simplified version that is described in the text, on account of the double-spin echo. The effective evolution time after the second gradient is $2\tau_{dq}$, as required.

Appendix A1.22 [73]

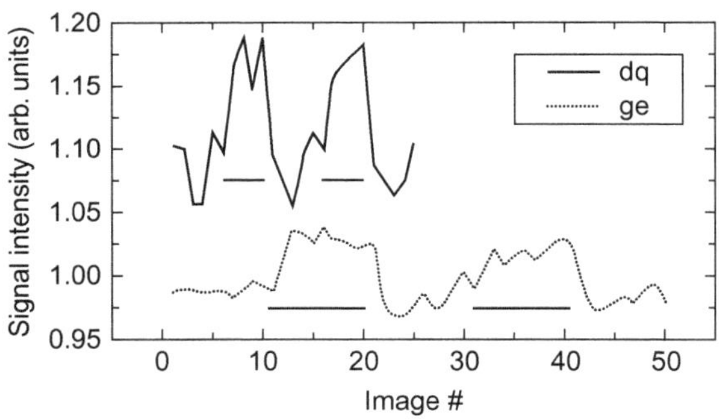

Figure A1.22 Time courses from activated pixels for BOLD (bottom) and iDQC (top) methods. Note that the signal intensity change is larger in the iDQC time course, as are the baseline fluctuations.

Appendix A1.23 [76]

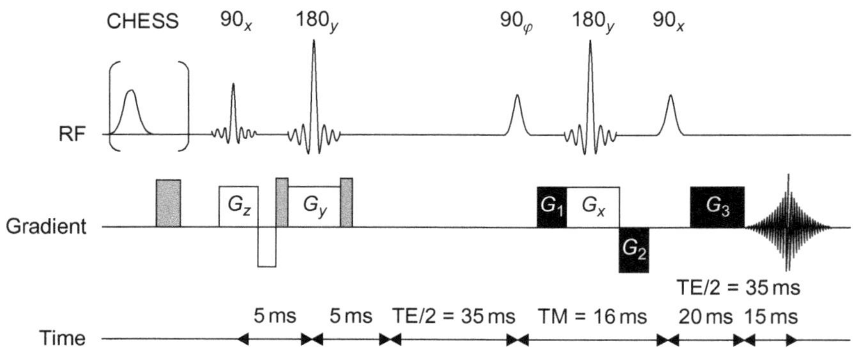

Figure A1.23 Volume-selective DQF with two frequency-selective 90° Gaussian pulses, preceded by CHESS. TM is 16.0 ms and t_1 is 5.0 ms. The gradients marked in black are used to elect the DQC. Spoiling gradients (gray) are applied along the first 180° pulse. The data acquisition starts at 20 ms before the echo. The duration of G_1 and G_2 are 1.0 and 4.0 ms, respectively. The amplitudes for G_1 and G_2 are both 22.0 mT/m. Phase-cycling scheme: first 90° Gaussian pulse, x, $-x$, x, $-x$; second 90° Gaussian pulse, x, x, $-x$, $-x$; ADC, x, $-x$, $-x$, x.

Appendix A1.24 [76]

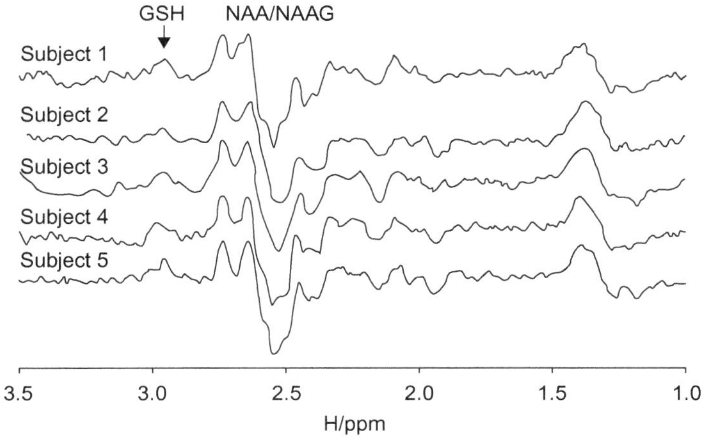

Figure A1.24 In vivo ^{1}H NMR spectra of GSH using the new DQC filter.

Appendix A1.25 [77]

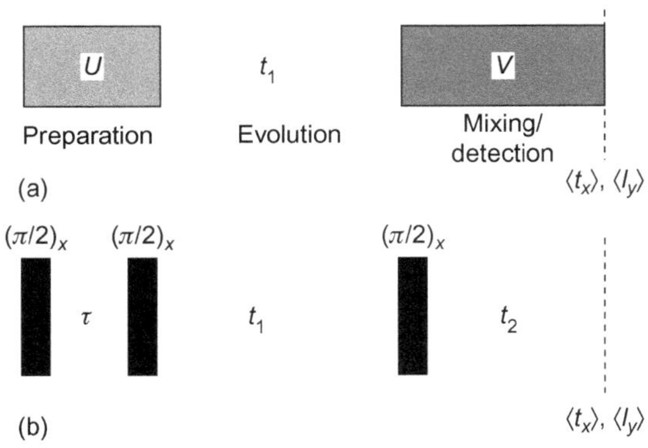

Figure A1.25 (a) Generic sequence for 2D experiments. (b) Simplest MQ pulse sequence. Fourier transformation of the signal with respect to t_1 can give frequencies that correspond to transitions with $[\Delta M] \neq 1$. Fourier transformation with respect to t_2 gives only frequencies with $[\Delta M] = 1$. This three-pulse sequence is a very simple example of the general structure in (a).

Appendix A1.26 [72]

The Equilibrium Density Matrix: Scalar Couplings

$\rho_{eq} = \exp(-H/kT)/\mathrm{Tr}\{\exp(-H/kT)$; $[i(h/2\pi)(\partial\rho/\partial t) = [H, \rho]$; $\rho(t) = U\rho(0)U^{+}] = 2^{-N}\Pi_i(1 - \Im I_{zi})$; $\Im = 2\tanh(h/2\pi(\omega_0)/2kT)$. *Chemical shift*: $\sigma_{eq} = \exp(-H/kT)/\mathrm{Tr}\{\exp(-H/kT)$; $[i(h/2\pi)(\partial\sigma/\partial t) = [H, \sigma]$, $\sigma(t) = U\sigma(0)U^{+}] = 1/2\{1-2\tanh[(h/2\pi(\omega_0))/2kT]I_z\}$.

Observed Signal

$\langle I_x\rangle = \mathrm{Tr}\{I_x(U\rho_{eq}U^{+}):\langle I_y\rangle = \mathrm{Tr}\{I_y(U\rho_{eq}U^{+})$

The operators $I_x = \Sigma_i x_i$ and $I_y = \Sigma_i y_i$, which represent the signal, are 1-spin, 1-quantum operators. All terms proportional to $\Im^2$ are 2-spin operators, and are not directly observable. In general, 2-spin operators can only be converted into 1-spin operators by a bilinear operator in the Hamiltonian (dipolar or scalar couplings). Any individual spin is scalar-coupled to far fewer than 10^4 other spins, so if dipole couplings can be ignored, the contribution to observable signals from the $\Im^2$ and higher terms is indeed small. The dipolar Hamiltonian can couple each spin to a very large number of other spins, albeit weakly, and this is what leads to the signal one needs to calculate.

Appendix A1.27 [72]

Double-Quantum (2Q) Crazed Sequence (Appendix A1.16/Figure A1.16)

During the first gradient pulse, the magnetization becomes dependent on position and dipolar couplings are reintroduced. To simplify the calculations, one assumes that the gradient pulses are short compared to the time in which dipolar couplings take effect. The dipolar couplings will not be introduced until after the second gradient pulse. The Hamiltonian during the gradient pulse is only the Zeeman Hamiltonian. However, the resonance frequency is spatially modulated by the gradient pulse. Starting from the equilibrium density matrix, the density matrix, after the second pulse $\pi/2$ and the gradient pulse, is $\rho(t_1, t_2 = 0) = 2^{-N}\Pi_i\{1i + \Im I_{zi}\cos(\Delta\omega t_1 + \gamma GTs_i) - \Im I_{yi}\sin(\Delta\omega t_1 + \gamma GTs_i)\cos(2\gamma GTs_i) + \Im I_{xi}\sin(\Delta\omega t_1 + \gamma GTs_i)\sin(2\gamma GTs_i)m$. An n-spin term leading to an observable operator has the form $I_{xi(yi)}\Pi_{j=1,\ j\neq 1}{}^{(n-1)}(I_{zj})$. Only the operators with a single transverse operator (I_x or I_y) produce an observable signal. Commutation with terms such as D_{ij}, I_{zi}, I_{zj} never strip out the transverse operators.

Appendix A1.28 [72, 81, 82]

Density matrix $= 2^{-N}\Im\Sigma_{i=1}{}^{N}[I_{xi}\sin(2\Delta\omega t_1) + I_{yi}\cos(2\Delta\omega t_1)](\Lambda_i)J_2(2\Lambda_i)$; $\Delta_s = [3(sz)^2 - 1]/2$ and $s = rs$; $i = \Im t_2(3/4)\Sigma_{j=1}{}^{N}D_{ij}\cos\gamma GT(s_j - s_i)$; J_n is Bessel function of order n; $\Sigma_{j=1}{}^{N} D_{ij}\cos[\gamma GT(s_j) = -[\mu_0(h/2\pi)\gamma^2/(4\pi)]\Sigma_{j=1}{}^{N}\{[3\cos^2\theta - 1]/(4r_{ij}{}^3)$ $\cos(\gamma GT)\}$.

Observed signal $= M^+(t_1, t_2) = \text{Tr}[\rho(t_1, t_2)\gamma(h/2\pi)(I_x + iI_y) = iM_0\exp[(-2i\Delta\omega t_1)$ $\exp(i\Delta\omega t_2)(2\tau_d/t_2\Delta_s)J_2(-t_2\Delta_s/\tau_d)$

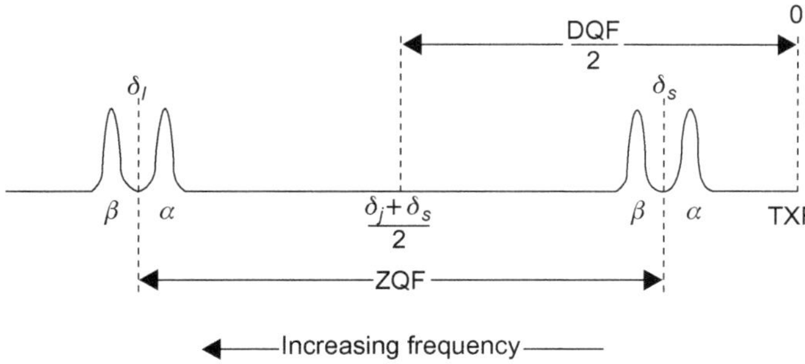

Figure A1.26 Schematic spectrum of a weakly coupled homonuclear 2-spin system (IS) showing the zero-quantum frequency (ZQF), which is independent of the transmitter frequency (TXR) and the DQ frequency. The TXR is set at an arbitrary point in the spectrum. We employ the "left-hand rule": a positive chemical shift corresponds to a clockwise rotation in the XY plane viewed from the $+Z$-axis, and the β vectors precess faster than the α vectors.

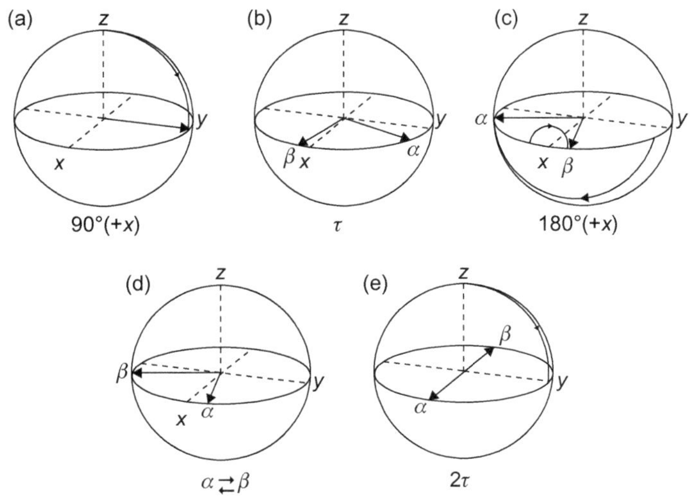

Figure A1.27 (a) Preparation of inverted antiphase S-spin magnetization aligned along the $\pm$ axes by means of a spin-echo sequence with the intervals t set equal to $1/(4J_{IS})$. (b) The α and β vectors evolve under chemical shift and spin–spin coupling until they subtend an angle of 90°. (c) The 180°(X) pulse rotates these vectors into the mirror-image configuration. (d) The interchange $\alpha \leftrightarrow \beta$ is achieved by a 180° pulse applied to the I-spin partner. (e) Further free precession refocuses the chemical shift and leaves the a vector along $+X$ and the β vector along $-X$.

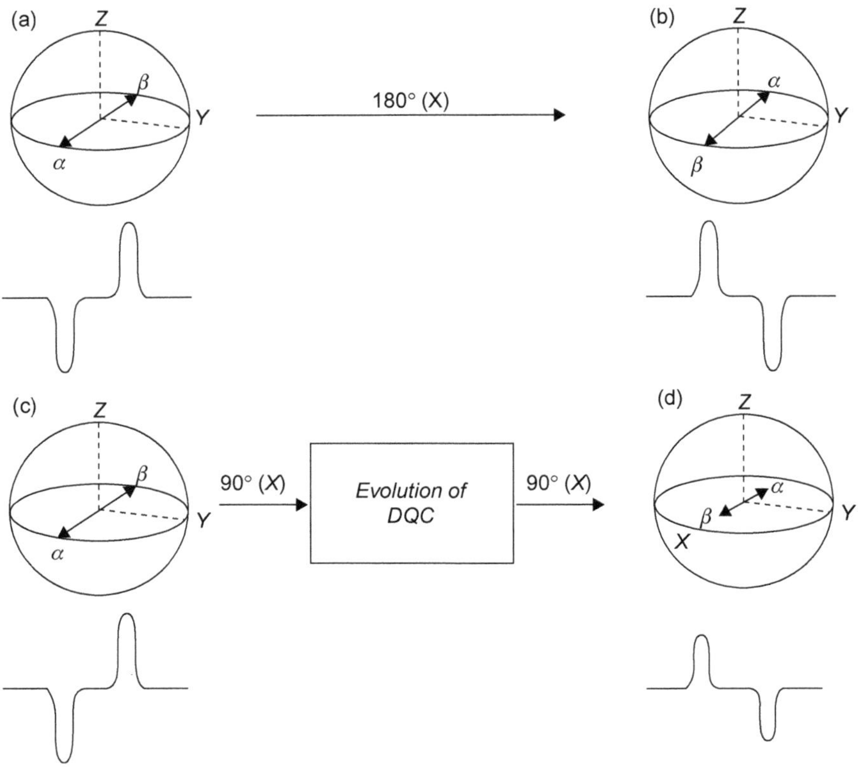

Figure A1.28 Antiphase S-spin magnetization (a) is inverted (b) by an 180°(X) pulse applied to the I spins. (c) When two 90° pulses are applied, separated by an evolution interval, an intermediate state of DQC is created. (d) After the second 90°(X) pulse, the overall result is similar to that in (b).

The principles of a density matrix treatment for iMQCs for a single-component sample and uncoupled S spins are easy to handle mathematically. This treatment can be extended to calculate the signal from the iMQCs in the presence of the J-coupling. The pulse sequence for the intermolecular CRAZED experiment is shown in Figure A1.16 [82]. For simplicity in notation, one considers the case of a $^{13}CHCl_3$ sample. Here ^{13}C and H spins are J-coupled. I and S denote two different kinds of spins, $I = H$, $S = {}^{13}C$. One assumes for simplicity that the proton signal is detected, with the detection period being t_2. The secular part of the Hamiltonian for this spin system can be represented as

$$
\begin{aligned}
H/\hbar &= \sum_{i}^{N}(\Delta\omega_I I_{zi} + \Delta\omega_s S_{zi} + 2\pi J I_{zi} S_{zi}) \\
&+ \sum_{i}^{N}\sum_{j}^{N}\left(3D_{ij}^{II} I_{zi} I_{zj} + 3D_{ij}^{SS} S_{zi} S_{zj}\right) \\
&+ \sum_{i}^{N}\sum_{j}^{N} 2D_{ij}^{IS} I_{zi} I_{zj}
\end{aligned}
$$

The first part is the contribution from the Zeeman effects and the J-coupling. The second and third parts account for the contributions from homonuclear and heteronuclear dipolar couplings, respectively, where the dipolar coupling constants are defined by

$$
D_{ij}^{IS} = \frac{\mu_0}{4\pi}\frac{\gamma_I \gamma_S \hbar}{4}\frac{1 - 3\cos^2\theta_y}{r_{ij}^3}
$$

$$
D_{ij}^{IS} = \frac{\mu_0}{4\pi}\frac{\gamma_I^2 \hbar}{4}\frac{1 - 3\cos^2\theta_y}{r_{ij}^3}
$$

Here r_{ij} is the separation between spins i and j. θ is the angle between the internuclear vector and the main magnetic field, and μ_0 is the vacuum permeability. The most useful type of heteronuclear coherence would be DQCs and ZQC. This involves SQC in the I-spin and the S-spin. For the ^{1}H and ^{13}C heteronuclear case, one can write $\gamma_s/\gamma_I = 1/4$. For the $n = \pm 5/4$ case, one gets the DQ peaks located at $(F_1, F_2) = [\pm(\Delta\omega_I + \Delta\omega_S), \Delta\omega_I]$.

$$
\begin{aligned}
M_{2Q}^{I+}(t_2, t_2) &= iM_0^I \exp(i\Delta\omega_I t_2)\exp\{\mp i(\Delta\omega_I + \Delta\omega_S)t_1\} \\
&\times J_1\left(-\frac{t_2\cos(\pi J t_1)}{\tau dI}\right) J_1\left(-\frac{2}{3}\frac{\gamma_I}{\gamma_S}\frac{t_2\cos(\pi J t_1)}{\tau dS}\right) \\
&\times \left[\cos(\pi J t_1)\cos(\pi J t_2)\left(\frac{\tau_{dI}}{t_2\cos(\pi J t_1)}\right) - \sin(\pi J t_1)\sin(\pi J t_2)\right. \\
&\left.\times \frac{\gamma_S}{\gamma_I}\left(\frac{3}{2}\frac{\gamma_S}{\gamma_I}\frac{\tau_{dS}}{t_2\cos(\pi J t_1)}\right)\right]
\end{aligned}
$$

The ZQ peaks obtained if $n = \pm 3/4$ are located at $(F_1, F_2) = \{\mp(\Delta\omega_I - \omega_S), \Delta\omega_I\}$, and the signal is

$$M_{ZQ}^{I+}(t_1, t_2) = iM_0^I \exp(i\Delta\omega_I t_2)\exp\{\mp i(\Delta\omega_I - \Delta\omega_s)t_1\}$$
$$\times J_1\left(-\frac{t_2\cos(\pi J t_1)}{\tau dI}\right) J_1\left(-\frac{2}{3}\frac{\gamma_I}{\gamma_S}\frac{t_2\cos(\pi J t_1)}{\tau dS}\right)$$
$$\times\left[\cos(\pi J t_1)\cos(\pi J t_2)\left(\frac{\tau_{dI}}{t_2\cos(\pi J t_1)}\right) + \sin(\pi J t_1)\sin(\pi J t_2)\right.$$
$$\left.\times\frac{\gamma_S}{\gamma_I}\left(\frac{3}{2}\frac{\gamma_S}{\gamma_I}\frac{\tau_{dS}}{t_2\cos(\pi J t_1)}\right)\right]$$

Appendix A1.29 [82]

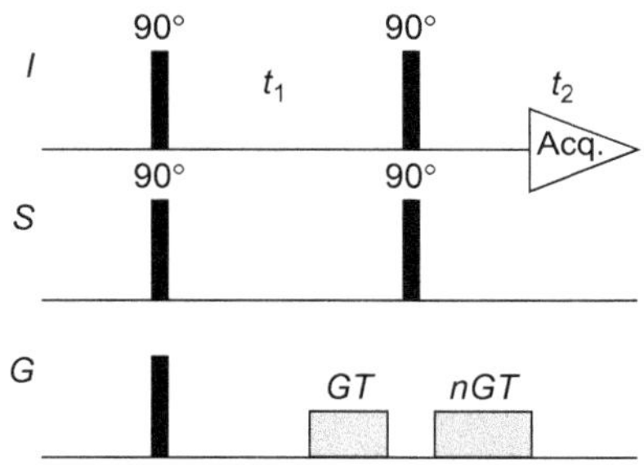

Figure A1.29 The heteronuclear CRAZED pulse sequence used to observe intermolecular heteronuclear MQCs in solution. A 90° pulse simultaneously given to both spins is followed by a free evolution time t_1 and a gradient pulse of strength G and length T immediately afterward; a second 90° pulse is given to both spins at once, followed by a second gradient pulse of area nGT. Detection of 1 spin occurs in t_2. In real experiments, a portion of the first gradient pulse is applied at the beginning of the t_1 period to prevent radiation damping from rotating the concentrated I-spin. For intermolecular $^{13}C-{}^{1}H$ MQCs, a ratio of ±3/4 selects ZQ terms and a ratio of ±5/4 selects DQ terms.

Appendix A1.30 [88]

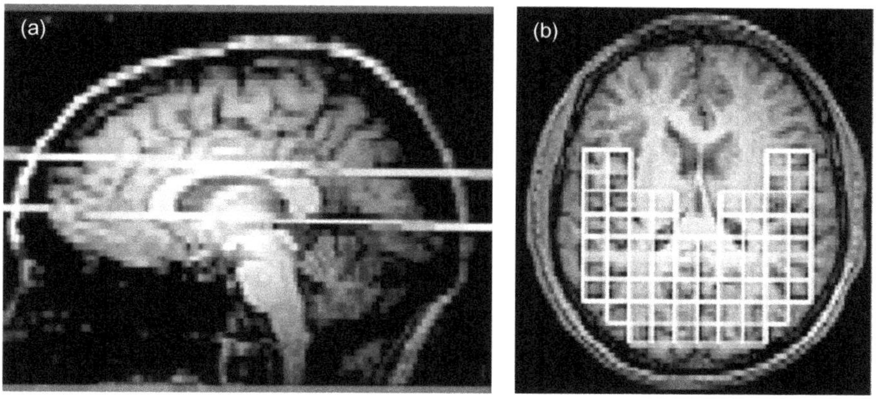

Figure A1.30 (a) Sagittal image at 4 T depicting the placement of the 30-mm-thick MRSI slab. (b) Oblique T_1-weighted image showing sampled 2D JMRSI voxels used to derive GM and WM estimates of brain GABA concentration.

Appendix A1.31 [88]

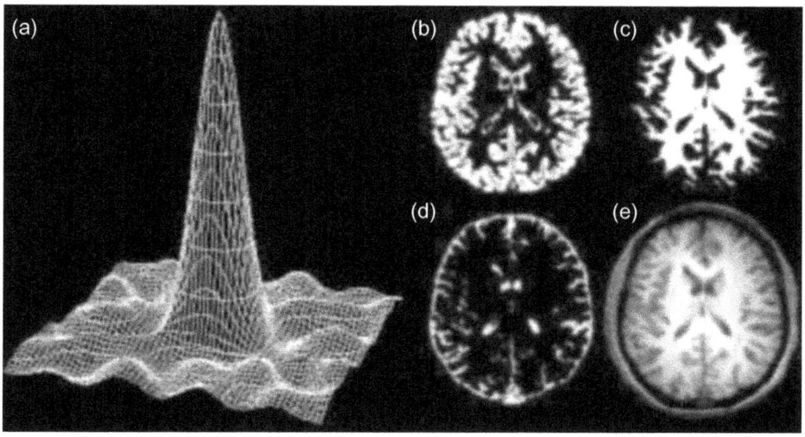

Figure A1.31 (a) Two-dimensional point-spread function of JMRSI acquisition, including the effects of *k*-space filtering. Segmented binary images of (b) GM, (c) WM, and (d) CSF. (e) T_1-weighted oblique image.

Appendix A1.32 [88]

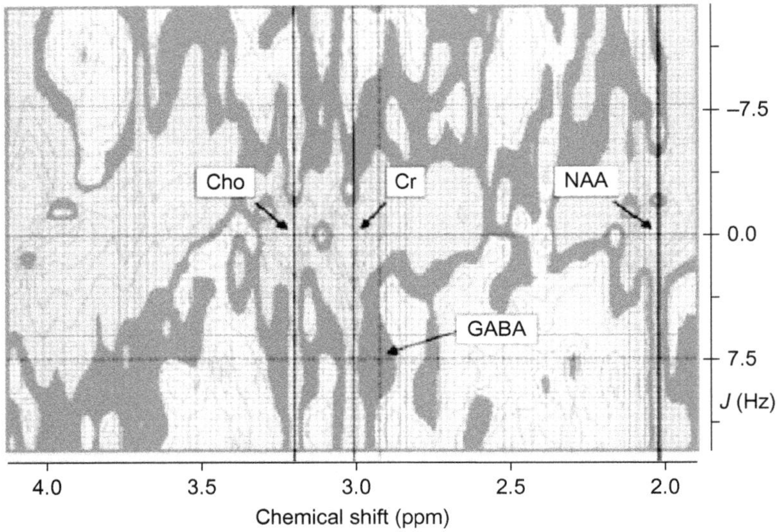

Figure A1.32 Two-dimensional plot of *J*-resolved spectrum from a voxel in the human brain. The main resonance lines of Cho, Cr, and NAA are clearly visible. The *J*-resolved GABA resonance at 2.95 ppm exists at the intersection of the $J = 7.5$ Hz and 2.95 ppm lines (dotted).

Appendix A1.33 [80]

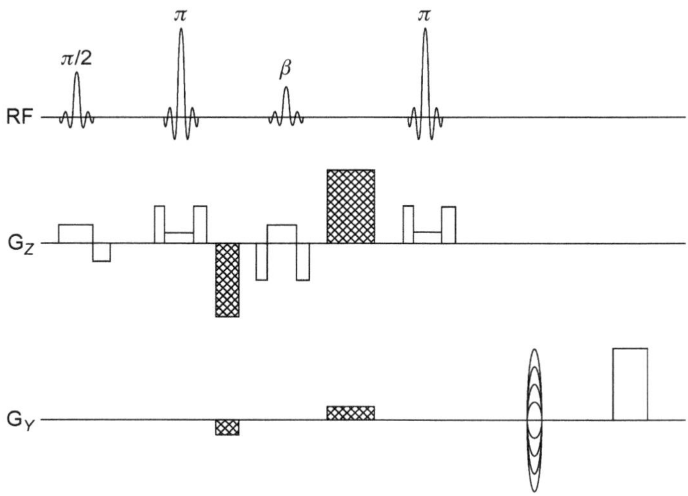

Figure A1.33 DQC imaging sequence used in this study. The locations of the DQC-encode gradients are for axial images, where Z is along the direction of the static field B_0. A pair of gradients along the y direction before phase-encode gradients is used to eliminate residual contamination from coherences other than DQCs. The polarity for the y-direction gradient pair is alternated for each RF excitation. In general, any image orientation can be chosen. The magic angle is always chosen to be formed with gradients along B_0 and perpendicular to it, and the other gradient perpendicular to B_0 of a small amplitude is used with two-step cycling.

Appendix A1.34 [80]

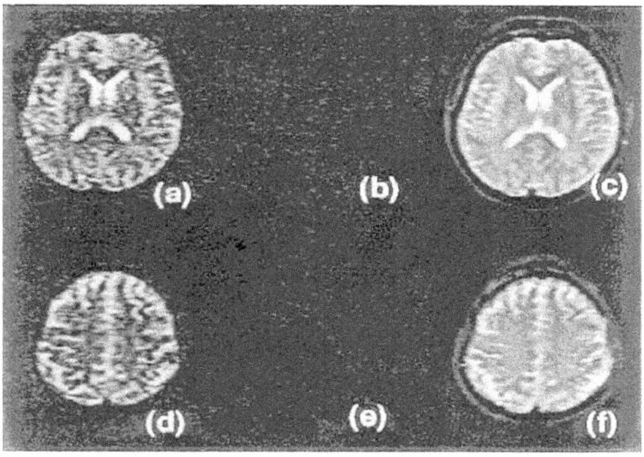

Figure A1.34 Images from two slices in a brain. (a, d) DQC images with DQC-encode gradients along B_0 direction; (b, e) DQC images with DQC-encode gradients along the magic angle; (c, f) T_2-weighted conventional SQC images. TR = 4000 ms, TE = 150 ms, matrix size 64×64, and 4 NEX is used for all images. All DQC images are displayed with the same window setting, but are different from the setting for the T_2-weighted images.

Appendix A1.35 [80]

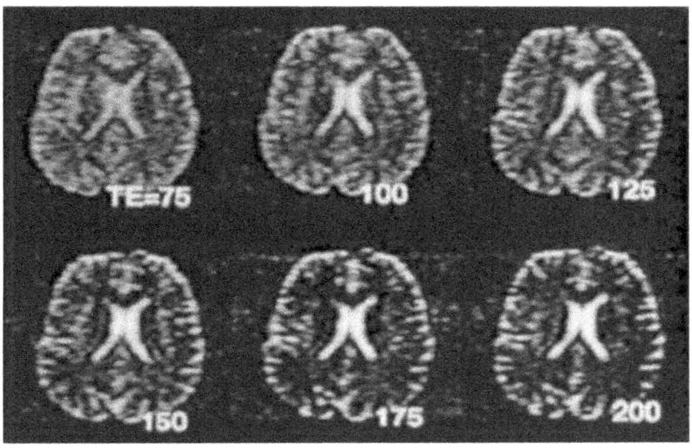

Figure A1.35 DQC images in a brain slice with different TE values as labeled on images. The DQC evolution time was 25 ms. TR = 4000 ms, matrix size is 64×64, and 4 NEX is used for all images. Images are displayed with the same window setting.

2 Magnetic Resonance Imaging of the Human Brain

Table of Contents

Curiosity and Education

It is very natural for human beings to look for the secrets of nature—it would be surprising if they did not. Curiosity constantly drives the human mind toward the unknown. Without curiosity, life would be very blank and the future very bleak.

Large nuclei (e.g., uranium) can be broken into smaller ones, through a process called *fission*. Smaller nuclei (e.g., hydrogen) can be fused to produce bigger ones (e.g., helium), in a process called *fusion*. Either way, a tremendous amount of energy is released. This cycle of make-and-break, e.g., helium to hydrogen and hydrogen to helium, has been active within our Sun, our star, for billions of years. Though this is only one part of the cycle of energy in the Sun, solar nuclear energy is the source of life on earth. No one knows what the beginning (the zero time) was, nor do we know when the end (i.e., the collapse of the star) will be. Driven by curiosity, humans have been able to recreate the fission source of energy (the nuclear energy we use) on earth. The second part of the Sun's mysteries is still unknown.

A secret is beautiful, and curious. Once a secret of nature is known, it is hardly a secret. However, the human mind can work out explanations for and mechanisms of more secrets, thereby permitting even further developments. This leads to more curiosity and further efforts to solve the mysteries of even more secrets.

Dissemination of knowledge is the key to further development. Carrying out research to understand the functions of the various organs and systems of the

Quantum Magnetic Resonance Imaging Diagnostics of Human Brain Disorders. DOI: 10.1016/B978-0-12-384711-9.00002-6

human body meets a very basic human desire. How various organs coordinate and perform a common function is as intriguing as it is puzzling. The precise correlation of body organs in their performance of routine bodily functions is still far from well understood. From its start as a matter of a physicist's curiosity, NMR technology has evolved, most recently with regard to biomedical materials. It has now become a gift of unparalleled value to the field of medicine. It was initially developed to answer the desire to understand the energetics of the orbital electrons in the phosphorous nucleus, as a part of research in semiconductor physics. As a dilute impurity in silicon lattice, phosphorous supplies free electrons to design a fast electronic computer (EC). Now there is a fundamental desire to understand the QM of the nuclear and electron spins in phosphorous, and many other elements, in the hope of creating a solid-state quantum computer that will work on the spins or charge of electrons or nuclei. As yet, this remains only a theoretical possibility.

An external applied magnetic field allows one to understand electron-spin resonance (ESR) and also the nuclear magnetic (spin) resonance. We live under the influence of Earth's magnetic field. The field is very small. The effects on our bodies are small too, but not too small for a curious mind to want to discover the interference effects on the various nuclei in the semiliquid/semisolid state present in the human body. In fact, NMR studies on the influence of Earth's magnetic field are in progress. We use NMR to unravel the intricacies of the physical, chemical, and mathematical (PCMcal) processes taking place in the human body round the clock. This can be done remotely, with little intervention inside the human body, i.e., noninvasively. Human minds figured out the PCM of the fission–fusion–fission cycle in the Sun, despite the Sun being millions of kilometers away. So far, though, the secrets of the human body have remained a mystery. It is time to value the desire to understand the human body. Treating it when it is sick is not good enough. For example, diabetes has been around since before the modern scientific civilization, and persists in the technological one that followed. Yet old treatments still continue, with little change or improvement. This is partly due to the lack of knowledge of science among the masses.

Science can be a powerful force—and even a preoccupation—of the human mind, and the chances are good that an invention or two will come from the curious mind. Investment in education about QM at the secondary level is a worthy initiative. Certainly, investment in newtonian mechanics (NM) has paid off. Undermining or underemploying mind is worse than pointless; it is destructive to both individuals and society at large. What price attaches to intellectual preoccupation of a human mind? Whatever the cost, it is better than self-destruction.

This chapter is dedicated to the researchers who, over decades, have been engaged in trying to unravel the science of the human brain. The PCM behind the dynamical processes that are going on all the time in the human brain is the key to understanding brain functions. The scope of this chapter is restricted to NMRI, in particular, to the application of NMRI to the human brain. NMR was discovered during the middle years of the twentieth century. Since then, perhaps millions of workers have contributed, in some respect or another, to the development of the field. Today, this field is nothing short of a miracle technology.

PCM is the basic language of science education and learning. English is the most commonly used medium through which science is propagated on a worldwide scale. However, it is a higher level of competence in PCM that can improve comprehension of the field of MRI and thus of medical diagnostics. Doctors' curiosity about application of MRI, as a noninvasive tool of diagnostics, is understandable. There is a pressing need to understand the function of the various PCM processes taking place in the human brain. MRI allows one to work out, qualitatively or quantitatively, disorders in the normal functions of the human brain. X-ray imaging provides easily obtained picture readouts of any mechanical faults in the rigid structure of the body (e.g., the bones). Unfortunately, the straightforward rules of thumb developed for diagnostics using X-ray imaging are not easy to replicate for diagnostics using MRI. Here one is dealing with the fast time dynamics of the nuclei, molecules, etc., in the body. In some ways, perhaps it is better that MRI is difficult. That difficulty may drive improvements in the quality of research and development in MRI, and enhance the development of a competent tool of diagnostics.

In the brain, the structures and artifacts are not as rigid or static as in the skeletal system. One is dealing, in the brain, with semiliquid, semisolid matter, plus millions of free atoms, electrons, and nuclei moving around and through this viscous, semirigid structure. The free entities (molecules, nuclei, etc.) exchange information (energy) among each other and with the surrounding structures. All this allows the brain work the way it does. NMR imaging is about the nonequilibrium nuclear dynamics in the brain; it is possible to visualize the nonequilibrium PCM processes occurring in the human brain through MRI.

To date we have only sketchy ideas, models, and theories, as we work to decipher the many mysteries of the brain. The memory and consciousness aspects of the brain, in particular, still remain an open question. There are many radical views on, and models of, the human brain; the quantum model of the brain is one extreme example. MRI research on the human brain can unravel some of these mysteries. It uses RF electromagnetic fields to simulate, experimentally, energetic events that may be close to the ones actually happening in the brain. The technique of MRI allows us to image the events, in situ and in real time, but lots of problems remain to be solved before an expert can really state, authoritatively, what exactly is going on behind the scenes. One cannot develop a plain English formula for the human brain arena—not yet, anyway.

The electrical synaptic events that take place within neuron networks propagate on the scale of nanoseconds ($\sim 10^{-9}$ s). This adds to the complexity of any model of the human brain. The energetic events happening in the brain occur on a much longer time scale, i.e., microsecond (10^{-6}) to millisecond (10^{-3}) scale. Still, MRI technology is not yet advanced enough to get a very precise picture. Maybe someone's curious mind will one day discover a way to improve the technology to such a degree as to enable us to visualize, analyze, and understand the curious functions of our brains.

MRI has now become an integral part of the hospital infrastructure. However, the field of MRI, which is now so critical to the practice of medical science, did not come into existence overnight. The background information presented here should be useful to many readers, medical or otherwise, and it is hoped that the reports of research trends and directions will also stimulate curiosity.

Many and diverse contributions have been made toward learning the basics of the PCM involved in MRI over the last century. This notably includes the contributions of QM. One of the benefits that we see today is that MRI has become a highly successful commercial venture in the service of humanity. We commonly see it in hospitals, research institutes, universities, etc. This is a typical example of how basic research eventually becomes intimately linked to the wealth and well-being of society. Today science has advanced to such a complex stage that one needs to be a PCM expert to make use of it or to advance it any further—and yet advancement is taking place by the hour. However, the dissemination of PCM knowledge and education in PCM in the community are far from closing the gap in knowledge developed about MRI and common literacy regarding it. This goes for the scientific community as well as for the community at large. The larger this knowledge-dissemination gap between the haves and have-nots, the slower the rate of progress in science, and consequently in any technology; MRI is no exception.

The amount of knowledge required to assimilate scientific developments and advances is getting bigger by the day. The number of people in the community who can understand the PCM is decreasing, and there are more and more people who are unwilling or unable to be educated about the higher levels of and developments in PCM. This becomes even more problematic as more information is generated. This phenomenon is widespread today: The level and amount of accumulated knowledge is getting bigger as more and more is being produced by intellectuals, but the number of professionals (and others) who can absorb and use this knowledge is getting smaller. The whole phenomenon of scientific development is becoming ever more restricted and localized among specific, specialized groups; likewise, the knowledge generated by research is being disseminated among fewer and fewer people. Consequently, fewer and fewer people understand it. Renewal and advancement of knowledge by the masses are becoming rare. This is a disaster being perpetrated on the future. One solution—probably the only one left—is to lure younger generations back to the fundamentals of PCM. At the moment, young people are fast going away from it, toward the apparently greener grasses of finance, business, and management, where employment opportunities are increasing by leaps and bounds.

Writing this chapter has done only a fraction of the job for a fraction of the people. Nevertheless, it is a step in the right direction toward a long-term goal. This chapter should bring a vast community of medical practitioners and the dwindling number of PCM experts closer together. We call these latter *dwindling* because many of them want to get away from the difficulties of coping with the pace of advancing knowledge, or would enjoy more material/financial advantages if they abandoned PCM. Nonetheless, basic science and fundamental research underlie all the technological developments and advantages that we enjoy today. Maybe the financial leaders of the world will heed the right message someday. Perhaps MRI technology, as an example, now has a better chance to convey that message.

This chapter is written for the particular benefit of those interested in the field of medical science, as well as for the intellectual development of the scientific community and humanity overall. Physicists, chemists, and mathematicians concentrate their efforts in specialized areas, as do most medical scientists. More often than not, though, medical scientists tend to learn through clinical interactions with their patients. After dyes, radioactive elements, etc., are intravenously or orally introduced into the body, an imaging test can be performed for a specific area of the body under investigation. The test only provides an overall physical picture in terms of the events that occurred in that area. A noninvasive recording of the events in the time-space examination, via spectroscopic energy transitions, excited remotely in a localized area of the body through NMRI, constitutes a giant step forward. However, many users possess only limited knowledge and understanding of the PCM involved in MRI. PCM forms an integral part of the use and development of this technology. Incompetence in PCM results in, at best, limited use of the technology. This puts a barrier between experts in PCM and the competent diagnostics that can be achieved by a medical scientist. Researchers in the area of MRI produce an enormous wealth of knowledge, but almost always, someone else must take on the tedious task of educating medical professionals' (MPs) and others about this knowledge. Broad visualizations and the associated vague interpretations of photographic images from clinical trials, in relation to the medical disorders being studied, are of limited value. The reality behind the technology is far beyond the comprehension of many MPs, and many questions remain unanswered. This leads to lots of speculation and ambiguities in diagnosis and treatment. This chapter tries to collect and simplify the information available on NMRI techniques and the PCM behind this technology. The information disseminated here, it is believed, would be useful to MPs who wish to know the field of NMRI from the ground up, and use it and develop it further. It is presented in a manner that we hope will enable MPs to understand the material, all the while recognizing that the ultimate objective is application for the benefit of patients.

Our purpose in writing this chapter is purely educational—a purpose that is more important today than ever. In years past, researchers were part of the teaching program in places like universities. This allowed researchers to pass on the knowledge they had acquired in their field of specialization to the next generation. Today researchers rarely do teaching and teachers rarely do research. The widening gap between the research performed and the knowledge passed on is becoming more and more obvious. For those who are interested, the work presented here will, it is hoped, lessen the gap between what they know and what they want to know. Much has yet to be done to enable MPs to benefit from the research and development being done in other scientific arenas. This chapter tries to bring together knowledge from diverse disciplines, generated through many different research avenues and approaches, to create common ground for solving the difficult and diverse problems facing medical science. The connections and sharing of information from and among various disciplines, as attempted here, is the key to success. This chapter is not written in the form of a textbook, or even a close substitute for a textbook. What is presented here is a summary—an abridged form of information—of the research and development efforts that have gone into building this field over half a century.

Readers may find that some subjective (personal) component is involved in making full use of this chapter. The threshold requirements may have been a part of the individual's development through education, opportunities, efforts, and more; certainly, one's academic training in one's own specialized discipline (e.g., medicine, physics, chemistry), and one's interaction with and knowledge of other disciplines will be very helpful in understanding the material in this chapter and making good use of it. The medical diagnostics that employ technologies such as NMR require many skills, including an understanding of the PCM of the underlying processes.

Given the diverse disciplines coordinated herein, the information presented in this chapter had to be acquired from many varied and widely different sources. Much of this information was drawn from publications in the scientific literature, including material in the form of illustrations, technical know-how, mathematical equations, etc. Though the information available in the literature is scattered, it is available today in a much more convenient way than was formerly possible, thanks to the developments in information science and technology that have taken place over the past decade or so. The information technology available today has accelerated the dissemination of knowledge, and thus the progress of science, for the benefit of everyone. The producers of research should rejoice that their work is being used for the benefit of humankind. Scientific advancement is possible only through free semination and dissemination of the knowledge produced. Those who are able to produce this knowledge should be happy that others use it. Some researchers may do enormous amounts of work without deriving much material benefit therefrom, but their sacrifices for the benefit of others may come back to them, in an enhanced form, on some later day. There is an Indian saying that those who got the spirit are better than those who got the city. Perhaps there is some wisdom in contentment with no material gain. Transmission of information in the form of a coordinated written work, such as this book, costs a lot in terms of time, effort, and resources. The only reward hoped for is overall progress in the field of science in general, and in particular progress in the field of medicine.

This chapter includes, at the end, a list of references citing the sources from which information, in one form or another, was drawn for use in this book. We express deepest gratitude to the sources of information (i.e., the scientific publishers, authors) that have in any way contributed to the content of this chapter, especially to all the authors of all those publications from which information was derived, advertently or inadvertently. Thanks are expressed once again to those whose work in the field of MRI has, directly or indirectly, been incorporated into this chapter. These researchers deserve much appreciation, in every respect, for the work they have produced. But for their work, our task would have been impossible to complete.

Sincerest gratitude is also expressed to those researchers who are working hard to take the field of medicine ahead, to the scientific community in general, and to those in the present civilization who want to progress to greater heights. A free mind is the mother of all creation. The pride of freedom lies in the coexistence of the human mind, body, and all the creations of nature, including plant and animal life. This is true for all. Coexistence is the key to the survival of all, and thus of each one. Therein lies the survival of all. God bless all; God bless one. There is no all without one; there is no one without all.

2.1 Introduction

The field of medicine has progressed a lot over the last century, but not as much as it could have, would have, or should have. As a comparison, over the past 50 years, progress in physics has been made by leaps and bounds. It started with the development of bulk semiconductors. This led to previously unimaginable micro-miniaturization of electronic circuitry on a single silicon (Si) chip. Smaller and smaller consumer goods (e.g., televisions, mobile phones) then followed. Without these inventions, our day-to-day lives would virtually come to a standstill. The high level of progress in the areas of PCM has enabled even greater technological developments. The visible results of this progress range from satellite communication technology, to nuclear technology, to putting humans into space, to the liquid crystal display in a laptop computer, to plasma TV (gas-state imaging), and so on, to mention only a few; the list could go on for pages. The technologies we enjoy today would have been impossible, though, without the research and development efforts of preceding decades. The disciplines of PCM have played a crucial role in producing these achievements.

When it comes to understanding the functions of the human brain, one looks to the field of neurology. Neuroscience is a young field, but its theories of complex neural systems are able to explain, to a great degree, the neuronal synaptic propagation of signals. These signals seem to propagate through firings on a timescale of nanoseconds (10^{-9} s). The quantum mechanical energy transitions in molecules in the human body environment can be observed by the technique of MRI, but it remains an open question as to how the molecular-rotational, electronic, or nuclear processes control our body functions. The transport of information, and the processes involved in performing a particular task within the human body is another area of interest and challenge. It is known that when a stimulus (mechanical, electrical, etc.) is applied to the brain, the brain responds within a millisecond's time. At the moment, it is believed that brain functions, including consciousness, are associated with the firing activity of a very large number of neurons spread all over the cortex and its associated satellites, such as the thalamus. Thus, any one conscious perception, or thought, must correspond to a widely distributed coalition of neurons firing together.

Lots of questions remain unanswered, and we need a practical demonstration of what really happens. One approach to understanding the human brain is to find out if QM is somehow involved in the functional operation of the brain. To do that, one can study multiple quantum coherences through NMRI of the brain. This approach could be very helpful in solving these unanswered questions. Some scientists question the need for QM: the neuron network model works fine in the analysis of synaptic functions [1], so why this quantum approach? This is a shortsighted view. The network model does not have the power to explain consciousness, as the human mind experiences it. The functions of the metabolites in synaptic activities remain an area of vital interest.

Successful applications of the power of LSNMR (liquid state) NMR are now being pursued analogously, in the SS, in the area of quantum computing [2]. As with NMR spectroscopy, a 2D visible-light spectroscopy that would seem to have far-reaching benefits is proposed. It is known that light is converted into chemical energy in a photosynthetic system. This is largely determined by the vibrational and electronic dynamics of the complex biological macromolecules involved. There has already been a demonstration of direct measurement of electronic couplings, combined with the dynamics of excitations, transferring between molecular energy levels in photosynthetic antennas [3]. Life in different forms, humans, animals, plants, etc., demonstrates QM processes occurring in their systems. Is there something common or something different among different forms of life? Whether the human brain performs quantum computing is a mystery yet to be solved. Though the human brain is certainly in the picture, at the moment the authors do not want this chapter to get bogged down in the question of whether the human brain is quantum or not quantum. Rather, the purpose of this chapter is to explore the benefits of quantum coherences among molecules, and whether one can produce real-time images of the processes taking place in the human brain, through NMR studies [4, 5]. It is hoped that, in the long term, MRI will produce a dependable technique with which to diagnose ailments of the human brain, and thus lead to better treatment of brain disorders.

2.1.1 Magnetic Dipole Moments

2.1.1.1 Atomic Moments [6]

It is now common knowledge that electrons move in stable, circular orbits around the net positive charge of the atomic nucleus (due to protons). This steady state is the result of a balance between the centripetal force, which tends to sling the electrons away from the center, and the attractive coulomb force between the electron and the nucleus (positive charge of the protons), which tends to do just the opposite. The result is a stable atomic orbit for the electrons. The charge circulating in a loop (periodic variation in time) gives rise to circulating electric current, $i = e/(2\pi r/v)$. Here r is the radius of the orbit and v the linear velocity (at an instant) along a tangent, at a point, on the orbit. The current in the loop, i, surrounds the area A covered by the loop. The two sides of the loop act like the north (N) and south (S) poles of a circular magnet. One can assume that the two fictitious poles N and S, as for an ordinary bar magnet, are placed at the center and on the opposite sides of the electronic orbit. The magnitude of the orbital magnetic-field strength of a dipole magnet, μ_1, expressed as an equivalent dipole in terms of microscopic parameters v and r, is

$\mu_l = iA = [e/(2\pi r/v)][\pi r^2] = (evr/2)$. The direction of the dipole moment would be along the N–S direction and perpendicular to the plane of the orbit. Magnetic lines of force, surrounding the atomic N–S magnet, would naturally be produced.

This is analogous to what an ordinary, small magnet does in the space surrounding around it. Another natural magnet example, though on a much larger scale, is the Earth, which behaves like a huge (N–S) magnet. The molten, ionized elements in the core of the Earth, with their associated protons, electrons, etc., produce large circulating currents. These currents generate a huge equivalent magnetic dipole, stationed at the center of the Earth. Because an electron has a negative charge, its magnetic dipole moment μ_l is antiparallel to its orbital angular momentum $L = mvr$, m being the electron mass, v the orbital velocity, and r the radius of the orbit. The ratio of the dipole magnetic moment to the orbital angular momentum L is called the *gyromagnetic ratio* (GMR) $= \mu_l/L = (e/2m)$. This ratio is conveniently written in terms of the basic atomic magnetic moment (created by electrons) unit, $\mu_b = (e\hbar/2m) = 0.927 \times 10^{-23}\ \mathrm{A\,m^2}$ ($\hbar = h/2\pi$, h is Planck's constant); μ_b is called the *Bohr magneton*. The final result for the GMR is $\mu_l/L = (\mu_b/\hbar)$.

Now, when this magnetic dipole is placed in an externally applied magnetic field H, it will experience a torque rotational force that tries to align it along the N–S magnetic-field direction (the reader is referred to an undergraduate textbook for the details). In the human body environment, before the application of a magnetic field, the magnetic moments would be randomly distributed in their spin directions. The random equilibrium is produced by thermal agitations and kinetic motions, including the vibrations and rotations of molecules. Application of the external magnetic field creates another equilibrium. This artificially created magnetic equilibrium is the means through which MRI gathers information about, for example, the metabolic activities in the human brain. This magnetic torque τ, in terms of the vector product of the vectors μ_l (the magnetic moment of a particular species of spins I) and H (when an atom forms part of a molecule placed in a medium such as a liquid), can be approximately represented by $\tau = \mu_l \times H = \mu_l H \sin\theta$. In the language of mathematics, this product is called the cross-product of the two vectors and θ is the angle between them.

Here the force generated is rotational, and is in the plane of the orbit, whereas the momentum (which here is the angular momentum) is normal to the plane of the orbit. This torque will try to align the dipole along the direction of the applied field. The energy consumed by the sample due to this orientation is $\Delta E = -\mu_l H$. The negative sign is to remind us that energy is absorbed (used) to perform the alignment. One should remember that these equations are valid only in vacuum. In real systems, like the human brain, corrections have to be applied to compensate for the viscous nature of the medium in which the nuclei are present. If there is no way for a system, consisting of a magnetic dipole moment μ_l in a magnetic field H, to dissipate energy, the orientational potential energy ΔE of the system must remain constant. In these circumstances, μ_l cannot align itself with H. Instead, μ_l will precess around H in such a way that the angle between these two vectors remains constant. At the same time, the magnitudes of both vectors remain constant. This precession motion is a consequence of the fact that the torque acting on the dipole is always perpendicular to its angular momentum. This is completely analogous to the case of a spinning top. It is easy to show that the magnitude of the angular frequency of precession of μ_l around H is given by $\omega_L = (\mu_b H)/\hbar$. This equation also indicates that the sense of the precession is in the direction of H.

This phenomenon is known as the *Larmor precession*, and ω_L is called the *Larmor frequency*. If the applied magnetic field is uniform in space, there will be no net translational (moving-sideways) force on the magnetic dipole (although there is certainly a torque). If the field is nonuniform (a gradient field), which is a more realistic situation, there will be a translational force in addition to the torque. An electron, moving with velocity v through a circular orbit in a region in which the H field is converging, feels a force proportional to $v \times H$. This force always has a component in the direction in which the field becomes more intense. One can show that the average force acting on the magnetic dipole is $Fz = (\partial H/\partial Z)\mu_{lz}$. Here z is the coordinate axis in the direction of increase of the field strength, and $\partial H/\partial Z$ is the rate at which the field strength increases. Thus, in a nonuniform magnetic field, a magnetic dipole experiences a torque that will cause precession. In addition, there is a force that will cause displacement.

This is as far as classical (Newtonian) mechanics goes. In the language of QM, the atomic magnetic moment can be written as $\mu = (-e/2m)\hbar I$, where I is a directed orbital quantum number. Because of the negative charge of the electron, μ and I are oppositely directed. In the presence of an external field, the system acquires magnetic energy, $-\mu \cdot H = (e/2m)\hbar(I \cdot H)$. Spatial quantization allows the vector I to take $2l + 1$ possible orientations between the limits $\pm l$. One should note that the negative values of l provide the lower energies. When both orbital and spin (magnetic) momenta are present, one can write, for a single electron, $\mu = g\mu_b j$ and $j = I$(orbital) + s(spin). For a whole atom (electron plus nucleus), $\mu = g\mu_b J$ and $J = L + S$. Here g refers to the mechanical ratio, the gyromagnetic ratio, and sometimes Lande's splitting factor. The spectroscopic state of an atom is determined by the quantum numbers L, S, and J. According to Hund's rule, the resultant state J in the ground state is obtained as follows: (1) S is maximized. (2) L is maximized subject to 1. (3) $J = |L - S|$, when a shell is less than half full, and $J = L + S$ when a shell is more than half full.

A nucleus also has a dipole magnetic moment and angular momentum, just as an electron circulating around the nucleus does. The protons and the neutrons in the nucleus are, together, the source of the net nuclear magnetic moment. It is the proton moment (because of its charge) that is directly relevant to the NMR. Figures 2.1–2.4 give a pictorial representation of the forces acting on the orbital

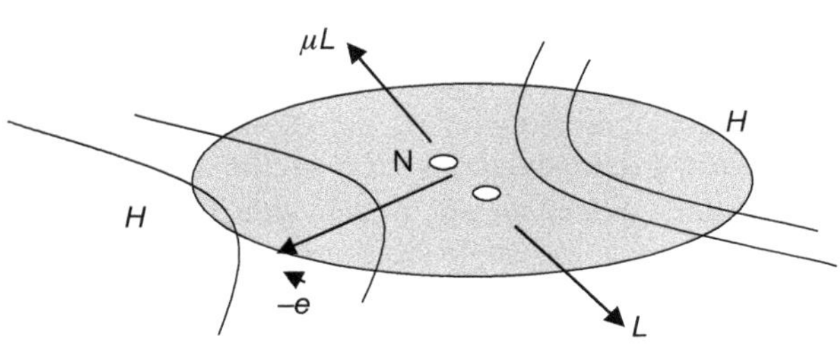

Figure 2.1 Orbital angular momentum (L) and the magnetic field (H) around it.

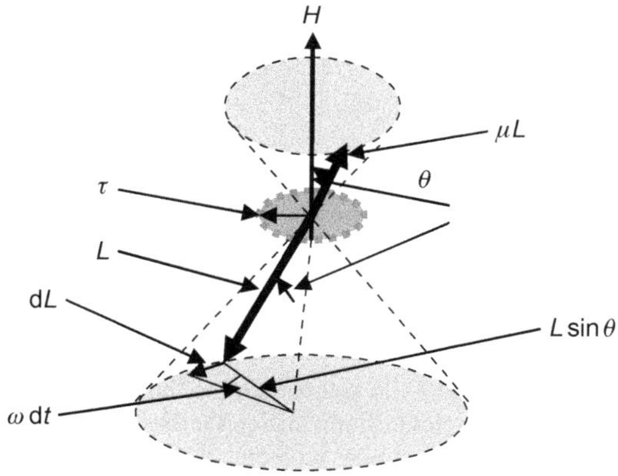

Figure 2.2 Dipole moment (μ_1), orbital angular momentum (L), and the magnetic field (H) around it. There is a torque, $\tau = \mu_l \times H = -(g_l\mu_b H/\hbar)(L \times H)$, on the electron in orbit. This torque gives rise to a change, dL, in the angular momentum in time, dt. According to Newton's Law, one can write $dL/dt = \tau$. However, the change dL causes L to precess through an angle $\omega\,dt$. Here ω is the precessional, angular velocity. At the same time, one knows that $dL = L\sin\theta\omega\,dt$. Also, it is known that $L\sin\theta\omega = dL/dt = \tau = (g_l\mu_b/\hbar) \times LH\sin\theta$. Here, $\omega = g_l\mu_b H/\hbar$. *Courtesy*: Ref. [6].

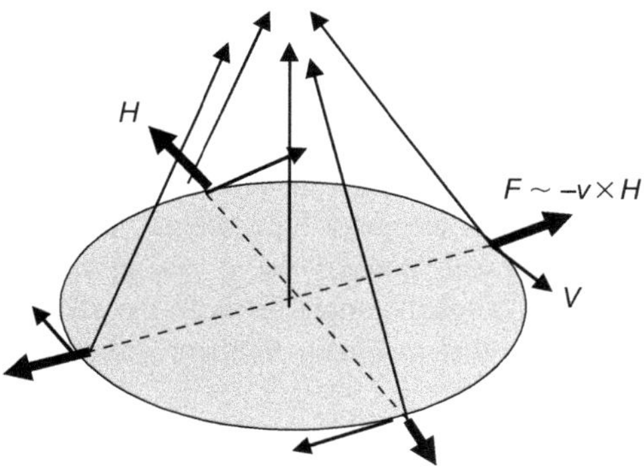

Figure 2.3 Forces on an electron in an orbit: There may be places where the applied field H is converging. There, the field exerts a directed force F on the electron moving in an orbit. The velocity v generates the circulating electric current. This force has a component that is radially outward. There is also a component in the direction in which H becomes more intense. Averaged over the orbit, all the radial components cancel out. Thus, there is an average force in the upward direction.

electrons, and the magnetic moments in a constant and gradient magnetic field. For a complete physical picture of the system, the reader is refered to the electricity and magnetism chapters of an introductory physics textbook.

2.1.1.2 J(L−S) Couplings, Energy Spectrum [7]

Electrons in atoms possess both orbital angular momenta L and spin S (precession around a magnetic field). There are also the resultant orbital and spin-angular momenta associated with a complete atom or an ion. These quantities are described in terms of the orbital quantum numbers, L, and the magnetic quantum number S. In Russell−Saunders coupling, L and S may be combined vectorially to produce J, the

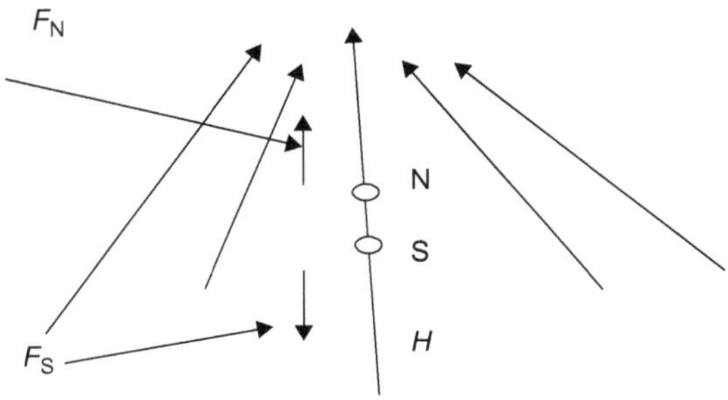

Figure 2.4 There are forces F_N and F_S in space (in a medium such as the human brain, for instance), due to the applied static field. Mathematically, these forces are expressed acting on the poles of the fictitious magnetic dipole moment. This moment is equivalent to (and as a result of) the circulating electron. The moment is located in a region where the applied field H is converging. But, because F_N is greater in magnitude than F_S, there is a net force on the dipole in the direction (upward) in which H becomes more intense.

total angular momentum of the atom or ion. It is known that L and S add up to a different J. They produce multiplet structures accurate and acceptable. The eigenvalues (degenerate energy spectrum) for the atomic orbital, atomic spin, and total atomic angular momenta are given as $\hbar[L(L+1)]^{1/2} = \hbar L'$; $\hbar[S(S+1)]^{1/2} = \hbar S'$; and $\hbar[J(J+1)]^{1/2} = \hbar J'$. Now consider the eigenvalues of the total angular momentum, $\hbar J'$. By using a weak magnetic field (which provides a convenient reference direction along the z-axis), one finds that the resolved components of $\hbar J'$ may take only the discrete values $-\hbar J$, $-\hbar(J-1)$, $-\hbar(J-2), \ldots, +\hbar(J-2), +\hbar(J-1), \ldots, +\hbar J$. This is called the *spatial quantization*. The eigenvalues of the z-component of the angular momentum are written as $\hbar m$, where m takes either integer or half-integer values in the range $-J \leq m \leq +J$. The magnetic moment parallel to J' is given as $\mu' = \mu_b L' \cos(L', J') + 2\mu_b S' \cos(S', J')$, where (L', J') is the angle between the vectors (L', J'), etc. However, it is preferable to express this magnetic moment directly in terms of J', so that $\mu' = g\mu_b J'$.

In spectroscopy, g is called the Lande's splitting factor, controlling the Zeeman effect (splitting of energy levels in the presence of a magnetic field). In the field of MRI, it is called the gyromagnetic ratio, although it is more correctly termed the magneto-mechanical ratio. An expression for g is easily obtained from the triangle of vectors, making use of the cosine rule, among L', S', and J'. In a magnetic or NMR context, we are more interested in the resolved components of the expectation value of $g\mu_b J'$, in the direction of the field (i.e., the z-direction). These resolved components are $-g\mu_b J$, $-g\mu_b(J-1), \ldots, 0, +g\mu_b(J-1), +g\mu_b J$. They are most conveniently written as $g\mu_b m$, where $-J \leq m \leq J$. The maximum value of the resolved magnetic moment is $g\mu_b J$, the atomic magnetic moment. One should emphasize that in the absence of a magnetic field, in an atom with $J > 0$, the atom will be in a degenerate state. All the levels of different m are equally represented in the ground-state configuration. The application of an external magnetic field removes this degeneracy through Zeeman splitting. The different m levels are now at different energies and they are occupied, in accordance with Hund's rule, to produce the magnetic moment. The large resultant dipole magnetic moment arises in the presence of the external field. This is because each moment tries to align itself along the field. This alignment takes place no matter how small the external field is. One should therefore expect a permanent atomic moment. Please refer to Figure 2.5 for a pictorial depiction of the $J-L-S$ couplings in atoms. The component of the total angular momentum in the field direction may take only discrete values determined by the quantum number m, where $-J \leq m \leq +J$. The vector representation of the total angular momentum and the corresponding geometry for the magnetic moment are roughly depicted in Figure 2.5 [7]. The resultant moment $M_{J'}$ is not parallel to J'. It precesses about J'. This leads to a resolved component of the magnetic moment $g\mu_b$. The mathematical expression for the total moment parallel to J' can be as follows: $\mu' = g\mu_b J' = \mu b L' \cos(L', J') + 2\mu b S' \cos(S', J')$. One can find the g value as $g = (3J'^2 + S'^2 - L'^2)/2J'^2 = 1 + [J(J+1) + S(S+1) - L(L+1)]/2J(J+1)$.

The proton does not have a magnetic moment equal to one nuclear magneton, but rather $2.79\mu_b$. Similarly, although the neutron has no net charge, it has magnetic moment $= -1.9\mu_b$, i.e., the nuclear magnetic moment is antiparallel to I. It is

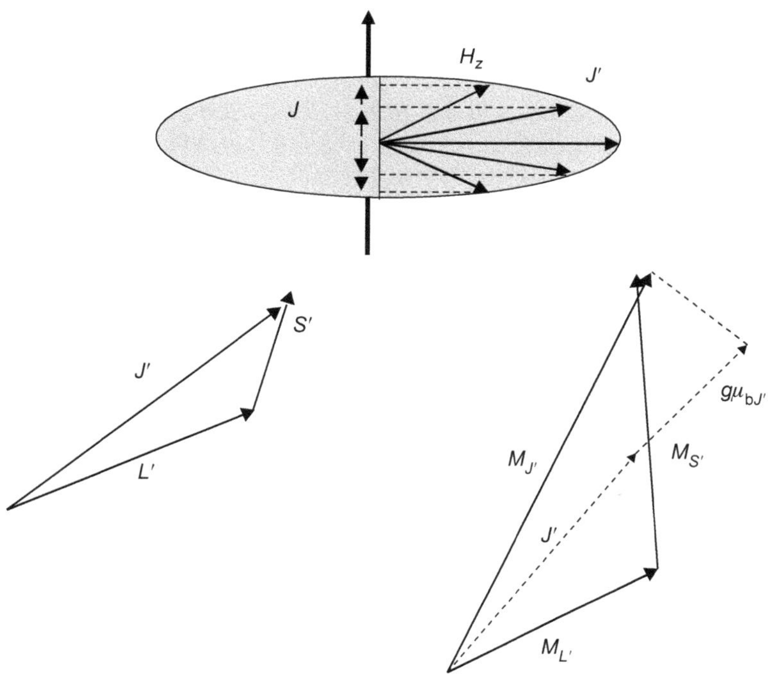

Figure 2.5 The left side of the figure represents the vector composition of the total angular momentum; the right side is the corresponding geometry for the magnetic moment. The resultant moment M_J' is not parallel to J' and therefore precesses about J'. However, its component $g\mu_b J'$ is parallel to J'.

expressed as $\mu = \gamma I$, γ being the nuclear magneton mechanical (gyromagnetic) ratio. It varies from nucleus to nucleus. Although this ratio is usually positive in sign, there are nuclei with negative values, e.g., for ^{107}Ag, $I = 1/2$, $\mu = -0.133\mu_b$. Owing to spatial quantization, a nucleus with quantum number I possesses $2I + 1$ levels, which, in the absence of an external field, are degenerate. An external magnetic field leads to the occurrence of $2I + 1$ discrete nuclear Zeeman levels, each of which is separated from its neighbors by an energy $\hbar\gamma H_0$. The frequency ω_L, also known as the Larmor frequency, defined by ω_L, is that with which the nuclear moment μ precesses around the direction of the applied field H_0. H_0 is assumed to be directed along the z-axis. For the proton in a magnetic field of 1 T, the value of $\hbar\gamma H_0 = 1.76 \times 10^{-7}$ eV, which, expressed as a frequency, corresponds to 42.5 MHz. In addition to nuclear Zeeman splitting, nuclear spin levels are also perturbed by electronic quadrupole moments that arise as a result of nonspherical electrostatic charge distribution. Quadrupole moments, however, are not part of the subject of this chapter.

2.1.1.3 The Nuclear Moment [7]

Many nuclei present in nature (e.g., in the human body) possess atomic angular momentum, as discussed in Section 2.1.1.2. This momentum is measured by a quantum number and has an associated magnetic moment μ. This nuclear moment allows interaction of the nucleus with the outer electron clouds (atomic orbits). Certain nuclear processes may therefore reflect changes in the electronic structure. The nucleus is an assembly of neutrons and protons. Each nucleon (i.e., each neutron or proton) has an effective radius of about 1.4×10^{-15} m and an intrinsic angular momentum, corresponding to a quantum number 1/2. In nuclei, the nucleons also possess orbit angular momentum, which when combined with their spin produces a resultant angular momentum, called the *nuclear spin* and denoted by I. As before, in the case of the nucleus, one can write the eigen (energy) values as $|I| = \hbar[I(I+1)]^{1/2}$. Here I is the appropriate quantum number. The resultant nuclear spin may take quite large values. The nuclear magneton $\mu_N = (e\hbar)/2M = 5.05 \times 10^{-27}$ J/T, M being the mass of the proton (Figures 2.6 and 2.7).

2.1.1.4 Magnetism of Atoms, Molecules, and Electrons [6, 7]

It is well known that the electrons in atoms always occupy orbits. This produces the maximum resultant spin, subject to Pauli's Quantum Principle (PQP). According to PQP, two electrons with the same spin direction cannot occupy the same energy level. Closed shells of electrons (refer to electronic structure of atoms in the Periodic Table of Elements) do not produce resultant orbital or spin-angular momentum. They therefore have no resultant atomic magnetic moment. In materials, they possess purely diamagnetic (inherent magnetic moment) properties. In contrast, any atom or ion with an incomplete shell or subshell of electrons (note the transition metal and rare earth metals) has an associated magnetic moment. In free atoms or ions, the magnetic state is preferred. This is because electrons with same spin structure avoid coming too close to one another (according to PQP). This avoidance reduces the coulomb (electron–electron repulsion) energy. So, for free

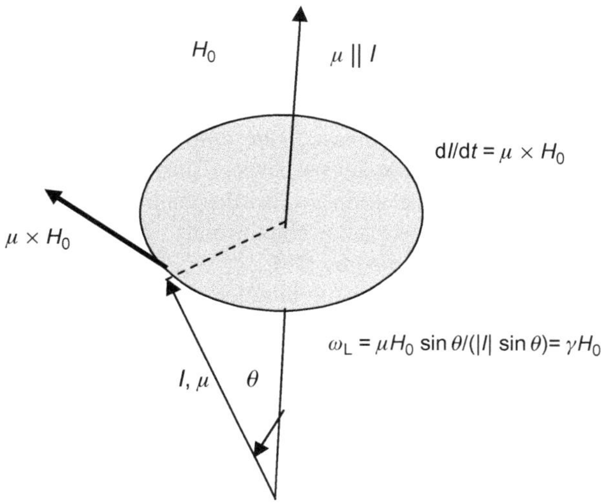

Figure 2.6 The precession of the nuclear spin I in an external field H_0. The rate of change of the angular momentum, $\omega_L I \sin\theta$, is equated to the torque, $\mu \times H_0$ (vector cross-product), produced by the magnetic field H_0.

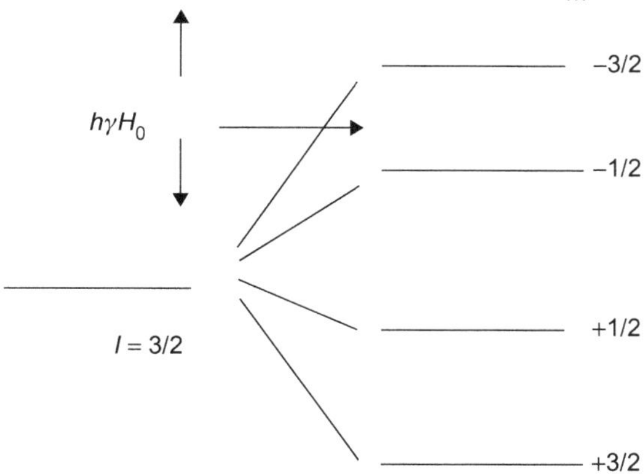

Figure 2.7 The Zeeman splitting of nuclear levels by an external magnetic field. *Courtesy*: Ref. [7].

atoms, the magnetic state is an ordinary state of matter. When atoms form aggregates, though, the atomic moment is usually lost. For example, in Na^+Cl^-, the magnetic moment is lost. Atomic hydrogen has a magnetic moment of one Bohr magneton (μ_b), but in its stable form in the hydrogen molecule, the electrons are paired and possess oppositely directed spins, and thus there is zero orbital angular momentum. The hydrogen molecule is therefore diamagnetic. This means that magnetism arises only from the application of an external field. By virtue of the energy added, an externally applied magnetic field changes the size of the electrons' orbit (diamagnetism) around the nucleus. Hence, induced magnetism arises. Diamagnetism is only induced by an external field, due to a reaction (screening) to the applied field. So, it is said to have a negative magnetic susceptibility, $\chi = -M/H$ (magnetic moment M generated per unit applied field). This is true for N_2 and for most dynamic molecular gases. The important exceptions are O_2 and

NO, which are both strongly paramagnetic (unpaired electron structure). Because the condensed form of many elements, like hydrogen and nitrogen, consists of aggregates of the molecules, these elements are purely diamagnetic.

In an ideal situation, as *H* (magnetic field) is increased, *M* should increase proportionally. In contrast, as the temperature is increased, the ability of the atoms and molecules to align with a magnetic field becomes weaker. Thus, to be useful, the susceptibility should be independent of the temperature (diamagnetic or paramagnetic). Of course, real systems in nature, e.g., the human body, are not ideal. This departure from the ideal can only be explained by QM.

In atoms, the odd number of valence electrons is the number of electrons in the outermost orbit. When they are isolated, they possess atomic magnetic moment, which is lost when a solid is formed. This is because the valence electrons lose their atomic character (association) and form an electron gas. This gives rise to (Landau's) diamagnetism and (Pauli's) paramagnetism among the free electrons in a solid. In the liquid state, the electrons partially belong to a specific atom but they have the fluidity to belong to others as well. Thus, one can say that atoms have magnetic moment associated with their electrons, but they generally lose it when they form molecules, compounds, or larger aggregates, because the valence electrons are, in various ways (ionic, covalent, or metallic bonds), shared with other atoms of the aggregate and lose their localized character. The electrons that provide the atomic moment in a free state maintain their atomic character in the liquid state only to a certain degree; hence, the condensate will be magnetic in a characteristic fashion. The pure transition metals are magnetic in the manner of an electron gas.

The strong magnetic properties of a minority of these electrons (in d transition metals, such as Mn, Fe, Co, and Ni) must be considered exceptional. In a real system, the term *interaction* is used to represent the complex, quantum mechanical exchange and correlation of forces. These forces govern the behavior of electrons and their magnetic properties. The classical magnetic dipole–dipole interactions between atomic magnets are extremely small. To observe them, one must create special environments, e.g., low temperatures. It may take only a curious mind to work out why this low-temperature requirement cannot be transformed into a room-temperature, spin-alignment enhancement in a laboratory situation. The human body seems to be such a laboratory. There the dipole magnetic moments produce some sort of coherence at the physiological temperature of 300 K. This may hold some clues about the mind, thought, and coherent brain function.

For the sake of completion, it is considered useful at this point to include the approximate mathematical expressions for magnetic susceptibility according to the classical (Newtonian) mechanic (NM) and QM models, respectively:

$$\chi(\mathrm{NM}) = (\mu_0 N \mu^2)/3kT$$

Here *N* is the number of atoms (Avogadro's number in case of a mole of a substance), μ the magnetic moment of each atom, μ_0 the permeability of the medium in which the dipole moments are present, *k* the Boltzmann constant, and *T* the temperature.

$$X(\mathrm{QM}) = [(\mu_0 N g^2 \mu_b^2)/3kT]J(J+1)$$

One should note that the quantum mechanical expression is the same as the NM one, except that the moment μ of the NM model has been replaced by the expression $\mu_b g[j(J+1)]^{1/2}$ or, say, by an effective value of $\mu_{eff} = \mu_b g[j(J+1)]^{1/2} = p\mu_b$. Here $p = g[j(J+1)]^{1/2}$ is called the effective Bohr magnetic quantum number.

Deviation from the ideal is what makes nonideal situations and adds to scientific researchers' curiosity about the secrets of nature. But the biggest beneficiary is humanity, which ultimately gets to use any new technology developed. So, who is the winner? Constructive or destructive use of a technology is a choice that rests with society.

2.1.2 Nuclear Induction

2.1.2.1 Observation of a Voltage Signal Through Nuclear Induction [8–11]

The nuclei of some atoms have a permanent magnetism, called *nuclear paramagnetism*. Macroscopically, its basics are very much like those of ordinary natural magnets, such as the ones children play with in school and at home. Another good illustration is Earth's magnetism. It is present, all the time and everywhere, on the Earth's surface, and is part of our day-to-day lives, even though the effect on a human body is considered of hardly any consequence. One uses a compass needle to find the rough direction of Earth's magnetic field, at any given position on Earth. Nuclear paramagnetism can be detected by applying an external magnetic field to the nucleus of an atom and detecting the voltage signal induced, in a copper coil placed nearby, as a response to the applied field.

Molecules are made of more than one atom; e.g., water has three atoms, H, H, and O, and thus three nuclei. These three nuclei can interact with each other and thus produce a fine structure, in the signal detected by an NMR study in a test tube. The signal can give us information about the bonds between the H, H, and O atoms. In contrast, when water is present in another environment—say, for example, in the human body—then the fine structure observed also reflects the other atoms and molecules present and the artifacts in the human body that the water surrounds. Protons (from, say, the nucleus of a hydrogen atom) in water can be used as a probe to image any chemical and physical processes that occur, including interactions with other molecules and the surrounding artifacts. Let us now come to the more specific area of interest here, i.e., the human brain. The distribution of gamma amniobutyric acid (GABA) in various parts of the brain can be studied by the response of the GABA to an externally applied field. It can reveal very useful information about any disorders in a brain, e.g., Alzheimer's disease. This is one of the reasons for our writing this chapter: ultimately, it is hoped that the information herein will help medical professionals to (MPs) better interpret the results of MRI of the brain and thus make better and more precise diagnoses of the diseases revealed. Understanding of the PCM of the processes occurring in the brain is the key that will help unlock useful developments in MRI of the human brain.

2.1.2.2 NMR Simple Theory and Concepts [8–11]

The phenomenon of NMR was first put into practice (experimentally), as predicted by theory, during the 1940s and 1950s. At that time, NMR was primarily a matter of mere curiosity; nobody predicted that it would develop into the powerful analytical tool that it has become today. This section reviews the development of the basic concepts of NMR, created through the foundational research [8–11]. This will allow readers to appreciate the intuitive power that successively led to the creation of NMR science, and our mostly unthinking acceptance of its current benefits. The science of the human brain is still in its infancy. It is hoped that it will progress as robustly as the other discoveries of physics. Another wonder of physics research is the development of QM, the applications of which are all around us today. Even the Nobel Prize, which those pioneer workers received, does not come anywhere near a sufficient appreciation of their contributions. Any invention in physics—and perhaps in any field—is much like a newborn baby: no one really knows what it will grow up to become and do. These days, many say that research should be pursued only in applied physics, not pure physics. These people apparently do not realize that one cannot have applied physics unless one has the pure physics on which to ground it. Applied comes out of pure: Pure physics provides the ideas on which to build applied physics. In turn, the developments in applied physics may illuminate some of the "pure" or theoretical physics that were not fully understood before. Pure physics research work on these aspects may then improve the applications or lead to the discovery of new ones. This cycle of pure-applied-pure has resulted in the technologies that we use and, indeed, rely on today. It is increasingly true that, in our current technological society, one cannot do anything useful for society without first learning the underlying PCM. Common sense dictates that we think (pure) first and build (apply) later.

Having noted that, let us now return to NMR. When an external, DC magnetic field is applied to a sample like the brain, an equilibrium between the nuclear magnetic moments in the solution and their semirigid environment is quickly established. The nuclear moments try to align themselves along the field, as that is the lowest energy configuration, but their interactions with their neighbors constantly deflect them away from the direction of the field. When a static magnetic field is applied along a chosen area of the body, such as the human brain, normally z is the chosen direction; the atomic and nuclear moments in the brain try to align themselves along the field during a short period of time. Similarly, there will be a time lapse if the field is removed, during which the nuclear moments return to their original positions. This time is called the *longitudinal* (parallel to the static field) *relaxation time*, denoted as T_1. The rate of change of the total magnetic moment (magnetic moment per unit volume) over a selected area of the object can be written as $dM/dt = \gamma[M \times H]$. This is the standard Bloch equation (SBE). M and H are the vectors (having components in three dimensions), M represents the magnetization (magnetic moment per unit volume) and H the applied (DC) magnetic field. The cross (product) between M and H represents a resultant vector, i.e., $M \times H = MH \sin\theta$, in magnitude, and perpendicular to both M and H in direction. The γ is the gyromagnetic ratio and is defined as M/A (for the whole sample). A is the total sum of the angular momentum components (per unit volume) due to all the nuclei and M is the sum of all the nuclear magnetic moments (per unit volume).

Just as the electrons precess (rotate) around the positive resultant charge of the protons of the nucleus, similarly the nuclear moment (vector) precesses around the applied field direction. The vector of this precision would ideally be parallel to H, but due to interactions with the environment it exhibits an angular frequency spectrum, which can be explored by the application of an externally applied oscillating magnetic field. It is applied in the perpendicular, x-direction with respect to the main Hz field. This in return results in inducement of an oscillating field in the second (y) perpendicular direction. It is this induced field in the y-direction that makes up the measured signal. H and M are now resolved into components as follows: $H_x = H_1 \cos \omega t$, $H_y = \pm H_1 \sin \omega t$, and $M_x = M \sin \theta \cos \omega t$, $M_y = M \sin \theta \sin \omega t$, and $M_z = M \cos \theta$. This holds under the restriction $\omega - \omega_0 \ll \omega_0$, where $\omega_0 = \gamma H_0$ is the resonance frequency of the ensemble of nuclei, to the incident RF electromagnetic radiation, and $H_1 \ll H_0$. The sign $\pm$ comes from the $\pm$ sign of γ. The SBE clearly implies that if θ is constant and chosen such that $\tan \theta = H_1/(H_0 - H^*)$, then $H^* = \omega/\gamma$, denoting the resonance field at frequency ω. The field H is that for which the Larmor frequency $\omega_L = \gamma H$. The direction of polarization starts to deviate noticeably from the z-direction as the difference $(H_0 - H^*)$ becomes comparable or small compared to the magnitude H_1 of the effective rotating field. It is perpendicular to the z-direction for $H_0 = H^*$, and for still further decreasing values of H_0 it turns toward the negative z-direction. Finally, it points in a direction opposite to H_0, for $(H^* - H_0) \ll H_1$. If one writes $\delta = (H_0 - H^*)/H_1 = \cot \theta$, the solution of the SBE then takes a very compact shape, as follows: $M_x = M \cos \omega t/(1 + \delta^2)^{1/2}$, $M_y = \pm M \sin \omega t/(1 + \delta^2)^{1/2}$, $M_z = M \sin \omega t/(1 + \delta_z)^{1/2}$. It clearly shows an increase of the rotating component of M upon approach to resonance ($\delta = 0$).

This has been just about the simplest model possible. It is nowhere close to an actual situation. Nevertheless, it introduces readers to some of the technical terms they may eventually encounter in applying the technique of NMRI to medical diagnostics. The above solution was formulated under the assumption that ω, H_0, and δ are constant. It can be shown that the theory is equally valid even if these parameters are not absolutely constant, provided that these quantities vary adiabatically, i.e., slowly enough that $\mathrm{d}\delta/\mathrm{d}t = \ll |\gamma H_1|$. For constant H_1 (i.e., for constant amplitude of the oscillating field), this variation of δ, and therefore of the components of M, takes place through two different procedures. In the first, the field H_0 in the z-direction is kept constant and the frequency ω of the oscillating field is slowly varied, thereby slowly varying the value H^*. The resonance field is given by $H^* = \omega/|\gamma|$. In the second, ω and therefore H^* are kept constant and H_0 is varied slowly. Whether a variation $\mathrm{d}H_0/\mathrm{d}t$ can be considered adiabatic depends on the half of the amplitude H_1 of the RF field. The more adiabatic any given variation is, the larger the H_1 will be. The condition for adiabatic variation can also be expressed by stating the z-field H_0. It has to pass through an interval comparable to the resonance width. This is during a time that is long compared to $t_1 = 1/|\gamma H_1|$.

This analysis, so far, has been oversimplified; the model requires further improvement. One needs to include the action of the atomic moments upon the nuclei. The importance of the atomic moments depends upon the substance under consideration. In some substances, such as water, the electronic spin moments will be paired off. The orbital moments (which may be present in free atoms and molecules) are quenched because of intermolecular interaction. One also needs to consider the presence of permanent atomic moments.

2.1.2.3 The Realistic Model [8–11]

The SBE introduced in the previous section will now be made to take a better shape, and will be called the first revised SBE (FRSBE). The revised phenomenological equation now is $d[M(t)]/dt = \gamma(M \times H) - i(M_x/T_2) - jM_x/T_2 - k(M_z - M_0)/T_1$. This represents the time dependence of the macroscopic atomic and nuclear polarization $M(t)$ under the influence of $H(t)$. The vectors i, j, k are the unit vectors in the x-, y-, and z-directions, and $\gamma = \mu/I\hbar$ ($\hbar = h/2\pi$, h being Planck's constant). Here γ is the gyromagnetic ratio, μ is the magnetic moment of a single nucleus (e.g., in the simplest nucleus, hydrogen, it will be the magnetic moment of a single proton), and $I\hbar$ is the nuclear angular momentum (I being the nuclear spin 1/2, 3/2, etc.). The magnetic field is assumed to have the form $H(t) = kH_0 + H_1(t)$, where H_0 is strong and constant, while H_1 is a relatively weak and arbitrary function of the time t. M_0 is the equilibrium polarization in the field H_0. The establishment of the thermal equilibrium is summarily described by the constants T_1 and T_2, called the longitudinal and TRTs, respectively. Starting with an arbitrary magnitude and direction, the z-component of M will, in the absence of the field H_1, reach the value M_0 with a time constant T_1, and the x- and y-components will vanish with a time constant T_2. Because of the large number of identical nuclei in the macroscopic sample, the behavior of the polarization is determined by the effect of the external field. This is also controlled by the molecules surrounding the neighboring nuclear moments, and by the orientation of the magnetic moment of each nucleus.

It is not considered appropriate to include a full mathematical treatment here. In practice, of course, one should analyze the complete dynamical theory of nuclear magnetization. The reader is referred to references [10, 11] for a view of the initial, conceptual development of the theory, and the later standard texts developed over the years.

2.1.2.4 NMR Adiabatic Pulses: Vector Representation (Appendices A2.1 and A2.2) [12]

When NMR was first created, resonance was achieved by sweeping the amplitude of the polarizing magnetic field H_0 in the presence of a perpendicular field H_1, which oscillates at a constant RF (in some references, the symbols B_0, B_1, etc., are used instead). This continuous-wave (CW) approach was later replaced by the pulsed NMR approach. This is performed in a static H_0 and uses a pulsed H_1 to excite the full band of spectral frequencies simultaneously. Typically, the carrier frequency of the pulse remains constant and is applied at the center of the spectral region of interest (ROI).

One can also follow an alternative approach, in which the carrier frequency varies with time during the pulse. These frequency-swept pulses are known as *adiabatic pulses*. The different spectral components are rotated, in succession, during the adiabatic frequency sweep. When the total sweep time is short relative to T_1, the transient response time of the spin system can be induced. This allows observation of the NMR phenomenon, e.g., the free induction decay (FID) or the echoes, related to the pulsed method. By rapidly sweeping the frequency of the adiabatic pulses, NMR experiments can be performed in the same manner as pulsed experiments. This means the length of the adiabatic pulses can be short enough (in time) to permit their use in most pulse sequences. In this way, the advantages of both classical CW and pulsed NMR approaches can be exploited.

In a sweep of either H_0 (the classical experiment) or the RF pulse-frequency adiabatic experiment, from one side of the resonance to the other, the net rotation of the magnetization M is highly insensitive to changes in H_1 amplitude. The adiabatic pulses offer a means to rotate M by a constant flip angle, even when H_1 is extremely inhomogeneous. Across the spectral bandwidth of interest, spins with different precession frequencies (*isochromats*) are sequentially rotated, as the frequency sweep $\omega_{RF}(t)$ approaches the resonance frequency ω_0, of each isochromat. With some types of adiabatic pulses, such as adiabatic full passage (AFP), the bandwidth $\Delta\Omega$ is dictated solely by the range of the frequency sweep, Appendices A2.1 and A2.2. For the spins precessing within this frequency band, the flip angle will be uniform, provided the orientation of the effective magnetic field changes slower than the rotation of M about the effective field. This requirement, which is known as the *adiabatic condition*, can be satisfied by using sufficiently high H_1 amplitude or by a slow frequency sweep. With the latter method, $\Delta\Omega$ can be arbitrarily wide, even when using the low peak RF power, provided that the pulse length Tp can be sufficiently long. The ability to achieve uniform flip angles over broad bandwidths with low H_1 amplitude is a unique feature of adiabatic pulses. With conventional constant-frequency pulses, $\Delta\Omega$ varies inversely to Tp. $\Delta\Omega$ and Tp are independent parameters in certain types of adiabatic pulses. So far, major efforts in NMR have focused on the design of a complex RF pulse to compensate for changes in H_1 amplitude and/or to increase the bandwidths.

A close relative of the adiabatic pulse is the composite pulse, which consists of a train of rectangular pulses of different phases. Although composite pulses can be derived to compensate for $>$10-fold variation of H_1, adiabatic pulses generally offer the greatest combined immunity to H_1 inhomogeneity and resonance offsets, for a given amount of RF power. Appendices A2.1 and A2.2 include diagrams of geometrical (vector) configurations in space of some adiabatic pulses.

2.1.3 Illustrative MRI: Technical Examples

2.1.3.1 Neurophysical Functional Magnetic Resonance Imaging [13]

Functional magnetic resonance imaging (fMRI) is a technique to measure BOLD hemodynamic changes. These changes might be, for example, alterations in blood flow, volume, or intravascular magnetic susceptibility. fMRI has limitations regarding the relationship between cerebral hemodynamic changes and actual neural activation. Active states of any brain site are characterized by the time-varying spatial distributions of action potentials (APs). These are superimposed on relatively slow-varying field potentials (FPs). In experimental functional imaging, an ideal situation would be microelectrode recordings of both of these potentials; this should include single-spike response and FPs. The latter relate not only to spike activity, but also to subthreshold integrative processes in areas such as dendrites. These are otherwise inaccessible.

An experimental technique that permits simultaneous electrophysiological measurement and imaging would be very useful. In particular, it could link the dynamics of the brain with neuronal synaptic activities. MRI applies standard magnetic-field gradients for image generation. One makes use of this during echo planar imaging (EPI) when one needs to locate the neural origin of the BOLD signal. This is done by examining the degree of correlation of the hemodynamic response to single-unit and multiunit activity (MUA), as well to the local field potentials (LFPs). For explanation of the results in fMRI studies, one has to keep in mind an important factor: neural activity and energy metabolism are tightly coupled. A quantitative relationship can actually be established between imaging signals and the cycling of certain cerebral neurotransmitters. Synaptic activity is tightly coupled to glucose uptake. NMR spectroscopy experiments have been carried out in this area. By studying such couplings, one sees that the energy demands of glutamatergic neurons account for most of the total cortical glucose usage. Some of the results in the study are included as an illustration (Figure 2.8).

2.1.3.2 MQC Imaging

2.1.3.2.1 Double-Quantum Coherence Filter [14]

GSH is a major antioxidant that plays a significant role in the detoxification of reactive oxygen species. GSH in the normal human brain varies from 0.8 to 3.1 mM. It is difficult to measure in vivo using proton MRS (PMRS) because of the severe overlapping of GSH resonances with more intense peaks from other metabolites, such as creatine (Cr). There are a few difficulties in the double-quantum coherence filter (DQCF) regime. Extra scans are required to achieve maximal editing yield. Because the slice-selective pulses in the sequence induce phase increments, the phase of the DQ excitation pulse has to be set correctly to eliminate the phase accumulation and maximize the SNR by additional phase-calibration scans. Another problem is the influence of the water presaturation pulses on GSH yield. With the proximity of the chemical shifts between the CH proton of GSH cysteinyl group (4.56 ppm) and water (4.70 ppm), the CH proton is generally assumed to be completely destroyed by water saturation pulses. The suppression of the CH proton of the GSH cysteinyl group eliminates its influence on detection of the signal arising from the CH_2 proton. However, if the CH protons are not completely suppressed by the water saturation pulses, the residual CH proton signal will be converted into the CH_2 proton at 2.95 ppm by *J*-couplings. Because this residual CH proton signal depends on the water saturation pulses, it may vary from

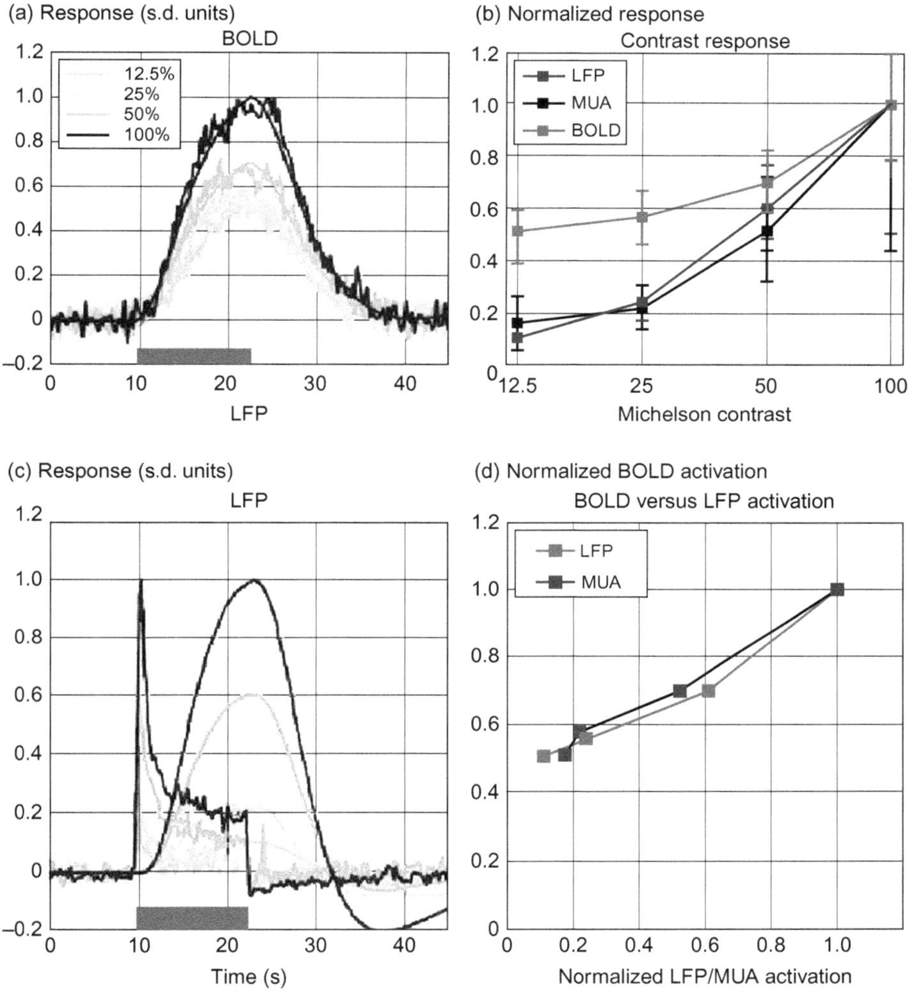

Figure 2.8 fMRI response to pulse stimuli at four different contrasts (12.5%, 25%, 50%, and 100%). (a) Mean fMRI superimposed with a model estimated with nonlinear curve fit. The scale parameter of the model was taken as the response amplitude. (b) Normalized response amplitude of LFP and MUA against contrast. Data from five sessions with a pulse duration of 12.5 s. (c) LFP response for four different contrasts. Smooth lines are the result of the neutral response, with the impulse response estimated by correlation analysis. (d) Normalized BOLD response as a function of LFP and MUA. Responses were normalized by dividing each response by the maximum response.

experiment to experiment; i.e., using water suppression may lead to variation in the detailed GSH signal.

A new DQF is introduced, and is combined with a PRESS. The distinguishing feature of this filter is that frequency-selective pulses, instead of nonselective pulses, are used both for generating DQCs and for converting them back into SQCs. A CHESS, consisting of three Gaussian pulses, each followed by a gradient crusher, is used for water suppression. In practice, a 90° excitation pulse and two 180° pulses, which form the PRESS, were used to achieve spatial localization. The signal stated from the CH proton of the cysteinyl group, which had been severely reduced by the water saturation pulses, was left in antiphase state by the first frequency-selective 90° Gaussian pulse, and not converted into observable CH_2 signal around 2.95 ppm by *J*-coupling. This minimized the potential contamination of the GSH signal by the CH proton signal, if the CH proton was incompletely suppressed by the water saturation pulses. Due to the strong couplings between the two CH_2 protons, signal detection starting from the full echo does not ensure a pure absorption *GS*; *H* in phase. G_1 and G_2 gradient magnetic fields were applied to crush magnetization that did not originate from the DQC. The onset of G_2, the magnetization of interest, was converted into SQCs by the second 90° Gaussian pulse. The last slice-selective 180° pulse was placed in the middle of the transverse-modulated (TM) period. It refocused the effects of the chemical shifts and the inhomogeneous H_0 field. The yield of the DQCF is defined as the ratio of the GSH amplitude from the DQCF to that from PRESS (TE, time of echo = 30 ms; TR, time of repeat = 2.0 s) in the same volume of interest (VOI) (Figures 2.9 and 2.10).

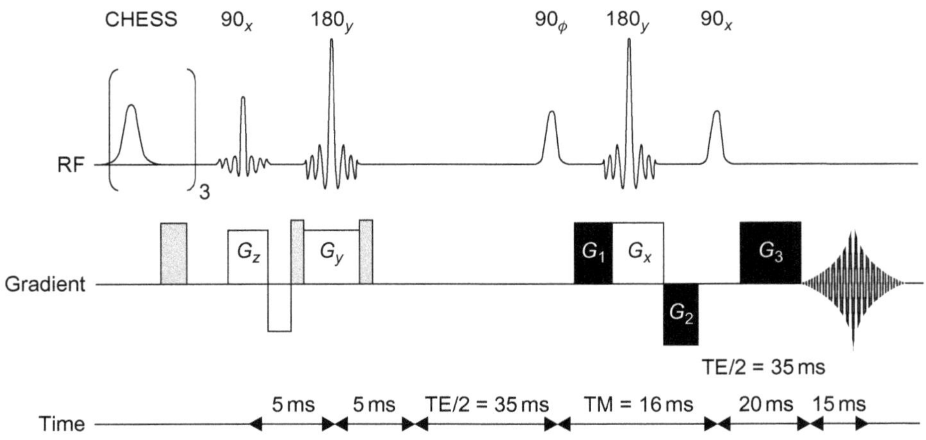

Figure 2.9 Volume-selective DQF with two frequency-selective 90° Gaussian pulses is preceded by CHESS. TM is 16.0 ms and t_1 is 5.0 ms. The gradients marked in black are used to select the DQC. Spoiling gradients (gray) are applied along the first 180° pulses. The data acquisition starts at 20 ms before the echo. The duration of G_1 and G_2 are 1.0 and 4.0 ms, respectively. The amplitudes for G_1 and G_2 are both 22.0 mT/m. Phase-cycling scheme: first 90° Gaussian pulse, $x, -x, x, -x$; second 90° Gaussian pulse, $x, x, -x, -x$: ADC, $x, -x, -x, x$.

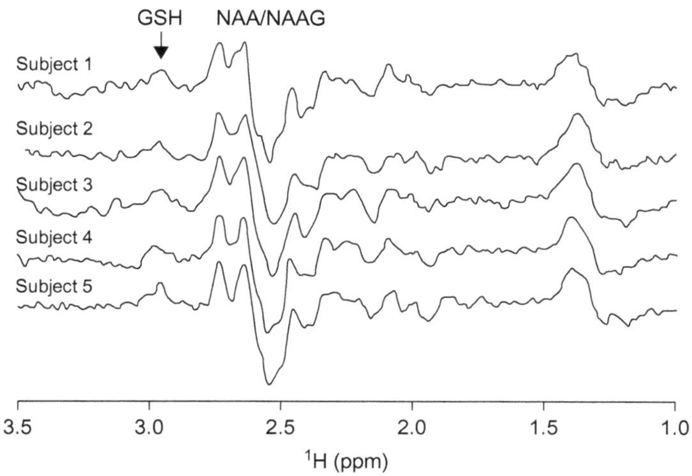

Figure 2.10 In vivo ^{1}H NMR spectra of GSH using the DQCF.

2.1.3.3 MRS of GABA-Neuropsychiatric Disorders [15]

GABA is the major inhibitory neurotransmitter in the nervous system and is integral to the management of brain excitability. GABA is synthesized in neurons from glutamine, with glutamate as an intermediate step. This is done via glutamic acid decarboxylate and the cofactor pyridoxal phosphate. In addition, released GABA is taken up by glial cells and metabolized by GABA-transaminase, and then enters the tricarboxylic acid cycle that will eventually yield glutamate, which will be converted into glutamine by glutamine synthetase. The glial glutamine can then be transported back into the neurons for synthesis of new GABA. There are at least two GABA pools within the neuron cytoplasmic vesicles. The synaptic effects of GABA are mediated through two major receptor subtypes, $GABA_A$ and $GABA_B$. GABA receptors inhibit neurons and are crucial to controlling brain excitability. They house recognition sites for benzodiazepines, which occur naturally in the central nervous system, as well as being synthesized as psychotropic mediations.

Because $GABA_A$ receptors also bind to other chemicals, including neuroactive steroids, ethanol, and anesthetics, the action of GABA receptors is modified by a number of endogenous and exogenous ligands. $GABA_B$ receptors are guanidine-nucleotide–binding proton-coupled receptors. Though little understood, $GABA_B$ receptors are thought to modulate and inhibit the generation of excitatory postsynaptic potentials and long-term potentiation in the hippocampus and in the mesolimbic dopaminergic neurons. In the healthy human brain, GABA concentrations range from 0.5 to 1.4 $\mu mol/cm^3$. GABAergic neurons are widely distributed and have been shown to represent 10–40% of nerve terminals in the cerebral cortex, substantia nigra, and hippocampus of nonhuman primates. It has been estimated that as many as 75% of all synapses within the central nervous system are GABAergic.

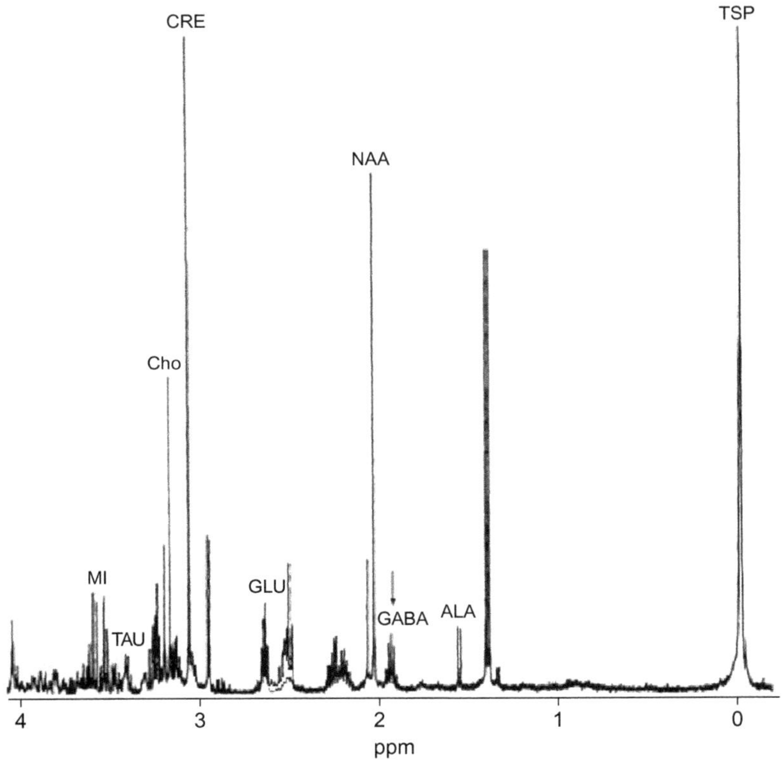

Figure 2.11 The spectrum on a 500-MHz (11.7 T) NMR spectrometer. The well-delineated GABA measurement is at 1.90−1.98 ppm. ALA: alanine, Cho: choline compounds, CRE: total creatine, GABA: γ-aminobutyric acid, GLU: glutamate, ^{1}H-MRS: proton magnetic resonance spectroscopy, MI: myoinositol, NAA: *N*-acetylaspartate, TAU: taurine, TSP: trimethylsilyl-propionic acid.

However, the human brain is chemically heterogeneous, and many metabolite concentrations, including GABA, vary by brain region, sex, and age. MRS provides a noninvasive means to identify and measure metabolites in vivo, ex vivo, or in vitro, especially metabolites that are present in relatively high concentrations in the brain (Figure 2.11).

2.1.3.4 PMRS/Point-Resolved Spectroscopy: DQF MRS [16]

To understand the functions of the brain, it is important to carry out a high-precision assessment of the concentrations of GABA+ (GABA plus homocarnosine) and GLX (glutamate plus glutamine). This also holds for the metabolites *N*-acetylaspartate plus, *N*-acetylaspartylglutamate (NAAGt), creatine and phosphocreatine (PCr), choline-containing compounds (Cho), and myoinositol (INS). It has been suggested that a decrease in GABA inhibition may be one of the causes of epileptic seizures. Antiepileptic drugs (AEDs) enhance GABA function. Identification of GABA via MRS is difficult owing to overlapping metabolite peaks. Methods such as spectral editing have now been developed that allow separation of the GABA signal, to allow quantification. The level of GABA and homocarnosine, a dipeptide of GABA and histidine, has been shown to increase with improved seizure control. This also happens following the administration of an AED that modifies GABA metabolism or function. It is common to see studies on subjects with idiopathic generalized epilepsy (IGE) for evidence of neuronal dysfunction (reflected by reduced NAA concentrations), or of cortical hyperexcitability (reflected by reduced GABA or increased GLX) (Figure 2.12).

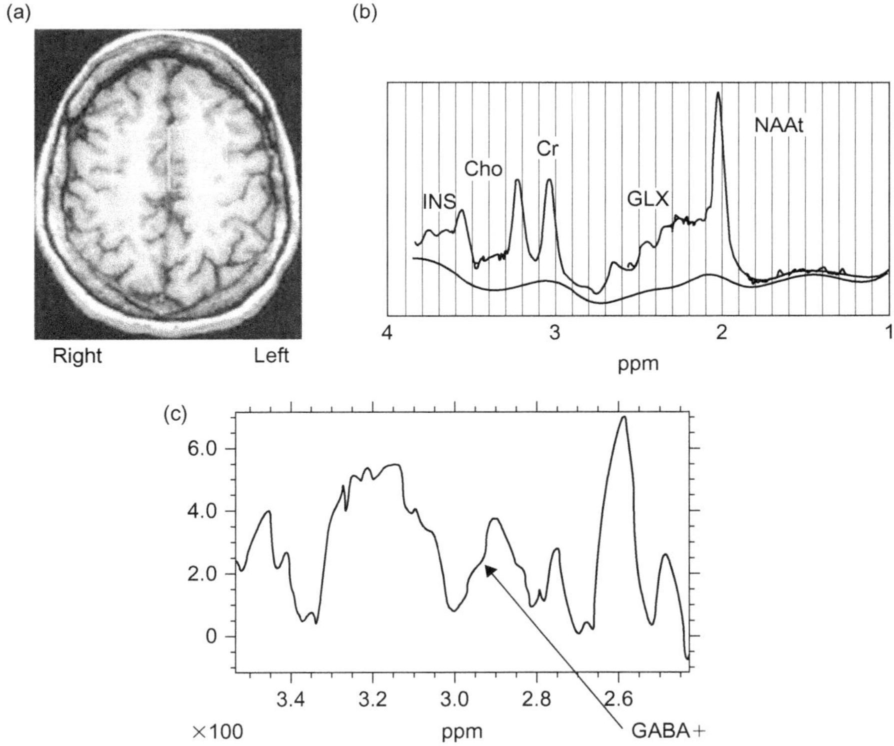

Figure 2.12 (a) Axial image showing the position of the prescribed frontal-lobe voxels. (b) A representative PRESS spectrum (echo time/repetition time 30/3000 ms) with superimposed LC model fit to metabolite concentrations and estimated baseline. (c) A filtered GABA spectrum from the right frontal voxel of a control. NAAt: *N*-acetylaspartate plus *N*-acetylasaparatylglutamate; GLX: glutamate plus glutamine; Cho: choline-containing compounds; Cr: creatine + phosphocreatine; INS: myoinositol; GABA+: GABA plus homocarnosine.

2.1.3.5 Spatial Localization of Point-Resolved Spectroscopy for GABA Measurement: DQ Excitation and Detection [17]

Dysregulation of the GABAergic system has been implicated in the pathophysiology of a wide range of neurological and psychiatric disorders, including epilepsy, schizophrenia, alcoholism, and stroke. GABA is present in the human cortex at a concentration of approximately 1 mM, and proton spectroscopy can be used to make localized in vivo measurements of GABA concentrations in the brain. One uses spin-echo difference techniques, in which successive scans are acquired with and without application of an inversion pulse. When the two spectra are subtracted, the unaffected singlet resonances are nulled and the γCH_2 triplet of GABA is detected. Spatial localization is achieved with either image-selected in vivo spectroscopy (ISIS) or the PRESS technique. The difference technique is, however, susceptible to cancellation errors from patient motion and other metabolites that produce slight variations between subsequent acquisitions.

Multiple-quantum filter (MQF) techniques can overcome the difficulties of the difference method, by achieving robust suppression in a single acquisition. This new technique uses the different phase-accumulation rate of the singlet and MQ states to selectively replace only magnetization that passes through MQC. This eliminates the MR signal from uncoupled metabolites such as creatine, choline, and the CH_3 group of NAA, as these resonances have no MQCs. A localized DQF sequence, with PRESS localization, for the in vivo detection of GABA in the human brain is shown in Figure 2.13. DQCs are excited in the coupled metabolites by the first three pulses ($\pi/2$, π, $\pi/2$). The SQCs excited by the initial $\pi/2$ pulse evolve, under the influence of the J-coupling interaction, into antiphase coherence which is in turn converted into an MQC by the second $\pi/2$ pulse. The MQCs are converted into observable SQCs by the third $\pi/2$ pulse. This is a frequency-selective

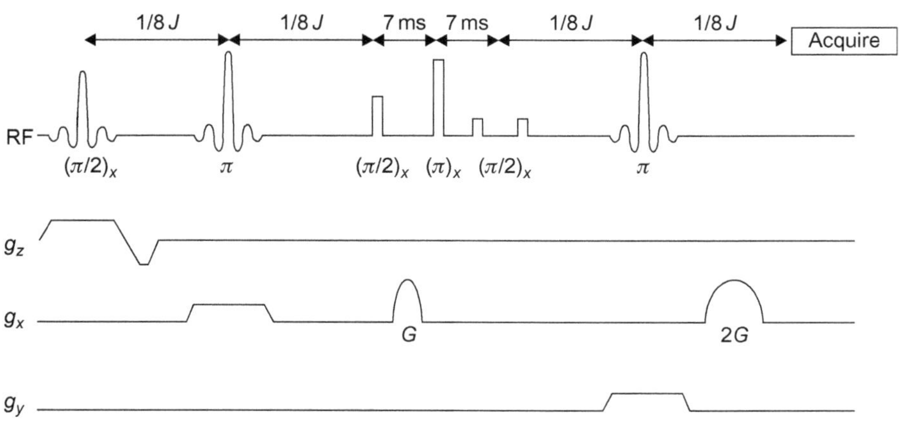

Figure 2.13 PRESS-localized DQF sequence for the in vivo detection of brain GABA. The initial CHESS water suppression pulses and the crusher gradients surrounding the slice-selective π pulses are not shown. The time period $1/8J$ was set to 17 ms.

binomial pulse with 7 ms spacing tuned to the βCH_2GABA resonances at 1.91 ppm. Use of this selective magnetization transfer pulse doubles the theoretical detection efficiency for the γCH_2 resonance at 3.02 ppm, by transferring the βCH_2 resonance to the desired γCH_2GABA resonances.

The π pulses in the sequence serve to refocus the effect of chemical shifts and provide slice selection for localization. The gradients G and $2G$ are used to crush magnetization that did not originate from a DQC. For a spin at location x, the gradient pulse of area G encodes the SQCs and DQCs with phase proportional to G_x and $2G_x$, respectively. At the time of the gradient pulse of area $2G$, all of the magnetization of interest has been converted back into an SQC. This magnetization acquires an additional position-dependent phase proportional to $2G_x$. The spins that passed through a DQC thus have no net phase difference across the voxel, whereas uncoupled metabolite resonances such as the 3.0-ppm creatine line acquire a net phase proportional to G_x, and contribute no net signal when averaged across the voxel. The delay between the pulses ($1/8J = 17$ ms) maximizes the production of antiphase magnetization at the time of the second $\pi/2$ pulse. The timing used is approximately half that of a conventional DQF sequence. The DQF detection efficiency is optimized when the first and second $\pi/2$ pulses have identical phases. Because one of these pulses is slice selective and the other is a hard $\pi/2$ pulse, the relative phase is, in general, dependent on the locality of the selected region. This position-dependent phase arises from phase accumulation during the brief delay between the direct digital synthesis of the transit frequency and the start of the slice-selective pulse. Figure 2.14 [17] shows the two-pulse sequence used to calibrate the relative phases of the initial excitation, and the DQ production pulse. In this sequence, the slice-selective RF and gradient pulses are identical to those of the DQF sequence. The relative phase difference between the initial $\pi/2$ pulse and the hard π pulse is obtained by adjusting the phase of the hard π pulse until the phase of the water signal is identical when measured with the two sequences.

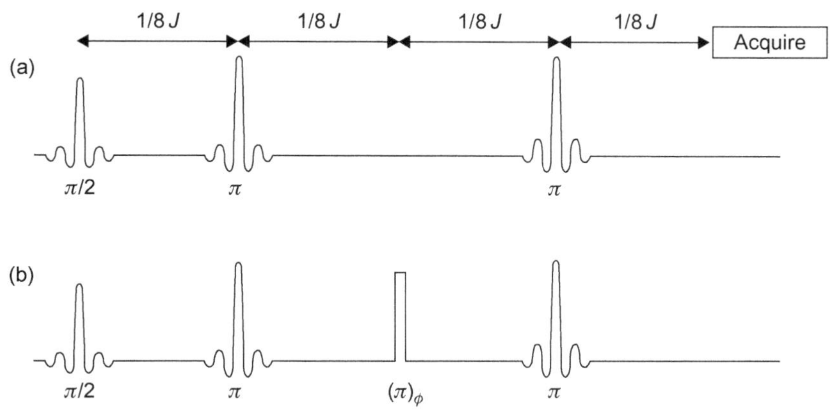

Figure 2.14 The two-pulse sequences used to calibrate the phase of the DQF sequence. The slice-selective RF and gradient pulses are identical to those of Figure 2.13.

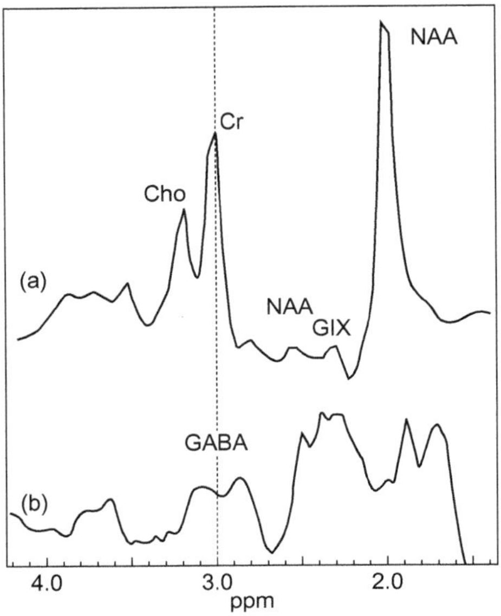

Figure 2.15 Localized in vivo spectra from human brain acquired with identical PRESS localization parameters. (a) SQ spectrum. (b) DQF with vertical scale expanded 12.5 times.

The spectra in actual cases were taken from a 27- to 55-ml voxel located in the midline parieto-occipital region. Conventional PRESS-localized single-quantum (SQ) spectra were acquired, to obtain a measure of creatine, by removing three hard pulses and associated time intervals from the sequence in Figure 2.13 [17]. DQ filtered spectra were obtained with this pulse sequence, after phase-shifting the second $\pi/2$ pulse by an amount determined by the phase-calibration procedure. Two hundred and fifty-six acquisitions were averaged with a spectral bandwidth of 2500 Hz, a transverse repetition (TR) time of 2 s, and a total acquisition time of 8.5 min (Figure 2.15).

2.1.3.6 *Localized ^1H (Proton) MRS J-Coupling: Human Brain (Occipital Lobe) GABA In Vivo [18]*

It has been shown that compounds with anticonvulsant properties act to reduce the uptake of GABA. This is what happens from the extracellular space: They enhance the activity of the GABA–benzodiazepine receptor complex and also block the degradation of GABA by GABA transaminase. Vigabatrin (4-aminohex-5-enoic acid) is a highly specific inhibitor of GABA transaminase. The drug binds to neuronal and glial GABA transaminase with high affinity, and irreversibly and markedly inhibits the enzyme. This raises GABA concentrations in vitro and in vivo. Vigabatrin has been shown to be a safe and effective antiepileptic medication in humans. It is well absorbed after oral administration and has predictable pharmacokinetics. The involvement of the GABAergic system in the pathogenesis and treatment of epileptic disorders suggests that an in vivo method of assessing GABA metabolism would be of value. ^1H NMR has the potential to provide regional measurements of brain

GABA concentrations, as well as the rate of GABA turnover. However, the proposed process is complicated by the proton resonance of GABA, which is overlapped by the larger resonance of creatine, glutamate, and NAA, as well as resonance from macromolecules.

In stronger magnetic fields, a homonuclear *J*-coupling-based editing-pulse sequence can be used to resolve the C4 and C2 GABA-proton resonances. The lower B_0 fields, and the need to perform spatial localization, in human brain studies have limited the application of spectral editing to lactate and glutamate. A multislice inversion recovery image, using an adiabatic hyperbolic secant pulse for H_1-insensitive contrast, was obtained to select the volume for the spectroscopy. For localized spectroscopy, a volume in the occipital lobe of approximately $2\,\text{cm}^3 \times 4\,\text{cm}^3 \times 4\,\text{cm}^3$ (x, y, z) was chosen from the image. In this instance, the y-dimension was parallel to the coil axis. Three-dimensional localization was obtained by using 3D ISIS. This is done in a sequence with 8-ms, phase-sweep hyperbolic secant inversion pulses (bandwidth = 2000 Hz). Maximum pulse power was 350 W. A curved-surface spoiler gradient with 1.5-cm wire separation was pulsed at 8 A, to provide x and y outer-volume suppression. Additional outer-volume suppression was achieved with a selective pulse in an x-gradient. A $\theta/3$ depth pulse, followed by a dephasing gradient, was used to reduce signal from high-flux regions near the surface coils. The surface-coil RF magnetic field (H_1) was sufficient to cut off outer-volume signal in the z-direction. A 5-ms five-lobe sin c (sin x/x) shaped pulse was truncated at the fifth zero crossing. Therein, a 2.35-mT/m (1000 Hz/cm for ^{1}H) y-gradient was used for spatially selective excitation. Crusher x-gradient pulses of 8-ms duration and 7.52-mT/m strength were used. This was in the spin-echo sequence to eliminate nonrefocusing magnetization. All gradient pulses had a 10-mT/m/ms rise rate. An 80-ms hyperbolic secant-selective inversion pulse and a semiselective refocusing pulse were calibrated to give a 90° pulse duration of 120 μs (used for water suppression). The power of the sin c and the refocusing pulse was calibrated by maximizing the signal between 3.5 and 3.7 cm from the coil center. This was done on a 1D profile of the pulse sequence along the coil center, with a z-cos c to form a 2D column. There was approximately threefold variation in H_1 strength from the center of the coil to the center of the localized volume. It is important to avoid excessive tissue heating from the surface-coil RF field. The RF holder kept all points of the head of the subject, at least 0.075, of a coil of diameter of 0.6 cm away from the coil wire. Regional RF heating was calculated by assuming a uniformly conducting medium. This is done using both a Faraday loop and a magnetic vector potential model. It is found to be <1 W/kg in the brain. In vivo data were acquired with an acquisition time of 410 ms, a sweep width of 2500 Hz, and a repetition rate of 3.39 s (Figure 2.16–2.18).

2.1.3.7 Two-Dimensional J-Resolved Spectroscopic Imaging: GABA in the Human Brain (In Vivo) [19]

The in vivo measurement of GABA using noninvasive MRS technique has been an area of considerable innovation. The primary inhibitory neurotransmitter in the

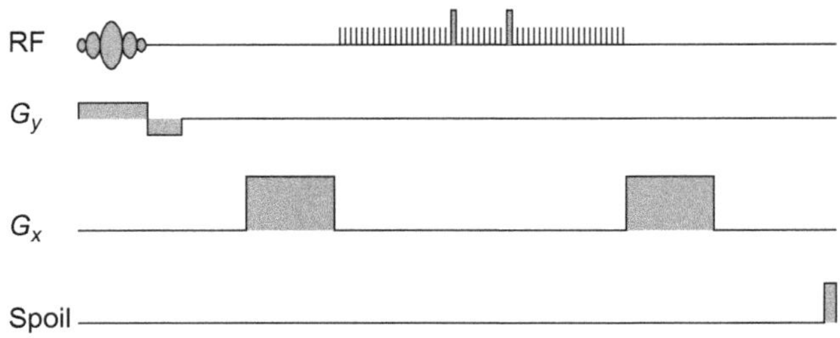

Figure 2.16 Pulse sequence used for the *J*-editing of GABA. Spectra are obtained either with or without a 26.5-ms DANTE inversion pulse applied to the 1.9-ppm C3 GABA resonance to induce *J*-modulation for the C4 resonance. The DANTE and 22 pulses are phase cycled as a single-pulse train. Subtraction of the spectrum obtained with the DANTE pulse from a spectrum obtained without the DANTE pulse gives the edited spectrum. Prior to the editing sequence, a 3D ISIS image with outer-volume suppression is used to localize the volume from which the signal is obtained. G_x: *x*-gradient, G_y: *y*-gradient, Spoil: spoiler gradient.

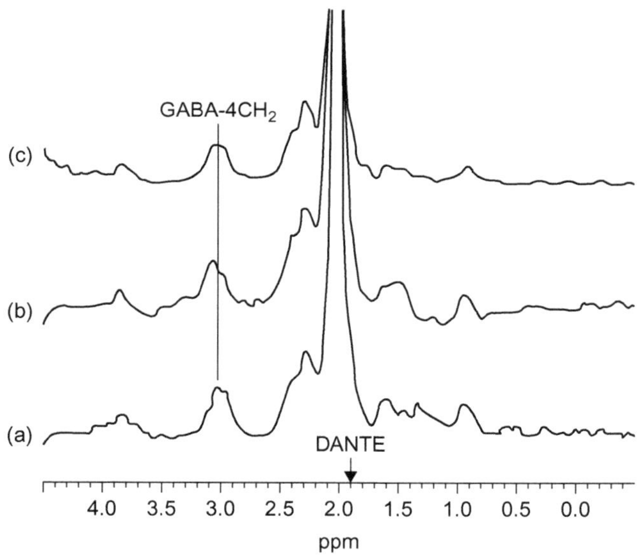

Figure 2.17 Localized GABA edited and control edited ^{1}H NMR spectra from the three nonepileptic subjects (a, b, and c). (Left) GABA-edited spectra obtained with the DANTE placed at 1.9 ppm. (Right) Control edited spectra obtained with the DANTE placed at 2.1 ppm. In the GABA-edited spectra, a resonance is present at 3.0 ppm, which corresponds in chemical and line width to the resonance C4 GABA (GABA-4CH_2). In the control edited spectra, no resonance intensity is observed at 3.2 ppm, indicating that the 3.0-ppm resonance in the GABA-edited spectra is selected on the basis of *J*-coupling at a resonance at 1.9 ppm. The large resonance at 2.0 ppm is from NAA, which is inverted by the DANTE pulse.

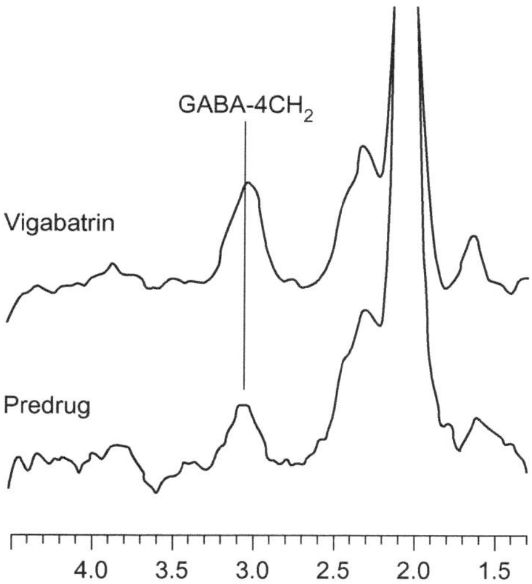

Figures 2.18 GABA-edited spectra measured in an epileptic subject before and during vigabatrin administration. Spectra: lower, before vigabatrin; upper, during treatment with 4 g/day of vigabatrin. The intensity of the edited C4 GABA resonance at 3.0 ppm (GABA human brain (occipital lobe)) is increased by 2.3 times over the intensity in the spectrum obtained before vigabatrin administration.

brain, GABA is of considerable interest in many neuropsychiatric and neurological disorders. It exists in low concentration in vivo (~1.0 μM, ml) and has a complicated, multiresonance spectrum that is overlapped by the dominant creatine (Cr), *N*-acetylaspartate (NAA), and GLX resonances.

Two-dimensional spectroscopic techniques have been developed to isolate and measure the weak and crowded signal from GABA, as well as the other *J*-coupled metabolites and macromolecules. One technique combines 2D *J*-resolved magnetic resonance spectroscopy (*J*-MRS) with chemical-shift imaging (CSI): 2D-JMRSI. The $^2CH_2$ GABA resonance situated at 2.97 ppm can be resolved from the neighboring creatine resonance at 3.0 ppm. The power of the 2D-JMRSI technique lies in its ability to obtain detailed 2D spectral information from multiple voxels. This allows one to perform a regional analysis of different GABA levels throughout the various components of the brain and to obtain a global estimate of brain GABA levels. Thus, 2D-JMRSI is a powerful tool for studying region- and tissue-specific drug effects on the GABA levels that appear to be the underlying neuropathological circuitry of psychiatric and neurological disorders. Numerous 1D MRS techniques have been developed to quantify all or part of the GABA triplet centered at 3.01 ppm in the human brain. In this spectral region, the GABA resonance is relatively unhindered by neighboring coupled and complex metabolite resonance groupings, such as glutamate and glutamine. Instead, it is almost centered on the dominant Cr resonance (3.03 ppm), an uncoupled peak that can be subtracted or filtered out leaving the GABA $^2CH_2$ group for measurement, as in *J*-editing and quantum filtering. 2D spectroscopic techniques have been explored in the quest to further isolate and measure the weak and crowded signal from GABA, as well as other *J*-coupled metabolites and macromolecules. The 2D *J*-resolved method exploits the 7.5-Hz *J*-coupling constant of the GABA resonance to isolate it from the dominant uncoupled resonance of Cr at 3.03 ppm, to facilitate the measurement of GABA and the other proton metabolites that fall within the *J*-coupled bandwidth. With this 2D method, an abundance of metabolite information is collected, including all *J*-coupled and uncoupled resonances within the sampled bandwidth, as well as metabolite T_2-decay information. Implementation of this technique is relatively easy and requires only echo time (TE) stepping, which can be incorporated into any spectroscopy sequence. Important issues that must be considered with this method are field strength, *J*-coupling, and contamination from underlying macromolecule resonances.

Two-dimensional JMRSI also permits adequate sampling of multiple TEs and *k*-space points to resolve GABA from neighboring Cr in a CSI sequence. One can address the issue of high field strength to resolve GABA with 2D-MRS, a technique that to date has only been demonstrated at lower fields. High field allows for more sensitive measurement and potentially shorter scan times or smaller voxels, but may theoretically pose a disadvantage due to the field independence of *J*-coupling. One can show that one can adequately resolve GABA from Cr at 4 T. Although the macromolecules at 3.0 ppm contribute significantly to the 1D spectral baseline at short TEs, they decay quickly and their signal is considerably attenuated through TE stepping. Following is the in vivo macromolecule data to illustrate this effect. One can demonstrate the method of reduced and optimized data acquisition to achieve the goal of implementing a novel 2D *J*-resolved spectroscopic imaging sequence that collects an entire 2D slab of high-resolution voxels, in a clinically usable scan time. 1D and 2D spectra plots of the pure GABA phantom are shown in Figure 2.19, and demonstrate the 2D structure of GABA at 4 T. The 1D spectra are coplotted with the 2D spectrum, and are extracted from $J = 0.0$ and 7.5 Hz for visual comparison. The $^2CH_2$ GABA resonance at $J = 7.5$ Hz sits at 2.97 ppm, and thus is clearly resolved from the $^2CH_2$ GABA resonance at $J = 0.0$ Hz centered at 3.01 ppm. Figure 2.20 [19] shows human in vivo spectra extracted from $J = 7.5$ Hz

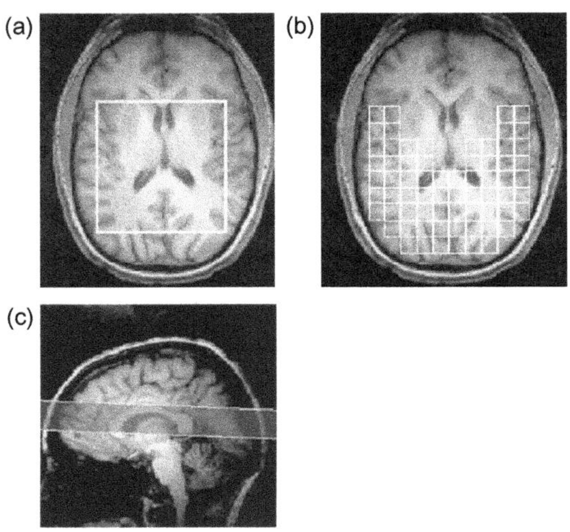

Figure 2.19 (a) Axial T_1-weighted image showing the position of the PRESS-localized/voxel used to acquire standard and metabolite-nulled 2D *J*-resolved proton spectra. (b) Oblique T_1-weighted image showing sampled 2D-JMRSI voxels used to derive global brain estimate of GABA concentration. (c) Sagittal image at 4 T depicting placement of the 30-mm-thick MRSI slab.

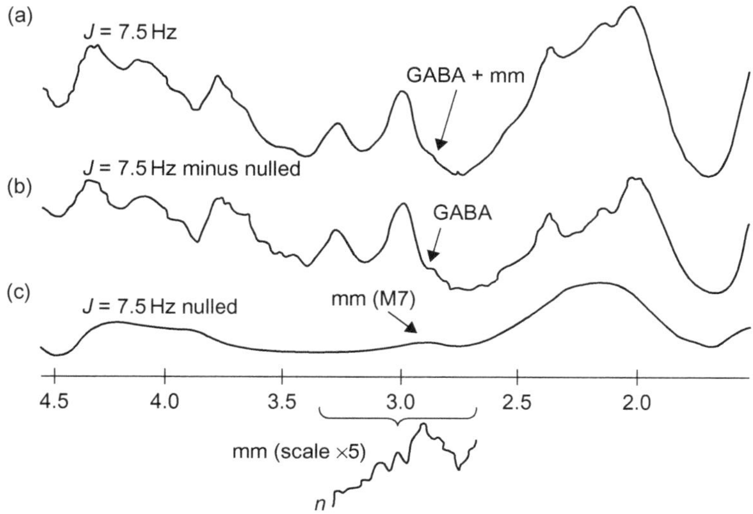

Figure 2.20 In vivo PRESS-defined 2D-JMRS human brain spectra (a) without and (c) with metabolite nulling, and the result of subtracting (b) a metabolite-nulled spectrum from (a) a standard spectrum (c): Residual component at 3.00 ppm is assigned to the macromolecule resonance M7. All three spectra are displayed with 1 Hz exponential filtering and on the same horizontal and vertical scales for comparison.

from the single-voxel 2D-JMRS experiment described above. The standard extraction reveals the combined GABA-residual macromolecule resonance at 2.97 ppm. Figure 2.20(b) [19] shows the result of subtracting the metabolite-nulled spectrum from the standard spectrum, revealing a slight improvement in resolution of the visible "macromolecule-free" GABA resonance at 2.97 ppm from neighboring Cr. The metabolite-nulled spectrum displays a low-SNR, broad signal at and around 3.0 ppm, which is assigned to the M7 macromolecule, with a J-coupling of 7.8 Hz. When the standard and macromolecule-free GABA spectra were fitted, the mean GABA level ($N = 6$) was found to be approximately 15% higher in the standard spectrum, indicating this amount of contribution to the standard GABA measurement from the M7 macromolecule resonance. When the results are plotted for the surrounding spectral extractions about $J = 7.5$ Hz, it can be seen that the macromolecule contribution to the standard GABA resonance area is minimal at $J = 7.5$ Hz and increases in the surrounding spectral extractions on either side, confirming that $J = 7.5$ Hz is the optimum 1D GABA spectrum. To indicate the global spectral quality resulting from the technique, Figure 2.21 [19] shows multiple 2D-JMRSI spectra obtained from several regions throughout the brain that contain different amounts of GM versus WM. The displayed GABA and TE-averaged spectra show good spectral quality and GABA resolution from each region at 4.5 ml nominal resolution, as well as a typical LC model fit. The global brain GABA level was determined to be 0.76–0.20 mM across subjects after correcting for 12% macromolecule contribution.

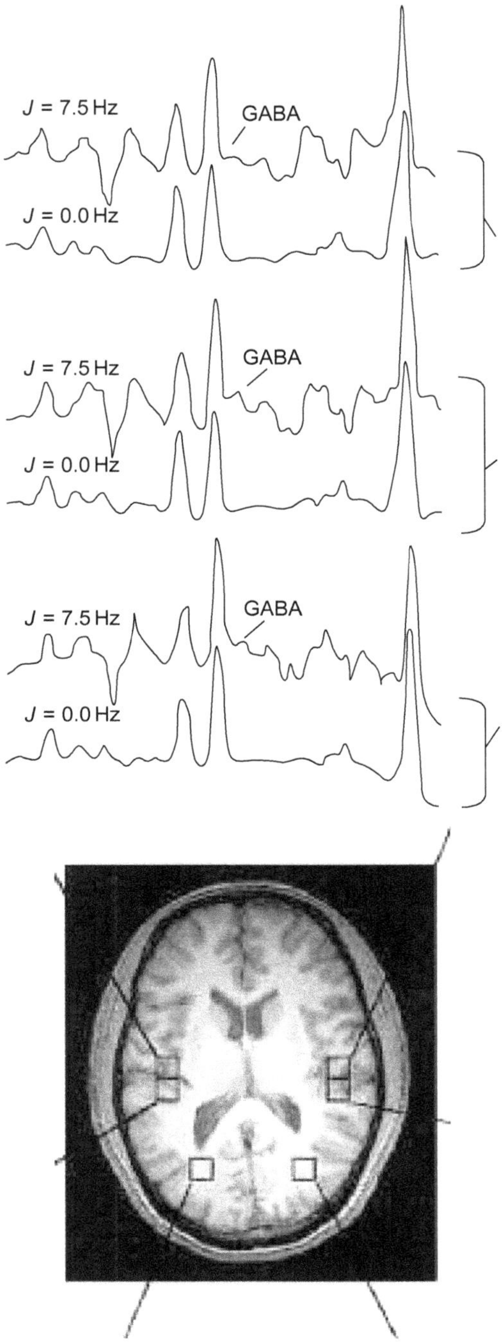

Figure 2.21 2D-JMRSI at 4 T on a healthy human volunteer, and 4.5-ml (nominal) voxels specified with $J = 0.0$ and 7.5 Hz 1D extractions. GABA spectra are scaled to that of TE-averaged spectra for visual comparison.

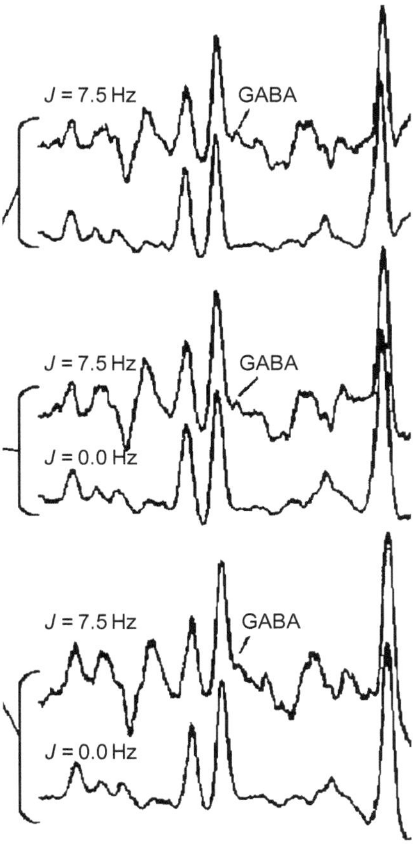

Figure 2.21 continued

2.1.3.8 Phosphorous MRS, 3D CSI, and Phospholipid Resonance Imaging: Human Brain [20]

Phosphorous MRS (^{31}P-MRS) can measure phosphomonoester (PME), phosphodiester (PDE), and brain membrane phospholipids (BMPs) implicated in the pathogenesis of psychiatric disorders and altered in tumor response to therapy. However, the large voxel sizes used in clinical ^{31}P-MRS at 1.5 T have so far prevented isolated studies of discrete brain areas such as the thalamus, hippocampus, cerebellum, and the interconnecting circuitry—areas of particular interest in disorders such as schizophrenia, but not accessible with surface coils. Higher field strengths (3 T), combined with more time-efficient phase-encoding schemes in 3D CSI, have reduced the effective voxel sizes obtainable in brain ^{31}P-MRS to as small as 12 cm^3 while still enabling quantification of PCr and adenosine triphosphate (ATP). Although the increased spectral dispersion at higher field strength should allow quantification of the individual metabolites, brain imaging research shows that, phosphorous (^{31}P) 3D CSI is a standard protocol for studying phospholipid metabolism in discrete regions of the human brain. This technique does not require ^{1}H-decoupling or nuclear Overhauser enhancement (NOE). The metabolites within the PME, phosphoethanolamine (PEth), phosphocholine (PCh), PDE, glycerophosphoethanolamine (GPEth), glycerophosphocholine (GPCh), and BMPs can be dealt with by this technique.

So far, no one has published the results of an attempt to quantify the individual PME/PDE metabolites at high field and measure the degree of in vivo quantification precision. There are three objectives. First, because SNR is crucial in maximizing spatial resolution and precision, part of this work focuses on developing and testing a ^{31}P-MRS sequence at 4 T, based on methods proposed in the literature, which produces high-quality ^{31}P in vivo brain spectra from very small voxels ($15\,cm^3$). Phantom experiments have compared a previously proposed "time-efficient" phase-encoding strategy and spherical encoding with random point omission to conventional phase encoding, attempting to find the best compromise between signal-to-noise ratio efficiency (SNReff) and degree of localization. Second, we need to optimize an in vivo spectral quantification model for precise quantification of the individual PME and PDE constituents (PEth, PCh, GPEth, GPCh, and MP), assessing in vivo precision with this method at $15\,cm^3$ effective volumes. Finally, millimolar concentrations (mM/brain tissue) of PEth, PCh, GPEth, GPCh, and methylene diphosphonic acid (MDP) must be derived and compared against other ^{31}P-MRS studies with ^{1}H-decoupling and NOE at lower fields and much larger voxels. This demonstration would show that the individual PEth, PCh, GPEth, GPCh, and MDP metabolites can be consistently detected and quantified from voxels as small as $15\,cm^3$ in the brain at 4 T without the aid of ^{1}H-decoupling or NOE (Figures 2.22 and 2.23).

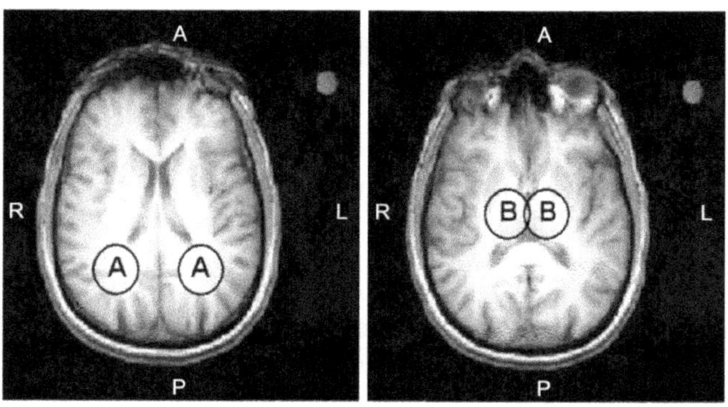

Figure 2.22 Magnetization-prepared rapid acquisition gradient echo (MPRAGE) transverse images of a healthy volunteer displaying effective voxel boundary (3.0 cm diameter) in the left/right parieto-occipital cortex (A) and left/right thalamus (B). MDP reference standard is visible in the upper-right area of the image (subject's left side).

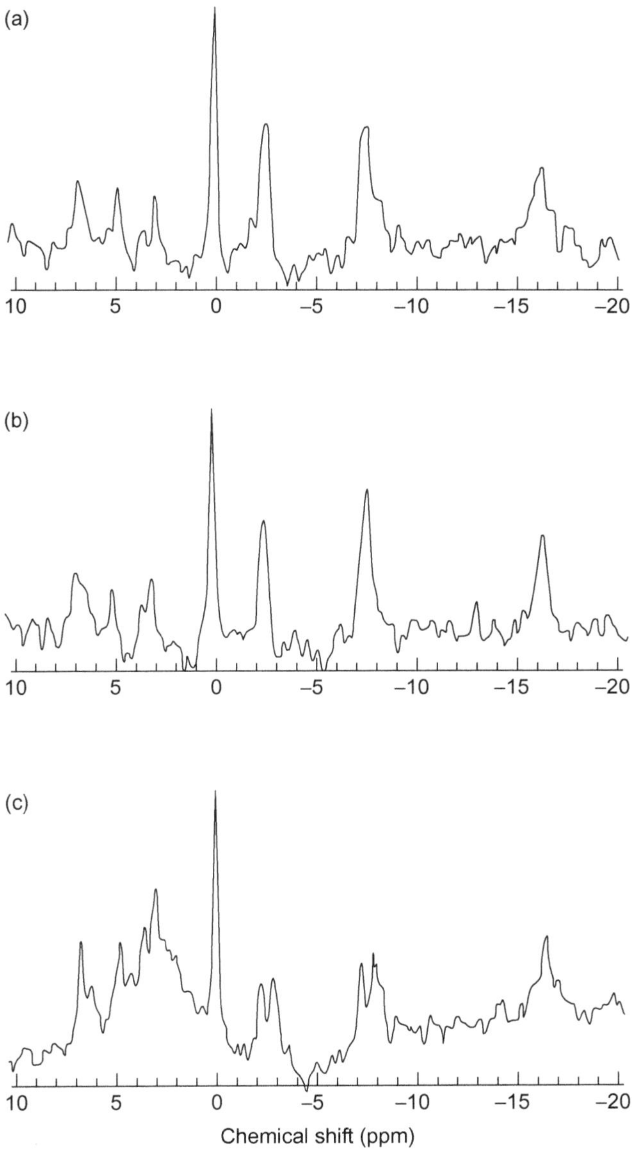

Figure 2.23 In vivo ^{31}P brain spectra from 15-cm^3 effective voxels at 4 T, in the parieto-occipital cortex (a), the thalamus (b), and a 60-cm^3 effective voxel at 1.5 T, ^1H with ^1H-decoupling and NOE (c) in the parieto-occipital cortex. Acquisition time was 45 min for (a) and (b), and 34 min for (c). 15-Hz exponential filtering is applied to the 4-T spectra (a, b) and 6-Hz exponential for the 1.5-T spectrum (c). The 1.5-T spectrum was acquired on a healthy volunteer, using a vision whole-body clinical MR scanner, a decoupling console, and a dual-tuned $^1H/^{31}P$ birdcage head RF coil operating in quadrature mode. Two-dimensional CSI was used, with ^{31}P-MRS parameters of: Tr, 1 ms; tip angle, 45°; spectral bandwidth, ±2 kHz; readout, 256 ms; complex points, 1024; FOV(x, y), 240 mm; phase steps (x, y), 8; averages/phase step, 8; slice thickness, 3 cm; k-space filter, 2D hanging (outermost four k-space points filtered in x and y). Decoupling was applied for the entire read duration (256 ms) using WALTZA sequence centered 1 ppm left of the water peak, and CW RF throughout the entire sequence (10% decoupling power) for NOE. Specific absorption rate (SAR) was approximately 0.6 W/kg.

2.1.3.9 PMRS: Human Cerebrum [21]

Advanced efforts are being made today in the pursuit of high-resolution analytical NMR spectroscopy, as this is a powerful tool for quantitative analysis of the mixture of compounds found in the human brain. Specimens of cerebral tissue from human subjects undergoing neurosurgical procedures are a potentially rich source of information on neurochemicals. The changes are associated with histopathological ones in the matched sample. Proton (^{1}H) NMR analysis of human brain extracts also complements MRS measurements made in vivo. This is done by providing spectra with better resolution. It enhances both the assignment of resonances and signal quantification. Specimens of the anterior temporal pole were obtained. Figure 2.24 [21] shows representative ^{1}H NMR spectra of perchloric acid extracts of human cortex; signals were assigned to the 19 compounds. Several major metabolites include lactate, creatine, NAA, inositol, and GABA. They have two or more resonances located in the uncrowded portions of the spectrum. These were measured independently. One needs to resolve the overlapping signals in crowded regions of the spectrum. *J*-resolved and shift-correlated spectra were used to clarify assignments. Quantification of in vivo spectra requires some concentration standard. This can be a large signal from a compound of known concentration. The compound may be an endogenous compound such as creatine or NAA. Both are abundant in the human brain.

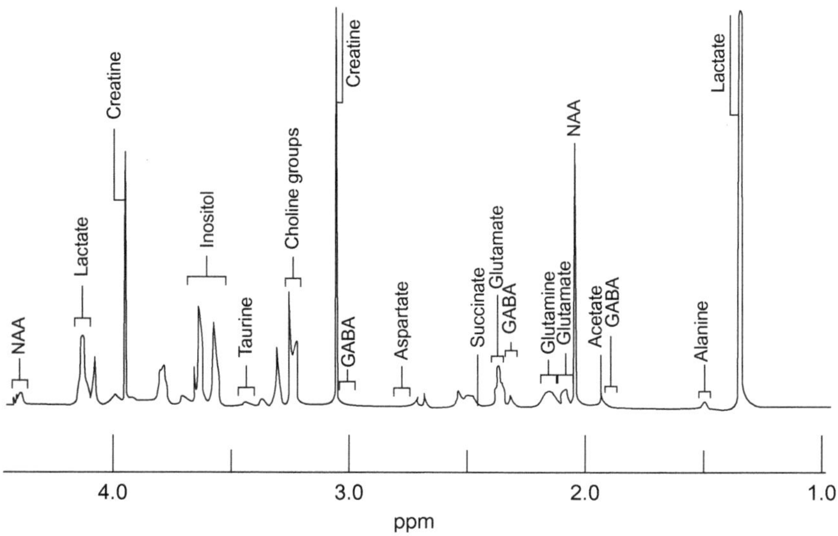

Figure 2.24 ^{1}H NMR spectrum (500 MHz) of perchloric acid extract made from the cortex of human brain tissue obtained during surgery for the amelioration of complex partial seizures. The amplitude of the signals is proportional to the number of hydrogen atoms producing the signal at that frequency. Quantitative measurements reveal the GM−WM difference.

2.2 Theoretical Concepts and Practical Realization of MRI

This section includes some examples to give readers first-hand awareness of some of the simple laboratory-stage concepts involved in MRI. These go far in the design and development of the technology of NMR. From its beginnings, rooted in scientific curiosity and a desire to perform experimental verification, the field has seen vast commercialization of NMR technology. Because of continuing developments and refinements, the techniques of imaging and analysis have now reached astounding stages of sophistication. However—lest we forget—the foundational concepts still remain as a basis for further research and development.

2.2.1 *Phase Localization In Vivo Using Surface-Coil Surface Spoiling: Inhomogeneous Magnetic-Field Gradient [22]*

A ferromagnetically generated surface magnetic gradient can be used for rapid dephasing of surface-lying spins. Such gradients differ from those generated electrically. They cannot be modulated readily in time (phase, frequency, amplitude). There is a drive toward ferromagnetically (and more recently, paramagnetically) generated surface-spoiling gradients and toward achieving the simplest possible localization scheme. One can explore the use of single-pulse-and-collect acquisition in concert with switched (on/off) and continuous (analogous to ferromagnetically generated) electrically driven surface gradients. Remarkably, one finds that very short switched-gradient periods, of as little as 0.5–1.0 ms, are sufficient to eliminate surface-layer spin contributions, with gradient-driving currents of approximately 0.5–1 A. Furthermore, because the surface gradient is essentially (switched) off during data-acquisition, line-width resolution (i.e., B_z, homogeneity) is not markedly degraded by partial penetration of the nonideal "surface gradient" into the deep-lying VOI. This is not likely to be the case with a continuous-surface gradient, where localization is achieved at the apparent expense of added line-width broadening.

Here we contrast the use of switched and continuous-surface gradients in concert with single-pulse-and-collect surface-coil detection for ^{31}P NMR localization in rat liver in vivo. The liver was chosen not only for its central metabolic importance, but also because of its lack of PCr. It allows the signal intensity of this compound to serve as a marker of muscle tissue contributions (the surface component) to the spectrum, as muscle is rich in PCr. In addition, the β-phosphate resonance of ATP (β-ATP) in the liver has a short T_2, and preservation of this resonance is an appropriate test of any T_2-dependent localization technique. Experiments were performed at 4.7 T in a vertically mounted magnet of 85-mm useable bore diameter. Anesthetized male Sprague–Dawley rats (ca. 200 g) were employed without surgical intervention. Figure 2.25 illustrates schematically (in cross-section) the relative orientation of the surface coil, surface magnetic-gradient coil, and rat body; the gradient coil can of course also be placed on the other side of the surface coil (see figure captions). The surface gradient-coil global geometry and current path are

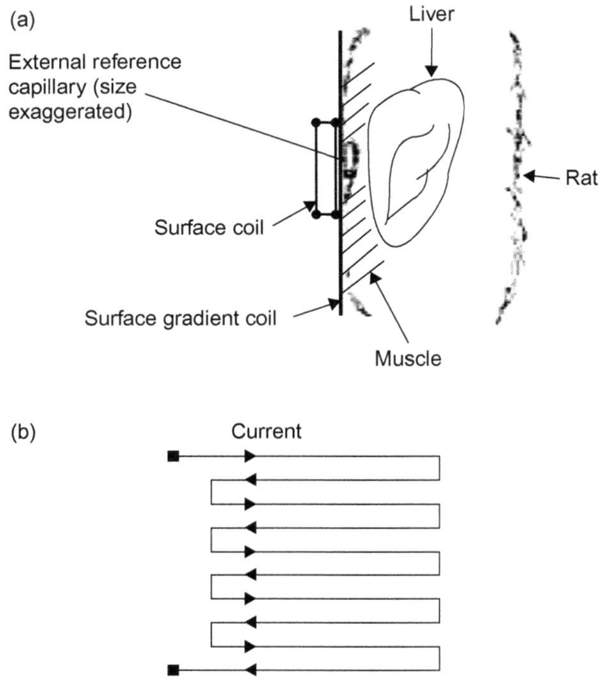

Figure 2.25 (a) Cross-sectional schematic illustration of the relative orientations of the surface-coil RF antenna, the surface magnetic-gradient generation coil, and the body of the rat in a vertical-bore magnet. The gradient coil could also be placed on the other side of the surface-coil antenna, and its overall geometry is not limited to a strictly planar (2D) configuration. As illustrated, the gradient coil's global dimensions ($l \times w$) are substantially greater than that of the surface-coil diameter. (b) Illustration of the gradient-coil geometry and current path employed in the experiments reported herein; numerous other global gradient-coil geometries and current paths are also likely to be effective. The gradient coil employed in the experiments reported herein had a 5 mm spacing between antiparallel current elements (which was close to the "surface depth"), was approximately square (7 cm × 7 cm), and was partially wrapped (halfway) about the sides of the rat.

also shown; numerous other global gradient-coil geometries and current paths are likely to be effective as well. As employed in the experiments reported herein, the gradient coil is partially wrapped halfway around the sides of the rat. Figure 2.26 presents the RF pulse and magnetic-gradient timing diagram, along with representative ^{31}P spectra taken at different continuous-gradient current levels. The RF pulse was set to a nominal 90° at the VOI (liver) and the surface gradient was turned on continuously, including during the entire data-acquisition period. The pulse-repetition period (3 s) was set to provide quantitative acquisition conditions for liver phosphorous metabolites (i.e., $\geq 3T_1$). Because the surface gradient employed herein is nonideal, in that it does not go identically to zero everywhere at

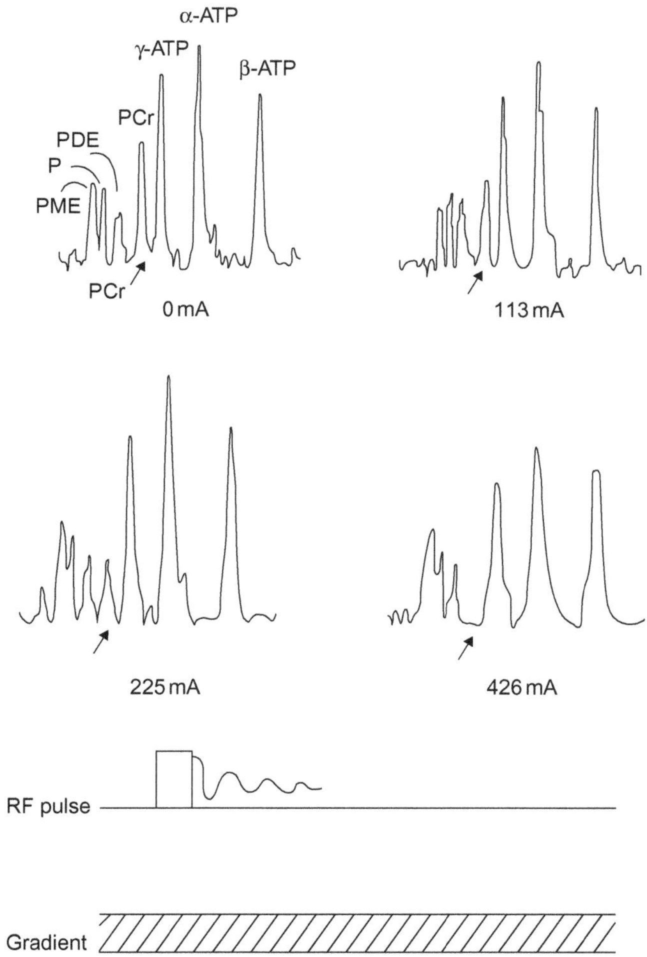

Figure 2.26 Continuous-magnetic-gradient and RF pulse-timing diagram and representative ^{31}P spectra of rat in vivo. Pulse-and-collect acquisition was employed with the surface gradient on throughout the entire acquisition period. Each spectrum is the result of 100 scans at a 3-s pulse-repetition period (TR) for a total acquisition time of 5 min. A nominal 90° pulse (~4.0 μs) at the liver was employed with a flat, 2-turn, close-wound surface coil of 16 mm internal diameter (i.d.) and 18 mm outer diameter (o.d.); no effort was made to apply a 180° surface nulling pulse or to fine-tune signal-to-noise optimization. The surface coil was placed between the gradient coil and the animal's body (i.e., directly against the animal). The gradient-driving current is given below the relevant spectrum in units of mA. The liver "baseline hump" has been removed via cubic-spline fit. Resonances are assigned as phosphomonoesters (PME), inorganic phosphate (P), phosphodiesters (PDE), phosphocreatine (PCr, a marker of muscle tissue), γ-phosphate of ATP (γ-ATP), and likewise for α-ATP and β-ATP.

the border to the liver volume, partial penetration of the magnetic gradient into the VOI during data acquisition results in inhomogeneous broadening of the resonance line widths. This is illustrated in Figure 2.26 [22], where the line widths broaden substantially as the gradient-driving current is increased from 0 to 0.43 A. Nevertheless, the overlying muscle contribution to the ^{31}P spectra, evidenced most clearly by the PCr resonance, is markedly reduced as the gradient baseline is corrected with a cubic-spline fit to remove the liver ^{31}P baseline hump that derives from the broad phospholipid head group resonances.

Figure 2.27 [22] is a similar presentation of the switched-gradient experiment. In this case, the gradient is switched on immediately following the RF pulse for a period of only 1 ms. The surface gradient is essentially off during the data-acquisition period. In place of postfunction data manipulation, a DANTE (delays alternating with nutations for tailored excitations) pulse sequence was employed to saturate the ^{31}P baseline hump. Quantitative acquisition for liver metabolites was used with a pulse-repetition period of 4 s. Consistent with the continuous-gradient experiment described above, a gradient-driving current of 0.5 A was found sufficient to eliminate the overlying muscle contribution to the ^{31}P spectrum (i.e., elimination of the PCr resonance and reduction of ATP resonance intensities contributed by the muscle). Importantly, however, the switched-gradient experiment caused no degradation in line-width resolution. This is a substantial advantage over use of a continuous gradient for surface spoiling, at least with the gradient-coil design employed herein. Furthermore, the short gradient period required to effect localization (periods as short as 0.5 ms have been successfully employed with multicompartment phantoms) allows detection of resonances with relatively short T_2 relaxation times, e.g., β-ATP. In summary, switched-gradient surface spoiling provides a convenient, effective, and robust means to eliminate spectral contributions from overlying tissues, and thus eliminate the deep-lying localization properties of surface coils. Although not necessarily intended to compete with localizing techniques requiring high-power linear gradients, surface spoiling does provide a routine means of eliminating surface tissue contributions to spectroscopic or imaging-based interrogation of deep-lying tissues.

2.2.2 Pulsed Field Gradients: Separation of the Different Orders of Multiple-Quantum Transitions [23]

There are several methods to generate coherent multiple-quantum transitions (MQTs). In one of them, it is done by means of two nonselective 90° pulses in a homonuclear (same species nuclei) coupled-spin system. These pulses are separated by a time interval τ. The coherence created thereby then evolves during a time t_1. After this, another 90° pulse partly transfers the MQT coherence back into observable SQC. This is detected during the time t_2 (Figure 2.28 [23]). Fourier transformation of the 2D time signal, with respect to t_1 and t_2, gives a 2D frequency spectrum containing information about the coherence that existed during the evolution period. This is presented along the ω_1 (frequency) axis. In general, along the

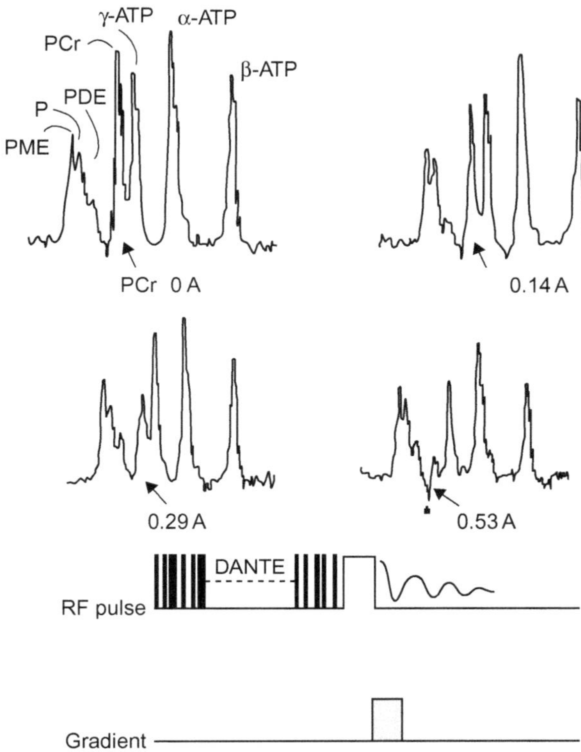

Figure 2.27 Switched-magnetic-gradient and RF pulse-timing diagram and representative ^{31}P spectra of rat in vivo. A different animal and surface coil were employed relative to those used to generate the data in Figure 2.26. Pulse-and-collect acquisition was employed with the gradient switched on for only 1 ms immediately following the interrogation pulse, a nominal 90° flip angle at the liver (~30 μs); no fine-tuning of signal-to-noise optimization was attempted. A solenoidal, 3-turn, close-wound surface coil of 18 mm o.d. was employed, with the gradient coil placed between the surface coil and the animal as illustrated in Figure 2.25(a). A DANTE sequence (4000 × 1 μs pulses with a 200-μs pulse-repetition period) preceded the interrogation pulse to suppress the liver "baseline hump"; the final spectrum identifies the frequency position at which DANTE hump suppression was applied. Such suppression can be applied effectively almost anywhere within the envelope of the hump, and in this case may partially obscure the PCr intensity and lineshape. Each spectrum is the result of 192 scans with a pulse-repetition period (TR) of 4 s for a 12.3 min total acquisition time. Other notations are as in Figure 2.26 [22].

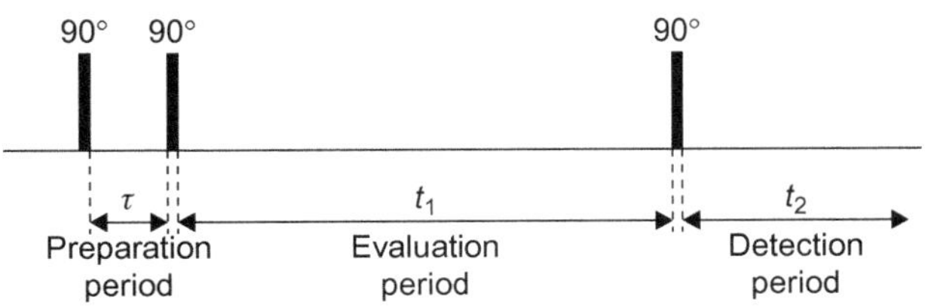

Figure 2.28 Pulse sequence used to create and detect MQC of all orders by way of an (2DFT) experiment.

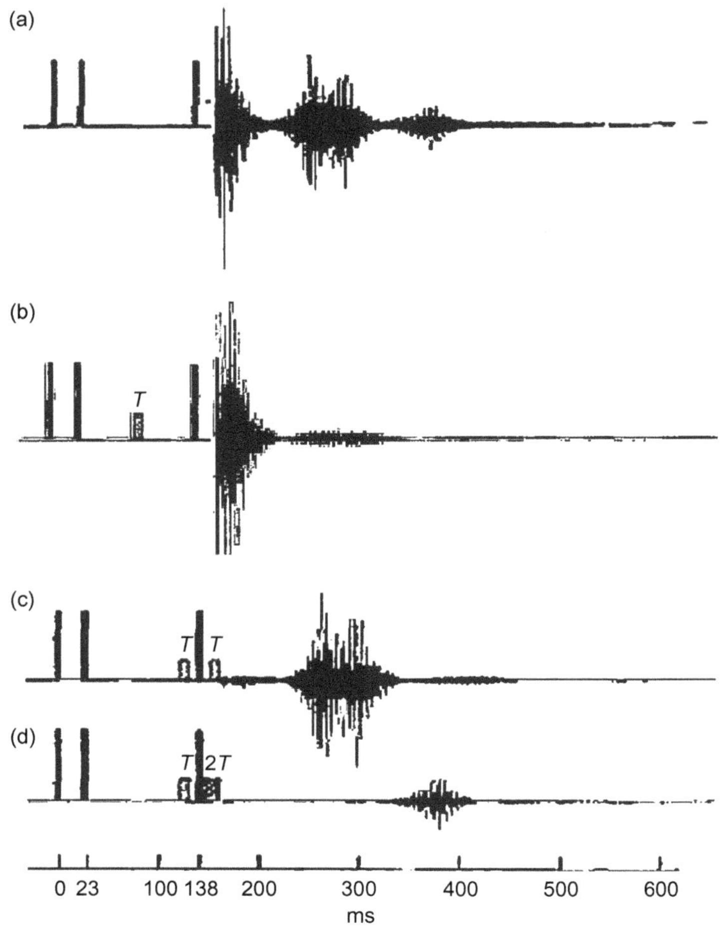

Figure 2.29 Signals from an *AX* (spin) system after a three-pulse sequence (a) without gradient pulses; (b) with a gradient pulse applied during the evolution period; (c) with equal gradient pulses at the end of the evolution and the beginning of the detection period; and (d) with the gradient pulse during the detection period doubled in length.

ω_1 axis, one finds resonance lines that correspond to all orders of transitions in the coupled-spin system. In contrast, along the ω_2 axis, only the single-quantum transitions (SQTs) observed during t_2 are visible (Figure 2.29 [23]).

2.2.3 NMR Spin-Echo Planar (Multiplanar) Image Formation [24]

This section provides readers with a brief exposure to the mathematics involved to explain 2D or 3D spin-density imaging. In this case, an initial level of comprehension of the classical mechanics (CM, Newton's calculus) is enough to explain the main features of the technology. In fact, QM is no more difficult than CM. The only difference is that CM is taught at the secondary level of education, all over the world; QM is not. That is what makes the difference: It is a matter of exposure of a human mind, at an early stage of the person's development, that creates assumptions about difficulty. The mathematics involved here exploits the properties of spin-echoes in time-dependent magnetic-field gradients. Operation of the EPI technique relies on the selective excitation of the specimen in switched magnetic-field gradients; essentially, it is an extension of a standard line-scan method. The method achieves spatial selectivity, either in part or fully, by using the properties of the spin-echoes. One employs time-dependent magnetic-field gradients. The idea uses determination of a continuous spin-density distribution when periodicity in the time domain is imposed. It produces high-speed, cross-sectional NMR images corresponding to the mobile spin-density variation throughout a living biological specimen.

To keep it simple, the discussion is limited to 2D imaging in one plane. The sample is placed in a large, uniform, static magnetic field B_0, which defines the equilibrium spin polarization axis. A linear magnetic-field gradient $G_x = \partial B_x/\partial x$ is also applied and, at the same time, a tailored 90° RF pulse excites the spins within a single slice of thickness Δx at x_0, giving an FID. Immediately following this excitation pulse, the gradient G_x is switched off and the FID is observed in a switched gradient $G_y = \partial B_x/\partial y$ and a steady gradient $G_z = \partial B_x/\partial z$. The effect of gradient switching is best understood if one studies what happens when $G_z = 0$. Suppose that, in a time τ_b, a steady gradient G_y has caused FID amplitude to decay completely to zero. If this decay alone were sampled and Fourier transformed, it would of course yield the projection profile of the spin distribution along the y-axis in the x_0 plane. However, by reversing the gradient direction (or by using a 180° RF pulse), the decayed FID signal can be made to grow into a spin-echo, in a further time period τ_b, which then decays again. Further gradient reversals can thus recall the signal. It is (frequency) times (2 here), provided that $t_b = 2\tau_b < T_2$. Here, T_2 is the spin–spin (transverse) relaxation time of the specimen. Recalling the signal in this way, and sampling the full spin-echo train, imposes a discreteness on the Fourier-transformed projection profile. The discrete frequency spacing is given by $\Delta\omega = \pi/\tau_b$. External shaping of the spin-echo train makes it possible to further broaden the discrete lines into rectangular or other desired profiles. In the full 2D experiment, signal sampling is performed with the additional steady gradient G_z. It broadens the individual discrete line, to yield (for a single echo train), using Fourier transformation, a complete set of resolved cross-sectional profiles. This is for the spin distribution across the thin layer in the specimen. The profiles can then be appropriately transformed into a rectangular array of data points within a computer memory. This digitized data can then be made to form a visual image, as an output in a monitor display.

In generalization of this experiment to three dimensions, a multiplanar selection process can be incorporated by modulating both G_x and G_y while maintaining G_z steady. In this case, the initial selective excitation pulse can be replaced by a conventional nonselective 90° pulse. The effect of gradient switching, together with

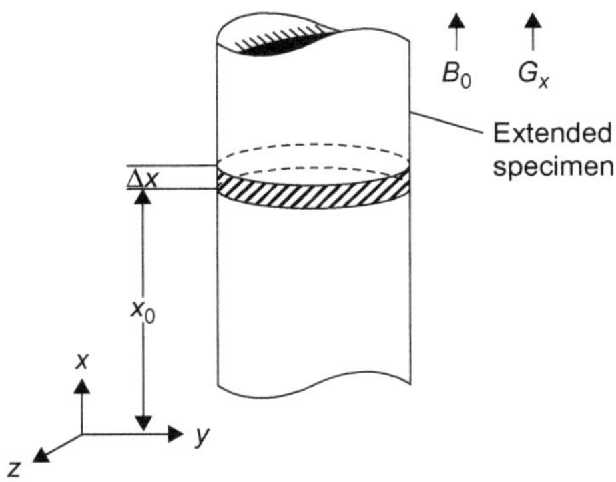

Figure 2.30 Sketch showing selectively excited layer of spins of thickness Δx in the plane x_0 within an extended specimen.

digital sampling of the signal, is to impose an otherwise continuous spin-density distribution, a discrete lattice-point distribution, with spatial periodicities a, b, and c. The FID signal in the rotating reference frame, at time t following the pulse, is given by $S(t) = \text{Re}(\text{the real part})\int\int\int \rho(x, y, z)\exp[\mathrm{i}\gamma \int_0^t (xG_x(t') + yG_y(t') + zG_z(t'))\,\mathrm{d}t']\mathrm{d}x\,\mathrm{d}y\,\mathrm{d}z$. Here $\rho(x, y, z)$ is the continuous spin-density distribution of the sample and γ is the gyromagnetic ratio. A detailed Fourier-transform treatment for digital sampling of $S(t)$, for a time τ_c (keeping G_z steady), introduces a discreteness, along the z-axis, in spin-density distribution. The points are spaced at $z = z_0 + nc$, where n is an integer and $c = z$, space coordinate along z-direction. This corresponds to an angular frequency interval $\Delta\omega_z = 2\pi/\tau_c = \gamma c G_z$. Thus, the $S(t)$ becomes $S(t) = \Sigma \rho_{lmn} \cos t(l\Delta\omega_x + m\Delta\omega_y + n\Delta\omega_z)\Delta v_{lmn}$, where $\Delta v_{lmn} = abc$ is the unit cell volume, the spins of which contribute to the signal at each lattice point. The modulation periods (and hence the gradients) are chosen so that $\Delta\omega_x/M = \Delta\omega_y = N\Delta\omega_z$, where M and N are the largest values of integers, m and n, respectively, in the imaging field. One can then see that all points in the distribution ρ_{lmn} are uniquely defined in the frequency domain. Fourier transformation of $S(t)$ will thus yield, in one calculation, the complete 3D spin-density distribution function ρ_{lmn} (Figure 2.30).

2.2.4 Two-Dimensional Spatially Resolved Spectroscopic Imaging [25]

A 2D spectroscopic imaging technique relies on collection of a series of acquisitions, where a time delay t_1 is varied from one excitation to the next and data are sampled during the acquisition period t_2. There has been emphasis on fast spectroscopic imaging, using chemical shift, in two or three spatial dimensions. In the imaging technique discussed here, one applies time-varying gradients during the readout period to acquire spatially resolved spectroscopic (chemical-shift) data rapidly for each t_1 step. The rapid spatial encoding of chemical shift allows several increments of t_1 to be acquired within a reasonable time. Thus, the collection of two spectral dimensions and two or three spatial dimensions becomes possible in a single study. This method does not suffer SNR loss as compared with a conventional single-voxel 2D method, as long as the voxel size, imaging time, and other timing parameters, such as echo time (TE) and the repetition time, are kept constant. In this technique of spatially resolved 2D spectroscopy, one applies readout gradients that trace spiral-based paths in k-space. The spiral trajectories make highly efficient use of the gradient hardware. Data collected with time-varying gradients cannot be reconstructed directly with fast Fourier transform (FFT), as the sampling points are not on a rectilinear grid. To overcome this problem, one applies a gradient reconstruction, a general algorithm that resamples nonuniformly spaced data onto a Cartesian grid prior to FFT. In essence, here one uses spiral-based k-space trajectories, on a standard clinical scanner, to collect data simultaneously for two spatial dimensions (x, y) and two spectral dimensions (f_1, f_2). The method is demonstrated with a J-resolved acquisition of a phantom, with separate bottles of 100 mM lactate in water and approximately 700 mM ethyl alcohol in water. The data are resolved into 18 × 18 pixels over a 24-cm field of view (FOV); resolution of 1.3 cm × 1.3 cm in plane A, 400 Hz spectral bandwidth with 3.8 Hz resolution in chemical shift (readout period of 262 ms), and 32 steps with 1.56 Hz resolution in f_1. Figures 2.31–2.34 provide the reader with more information about the technique of CSI.

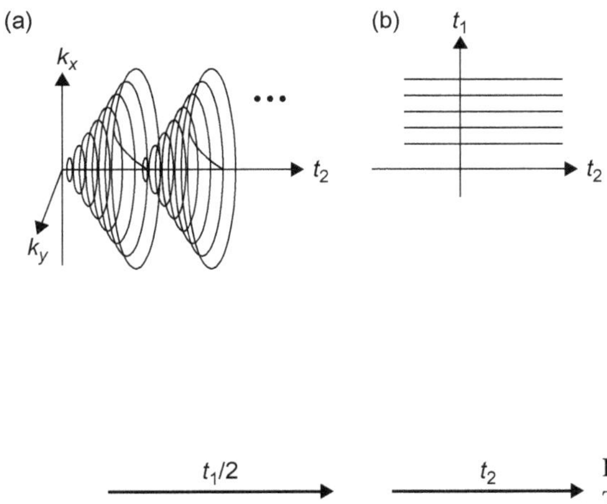

Figure 2.31 Schematic illustration of the k-space trajectory. (a) (k_x, k_y, t_2)-space is sampled on spiral trajectories repeated throughout the readout period. (b) The trajectory in (a) is repeated for each value of t_1, thus sampling (t_1, t_2)-space as shown. The asymmetry around the origin of t_2 is due to asymmetric sampling of the spin-echo during the readout period.

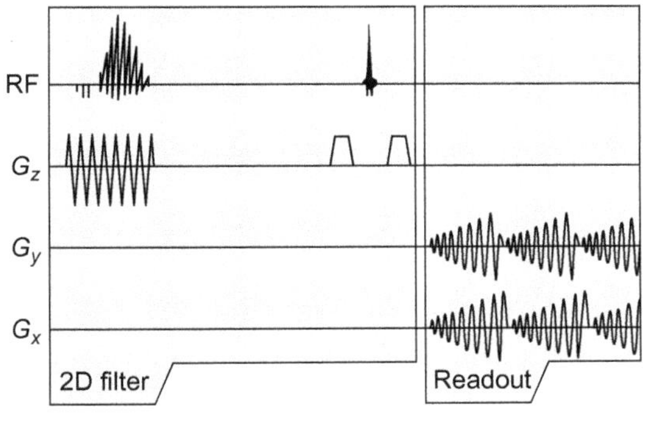

Figure 2.32 Pulse-sequence diagram. The 2D spectroscopy filter is a spin-echo sequence with $t_1 = TE$. During the readout period, time-varying gradients of the x- and y-gradients trace the spiral trajectories in Figure 2.31.

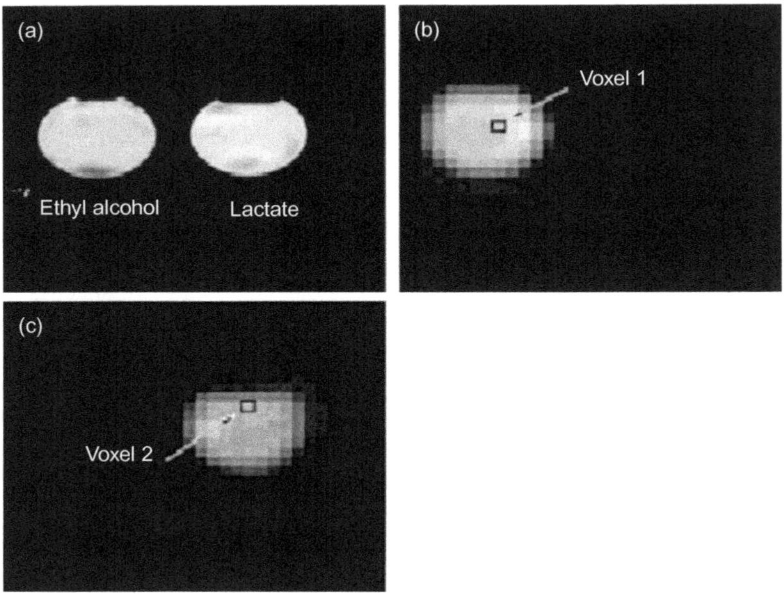

Figure 2.33 Spectroscopic images. (a) A gradient-recalled echo image of a phantom that contains ethyl alcohol in water (bottle on left), along with a separate bottle of lactic acid in water (bottle on right). The alcohol concentration is approximately 700 mM, and the lactate concentration is approximately 100 mM. (b) A spectroscopic image, obtained with a least-squares estimate of the alcohol peaks in the spectroscopic data. Each of the voxels in the 32×32 pixel image contains a 2D spectrum. (c) A spectroscopic image obtained with a least-squares estimate of the lactate doubled in the spectroscopic data. Spectra from voxels 1 and 2 are shown in Figure 2.34 [25].

2.2.5 CSI of GABA in the Human Brain: MQ Filtering [26]

GABA is the major inhibitory neurotransmitter in the mammalian brain, with a concentration of about 1.1–1.2 mmol/ml in normal human cortex. It has been estimated that 30–50% of all brain synapses are GABAergic. GABA is essential to normal brain function. Dysfunction of GABAergic neurotransmission has been implicated in many neurologic and psychological diseases, including epilepsy, alcoholism, Huntington's disease, depression, and schizophrenia. Recently, several ^{1}H NMR spectroscopy methods have been proposed, based on *J*-modulation inhibition, double- or triple-quantum filtering, *J*-refocused coherence transfer, or *J*-resolved spectroscopy, to measure GABA from brain extracts or in vivo animal or human brain. Single-voxel GABA measurements have demonstrated lowered GABA concentration in adult and pediatric epilepsy, as well as dynamic changes with pharmaceutical treatment. Elevated GABA has been found to be correlated with improved seizure control. The ability to measure the spatial distribution of GABA using CSI should be of great value for understanding the role of GABA metabolism in normal

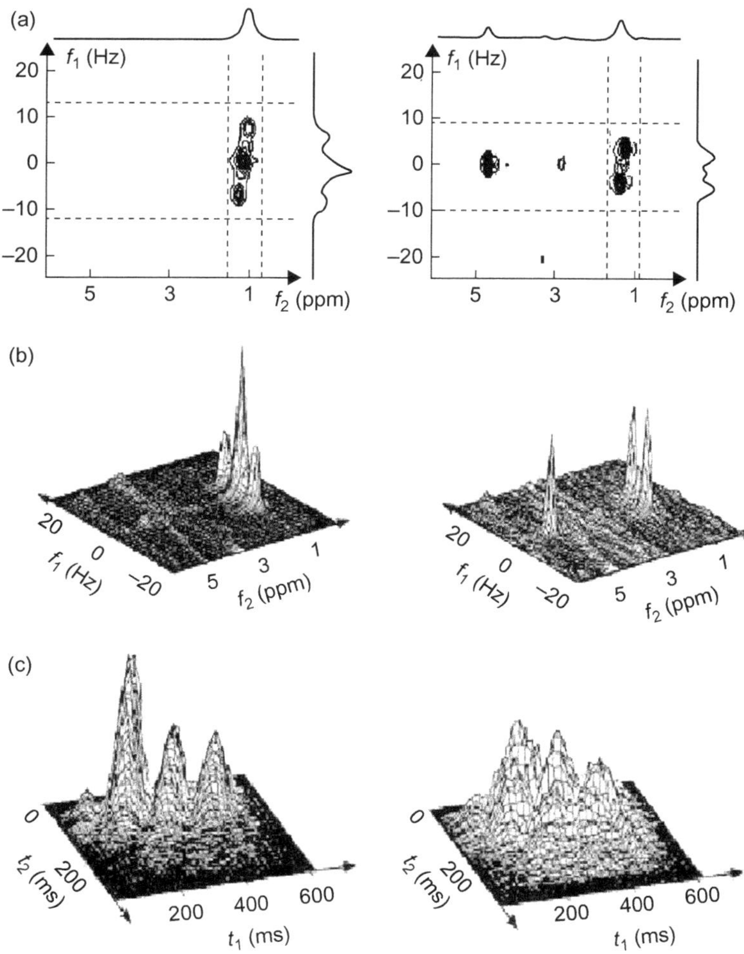

Figure 2.34 Magnitude spectra and time-domain data from voxel 1 (left column) and voxel 2 (right column). (a) Contour plots with 10 contour lines of the 2D spectra in voxels 1 and 2, along with projections on the f_1 and f_2 axes over the frequency ranges indicated by the dashed lines. The projection on the f_1 axis shows a well-resolved multiplet structure, whereas limited chemical-shift resolution due to T_2* decay limits the resolution of the projection on the f_2 axis. (b) A 3D surface plot of the same spectra as depicted in (a). The mesh plot shows the alcohol triplet structure as well as residual unsuppressed water. (c) Magnitude time-domain data from voxels 1 and 2. Modulation of the data, as a function of both t_1 and t_2, is apparent.

and pathologic conditions and/or in pharmaceutical treatment. In epilepsy, CSI studies of creatine, choline, and NAA have shown regional alteration of tissue biochemistry in areas around the seizure focus. Regional variations in GABA metabolism would be anticipated in this condition, and would provide a novel parameter for identification and characterization of epileptogenic tissue.

The application of in vivo GABA CSI has been limited by the requirement of high sensitivity, due to the relatively low concentration of GABA, and the requirement to suppress the intense overlapping creatine signal at 3.0 ppm by approximately 100- to 1000-fold. In CSI, achievement of this suppression complicates the usage of *J*-modulation, inhibition due to the requirements for spatial insensitivity to B_0, B_1 inhomogeneity and for the subject to have minimal movement for 30–60 min. An alternative to *J*-modulation inhibition methods is the use of MQ filtering, which allows the uncoupled resonance of creatine to be minimized by gradient dephasing. However, in vivo application of these methods, for both single-volume spectroscopy and CSI, has been limited by loss of sensitivity or selectivity relative to *J*-modulation inhibition-based methods. It is proposed that a robust method based on "single-shot" selective MQ filtering, at 2.1 T with $G_{max} = 0.9$ G/cm, to obtain chemical-shift images of GABA from a human brain in vivo, is possible. Product operator calculations and coherence pathway analysis are used for pulse-sequence design. The method has been tested extensively on phantom samples. In vivo GABA imaging from the occipital cortex of a healthy volunteer is presented. GABA ($H_2NCH_2CH_2CH_2COOH$) is an $I_2S_2W_2$ spin system. Here GABA-4 (I) is resonating at 3.01 pm, GABA-3 (S) at 1.91 ppm, GABA-2(W) at 2.30 ppm, and $J = 7.3$ Hz. At 2.1 T (Tesla is the unit for magnetic field), $\Delta\nu$ (89.43 MHz)/J = 13.5 Hz for GABA-3 and -4 and 4.8 Hz for GABA-2 and -3, respectively (Figures 2.35 and 2.36).

2.3 Latitudes and Longitudes in Various Techniques of MRI in NMR

2.3.1 CSI: Echo Planar Shift Mapping [27]

The EPI method is known for its high-speed imaging capability. It also produces real-time movie images. One can adapt it to provide spatial chemical-shift information. There are two basic variants of EPI, both of which allow spin density and isotropic chemical shift to be measured. This is done in a defined spatial region, to be simultaneously observed and unambiguously extricated. One of these is echo planar shift mapping (EPSM). This is a very fast method. It takes only four separate experiments to produce a 3D chemical-shift map. The other variant is a hybrid projection reconstruction echo planar (PREP) method. It is a relatively slower method, but is also a simpler technique. Basically, one has to produce mapping of the isotropic part of the chemical shift $\delta(x, y, z)$. Initial excitation of the spin system is done by using a nonselective pulse that excites all the spins within the volume covered by the transmitter coil. One then produces a 4D map (x, y, z, δ).

A prerequisite for application of much of the theoretical analysis is a mechanism for accurate slice definition. This is done with a modified selective irradiative sequence. In the more usual selection process, an extended object, which is placed in a polarizing magnetic field B_0 $(0, 0, B_0)$ and a magnetic-field gradient G,

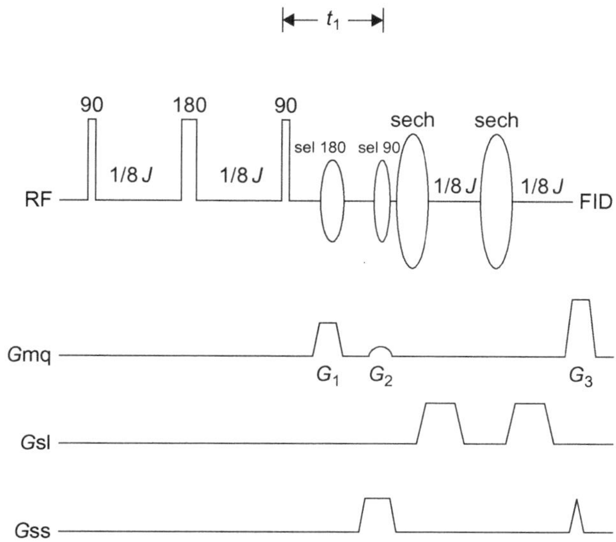

Figure 2.35 The sequence shown is preceded by a spatially selective inversion pulse to select a 2-cm y slice followed by a nulling delay of 173 ms for surface lipid suppression, and a standard 2D ISIS scheme to prelocalize a 6 cm × 6 cm x, z plane for further outer-volume lipid suppression. Both spatially selective inversion and the 2D ISIS used hyperbolic secant pulses (4.0 ms, $\mu = 5.0$, 1.0% truncation). The basic pulse sequence is proposed for CSI of GABA with all RF pulses applied along the x-axis. The GABA MQ state is prepared by the first two 90° pulses and an evolution period of $1/4J$, followed by a gradient filter that selects the DQ = ZQ pathway and dephases creatine single quantum. B_1-insensitive slice selection is achieved by a pair of identical hyperbolic pulses, which also refocus chemical-shift evolution during the second $1/4J$ period. Surface spoiler gradients (Gss) were added during the ZQ period and before acquisition to enhance creatine and surface lipid suppression. RF and gradient pulses for outer-volume suppression, as well as phase-encoding gradients for CSI, were not drawn, so as to preserve clarity.

is irradiated with a tailored selective RF pulse $B_1(t)$ applied along the x-axis in the rotating reference frame so as to excite essentially only those spins that lie in the range Az at displacement z. The corresponding range of Larmor frequencies excited is given by $\Delta w = \gamma G \Delta z$. This condition is only approximate, however, because such tailored pulses automatically introduce some dispersive element of magnetization. One can produce a net magnetization along the y-axis with, ideally, zero magnetization along both the positive and negative x-axes. This situation may be approached by using a focused selective pulse in which either the select gradient G is temporarily reversed following excitation, or G remains positive and a nonselective 180° RF pulse is applied (Figure 2.37). Both arrangements produce an echo-like signal that maximizes at time $3t_wJ/2$, where t_w is the RF pulse length. This is the case when G_z is constant. The unwanted x-component of the signal may be

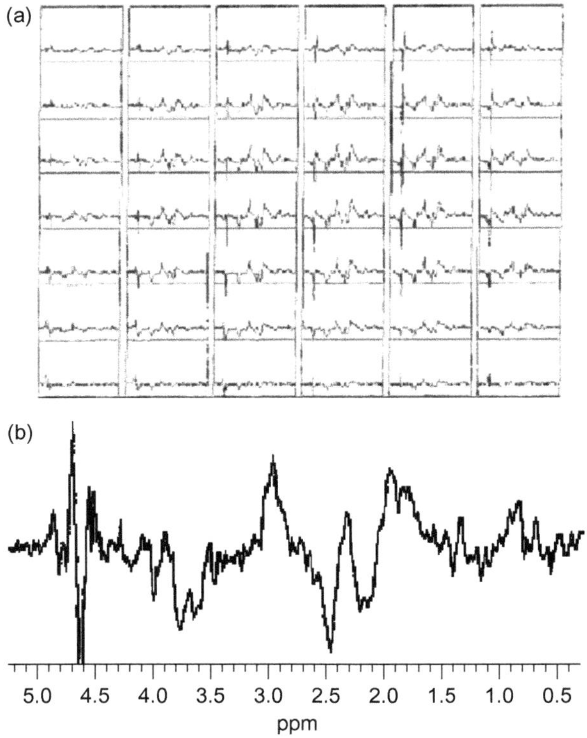

Figure 2.36 Coronal chemical-shift image (a) ($y = 0$) of GABA from the occipital lobe of healthy volunteer using the selective MQ filtering method shown in Figure 2.35. The total acquisition time of the GABA image was 35 min. The slice thickness was 3 cm. 8 × 8 phase-encoding steps were used and zero-filled to 16 × 16. A 2-Hz exponential line broadening was applied to the chemical-shift dimension, whereas a Hamming filter was applied to the two spatial dimensions before FT. The nominal voxel size is 1.5 cm × 3 cm × 1.5 cm. Spectrum from a voxel at isocenter (b). GABA peak at 3.0 ppm was clearly resolved. Multiplets from NAA, glutamate, glutamine, and inositol also passed the MQF. A 2D ISIS scheme was used to obtain 3D localization for single-voxel GABA measurement using MQ filtering.

minimized by subtle changes in the RF pulse shape. When chemical shifts are present, some spins outside the confines of the slice are well defined, and in general contribute to the observed signal, whereas the contributions of other spins within the slice may be lost. This effect may be minimized by increasing G and satisfying the condition $\gamma B_1 > \delta_{max}$. Here δ_{max} is the largest chemical shift present.

A more serious problem is the failure of chemically shifted spins to properly refocus in a gradient reversal. This means that each chemical species has a different starting phase, thus introducing a mixture of absorptive and dispersive components into the FT. This problem may be overcome by using a nonselective 180° RF pulse

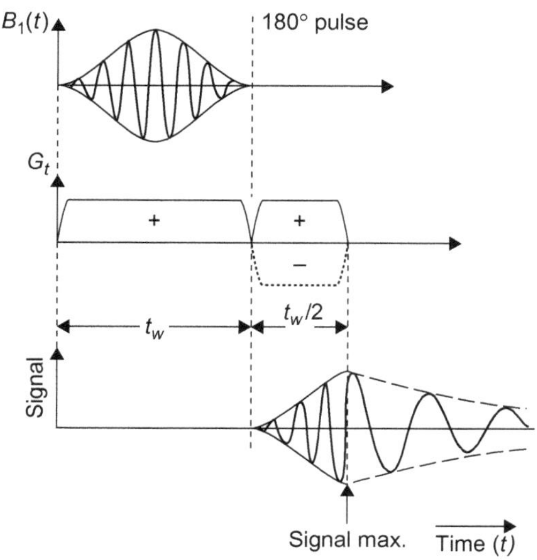

Figure 2.37 A sketch of a focused-selection pulse sequence suitable for exciting spins in a thin slice of material. $B_1(t)$ is the suitably shaped RF pulse lasting a time t_w. During this time, the select gradient, G_z, is pulsed +ve as shown. No NMR signal is observed in this phase. At $t = t_w$, the gradient is reversed (dotted line) for a time $-t_w/2$, during which the nuclear signal grows as shown to its maximum value. At this point, G_z is switched off. With no other gradients on, the signal would decay away more slowly with T_2 or due to static field inhomogeneity. An alternative to gradient reversal is also shown (solid line). Here the gradient is reduced to zero while a nonselective 180° RF pulse is applied, then restored to a positive level for $t_w/2$.

instead of gradient reversal. Alternatively, gradient reversal may be used together with power FTs to remove phase artifacts, although this procedure would almost certainly result in a broadening—and hence loss of resolution—of the chemical-shift spectra. Because slice selection does not unduly perturb adjacent layers of material, rapid scanning of several layers may be achieved. This can be done without the necessity of waiting time T_1, the spin-lattice relaxation time, between different plane scans. Figure 2.38 shows the pulse-timing diagram for a basic echo planar experiment (unbroken lines). Slice selection in the excitation period T_E is followed by G_y, modulation and complete signal and echo sampling in time T_S, which in turn is followed by a delay period T_D. The experimental cycle time T_{cycle} is therefore given by $T_{cycle} = T_E + T_S + T_D$. Fourier transformation of the complete spin-echo sequence gives a stick spectrum. When we have a sample with several spin species, or more practically, one species with several different chemical environments giving rise to several chemically shifted spin sites, the chemically shifted nuclei will modify the stick spacing, provided all resonances are excited in the initial selective excitation sequence. For a general nonhomogeneous and irregularly shaped object that is arbitrarily placed in the object field, the projection in $+G_y$ is not the same as that in $-G_y$. They are mirror images. This point is embodied in the reverse sign in the second delta function.

2.3.2 Phase-Encoded Echo Planar Mapping [28]

The timing sequences for phase-encoded echo planar (PEEP) mapping are shown in Figure 2.39 [28]. The gradient field applied along the y-direction is rapidly switched between positive (+) and negative (−) values. This is common to EPI

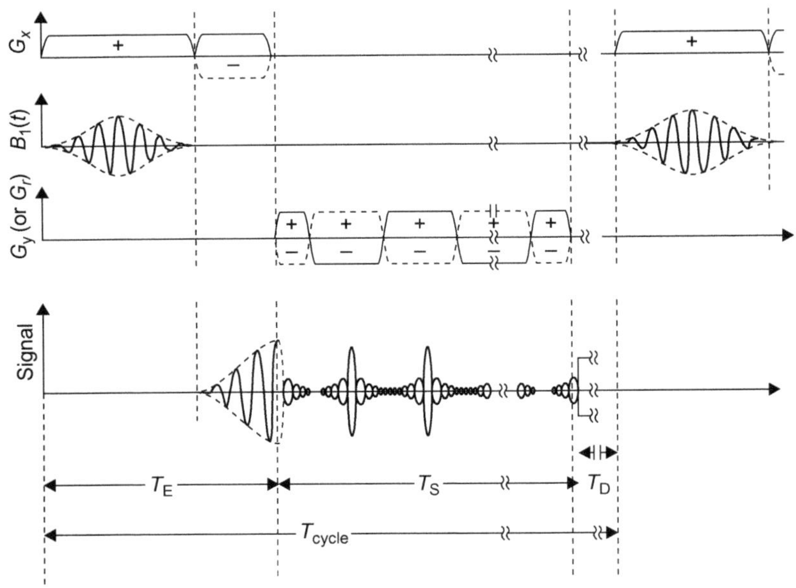

Figure 2.38 Timing diagram of the part EP sequence in which $G_x = 0$ (not shown), with modification of the alternating initial phase of the G_y modulation gradient (dotted). Also shown are the cycle time T_{cycle}, the excitation time T_E, the sampling period T_S, and the delay T_D. This is the sequence necessary for the PREP technique.

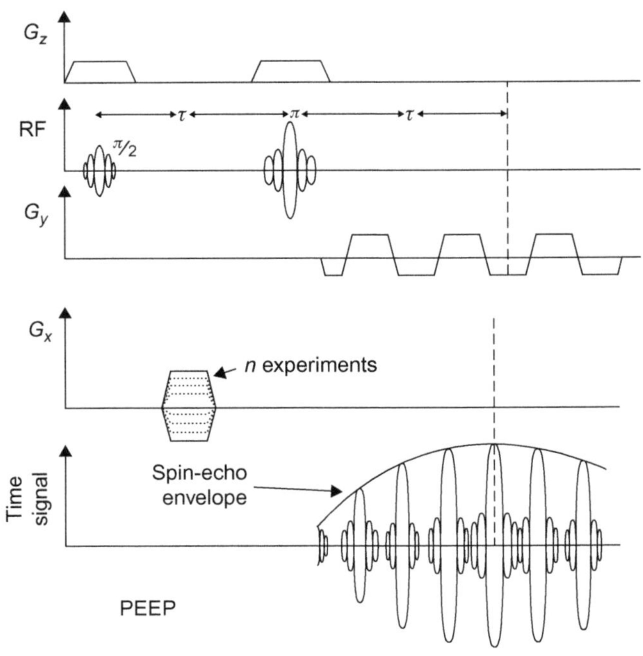

Figure 2.39 Gradient and RF timing diagram for the PEEP sequence.

techniques. The gradient applied along the x-direction is a phase-encoding gradient common to two-dimensional Fourier transform (2DFT) imaging experiments. The signal is modified in amplitude by a $90^\circ - \tau - 180^\circ$ spin-echo envelope, which ensures that both sides of k_δ (reciprocal space in the chemical-shift space) are sampled and, at the same time, minimizes signal loss caused by inhomogeneities in the main field. The presence of different chemical species provides continuous phase encoding. It is done in much the same manner as in a continuous x-gradient, in the case of the original EPI sequence. This enables complete mapping of the k_y, k_δ plane from a single RF excitation. Consecutive experiments with increased amplitude of the phase-encoding gradient cause increased dephasing of the time signal, in a manner similar to 2DFT. Figure 2.40 [28] depicts the k-space coverage of the experimental sequence shown in Figure 2.39. There are n experiments, the data of which constitute n points along the x-axis, running from $(-n/2 + 1)$ to $n/2$, i.e., full negative to full positive phase encoding. Figure 2.40(a) [38] shows the presample evolution in the k_x, k_δ plane. The sampling then begins in the presence of the switched gradient along the y-axis. The resulting trajectory in the k_y, k_δ plane at a position $n/2$ is shown in Figure 2.40(b) [28]. This plane is orthogonal to the k_x, k_δ plane (Figure 2.40(a) [38]). The sequence is then repeated with a phase-encoding G_x pulse, the amplitude of which is incremented so that the sampling starts at the next k_y, k_δ plane at a position $(n/2 - 1)$ along the k_x axis. This is repeated until all k_y, k_δ planes along the k_x axis have been sampled. A full 3D dataset k_δ, k_x, k_y, suitable for a 3D modulus FT, is thus acquired.

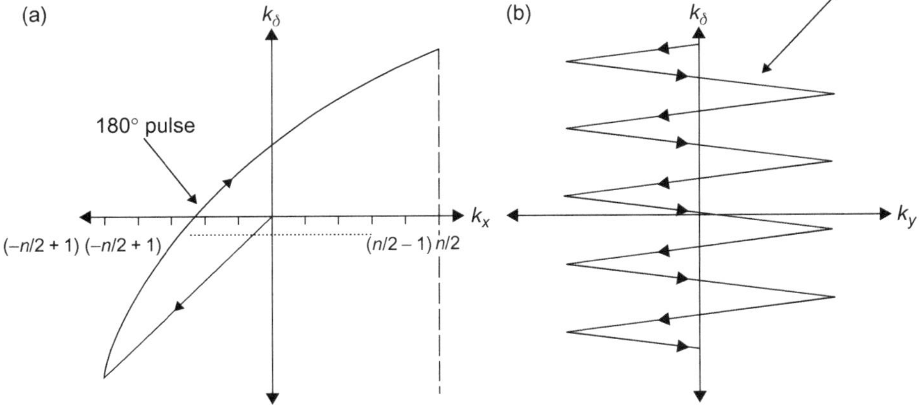

Figure 2.40 The k-space coverage resulting from the PEEP sequence. (a) The presample evolution in the k_x, k_δ plane caused by the phase-encoding gradient and the 180° RF pulse. The k_y axis is perpendicular to this plane. The dotted line represents the k_y, k_δ plane shown in (b) where the data sampling begins. This plane is moved along the k_x axis as the phase-encoding gradient is incremented.

2.3.3 Spatial (Volume) Selectivity: Time-Varying Gradients [29]

Since the early conceptions of NMRI using gradient fields for spatial selectivity, a wide variety of novel imaging approaches have been studied. Some have exhibited remarkable performance levels and have therefore been incorporated into commercial machines. Despite this success, a number of important problems remain. Techniques that provide the desired resolution levels have relatively long acquisition time, such that respiratory, cardiac, and peristaltic motions will degrade the image. In addition, magnets that provide the requisite degrees of (very high) uniformity are startlingly expensive. Another problem is the inability of existing localization approaches to do justice to the potential for in vivo NMR spectroscopy; present methods are time-consuming and have poor spatial resolution. This analysis presents both general and specific techniques that attempt to overcome these problems.

In an ideal system, acquisition time would be based solely on considerations of resolution and SNR, with these being readily traded off as desired. We could acquire all the data in an arbitrarily short interval, as with an X-ray image. The resultant data could be presented at low resolution with relatively high SNR, and vice versa. The only motivation for increasing the acquisition time would be the need for greater SNR at a given spatial resolution. Presently, one attempts to accomplish this through the use of time-varying gradients, which have been used in NMR for various imaging approaches. However, these approaches have not been able to tap the enormous potential of this modality. For example, the sensitive-point method has, thus far, been limited to the sequential acquisition of one point at a time, and therefore has not been competitive with other imaging approaches. However, the sensitive-point method does represent an interesting approach to spatial localization for in vivo spectroscopy. Another current use of time-varying gradients is the echo planar system, in which, following a single selective excitation pulse, a sequence of echoes is produced using either a periodically reversed gradient signal or 180° excitation pulses. The effective net result is a modulated function with its associated sidebands, where each sideband represents a specific coordinate point in space. Although this approach provides remarkably rapid data acquisition in the 10–100 ms range, it restricts the number of resolvable elements to the number of available sidebands. Thus far, images with 32×32 elements per plane have been produced. In a variation of the echo planar technique, an improved decoding method was introduced. Although one of the techniques discussed herein is related to the method proposed, the processing operation is quite different. The data-acquisition approach is the same as in the echo planar system: namely, the excitation of a selected slice with a static gradient on one axis and a periodic gradient on the other. The concept of multiplying the received signal by the integral of the gradient modulation is introduced and expanded here. Several different variants of EPI are being used in the field of MRI. In the simplest, a sequence of echoes is produced following a single excitation pulse, using either a periodically reversed gradient signal or a 180° excitation pulse. The net result is a modified function with its associated sidebands. Each sideband represents a specific coordinate point in space. This approach provides rapid data acquisition in the 100 ms range, but it too restricts the number of resolvable elements to the number of available sidebands.

For purposes of discussion, assume an NMR system with the static field in the z-direction and a time-varying gradient field $G(t)$ in the x-direction. Following a 90° excitation pulse at the resonant frequency ω_0, the time-varying magnetic moment from the Bloch equations is given by $\Delta m/\delta t = -m\{\mathrm{i}\omega_0 + \mathrm{i}yxG(t) + t/T_2\}$, where $m = m(r, t) = m_x(r, t) + \mathrm{i}m_y(r, t)$. Upon integration, one obtains $m = m_0\exp\{-\mathrm{i}\omega_0 t - \mathrm{i}\gamma xD(t) - t/T_2\}$, where $D(t) = \int_{t1}^{ti+t}(G(t)\mathrm{d}t)$. Following phase-sensitive synchronous detection in which the real and imaginary parts are preserved using multiplication by $\sin\omega_0 t$ and $\cos\omega_0 t$, followed by low-pass filtering, one has the baseband signal $m(r, t) = m_0(r)\exp(-T/t)$ $\exp[-I\gamma xD(t)]$. A long-term integration of $m(t)$, using a time $T \gg T_2$, yields a spatially localized $m(r) = m_0(r)s(x)$; $s(x)$ is the localization function in the x-direction given by $s(x) = \mathrm{Re}[1/T\int\exp[-\mathrm{i}\gamma xD(t) - t/T_2]]\mathrm{d}t$. The volume integral of $m(r)$ thus isolates the region of the origin. This is essentially the sensitive-point method, in which $G(t)$ is cosinusoidal and $D(t)$ sinusoidal, in which case the localization function $s(x)$ is (T_2/T) $J_0(\gamma G_x/\omega_m)w$. Here G is the peak amplitude of the gradient modulation and ω the angular frequency. Use of this method of localization using time-varying gradients, in which the signal is received followed by synchronous detection and integration, has been limited because of its apparent inefficiency. Sequencing through each coordinate point to derive the local moment density would indeed be very inefficient. This problem is removed, however, when we realize that $m(r, t)$, in the presence of time-varying gradients, contains the desired information about *every* point in the volume. This can be simply illustrated by using the same 1D model.

Another potentially exciting area in NMR is in vivo spectroscopy, where biochemical volumetric NMRI information is derived from specific regions of the body. Thus far, this has been done primarily through surface coils. These have limited spectral localization capability and do not enable acquisition of images of interior regions. More importantly, this method does not enable electronic selectivity of all regions of the body that could provide images based on spectroscopic information. In vivo spectroscopy has been attempted using 3D and 4D Fourier-transform approaches. These techniques are very time-consuming and require a very high degree of homogeneity of the magnetic field, along with a high degree of linearity of the gradient fields. The sensitive-point method has been successfully applied to spectroscopy. However, despite the reduction in acquisition time and its relative immunity to inhomogeneity, this technique has been limited by the apparent acquisition of one region at a time. However, with the precession in methods, this clearly becomes a very worthwhile approach. Using time-varying periodic gradients at each axis, data can be acquired representing the spectroscopic information of all regions of the body. These can then be processed to provide the desired spectral localization. In spectroscopy, $m(r, t) = m(r)\Sigma_n\exp\{-\mathrm{i}\omega,t - \mathrm{i}\gamma[D(t)r] - t/T_2\}$, where ω represents the spectral components of amplitude; in general, T_2 will be a function of both r and n.

2.3.4 CSI: Multiple Frames [30]

MRSI has employed many techniques to obtain the desired imaging results from the spatial and spectroscopic information derived from any object. In a clinical situation, it is important to minimize the time spent in acquisition of the imaging data. In a typical case, one may use 128×128 spatial points (pixels) over a small area. This means 16,384 FID signals from the chemical-shift information, which must be translated into imaging. The more variants there are in chemical-shift species, the more variation in chemical shift there will be to be collected. This is necessary to achieve better-resolution imaging. To reduce the total imaging time, the number of FIDs for a given spatial resolution has to be kept to a minimum. MRI techniques map the spatial distribution of NMR signal intensities. CSI techniques and MRS techniques, used together, have the potential to bring additional chemical-shift information to bear on clinical diagnostic problems. However, some practical problems remain in obtaining both spatial and chemical-shift information using clinical MRI systems. The following are some new acquisition techniques that produce either localized spectra or multiple-frame chemical-shift images, which can be implemented using conventional whole-body MRI systems (with no hardware modifications) and can be performed in less than 10 min. The technique uses the slice selection and phase encoding commonly used in MRI to define two of the three spatial directions, but applies a cyclic trapezoidal read gradient in the third direction. The signal produced during this cyclic read gradient reflects both the spatial distribution in the third direction and the evolution of the FID due to the chemical-shift distribution.

MRSI actually refers to a family of techniques for obtaining spatial and spectroscopic information simultaneously. MRSI can produce a set of chemical-shift images, each representing a distinct section of the chemical-shift spectrum. CSI techniques include the fat- and water-imaging technique and similar two-frame techniques, as well as multiple-frame techniques. MRSI is also a flexible method for localized NMR spectroscopy: Spectra from an array of spatial locations (pixels) are obtained in one acquisition. Unlike other imaging techniques for CSI and other spectroscopic techniques for localized MRS, spectroscopic imaging produces a rather large 3D or 4D dataset that can be sliced into chemical-shift images or taken apart longitudinally into separate NMR spectra from specific locations.

This flexibility is attractive, but it requires acquisition of a large number of individual data points. To make the necessary N-independent measurements in an $N \times N$ array for spectroscopic imaging, it is necessary to produce $N \times N$ different degrees of phase encoding. In the simple phase-encoding form of spectroscopic imaging, the number of FIDs acquired must be at least the number of independent spatial locations (in this case $N \times N$). If N is 128, corresponding to a modest spatial resolution, at least $128 \times 128 = 16,384$ FIDs must be acquired to obtain a complete dataset. At a reasonable repetition time, the experiment times required would be too long for a clinical study. The rapid spectroscopic imaging techniques, including the technique presented here, overcome this difficulty by reducing the number of FIDs required for a given spatial resolution. They do this by applying gradient fields during acquisition, so that spatial information is "multiplexed" into the FIDs. The multiple-output chemical-shift imaging (MOCSI) technique was developed to provide a large set, rather than just a pair, of chemical-shift images, using an unmodified commercial MRI system. MOCSI is similar to certain other techniques for rapid spectroscopic imaging in that it uses a single periodic gradient during acquisition to encode spatial information rapidly. However, the other techniques use the rapidly switched, nearly rectangular gradient waveforms of the EPI method, which require more intense and more rapidly switched gradients than are achievable in a typical commercial MRI system. MOCSI was developed as a technique suitable for commercial clinical MRI systems, with gradient intensity and slew rates insufficient for the EPI-based acquisition methods. As implemented on the 2-T Philips imaging system at Philips (Forschungslaboratorium Hamburg, Hamburg, West Germany), the MOCSI acquisition technique relies on phase-encoding y-gradients, with periodically varying x-gradients applied during data acquisition to scan back and forth in k. For each of the 128 steps in phase encoding, this periodic gradient was repeated with four different delays in its beginning time, so as to sample more frequently in time and to achieve a spectral bandwidth (see Figure 2.41 [30]). The trapezoidal gradient waveform had 2 ms plateaus at maximum positive and maximum negative gradient levels, with 2 ms ramps between positive and negative. This 8-ms period, when interleaved 4 times, gave a 2-ms sampling time, leading to a 5.9-ppm spectral bandwidth, sufficient to include the water and lipid features dominating the in vivo proton spectrum. More rapid switching and larger gradients would have made it possible to achieve an even greater bandwidth in fewer acquisitions, but the acquisition (8.5 min at TR = 1 s) was sufficiently brief for clinical use.

As implemented, the method acquired an FID signal (including, of course, "gradient echoes"), but a Hahn spin-echo could be used to produce T_2-weighted images. An important feature of MOCSI acquisition is that data are acquired continuously throughout the gradient switching periods, and these data are used in the

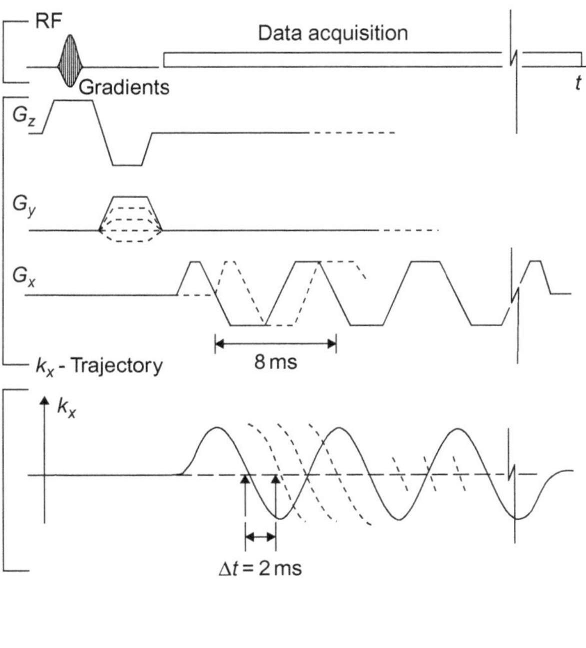

Figure 2.41 The pulse sequence for MOCSI. A single slice-selective RF pulse is applied during the selection gradient (in this example, in the z-direction). Following excitation of the slice, there is a conventional phase-encoding gradient pulse and the periodic trapezoidal MOCSI waveform, applied during data acquisition.

reconstruction. This leads to efficient use of acquisition time, although it requires special reconstruction measures. In any spectroscopic imaging technique using periodic gradients, the data may be reorganized after acquisition into "*k*-FIDs." A *k*-FID is simply the FID that would be obtained from the object in the absence of gradients, if it were first phase-encoded to *k* cycles/cm by radiant pulses (*k* is a 3D vector representing phase encoding in *x*, *y*, and *z*). In the MOCSI technique, as in the other rapid MRSI techniques using periodic gradients, each *k*-FID has two distinct sets of samples uniformly spaced in time, one corresponding to positive gradient segments and the other corresponding to negative gradient segments (see Figure 2.42, in which these are indicated by dots and circles, respectively). In the reconstruction used here, the data are first reorganized into *k*-FIDs from the positive segments (the odd echoes). Figure 2.42(a) shows a portion of the sampling trajectory and sample points in the (*k*, *t*)-plane for the case of ideal rectangular gradient waveforms, which can be closely approximated in small-bore systems and with special-purpose gradient-coil systems. For rectangular gradient waveforms, sample points fall on straight line segments in the (*k*, *t*)-plane. Figure 2.42(b) shows a portion of the sampling trajectory and sampling points in the (*k*, *t*)-plane for the periodic trapezoidal gradient waveforms used in MOCSI. These trapezoidal gradient waveforms are a concession to the limited gradient capabilities of the typical clinical MRI system, which can produce full-on straight lines in the (*k*, *t*)-plane only during the 2-ms intervals of constant gradient. During the intervening 2-ms periods, the gradient is either increasing or decreasing linearly, so that the sample points fall upon a parabolic sample path in the (*k*, *t*)-plane. As a result of these nonlinear sampling trajectories, the correct reconstruction of chemical-shift images and localized spectra from MOCSI data demands two measures not needed in other methods, in which sample points always fall along straight lines in the (*k*, *t*)-plane.

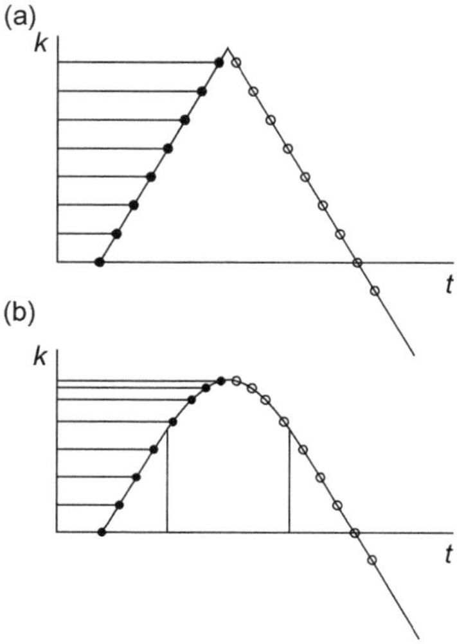

Figure 2.42 Schematic of sections of the sampling trajectories in the (*k*, *t*)-plane for (a) a technique using periodic rectangular gradient waveforms and (b) the MOCSI technique, which uses periodic trapezoidal gradient waveforms achievable in a typical clinical MRI system. Samples are taken at uniformly spaced time intervals. For clarity, only some of the sample points are shown. The *k*-FIDs are sampled along the horizontal lines at the points indicated by dots (odd echoes) and circles (even echoes). During the 2-ms periods of gradient switching (i.e., between the vertical lines in (b)), the *k*-FIDs have delays that are nonlinear in *k*, and the *k*-values sampled during these periods are not uniformly spaced. Because of this, accurate reconstruction of MOCSI data requires special processing techniques.

2.3.5 MQTs: 2D (Homonuclear) Spectroscopy [31]

This detection technique is about forbidden quantum transitions in 2D spectroscopy. It starts with a preparation period $t < 0$. This is required for the population of the corresponding off-diagonal matrix elements, of the density operator $\sigma(0)$. During the following evolution period $0 < t < t_1$, the MQT matrix elements are permitted to evolve under the influence of the unperturbed Hamiltonian H. At $t = t_1$, a mixing $P(\alpha)$, with rotation angle α, is applied to transform the unobservable MQT matrix elements into observable SQT (1QT) matrix elements. During the detection period $t > t_2$, the transverse magnetization is observed as a function of $t_2 = t - t_1$. This experiment is repeated for different systematically varied values of t_1. The resulting 2D signal function $S(t_1, t_2)$ is Fourier transformed in two dimensions to obtain a 2D spectrum $S(\omega_1, \omega_2)$, with the desired MQT information in the ω_1 direction. Several techniques exist for the creation of MQT matrix elements in the density matrix (Figure 2.43 and Table 2.1).

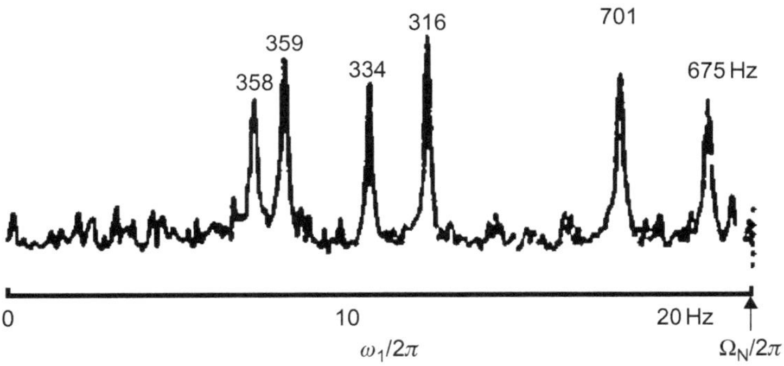

Figure 2.43 MQT spectrum of 2-furancarboxylic acid methyl caster obtained by applying a strong field gradient during the evolution period. The projection of the 2D spectrum on the ω_1 axis is shown. Note that folding at Nyquist frequency, $\Omega_N/2\pi = 22.0\,\text{Hz}$, has been utilized to enhance the digital resolution. The indicated true line positions can be obtained by computing $2\Omega_N - \omega_1$ and $4\Omega_N - \omega_1$, respectively.

Table 2.1 Simple Possibilities for the Selective Detection of MQ Transitions (for the Barred-Phase Values, the Resulting Signals Must Be Subtracted)

Phase Values for the Coded Experiments						Observed Values of the MQTs									
0°						0	1	2	3	4	5	6	7	8	9
0°	180°					0		2		4		6		8	
0°	180°						1		3		5		7		9
0°	90°	180°	270°			0				4				8	
0°	90°	180°	270°					2				6			
0°	60°	120°	180°	240°	300°				3						9

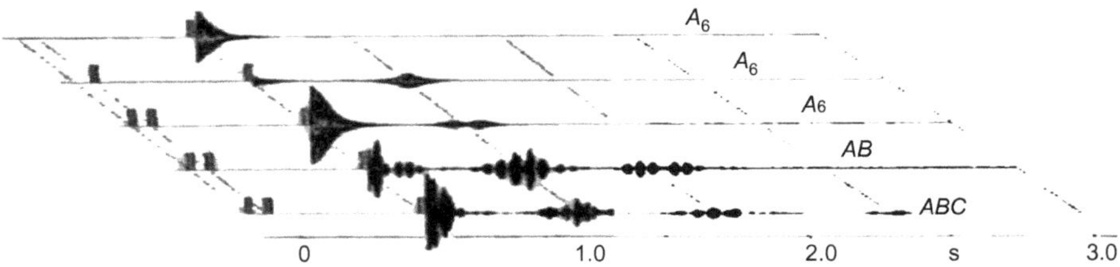

Figure 2.44 Illustration of MQ transfer echoes. The first three traces are obtained on Appendices A2.10–A2.12 spin system of benzene for three different pulse-excitation schemes. The last two traces are obtained from the *AB* spin system of 2,3-dibromothiophene and from the spin system of acrylonitrile, respectively, at 60-MHz proton resonance. The separation τ of the first two pulses is 35.8 ms; the delay t_1 between the second and third pulses is 0.6000 s.

2.3.6 MQTs: Coherence Transfer Echoes (Heteronuclear) Spectroscopy [32]

In this technique, the coherence is transferred, between defocusing and focusing, from one transition to another transition. The two transitions involved may be differently sensitive to inhomogeneous interactions, and thus defocusing and refocusing will proceed with unequal speed. This results in defocusing and refocusing periods of different lengths. Allowed and forbidden transitions may be involved in the echo formation; these are the *coherence transfer echoes*. A particular type of such echoes is magnetization transfer echoes, which can be observed in NMR of coupled-spin systems. A magnetization transfer between transitions belonging to different nuclear species leads to heterogeneous transfer echoes. The transfer of coherence from an MQT to an SQT may create MQ transfer echoes (Figure 2.44).

2.3.7 Spatially Resolved NMR Spectroscopy: A Combined Imaging Technique [33]

One can apply a periodically oscillating or inverting field gradient on an object during observation of an FID. The resultant signal takes the form of a spin-echo train. The external field-gradient effect on the signal may be periodically canceled at each echo maximum. Thus, it is possible to obtain a spectroscopic FID from the envelope of the echo train. Because the signal is also modulated by the external field-gradient effect, 1D spatial information can simultaneously be extracted from the echo train. It is possible to obtain 2D, spectroscopic-spatial information from a 1D echo train (Figures 2.45 and 2.46).

2.3.8 PCM: The Basics of MRI

This section is included to make readers aware of the basic mathematics involved in the techniques of imaging. It is hoped that this may induce the mathematically weaker sections of the scientific community to try to develop the mathematical skills

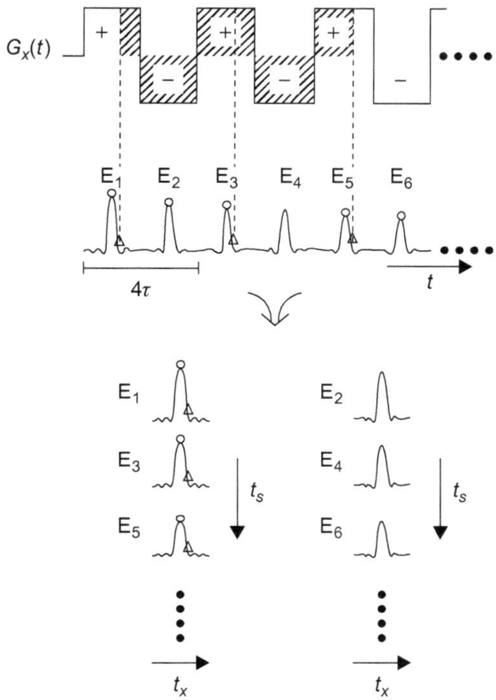

Figure 2.45 Echo rearrangement. Odd- and even-numbered echoes are separately rearranged to give two sets of 2D echo arrays. Note that the signal time development along the t_s axis is determined purely by the internal spectroscopic interaction, because along the t_s axis the external field-gradient effect on the signal is quenched as long as $G_x(t)$ is periodic. In contrast, time development along the t_x axis (or each echo shape) is dominated by the temporarily time-independent external local interaction.

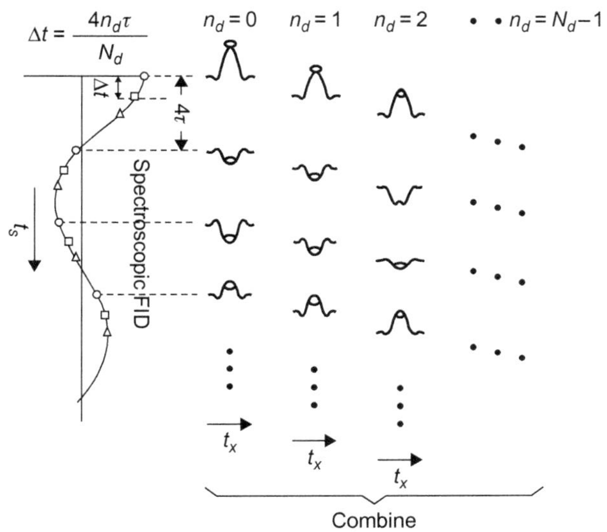

Figure 2.46 Combination of echo arrays obtained from delayed spin-echo trains. Each of the measured N_d sets of echo trains is separated and rearranged as in Figure 2.45. The resultant echo arrays are then combined in accordance with the spectroscopic time order to provide a similar but larger echo array. This results in an N_d-fold expansion of the spectral bandwidth. *Note*: The technique provides the data combination necessary for the expansion. Δt is chosen for Fourier transformation of the combined data.

needed to understand the field of MRI. Efforts to become familiar with the intricacies of the science of NMR, and its applications in medical imaging technology, will be most worthwhile, especially if MPs become part of the development. Newtonian mechanics (NM), e.g., the differential calculus, form part of the learning process of most science undergraduates, up to a certain level. Therefore, as far as possible in writing this chapter, we have adopted NM as the medium of exposure. In the particular case of imaging based on distant dipole–dipole quantum mechanical correlations, however, one needs a good comprehension of the field of QM. This QM knowledge enables one to fully understand the energy transition processes occurring within nuclei, and the quantum cross-correlations among the nuclei. In a situation like the human brain, the correlations of the energy transitions of the nuclei are spread over a wide space. These can be technically collected, so as to get the required quantum imaging of the brain. The application of mathematical concepts to an imaging technique is the key to development of an imaging technology. The idea underlying this section is to enable readers to acquaint themselves with various classical and quantum physics concepts.

These concepts are the building blocks of MRI technology. Our intention in writing this chapter is to keep the mathematics involved to the NM level, as much as possible. One must realize, though, that if one really wants to master the field of NMRI, one cannot avoid QM. In fact, QM is much more intuitive than NM, and certainly is no more difficult to learn than NM. It is an ideal tool to quantify logical deductions and solve problems at the nuclear and atomic levels. This chapter is in fact an effort to bridge the gap between the two approaches. QM concepts, which have evolved over more than half of the last century, are the backbone and foundation of MRI. Appendices at the end of this chapter give practical illustrations of both the prevalent and the progressing imaging techniques, through the lens of mathematics, both classical and quantum. (Presentation of the information in this manner makes the task of writing this chapter a little bit easier.) An explanation of the phenomenological theory of NMRI is beyond the scope of this chapter. The appendices are not arranged in any particular order, nor are they intended to act as a textbook. One may find that the information in different sections overlaps to some extent. This is intentional, because one should realize that exposure to diverse approaches and concepts is very helpful in learning something that is new and possibly difficult to comprehend. In some situations, this process of learning may itself be an essential part of one's education; exposure to possibly somewhat repetitive information, from various viewpoints in the field, is a good way of learning. This is particularly true where formal means of learning are limited or unavailable.

2.3.8.1 Heteronuclear 2DFT: MRS [34]

This technique has two basic elements in it. First, there can be indirect detection by the transfer of longitudinal magnetization between the subsystem S, to be investigated indirectly, and the subsystem I, the resonances of which are being observed. This transfer proceeds through cross-relaxation mechanisms that couple the two subsystems and permit an incoherent transfer of energy. It can be utilized in the laboratory or in the rotational frame. The energy transfer requires a finite amount of time, determined by the relevant relaxation parameters governing cross-correlation. Second, the frequency dependence of the effect permits an indirect, point-by-point measurement of the spectrum of the subsystem S, by observing the resonances of the I subsystem. The mechanism requires a coherent coupling between the two subsystems leading to a resolved fine structure in the observed spectrum. It is characterized by an immediate response of the detection system, as it does not involve relaxation processes. In an alternative approach, a mechanism is devised whereby one relies on a coherent transfer of transverse magnetization between the two subsystems. A coherent coupling between the two subsystems must exist (Appendix A2.3).

2.3.8.2 Spin-Echoes: Pulsed Nuclear Magnetic Double Resonance (Appendices A2.5 and A2.6) [35]

In an ensemble of atoms and molecules, the electrons precess around their nuclei in circular orbits. The plane of this circulating current gives an intrinsic magnetic dipole character to the atoms and molecules. These planes are randomly oriented in space. The resultant intrinsic dipole magnetization (magnetic moment per unit volume) of the substance is thus zero. When an external magnetic field is applied to the substance, the planar orbits tend to align themselves perpendicular to the applied field. The result is a net magnetization in the substance. However, the net magnetization is not perfectly aligned along the field as it should be: There are distractions to perfect alignment. These distractions are due to interactions among the nuclei (e.g., due to the surrounding cloud of electrons), called the *chemical shift*, and the spin–spin coupling, termed the J-coupling. Other disturbances are interaction with the environment in which the nuclei are confined and move. When the external field is removed, the magnetization returns to zero, that is, to its equilibrium value. There is a continuous Larmor precession of the spins in the ensemble, due to the static magnetic field applied in the z-direction. The time this precession keeps, phase memory, is called as the spin–spin relaxation time T_2. A driving RF pulse of duration t_w ($<T_2$) creates a nonequilibrium macroscopic magnetization, M. Transient nuclear induction can be used, in between RF pulses, to create spin-echoes that are spontaneous induction signals. They appear due to constructive interference of macroscopic magnetic moments after more than one RF pulse has been applied.

2.3.8.3 Nuclear Dipole–Dipole Coupling Through Chemical Shift: Intranuclear and Internuclear Orbital Electrons (Appendices A2.7 and A2.8) [36]

A local magnetic field, at the position of the nucleus, is caused by the precession (called the *Larmor precession*) of the outer-nuclear electrons. This is seen in the presence of an externally applied magnetic field. In the first order, the effect, due to a spherically symmetric electron distribution, is given by the diamagnetic correction. In the second order, there are electrons that undergo attraction through two or more nuclei in molecules. This induced paramagnetism, in many cases, is comparable to or larger than the diamagnetic correction that arises from a perturbation.

In this perturbation, the ground state mixes with the paramagnetic, excited state of the molecule. It causes the observed large chemical shift that occurs when the energy level of the paramagnetic state lies very close to that of the ground state. The analysis can be understood, in the case of two-proton coupling, where J-splitting appears in the transitions between the pure triplet states, which are linear combinations of singlets, and the triplet terms. Normally, transitions to the singlet state are forbidden, but may occur when the identity between the two protons is removed by a difference in chemical shift.

2.3.8.4 Pulsed NMR Double Resonance: Dipole–Dipole Interactions (Appendix A2.9) [37]

The effect of pulsed double resonance is clearly indicated by the study of proton (A spin) echoes coupled to phosphorous nuclei (B spins). In an aqueous solution of phosphorous acid, $HPO(OH)_2$, the indirect spin–spin interaction occurs between H and P nuclei in the HPO group. It is described by $J(I_H I_P)$ in frequency units. Here I is the spin operator, and $J/2\pi = 708$ Hz. The H concentration involved in the coupling is about one-eighth of the total proton concentration in the sample used. A 90–180° double sequence is applied to obtain the proton echo. A third pulse is applied at the P resonance, at t', with respect to the first 90° proton pulse. One considers only the macroscopic magnetic moment M_0 due to those protons coupled to the phosphorous. They have an off-resonance angular frequency $\Delta\omega$ with respect to the frame of reference rotating at a Larmor frequency ω_H in the absence of J-coupling.

2.3.8.5 DQ Transitions (Appendices A2.10–A2.12) [38]

The technique of double resonance proceeds in a more direct manner. It identifies transitions that have an energy level in common. It is convenient to redefine two such transitions, called *connected transitions*. One can divide them into two categories. For any two connected transitions, the two energy levels that are not common to both transitions must have the same magnetic quantum number. These transitions are referred to as *regressive*. For quantum numbers that differ by two units, the transitions are referred to as *progressive*. Using these definitions, the double-resonance behavior may be simply stated in the form of three rules. (1) If a nondegenerate transition is subjected to a weak RF field, all transitions that are connected with it will split into doubles. (2) If line widths are determined by magnetic-field inhomogeneity, and if all the nuclei have the same gyromagnetic ratio, the irradiation of one transition will split regressive connected transitions into well-resolved doubles, but will split progressive connected transitions into broadened doubles. (3) The magnitude of the splitting (which is best measured for the well-resolved doubles) is proportional to the RF field H_2 and to the square root of the intensity of the irradiated line. A second possible way of recognizing progressive connected transitions is through observation of DQTs. It is well known that the normally forbidden $\Delta m = 2$ transitions involving two quanta can be made to appear in an NMR spectrum if the RF power is increased above the level set by the normal requirement that the SQ lines not saturate. They are characterized by their sharpness (compared with the power-broadened SQ line), and by the dependence of signal strength on H_1 in the region below saturation. A pair of progressive transitions is readily determined by recognizing that the average of the two frequencies of the pair must closely correspond to the transition frequency of a DQ line. A DQ line will define a group of two or more pairs of progressive transitions. In the simplest case, that of two coupled spin-½ nuclei, there is just one DQ line, and it is a simple consequence of the frequency sum rules that two transitions constituting a progressive pair must be symmetrically spaced on each side of the line. If the transition frequencies *g*, *h*, *i*, *j*, Appendix A2.11 (Figure A2.11(a) [38]), are ordered by increasing (or decreasing) frequency, the energy-level arrangement is that shown in Appendix A2.11 (Figure A2.11(b) [38]). An inversion of this energy-level diagram corresponds to a reversal of the spin-coupling constant. The two progressive pairs are *g*, *j* and *h*, *i*. The ordering is such that *g*, *i* and *j*, *h* are regressive pairs. This ordering also applies to the *AB* or *AX* part of a more complex spectrum in systems where all other spins are loosely coupled to each other and to the particular pair under consideration. For example, a part of a more complex energy-level diagram is sketched in Appendix A2.11 (Figure A2.11(c) [38]). Here, if the lines *g*, *h*, *i*, *j* correspond to transitions of the *AB* spins, then either line *f* or line *k* must be a "combination" line and therefore be weak or unobserved, as indicated in Appendix A2.11 (Figure A2.11(d)).

This observation can be a great aid in assignment of these more complex systems, as a complete assignment can be made from only a small number of DQTs. The observation of MQTs is limited to transitions involving spin changes of nuclei with the same gyromagnetic ratio. The dependence of the maximum intensity of a DQ line on the chemical shift, δ, between two closely coupled spin-½ nuclei can be calculated. The analysis shows that the optimum value of the RF field H_1 is proportional to $\delta/J^{1/2}$ and the maximum height of the DQT is proportional to $J^{1/2}/\delta$. The frequency difference δ between two different isotopes is so large, at normal magnetic-field strengths, that such transitions would be unobservable above the noise. Systems that consist of a group of nuclei tightly coupled to each other, but only loosely coupled to other nuclei, may be treated in the same manner as they are in the analysis of normal NMR spectra. The transitions of the tightly coupled group are treated as a superposition of several spectra, each corresponding to a different orientation, of the loosely coupled spins. Thus, an ABCX spectrum can be treated as a superposition of two ABC spectra corresponding to two orientations of the *X* spin. This technique can be illustrated by treating the proton spectrum in trivinyl phosphine as a superposition of two ABC spectra.

It is instructive to study the frequency shifts in such a system. For this purpose, it is convenient to consider separately terms with the different values that the magnetic quantum number M_k can take in this expression. Let *r* designate all terms with $M_r = M_s + 1 = M_q - 1$, terms with $M_t = M_s - 1$, and *p* terms with $M_p = M_q + 1$, as is illustrated in Appendix A2.12 (Figure A2.12). The normal SQT frequencies ω_{kj} are defined by $\omega_{kj} = E_k - E_j$. The DQ frequency condition may be expressed in the form $\omega_1 = 1/2(\omega_{sr} + \omega_{rq}) + d_{sq}$, where d_{sq} is the shift in frequency from the center of the pairs of SQ lines ω_{sr} and ω_{rq}. The observation of MQTs is limited to transitions involving spin changes of nuclei with the same gyromagnetic ratio. The dependence of the maximum intensity of a DQ line on the chemical shift δ, between two closely coupled spin-½ nuclei, can be calculated.

2.3.8.6 EPSM: Projection Reconstruction and Echo Planar Mapping (Appendix A2.13) [39]

This technique allows both spin density and chemical shift, in a defined plane, to be simultaneously observed. In many biological systems, chemical-shift studies of ^{31}P and ^{13}C can be useful means of studying metabolic pathways in vivo. The information is of great clinical value in the diagnosis of disease and in the study and monitoring of the effects of drug treatment. There are two variants: the EPSM and the PREP methods. Both methods have the advantage of speed over the point-by-point tomographical magnetic resonance (TMR) approach. The FID signal in the rotating frame of reference at time t, following a spin excitation, is $S(t) = \mathrm{Re(real\ part)} \iint \Delta\rho_z(x,\ y)\exp[\mathrm{i}\delta(x,\ y)t]\exp[\mathrm{i}\gamma \int (xG_x(t') + yG_y(t)\mathrm{d}t']\mathrm{d}x\,\mathrm{d}y$. Here one assumes that the RF pulse is selective to plane Z with slice thickness Δz. The average spin density over the selected slice is $\rho_z(x, y)$; γ is the gyromagnetic ratio, considered the same for all the spins. $G_x(t')$ and $G_y(t')$ are the time-dependent magnetic-field gradients, and $\delta(x, y)$ is the chemical shift, assumed to be independent of the thickness z. Appendix A2.13 (Figure A2.13(a) [39]) shows the principal spin projections in the static gradients for an echo field comprising three disks of material with chemical shifts δ_1, δ_2, and δ_3. Figure A2.13(b) [39] (Appendix A2.13) is the discrete projection of the object when G_y only is modulated, and is the basis of the PREP experiment. Figure A2.13(c) shows the further splitting of each stick that occurs when a modulated gradient G_x is added. This is the data form produced in an EPSM experiment. In the EPSM technique, only four separate experiments are required to produce an x, $y - \delta$ map. These may be combined into two pairs or possibly one single experiment, for snapshot studies. This form could be particularly useful for time-course studies of metabolic processes, and also for rapid magnetic-field alignment. Data processing requires a single 1DFT. The PREP method requires $2m$ separate experiments, which may be constructed to m pairs, where m is the ray size. Data is processed initially by a 1DFT, and then by a projection reconstruction algorithm. Both methods (EPSM and PREP) have the ability to produce spin-density images that are free of spatial distortions arising from static field inhomogeneities.

2.3.8.7 Multiple- (Double-)Quantum MRI (Appendices A2.14–A2.16) [40]

For a description of nonlinear properties, it is not sufficient to consider just the amplitude of the signals. One should include phase as well, as a function of frequency. It is necessary to include at least one further parameter, such as the RF field strength or a second frequency. This is done, for example, in DQ resonance. A graphic representation of such a set of data naturally leads to 2D or, more generally, to multidimensional spectroscopy. Normally used nonlinear techniques to note in this context are illustrated in Appendix A2.14. A brief description is as follows. (a) *Frequency space experiment*: A simultaneous application of two frequencies, and measurement of the response as a function of both frequencies, leads directly to a 2D spectrum. This is the principle of conventional double resonance. (b) *Mixed-frequencies space-time experiment*: A system perturbed by a strong RF field with frequency ω_2 can be investigated by applying an RF pulse and measuring its response. Fourier transformation of the FID and repetition of the experiment for various perturbing frequencies leads also to 2D spectra with properties very similar to conventional double resonance. (c) *Time-space experiment*: A 2D experiment done completely in time-space requires two independent time variables, as a function of which a signal amplitude can be measured. A 2DFT of the 2D time signal then produces a 2D spectrum. (d) *Stochastic resonance experiment*: From the response of a nonlinear system to a Gaussian random perturbation, it is also possible to compute a 2D spectrum by calculating higher cross-section-correlation functions between input and output noise and Fourier-transforming them.

This subsection is restricted to evaluation of the basics of time-space experiments. In 2D Fourier-transform spectroscopy (FTS), the 2D spectrum is obtained by Fourier-transforming a signal $S(t_1, t_2)$, which depends on two independent time variables t_1, t_2. For the introduction of the two time variables, it is necessary to mark out two positions on the time axis and to partition the experiment time into three periods. It is convenient to let t_1 be the duration of the second period and t_2 the running time in the third period, as shown in Appendix A2.15. The signal $S(t_1, t_2)$ is then measured in the third period as a function of t_2 with t_1 as a parameter. The three phases, which are characteristic of all 2D FTS experiments, are named according to their physical significance: (1) $t<0$ Preparation Period: The system is prepared in a suitable initial state, described by the density operator $\sigma(0)$. (2) $0<t<t_1$ Evolution Period: The system evolves under the influence of the Hamiltonian (the steady-state energy of the system) H_1. One assumes that at the end of this interval, a particular state which depends on H_1 and on the elapsed time t_1 is created. (3) $t_1<t_2$ Detection Period: The system develops further under the influence of Hamiltonian H_2 (H_1 changes into its new value H_2). During this time, the transverse magnetization $M_y(t_1, t_2) = S(t_1, t_2)$ is detected as a function of t_2. A large number of experiments, using different durations t_1 of the evolution period, have to be performed to obtain a sufficiently dense sampling of the 2D time function $S(t_1, t_2)$.

In a multidimensional extension of 2D spectroscopy, the system has to go through several evolution periods, each of which must be varied systematically in length. Many homonuclear and heteronuclear experiments are possible. A two-pulse experiment is a good illustration. The preparatory phase ends with a non-selective RF pulse at time $t=0$ (called the *preparatory pulse*). A flip angle of 90° is usually employed to generate off-diagonal elements of the density operator, which evolves under the influence of H_1 during the evolution period. This period is ended by a second RF field (called the *mixing pulse*) at time $t=t_1$. It mixes the various magnetization components and enables them to be measured during the detection period. This experiment permits evolution of the energy-level schemes of the coupled-spin systems.

Another very simple application is the distinction of resonance lines belonging to different molecules in a mixture. In another experiment, Appendix A2.15 (Figure A2.15 [40]), an RF field is applied during the evolution period. Separate preparation and mixing pulses are not required. During the evolution period, Torrey oscillations will develop. They can be associated with the various resonance transitions. A modification of this experiment, in which a magnetic-field gradient is added during the detection period, may serve as a means to measure spatial inhomogeneity of the RF field strength. One can use an experiment where a non-equilibrium state $\sigma(0)$ is created during the preparatory period. Application of the first two pulses permits the population of all matrix elements of the density operator. The behavior during evolution and detection periods then completely characterizes the initial nonequilibrium state. Here, 2D spectroscopy is used as a means to measure the instantaneous state $\sigma(0)$ of a perturbed system, by elucidating the matrix elements responsible for the higher-order transitions. One can also use a preparatory pulse that generates the required transverse magnetization. During the following time periods, it precesses in the presence of two different field gradients. It permits measurement of the 2D or 3D spatial spin density of macroscopic objects.

There is also an experiment that has been used to measure off-diagonal elements of the density operator $\sigma(0)$, created during a chemical-reaction-induced dynamic nuclear polarization. The heteronuclear experiment leads to 2D-resolved carbon-13 spectroscopy, a promising method for unraveling complicated uncoupled C-13 spectra. During the evolution period, the C-13 spin precesses in the absence of proton–carbon couplings, while the complete Hamiltonian determines the evolution during the detection period. This permits separation of the multiplets that originate from different carbon spins. Yet another experiment, i.e., a typical cross-polarization experiment, is used in solids to detect rare nuclei. The evolution period here is identical to the cross-polarization time. During this period, transient oscillations have been observed. They are caused by the coherent dipolar interactions of directly bound nuclei. A 2D representation of these phenomena permits one to obtain structural information on solid samples, for single crystals as well as for powders.

One can perform an experiment that is a heteronuclear modification of the basic two-pulse experiment. It permits one to unravel the multiplet structure of heteronuclear spin systems. In the experiments described previously, a 2DFT is required to generate the desired complete 2D spectrum. The information obtained, which is analogous to that obtained from double resonance, is provided by the rotation superoperator R' that couples the various transitions. One also obtains information regarding the flip-angle dependence of the intensities and phases. 2D spectra contain a wealth of information on the topology of the energy-level diagram. *Information on transverse relaxation processes*: The line shapes are determined by the TRTs T_{2kl}. They can also be determined even in the presence of field-inhomogeneity broadening. *Information on the initial state $\sigma(0)$ of the spin system*: In conventional spectroscopy, allowed transition can be detected only in 2D FTS. In contrast, it is possible to measure all matrix elements of $\sigma(0)$ in a unique manner. In particular, it is possible to observe matrix elements responsible for zero-, double-, and multiple-quanta transitions. For weakly coupled nonequivalent spins-½, it is possible to distinguish three classes of connectivity patterns. Their definition follows the well-known terminology used for directly observed transitions, distinguishing regressive and progressive transitions (Appendix A2.16).

2.3.8.8 DQ Transitions

2.3.8.8.1 DQT in Deuterium (Appendix A2.17) [41]

This illustration demonstrates a case involving a combination of spin double resonance and the coherence properties of MQTs. In high-resolution SS NMR, for any but spin-½ nuclei, there is a problem of spectral splitting and broadening, induced by the interaction of the nuclear quadrupole moments with electric-field gradients. Appendix A2.17 shows an example where this particular case was studied in reference to deuterium.

2.3.8.8.2 Nuclear Dipole–Dipole Interaction: Quadrupole Hyperfin Interactions (Appendices A2.18–A2.21) [42, 43]

One can use nuclear dipole–dipole interaction (NDDI), as part of NMR spectroscopy, to deduce atomic positions in both the LS and SS of matter. This NDDI, which is used for the determination of structural parameters, has special appeal because the interactions are unlike any other perturbations experienced by nuclear spins. The strength of NDDI is directly specified by interatomic distances and orientations. One needs to consider the effect of neighboring dipoles on a particular reference spin, depending on the structural environment. To observe NDDI, strong rotating fields of amplitudes $H_{1S} = \omega_{1S}/\gamma_S$ (rarer species) and $H_{1I} = \omega_{1I}/\gamma_I$ (abundant species) are applied at frequencies ω_S and ω_I. These are near the resonance frequencies $\omega_{0S} = \gamma_S H_0$ and $\omega_{0I} = \gamma_I H_0$, respectively. Eventually, under the influence of $\hat{H}_{\mathrm{ff}}(t)$, the two species come to a common temperature.

2.3.8.9 DQ Resonance: High-Resolution NMR (Appendices A2.22–A2.24) [44]

In this technique, one needs two transverse RF fields H_1 and H_2. They should be either of the same order or greater than the perturbed spin-coupling constant J. However, the spectra of strongly coupled spin systems generally have transitions separated by frequencies much less than J. If simple double-irradiation spectra are required, considerably weaker fields would have to be employed. One essential step in double-resonance experimentation consists of identifying transitions that have an energy level in common. This allows the network of transitions between energy levels to build up. It may then be correctly oriented in terms of the increasing energy. This is done by observing the details of the line profile obtained in the double-resonance spectra. In this selected case, the analysis includes strongly coupled (highly perturbed) spin systems, such as are often found in high-resolution NMR spectra. The analysis is normally restricted to the condition that the perturbing RF field be near a single nondegenerate transition frequency. Nevertheless, it can be extended to include two transition frequencies, where the two transition frequencies are close together but have no common energy level. Examples of degenerate transition frequencies with a common energy level occur in loosely coupled systems. The spectra observed in high-resolution NMR can be described by the Hamiltonian, $H = -\Sigma_i \omega_i I_i^z + 2\pi \Sigma_{i<j} J_{ij} I_i I_j$. Here $\omega_i = \gamma_i(1 - \rho_i)H_0$, and ρ = the magnetic shielding parameter. The quantity ω_i represents the resonance frequency of the *i*th nucleus, in the absence of any coupling between nuclei. I is the vector spin operator of the nucleus I, with the Z-component I_i^z, and J_{ij} is the spin coupling between the nuclei i and j measured in units of Hz.

For double-resonance experiments, it is necessary to consider the interactions of the spin system with two RF fields. The first is the usual RF amplitude $2H_1$ and angular frequency ω_1 applied along the x-axis. The spectrometer receiver is tuned to accept signals created in a narrow band of frequencies around ω_1. The second field, referred to as the *perturbing field*, has amplitude $2H_2$ and frequency ω_2, and is also applied along the x-axis. Both fields may be broken down into two equal counter-rotating fields; only the component rotating in the same sense as the nuclear precession appreciably perturbs the nuclei. There are two possible arrangements of the three energy levels p, r, and s, and they must be clearly distinguished because they lead to important differences in the double-resonance spectra, Appendix A2.22. One energy level must always be common to both transitions; the other two energy levels may then either have the same spin quantum number ($\Lambda = 0$) or spin quantum numbers that differ by two units ($\Lambda = 2$). Consider first the case $\Lambda = 0$. In the limit $H_2 = 0$, ω_{rp} is an allowed transition with $\Delta M = \pm 1$, whereas ω_{ps} is a forbidden transition with $\Delta M = 0$. However, the RF field H_2 causes the states formerly defined by Ψ_r and Ψ_s to be mixed, to give new states Ψ_r' and Ψ_s'. In this case the transitions with $\Delta M = 0$ become allowed. The frequency of the RF field, H_1 in the rotating coordinate system, is $\omega_1 - \omega_2$, and it must satisfy the Bohr condition, $\omega_1 - \omega_2 = \varepsilon - (Ep + \omega_2 Mp)$, with $t = r$ or s. The second case of interest is that with $\Lambda = 2$, Appendix A2.22. In the limit $H_2 = 0$, ω_{sp} is an allowed transition with $\Delta M = \pm 1$, whereas ω_{rp} is a forbidden transition, with $\Delta M = \pm 2$. A molecule containing two nonequivalent nuclei is the simplest model for such double-irradiation experiments. To make the point that strongly coupled systems may be investigated by use of this technique in exactly the same way as weakly coupled systems, the *AB*-type molecule 2-bromo-5-chlorothiophene was chosen. It has the simplest set of energy levels.

2.3.8.10 MQTs: NMR (Appendix A2.25) [45]

The system considered here is one that consists of a group of nuclei of the same species. The nuclei are exposed to a strong and constant magnetic field H_0, and to an RF field whose frequency is in the vicinity of the Larmor frequency. This is the frequency of the magnetic moments. There exist chemical shifts, which have the same value for certain subgroups of equivalent moments. Further, two such groups, s and t, are coupled to each other. This coupling is achieved by an interaction energy proportional to the scalar product of the total spin vectors I_s and I_t, respectively.

2.3.8.11 Spin Decoupling in Abundant Isotope While Observing Resonance in Another (Appendices A2.26 and A2.27) [46]

NMR lines, in diamagnetic solids, are dominated by magnetic dipole–dipole (D–D) interactions between spins. Hidden in these lines is a wealth of information, derived from much smaller interactions of interest. One can apply either of two basic approaches. The DQ process is inherently less probable than the allowed SQ processes (if both are induced on resonance). It is resonant for all deuterium spins, but the allowed ones are not. The same reasoning can be invoked for the spin decoupling of the two strongly dipolar-coupled, line spin-½ nuclei from a third unlike nuclear spin.

2.3.8.12 Rotating Coordinate Frame of Reference: MQTs

2.3.8.12.1 Single Nuclear Species: Double Resonance (Appendix A2.28) [47]

In the situation of an oscillating (rotating), i.e., time-varying, magnetic field, one can write the equation $dL/dt = \partial L/\partial t + \omega \times L$. Here one notices that L, the angular momentum, appears on both sides of the equation. It is as if measured by a stationary observer. The symbol $\times$ stands for a cross (vector product) between ω and L. This equation can be applied to interpret the effect of the rotating field used, and this is in fact done in various NMR experiments. In most of these experiments, there is a constant field H_0 along the z-direction. In addition to this field, another (usually much weaker) field H_1, perpendicular to H_0 along the x-direction, is also applied. It rotates with angular velocity $-\omega$. From the point of view of a coordinate system rotating with H_1, none of the magnetic fields is changing as a function of time. The method can be used to demonstrate the criterion for the rate of change of a field to be "adiabatic." It is such that a nuclear moment preserves its magnetic quantum number (classically its angle) relative to the field as the field is varied.

2.3.8.12.2 Rotating Coordinate Frame of Reference: Two Different Nuclear Species (Double Resonance) (Appendix A2.29) [48]

A spin sample is chosen in which the nuclei to be detected, the "b" nuclei, have "a" nuclei as abundant neighbors in the same crystal. The "a" nuclei yield a large signal. The dipolar field due to b nuclei determines a portion of the local field h_{local} acting upon the "a" nuclei. The spin-echo formed by the "a" ensemble depends on the persistence of this local field. The RF pulse is applied to the "b" system within the echo memory time T_2 of the "a" spins. The "a" echo is attenuated by a certain amount because the local field due to the "b" spins is scrambled. The "b" resonance is therefore indicated.

2.3.8.12.3 Spin Temperature Theory: Quadrupole–Dipole Interactions [49]

Dynamic orientation experiments, on partially deuterated, organic materials doped with paramagnetic materials (e.g., Cr^{5+} complexes), are a good source of useful information. Very high spin vector polarization can be attained in these materials.

Such high polarizations are obtained by irradiating the electron-spin system near the MR frequency of the Cr^{5+} complexes (69.5 GHz in 25-kG magnetic fields). This causes a cooling of the "electron spin–spin interaction reservoir."

2.4 Summary and Conclusion

NMR technology has provided a complete method for analytical science in the field of medicine, both in vivo and in vitro. It has the versatility to spectroscopically determine the dynamic functioning of the various chemical elements present in the human brain that direct and enable the physical performance of various organs, as well as the brain itself. Still, the technology has limitations. It is partially invasive, albeit within allowable limits. The method carries out initial polarization of the chemical nuclei in a shallow depth through the surface of the brain. This is done by the application of a static (normally <4 T) magnetic field. The dipole magnetic moments of the polarized nuclei are subsequently excited using much smaller additional transverse RF magnetic fields. As the nuclei relax to their initial state (i.e., the static polarized direction), they emit electrical signals from their various locations in the brain. The magnitude of the signal from different nuclei is quantized and is space- and time-dependent. This forms the basic imaging source from the subsurface of the brain. The imaging is limited in depth from the surface, because radio waves can penetrate to only a limited depth in the brain. This limitation comes from the reaction (a natural shielding) by the fluids in the brain to the applied RF field. Isn't it lucky that the human body knows how much it can take?

Any excess can be very destructive. Bringing out the response in a tangible form requires sophisticated electronics technology. Different nuclei will require different strengths of field, because of the different quantized states the excited magnetic moments can take on. Success depends on the sophistication of the mapping and imaging techniques developed. The polarized magnetic moments (spins) precess around a local field axis (fixed by the applied static magnetic field) because of the angular momentum of the spins. The spins can absorb the RF electromagnetic energy only in quantum jumps, so they remain in an excited state for only a brief time. Interaction of the precession of the nuclear spins with other spins and the environment through which they move is mapped by the RF pickup coils placed close to the brain. In one imaging technique, one uses the echoes received from the RF fields applied. These have to be of suitable amplitude, frequency, and phase, and mapped over the surface of the brain. EPI is a prominent technique employed in MRI. Its several variants are used to improve the intensity and/or contrast of the images. The interactions of the applied RF fields, which result in the natural FID of the nuclei, produce the desired, detectable signals. The influence of the electron clouds present around various species of nuclei produce the chemical shift used in imaging. The nuclear and atomic magnetic dipole interactions (quantum correlations) create the variable parameters, along the surface and in the depths of the brain, that are used in imaging.

The imaging process involves interaction of magnetic fields, no matter how small, with the body and is therefore somewhat invasive. However, one can reduce dependence on the size and diversity of various technical requirements in MRI to a large extent and make the system manageable and reliable. Reducing the volume of the infrastructure requires huge input from and understanding of the PCM of the various techniques involved. There is a great need to renew the basic support structure of PCM from the grass roots up; this includes sound education in quantum science of a much wider community, at both the secondary and tertiary levels. A firm foundation of PCM knowledge is required to support and develop the specialist knowledge needed for continued progress in MRI; without this, the task is likely impossible.

Even as the field of MRI progresses, it is wise to consider alternative routes as well; one must keep in mind that MRI is not a perfectly noninvasive technique. Some parallel technologies are at various stages of research and development, but only in certain restricted parts of the world. A noninvasive method of imaging and analysis, such as magnetoencephalography (MEG), is a possible alternative direction [50–53]. A lot depends on the development of sensors that can detect the very weak electrical and magnetic signals from the naturally occurring metabolic activities in the brain. It will be a great achievement to devise a technique in which there is no need to apply strong external magnetic fields. Recent advancements in superconducting quantum interference devices (SQUIDs) enable one to detect very weak magnetic signals of pico (10^{-12}) to femto (10^{-15}) Tesla strength. This has made it possible to develop MEG as a tool of a totally noninvasive character. The small space it occupies is an added advantage. All one is after in this approach is measurement of the small magnetic fields generated as a result of the motion of metabolites through various channels in the brain. As the ions move, the natural dipole magnetic fields associated with the spins of the nuclei are in a dynamic state; so are the electrical fields generated by the channeled ions moving through the synaptic trees. They interact with each other and with the environment through which they move. A totally noninvasive brain imaging probe would enable the remote detection of interactions by sensors near the surface of the brain. Current and future advances in high-temperature superconductor (HTSC) research will allow HTSC SQUIDs, operated at 77 K instead of low-temperature superconductor (LTSC) SQUIDs operated at 4 K, to be used as sensors. Due to their better operating efficiency, the use of liquid-helium-cooled LTSC sensors is mandatory at present. The use of liquid nitrogen (boiling point 77 K) for HTSC SQUIDs, instead of liquid helium (boiling point 4 K), in the future would make MEG a very mobile, miniaturized technology.

HTSC research, it is hoped, will also pave the way to help reduce the size of the infrastructure needed for present NMR technology. This may be done by replacing LTSC magnets with HTSC magnets, when they are developed to the required standard. At present, liquid helium must be used to cool LTSC coils enough to create the huge DC magnetic fields required for NMR. It defeats the purpose for affluent nations to assume that someone else will do the research (HTSC research is only a typical example). The attitude is that we will simply go and buy the equipment when and wherever it becomes ready and available. From a strictly economic theory viewpoint, this idea at first appears clever, but something is missing here: The education of the young minds of a nation, which can be part of the process of research and development, is considered of negligible importance. What can be more important than the future of the coming generations? Lest we forget, ignorance has its own cost—a cost that becomes prohibitive as time goes on. The use of sensitive and miniaturized HTSC SQUIDs as sensors will open new directions for methods of imaging. Parallel and competitive research and development efforts by several independent workers create an environment of learning, competition, and advancement. This is essential for the constructive and healthy occupation of many minds, both young and old, in a healthy society. It has already been shown that it is possible to use very small DC magnetic fields (e.g., the Earth's own magnetic field, at ~micro-Tesla strength) as a polarizing field and perform NMR analysis using very sensitive sensors (e.g., SQUIDs) [54–60]. The strong magnetic fields used in conventional NMR enable high-intensity images, but that same high intensity becomes a drawback with regard to resolution of fine details. The overlapping of strong signals from adjacent regions results in poor resolution. The naturally occurring, very small, homogenous magnetic-field gradients (10^{-9} T/m) in Earth's field provide an opportunity for a technique with good resolution, but in a laboratory situation, achievement of a high degree of resolution using a small homogenous magnetic-field gradient adds to the complexity of MRI technology.

In this regard, another kind of technology deserves mention here. Already in use in the field of medical diagnostics, this decades-old technology is based on electrical fields rather than magnetic ones. Electroencephalography (EEG) is purely an electric-field arena technique, and provides a good means of understanding the results. Electric current propagates via neuronal synapses in the human brain. The stimuli used in this technique can be either noninvasive (e.g., audio, video) or invasive (e.g., a chemical tracer, an external electric field). As noted, EEG is a clinical diagnostic tool routinely used in the medical field at present. MEG is a completely noninvasive tool which detects the emitted magnetic signals from the brain in response to audio, visual, etc., stimuli. Complementary studies of human brain activities, through EEG and MEG, can provide valuable information about brain disorders as well as about the cognitive powers of the brain [61, 62].

The quantum model of the brain provides a powerful tool for understanding the connection between the cognitive powers of the mind and the associated chemical functioning [63, 64]. In this context, one should remember that it is the diversity and interaction of varied human minds that matters. No matter which direction the right concept comes from, it advances us toward the desired goal. That is all that matters. The breakthrough idea may be purely a mathematical one. The mathematics involved requires a good knowledge of classical and QM. Any contribution, no matter how small, can make a difference in understanding the intricacies that will lead to an ultimate technology [65–67]; we should be careful not to overlook the unnoticed contribution that could make all the difference. In the next few sections we have elaborated useful illustrations as guidance for readers. We hope that these may also act as motivation for readers to educate themselves about possible future developments in the field of NMR, which has so much potential to bring us closer to understanding the mysteries of the human brain.

2.4.1 Magnetoencephalography (Appendices A2.32–A2.35) [50]

The number of neurons engaged in active processing of information in a human brain is roughly 10^{11}. Approximately half of these are present in the cerebellum and cerebral cortex regions. This assembly of the cells is massively interconnected. By itself, the cerebral cortex, a structure essential to higher cognitive function in humans, supports approximately 10^{14} individual connections; an average of roughly 10^4 per neuron. Each neuron is an exceedingly complex structure, with the capacity to undergo plastic changes in function and also in connectivity within any given network of cells. All of human behavior emerges from the moment-to-moment interaction of the elements of this complex system. Communication between neurons occurs chemically through the transfer of neurotransmitters at neuronal synapses and, less commonly, through direct electrical contact. Currents flow within the extensive dendrite tree that supports most synaptic connections in the cortex. This dendrite mass mediates a substantial part of the information-processing capacity of the human brain. One can see an activation of localized regions of cortex following a sensory stimulation. This happens prior to the generation of movement during ongoing cognitive processes. These processes elicit current flow within the assembly of dendrites that can be detected at the surface of the scalp.

2.4.2 NMR in Weak Magnetic (Micro-Tesla) Fields

2.4.2.1 NMR in Earth's Magnetic Field (Appendices A2.36 and A2.37) [57]

The effective magnetic field on a nucleus, and consequently the observed NMR frequency, is subtly changed by the effects of orbiting electrons, the chemical shift, and the interacting spins of the neighboring nuclei (the *J*-coupling). Compared to the fields that can be attained with the superconducting magnets required for conventional NMR, the Earth's magnetic field is very weak. It varies from 25 micro-Tesla (μT) at the equator to 75 μT at the poles. Before an NMR measurement using the Earth's field can be made, the spins generally have to be polarized by a high-strength magnetic field, Appendix A2.36. The polarization can also be transferred from more readily polarizable electrons, using a method known as *dynamic nuclear polarization*. Such techniques have been used to measure the precise Larmor precession frequency of protons. They can also be used to measure the relaxation time that spins require to return to their normal state following polarization. This relaxation includes the *J*-coupling between orbital momentum (angular momentum L of the electrons) and spin (magnetic moment S) in a nucleus, and nucleus-to-nucleus couplings. The disadvantage of low-field NMR for chemical analysis is that the chemical shift (nuclei-orbital electron interaction) of the spectral lines is smaller than the magnetic field created in dipole–dipole interaction. This renders direct information about a nucleus's chemical environment, gained from the chemical shifts, effectively unobservable. There is overwhelming screening of the applied magnetic field by surrounding electrons. This is not a problem in dipole–dipole interaction. The *J*-coupling between nuclear spins that is caused by chemical bonds (not involving electrons) is almost independent of the magnetic field. Its effect is proportionally greater when the field is low. In *J*-coupling, the quantum states of shared electrons impart information about the chemical environment through indirect dipolar coupling of electrons that are relatively close to one another. The effect of the resultant *J*-coupling on the NMR frequency is at an observable level of between a few tenths and a few tens Hz. The narrow ^1H lines observed in the Earth's field allow accurate measurement of the heteronuclear *J*-coupling constants of the molecules. The *J*-coupling (*J*-C) constant is a measure of the strength of the electron-mediated, indirect nuclear spin–spin coupling.

2.4.2.2 Nuclear Magnetism in Earth's Magnetic Field (Appendices A2.38 and A2.39) [59]

The molecular motions in liquids reduce the effect of the magnetic dipolar and electric quadrupole interactions to that of broadening, not splitting. This is in contrast to solids, where splittings are observed instead. In ordinary liquids, only one kind of interaction leads to an observable decomposition of the resonance: the indirect J interaction between the spins. It decomposes the resonance line of a nucleus, when interacting with another nucleus of spin I, into $2I + 1$ components, which often are easily resolvable. In the case of interaction between two spin-½ separated by an interval of frequency $\Delta \upsilon$, one has a multiplet, for which relative frequencies and intensities are given for different relative values of J and $\Delta \upsilon$. This is when J and $\Delta \upsilon$ are smaller than υ_0, the Larmor frequency. In the case of protons, the known Js are on the order of 10 Hz, and in Earth's field, chemical shift δ is on the order of 10^{-3} or 10^{-2} Hz. The multiplet is thus not resolvable. What holds in the case for two nuclei of spin-½ essentially also holds for more complex systems, such as hydrocarbons, alcohols, etc. The situation is completely different for the indirect interaction between nuclei A and B of different species.

2.4.2.3 Scalar (J) Couplings in Micro-Tesla Magnetic Fields (Appendices A2.40 and A2.41) [55]

In high-field NMR, the information content of the spectrum is often limited by spectral resolution. The spectral resolution is determined by the width of the spectral lines, which is directly proportional to the strength of the applied field. The width of the lines thus masks fine structure in NMR. To obtain the resolution necessary to distinguish different resonance lines, a very high field homogeneity (a few parts per billion) is required. Both these stringent requirements are easily met in Earth's magnetic field. The field there is very small and very uniform. Spatial gradients can be approximately 1 nT/m, and time variation can be approximately 1 nT/s, at a particular spot on the Earth where one experiences a magnetic field of about 50 μT. In normal NMR equipment, this is achieved only at very high cost. We human beings live in Earth's magnetic field and perform day-to-day functions under its influence. It is therefore worth learning how to image the motion of the magnetic dipoles present in the human brain in its most natural environment. The cognitive activities of the brain use very small amounts of energy and produce magnetic signals in the 10^{-12}T range. These signals can be detected by the recently developed SQUID sensors. Such sensors are sensitive to magnetic flux rather than to the rate of change of the flux, in contrast to the surface coil in conventional NMR. Furthermore, the SQUIDs function independent of the frequency of the signal, thus providing a broadband system. As the SQUID magnetometer is untuned, the NMR signal strength is independent of the Larmor frequency for a given sample magnetization. Thus, the area under the NMR line is conserved. For a given magnetic-field inhomogeneity, the width of the NMR line scales linearly with the magnetic field. The peak height and the SNR are enhanced as the measurement field is reduced. For measurement, a much lower magnetic field is applied in a direction orthogonal to the polarizing field. When the polarizing field is reduced nonadiabatically to zero, the sample magnetization precesses in the measurement field.

2.4.3 Brain Function Modeling

2.4.3.1 Electrophysiological Model of the Brain (Appendices A2.42 and A2.43) [61, 62]

It is possible to understand a wide variety of brain phenomena by correlating the electrophysiological model of the brain (EMB) with electroencephalograms (EEGs) of brain functions. Both EMB and EEG are widely used in diagnostics, but these techniques are also an important source of basic knowledge. This is how the close correlations of the dynamics and cognition within the human brain are inferred. EEG alone cannot provide all the answers needed to develop an overall model of the brain; input from other studies, such as MRI of the brain, is equally essential. MRI analysis fills this gap. EEG results from cortical activity are aggregated over scales much larger than the individual neurons. Appendices A2.42 and A2.43 illustrate the results of an exemplary mathematical simulation model.

2.4.3.2 Quantum Model of the Brain

2.4.3.2.1 Mind–Brain Problem [63, 64]

The functioning of the human brain involves some basic related processes. The first is the deterministic evolution of the state of the physical processes occurring in the human body. Second is the interactions with the surroundings (the physical universe) in which we live. The process is a local dynamical process, with all the causal connections. These arise initially from the interactions between neighboring (localized) microscopic elements inside the body and secondarily from the macroelements outside the body. The local process holds only during the interval between deterministic quantum states. These quantum events involve two basic choice-making processes. The first is a choice of "yes or no" to a question posed by the mind–brain system. The second is a choice of an answer to a question posed by nature (the external environment and the internal chemical events in the body). The answer to the first question is either yes or no. This second choice, which rests with nature, is partially free (you may or may not pose a question). It is a random choice, subject to the statistical rules of quantum theory. In real life, the first choice is made by an experimenter (or a querying mind) as to which aspect of nature is going to be probed. This is the essential element of the quantum theory. The quantum rules cannot be applied until and unless some specific question is first selected. Only yes-or-no questions are permitted. All other possibilities can be reduced to these. Each answer, yes or no, injects one bit of information into the quantum universe. These bits of information are stored in the evolving, objective quantum state of the universe, which is a compendium of these bits of information. In the body, this information (through the to-and-fro metabolic motion of chemical elements) evolves in accordance with the laws of atomic physics.

The quantum state has an ontological character that is in part matter-like. It is expressed in terms of the variables of atomic physics. It evolves between events, under the control of the laws of atomic physics. Each event injects the information associated with a subjective (mind) perception, made by some observing system, into the objective state of the universe (body and its environment). The conceptualization of the natural process arises not from some preconceived speculative intuition (unless corrupt). Instead, it follows directly from an examination of the mathematical structure injected into science (the thinking process). This is done by our study of the structure of the relationship between our experiences (e.g., the thought, the apple falling to the ground—a symbol of gravity). The quantum state of the universe is rooted in the atomic properties of nature. It is an information structure that interacts with, and carries into the future, the information content of each mental event. This (mental) state has causal efficacy. Via statistical laws, it controls the properties governing the occurrence of subsequent events.

Even thinking costs energy. It is not free. You have to be alive and of sound mind to be able to think. The physical world is thus conceived of similarly, as an objectively stored compendium of locally efficacious bits of information. The instantaneous transfer of the information along the preferred outlets can now be understood to be changes, not in just human knowledge, but in an abstract of the objective information. In a physical world, a key microproperty of the human brain pertains to the migration of calcium ions through microchannels. Through these channels, the ions enter the interior of nerve terminals, reaching a site where they trigger the nerve to release the contents of a vesicle of neurotransmitter. The quantum mechanical rules mandate that each release of the contents of a vesicle of neurotransmitter generates a quantum splitting of the brain into different, classically describable, components or branches. Evolutionary considerations dictate that the brain must keep the brain–body functioning in a coordinated way. More specifically, the brain–body unit must plan and effectuate, in any normally encountered situation, a single coherent course of action that meets the needs of the individual.

But, due to quantum splitting, the quantum brain will tend to decompose into components that specify alternative possible courses of action. Thus, the purely mechanical evolution of the state of the brain will, in accordance with quantum mechanical mathematics, normally cause the brain to evolve into a growing ensemble of alternative branches. Each of these branches is essentially an entire, classically described brain, which specifies a possible plan of action. This ensemble, which constitutes the quantum brain, is mathematically similar to the ensemble that occurs in a classical model of the brain. One takes into account the uncertainties in our knowledge of the initial conditions of the particles and fields that constitute the classical representation of the brain. There is a close connection between what QM theory predicts and what classical physics and NM derive.

2.4.3.2.2 Brain Quantum Correlations: Double- and Zero-Quantum Transitions (Appendix A2.44) [66]

PMRS is an important tool for measuring in vivo, dynamic activities of metabolites. The spectra obtained thereby, however, contain many overlapping lines, with a large variation in peak sensitivity. It is very important to be able to develop techniques so that only a selected range of metabolites is measured. This required editing can be obtained by measuring signals from a certain order of MQCs. MQCs can in principle enable high-resolution spectroscopy. Magnetic-field gradients are often used to achieve this aim, although water and lipid suppression and static (B_0) and varying (B_1) field inhomogeneity are still considerable obstacles to achieving high resolution. It is possible to select signals from different compounds using differences in relaxation constants. This can be done with a spin-echo experiment, using long and short echo times. Other techniques are based on differences between scalar-coupling constants and differences in chemical shifts.

Two basic methods can be distinguished. The first is a two-scan spin-echo experiment. In the second, scan-selective decoupling is used, or the 180° pulse is made frequency selective. The second method is based on homonuclear (same kind of nuclei) polarization transfer. It is a good illustration because for lactate detection, the α-proton polarization is transferred to the CH_3 group. One can use an editing method in which one uses three frequency-selective 90° pulses to create correlated z-order magnetizations. This method yields 100% of the signal intensity, but it has two disadvantages: It does not differentiate between similar spin systems, e.g., lactate and alanine. Also, the sequence is sensitive to the frequency of the second selective pulse. In addition to the methods just mentioned, MQCs have proven to be useful in editing signals from certain spin systems. These techniques can use scalar coupling, chemical-shift differences, and/or MQ frequencies to obtain editing capabilities. The use of MQ frequencies is of special importance when two metabolites with similar and severely overlapping spectra (e.g., of lactate and aniline) must be separated. The main disadvantage of the MQC technique is that in most cases, a reduction of the SNR (intensity) occurs. One of the advantages of the MQC technique is that an additional parameter—the modulation due to MQ frequencies—can be used to distinguish different compounds by MRS. For an AX_3 (e.g., CH_3) system, the lactate and alanine MQCs can be created with the first three pulses of the sequences shown in Appendix A2.44. Because of the 180° pulse, the creation of the MQCs does not depend on the chemical shifts. This pulse also removes the effects of static field inhomogeneity. MQCs evolve during t_1 and are partly converted into antiphase magnetizations by the third 90° pulse. During τ_1 and τ_2, they evolve into detectable transverse magnetizations. If $\tau_1 = \tau_2$, the effects of the chemical shift during τ_1 are compensated for in τ_2 because of the last 180° pulse. The effects of the sequences of a weakly coupled spin system can be calculated, with operator formalism.

The X_3 (i.e., the hydrogen H_3) spins are denoted by I_1, I_2, and I_3 and have frequencies in the rotating frame of ω_1, ω_2, and ω_3. The A (i.e., the carbon C) spin is denoted by I_4 with frequency ω_4. Weak scalar coupling (coupling constant J) is assumed between A and X spins. Signal intensities are given relative to the intensity of an FID signal. Effects of relaxation are omitted. At $t_1 = 0$, the density matrix σ is $\sigma\ (t_1 = 0) = 2I_{1x}I_{4y} + 2I_{2x}I_{4y} + 2I_{3x}I_{4y} - 8I_{4x}I_{3y}I_{2y}I_{1y}$. Some studies have found the optimal value of τ_0 (scalar coupling) to be $1/4J$. Magnetizations of the last term contribute only to the signals at the CH frequency, which have a low intensity. They give the spectrum peaks close to that of water resonance. Because the signal at the CH_3 frequency can be detected more easily, the last term can be omitted. In contrast, the first three terms, which finally give signals at the CH_3 frequency, behave identically, so only the first term needs to be described.

Only coherences (σR) that contribute to the signal detected during t_2 will be analyzed. The double- and zero-quantum (ZQ) frequencies are $\omega_D = \omega_1 + \omega_4$ and $\omega_Z = \omega_1 - \omega_4$, respectively. To detect only signals corresponding to a certain order of MQCs, all other coherences (including those of water and lipids) must be eliminated. This can be done with gradient pulses or phase-cycling schemes. Phase-cycling methods are unfavorable for in vivo studies, because subtraction errors due to motion may be large. For in vivo applications, the MQC orders are most conveniently separated by B_0 gradients. A detailed mathematical analysis of the system shows that the signal measured during t_2 is phase-modulated with $\pm\omega_D t_1$. The signal intensity is only 1/4 of the total magnetization of the observed spins. This is understandable, because half of the magnetization at $t = \tau_0$ is converted into ZQCs and half into DQCs. Because of the gradient pulses, signal is detected only from the echo or the antiecho, which again halves the signal intensity. An advantage of applying the second gradient pulse after, rather than before, the 180° pulse is that the unwanted magnetizations resulting from this pulse are destroyed. A disadvantage is more risk of signal loss due to motion artifacts, because the time between G_1 and G_2 is larger. The best solution is spoiler gradient pulses before and after the last 180° pulse, and application of G_1 and G_2 immediately before and after the read pulse.

Besides editing via selection of certain MQC order, further editing is possible using the MQC frequencies to separate (nearly) overlapping signals, e.g., the CH_3 groups of alanine and lactate. This can be done with two-scan measurement, with two values for t_1, t_{11}, and t_{12}, such that $(\omega_D)_{lac}(t_{11} - t_{12}) = 2\pi k_1$ and $(\omega_D)_{ala}$ $(t_{11} - t_{12}) = (2k_2 + 1)\pi$; k_1, k_2, being integers N. The lactate signal is measured twice with the same sign, whereas the alanine signal is inverted in the second measurement. By adding or subtracting, spectra are obtained with respectively, only lactate or alanine doublets. If the difference $t_{11} - t_{12}$ is not much smaller than 1/*J*, a correction for modulation due to coupling effects is necessary. It is also possible to modulate the lactate CH_3 group through an odd number of half-cycles of the DQ frequency during the period $t_{11} - t_{12}$. In that case, the alanine CH_3 group should modulate an even number of half-cycles. Further editing can also be achieved with a two-scan measurement using an inverted gradient (G_1 or G_2) in the second scan. In this case the $I_{1x}\sin(\omega_D t_1)$ terms, in the density matrix σ_R, cancel after addition of the signals. The result is a signal that is amplitude-modulated with $\omega_D t_1$. If t_1 is chosen such that $(\omega_D)_{lac}t_1 = k_1\pi$ and $(\omega_D)_{ala}t_1 = (2k_2 + 1)\pi/2$; k_1, k_2, integers *N*; the two measurements yield a spectrum with the lactate doublet but without the alanine doublet. The latter method is sensitive to eddy currents. This can be circumvented by optimizing G_2 in both scans to compensate for the effects of the eddy currents.

It is possible to increase the intensity of the DQCs. This is done by using a frequency-selective pulse (on the *A* spin) to convert the DQCs into antiphase magnetizations, Appendix A2.44, Figure A2.42(b) [66]. With the sequence of Figure A2.42(d) [66], Appendix A2.44, during t_1, all coherences are dephased by gradient pulse G_1. Only the ZQCs and the longitudinal magnetizations are not affected by this pulse. To detect transverse magnetizations during t_2 resulting from ZQCs, the gradient pulse G_2 during τ_1 or τ_2 should be as in Figure A2.42(b) [66], Appendix A2.44. There is suppression of the longitudinal magnetizations (present during t_1). The density matrix at $t_2 = 0$ is given by $[\tau_1 = \tau_2 = (2k + 1)/4J$ with k integer $N]\sigma_R(t_2 = 0) = 1/2I_{1y}\cos^2(\pi J t_1)\cos(\omega_Z t_1)$. From this expression, it can be seen that the maximal ZQ signal is detected in a single-scan measurement—which, however, still results in 50% loss of signal intensity. Furthermore, the measured signal is amplitude-modulated. This again makes it possible to selectively detect, for example, the alanine or lactate in a single scan. If t_1 is chosen such that $(\omega_Z)_{lac} = k_1\pi$ and $(\omega_Z)_{ala} = (2k_2 + 1)\pi/2$, k_1, k_2 integers *N*, the lactate doublet signal is measured and the alanine signal is suppressed.

The signal intensity can be further improved. Signals corresponding to the ZQCs and the DQCs are measured simultaneously during t_2. This is done with the sequence in Figure A2.42(e) [66], Appendix A2.44. During $t_1 = t_a + t_b$, the first gradient pulse dephases the DQCs. A selective 180° pulse (on the *A* spin) in the middle of t_1 converts the ZQCs into DQCs and the DQCs into ZQCs. It is also possible to use a selective pulse on the *X* spins. The second gradient pulse dephases the DQCs then present. At the end of t_1, the ZQCs and DQCs at $t_1 = 0$ are dephased to the same extent. After the selective read pulse, which has the effect described earlier, the magnetizations resulting from both the ZQCs and the DQCs are rephased by the last gradient pulse. The density matrix at $t_2 = 0$ is given by $[2G_1 = 2G_2 = G_3$ and $\tau_1 = \tau_2 = (2k + 1)/4J]$: $\tau_R = -1/2I_{1y}[\cos(\omega_D t_a + \omega_Z t_b) + \cos(\omega_Z t_a + \omega_D t_b)] + 1/2I_{1y}[\sin(\omega_D t_a + \omega_Z t_b) + \sin(\omega_Z t_a + \omega_D t_b)]$. The signal measured during t_2 is neither amplitude- nor purely phase-modulated. If $t_a = t_b$ and $\omega_4 = 0$, the ZQ and DQ frequencies are equal and the signal is phase-modulated with $2\omega_1 t_a$. Note that the signal intensity is maximal (100%), which is very important for in vivo studies. Because of the selective 180° pulse, the coupling during t_1 does not modulate the signal.

As described, gradient pulses are used to select the DQCs and the ZQCs, and a frequency-selective read pulse is used to convert the MQCs into antiphase magnetizations. For the DQCs, water and lipid suppression, which is essential for in vivo studies, is relatively simple if the gradient pulses are used to select the DQCs. In that case, the gradient G_2, Appendix A2.44, Figure A2.42(a) and (b), dephases the unwanted magnetizations from uncoupled protons. One can understand this easily if one realizes that the coupled protons of the lipids are not affected by the selective pulse. This means that the MQCs are not converted into transverse magnetizations. To suppress the water signal, one may use a frequency-selective adiabatic pulse to invert the water resonances. The sequence starts at a time t_0 after this pulse; t_0 is chosen to minimize the remaining water signal. This method can also be used as an additional water suppression technique for the DQ pulse sequences. The DQ sequence (FigureA2.42(b) [66], Appendix A2.44) and the combined Z-plus-DQ sequences (Figure A2.42(e) [66], Appendix A2.44) are, during the MQ evolution time t_1, sensitive to B_0 inhomogeneity. It is possible to correct for this by shifting the last 180° pulse in time.

2.4.4 Three-Dimensional Imaging

2.4.4.1 The k-Trajectory Image Formulation (Appendices A2.45–A2.49) [65]

In NMR, one causes the object being imaged, after it is irradiated with an RF field over a certain volume, to emit RF signals. These signals contain information about the nuclear energy transitions and their interaction with the environment. The information concerns the spatial distribution of the nuclear magnetization within the object. More precisely, it concerns the magnitude of the transverse component of the magnetization, which in general reflects spin density and the relaxation times T_1 (longitudinal magnetization)) and T_2 (transverse magnetization) in combination. The transverse component of the magnetization terms (measured in terms of T_2) is the quantity whose distribution is represented in the RF signal. Each imaging method must also have a strategy for a decoding procedure, which extracts the desired information from the noisy signal. The *k*-trajectory (*k*-T) methodology of imaging is a general approach, and is applicable to NMR as well. The essence of the *k*-T formulation is that the time-varying magnetic-field gradients map the spatial-frequency-domain content of the object directly into the FID signal. Mapping may be visualized in terms of a trajectory in the spatial-frequency space, referred to as the *k*-T domain. The *k*-T formulation applies only to methods in which an FID (with or without echoes) is produced. The signal suggests that one can see the continuous-time FID, $s(t)$, as a weighted observation of the spatial distribution $\mathbf{F(k)}$, corresponding to the original object. The encoding process then consists of sampling along a set of *k*-trajectories determined by the time-varying field gradients. Subsequently, decoding consists of processing the samples (usually by means of the FT) to obtain a discrete image, which is an estimate of the original spatial distribution.

2.4.4.2 Various k-Trajectories in Imaging (Appendices A2.55–A2.56) [67]

The *k*-trajectories are designed to shorten the scan time by reducing the number of excitations. They differ from traditional 3D trajectories in that they employ time-varying gradients. The sampling does not occur in a single straight line. Curved and echo planar trajectories in general allow longer readouts and in turn reduce the scanning time. These trajectories provide flexibility for trading off SNR against scan time. If SNR is important, it can always be recovered by averaging. To achieve efficient acquisitions, the trajectories must sample *k*-space in the most uniform way possible. The design is restricted by hardware limitations. The maximum gradient amplitude (the speed along the trajectory) and the maximum slew rate of the gradients (vectorial acceleration of the trajectory) are required. The reconstruction procedure does not impose any further restriction. The data can be sampled in any fashion provided sufficient *k*-space is used: for example, it is not possible to place a sphere of diameter equal to 1/(FOV) in *k*-space without including at least one sample. To obtain isotropic resolution or, equivalently, a spherically symmetrical impulse, a sphere in *k*-space must be sampled. In 3D, the time saved by sampling a sphere instead of a cube is approximately 48%, compared with 22% for sampling a circle instead of a square in 2D.

An auxiliary goal is for the trajectories to achieve a certain level of insensitivity to flow and motion artifacts. This is especially important for cardiovascular applications. The motion/flow properties of the trajectories are compared, on the basis of these two criteria, in the *k*-space formalism. First, to have good flow properties, the gradient's first moment should be small near the origin of *k*-space. Second, the gradient should vary smoothly as a function of *k*-space position. Some prevalent *k*-Ts are compared, as to preparation time, scan time, etc., in (Appendix A22.50). Typical background technical parameters of the trajectories are as follows: The FOV is 20 cm × 20 cm × 20 cm; resolution is 100 × 100 × 100 pixels (2 mm × 2 mm × 2 mm); readout duration is 14,336 ms; repetition time (TR) = 50 ms; maximum field gradient is $dG/dx|_{max} = 1$ G/cm; and maximum time of variation of the field gradient is dG/dt [$dG/dx|_{max} = 2$ G/cm/ms]. Other characteristic properties of the trajectory are given in Table 2.2.

The idea behind gridding is to convolve the sampled data with a kernel such that it is possible to obtain the value of the smoothed data at the grid points. The smoothing operation in the frequency domain can be undone with a division in the object domain, after the FFT operation. One can employ the spherical stack of spirals and the density-compensated cones trajectories experiments. The spherical stack of spirals uses the shortest imaging time. It can also be modified for anisotropic fields of view.

The cones trajectory is the fastest of the trajectories that do not require preparation time and possesses good flow properties in all directions. The sequences are implemented with static gradient $|G| \leq 1$ G/cm and time-varying gradient $|dG/dt| = \leq 2$ G/cm/ms. Three orthogonal slices taken from a 3D dataset of the head, using the spherical stack of spirals as *k*-T, are shown in Appendices A2.53–A2.55.

Table 2.2 Characteristic Properties of the Trajectory

Number	Technical Nature	Advantages/Disadvantages
1	*A conventional 3DFT*: 3D discrete FT: After each excitation, a constant gradient is applied, such that a straight line is acquired in one *k*-space dimension. It serves as a reference trajectory (Appendix A2.51 (Figure A2.47 [67])).	The advantages of this trajectory are its simplicity, ease of reconstruction, and short readouts. The main disadvantage is its long scan time. The 3DFT trajectory is sensitive to flow and motion of artifacts. The magnitude of the gradient's first moment is significant near the origin of the *k*-space. Flow compensation reduces the sensitivity at the expense of increasing the echo time. The palatial motion produces artifacts due to phase discontinuities from readout to readout.
2	*Cylindrical 3DFT*: The simplest modification to the 3DFT trajectory is to make the resolution more isotropic by restricting the *k*-space. The space it covers converges to a cylinder.	This trajectory has the same advantages as the regular 3DFT. The number of required excitations is reduced. It still has the same minimum TE (excitation time) requirement and sensitivity to flow artifacts (Appendix A2.50 (Figure A2.48 [67])).
3	*3D echo planar*: If more than one line in *k*-space is acquired in each readout, the number of excitations is reduced. This can be done by extending the idea of echo planar to three dimensions.	It has the advantage of being easy to reconstruct and the disadvantage of not having good flow properties, due to the phase-encoding steps (Appendix A2.50 (Figure A2.49 [67])).
4	*Cylindrical 3D echo planar*: As with the 3DFT trajectory, the 3D echo planar trajectory can also be modified, such that only a cylinder in *k*-space is required.	This trajectory has the same advantages as the regular 3D echo planar, but the number of required excitations is reduced (Appendix A2.50 (Figure A2.50 [67])).
5	*Stack of spirals*: To reduce the number of excitations, a time-varying gradient trajectory based on spirals is considered. The stack of spirals is formed by placing each planar spiral acquisition on top of another in *k*-space (Appendix A2.51 (Figure A2.51 [67])).	The main advantage of this trajectory is that it reduces the number of excitations. Although the flow properties are relatively good in-plane, a phase encoding in the third dimension still produces ghosting artifacts under oscillatory motion. Because the sampling is not on a Cartesian grid, the construction is more computationally intensive. A 2D gridding algorithm is employed for each plane in *k*-space.

6	*Spherical stack of spirals*: By reducing the circle diameter, and the number of interleaves of planes farther away from the origin, the stack of spirals is modified such that only a sphere is acquired.	It has the same advantages and disadvantages as the stack of spirals, but reduces the number of excitations needed (Appendix A2.51 (Figure A2.52)).
7	*Shells*: Instead of acquiring planes in *k*-space, it is also possible to acquire spherical surfaces. After a phase-encoding step that moves the *k*-space location from the origin to the plane of one of the concentric spheres, the trajectory spirals down on the surface of the sphere (Appendix A2.51 (Figure A2.53 [67])).	As with the spherical stack of spirals, this trajectory needs a phase-encoding step that requires a preparation time of 1.42 ms and increases the sensitivity to pulsatile flow artifacts in one direction. In the other directions it possesses good flow properties. The advantage of this trajectory is the reduced number of excitations required.
8	*3D projection reconstruction (PR)*: This is a conventional trajectory for 3D *k*-space that does not have a minimum TE requirement. The acquisition starts at the origin of the *k*-space and moves outward along a straight line (Appendix A2.52 (Figure A2.54 [67])).	This trajectory is simple to implement and readouts may be very short. It does not have a minimum TE requirement (the origin of the *k*-space is acquired immediately after the excitation), and its flow properties are excellent. The big disadvantage of this trajectory is long total scan duration, because it requires a large number of excitations.
9	*Spiral-PR hybrid*: A trajectory that also avoids the minimum TE, as well as the phase-encoding step. Planar spiral trajectories are rotated around a common axis in their planes (Appendix A2.52 (Figure A2.55 [67])).	The trajectory possesses all the advantages of planar and spiral: good flow properties; allows tradeoff of scan time for SNR. In addition, it does not have a minimum TE requirement, nor does it require phase-encoding steps. As compared with the spherical stack of spirals, the price paid for these better properties is an increase in the number of excitations.

(Continued)

Table 2.2 (continued)

Number	Technical Nature	Advantages/Disadvantages
10	*Yarn*: The yarn trajectory is inspired by the way yarn is rolled into a ball. The natural tendency of rolling the yarn with the least possible curvature suits the hardware constant of slew rate. It is defined as follows: Concentric circles are drawn on a plane. The radii differences are given by the desired FOV. The circles are divided into an integer multiple of the number of interleaves. Starting from the center, semicircles perpendicular to the plane are drawn connecting opposite (as opposite as possible) points in the circles.	Because this trajectory can be made with arbitrary readout lengths, it allows a tradeoff between SNR and scan time. The minimum TE is zero and it has good flow properties. The yarn trajectory is not simple to implement and care is required to make the sampling of *k*-space as uniform as possible (Appendices A2.54 and A2.55 (Figure A2.56 [67])).
11	*Cones*: Another alternative is to use the cones trajectory. The surface of each concentric cone is traversed with a spiral-like trajectory. A variant of this trajectory, called density-compensated cones, reduces oversampling by varying the pitch of the spiral, as it gets farther away from the origin, to compensate for the nonuniform separation between cones. It has all the advantages of the cones trajectory.	It has good flow properties and the minimum TE is zero. As with all nonplanar and nonuniform trajectories, it requires a 3D gridding reconstruction algorithm. There is inefficiency in sampling *k*-space because there is oversampling near the origin (Appendices A2.54 and A2.55 (Figure A2.57 [67])).

To use the FFT algorithm to reconstruct the data, it is necessary to map the nonuniformly sampled points to a Cartesian grid. This is done with a gridding algorithm.

2.4.5 The Outlook

2.4.5.1 High-Definition MRI (Appendices A2.57 and A2.58) [68]

The EPI method relies on the FT of the signals to extract the desired correlation (chemical shift) between intervening Bohr (bound electrons to nuclei) precession frequencies. For conventional 2D NMR spectroscopy, the frequencies to be correlated occupy separate portions of the experiment. In 2D EPI, one alternates between the relative contributions of the frequencies to define the spin's evolution. This is intended to deliver a complete 2D, time-domain interferogram following a single excitation of the spins. This *k*-space walk principle can be carried out by switching the intervening magnetic-field gradients. This can also be done by modulating them concurrently, leading to a variety of related, fast-acquisition modes. In a different approach, one can also use spatial acquisitions in conjunction with the typical advantages of EPI. These alternatives open up vast new horizons in multidimensional imaging. This kind of imaging is done by applying an external, magnetic-field gradient, which uses spin located at different positions, with individually addressable frequencies. It can be applied in conjunction with a frequency-incremented excitation or inversion of the spins. The gradients lead to the possibility of spatial encoding of the NMR interactions to be measured. They allow one to encode an interaction Ω_1 with a phase $\Phi(t)$ as approximately $\Omega_1(r - r_0)$, rather than with the usual temporal encoding, as $\Phi(t) \sim \Omega_1$. Here r and r_0 represent some arbitrary position coordinates, as an illustration.

2.4.5.2 Imaging in Low Fields

2.4.5.2.1 Dynamic Nuclear Polarization: Overhauser-Enhanced Magnetic Resonance Imaging (Appendix A2.59) [69]

Overhauser-enhanced magnetic resonance imaging (OEMRI) has been shown to be a viable alternative to conventional NMRI. It offers possible reduction in the complexity and complications of the hardware and the software involved. The advantage of OEMRI is that both physiologic and anatomic information can be extracted simultaneously. This technique is based on the Overhauser effect. In OEMRI, the electron-spin resonance (ESR) of a free radical is irradiated during acquisition of an MR image. It is followed by transfer of polarization from unpaired (free from bonding) electron spins to coupled proton spins. This results in enhancement of the NMR signal in the regions of the sample containing free radicals. The result reveals the spatial distribution of the signal in the final image. It leads to the possibility of NMRI with very low magnetic fields. OEMRI offers better sensitivity than conventional MRI. Moreover, its spatial resolution is not limited by the line width of the free radical, in contrast to ESR imaging. Both the temporal and spatial distributions of spin probes in vivo can be simultaneously monitored with reasonably good resolution. Nitroxyl radicals have been widely used as spin probes for low-frequency in vivo ESR imaging. These act as contrast agents for OEMRI.

2.4.5.2.2 NMR Spectroscopy in Earth's Magnetic Field (Appendices A2.60 and A2.61) [70]

NMR spectroscopy allows a multidimensional (space and time) representation of various internuclear and intranuclear interactions. These may be between different atomic sites in molecules; these molecules may be dispersed over a relatively large area. A human brain is a good example. Earth's field NMR (EFNMR) can provide high resolution in space because the magnetic field ($\sim 60\,\mu T$) is very uniform. This holds for both spatial and time gradients over the surface of the Earth. EFNMR is of special interest regarding the function (and imaging) of the human brain. Earth's magnetic field is portable and requires no magnet. Furthermore, all of nature's organisms (including humans) are immersed in Earth's field and subject to its interference. Using EFNMR, it is possible to measure heteronuclear *J*-couplings (spin–spin couplings) that are field independent. Current NMR technology allows spin ensembles to evolve under the influence of local interactions. This is achieved by application of the trains of RF pulses to the surface of the brain.

2.4.5.3 QM of the NMRI

2.4.5.3.1 Quantum Magnetic Resonance Imaging (Appendix A2.62) [71]

To fully comprehend the field of quantum magnetic resonance imaging (QMRI), one needs to read a basic text on QM; several such textbooks are available. This and the next subsection may be able to create reader awareness of the magnitude of QM knowledge required. The example chosen for discussion is the simplest one possible: the case of a 2-spin, I and S ($=1/2$), system. It only underscores the requirement of learning advanced mathematical skills needed to be able to produce a machine such as the MRI apparatus.

To describe the 2-spin system, I and S (each $= 1/2$), one first creates a suitable mathematical basis. The chosen basis is the two functions, as the product of the two single-spin-system wave functions m_i and m_s. These are $|\varphi_i\rangle = |m_i, m_s\rangle$. On expansion, these are $\varphi_1 = \alpha\alpha(1/2, 1/2)$, $\varphi_2 = \alpha\beta(1/2, -1/2)$, $\varphi_3 = \beta\alpha(-1/2, 1/2)$, and $\varphi_4 = \beta\beta(-1/2, -1/2)$. These are the four possible spin combination wave functions. An algebraic approach to the problem, using operator representation of the wave functions, is more efficient than finding the solution of the wave functions using the famous Schrödinger's wave equation. The components of the operators in the Cartesian coordinate frame of reference, for a two-spin system, can be formed from the 1-spin operators by taking the products of the set $\hat{E}, \hat{I}_x, \hat{I}_y, \hat{I}_z$ with the set $\hat{E}, \hat{S}_x, \hat{S}_y, \hat{S}_z$. Here $\hat{E}$ is an identity operator, which has unity (1) as the four diagonal elements, and all others being zero. By its nature, when multiplied with a matrix, it does not change the other matrix. This leads to 16 operators, as given in Appendix A2.62. Matrix representation of the operators provides a suitable approach, keeping in mind the laws of QM.

2.4.5.3.2 Two-Dimensional Quantum Coherences (Appendices A2.63 and A2.64) [72]

The discrimination between different types of spin systems, using ZQCs and MQCs, can be understood by considering an energy-level diagram, as in Appendix A2.63. Noncoupled (isolated single) spin-½ nuclei in a magnetic field undergo SQ ($\Delta M = \pm 1$) transitions (Figure A2.63(a)) exclusively, as there are only two available spin states. Coupled-spin systems, however, can participate in ZQ ($\Delta M = 0$), SQ, and higher-order ($\Delta M = <-1,\ = >1$) transitions. The energy-level diagram in Appendix A2.63, Figure A2.63(b) [72], for a weakly coupled *AX* system, shows 4-spin states. It allows six possible transitions. ZQCs and MQCs (occurring at certain frequencies) can be employed to discriminate against singlets (isolated relatively) arising from solvent peaks, e.g., water. Furthermore, differences in ZQCs and MQCs arising among various types of coupled-spin systems can provide additional spectral capability. For example, the DQ frequency is the sum of the chemical-shift offset, with respect to the transmitter position, for each resonance in a coupled-spin system. Therefore, the DQ frequency will usually be unique for a particular group of coupled spins. It can be used to distinguish between different coupled-spin systems. In addition, the relative DQ frequencies among different coupled-spin systems can be favorably manipulated by judicious choice of a transmitter position.

2.4.6 Medical Diagnostics: MRSI (Appendices A2.65 and A2.66) [73]

Spatial localization techniques for proton brain spectroscopy fall into two general categories: either single-voxel techniques (i.e., where a spectrum is recorded from a single brain region), or multi-voxel techniques (where multiple regions are acquired simultaneously). The latter is called MRSI; it is also known as CSI. The two most commonly used single-voxel methods are stimulated echo acquisition mode (STEAM) and PRESS. In spiral MRSI, a gradient waveform in two dimensions is applied. It traverses a spiral trajectory in two directions of *k*-space, again concurrent with the evolution of the readout time. The *k*-T space can be mapped out, using multiple shots, until all data points are collected.

Another approach, in which one applies reduced phase-encoding strategies, is used to reduce scan time in 2D or 3D MRSI. This is similar to approaches that have been developed for parallel MRI. The basic principle is called sensitivity-encoding MRSI (SENSE MRSI). The idea is to use the inhomogeneous B_1 fields of multiple phased-array coils to encode some of the spatial information, thereby allowing fewer conventional phase-encoding steps to be used and hence reducing scan time. PMRS of the brain gives useful information on a number of different neurological and psychiatric disorders.

2.4.7 Quantum Horizons

2.4.7.1 MQTs: Spin Counting (Appendices A2.67 and A2.68) [74]

In the LS, *J*-coupling networks can be exploited to yield knowledge regarding molecular connectivity. In contrast, in the solid state, in which there is much wider uniformity in space, spatial proximity can be studied using dipole spin (*J*—*J*) interactions. A magnetization state of MQC with quantum order $\pm p$, in a system of nuclei, with spin *I* is indicative of the presence of at least $p/2I$ spins in a coupled network. Thus, by establishing the maximum possible MQC order that can be achieved, a count of the number of spins in a coupled network can be made. The excitation of a magnetization state of MQCs, with order $\pm p$, does not exclude the presence of more than $p/2I$ spins. Therefore, such a count can only be considered to place a minimum bound on the number of spins, unless it can be established that higher coherence orders cannot be excited and are not absent due to experimental restrictions. These generic sequences are shown in Appendix A2.67. They differ in the way in which 90° RF pulses are combined. This is in reference to the periods during which ZQ, SQ, or DQ Hamiltonians are applied. Other sequences are also possible, such as those that commence with cross-polarization rather than a simple 90° pulse. The classification of Hamiltonian as ZQ, SQ, or DQ indicates the change in coherence order; i.e., they are interconverted between the states of magnetization. Bonded silica phases have been investigated using spin counting; these consisted of silane groups (SiR3, *R* is an alkyl group) bound through a Si—O linkage to a silica surface, Appendix A2.68.

2.4.7.2 MQTs: Selection Rules (Appendix A2.69) [75]

MQ spectroscopy with ^{13}C may prove to be a useful probe of the conformations and oligomerization states of ^{13}C-labeled biopolymers such as peptides and proteins. An important aspect of MQ spectroscopy is the use of pulse sequences to create effective dipole—dipole coupling Hamiltonians that are time-reversible, i.e., they can be reversed in sign by some experimental manipulations, such as a shift of the phase of the RF pulses. The property of time-irreversibility is logically distinct from the property of time-reversal invariance. The latter property is shared by all effective Hamiltonians that are bilinear or even-power functions of spin-angular momentum operations, whereas the former property is obtained only in special cases. Suppose the effective coupling Hamiltonian (ECH), in the preparation period of an MQ experiment, is opposite in sign to the ECH in the mixing period of the experiment. All the MQ transitions, which would otherwise drastically reduce the MQ signal amplitude, are thereby eliminated.

2.4.8 Summary and General Outlook

In one way or the other, MRI of the human brain involves using quantum data obtained from a scan of a human brain in response to the static and dynamic (RF) magnetic fields applied to the surface of the brain. The FID of the nuclear spins, and their interactions with the orbital electrons (chemical shift), other nuclear (heteronuclear) spins (J-couplings), metabolites, etc., are converted into a time-spatial image map of the brain. This FID time-space image basically gathers information of a statistical average (time-space) of the interactions. To improve the intensity of the picture, which is basically the SNR, the data are collected over several periods. An optimum has to be struck, so as not to increase scan time beyond the limits of patient comfort, normally not more than about 10 min. Resolution of the image is determined by how finely the RF probe can excite the interactions in space and time, and how closely in space and time the sensors can pick up the signal and discriminate the various interactions.

For an MP, it would be ideal to know the details about a particular spot, as compared with the adjacent area, to better diagnose a particular disease. MPs would be even happier if they could also see a movie (millisecond to second in duration) of the dynamics of the spot. This would add great power to the diagnostic weaponry. However, this places more and more demands on the technology (commerce) and ultimately on the science (the curiosity) that permit such improvements and innovations. There are various problems of basic science yet to be solved to make NMR a cheaper and simpler technology. As an example, inhomogeneities of the magnetic field used add to the noise in the picture. At the same time, precision in the inhomogeneity, i.e., the magnetic-field gradients that can be achieved, allows various quantum reactions going on in the brain to be accurately imaged. It is analogous to creation of antivenins: one needs a snake's venom in order to develop an antidote for it. Even to understand the quantum reactions in NMR and MRI, one needs a good knowledge of QM, let alone to improve the quality of imaging and thus the diagnostic use of such imaging technologies.

QM is no more difficult to understand than Newtonian mechanics; the difference has arisen from the fact that NM is introduced at the secondary level of education (and has been taught at that level for more than a century), whereas QM has remained at the tertiary level of education. It does not take an Einstein to create an Einstein. A wider knowledge and circulation of science, introduced at a much earlier level of education, may bring out more geniuses to solve more complex problems. The technologies that society uses—and takes for granted—today are the product of the science education pursued globally during the past century. In the context of MRI of the human brain, the discussion of the QM involved should have been stimulating, at least for some.

Scientifically, the human brain can be treated as a nuclear reactor, with a mixture of atoms, molecules, nuclei, and electrons in a dynamic state of operation, round the clock. In the language of QM, it is a *dirty* mixture (different species of nuclei, atoms, etc., performing different and complex tasks). In QM, the solution of a problem involving a pure mixture is easy and straightforward [76]. An impure mixture, like the one we have in our brains, is a problem of immense mathematical complexity. Nonetheless, the future is not entirely bleak. Rewards or no rewards, there are curious minds that are eager to learn about and solve this difficult problem of the "dirty" brain. By gathering and summarizing information from the experimental and theoretical studies conducted to date, this chapter has tried to bring to light the difficulties involved, as well as the solutions being tested and the approaches being followed. May it motivate more people to enter the world of QM, enjoy its secretive nature, and eventually produce the desired methods of diagnosing the disorders prevalent in the human brain.

References

[1] C. Koch, K. Hepp, Nature 440 (2006) 611–612.
[2] R. Tycko, Nature 434 (2005) 966–967.
[3] R.M. Hochstrasser, Nature 434 (2005) 570–571.
[4] T. Brixner, S. Jens, H.M. Vaswani, M. Cho, R.E. Blankenship, G.R. Fleming, Nature 434 (2005) 625–628.
[5] S. Stapf, Nature 437 (2005) 488–489.
[6] R. Eisberg, R. Resnick, Quantum Physics of Atoms, Molecules, Nuclei, and Particles, John Wiley & Sons, New York, USA, 1974, ISBN 0-471-23464-8.
[7] H.P. Myers, Introductory Solid-State Physics, Taylor and Francis Ltd., London, UK, 1990.
[8] F. Bloch, Phys. Rev. 70 (1946) 460–474.
[9] F. Bloch, W.W. Hansen, M. Packard, Phys. Rev. 70 (1946) 474–485.
[10] R.K. Wangsness, F. Bloch, Phys. Rev. 89 (1953) 728–739.
[11] F. Bloch, Phys. Rev. 83 (1956) 104–135.
[12] A. Tannus, M. Garwood, NMR Biomed. 10 (1997) 423–434.
[13] N.K. Logothetis, J. Paulis, M. Augath, T. Trinath, A. Oeltermann, Nature 412 (2001) 150–157.
[14] T. Zhao, K. Heberlein, C. Jonas, D.P. Jones, X. Hu, Magn. Reson. Med. 55 (2006) 676–680.
[15] L. Chang, C.C. Cloak, T.J. Ernst, Clin. Psychiatr. (64) (2003) 7–14.
[16] J. Robert, M.R.C.P. Simister, M.A. McLean, G.J. Barker, J.S. Duncan, Neurology (61) (2003) 897–902.
[17] J.R. Keltner, L.L. Wald, B.deB. Frederick, P.F. Renshaw, Magn. Reson. Med. 37 (1997) 366–371.
[18] D.L. Rothman, O.A.C. Petroff, K.L. Behar, R.H. Mattson, Proc. Natl. Acad. Sci. USA (90) (1993) 5662–5666.
[19] J. Eric Jensen, B.deB. Frederick, L. Wang, J. Brown, P.F. Rnshaw, Magn. Reson. Med. 54 (2005) 783–788.
[20] J. Eric Jensen, D.J. Drost, Ravi Menon, P.C. Williamson, NMR Biomed. 15 (2002) 338–347.
[21] O.A.C. Petroff, D.D. Spencer, J.R. Alger, J.W. Prichard, Neurology (39) (1989) 1197–1209.
[22] Chen Wand, J.J.H Ackerman, NMR Biomed. 4 (1989) 205–207.
[23] A. Bax, P.G. De Jong, A.F. Mehlkopf, J. Smidt, Chem. Phys. Lett. 69 (1980) 567–570.
[24] P. Mansfield, J. Phys. C Solid State 10 (1977) L55–L58.
[25] E. Adalsteinssen, D.M. Spielman, Magn. Reson. Med. 41 (1999) 8–12.
[26] J. Shen, D.C. Shungu, D.L. Rothman, Magn. Reson. Med. 41 (1999) 35–42.
[27] P. Mansfield, Magn. Reson. Med. 1 (1984) 370–386.
[28] D.N. Gulfoyle, A. Blamire, B. Chapman, R.J. Ordidge, P. Mansfield, Magn. Reson. Med. 10 (1989) 282–287.

[29] A. Macovski, Magn. Reson. Med. 2 (1985) 29–40.
[30] D.B. Twieg, Magn. Reson. Med. 12 (1989) 64–73.
[31] A. Wokaun, R.R. Ernst, Chem. Phys. Lett. 52 (1977) 407–412.
[32] A.A. Maudsley, A. Wokaun, R.R. Ernst, Chem. Phys. Lett. 55 (1978) 9–14.
[33] S. Matsui, K. Sekihara, H. Kohno, J. Magn. Reson. 67 (1986) 476–490.
[34] A.A. Maudsley, R.R. Ernst, Chem. Phys. Lett. 50 (1977) 368–372.
[35] E.L. Hahn, Phys. Rev. 15 (1950) 580–594.
[36] E.L. Hahn, D.E. Maxwell, Phys. Rev. 88 (1952) 1070–1084.
[37] M. Emshwiller, E.L. Hahn, D. Kaplan, Phys. Rev. 118 (1960) 414–424.
[38] W.A. Anderson, R. Freeman, C.A. Reilly, J. Chem. Phys. 39 (1963) 1518–1531.
[39] P. Mansfield, J. Phys. D Appl. Phys. 16 (1983) L235–L238.
[40] W.P. Aue, E. Bartholdi, R.R. Ernst, J. Chem. Phys. 64 (1976) 2229–2246.
[41] S. Vega, T.W. Shattuck, A. Pines, Phys. Rev. Lett. 37 (1976) 43–46.
[42] R.K. Hester, J.L. Ackerman, V.R. Cross, J.S. Waugh, Phys. Rev. Lett. 34 (1975) 993–995.
[43] L. Muller, A. Kumar, T. Baumann, R.R. Ernst, Phys. Rev. Lett. 32 (1974) 1402–1406.
[44] R. Freeman, W.A. Anderson, J. Chem. Phys. 37 (1962) 2053–2073.
[45] S. Yatsiv, Phys. Rev. Lett. 113 (1959) 1522–1537.
[46] A. Pines, D.J. Ruben, S. Vega, M. Mehring, Phys. Rev. Lett. 36 (1976) 110–134.
[47] I.I. Rabi, N.F. Ramsey, J. Schwinger, Rev. Mod. Phys. 26 (1954) 167–171.
[48] S.R. Hartmann, E.L. Hahn, Phys. Rev 128 (1962) 2042–2053.
[49] W. de Boer, Phys. Rev. B 12 (1975) 828–838.
[50] M. Hamalainen, R. Hari, R.J. llmoniemi, J. Knuutila, O.V. Lounasmaa, Rev. Mod. Phys. 65 (1993) 413–497.
[51] R. Salmelin, R. Hari, O.V. Lounasmaa, M. Sams, Nature 368 (1994) 463–465.
[52] O.V. Lounasmaa, M. Hamalainen, R. Hari, R. Salmelin, Proc. Natl. Acad. Sci. USA (93) (1996) 8809–8815.
[53] C.D. Tesche, J. Low Temp. Phys. 135 (2004) 773–791.
[54] O.A. Shushakov, Magn. Reson. Imag. 14 (1996) 959–960.
[55] R. McDermott, A.H. Trabesinger, M. Muck, E.L. Hahn, A. Pines, J. Clarke, Science 295 (2002) 2247–2249.
[56] J. Stepisnik, Nature 439 (2006) 799–801.
[57] S. Appelt, H. Kuhn, F. Wolfgang Hasing, B. Blumich, Nature Phys. 2 (2006) 105–109.
[58] R. Mcdermott, N. Kelso, S.-K. Lee, M. Moßle, M. Muck, et al., J. Low Temp. Phys. (135) (2004) 793–821.
[59] G.J. Bene, Phys. Rep. 58 (1980) 213–266.
[60] J. Stepisnik, V. Erzen, M. Kos, Magn. Reson. Med. 15 (1990) 386–391.
[61] A. Robinson, C.J. Rennnie, D.L. Rowe, Phys. Rev. E 65 (2002) 0 41924-1-9
[62] M Breakspear, J.A. Roberts, J.R. Terry, S. Rodrigues, N. Mahant, P.A. Robinson, Cerebral Cortex 16 (2006) 1296–1313.
[63] H.P. Stapp, Found. Phys. 31 (2001) 1465–1499.
[64] R.A. Mould, Found. Phys. 33 (2003) 591–612.
[65] D.B. Tweig, Mod. Phys. 10 (1983) 610–621.
[66] J.E. Van Dijk, A.F. Mehlkopf, W.M.M.J. Bovee, NMR Biomed. 5 (1992) 75–86.
[67] P. Irarrazabal, D.G. Nishimura, Magn. Reson. Med. 33 (1995) 656–662.
[68] A. Tal, L. Frydman, J. Magn. Reson. 182 (2006) 179–194.
[69] A. Milton, F. Benial, K. Ichikawa, R. Murugesan, K.-i. Yamada, H. Utsumi, J. Magn. Reson. 182 (2006) 273–282.

[70] J.N. Robinson, A. Coy, R. Dykstra, C.D. Eccles, M.W. Hunter, P.T. Callaghan, J. Magn. Reson. 182 (2006) 343–347.
[71] F.J.M. Van De Ven, C.W. Hilbers, J. Magn. Reson. 54 (1983) 512–520.
[72] C.H. Sotak, D.M. Freeman, R.E. Hurd, J. Magn. Reson. 78 (1988) 355–361.
[73] P.B. Barker, D.D.M. Lin, Prog. NMR Spectr. 40 (2006) 99–128.
[74] C.E. Hughes, Prog. NMR Spectr. 45 (2004) 301–313.
[75] R. Tycko, J. Magn. Reson. 130 (1999) 302–307.
[76] A. Goswami, Wm. C. Quantum Mechanics, Brown Publishers, Dubuque, IA, USA, 1992, ISBN 0-697-11811-8.

Appendix A2.1

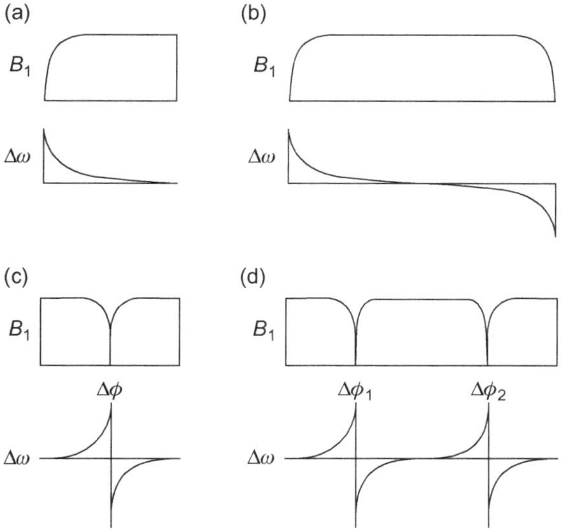

Figure A2.1 Examples of four adiabatic pulses. Adiabatic passages: (a) adiabatic half passage (AHP) and (b) adiabatic full passage (AFP). Class of composite pulses: B_1-insensitive rotations (c) BIR-1 and (d) BIR-4.

Appendix A2.2

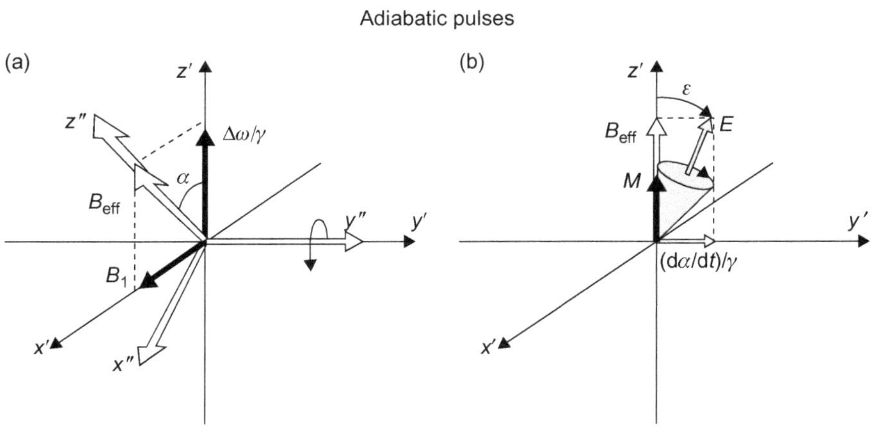

Figure A2.2 Vector diagrams showing the effective field and its components in two rotating frames of reference. (a) Relationship between the frequency-modulated (FM) frame, x', y', z' (thin axes), and the B_{eff} frame, x'', y'', z'' (thick axes). (b) Magnetic-field components and evolution of the magnetization vector (M) in the B_{eff} frame.

Appendix A2.3

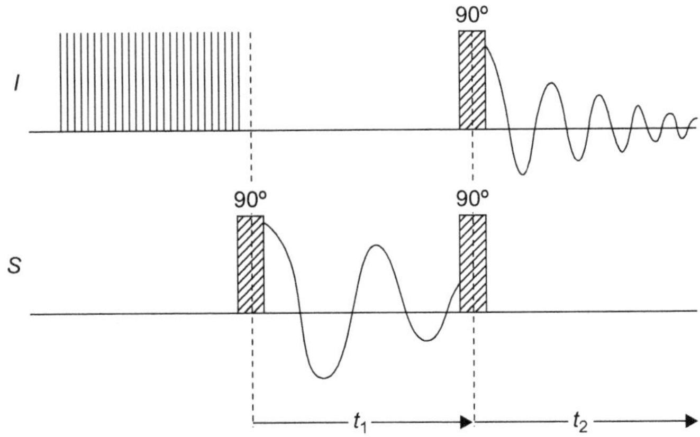

Figure A2.3 Basic pulse sequence for 2D heteronuclear spectroscopy.

Appendix A2.4

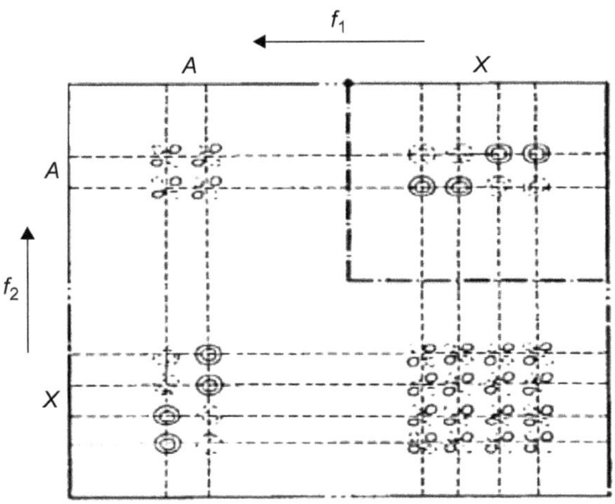

Figure A2.4 Schematic diagram illustrating the relationship to 2D auto-correlated spectroscopy for an A_3X spin system. Only the off-diagonal block framed by dashed lines is observed in a 2D heteronuclear experiment.

Appendix A2.5

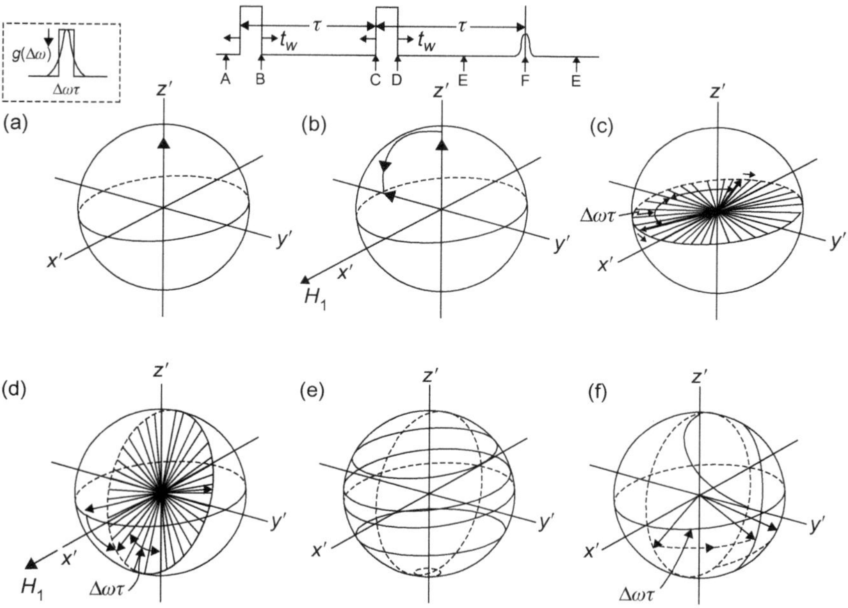

Figure A2.5 The pulse condition $\omega_1 t_\omega = \pi/2$. One notices formation of the eight-ball echo pattern in the coordinate system rotating at angular frequency ω. The moment vector monochromats are allowed to travel completely in a time $\tau \gg 1/\Delta\omega_{1/2}$. This is before the second pulse is applied. The echo gives the maximum available amplitude at $\omega_1 t_w = 3\pi/2$. $\omega_1 \gg \Delta\omega_{1/2}, t_w \ll \tau \ll T_1, T_2, \omega_1 t_w = \pi/2$.

Appendix A2.6

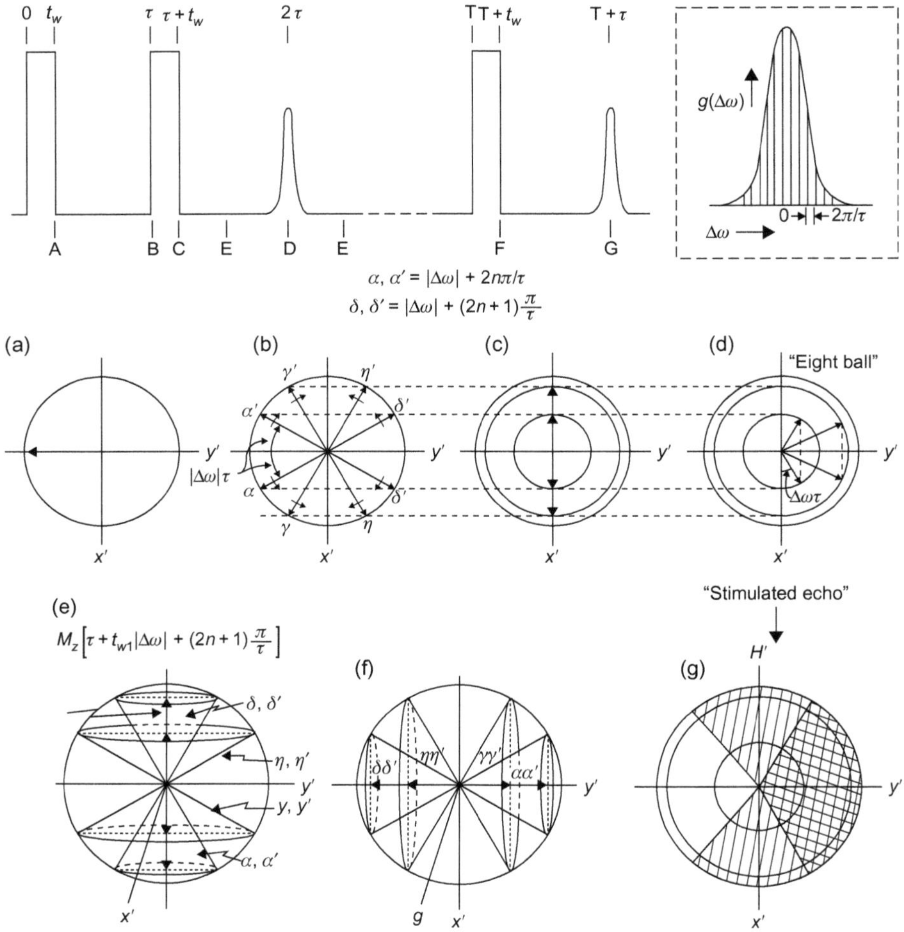

Figure A2.6 The simulated echo is at $t = T + \tau$. The conditions of the special case, for the primary echo model, are as in Figure A2.5. For a given $|\Delta\omega|$, the symbols α, α', and δ, δ' denote the moments. They have Larmor frequencies such that the precess angles are $\Delta\omega\tau + 2n\pi$ and $\Delta\omega\tau + (2n+1)\pi$, respectively. It happens in time $t = \tau$. The n is any integer. It applies to frequencies within the spectrum, which will lie in a pair of cones corresponding to specific $\Delta\omega$. These cones provide M_z components after the pulse at τ. These are available for simulated echo formation after the pulse at T. The shaded area in (g) indicates the density of moment vectors. The absence of vectors on the $-y'$ side leaves a dimple in the unit sphere.

Appendix A2.7

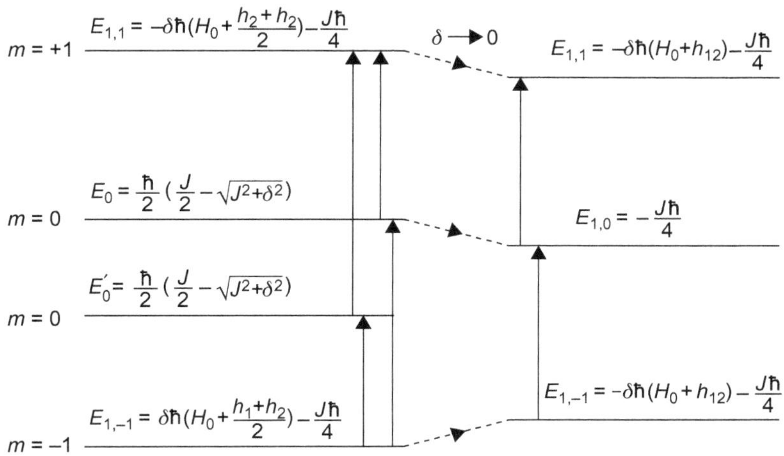

Figure A2.7 Zeeman energy-level diagram for a nonequivalent 2-spin system ($I = 1/2$, for each spin). For $\delta = 0$, the magnitude of $h_{12} = h_1 = h_2$ chosen is arbitrary.

Appendix A2.8

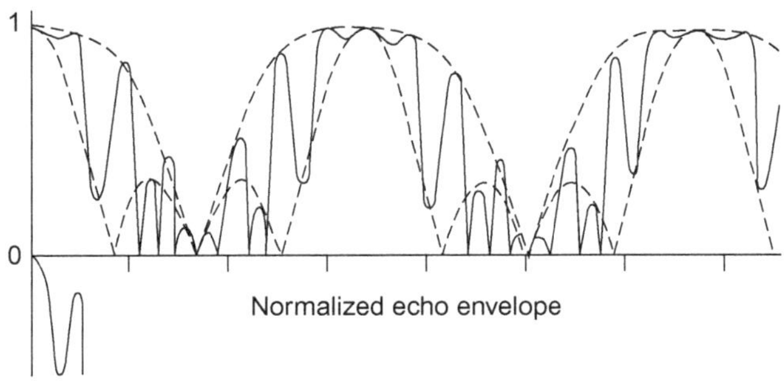

Figure A2.8 Experimental echo-envelope plots for protons in 1-dichloro-2-chloroethane. The dotted lines plotted from the theory trace out the envelope of the upper and lower limits of the echo-envelope modulation plot. The periodic doubling of the modulation frequency appears because the echo experiment reveals only the absolute magnitude of the modulation. The region of modulation doubling (as indicated within the small dotted lobes that meet the abscissa) signifies in the theory that the sinusoidal plot changes sign for a few cycles; that is, with respect to a reference axis, the nuclear magnetization vector of the echo reverses direction by 180°.

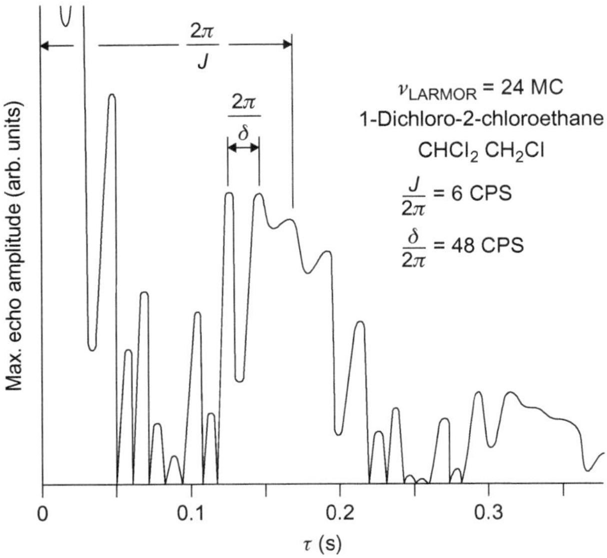

Figure A2.8 continued

Appendix A2.9

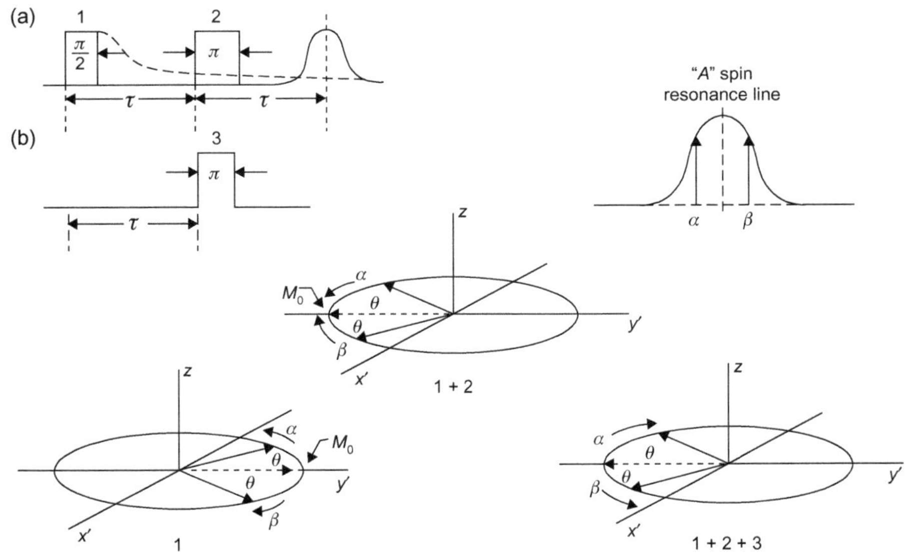

Figure A2.9 Double-resonance pulse effect of spin-echo attenuation.

Appendix A2.10

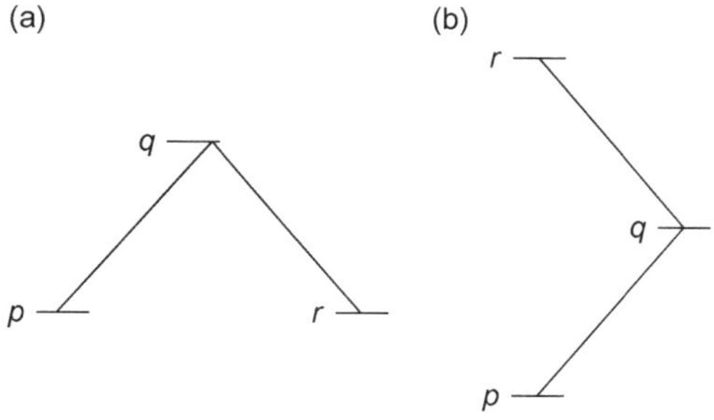

Figure A2.10 Two transitions that have a common energy level (q) are here defined as regressive if the initial and final levels (p) and (r) have the same magnetic number, as in (a); and as progressive if the initial and final levels differ in magnetic quantum number, as in (b).

Appendix A2.11

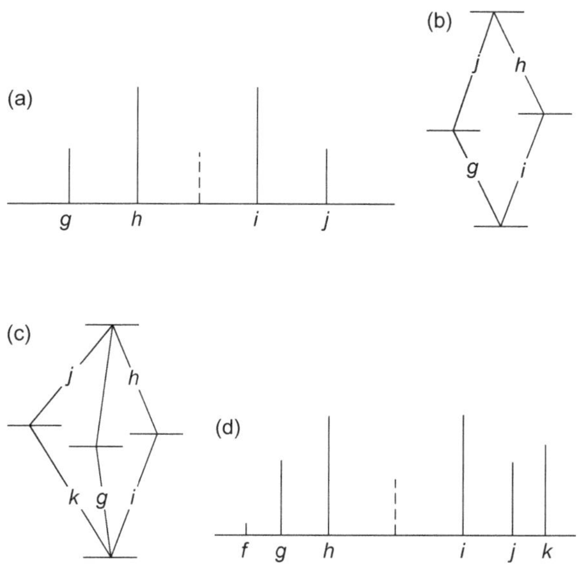

Figure A2.11 (a) The labeling of the normal transition frequencies found in a system of two coupled spin-½ nuclei. (b) The corresponding assignment of the transition frequencies to the energy-level diagram. Note the relationship of the progressive pairs g, j and h, i and the regressive pairs g, i and h, j. The dotted line represents the position of the DQT. (c) Part of an energy-level arrangement of a more complex spin system involving additional spins. The relationship of the progressive and regressive pairs of the transitions g, h, i, and j is preserved if all additional couplings are sufficiently weak. (d) A hypothetical spectrum obtained from the transition shown in (c). Line f is a "combination" line.

Appendix A2.12

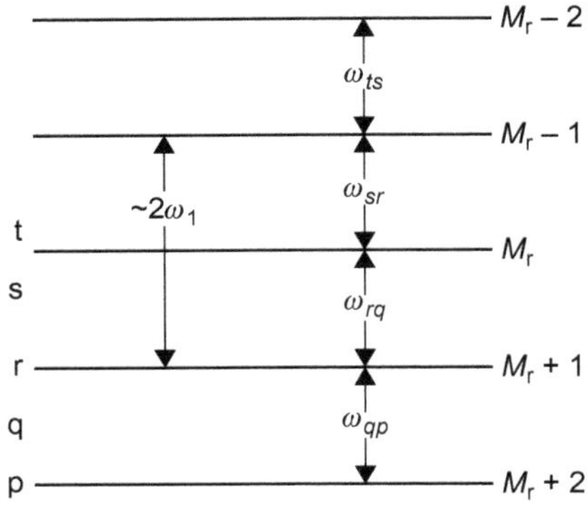

Figure A2.12 The energy levels that must be considered when calculating the frequency shift of a DQT $\omega_{rq}/2$.

Appendix A2.13

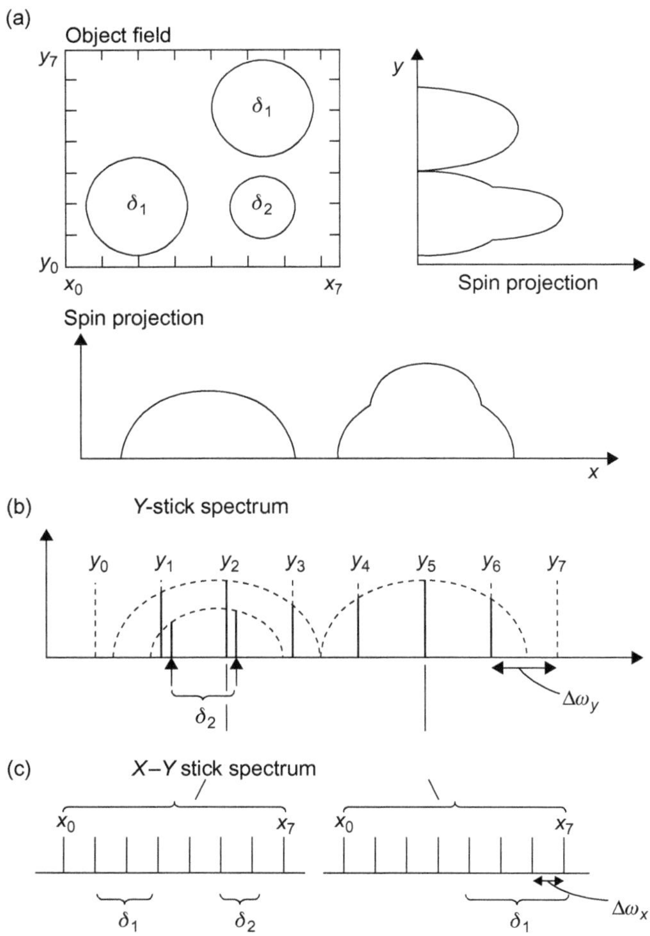

Figure A2.13 Sketches showing (a) object comprising disks of material with different shifts and typical projections in static gradients; (b) discrete stick spectrum of object in a square-wave-modulated y-gradient; and (c) further splitting of y-sticks when a square-wave-modulated x-gradient is added for the case when $L = M = 8$.

Appendix A2.14

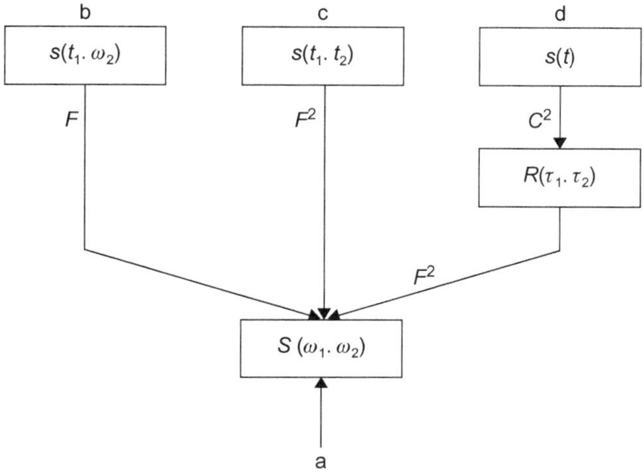

Figure A2.14 Normally used nonlinear techniques.

Appendix A2.15

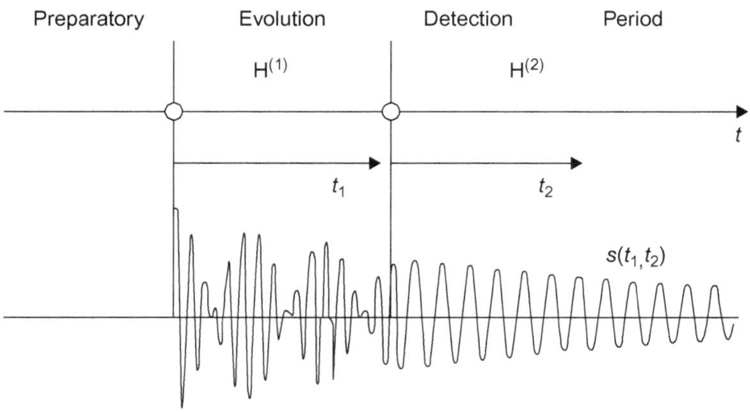

Figure A2.15 Partitioning of the time axis in a 2D FTS experiment.

Appendix A2.16

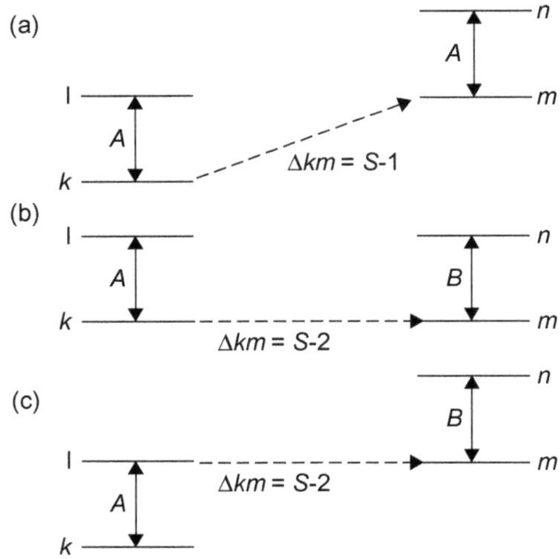

Figure A2.16 Definition of connectivity classes for nonequivalent spins-½. (a) Parallel pair, two *A* spin transitions; along the broken line, *S*-1 spins (≠*A*) change their polarization. (b) Regressive pair, one *A* and one *B* transition; along the broken line, *S*-2 spins (≠*A*, *B*) change their polarization. (c) Progressive pair, one *A* and one *B* spin transition; along the broken line, *S*-2 spins (≠*A*, *B*) change their polarization.

Appendix A2.17

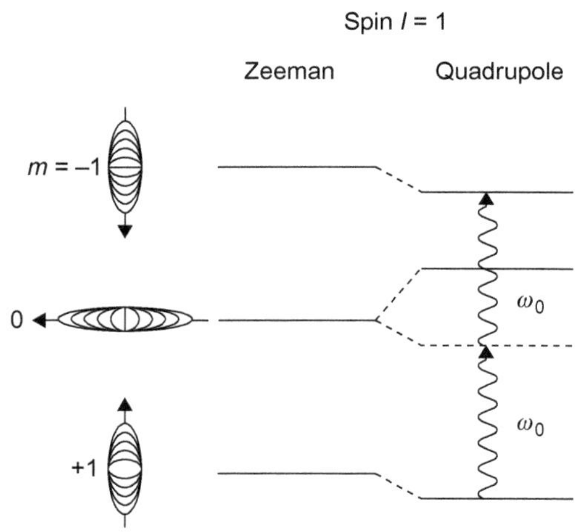

Figure A2.17 Deuterium in single crystal of oxalic acid dihydrate. Deuterium spin-1 energy levels are shifted by $-1/3\omega_Q$ and $+2/3\omega_Q$ by interaction of the electric quadrupole with an axially symmetric electric-field gradient. The transition $m = 1 \rightarrow -1$ is induced by a DQT at the unshifted frequency ω_0 with an RF field of intensity ω_1 and duration t_{P1}, where $\omega_1 \ll \omega_Q$ and $(\omega_1{}^2/\omega u_Q)t_{P1} = 1/2\pi$.

Appendix A2.18

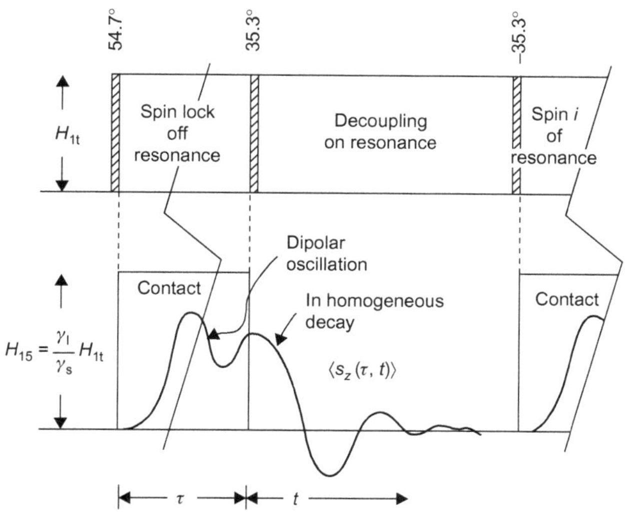

Figure A2.18 Schematic diagram of experimental procedure.

Appendix A2.19

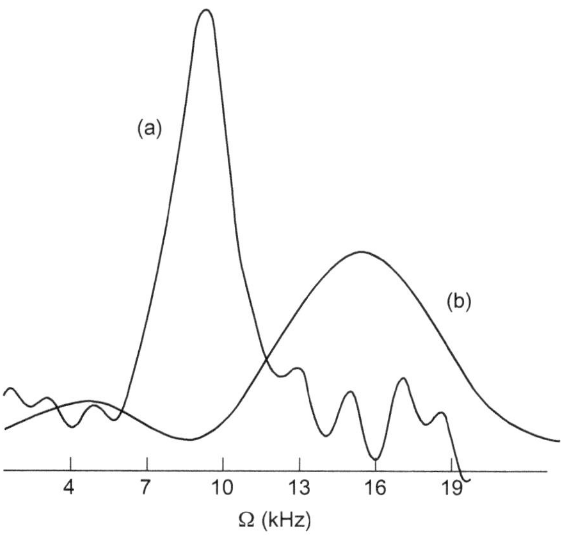

Figure A2.19 Dipolar oscillation spectra of the lower-frequency ^{13}CHOH line for (curve a) off-resonance and (curve b) resonant spin locking of the protons.

Appendix A2.20

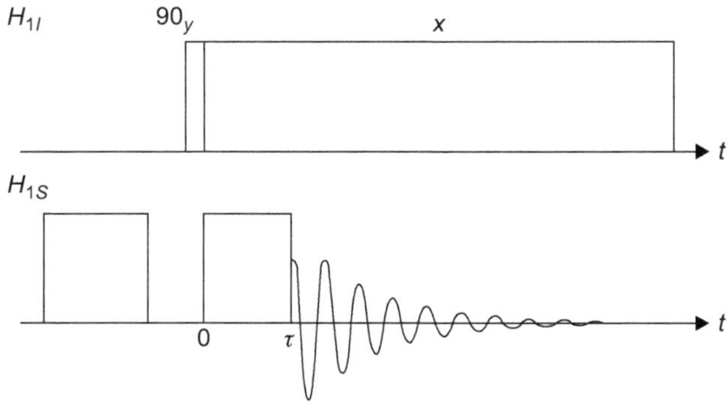

Figure A2.20 Pulse sequence of the cross-polarization experiments.

Appendix A2.21

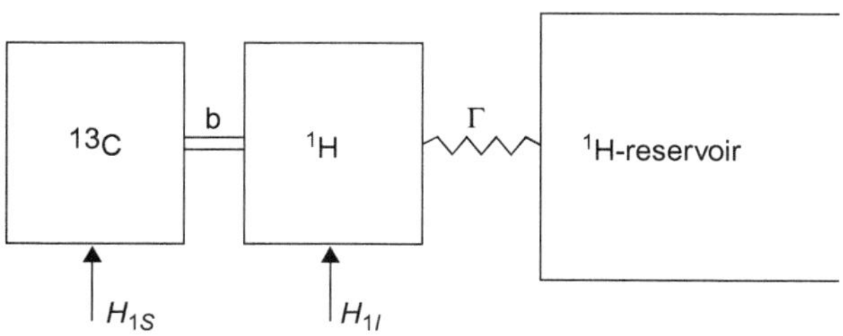

Figure A2.21 Schematic representation of the theoretical model used in the computations.

Appendix A2.22

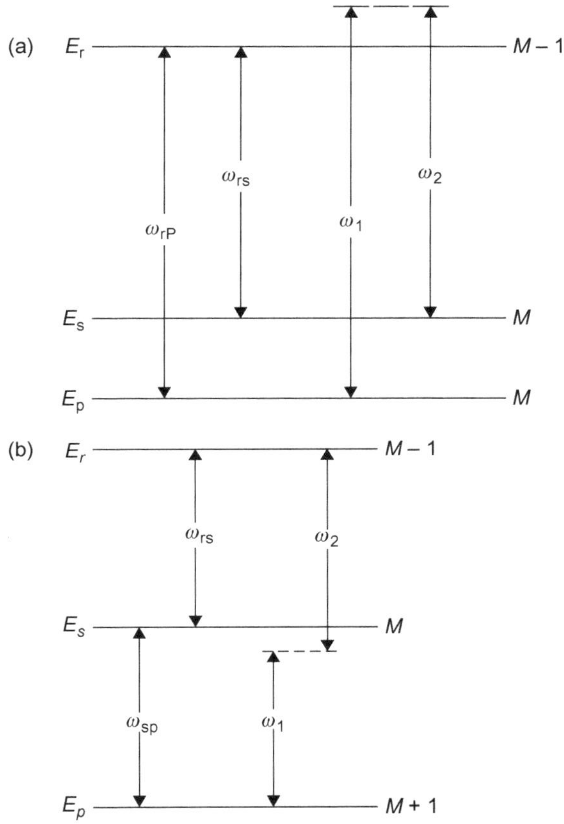

Figure A2.22 The two possible arrangements of energy levels for nuclear magnetic double-resonance experiments. In (a), the energy level common to both transitions is E_r. The other two levels then have the same magnetic quantum number M, and this situation is denoted by $\Lambda = 0$. In (b), E_s is the common energy level and the other two levels differ in spin quantum number by two units, a situation denoted by $\Lambda = 2$.

Appendix A2.23

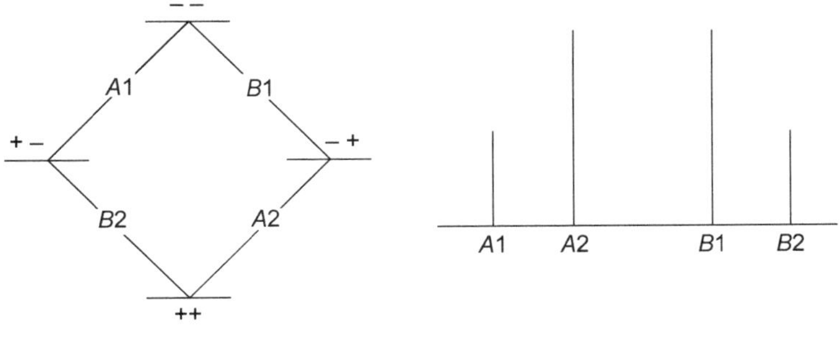

Figure A2.23 The energy-level diagram and transition frequencies for a two-spin system AB.

Appendix A2.24

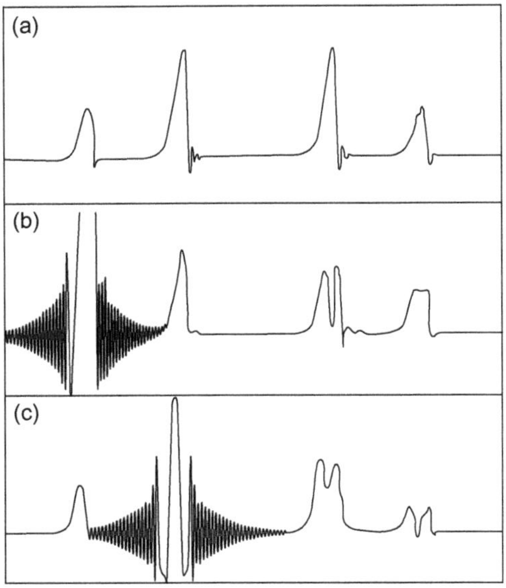

Figure A2.24 The AB spectrum of 2-bromo-5-chlorothiophene recorded by sweeping the frequency ω_1 (a) unperturbed; (b) with a second RF field H_2 at resonance on the line A1; (c) with the same RF field at resonance on line A2. The beat frequency is caused by the passage of ω_1 through ω_2 during the scan. Note the sharpness of the components of the B1 doublet in (b).

Appendix A2.25

$$S_a = \frac{16N\hbar\omega\gamma^4 J^2\delta^{-4}H_1[\sigma_0(a)-\sigma_0(b)]\Gamma_{ab}}{\theta^2+\Gamma_{ab}+4\Gamma_{ab}T_{ab}(\gamma H_1)^4J^2\delta^{-4}}\,\mathrm{e}^{-\mathrm{i}\omega t}$$

Table A2.1

I_A	I_B	Initial state M_A	M_B	Frequency[a]	Degeneracy	Relative Intensity	Grouping of Unresolved Lines	
3/2	1	−3/2	−1	$\frac{1}{2}(\omega_A+\omega_B)-\frac{3}{4}J+\rho$	1	3	3	α_1
3/2	1	−3/2	0	$\frac{1}{2}(\omega_A+\omega_B)-\frac{1}{4}J+\rho$	1	3	9	
3/2	1	−1/2	−1	$\frac{1}{2}(\omega_A+\omega_B)-\frac{1}{4}J+2\rho$	1	4		α_2
1/2	1	−1/2	−1	$\frac{1}{2}(\omega_A+\omega_B)-\frac{1}{4}J-\rho$	2	2		
3/2	1	1/2	−1	$\frac{1}{2}(\omega_A+\omega_B)+\frac{1}{4}J+\rho$	1	3		
3/2	1	−1/2	0	$\frac{1}{2}(\omega_A+\omega_B)+\frac{1}{4}J+2\rho$	1	4	9	α_3
1/2	1	−1/2	0	$\frac{1}{2}(\omega_A+\omega_B)+\frac{1}{4}J-\rho$	2	2		
3/2	1	1/2	0	$\frac{1}{2}(\omega_A+\omega_B)+\frac{3}{4}J+\rho$	1	3	3	α_4

[a] $\rho = J^2/[4(\omega_B-\omega_A)]$.

Appendix A2.26

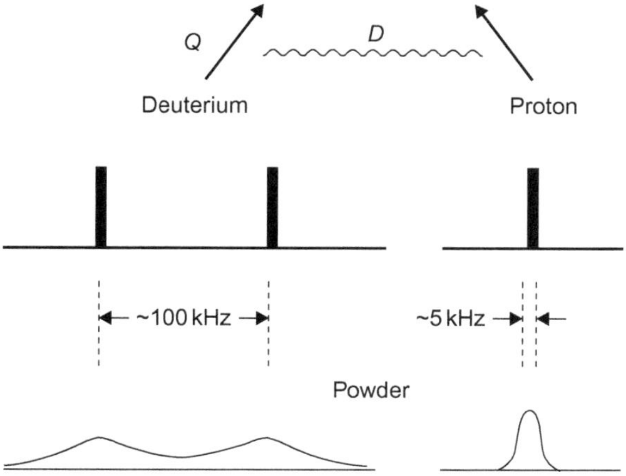

Figure A2.25 Schematic representation of dipolar-interacting proton and deuterium spins and NMR spectra. The dipolar coupling is D and the deuterium is characterized by quadrupolar splitting Q from an axially symmetric electric-field-gradient tensor. The problem with spin-decoupling deuterium comes from the large spread of frequencies (~200 KHz).

Appendix A2.27

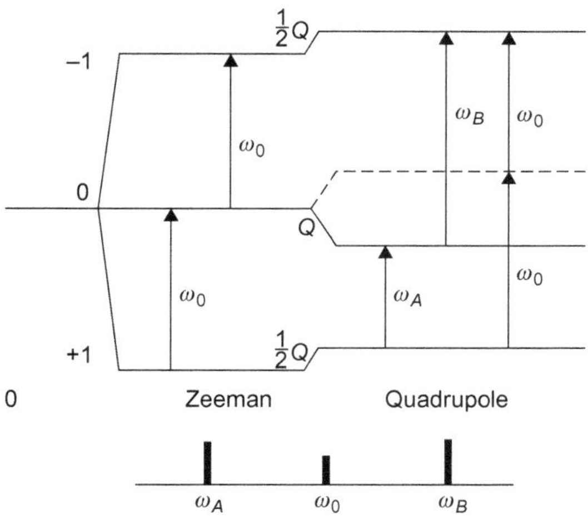

Figure A2.26 Energy-level diagram for discussion of deuterium transitions. A DQT $1 \rightarrow -1$ occurs at the unshifted central frequency ω_0.

Appendix A2.28

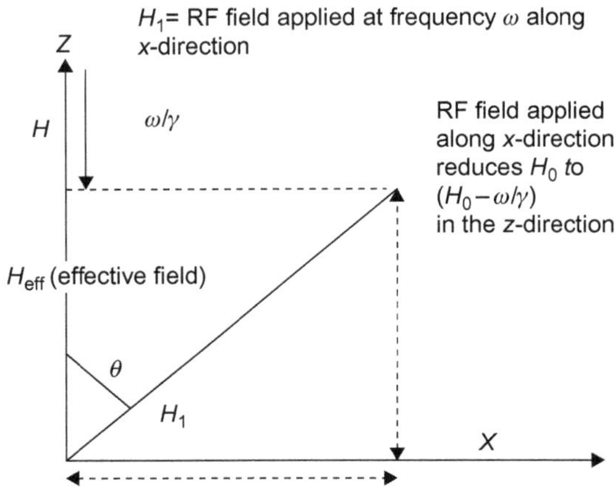

Figure A2.27 $(dL/dt) = [\mu \times H] = [\gamma(L \times H)]$; $(dL/dt) = [\partial L/\partial t + \omega \times L]$; L = nuclear orbital angular momentum; μ (magnetic moment) $= L\gamma$ (gyromagnetic ratio) $= L(\mu_b/\hbar)$; μ_b = one Bohr magneton $= (e\hbar/2\,m)$ (m = mass of the electron) $= 0.927 \times 10^{-23}\ \mathrm{A\,m^2}$; $H_{eff} = [(H_0 - \omega/\gamma)^2 + {H_1}^2]^{1/2}$.

Appendix A2.29

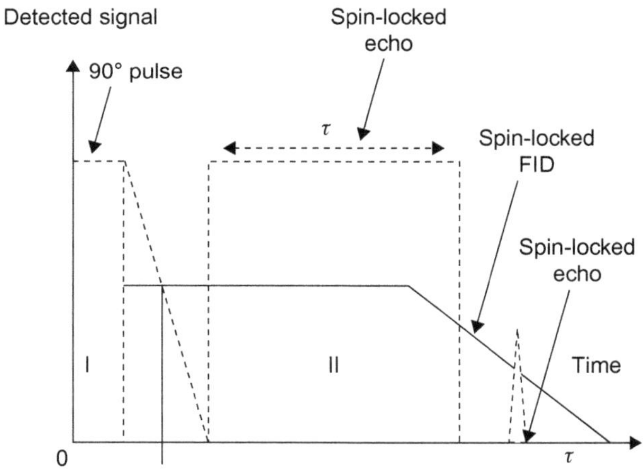

Figure A2.28 In a simple model, one assumes that there are uniform spin temperatures, T_a and T_b. This is done in the rotating frame of reference for each ensemble a (abundant species) and b (minority species). One writes the expressions for the rate of cross-coupling of spin populations thus: $\partial(n_a)/\partial(t) = -\partial(n_b)/\partial(t) = W_0(n_b N_a - n_a N_b)$, where N_a and N_b are the numbers per ml of a and b spins, respectively, and n_b, n_b are the numbers of the a and b spins in their magnetically excited states (Zeeman splitting) at any instant of time. W_0 represents the average rate of interaction per a–b pair. The dephased components of magnetization develop in a time τ, after a 90° pulse. They are locked in the rotating frame for a time T. For this time, the magnetization has x-axis memory along H_1. This gives rise to an echo after the locking pulse is removed. The echo performs, in its amplitude dependence, like a stimulated echo formed after a 3-90°–90°–90° pulse sequence. This is when τ is the interval between the first two pulses, and T is the interval between the second and third pulses. The stimulated echo is seen at $T + 2\tau$.

Appendix A2.30

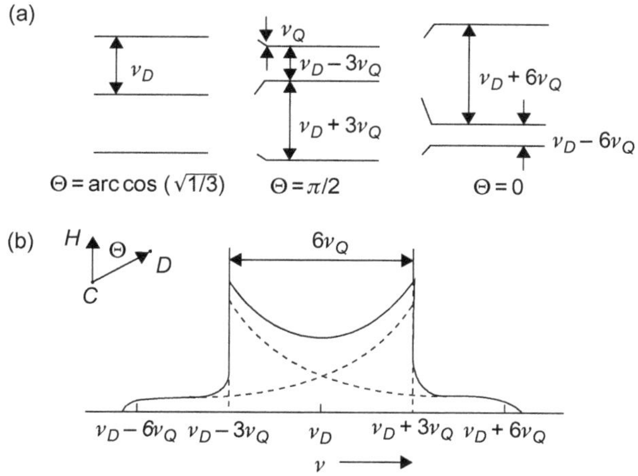

Figure A2.29 (a) Energy-level diagram of the deuteron-spin system for three values of the factor $3\cos^2\theta - 1$: namely, 0, -1, and 2, which correspond to the center, the peaks, and the pedestals of the deuteron lines, respectively. θ is roughly the angle between the direction of a $C-D$ bond and an external magnetic field. (b) Theoretical deuteron line shape, which is the sum of the two possible transitions $m = 0$ to $m = 1$ and $m = -1$ to $m = 0$ (dashed lines). Some line broadening due to spin–spin interactions has been taken into account, which makes the line smoother. Otherwise the peaks would tend to infinity.

Appendix A2.31

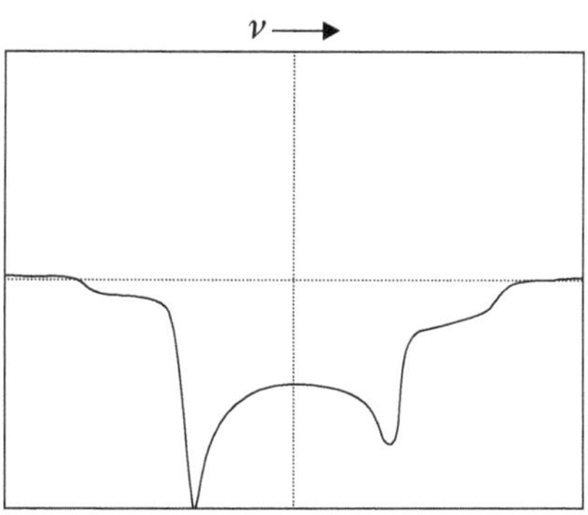

Figure A2.30 Deuteron MR line in propanediol $-D6$, measured in a 25-kG magnetic field. The distance between the peaks is 117 kHz ($6\nu_Q$). The deuterons are dynamically polarized to -43%, which causes the asymmetry.

Appendix A2.32

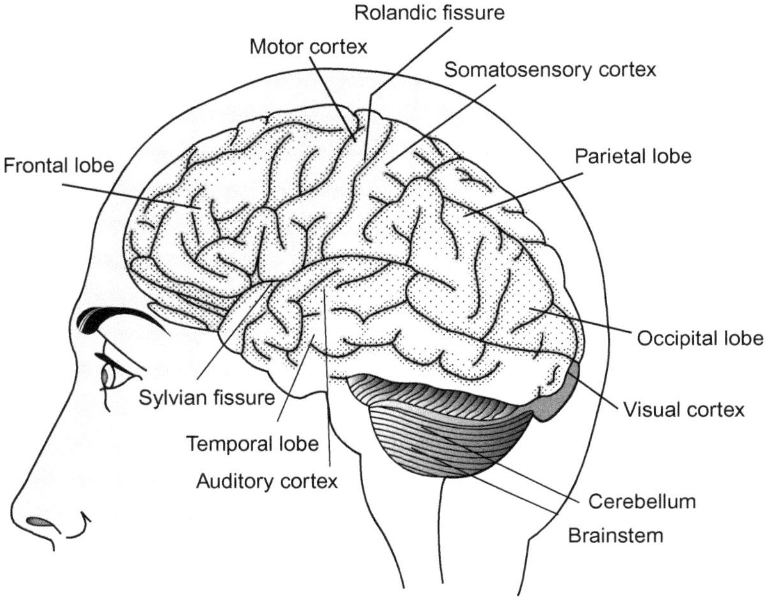

Figure A2.31 Human brain: Some of the important structural landmarks and special areas of the cerebral cortex are indicated.

Appendix A2.33

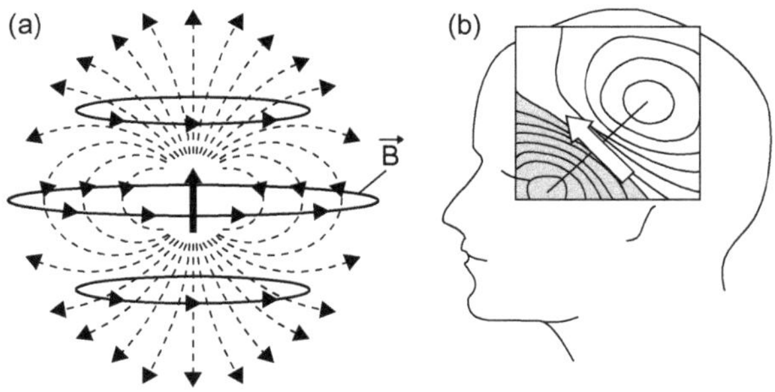

Figure A2.32 Magnetic field of a current dipole. (a) Current dipole (large arrow) in a homogenous conducting medium. Examples of volume currents (dashed curves) and magnetic-field lines *B* (solid curves) produced by the primary current are shown as well. (b) Examples of the topographic field map calculated from the measured MEG signals. The simple geometrical construction for locating the equivalent current dipole in the brain is also illustrated: the dipole is midway between the field extrema (compared with Figure A2.33).

Appendix A2.34

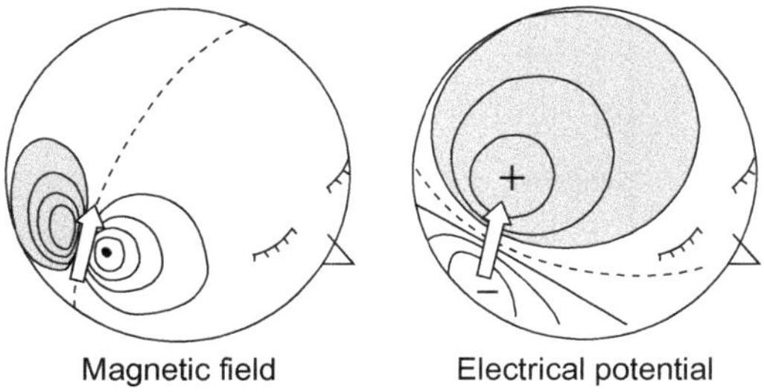

Figure A2.33 Schematic illustration of idealized magnetic field and electric-potential patterns produced by a tangential dipole (white arrow). The head was approximated with a four-compartment sphere consisting of the brain, the CSF, the skull, and the scalp. From noninvasive measurements of the MEG or EEG field distributions, the active area in the brain can be determined by a least-squares fit to the data.

Appendix A2.35

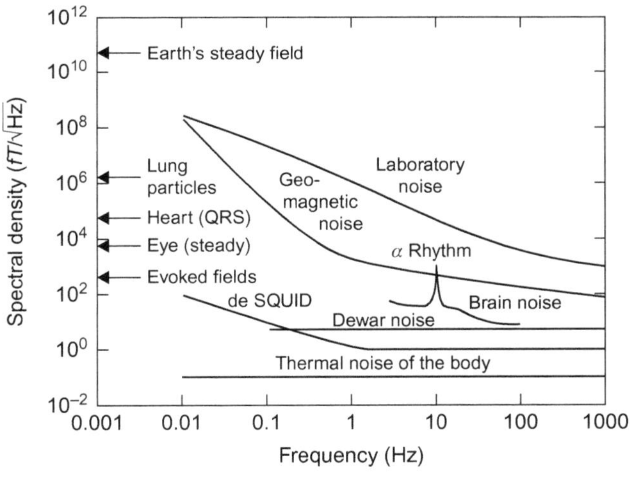

Figure A2.34 Peak amplitudes (arrow) and spectral densities of fields due to typical biomagnetic and noise sources.

Appendix A2.36

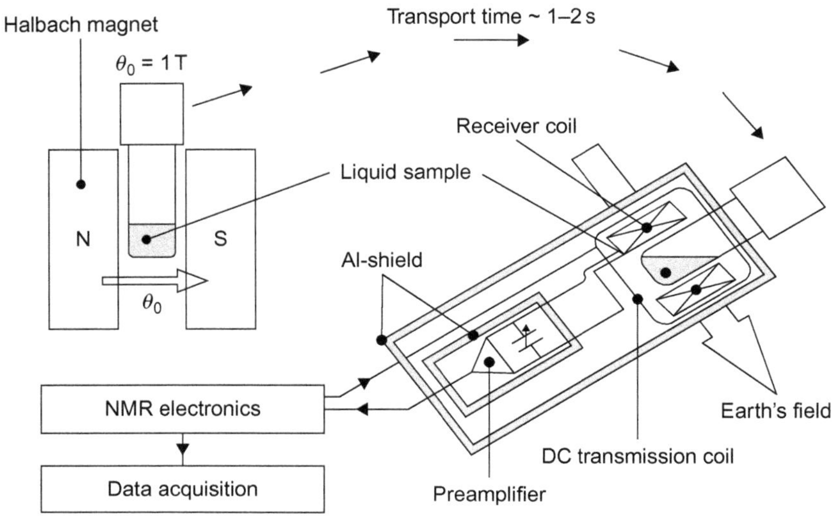

Figure A2.35 Setup of mobile, ultrahigh-resolution, EFNMR. The nuclear spins of a liquid sample are first prepolarized, using a Halbach magnet, and then transported to the NMR probe of the spectrometer. After a 90° DC pulse excitation, the FID of the nuclear spins in the Earth's magnetic field (~50 mT) is acquired and processed.

Appendix A2.37

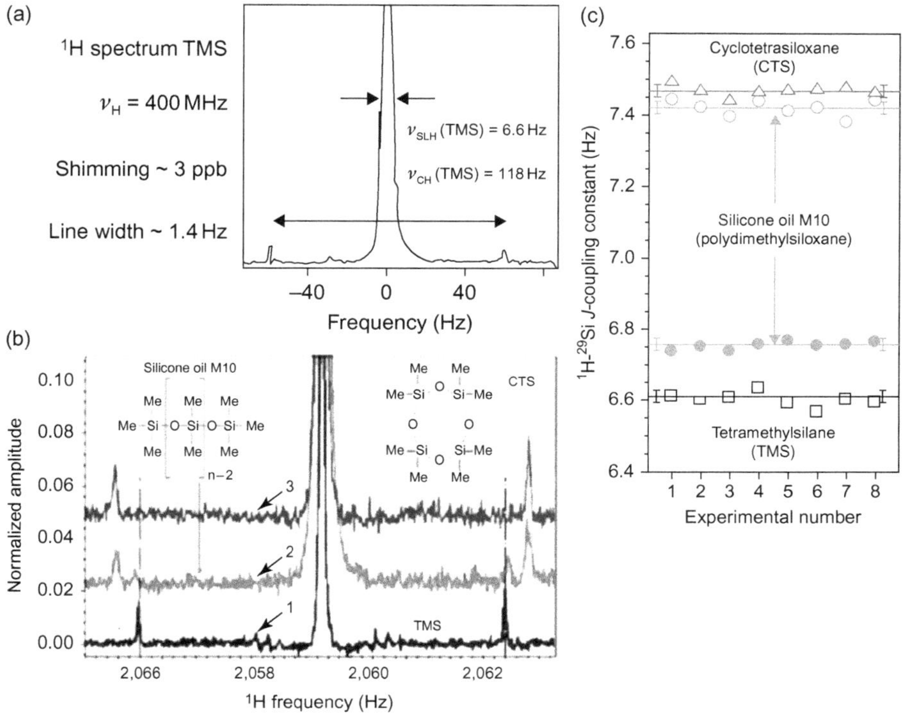

Figure A2.36 Precision measurements of $^1H-^{29}Si$ J-couplings for different silicon compounds. (a) ^1H-reference spectrum of tetramethylsilane (TMS) measured at the ^1H Larmor frequency, $\nu_H = 400$ MHz, with a superconducting magnet shimmed to a few ppb. The original ppm scale is rescaled to the frequency scale. (b) Single-scan ^1H EFNMR spectra of TMS (1→Dark Gray), silicone oil M10 (2→Gray), and cyclotetrasiloxane (CTS; 3→Light Gray). All three spectra show different values of the $^1H-^{29}Si$ J-coupling constants. The amplitude of the central peak (which corresponds to 95.3% of the protons without J-coupling) is normalized to 1. (c) Statistics over eight scans of all measured $^1H-^{29}Si$ J-coupling constants. The squares correspond to TMS, the circles to silicone oil, and the triangles to CTS. The fluctuations arise mainly from the imperfect phasing of the spectrum. The error bars indicate the standard deviation of the mean J-value (solid lines).

Appendix A2.38

Table A2.2 Characteristics of Some Useful Nuclei

Nuclei	I ($\hbar$)	γ	μ (n.m.)
$^{1}_{1}H$	$\frac{1}{2}$	42.577	+2.79
$^{2}_{1}D$	1	6.536	+0.86
$^{7}_{3}Li$	$\frac{3}{2}$	16.547	+3.26
$^{14}_{7}N$	1	3.076	+0.40
$^{19}_{9}F$	$\frac{1}{2}$	40.055	+2.63
$^{31}_{13}P$	$\frac{1}{2}$	17.235	+1.13

I = angular momentum in the unit $\hbar$; γ = resonance frequency in MHz in a field of 1 T; μ = nuclear magnetization.

Appendix A2.39

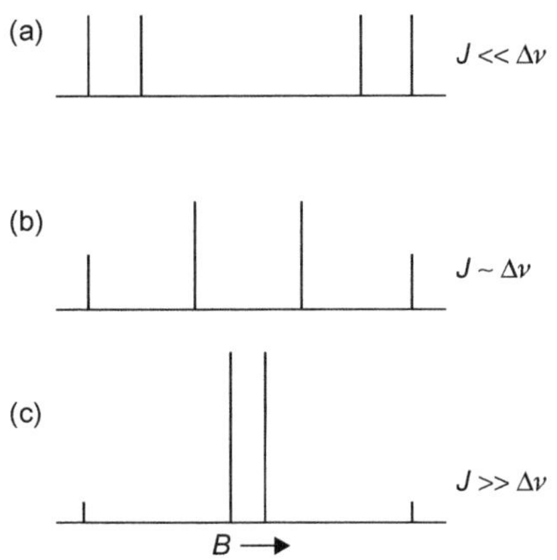

Figure A2.37 The spectra of two interacting nuclei of spin-½.

Appendix A2.40

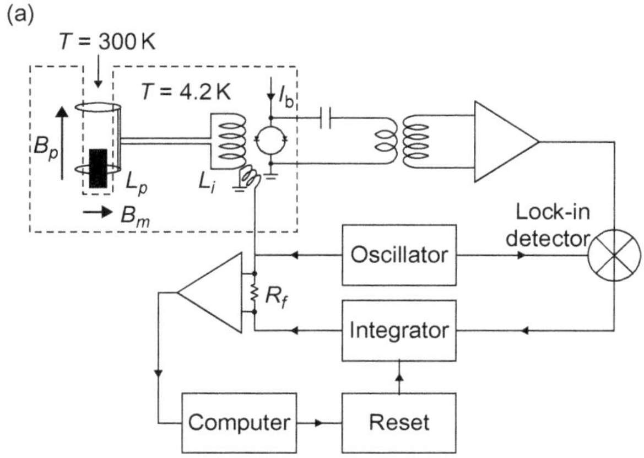

Figure A2.38 (a) Schematic of SQUID spectrometer for micro-Tesla-field NMR. The detector used was an Nb–AlOx–Nb DC SQUID. The DC SQUID consists of a superconducting loop interrupted by two Josephson junctions. When biased with a current I_b slightly above the critical current of the junctions, the SQUID acts as a flux-to-voltage transducer. To enhance its sensitivity to magnetic fields, the SQUID is often operated with a superconducting flux transformer consisting of a pickup coil tightly coupled to the SQUID loop. The flux transformer operates on the principle of flux conservation in a superconducting loop, which involves no frequency dependence. Thus, the SQUID magnetometer can detect broadband at arbitrarily low frequencies with no loss in sensitivity. In these experiments, the input coil (with inductance L_i) of the transformer was integrated onto the SQUID chip; the niobium wire pickup coil (with inductance L_P) was wound in a gradiometric fashion around the tail section of a cryogenic insert, made of Pyrex, which was surrounded by liquid sample that was lowered into the tail of the insert; a resistive heater maintained the sample temperature at around 300 K. A single-layer solenoid of copper wire wound directly on the sample cell produced the polarizing field. A set of coils located in the helium bath provided the measurement field. The belly of the helium dewar was lined with a superconducting lead (Pb) sheet, and the dewar was surrounded by a single-layer mu-metal shield to attenuate both the static magnetic field of the Earth and external magnetic fluctuations. The SQUID was amplified, integrated, and fed back to the SQUID as a magnetic flux. The voltage across the feedback resistor R_f was thus proportional to applied flux. In this way, the SQUID acted as a null detector of magnetic flux. (b) Pulse sequence for micro-Tesla-field NMR. The sample was polarized in a field B_p of about 1 mT for a time that is long compared to the spin-lattice relaxation time of the sample (typically several seconds). In addition, a measurement field B_m of a few micro-Tesla was applied in an orthogonal direction. When the polarizing field was removed nonadiabatically, the spins precessed in the measurement field. As the spins precessed, they lost phase coherence due to the poor homogeneity of the measurement field, which was abruptly reversed, causing the spins to reverse the sense of their precession. During the interval from τ to 2τ, phase coherence was restored, and at a time 2τ the echo amplitude was a maximum. This field-inversion spin-echo is to be contrasted with the conventional Hahn spin-echo, in which the phase of the spins is inverted by means of RF pulses, but the sense of precession is preserved.

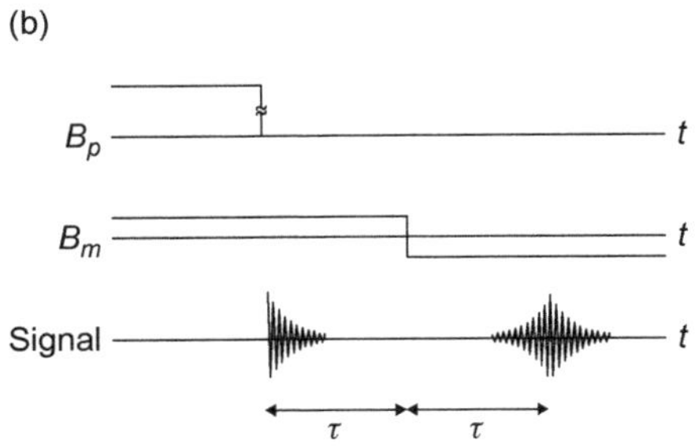

Figure A2.38 continued

Appendix A2.41

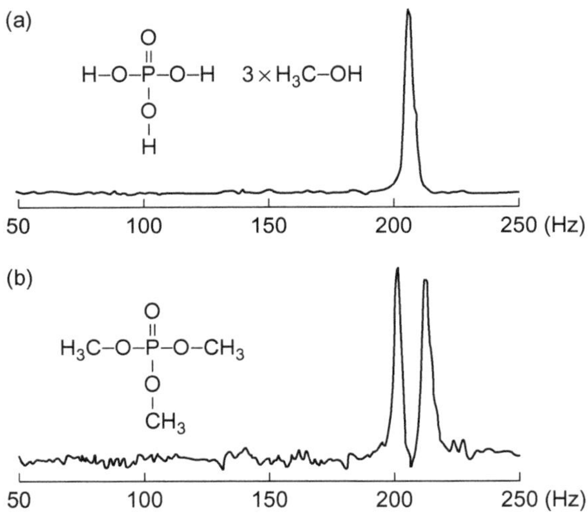

Figure A2.39 (a) NMR spectrum of 5 ml of 3 parts methanol, 1 part phosphoric acid (85% in water) measured in a field of 4.8 μT. The spectrum is the average of 100 transients. Rapid spin exchange with the protons in water obscures the proton–phosphorous scalar coupling in phosphoric acid, and the proton spectrum consists of a sharp singlet. (b) NMR spectrum of 3 ml of neat trimethyl phosphate (Sigma-Aldrich) measured in a field of 4.8 μT. The spectrum is the average of 100 transients. Electron-mediated scalar coupling of the nine equivalent protons to the ^{31}P nucleus splits the proton resonance into a doublet, with a splitting that is determined by the coupling strength *J*. For this particular coupling, of three covalent bonds, $J_3[P,H] = 10.4 \pm 0.6$ Hz. Scalar coupling to the nine protons splits the ^{31}P resonance into a decouplet; those lines are below the noise level. The reduction in SNR is due to the lower sample volume and filling factor.

Appendix A2.42

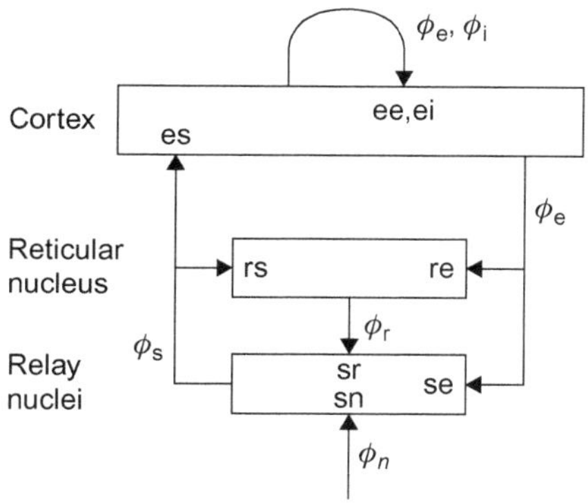

Figure A2.40 Schematic of corticothalamic interactions, showing the locations at which ν_{ab} and G_{ab} act.

Appendix A2.43

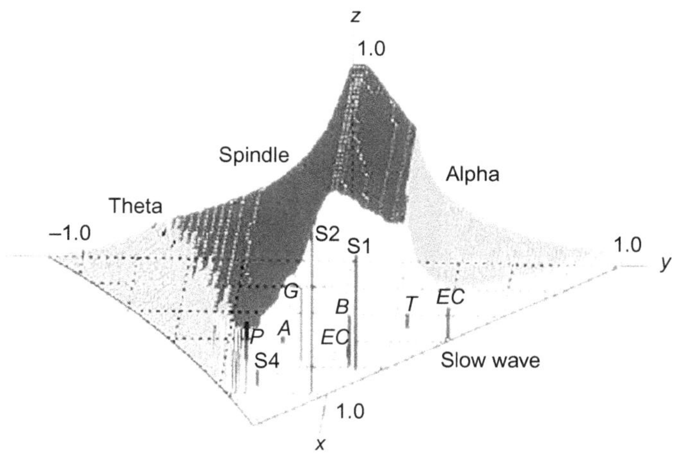

Figure A2.41 Stability zone for nominal parameters, except $\alpha = 60\,s^{-1}$. The surface is shaded according to instability, as labeled (dark gray for spindle, light gray at right for alpha, light gray at left for theta), with the front right face left transparent as it corresponds to a zero-frequency instability. Approximate locations are shown as eyes open (EO), eyes closed (EC), sleep stage 1 (S1), S2, S4, deep anesthesia (A), an onset of petit mal (P) seizure, with each state located at the top of its bar, whose x and y coordinates can be read from the grid. Irregularities in edges and shading are numerical artifacts. A very narrow, rapidly tapering extension of the stable zone continues to larger negative y for $x \approx z \approx 0$ (not shown).

Appendix A2.44

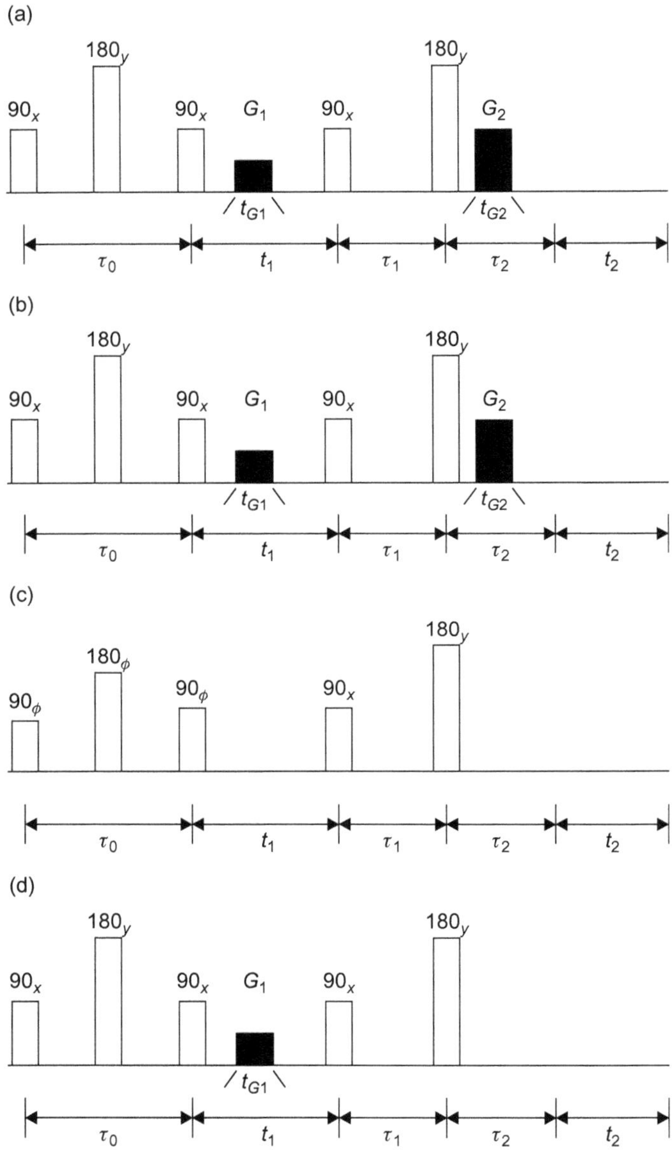

Figure A2.42 MQ editing sequences. (a) DQ sequence selecting the DQCs with gradient pulses. All RF pulses are nonselective. (b) DQ sequence as in (a), but with a read pulse that is frequency selective on the CH resonance. (c) MQ sequence using phase cycling to select a certain order of coherences. (d) ZQ sequence selecting the ZQCs with a gradient pulse. (e) Another sequence: ZQCs and DQCs are selected with gradient pulses $t_1 = t_a + t_b$. The selective 180° and 90° pulses excite only the CH resonances.

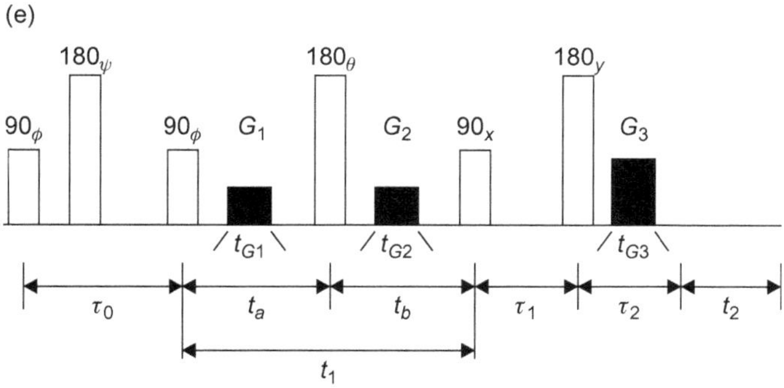

Figure A2.42 continued

Appendix A2.45

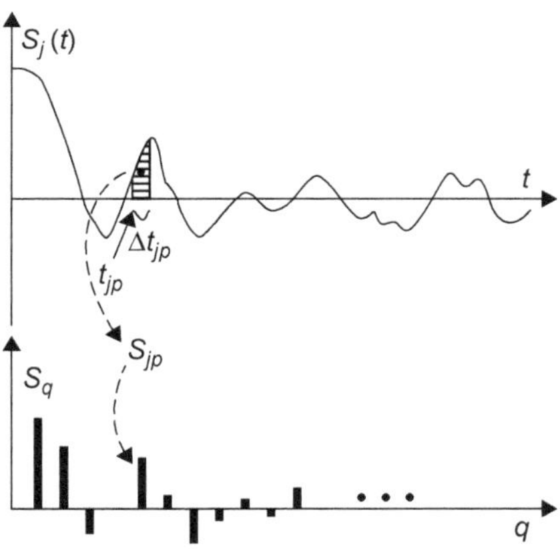

Figure A2.43 The datum S_{jp} results from sampling the jth FID $S_j(t)$ during the pth time interval $[t_{jp}, t_{jp}, +\Delta t_{jp}]$, and corresponds to a datum S_q in the discrete spatial-frequency coordinate q.

Appendix A2.46

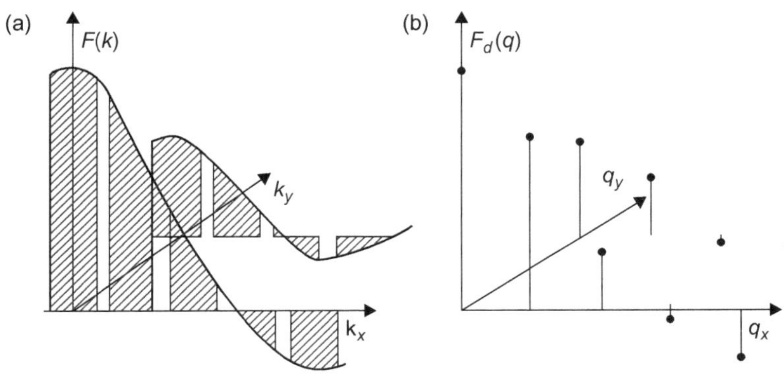

Figure A2.44 The discrete spatial-frequency $F_d(q)$ corresponds to samples of the continuous-variable distribution $F(k)$. For methods in which the k plane is scanned parallel to the k_x axis, sampling is impulsive in k_y, and integrated in rectangular in k_x, as indicated here by the shaded areas.

Appendix A2.47

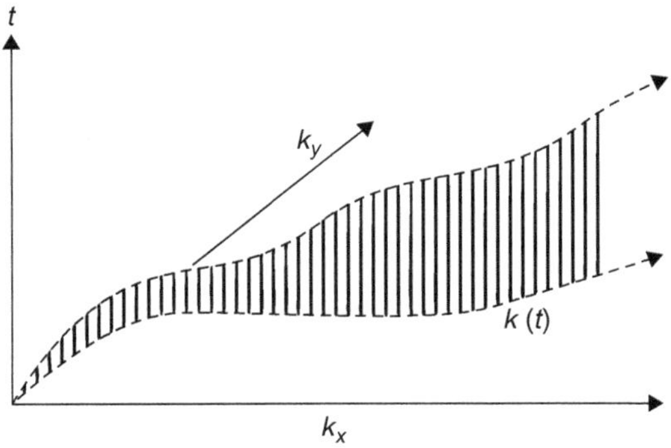

Figure A2.45 An arbitrary k-trajectory passing above the spatial-frequency [k] plane, with sampling path on the k plane indicated by the dashed line. At time t, the complex FID $s(t)$ is a sample of the k-domain distribution $F(k)$ at $k(t)$, but weighted by the exponential decay $\exp(-t/T_2)$. Where $d[k(t)]/dt$ (equivalently, the gradient magnitude) is small, the trajectory "rises" more rapidly, devoting a larger segment in time to sampling the corresponding segment of k-space and thereby sampling that k location more accurately.

Appendix A2.48

Table A2.3 Parameters of Approximate Discrete Transfer Functions for Generalized Fourier NMRI Methods

K_f = **highest k-value sampled**
q_x, q_y = **discrete spatial-frequency coordinates**
N = **dimension of (square) sampling array**
T_d = **delay for "ring-down" period**
G_x, G_y = **constant x- and y-gradient magnitudes**
$\{T_y\}$ = **fixed duration of phase-encoding y gradient** $H(q) = A\Delta t_q \exp(-t_a/T_2)$

Method	t_d	t_a
Original Fourier[a]	$(q_x/G_x + q_y/G_y)k_f/N$	$(T - T_d - q_y/G_y)/N)$
Spin warp	$(T_y) + T_d + (q_x/G_x)k_f/N$	$(T - T_d - (T_y))/N$
Rotating Frame Projection	$(q_x^2 + q_y^2)^{1/2}k_f/GN$	$(T - T_d)/N$
Rectilinear, echo planar		
	$[q_x + N(q_y - 1)]T/N$	T/N, q_y, even
	$[(N - q_x) + N(q_y - 1)]T/N$	T/N, q_y, odd

[a] The form for rotating-frame zeugmatography is essentially similar.

Appendix A2.49

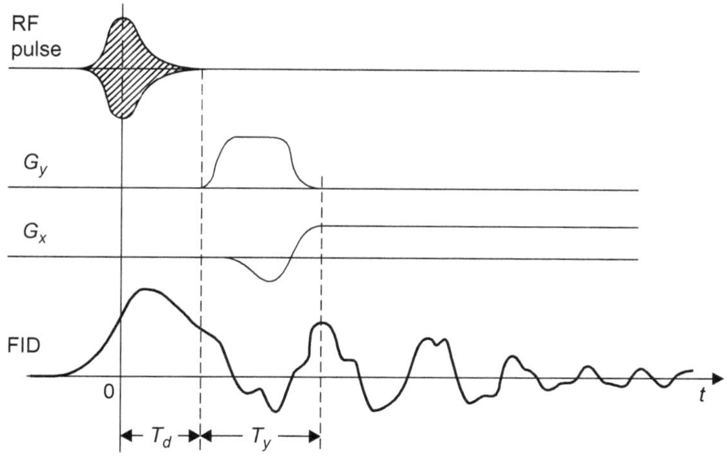

Figure A2.46 Timing relationships during a single FID of a simplified spin-warp method in which selective irradiation for plane selection is not used. G_x and G_y denote x- and y-gradients, respectively; the pulsed G_y accommodates phase encoding. A preliminary ring-down period T_d and phase-encoding period T_y precede the period devoted to FID sampling, which extends to a time T.

Appendix A2.50

Table A2.4 Some Prevalent *k*-Trajectories Compared as to Preparation Time, Scan Time, etc.

Trajectory	Total Number of Excitations	Scan Time[a] (min:s)	Prep. Time[b] (ms)	Min. *TE*[c] (ms)
3DFT	10,000	8:20	1.43	2.29
Cylindrical 3DFT	7854	6:33	1.43	2.29
3D echo planar	1667	1:23	1.43	2.29
Cylindrical 3D echo planar	1310	1:05	1.43	2.29
Stack of spirals	900	0:45	1.42	0
Shells	629	0:31	1.42	0
Spherical stack of spirals	628	0:31	1.42	0
3D RR	31,416	26:10	0	0
Spiral–PR hybrid	1413	1:10	0	0
Cones	1152	0:57	0	0
Yarn	1044	0:52	0	0
Density-compensated cones	800	0:40	0	0

The trajectories are presented in order of decreasing number of excitations, first for trajectories with a preparation time, then for trajectories without preparation time.
[a]Scan time is computed for TR = 50 ms and each acquisition is 14.336 ms long.
[b]Preparation time is the maximum of the times required to go from the origin of *k*-space to the start of the acquisition.
[c]Minimum TE is the minimum time at which the origin of *k*-space is read.

Appendix A2.51

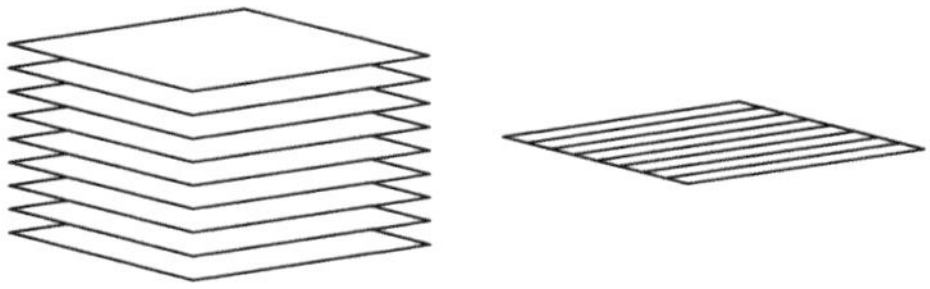

Figure A2.47 DFT trajectory.

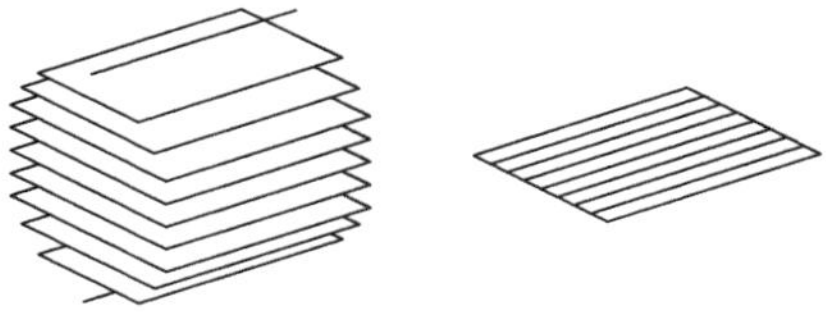

Figure A2.48 Cylindrical 3DFT trajectory.

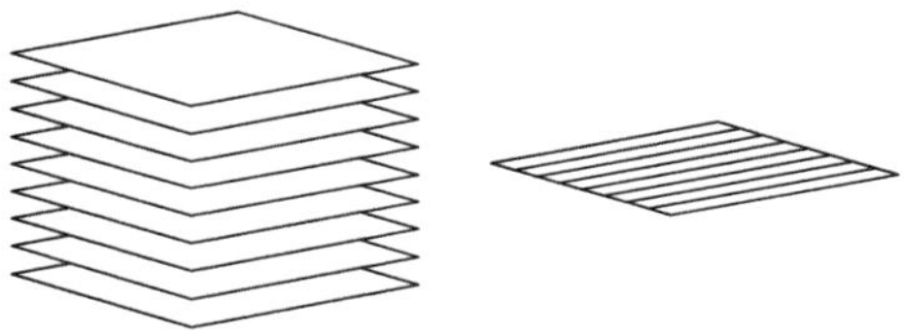

Figure A2.49 3D echo planar trajectory.

Appendix A2.52

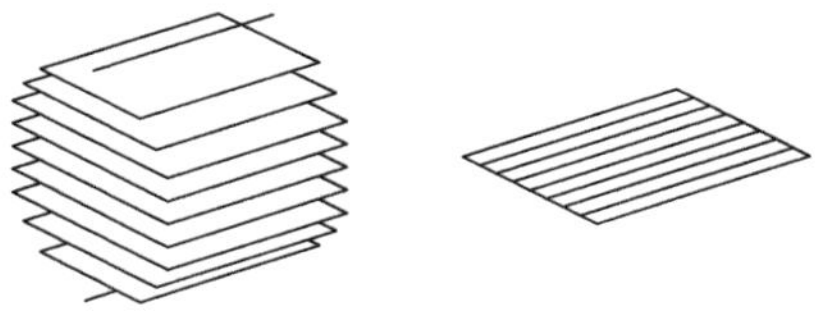

Figure A2.50 Cylindrical 3D echo planar trajectory.

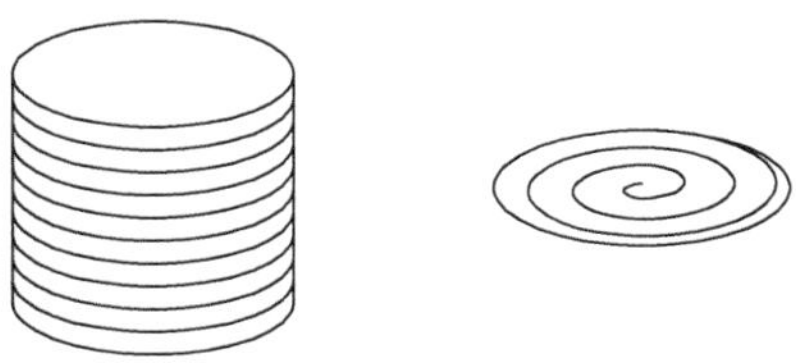

Figure A2.51 Stack of spiral trajectory.

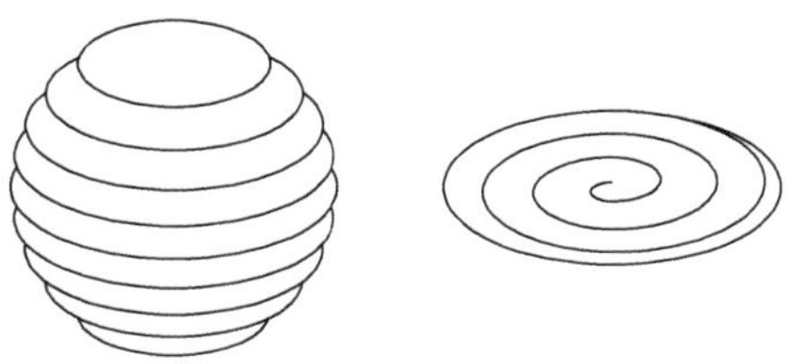

Figure A2.52 Spherical stack of spirals trajectory.

Appendix A2.53

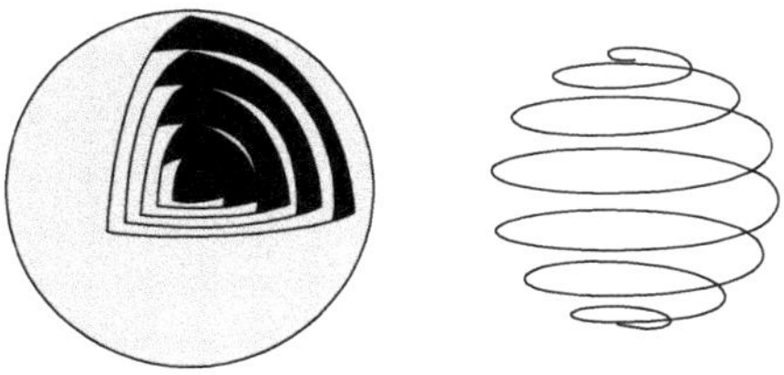

Figure A2.53 Shells trajectory.

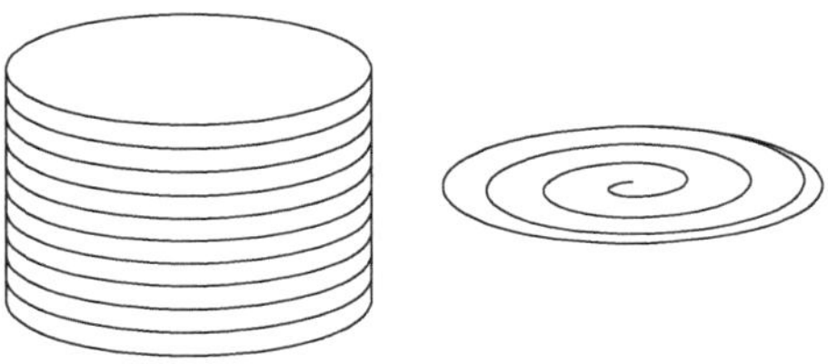

Figure A2.54 Three-dimensional projection–reconstruction trajectory.

Appendix A2.54

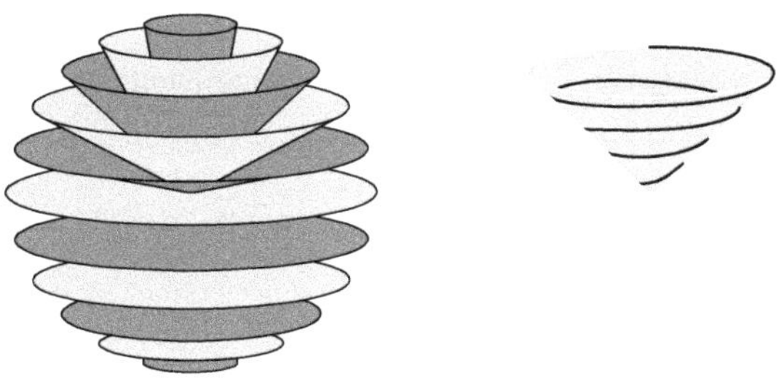

Figure A2.55 Spiral–PR hybrid trajectory.

Appendix A2.55

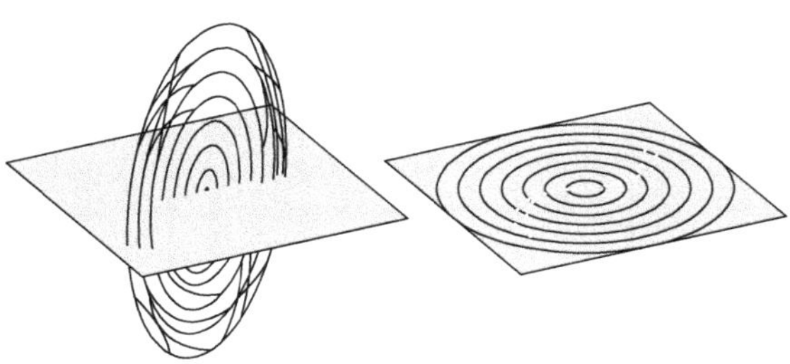

Figure A2.56 Yarn trajectory.

Appendix A2.56

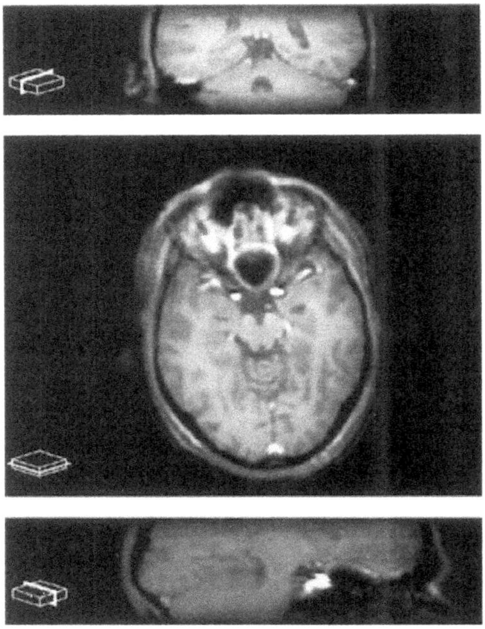

Figure A2.57 Coronal, axial, and sagittal slices of a 3D dataset of the head acquired with a spherical stack of spirals in 55.7 s.

Appendix A2.57

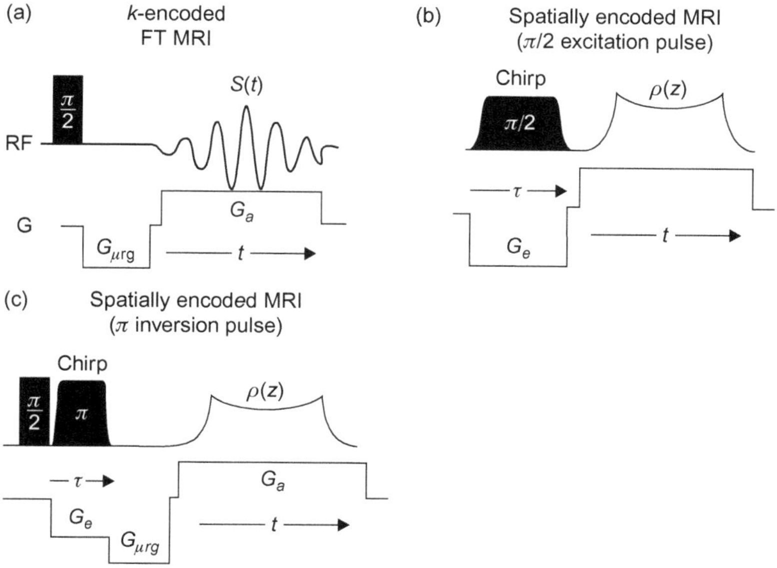

Figure A2.58 Comparison of schemes capable of yielding MR profiles $\rho(z)$ from a 1D sample. (a) Conventional scheme in which a gradient-driven k-domain $S(t)$ echo is monitored following a hard pulse, leading after FT to the 1D profile. (b) Non-FT scheme in which positions are initially encoded via the application of a frequency-swept excitation pulse, and subsequently read out via the application of a decoding gradient. (c) Same item as (b), but with the encoding phase profile resulting from an adiabatic p-sweep that follows the initial hard-pulse excitation of the spins.

Appendix A2.58

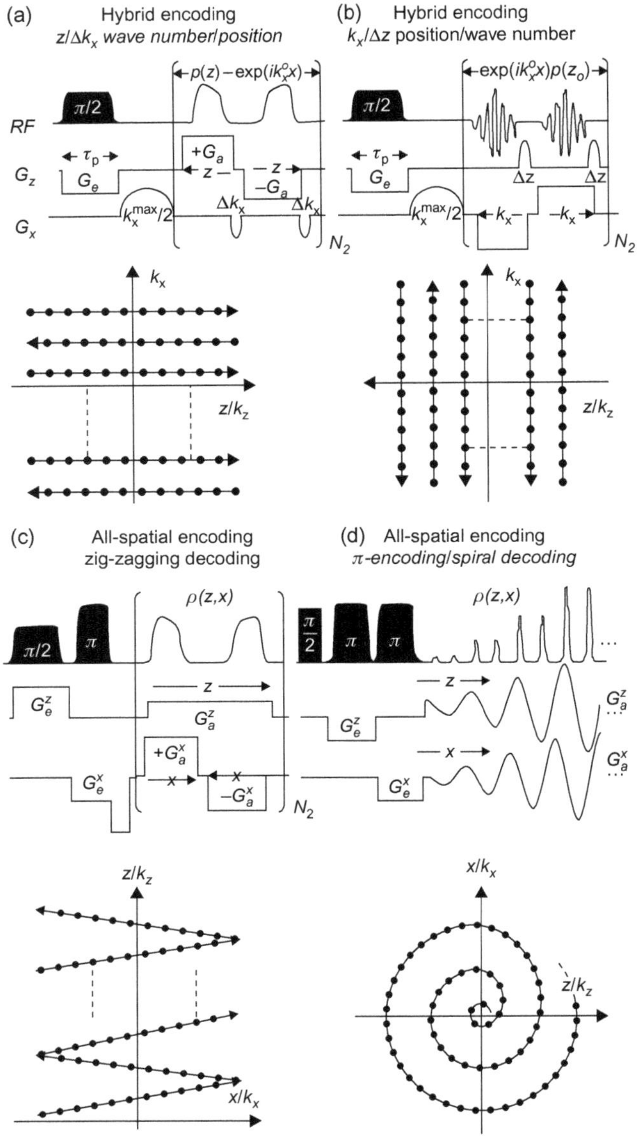

Figure A2.59 Examples of different schemes that, with the aid of spatial-encoding principles, can afford 2D MRI profiles within a single transient. (a) Hybrid scheme, relying on an initial Ge-driven encoding along one axis coupled to repetitive decoding echoes separated by small-phase incrementations along the orthogonal direction. (b) Hybrid scheme in which the roles of (a) have been reversed, and a single spatial decoding is interrupted by numerous repetitive *k*-domain acquisitions. (c) Purely spatial-encoding alternative, in which, following an initial $\pi/2-\pi$ encoding along orthogonal axes, the image is rasterized along a zigzag trajectory. (d) Spatial-encoding alternative whereby the profile is rasterized by an outward-expanding spiral trajectory. Shown for clarity are the *k*/*r*-space trajectories executed in each case, with dots representing the digitized data points.

Appendix A2.59

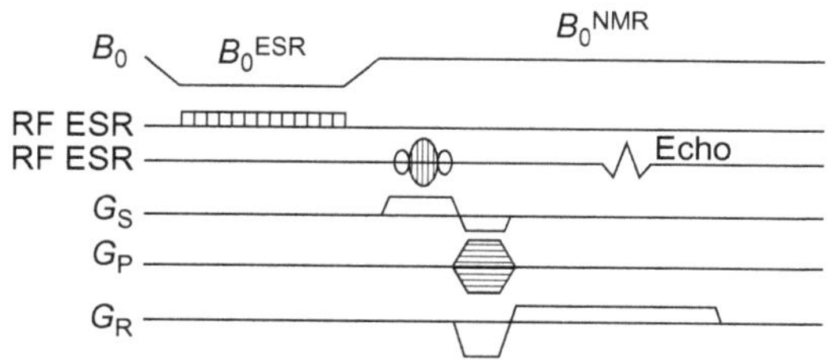

Figure A2.60 Field-cycled OEMRI pulse sequence starts with the ramping of the B_0^{ESR} field to 6.089 mT for ^{14}N-labeled nitroxyl radical or 6.569 mT for ^{15}N-labeled nitroxyl radical, followed by switching on of the ESR irradiation (220.6 MHz). The B_0 is ramped up to 14.529 mT before the NMR pulse and the associated field gradients are turned on. The field settling time was set at 18 ms.

Appendix A2.60

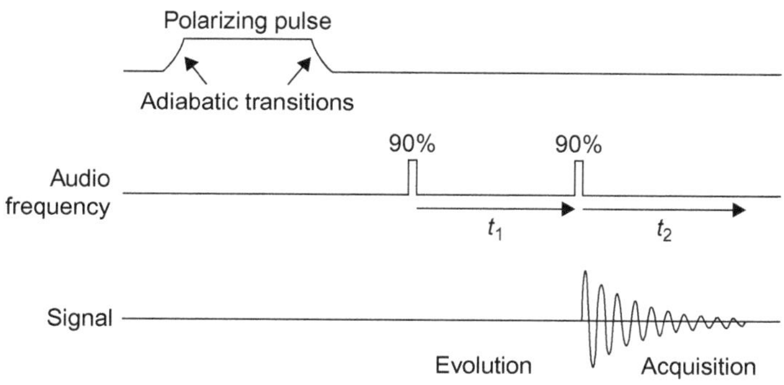

Figure A2.61 Timing sequence showing the prepolarizing pulse (duration around 6 s), and the two phase-locked audio frequency excitation pulses. The evolution period is t_1 and the acquisition time domain is t_2.

Appendix A2.61

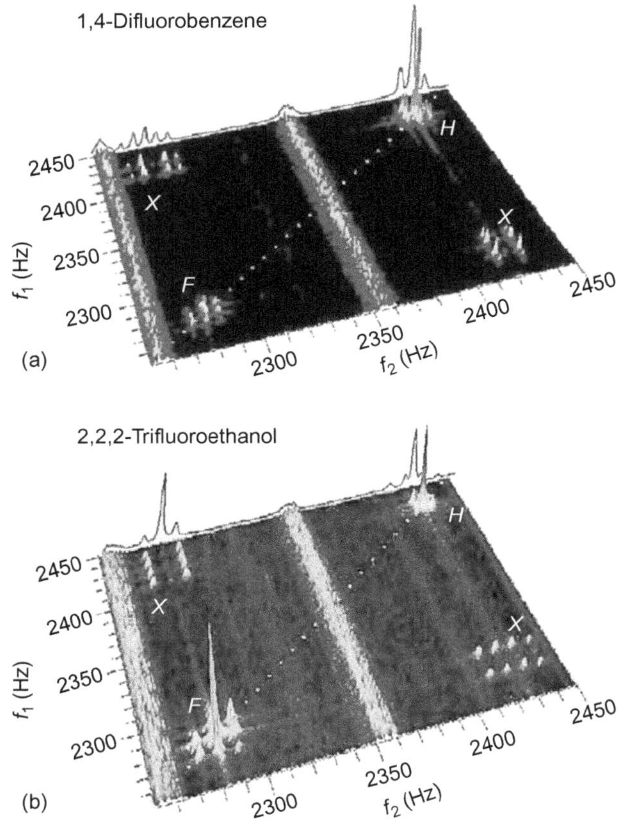

Figure A2.62 Experimental 2D COSY NMR spectra for (a) difluorobenzene and (b) trifluoroethanol. Both were obtained at audio frequencies 2.28 kHz (^{19}F) labeled *F*, and 2.43 kHz (^{1}H) labeled *H*. *X* represents the cross-peaks. The streaks along f_1 are from low-frequency interference.

Appendix A2.62

Table A2.5 Operator Matrices for a 2-Spin System ($I = S = 1/2$)

(1) Diagonal

$$O_1 = E = \begin{bmatrix} 1 & 0 & 0 & 0 \\ 0 & 1 & 0 & 0 \\ 0 & 0 & 1 & 0 \\ 0 & 0 & 0 & 1 \end{bmatrix} \quad O_2 = I_I = \frac{1}{2}\begin{bmatrix} 1 & 0 & 0 & 0 \\ 0 & 1 & 0 & 0 \\ 0 & 0 & -1 & 0 \\ 0 & 0 & 0 & -1 \end{bmatrix} \quad O_3 = S_z = \frac{1}{2}\begin{bmatrix} 1 & 0 & 0 & 0 \\ 0 & -1 & 0 & 0 \\ 0 & 0 & 1 & 0 \\ 0 & 0 & 0 & -1 \end{bmatrix} \quad O_4 = \mathrm{I}_g S_z = \frac{1}{4}\begin{bmatrix} 1 & 0 & 0 & 0 \\ 0 & -1 & 0 & 0 \\ 0 & 0 & -1 & 0 \\ 0 & 0 & 0 & 1 \end{bmatrix}$$

(2) Transverse *I* (SQCs)

$$O_5 = I_x = \frac{1}{2}\begin{bmatrix} 0 & 0 & 1 & 0 \\ 0 & 0 & 0 & 1 \\ 1 & 0 & 0 & 0 \\ 0 & 1 & 0 & 0 \end{bmatrix} \quad O_6 = I_y = \frac{1}{2}\begin{bmatrix} 0 & 0 & -1 & 0 \\ 0 & 0 & 0 & -1 \\ 1 & 0 & 0 & 0 \\ 0 & 1 & 0 & 0 \end{bmatrix} \quad \mathrm{O}_7 = \mathrm{I}_x S_z = \frac{1}{4}\begin{bmatrix} 0 & 0 & 1 & 0 \\ 0 & 0 & 0 & -1 \\ 1 & 0 & 0 & 0 \\ 0 & -1 & 0 & 0 \end{bmatrix} \quad O_8 = \mathrm{I}_y S_z = \frac{1}{4}\begin{bmatrix} 0 & 0 & -1 & 0 \\ 0 & 0 & 0 & 1 \\ 1 & 0 & 0 & 0 \\ 0 & -1 & 0 & 0 \end{bmatrix}$$

(3) Transverse *S* (SQCs)

$$O_9 = S_x = \frac{1}{2}\begin{bmatrix} 0 & 1 & 0 & 0 \\ 1 & 0 & 0 & 0 \\ 0 & 0 & 0 & 1 \\ 0 & 0 & 1 & 0 \end{bmatrix} \quad O_{10} = S_y = \frac{1}{2}\begin{bmatrix} 0 & -1 & 0 & 0 \\ 1 & 0 & 0 & 0 \\ 0 & 0 & 0 & -1 \\ 0 & 0 & 1 & 0 \end{bmatrix} \quad O_{11} = I_x S_x = \frac{1}{4}\begin{bmatrix} 0 & 1 & 0 & 0 \\ 1 & 0 & 0 & 0 \\ 0 & 0 & 0 & -1 \\ 0 & 0 & -1 & 0 \end{bmatrix} \quad \mathrm{O}_{12} = \mathrm{I}_z S_y = \frac{1}{4}\begin{bmatrix} 0 & -1 & 0 & 0 \\ 1 & 0 & 0 & 0 \\ 0 & 0 & 0 & 1 \\ 0 & 0 & -1 & 0 \end{bmatrix}$$

(4) ZQCs and DQCs

$$O_{13} = I_x S_x = \frac{1}{4}\begin{bmatrix} 0 & 0 & 0 & 1 \\ 0 & 0 & 1 & 0 \\ 0 & 1 & 0 & 0 \\ 1 & 0 & 0 & 0 \end{bmatrix} \quad O_{14} = I_y S_y = \frac{1}{4}\begin{bmatrix} 0 & 0 & 0 & -1 \\ 0 & 0 & 1 & 0 \\ 0 & 1 & 0 & 0 \\ -1 & 0 & 0 & 0 \end{bmatrix} \quad O_{15} = I_x S_y = \frac{1}{4}\begin{bmatrix} 0 & 0 & 0 & -1 \\ 0 & 0 & 1 & 0 \\ 0 & -1 & 0 & 0 \\ 1 & 0 & 0 & 0 \end{bmatrix} \quad O_{16} = \mathrm{I}_y S_x = \frac{1}{4}\begin{bmatrix} 0 & 0 & 0 & -1 \\ 0 & 0 & -1 & 0 \\ 0 & 1 & 0 & 0 \\ 1 & 0 & 0 & 0 \end{bmatrix}$$

Appendix A2.63

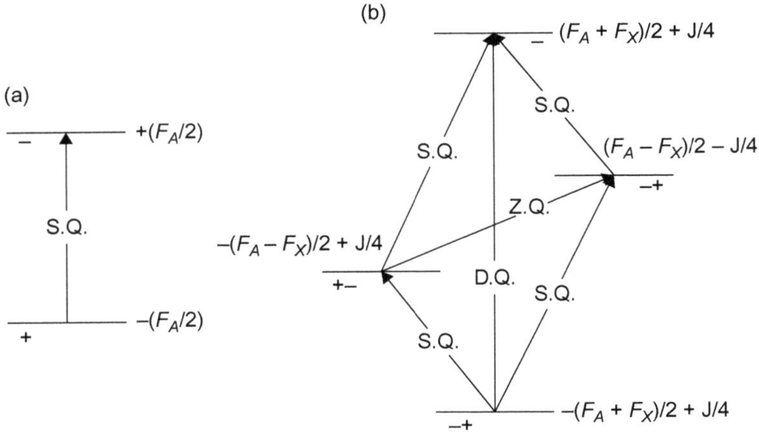

Figure A2.63 (a) The energy-level diagram for an isolated spin-*f* system in a magnetic field. (b) The energy-level diagram for a weakly coupled *AX* spin system in a magnetic field.

Appendix A2.64

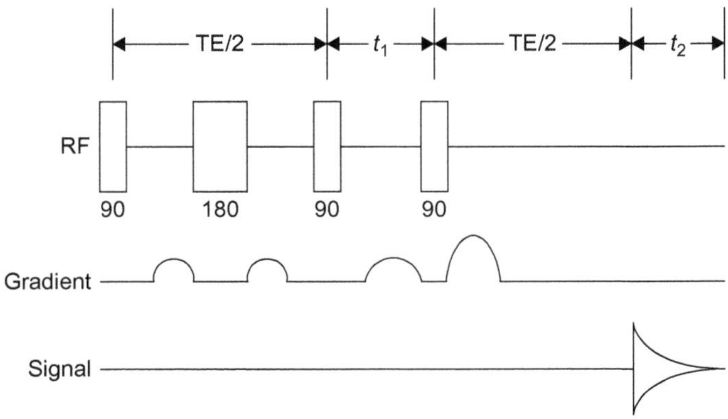

Figure A2.64 DQC transfer pulse sequence.

Appendix A2.65

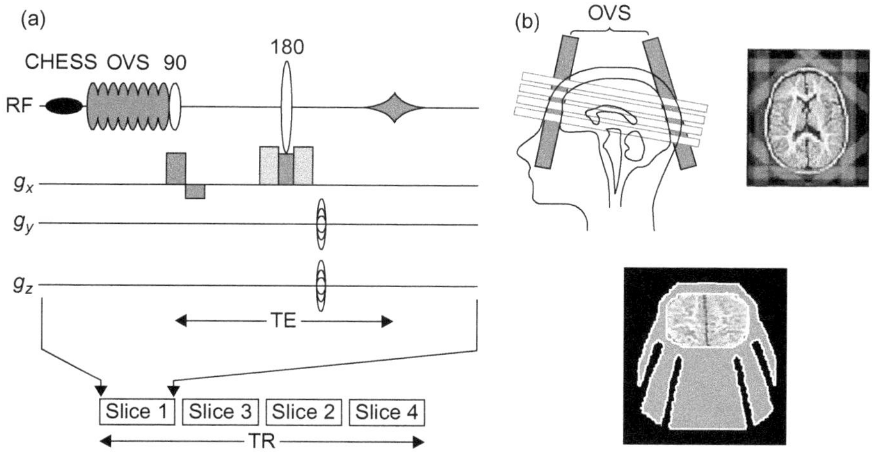

Figure A2.65 (a) Multislice 2D MRSI pulse sequence; a slice-selective spin-echo is preceded by eight outer-volume suppression (OVS) pulses (arranged to saturate the lipid signals in the scalp) and a CHESS water suppression pulse. Two-dimensional phase-encoding gradients are applied on g_Y and g_Z. In this example, four slices are interleaved within one repetition time (TR). Gradients associated with the OVS and CHESS pulses are omitted for clarity. (b) Schematic representation of the location of the OVS pulses, forming an octagonal cone in order to conform to the contours of the skull. The four oblique axial MRSI slices are also represented on the sagittal schematic.

Appendix A2.66

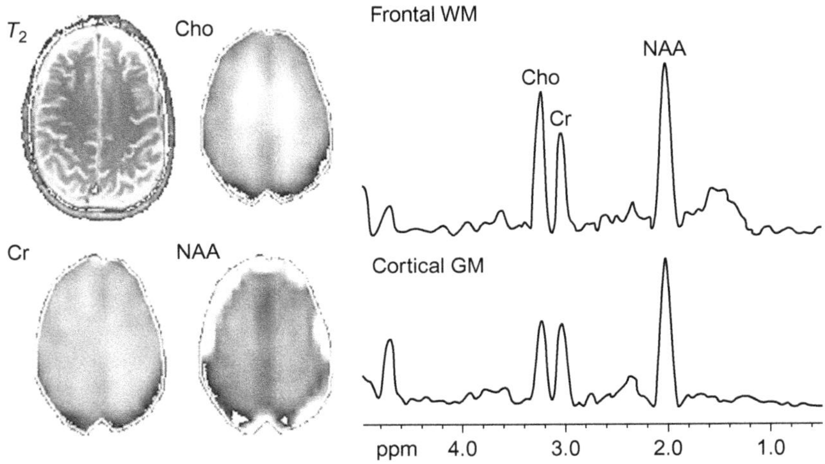

Figure A2.66 SENSE MRSI: sample data from a patient with HIV infection and dementia recorded at 3.0 T. The conventional T_2-weighted MRI scan is unremarkable except for the presence of atrophy, whereas MRSI shows an elevated choline signal bilaterally in the frontal-lobe WM, believed to be due to gliosis. The rapid data acquisition using SENSE MRSI allowed this relatively uncooperative patient to be scanned without using sedation or anesthesia.

Appendix A2.67

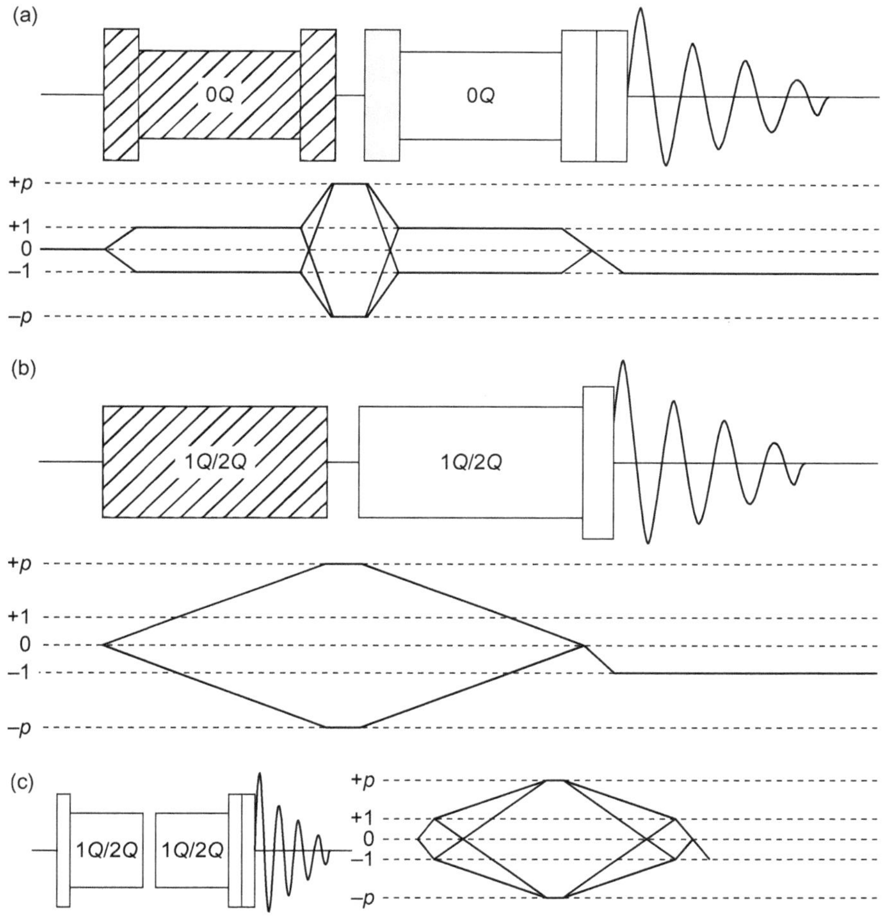

Figure A2.67 Schematics of spin-counting experiments. The taller rectangles indicate 90° pulses. (a) A ZQ scheme, in which transverse magnetization evolves under a ZQ operator to generate antiphase magnetization from which MQ states may be excited. The phases of the shaded 90° pulses were chosen to select particular parities of the coherence order. (b) A spin-counting experiment commencing from longitudinal magnetization, using SQ or DQ operators to directly excite MQ states. (c) A spin-counting experiment commencing from transverse magnetization, using SQ or DQ operators to excite MQ states. In all three experiments, the hatching indicates those parts of the pulse sequence to which a phase ϕ was applied. This phase can be cycled to achieve MQ filtering or incremented to generate a single dataset showing all excited coherence orders.

Appendix A2.68

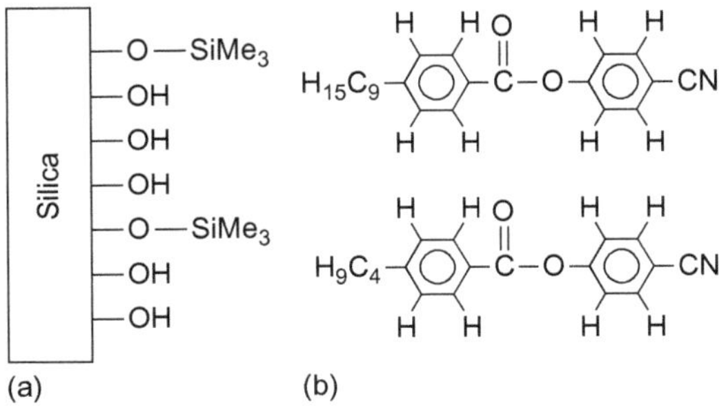

Figure A2.68 Trimethylsilane bound to silica, showing isolated trimethylsilane groups as indicated by spin counting. (b) The components of a liquid crystal on which others carried out spin-counting studies.

Appendix A2.69

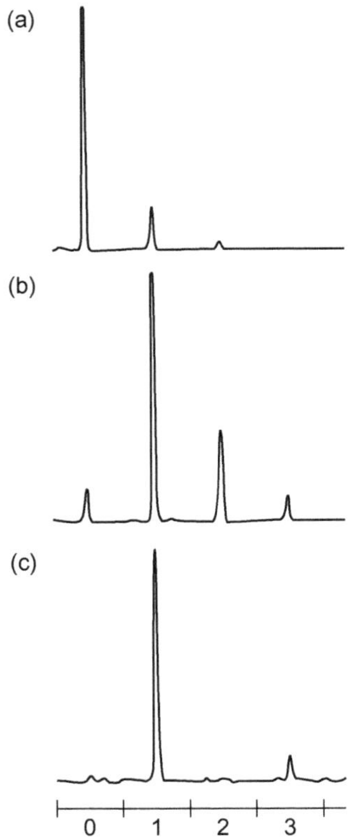

Figure A2.69 Experimental ^{13}C MQ NMR spectra of the 20-residue peptide helix 20, with ^{13}C labels at methyl carbons of Ala5, Ala8, Ala11, Ala14, and Ala17, in lyophilized form. The effective dipole–dipole coupling Hamiltonian for MQ excitation is the SQ operator HSQ in (a) and (b), and the DQ operator HDQ in (c). The initial condition and detected operator are I_z in (a) and I_x in (b) and (c). Excitation spectra for MQ orders 0 through 5 are plotted side-by-side, with a spectral width of 50 kHz in each order.

What Is Multiple-Quantum Coherence?

In the brain, a macro-molecule has a local order due to its own internal magnetic field.

This local order is created by the multiple-quantum coherences (MQCs), between electrons, atoms, nuclei, etc., within the hetero-nuclear character. The application of an external magnetic field in the laboratory creates imposed multi-quantum structure over larger distances.

It becomes the source of intermolecular multiple-quantum coherence imaging (iMQCI).

This is the concept behind the new technology, now emerging, called the quantum magnetic resonance imaging (QMRI). Double-quantum coherence (DQC) and the zero-quantum coherence (ZQC) are the two typical coherences used in imaging.

The physical picture of the two coherences is illustrated below.

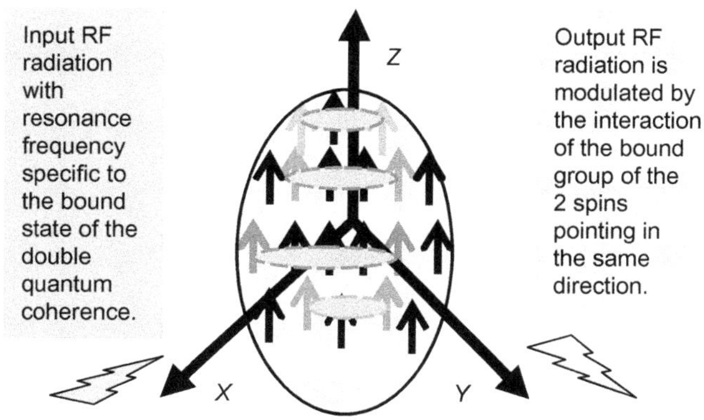

DQC: Two spins pointing in the same direction, near and far, act in quantum coherence, i.e., share energy in quantum jumps.

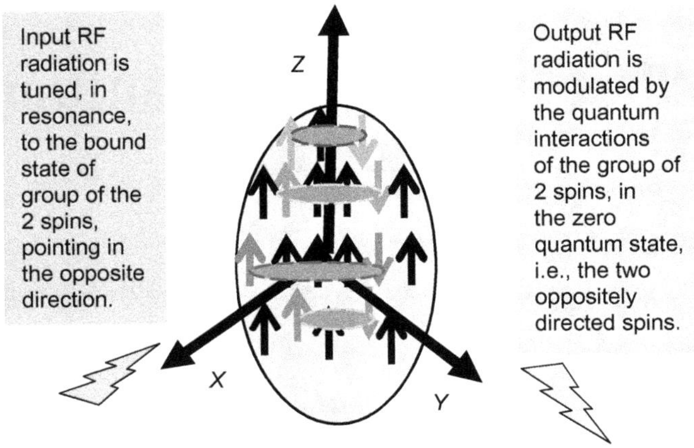

ZQC: Two spins pointing in the opposite direction, near and far, act in quantum coherence, i.e., share energy in quantum jumps.

3 Magnetic Resonance Imaging Diagnostics of Human Brain Disorders

Table of Contents

3.1 Introduction

3.1.1 *Education and Training in Radiology [1–4]*

This section illustrates some of the educational directions being pursued in the radiology subfield of MRI. Professional and governmental initiatives for regulation and standardization of procedures and education in nuclear medicine, as well as other radiological practices and procedures, are in place, but there is still a long way to go. Currency in education is made difficult by the rapid changes in technologies and the constant proposal of new concepts.

Through proper practical training and theory courses, medical professionals (MPs) can be brought to a required level of competence, so as to establish a uniform standard regarding procedures for evaluation of data, e.g., of magnetic resonance spectography (MRS), and of the images produced through MRI, magnetic resonance angiography (MRA), and other such technologies. MPs should have adequate education and training in areas like radiation biology, physics, instrumentation, and clinical aspects of radiology, among many others.

3.1.2 *Diagnostic Radiology*

3.1.2.1 *Vessel Tortuosity: Brain Tumor Malignancy [5]*

It has been observed that cancers express growth through alteration in the surrounding vasculature and blood flow, physiologically and morphologically. One searches,

Quantum Magnetic Resonance Imaging Diagnostics of Human Brain Disorders. DOI: 10.1016/B978-0-12-384711-9.00003-8

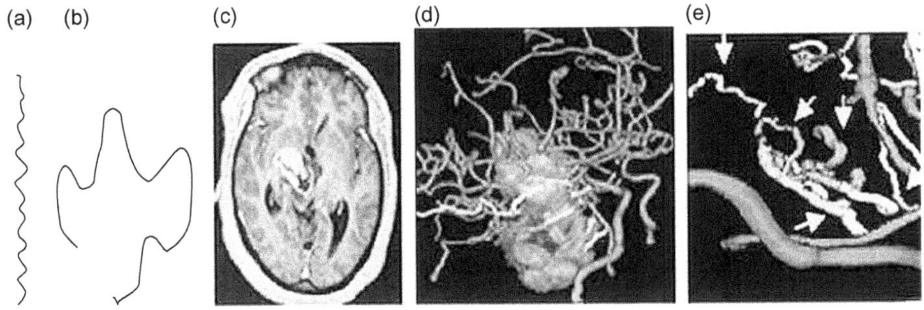

Figure 3.1 Examples of tortuosity types. (a) High-frequency wiggles resulting in high curvature per unit distance. This type of tortuosity is characterized by elevated sum-of-angles metric values. (b) Larger-amplitude curve with frequent changes and a short distance between endpoints. This type of tortuosity is characterized by elevated inflection-count metric values. (c) T_1 gadolinium-enhanced slice of a malignant tumor. (d) Three-dimensional rendering of segmented tumor and vessels. (e) Magnification of a cluster of tumor-associated vessels. The smooth vessels in the foreground lie far outside the lesion and have a normal shape. The vessels within or traversing the tumor exhibit abnormal tortuosity by both metrics (arrows).

for example, for foci of neoangiogenesis or for abnormal vascular permeability. One often finds malignancy-induced alterations of vessel shape affecting the larger vessels visualized directly by MRA. Because cancer and many other diseases alter vessel morphology, a quantitative measurement of vessel shape offers potentially valuable information complementary to that provided by perfusion and other imaging methods.

Figure 3.1 is a pictorial illustration of various tortuosity types.

3.1.2.2 Diffusion-Weighted Images: Intracranial Cyst-Like Lesions [6]

It is possible to design MR sequences to evaluate the diffusion movements of protons, through diffusion-weighted images (DWI). In these images, a bright signal identifies a region where the diffusion along a spatial axis is restricted. The contents of a cystic lesion frequently have the signal intensities of a generic homogenous hyperproteinic fluid, i.e., hypointensity in T_1-weighted images (longitudinal relaxation time in NMR), and hyperintensity in T_2-weighted images.

DWI can also give information about the microscopic organization of these fluids. Differential diagnosis of cystic lesions is mainly based on location, borders, presence of mural nodules, calcifications, and satellite or multiple lesions. The content of the cyst usually appears as a homogenous hyperproteinic fluid in MRI studies, giving limited adjunctive information. The signal in MRI sequences depends on chemical composition of tissues; cellular organization and metabolism are of little relevance. DWI is designed to evaluate diffusion movements of the molecules on a spatial axis. These incoherent microscopic molecular movements cause loss of phase of the protons and reduce signal intensity. Diffusion may be evaluated on a different spatial axis at different time/space scales, depending on the direction,

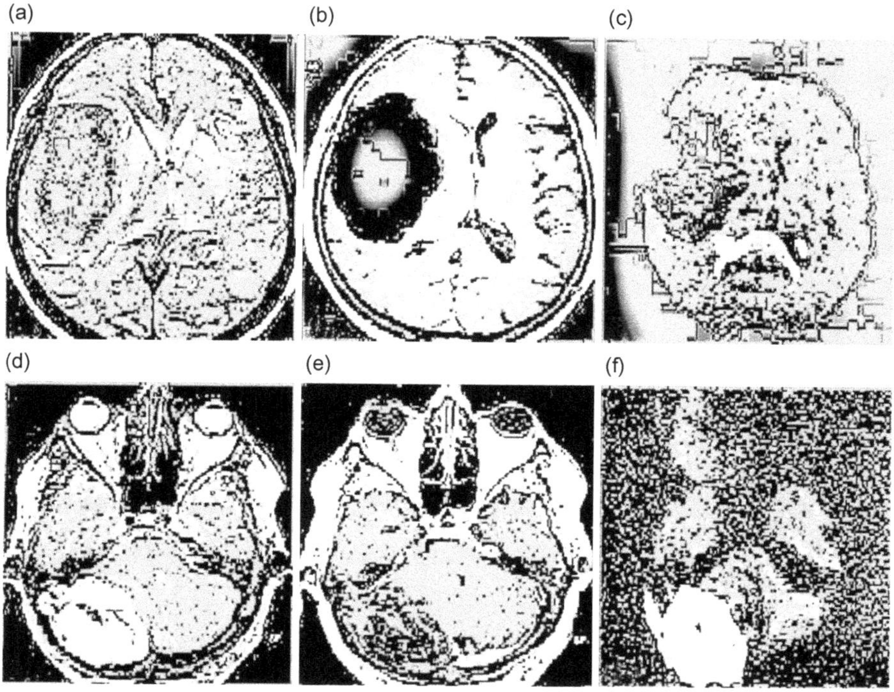

Figure 3.2 Arachnoid cyst (patient 20). T_1, T_2 and DWI (diffusion weighted images). (a) T_2-weighted turbo spin-echo axial slice (2800/12; echo train length 5). (b) T_1-weighted spin-echo axial slice (600/12). (c) DWI with echo planar sequences axial slice (6600/160). Signal intensities of the cyst in all the sequences, including DWI, are the same as for the CSF. Cholesteatoma (patient 24) (d, e) T_2-weighted turbo spin-echo axial slice (2800/98; echo train length). (e) T_1-weighted spin-echo axial slice (600/12). The lesion appears isointense to the CSF in all the sequences. (f) DWI with echo planar sequences axial slice (6600/160). The lesion is strongly hyperintense and is clearly different from the CSF.

timing, and strength of the applied magnetic field gradients. Signal may simply be converted into an image or may be further elaborated to obtain a quantitative map of the ADC. A region of restricted diffusion on a given spatial axis will appear hyperintense on the DWI and hypointense on the ADC maps. Diffusion in the tissues depends on capillary flow and on the intracellular movements requiring active metabolism. It is strongly limited by cell membranes and is preferentially oriented along the direction of the fibers. DWI can give information on the cellular organization and metabolism of a tissue that cannot be obtained with other techniques. It can differentiate tissues, with the same signal on T_1- and T_2-weighted images, based on their microscopic organization and metabolism. DWI is highly sensitive to artifacts—in particular, to macroscopic movements. This may be useful in some cases because it increases the contrast between stationary and mobile tissues, but it usually worsens SNR (Figure 3.2).

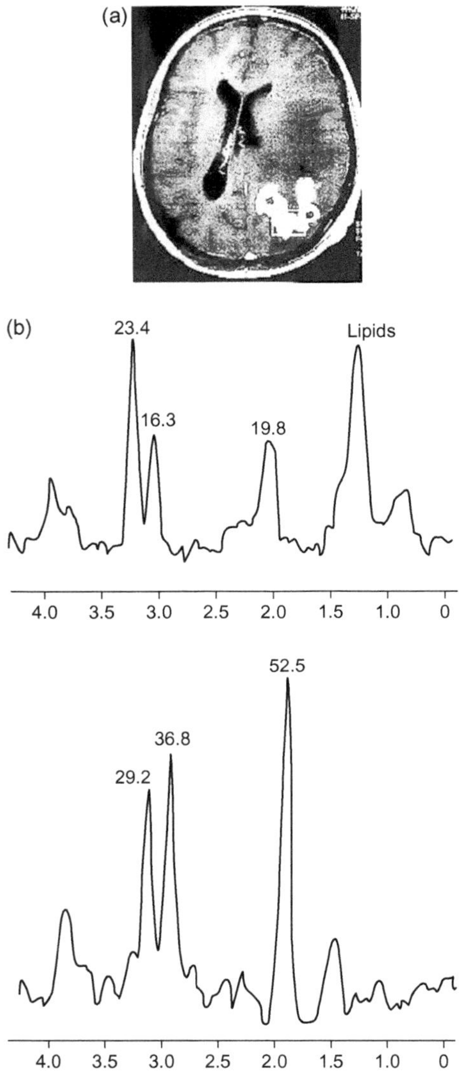

Figure 3.3 Brain abscess. (a) Location. Left Spectrum: mass lesion. Right spectrum: contralateral brain. (b) Values at top of NAA (2.0 ppm), Cr (3.0 ppm), and Cho (3.2 ppm) reflect signal peak integrals (arbitrary units). Metabolite levels relative to contralateral brain: NAA 38%, Cr 44%, Cho 80%. Resonances at 1.3 and 0.9 ppm are due to lipids. In contrast to metastases, high-grade gliomas, and other neoplastic lesions, Cho levels are decreased in inflammatory brain disease.

3.1.2.3 PMRS: Diagnosis of Intracranial Mass Lesions [7]

PMRS provides useful information, complementing that gained through MRI, with regard to cell membrane proliferation, neuronal damage, energy metabolism, and necrotic transformation of brain or tumor tissues. One needs to be able to interpret both images and the spectroscopy data accurately to derive information about the location, edema, mass effect, calcification, cyst formation, visualization, and contrast enhancement, given the patient's age and clinical presentation.

MRI, computer tomography (CT), and angiography methods provide what is basically structural data. PMRS elucidates different information, related to neuronal integrity, cell proliferation or degradation, and energy metabolism. In conjunction with structural imaging modalities, it also reveals necrotic transformation of brain and tumor tissues. PMRS plays an increasingly important role in a number of common neurological disorders, such as stroke, epilepsy, multiple sclerosis (MS), HIV, dementia, head injury, and near-drowning. The great clinical utility of PMRS lies in its ability to help solve key diagnostic problems; e.g., differentiation between neoplastic and nonneoplastic lesions; or between low- and high-grade tumors, infarction. However, methods of separating low-grade gliomas and metastases from primary brain tumors and abscesses have yet to be properly established (Figures 3.3 and 3.4).

3.1.2.4 Copper Metabolism in Wilson's Disease: Acute Putaminal Necrosis and WM Demyelination [8]

Neurodegenerative Wilson's disease (NWD) is a genetic disease caused by mutations in a P-type ATPase; specifically, Wilson's disease (hepatolenticular degeneration) is an autosomal recessive disease, caused by mutations in the ATPFB (copper-binding ATPase) gene, located at band q14.3 of chromosome 13. This defect leads to abnormal copper transport and metabolism in mitochondria and causes apoptotic and necrotic cell death by oxidative damage after excess accumulation of intracellular copper. Neurological symptoms, when present, usually follow the typical course of hepatolenticular degeneration. This is accompanied by clinical presentations of overwhelming hepatic and ocular copper accumulation, i.e., hepatic dysfunction and the typical Kayser–Fleischer ring of the cornea.

The impairment of copper transport across cell membranes leads to accumulation of copper in the liver, brain, cornea, and kidney, and causes toxicity to those organs. The damaged copper transport mechanism secondarily causes a low serum ceruloplasmin level (Figure 3.5).

3.1.2.5 Brain Metabolite Changes: Children with Poorly Controlled Type 1 Diabetes Mellitus/PMRS [9]

Type 1 diabetes mellitus (DM) is a complex metabolic disorder, and is often associated with complications secondary to central nervous system (CNS) involvement. In addition to acute changes that develop during episodes of hypoglycemia or hyperglycemia, there is long-term impairment of cognitive functions in type 1 diabetes mellitus patients who have poor glycemic control. These cerebral complications may be referred to as *diabetic encephalopathy*, which is diagnosed by clinical, psychological, and electrophysiological tests. The underlying mechanisms that cause brain damage in diabetes are not clear. Fluctuations in blood glucose level, secondary to exogenous insulin use, and acute and/or chronic metabolic and vascular impairment may cause functional and structural changes in diabetic patients.

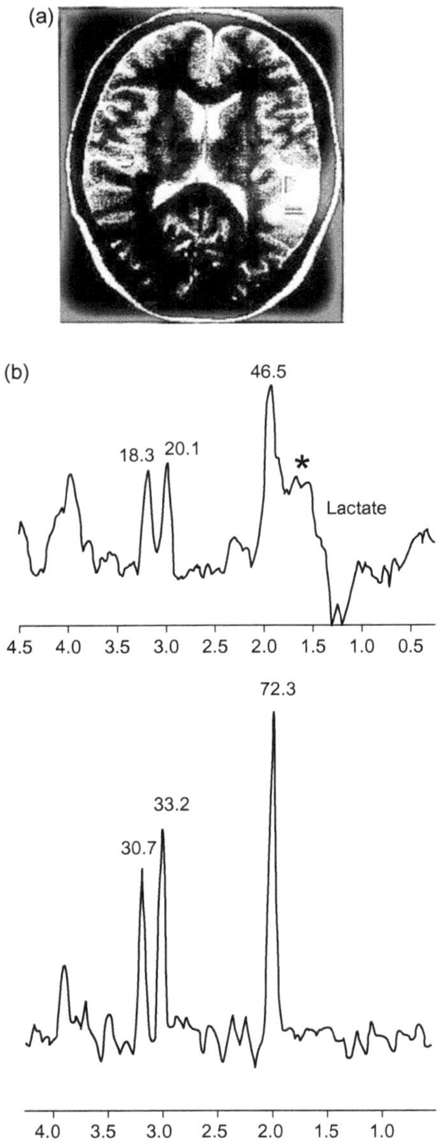

Figure 3.4 Cerebral infarction. (a) Location. Left spectrum: lesion. Right spectrum: contralateral brain. (b) Values at top of NAA (2.0 ppm), Cr (3.0 ppm), and Cho (3.2 ppm) reflect signal peak integrals (arbitrary units). Metabolite levels relative to contralateral brain: NAA 64%, Cr 61%, Cho 59%. All brain metabolites (including Cho) are decreased. Note the prominent phase-reversed lactate resonances at 1.3 ppm and the acetate peak at 1.8 ppm.

In vivo PMRS is a noninvasive imaging method for monitoring chemical and metabolic changes in areas of interest within the brain. PMRS can provide information about the concentration and relative levels of proton-including metabolites, and may assist in distinguishing between normal and pathological tissues.

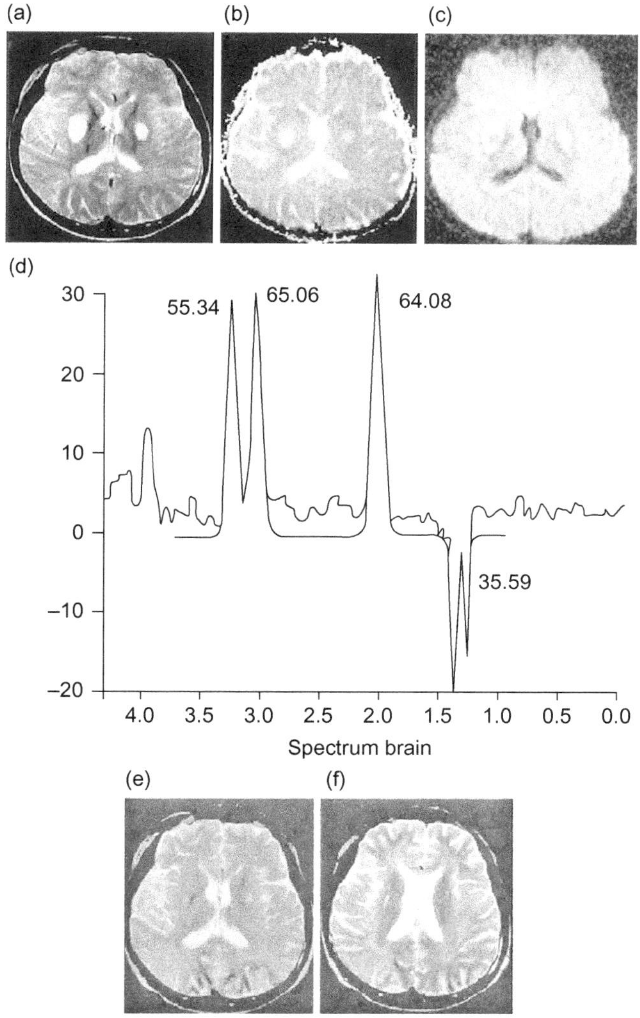

Figure 3.5 A 12-year-old child with NWD and subnormal copper metabolism. Initial MRI study at the level of the basal ganglia shows asymmetrical putaminal hyperintense lesion on T_2WI (a). The lesions are hyperintense on ADC maps (b) and hypointense on DWI ($b = 1000$) (c). ^{1}H proton MRS shows a decreased NAA/Cr ratio, normal Cho/Cr ratio, and elevated Lac/Cr ratio (d). A repeated MRI study 6 months later shows collapse of the bilateral putamina and diffuse WMH over the bilateral cerebral hemispheres on T_2WI (e, f). ADC values of the locations corresponding to the previous putaminal lesions were slightly higher on the right and normal on the left compared with the control group ($0.89 \pm 0.11 \times 10^{-3}$ mm^2/s on the right and $0.78 \pm 0.09 \times 10^{-3}$ mm^2/s on the left). The WMH became more extensive on T_2WI, as seen from adjacent slices such as that shown in (f), indicating more generalized WM demyelination.

3.1.3 Basics of QM [10]

3.1.3.1 Quantum Radiation (Light Waves)

The light radiation received from a source (e.g., our Sun) is received as quantums of photons. These photons have discrete energy spectra, in both wavelength and frequency; it is not a continuous spectrum. This is now common knowledge. Planck developed the following analytical equation of the energy a bundle (quantum) of photons of a particular frequency, f, receives. It is expressed as $E = hf/(e^{hf/kT} - 1)$. Here k is the Boltzmann constant ($=1.3806 \times 10^{-23}$ J/K), h is Planck's constant ($=6.6260755 \times 10^{-34}$ Js), and T is the temperature of the source of the photons in °K (Kelvin). At the secondary level of education in science, students are introduced to the electronic quantized orbits around a nucleus, and the associated quantum jumps between orbits, that is responsible for emitting light radiation of different frequencies and wavelengths. This process has been verified for all the chemical elements in nature (i.e., all the elements on the periodic table). In fact, the periodic table of elements has been arranged and classified according to the electronic orbit configurations of the different elements.

What can be more convincing evidence that we live in a quantum world? Bohr developed the following mathematical relation, for frequency f of radiation, to depict the quantum nature of light: $f = (2\pi^2 me^4/h^3)(1/n_j^2 - 1/n_i^2)$. Here, e is the electronic charge (1.602177×10^{-19} C), m the electronic mass ($=9.10939 \times 10^{-31}$ kg), and n_j, n_i the lower- and higher-orbit quantum numbers (1, 2, 3, . . .), respectively.

The radius of a quantized electronic orbit is given as $r = (1/4\pi^2)(n^2h^2/e^2m)$, n being the quantum number corresponding to a particular orbit. For $n = 1$, the closest orbit, $r = a_0$, called the Bohr radius, is $a_0 = 0.529 \times 10^{-10}$ m. The simple electronic transition theory mentioned earlier and the associated radiation emission have been experimentally verified for all sources of electromagnetic radiation, not just visible light. In fact, most of the communication gadgets that human civilization uses today are based on the use of some form of electromagnetic radiation. Some examples are the use of optical (visible) light, in television, and invisible light (e.g., X-rays), for human skeleton imaging, and so on. These technologies are based, directly or indirectly, on the lowest level of knowledge of QM. As one goes a bit higher up in technology level (e.g., to lasers, MRI machines), one finds that the QM involved has also been developed further. Basic QM is well suited to creating and forecasting the performance of technological applications (e.g., of MRI machines) in diverse areas. QM is a fundamental tool, which forms the basis for research and development, and for exploration of further applications, in fields like materials analysis and medical diagnostics. Yet education in QM, in the wider community, has never changed. The only people who study QM as a discipline are physicists, mathematicians, and chemists, as part of their tertiary education, and perhaps to a lesser extent electrical and electronic engineers. There is no reason not to teach QM at the high school level; QM is no more difficult to learn than Newtonian mechanics (NM). NM is already a standard part of study at the secondary level. Once QM is integrated into the school education system, as a normal secondary-level subject, then more and more students will choose to take it up at the tertiary level, and thus QM can take its rightful place as an essential part of all the disciplines of science.

3.1.3.2 *Transition from Light Waves to Matter Waves*

It is now common knowledge that not only photons, but also particles such as electrons and protons, behave as waves in some circumstances and particles in others. Electron microscopes work on the principle of wave phenomena, using diffraction of electron waves. Hence, the photon equation E (energy) $= \hbar\omega$, where $\hbar = h/2\pi$ and $\omega = 2\pi f$ (f is linear frequency expressed as c/s and ω is angular frequency expressed as rad/s), can also be applied to an electron, in the particle sense. Now, a photon has zero mass, and according to the theory of relativity, one can write the energy of radiation of the photon as $E = pc$. Here p is the momentum and c is the velocity of light. For waves, one can write the relation for the linear frequency of the radiation emitted as $f = ck$. Here k is the linear wave number (numbers/meter) $= 1/\lambda$ (it can also be expressed as an angular wave number, by multiplying with 2π), and λ is the wavelength of the wave, expressed in linear units, in nm (10^{-9} m). Combining these two, one gets $p = hk$. On writing, k, in angular units (rad/m), one can write, $\lambda = \hbar/p$. What this means is that the wavelength of the wave of a particle of matter is inversely proportional to the particle's momentum; the constant of proportionality is Planck's constant. It took de Broglie's great insight to suggest that this relation holds for all matter and all particles. The constant h thus fixes the order of scale (time and space) of all things.

There are amazing things happening in the quantum world of matter, which is all around us. The human body experiences and is subject to the quantum nature of matter as well, all the time.

3.1.3.3 *Wave—Particle Duality*

3.1.3.3.1 The Schrödinger Wave Equation

About a century ago, Schrödinger, using his intuition, developed the following equation: $i\hbar[\partial\Psi(x, t)/\partial t] = (-\hbar^2/2m)(\partial^2\Psi(x, t)/\partial x^2)$. It is fittingly called the *Schrödinger equation.* This seemingly strange equation provided a suitable alternative to the world of Newtonian mechanics. NM is suitable only for rigid, noninteracting bodies. Thus, the advent of the Schrödinger equation created a new world: the quantum world. One can certainly wonder whether civilization would be the same as it is today without this development.

This equation is really no different from any differential equation used in NM. The only difference is that, instead of analyzing macroscopic quantities like velocity and momentum, as an unknown quantity, as is done in NM, Schrödinger's equation examines the wave function of a microscopic particle. It describes the energy transitions within the particle, as well as the particle's interaction with other particles in the surrounding environment. This equation has enabled us to redraw the entire atomic periodic table based on the single unique footing of electronic orbit structure;

that is, of shells' orbital quantum numbers, $n = 1, 2, 3$, and their subshells, s, p, d, f, corresponding to angular-momentum quantum number $l = 0, 1, 2, \ldots$.

Schrödinger's formulation was, as noted, a product of his intuition. Perhaps the environment in those days was very congenial for such intuitive leaps; despite being a virtue that almost every human being has, no comparable intuition-based breakthroughs have been made in the decades since. The world today is a much smaller place; better means of communication, better research facilities, etc., are available. Why, then, has there been such delay in the next "quantum leap" in science?

Now let us put philosophy aside and do some more work in exposing the wonders of QM. One can consider the solution of the above equation as a plane wave, $\Psi = \exp(ikx - iwt) = \exp[i(px - Et)]/\hbar$ (E is the energy and t the time). By substituting in the above equation, the result $E = p^2/2m$, using $p = mv$, one gets $E = 1/2mv^2$. This is the familiar NM particle (rigid body) equation, called the *kinetic energy* of a moving rigid body. Thus in the limit, the wave equation of an SQ particle and NM give the same result. It should be pointed out here, though, that NM cannot be straightforwardly applied to an ensemble of atoms, molecules, nuclei, etc.

The particles in an ensemble such as the human brain react with each other, exchange quantums of energy, and perform metabolic activities, etc. To more closely model reality, one need to define a physical aspect of the wave function, analogous to that of a particle, so that the velocity, momentum, etc, of a particle can be measured.

3.1.3.3.2 Probability of Measurement of a Wave Packet

A single plane wave, $\varphi(x, t) = \exp[i(kx - \omega t)] = [\cos(kx - \omega t) + i\sin(kx - \omega t)]$ (sine or cosine wave) is used to describe a particle. It is infinite in its extent in space. The mathematical method of representing such a wave (sine or cosine) makes mathematical treatment of a wave much easier to handle in a real situation. The symbol $i = \sqrt{-1}$ takes the mathematical formalism into a complex domain treatment of trigonometry. Once a solution of the wave performance—say, a solution through a differential equation—is obtained, then one simply looks in the solution for cos wave in the real part, and for the sin wave in the imaginary part, of the expression. (The mathematical formalism of complex trigonometry is presently part of tertiary-level coursework in the mathematics discipline, but it could easily become part of secondary-level education.) One needs to use a superposition of waves to localize a wave, and therein lies the particle aspect of the wave. A good example is the interference of light waves through a diffraction grating. Superposition of waves results in the production of maximas and minimas, resulting in bright and dark fringes and ultimately a localized intensity of light. This makes measurement of the amplitude and phase at a particular point much easier, because there are more waves (particles) reinforcing each other at bright spots than there are at dark spots (where they are destroying each other).

This enables a greater probability of finding a particle at a particular position. The modulus square of the wave function $\varphi(x, t)$, at a point, gives the probability,

$P(x, t) = |\varphi(x, t)|^2$, of finding the particle at a particular point. The total probability over space is $\int_{-\infty}^{+\infty} P(x, t)\mathrm{d}x = 1$. A wave function satisfying this condition is said to be *normalized*. For a plane wave, $P(x, t) = |\exp[\mathrm{i}(kx - \omega t)]|^2 = 1$.

One needs to use a wave packet (superposition of waves) to precisely define the position of a particle, as the particle may be anywhere, at any particular instant of time. One should realize that the Schrödinger equation is a linear equation, so the superposition of waves (wave packet) is also its solution. This is the core of quantum mechanical science: It is called the *superposition principle*.

3.1.3.3.3 A Familiar Wave Packet—The Gaussian Wave Packet: Localization of a Particle

It is mathematically convenient to choose the well-known Gaussian wave packet $(\mathrm{GWP}) = \exp[-(k - k_0)/2(\Delta k)^2]$. Here k is the wave number $= 1/\lambda$, λ being the wavelength of the wave. The modulo square $|\mathrm{GWP}|^2$ of this wave packet is the familiar bell (random distribution) curve. The particle localization procedure is carried out by using familiar FT mathematics. A wave packet, centered at position x_0 (real space), can be represented as $\varphi(x, x_0) = \int_{-\infty}^{+\infty} \mathrm{d}k \exp[\mathrm{i}k(x - x_0)F(k, k_0)$. This is called the FT of $\varphi(k, k_0)$. One should notice here that when $x = x_0$, the exponential term $= 1$ and the contributions to the integral from different ks add up in phase (constructive interference). If $x - x_0$ is large, though, the contributions from different ks oscillate and produce destructive interference. One can choose $F(k, k_0) = \mathrm{GWP}$.

One can then calculate the probability (x) of the normalized wave packet, i.e., by following the condition $\int_{-\infty}^{+\infty} \varphi^* \varphi\, \mathrm{d}x = 1$. This is seen to lead to the condition $\Delta x\, \Delta k \geq 1$. Here φ^* is the complex conjugate of φ (i.e., one replaces i in φ by $-$i, to get φ^*). By using the quantum relation $p = \hbar k$, the above uncertainty condition becomes $\Delta x\ \Delta p \geq \hbar$. This is the familiar uncertainty principle of QM. The uncertainties in Δx and Δp are the known, standard deviations in measurement of a quantity: $(\Delta x)^2 = \langle x^2 \rangle$ (average of the squares) $- \langle x \rangle^2$ (square of the average), $(\Delta p)^2 = \langle p^2 \rangle$ (average of the squares) $- \langle p \rangle^2$ (square of the average).

3.1.3.4 Position Momentum and Energy Operators in QM: The Human Brain Model

In NM, the law of motion states that momentum p is related to mass and velocity as $p = mv$. In QM, p becomes an operator: $p = -\mathrm{i}\hbar(\partial/\partial x) = \hbar/\mathrm{i}(\partial/\partial x)$, $\mathrm{i} = \sqrt{-1}$. Similarly, the position x becomes an operator: $x = \mathrm{i}\hbar(\partial/\partial p)$. In classical physics (NM), physical quantities x and p commute. In QM, however, they do not necessarily commute. The first basic law of QM is $[xp - px] = \mathrm{i}\hbar$. The second law is that the operators have a real expectation value. Mathematically, it means, for an operator, $\langle O \rangle_\psi = \langle O \rangle^*_\psi$.

The symbol * denotes the complex conjugate of a quantity (in a matrix representation of an operator, this amounts to changing rows into columns and columns into rows, and i into $-$i). This means that an operator in QM is equal to its complex conjugate. The operators with real expectation value are called *Hermitian* *operators*. In QM, the energy operator is written as $H = p^2/2m + V(x)$. This is clearly analogous to NM, in the sense that the first term corresponds to the familiar kinetic energy (KE), and the second term to the potential energy (PE). In QM, p^2 is also an operator and $= [-i\hbar(\partial/\partial x)][-i\hbar(\partial/\partial x)] = -(\hbar^2\partial^2/\partial^2 x)$ (p is an operator). The potential energy $V(x)$ of a particle—say, an electron or a proton—in a confined environment (potential), such as the restricted environment of a human brain, can be modeled mathematically. One can work out the possible energy levels in an externally applied RF field (additional external source of potential energy). These can be compared with experimental NMR spectroscopy results of the human brain, and the discrepancies can be used to further improve the brain model. The Schrödinger equation energy-level solutions for the free atoms of the chemical elements in the periodic table have been worked out. Their agreement with experimental results is well documented, and the results are tabulated in chemistry and physics handbooks. The basics that apply to electron spectroscopy of elements also apply to the nuclear (proton) spectroscopy of the elements. This bank of prior knowledge is a valuable tool for the new breed of researcher. It provides a starting point from which they can build magnetic resonance imaging of the human brain (MRIOHB) into a versatile tool for diagnosis of brain disorders in the future.

3.1.3.5 The Concept of Measurement in QM

The effect of an operator O operating on a set of eigen (energy) functions ψ_i is to generate eigenvalues o_i, such that $O\psi_i = o_i\psi_i$. The set of functions ψ_i are the eigenfunctions of the eigenvalue equation (e.g., the Schrödinger equation) of the operator O. This constitutes an infinite set of linearly independent, orthogonal functions. They are chosen to be a complete orthonormal set, such that $\int \psi_{i*}\psi \, dx = \delta_{ij}$ ($=1$ if $i = j$ and $=0$ if $i \neq j$). Using this concept, a wave function in QM is expressed as superposition of the complete orthonormal set, as above, i.e., $\psi = \Sigma_j c_j \psi_j$. One can show that the expectation value of the operator O, in the orthonormal set, is $\langle O \rangle = \Sigma_j |c_j|^2 o_j$. Thus one can say that if one measures a physical quantity, controlled by the operator O, with eigenvalue o_j, then there is a possibility of measuring, with probability 1, if $|c_j|^2 = \delta_{ij}$. This means that the system is in the state $\psi_i(i = j)$. To ensure reproducibility of the measurement, one must make certain that a coherent superposition reduces to an eigenfunction upon measurement. One calls this the *reduction postulate*.

3.1.4 Matrix Mechanics Approach to QM: The MM-QM

When there is an ensemble of nuclei (say, for example, protons), as in a situation such as the human brain, one must write the Schrödinger equation for each particle, and solve the equations in a collective manner, as an interacting group. This leads to the impossible task of formulating a single Schrödinger equation for the ensemble, with many different nuclei (in different internal and external potential fields, different coordinate points in time and space, etc.), and then solving it. Readers are referred to a standard text to learn how the complexity of the problem multiplies as one goes from a simple atom like hydrogen (single electron and single proton) to a slightly more involved atom, such as helium (two electrons, two protons, and two neutrons). One quickly realizes that solution through the Schrödinger equation of a multiparticle system becomes an impossible task.

During early part of the past century, some ingenious physicists were able to invent a mathematical formalism that can handle such large and diverse systems. This is called matrix mechanics (MM) formalism. Its concepts are simple, and can easily be extended to systems with many different particles, and of different nature, as in the human brain. Because matrices, like QM operators, do not commute (i.e., for matrices A and B, $AB \neq BA$), it was realized that they offer the right choice for analyzing QM operations. The MRI technique, basically, involves exciting quantum mechanical states of spins of nuclei; this is achieved by applying a static magnetic field in one (z) direction, normally followed by application of two transverse (x-, y-directions) RF fields. The RF excitation fields are tracked by RF measurement sensor (coils), which scan the output over a selected area; the scan results are then electronically converted into an image. All the quantum excitations of the nuclei and their interactions can be mathematically worked out through MM, and the mathematical results can then be compared with experimental results. Such spectroscopic studies, in vitro and in vivo, have enabled scientists to construct standard tables, graphs, and other references. This standard reference material can be very valuable for MPs in developing methodologies for medical diagnosis of various brain disorders.

3.1.4.1 *Basic Rules of MM*

A wave function $\varphi(x)$, describing a quantum particle in space, can be expressed as a linear combination of another set of functions. This new set of functions is composed of eigenfunctions Φ. They form a complete set of representations for an operator O (which will be taken as a matrix). Thus we say $\psi(x) = \Sigma_i c_i \Phi_i$. Now, as with any mathematical formalism, MM has to be developed in such a way that the whole system is mathematically homogenous and consistent. This is like building a pyramid: Even as you build higher levels of the pyramid, the basic building block (i.e., the foundation or smallest block) remains the same. In QM, to achieve valid multiplicity of the system, we use coordinate space, called *Hilbert space (HS)*. This space can be built to any dimension. The basis (the lowest and extendible blocks) of the states comes from 2^N. N in our case has two values, i.e., spin up or spin down. So, the dimension of HS becomes $2^2 = 4$. If we have 2 spins, with 4 states (two of each spin), the space becomes $4 \times 4 = 16$ dimension in structure. The mathematical homogeneity of the system here is guaranteed by taking the expansion set to be an orthonormal set. This is done by taking $\Sigma_i |\Phi_i\rangle\langle\Phi_i| = 1$. The $|\Phi_i\rangle$ and $\langle\Phi_i|$ are complementary vectors in HS. They have to satisfy the above condition.

This meaning of complementarity will become clear shortly; first, though, we explain the meaning of the symbols bras $\langle|$ and kets $|\rangle$. This notation of representations for the eigen (energy) functions (in fact, all the basics of MM) were developed by the physicist Dirac. Bras and kets are the basis states (vectors) of HS, and HS is the quantum mechanical vector space of the QM operator. This is like the x-, y-, z-, t-space of NM. The bras $\langle|$ and kets $|\rangle$ are part of a complete set called the *basis set*. One is the complement (the other half or partner) of the other. The partnership is such that $\langle\Phi_i|\Phi_j\rangle = \delta_{ij}$ (called the Dirac delta function) $= 1$, if $i = j$, $= 0$, if $i \neq j$.

This is what makes the set complete, sometimes called a *normalized set*. Now we must define the coefficient c_i in the expansion $\psi(x) = \Sigma_i c_i \Phi_i$. We take $c_i = \langle\Phi_i|\psi\rangle$, which is the probability amplitude of the wave function ψ. The coefficient c_i can be interpreted as the scalar (sometimes called the *direct*) product of the two vectors Φ and ψ. This scalar product can be interpreted as the probability amplitude, such that we start with a ket $|\psi\rangle$ and with the bra $\langle\Phi_i|$. Now the difference between ket and bra becomes clear. A *ket* is the initial state; in other words, a starting position. A *bra* is a final state. There are two possible resultant effects of an operation. Also, $|c_i|^2$ is called the probability of finding the system in the eigenstate $|\Phi_i\rangle$, starting from the state $|\psi\rangle$. If ψ is normalized, the sum of all probabilities must add up to 1.

3.1.4.2 Operator Spins and Matrices

Let us consider the following algebraic equation in the ket–bra notation just introduced: $\langle j|\Phi\rangle = \Sigma_i\langle j|\acute{A}|i\rangle\langle i|\psi\rangle$. All the quantities in it are numbers, including $\langle j|\acute{A}|i\rangle$. One can say $\langle j|\acute{A}i\rangle$ is an algebraic representation of the abstract operator $\acute{A}$, in the basis $|i\rangle$. It is referred to as the *matrix element* of $\acute{A}$ between the states $|i\rangle$ and $|j\rangle$. Also note that in the matrix element $\langle j|\acute{A}|i\rangle$, $\acute{A}$ operates to the right, giving a new ket; one then takes the scalar product of this new ket by multiplying the bra from the left. One should remember that the quantum mechanical operators represent observables. Also, they are linear and Hermitian. The Hermitian conjugate of the operator $\acute{A}$ is denoted as $\acute{A}\dagger$, and is determined from the relation $\langle\Phi|\acute{A}\dagger|\psi\rangle = \langle\Phi|\acute{A}|\psi\rangle^*$. It is now clear why one tacitly assumes that an operator sandwiched between a bra and a ket is operating on its right: because on the left (i.e., on the bra) it is really its Hermitian conjugate, $\acute{A}\dagger$, that operates. In this definition, $[\acute{A}\dagger]\dagger = \acute{A}$ operates on the bra to the left, and then operates on the right side, as a result of the (forward)* = backward rule. An operator $\acute{A}$ is said to be Hermitian if $\acute{A}\dagger = \acute{A}$. Thus, for a Hermitian operator, $\langle\Phi|\acute{A}|\psi\rangle = \langle\psi|\acute{A}|\Phi\rangle^*$. One should further note that eigenvalues of Hermitian operators are real.

3.1.4.3 Spins and Matrices

Particles such as nuclei of atoms, protons, electrons, etc., have intrinsic (natural) spin. This is basically a dipole magnetic moment. In electrons, this moment (spin) is the result of the circular orbits of the electrons around the nucleus. The orbiting of the electrons (*charge*) generates a stable circular current, due to their quantized orbits, and thus the magnetic moment. At a particular instant of time, if the system is free of any external force, the spins of particles could be oriented in any direction. Thus the overall spin of the system, which is measured as magnetic moment per unit volume (called the *magnetization M*) is zero. However, by the application of an external static magnetic field—say, in the z-direction—almost all the spins can be polarized (oriented) in the z-direction.

The magnetization M (total magnetic moment per unit volume of the particles in an ensemble), in terms of the intrinsic moment of a particle, can be worked out as follows: $M = \Sigma_i Mi$. For one particular species of nuclei, M is given as iA. Here, i is the electron current in an orbit and A is the area of the orbit. The current, i, is the charge rotating per second $= e/T$. T is the time period per orbit, and is given as $(2\pi r/v)$, v being the linear (tangential) velocity of the electron at any instant of time, and r being the radius of the orbit. Then $M = (evr/2)$. This expression can be rewritten in terms of the fundamental unit of magnetic moment, the Bohr magneton, $\mu_b = (e\hbar/2m)$, and the angular momentum of the orbit $L = mvr$. Upon simplification, the magnetic moment for a single electron $M = iA$ is given as $M = N(\mu_b/\hbar)L$ (refer to Section 3.1.4.7). Here N is the particle density (number of particles per unit volume), $\hbar$ is Planck's constant, and e is the charge of the particle. One takes μ_b as the fundamental unit of the magnetic moment. The ratio $= \mu_l/L$ (the ratio of the magnetic moment to the angular momentum of the particle in the orbit) is called the *gyromagnetic ratio.* The linear (tangential) velocity v of the particle in the orbit equals ω (angular velocity) r. In this way the magnetization is expressed in terms of the fundamental constants of the particle. The Hamiltonian H of the system, which is the interaction energy between the applied field B_z and the intrinsic spin of the system (total magnetic moment), can be written as H (energy) $= -M_zB_z = \omega_L\hbar$. Here $\omega_L = (eB_z/2m)$ is called the *Larmor precession frequency.* The reader is reminded that what applies to charged particles (electrons) also applies to the charged particle proton (e.g., the nucleus of hydrogen). The difference is only in their charge (proton has positive charge) and mass (the proton is much heavier than an electron). In the absence of the magnetic field, the spin-up (+) and the spin-down (−) states of the proton are degenerate. The magnetic field removes this degeneracy.

In a tabular form, one can write the energy of the states as follows:

Table 3.1 Energy of the Spin States

	$\|\mathbf{I}_z = +\rangle$	$\|\mathbf{I}_z = -\rangle$
$\|I_z = +\rangle$	$-\mu_p B_z$	0
$\|I_z = -\rangle$	0	$+\mu_p B_z$

Here $\mu_p = (eg_l\hbar/4m_p) \approx 2.79$ nuclear magnetons, where m_p is the mass of the proton. The significance of g_l, which is called the orbital g factor, will become clear a little later. The quantity μ_p is the magnetic moment of the particle (proton or electron).

3.1.4.4 The z Spins I_z (Figure 3.6(a))

Now one needs to translate the basic mathematical formulation (available in any introductory textbook on QM) into the language of MM. In the absence of an external static field, there is random orientation of spins. In the presence of such a field, there is alignment along the z-axis. This can be interpreted as a transition from an initial state to a final state: $\langle f|A|i\rangle$. A is an operator that operates on the initial state of the system to create the final state. One has to make a conscious choice of an A that will perform the desired operation. The base set chosen in MM is $I_z = \pm\rangle$. One can think of the whole operation as passing through an I_z filter (the I_z operator). The table can be rewritten in the language of MM as:

Table 3.2 The I_z Spin Operator

	$\vert I_z = +\rangle$	$\vert I_z = -\rangle$
$\vert I_z = +\rangle$	$\langle I_z = + \vert A \vert I_z = +\rangle$	$\langle I_z = + \vert A \vert I_z = -\rangle$
$\vert I_z = -\rangle$	$\langle I_z = - \vert A \vert I_z = +\rangle$	$\langle I_z = - \vert A \vert I_z = -\rangle$

One should notice that one is arranging the columns and rows in descending order of the component of the spin. This is an important convention. Once the matrix A is known, one can determine the matrix elements of A by sandwiching it between any two arbitrary states, using $\langle f|A|i\rangle$. For the I_z operator, one notices that off-diagonal elements of the matrix A are zero. Thus, the matrix is

$$\begin{pmatrix} (1/2)\hbar & 0 \\ 0 & (-1/2)\hbar \end{pmatrix}$$

The matrix of I_z is its own basis. It is a diagonal matrix. Sometimes an operator is made of the product of two matrices, e.g., $C = AB$. Suppose, for simplicity, that $|i\rangle$ and $|j\rangle$ belong to the same basis that is used to define the matrices A and B. One should note here that $AB \neq BA$. This is what one needs in QM, because matrices should not commute. When they do not equal each other, they are suitable as quantum operators. The ket $|i\rangle$ and the bra $\langle f|$ vectors can be conveniently represented as column and row vectors, as

$$\begin{pmatrix} I_z = +|i\rangle \\ I_z = -|i\rangle \end{pmatrix} \quad \text{and} \quad \begin{pmatrix} \langle f|I_z+\rangle \\ \langle f|I_z-\rangle \end{pmatrix}$$

One should note that whereas the operators are represented as square matrices, kets and bras are represented by column and row matrices, respectively. The two-component column matrix under consideration here for the spin is called a *spinor*. So, for the initial two-states (kets, $|I_z = \pm\rangle$) of spin, the spinors are $\chi_+ = \begin{pmatrix} 1 \\ 0 \end{pmatrix}$ and $\chi_- = \begin{pmatrix} 0 \\ 1 \end{pmatrix}$, respectively. For the final states (bras, $\langle I_z = \pm|$), the spinors are $\chi_+\dagger = \begin{pmatrix} 1 & 0 \end{pmatrix}$ and $\chi_-\dagger = \begin{pmatrix} 0 & 1 \end{pmatrix}$. Here the symbol † denotes the Hermitian conjugate of a matrix. This is the matrix obtained from the original by interchanging rows and columns, and taking the complex conjugate. Hermitian conjugation changes a column matrix into a row matrix and a row matrix into a column matrix. Consider now the multiplication

$$\hbar/2\begin{pmatrix} 1 & 0 \\ 0 & -1 \end{pmatrix}\begin{pmatrix} 1 \\ 0 \end{pmatrix} = \hbar/2\begin{pmatrix} 1 \\ 0 \end{pmatrix}$$

This is an eigenspinor (vector), eigenvalue (vector) equation. The spinor $\begin{pmatrix} 1 \\ 0 \end{pmatrix}$ is an eigenspinor of I_z corresponding to eigenvalue $\hbar/2$. It is the matrix representation of the eigenstate of S_z denoted by $|I_z = +\rangle$.

3.1.4.5 The Spins I_x and I_y (Figures 3.6(b–e))

Starting with the basis of I_z, one can further develop the basis of I_x and I_y. It is instructive to relate the mathematics to some actual physical phenomena. We know that electromagnetic radiation is a transversely polarized radiation; i.e., if the radiation progresses (in time) in the z-direction, the amplitude of the radiation oscillates in the x–y plane only. Let us start initially with unpolarized radiation. This radiation would propagate in any random direction in space (unpolarized radiation). We now pass it through an X-polarizer (filter): if it is a perfect polarizer, the radiation coming out of the filter has no oscillating amplitude in the y-direction ($b = 0$), but has full amplitude (a) in the x-direction. One should remember that these are probability amplitudes, with values between 0 and 1. Now, if we place a Y-polarizer next to the X-polarizer, then $a = 0$, as it allows amplitude only in the y-direction. Thus, it completely quenches the radiation, and no radiation comes out after application of the Y-polarizer. Suppose now that we place a 45° polarizer between the X- and Y-polarizers. Then, after the Y-polarizer, one should expect some radiation to come out. One can then write the base kets (initial state) $|I_x \pm (\text{up and down})\rangle$ as follows: $|I_x = +\rangle = (1/\sqrt{2})[|I_z = +\rangle + |I_z = -\rangle]$ and $|I_x = -\rangle = (1/\sqrt{2})[-|I_z = +\rangle + |I_z = -\rangle]$. The factor for each amplitude $(a, b) = (1/\sqrt{2})$ arises because the total amplitudes have to be such that $a^2 + b^2 = 1$. In the spinor representation, the eigenspinor of I_x becomes, for the spin-up state,

$$\begin{pmatrix} \langle I_z = +|I_x = +\rangle \\ \langle I_z = -|I_x = +\rangle \end{pmatrix} = \begin{pmatrix} 1/\sqrt{2} \\ 1/\sqrt{2} \end{pmatrix}$$

and, for the spin-down state $|I_x = -\rangle$, the eigenspinor is

$$\begin{pmatrix} \langle I_z = +|I_x = -\rangle \\ \langle I_z = -|I_x = -\rangle \end{pmatrix} = \begin{pmatrix} -1/\sqrt{2} \\ 1/\sqrt{2} \end{pmatrix}$$

Now, because the y-spin states have to be distinguishable from the x-spin states (another postulate of QM), one has to use a different basic phenomenon to work out these states. This other phenomenon is that of circularly polarized radiation (light). This can be created if one passes linearly polarized light through a quarter-wave plate. The circularly polarized light is very different from the 45° polarized light. Mathematically, circularly polarized light is also a linear combination of x- and y-polarized light, but there is a difference: the amplitude of the y-polarized light is 90° out of phase with that of the x-polarized light. In the basis chosen here, one can write $I_y = \pm\rangle(1/\pm\sqrt{2})[|I_z = +\rangle \pm \mathrm{i}|I_z = +\rangle]$. These are distinct from

$|I_x = \pm\rangle$, as was desired. These are the eigenstates of the operator I_y. The spinor notation for the eigenspinors of I_y are thus

$$\begin{pmatrix} 1/\sqrt{2} \\ i/\sqrt{2} \end{pmatrix} \frac{\pi}{4} \quad \text{and} \quad \begin{pmatrix} 1/\sqrt{2} \\ -i/\sqrt{2} \end{pmatrix}$$

respectively, for the I_y spin-up and spin-down states.

One can make use of the fact that spins obey angular-momentum algebra. Thus, one can write $I_xI_y - I_yI_x = [I_x, I_y] = \mathrm{i}\hbar$, x, y, z cyclic (the standard anticommutation relation for two QM operators), because $I_\pm = I_x \pm \mathrm{i}I_y$. So $I_x = (I_+ + \mathrm{i}I_-)$ and thus $I_y = (1/2\mathrm{i})(I_+ - \mathrm{i}I_-)$. Now only two states, $\langle +1/2|I_+|\langle -1/2\rangle$ and $\langle +1/2|I_+|\langle -1/2\rangle$ are permissible. Consequently, the matrices are

$$I_x = (\hbar/2)\begin{pmatrix} 0 & 1 \\ 1 & 0 \end{pmatrix} \quad \text{and} \quad I_y = (\hbar/2)\begin{pmatrix} 0 & -i \\ i & 0 \end{pmatrix}$$

3.1.4.6 The Basis of Four (2 × 2) Matrices for 1-Spin System: Build 4 × 4 Matrices (2-Spin System)

Now one can write $I_x = 1/2\hbar\sigma_x$, $I_y = 1/2\hbar\sigma_y$, and $I_z = 1/2\hbar\sigma_z$. The matrices σ_x, σ_y, and σ_z are called *Pauli matrices* and are as follows:

$$\sigma_x = \begin{pmatrix} 0 & 1 \\ 1 & 0 \end{pmatrix}, \quad \sigma_y = \begin{pmatrix} 0 & -i \\ i & 0 \end{pmatrix}, \quad \text{and} \quad \sigma_z = \begin{pmatrix} 1 & 0 \\ 0 & -1 \end{pmatrix}$$

The fourth matrix (identity matrix) is

$$I = \begin{pmatrix} 1 & 0 \\ 0 & 1 \end{pmatrix}$$

They form a complete set of four 2 × 2 matrices. Any 2 × 2 Hermitian matrix (HM) can be expressed in terms of these four matrices. The Pauli matrices satisfy the following relations: $\sigma_i^2 = 1$, $i = x, y, z$; $\sigma_i\sigma_j + \sigma_j\sigma_i = 2\delta_{ij}$, $i, j = x, y, z$; $[\sigma_x, \sigma_y]$ (anticommutation) $= 2\mathrm{i}\sigma_z$, x, y, z cyclic. One can also write $\mathbf{I}$ (vector) as $\mathbf{I} = \hbar\boldsymbol{\sigma}/2$. Also, $\sigma^2 = \sigma_x^2 + \sigma_y^2 + \sigma_z^2 = 3$; $I^2 = \hbar^2\boldsymbol{\sigma}^2/4 = 3\hbar^2/4 = s(s+1)\hbar^2 (s = 1/2)$.

3.1.4.7 The Transformation Matrix: The Similarity Transformation

As an example, the product UI_yU^{-1} is called the *similarity transformation* of I_y. I_y is diagonalized by the similarity transformation. This is done by using the unitary matrix A. A similarity transformation with a unitary matrix is called a *unitary transformation*. This is the transformation matrix, e.g., between the I_z-basis and the I_y-basis. The transformation to the new basis representations is linked with the diagonalization of the matrix in the previous basis. In the example chosen, I_y, in its home basis representation, should be such that

$$\begin{pmatrix} 1 \\ 0 \end{pmatrix} = U\begin{pmatrix} 1/\sqrt{2} \\ i/\sqrt{2} \end{pmatrix}$$

and

$$\begin{pmatrix} 0 \\ 1 \end{pmatrix} = U\begin{pmatrix} 1/\sqrt{2} \\ -i/\sqrt{2} \end{pmatrix}$$

Now the ordering of the matrices should be from the highest (1, 0) to the lowest (0, 1) eigenvalue and there must be normalization and orthogonality of spinors. One thus makes the choice of U as follows:

$$U = \begin{pmatrix} 1/\sqrt{2} & -i/\sqrt{2} \\ 1/\sqrt{2} & i/\sqrt{2} \end{pmatrix}$$

The transformation matrix U from the I_z-basis to the I_y-basis is given as follows:

Table 3.3 The Transformation Matrix

	$\vert I_z = +\rangle$	$\vert I_z = -\rangle$
$\vert I_y = +\rangle$	$\vert I_y = +\rangle \vert I_z = +\rangle$	$\vert I_y = +\rangle \vert I_z = -\rangle$
$\vert I_y = -\rangle$	$\vert I_y = -\rangle \vert I_z = +\rangle$	$\vert I_y = -\rangle \vert I_z = -\rangle$

Let us now define the inverse of U, which is U^{-1}, such that

$$U^{-1}U = UU^{-1} = I = \begin{pmatrix} 1 & 0 \\ 0 & 1 \end{pmatrix}$$

Here **I** is the unitary matrix. So

$$U^{-1} = \begin{pmatrix} 1/\sqrt{2} & 1/\sqrt{2} \\ i/\sqrt{2} & -i/\sqrt{2} \end{pmatrix}$$

One can verify that U^{-1} is the Hermitian conjugate (HC) of the matrix U, i.e., $U^{-1} = U\dagger$. Now finally

$$\hbar/2\begin{pmatrix} 1/\sqrt{2} & -i/\sqrt{2} \\ 1/\sqrt{2} & i/\sqrt{2} \end{pmatrix}\begin{pmatrix} 0 & -i \\ i & 0 \end{pmatrix}\begin{pmatrix} 1/\sqrt{2} & 1/\sqrt{2} \\ i/\sqrt{2} & -i/\sqrt{2} \end{pmatrix} = \hbar/2\begin{pmatrix} 1 & 0 \\ 0 & -1 \end{pmatrix}$$

One can see that the result is a diagonal matrix that gives the eigenvalues of I_y. By using this technique, one can find the eigenstates of a Hamiltonian (the energy matrix) in some suitable starting basis. The diagonal matrix of the Hamiltonian in the new basis gives the eigenvalue spectrum, and the transformation matrix to the new basis determines all the probabilities for physically measurable quantities. In Cartesian (NM) space, **L** has three components: $L_x = yp_z - zp_y$, $L_y = zp_x - xp_z$, and $L_z = xp_y - yp_z$.

In QM, in the same space, p and x do not commute, i.e., $[x, p_x] = xp_x - p_x x = i\hbar$. But when the combination is from two different spaces, they do commute, e.g., $[x, p_y] = xp_y - p_y x = 0$. Following this rule of algebra, one can work out the following cyclic-in-order, anticommutation rules for the components of L: $[L_x, L_y] = i\hbar L_z$, $[L_y, L_z] = i\hbar L_x$, and $[L_z, L_x] = i\hbar L_y$. Because the results are cyclic in order, it is not difficult to remember. We remind the reader that this cyclic, symmetrical algebra is not accidental. It represents a clever choice of the spaces and vectors, and the mathematical ingenuity of the inventors of the coordinate transformation (e.g., orthogonal) from one space to another. Now the square of the angular-momentum, $\mathbf{L}^2 = L_x^2 + L_y^2 + L_z^2$, can be shown to commute with individual components L_x, L_y, and L_z, i.e., $[\mathbf{L}^2, L_i] = 0$, $i = x, y, z$, in cyclic order.

To learn about the angular-momentum algebra approach to QM, readers are referred to a standard textbook, e.g., Ref. [10]. This section summarizes only the required results useful in reference to QMRI. Now, because $\mathbf{L}^2$ and all the components L_i commute, in the language of QM this means that one can simultaneously measure eigenstates of $\mathbf{L}^2$ and any of the three components of L (i.e., L_x, L_y, and L_z). Hence, out of the four operators of $\mathbf{L}^2$—L, L_x, L_y, and L_z—the eigenvalues of only two can be used to label angular-momentum eigenvalues; others will be dependent on each other. One can alternatively say as follows: As the particle moves on the sphere, R ($r = R$), the radius of the sphere remains fixed; thus, only two coordinates, θ and φ, are enough for the position of the particle.

The spaces spanned by the angular-momentum operators must be 2D (cylindrical) only.

One chooses $\mathbf{L}^2$ and L_z as two operators whose eigenvalues label the angular-momentum eigenstates.

Now one again uses the ket and bra notation, which is basically the standard format used in the mathematical language of QM, to write the following useful operator relation: $\mathbf{L}^2|\lambda \mathrm{m}\rangle = \lambda|\lambda m\rangle$; $L_z|\lambda m\rangle = m|\lambda m\rangle$. The operator on the left-hand side of the equation operates on the eigenvector (eigenstates) $|\lambda m\rangle$, and produces its eigenvalue (energy value) followed by the original eigenvector. In the language of the coordinates θ, φ, one can write the following relations: $\langle\theta, \varphi|\lambda m\rangle = \psi_m(\theta, \varphi)$. This is the wave function representing the position of the particle with θ, φ coordinates and having quantum number values λ and m. Further, $\langle\theta, \phi|\mathbf{L}^2|\lambda m\rangle = \lambda\langle\theta, \varphi|\lambda m\rangle$ and $\langle\theta, \varphi\ |L_z|\lambda m\rangle = m\langle\theta, \varphi\ |\lambda m\rangle$. For the sake of completion one needs to write $\mathbf{L}^2$ and L_z in terms of the coordinates θ, φ (spherical coordinate system). See Ref. [10] for details.

3.1.4.8 The J*-Operator [10, 19]*

The J-operator follows the same commutation and anticommutation rules, plus some thing more as follows: $J_xJ_y - J_yJ_x = [J_x, J_y] = i\hbar J_z$, x, y, z cyclic; $J^2 = J_x^2 + J_y^2 + J_z^2$; $[J^2, J_i] = 0$, $i = x, y, x$ cyclic. As below, one chooses the eigenstates to be simultaneous eigenstates of J^2; J_z: $J^2|\lambda_J m\rangle = \lambda_J|\lambda_J m\rangle$ and $J_z|\lambda_J m\rangle = m\hbar|\lambda_J m\rangle$. Further, one defines the following mathematical operators as an addition set, i.e., $J_\pm = J_x \pm J_y$. Now, following the algebra shown in Figure 3.6(c) one can establish $[J_z, J_+] = \hbar J_+$; $[J_z, J_-] = -\hbar J_-$; and $[J_z, J_-] = 2\hbar J_+$. In addition, one has $J_+J_- = J^2 - J_z^2 + \hbar J_z$ and $J_-J_+ = J^2 - J_z^2 - \hbar J_z$. Following the individual $[J_i, J_i]$ cyclic anticommutation rules, as noted earlier, one can write $J_zJ_+ = J_+(J_z + \hbar)$; $J_zJ_+|\lambda_J m\rangle = J_+(J_z + \hbar)|\lambda_J m\rangle = (m+1)\hbar J_+|\lambda_J m\rangle$. Thus, $J_+|\lambda_J m\rangle$ is an eigenstate of J_z belonging to the eigenvalue $(m+1)\hbar$. One can interpret this as saying that it is a *raising operator*, as it raises the eigenvalue of J_z by one unit of $\hbar$. Notice that the operation by J_+ has no effect on λ, the eigenvalue of J^2, because $[J^2, J_+] = 0$. Logically, one should also have (one can prove this, though) $J_zJ_- = J_-(J_z - \hbar)$; $J_zJ_-|\lambda_J m\rangle J_-(J_z - \hbar)|\lambda_J m\rangle = (m-1)\hbar J_-|\lambda_J m\rangle$. Thus, J_- is a lowering operator. One can summarize as follows: $J_+|\lambda_J m\rangle C_{\lambda J}m|\lambda_J m{+}1$ and $J_-|\lambda_J m\rangle D_{\lambda J}m|\lambda_J m-1\rangle$.

One should realize that $(m\hbar^2) \leq \lambda_J$; for a fixed λ_J, the value of m is bounded, say between $m_{\max}$ and $m_{\min}$. In fact, one can show that $m_{\max} = -m_{\min} = j$ (another quantum number) and also that $\lambda_J = j(j+1)\hbar^2$. Furthermore, one can show that j can only be an integer, 0, or a half integer. One important thing to notice here is that J_i algebra permits half-integer values. This half-integer quantum number is the zest of the spin S, and $J = L \pm S$. Now, using further algebraic manipulations, one can show that the coefficients C_{jm} and D_{jm} are $C_{jm} = \hbar[j(j+1) - m(m+1)]^{1/2}$ and $D_{jm} = \hbar[j(j+1) - m(m-1)]^{1/2}$. The effect of the raising and lowering operators, J_+ and J_-, on angular-momentum states is thus $J_+|\lambda_J m\rangle = \hbar[j(j+1) - m(m+1)]^{1/2}|jm+1\rangle$ and $J|\lambda_J m\rangle = \hbar[j(j+1) - m(m-1)]^{1/2}|jm-1\rangle$.

3.1.5 Addition of Orbital and Spin Angular Momentum

3.1.5.1 Single Particle: J = L + S

The angular-momentum algebra illustrated in Section 3.1.4.8 will be applied here to MRI. In QM, using a spherical coordinate system, the coordinate θ is represented by the quantum number (QN) and the coordinate φ is replaced by the QN, denoted m. The eigenfunction of a single particle is represented by $|lm\rangle$ in space.

The effect of operation due to Lz gives the eigenvalues for Lz (operator) $|lm\rangle m\hbar$ (eigenvalue)$|lm\rangle$, $m = 0, \pm1, \pm2, \ldots$. Similarly, for $\mathbf{L}^2$, $|lm\rangle = l(l+1)\hbar^2$, with $l = 0, 1, 2, \ldots$, and $m = -l, -l+1, -l+2, \ldots, l-1, l$. It was shown that in another notation, called J-operator notation, l is replaced by j. The two new operators J_+ and J_- allowed the quantum numbers j and m to be raised and lowered by 1, upon operation. It is seen that in J notation l is replaced by $\lambda_j = j(j+1)\hbar^2$. In J-operator notation, the eigenfunction for a single particle, including space and spin coordinates, can be written as $|jm\rangle = \alpha|lm - 1/2, 1/2\rangle + \beta|lm + 1/2, -1/2\rangle$. The +1/2 and −1/2 correspond to the spin-up and spin-down quantum (coordinate) numbers of the particle. One should remember that, physically, spin (magnetic) and orbital (angular) momentum are closely related, through a constant, namely the gyromagnetic ratio of the nucleus of the particle. It is as though one were the cause of the

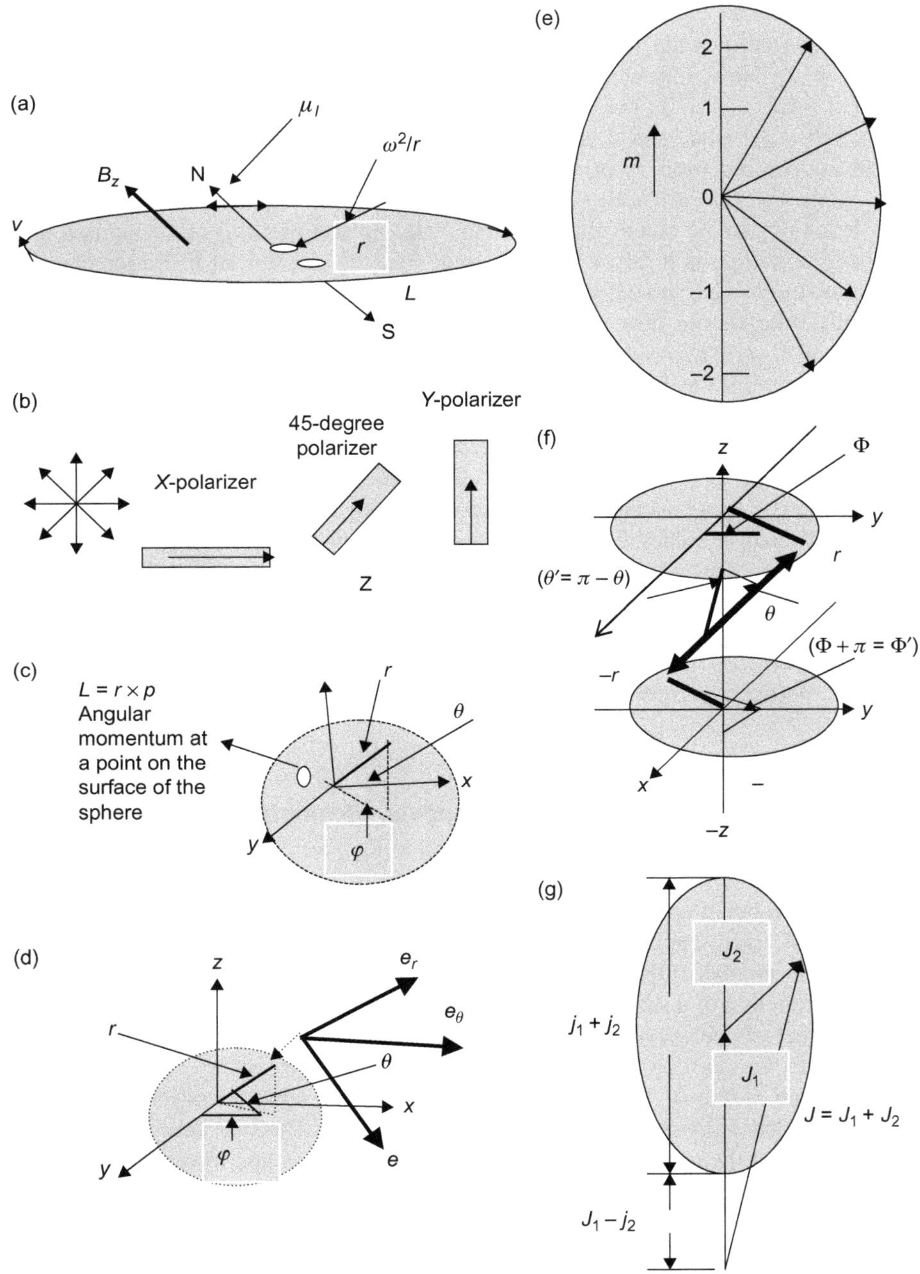

Figure 3.6 (a−g), Sections 3.1.4−3.1.5. The GMR = μ_l/L; μ_l = dipole magnetic moment; i = current = ev; $A = \pi r^2$; L = angular momentum; m = mass of the particle; v = tangential velocity; r = radius of the orbit; e = charge of the particle; $\mu_l/L = e/2m = (e\hbar/2m)/\hbar = \mu_b/\hbar$; μ_b = Bohr magneton = 0.927×10^{-23} A m^2 = the fundamental unit of magnetic moment.

other. Hence, for a single particle, $\mathbf{J}^2 = (L + \mathbf{S})^2 = L^2 + S^2 + 2\mathbf{LS} = L^2 + S^2 + 2L_zS_z + 2L_+S_- + 2L_-S_+$. One can use the $L_\pm$ and $S_\pm$ operator relations as if they were $J_\pm$ operators (logically, the effect has to be the same).

Our intention in exposing readers to the angular-momentum algebra of the **J**, **L**, **S** operators is basically to get them familiar with the terms *J*-coupling, *S*-coupling, etc., which are often used or referred to in the MRI literature. The reader who wants to learn all about the mathematics of the couplings is referred to a standard text on QM.

3.1.5.2 Two-Particle Spin System: Adding Spins Alone

Because the degrees of freedom of two particles are independent, one can write the commutation relation $[\mathbf{S_1}, \mathbf{S_2}] = 0$. The total spin is $\mathbf{S} = \mathbf{S_1} + \mathbf{S_2}$. The components of **S** will follow the usual angular-momentum commutation rules. One would be interested in the eigenvalues and eigenvectors of S^2 and S_z, which define the coupled representation. In the uncoupled basis (for two particles), S_z is as follows:

Table 3.4 The Spin S_z

	$\|1/2, 1/2\rangle$	$\|1/2, -1/2\rangle$	$\|-1/2, 1/2\rangle$	$\|-1/2, -1/2\rangle$
$\|1/2, 1/2\rangle$	$\hbar$			
$\|1/2, -1/2\rangle$				
$\|-1/2, 1/2\rangle$				
$\|-1/2, -1/2\rangle$				$-\hbar$

The matrix of $\mathbf{S^2}$, in the uncoupled basis, is $= \mathbf{S^2} = (\mathbf{S_1} + \mathbf{S_2})^2 = S_1{}^2 + S_1{}^2 + 2\mathbf{S_1S_2} = S_1{}^2 + S_1{}^2 + 2S_{1z}$.

$$S_{2z} + S_{1+}S_{2-} + S_1 - S_{2-} + = \hbar^2 \begin{pmatrix} 2 & 0 & 0 & 0 \\ 0 & 1 & 1 & 0 \\ 0 & 1 & 1 & 0 \\ 0 & 0 & 0 & 2 \end{pmatrix}$$

One should note here that S_z is the total component of the z-spin. It commutes with $s_1{}^2$, $s_2{}^2$, s_{1z}, and s_{2z}, so S_z should be diagonal in the uncoupled state (above). It is important to note that although the base states $|1/2, 1/2\rangle$ and $|-1/2, -1/2\rangle$ are eigenstates of S^2 (eigenvalue in absolute magnitude, $S(S+1)\hbar^2 = 2\hbar^2$, for $S = 1$), but $|1/2, -1/2\rangle$ and $|-1/2, 1/2\rangle$ are not. The preceding matrix is block diagonal in the center.

This part can be written as

$$\begin{pmatrix} 1 & 1 \\ 1 & 1 \end{pmatrix} = I(\text{identity matrix}) + \sigma_x(\text{single-spin x-matrix}) = \begin{pmatrix} 1 & \\ & 1 \end{pmatrix} + \begin{pmatrix} 0 & 1 \\ 1 & 0 \end{pmatrix}$$

3.1.5.3 Symmetry of Spins under Exchange (The Parity Principle) [10] (Figure 3.6(f))

When there are two particles of different species in an ensemble, both space and spin labels may be exchanged. This is called *Pauli's exclusion principle*. Using the spherical coordinate frame of reference (r, θ, φ), which describes space and spin of the two particles, one can break the state function in space. One may say either $r = -\mathbf{r}$, or $\mathbf{r} = \mathbf{r}$, but $\theta = \pi - \theta$ and $\varphi = \pi + \varphi$. Therefore, under exchange, the radial part remains unchanged; in contrast, the angular-momentum (spin) part becomes antisymmetric under exchange.

3.1.5.4 Generalized Angular-Momentum Coupling (J-Coupling): Glebsch–Gordan Coefficients [10] (Figure 3.6(g))

One can write $[\mathbf{J_1}, \mathbf{J_2}] = 0$ (commutation). Total angular momentum $\mathbf{J} = \mathbf{J_1} + \mathbf{J_2}$. The two sets of angular-momentum operators J_1^2, J_{1z}, J_2^2, J_{2z} and J_1^2, J_{1z}, J_2^2, J^2 and J_z commute. The two sets are connected through the following unitary transformation matrix elements (also called the Glebsch–Gordan coefficients (GGCs)): $|j_1j_2JM\rangle = \Sigma m_1,\ m_2\langle j_1m_1j_2m_2|j_1m_1j_2m_2\rangle$. The GGCs are $\langle j_1m_1j_2m_2|j_1j_2JM\rangle \equiv \langle j_1m_1j_2m_2|JM\rangle$. The inverse transformations can be written as $|j_1m_1j_2m_2\rangle = \Sigma_{JM}\langle j_1j_2JM|j_1m_1j_2m_2\rangle\,|j_1j_2JM\rangle$. One should write a transformation matrix as a real and unitary matrix–orthogonal, i.e., $\langle j_1j_2JM|j_1m_1j_2m_2\rangle = (\langle j_1m_1j_2m_2|j_1j_2JM\rangle)^* = (\langle j_1m_1j_2m_2|j_1j_2JM\rangle) \equiv (\langle j_1m_1j_2m_2|JM\rangle)$. Thus, $|j_1m_1j_2m_2\rangle = \Sigma_{JM}\langle j_1m_1j_2m_2|JM\rangle\,|j_1j_2JM\rangle$. The orthogonal relations for the GGCs then automatically follow as under $\Sigma_{JM}(\langle j_im_1j_2m_2|JM\rangle\langle j_im_1'j_2m_2'|JM\rangle) = \delta m_1m_1'\delta m_2m_2'$ and $\Sigma_{m1m2}\langle j_im_1j_2m_2|\mathrm{JM}\rangle\langle j_im_1j_2 m_2|J'M'\rangle = \delta_{JJ}'\delta_{MM}'$. Thus, GGCs must vanish unless $m_1 + m_2 = M$ (summations do not run over both m_1 and m_2). In addition, one has $\langle j_im_1j_2m_2|JM\rangle \neq 0$ only if $|j_1 - j_2| \leq J \leq j_1 + j_2$ (the triangle inequality). It tells one (Figure 3.6) the possible range of values of total J, when one adds the two angular momenta j_1 and j_2, whose values are fixed. In the language of GGCs, one can now write, for 2-spin-½ particles, $j_1 = 1/2, j_2 = 1/2$. The coupled state becomes

$$|SM\rangle = \Sigma m_s m_{s'}\langle 1/2m_s1/2m_{s'}\mid \mathrm{SM}\rangle|m_sm_{s'}\rangle\mid \mathrm{SM}\rangle\mid m_sm_{s'}\rangle$$

$$\equiv \begin{pmatrix} 1 & 1 \\ 1 & 0 \\ 1 & -1 \\ 0 & 0 \end{pmatrix} = \begin{pmatrix} 1 & 0 & 0 & 0 \\ 0 & 1/\sqrt{2} & 1/\sqrt{2} & 0 \\ 0 & 0 & 0 & 1 \\ 0 & 1/\sqrt{2} & -1/\sqrt{2} & 0 \end{pmatrix} = \begin{pmatrix} |1/2 & 1/2\rangle \\ |1/2 & -1/2\rangle \\ |-1/2 & 1/2\rangle \\ |-1/2 & -1/2\rangle \end{pmatrix}$$

Here the column represents the basis vectors themselves, not their components.

3.2 QM Applied to MRI

This section gives readers a brief exposure to the basics of QM as applied to an ensemble containing more than one kind of spins (heteronuclear system).

3.2.1 QM: MM

3.2.1.1 Density Matrix and NMR: 1-Spin (1/2, −1/2) System [11]

Density matrix technique is very usefully applied to NMR spectroscopy and imaging. Let us for the sake of argument suppose that there are two different (species) spins α (1/2, −1/2) and β (1/2, −1/2) in the system. The two states 1/2 and −1/2 (i.e., up and down spins) can have the following possibilities:

Table 3.5 The Up and Down Spins

Functions	***A***	***B***	**States**
Φ_1	1/2	−1/2	$\lvert\alpha\alpha\rangle$
Φ_2	−1/2	1/2	$\lvert\alpha\beta\rangle$
Φ_3	1/2	1/2	$\lvert\beta\alpha\rangle$
Φ_4	−1/2	−1/2	$\lvert\beta\beta\rangle$

One can describe the ensemble by using a typical state function, Ψ, written as a linear combination of basis states (functions), Φs, as $\Psi = \Sigma_n c_n \Phi_n$. The expectation value (average over space and time) of an observable of the system (e.g., the x component of the magnetization M_x, from the oriented component of the magnetization produced by a DC field applied along the z-direction) can be written mathematically as $M_x = \langle \Phi | M_x | \Phi \rangle$, where the Φs are four possible states (table above) of the spin combinations.

This can be expanded as $M_x = \langle \Phi | M_x | \Phi \rangle = \Sigma_{n,m} c_m c_n \langle m | M_x | n \rangle$. Here m and n are numbers 1−4, as above. Thus M_x can have 16 (4 × 4) values $M_{11}, M_{12}, \ldots, M_{21}, \ldots, M_{24}, \ldots, M_{44}$. M_{44} has 16 different coefficients $c_m c_n$, one for each different value of M_x. The ensemble average value can be written as $\langle |M_x| \rangle = \Sigma_{n,m} c_m c_n \langle m | M_x | n \rangle$. The basis functions, Φ_n, and matrix elements, such as $\langle m | M_x | n \rangle$, are time-dependent. Product terms of the coefficient, such as $c_m c_n$, appear so frequently that one can collect them together in a matrix, known as the density matrix σ, with matrix elements $\sigma_{nm} = c_m c_n$. The general rule follows: The ensemble average expectation value for any general operator OP can be written as $\langle OP \rangle = \text{Tr}$ (trace) $\{[\sigma][OP]\} = \text{Tr}\{[OP][\sigma]\}$ (that means the operators σ and OP commute).

The trace of an operator is equal to the sum of the diagonal elements of the matrix that represents the operator. If one neglects for a moment the effects of relaxation (moving of the spins to their equilibrium positions once the applied magnetic field is removed), one can follow the course of any NMR experiment with the use of a density matrix. One also needs two other simple tools: the rotation operator $U_i(\Phi)$, ($i = x, y, z$), or its matrix representation; and the time-evolution operator U_H. One can show that for a single-spin-½ system, in the equilibrium state, the diagonal elements of the density matrix are proportional to the energy levels and the off-diagonal elements are zero. The equilibrium density matrix is given by

$$\sigma(\text{average/expectation value}) = \begin{pmatrix} 1/2 + p/2 & 0 \\ 0 & 1/2 - p/2 \end{pmatrix}$$
$$= 1/2\begin{pmatrix} 1 & 0 \\ 0 & 1 \end{pmatrix} + 1/2\begin{pmatrix} p & 0 \\ 0 & -p \end{pmatrix}$$

the identity matrix part I (first part of RHS) is time-independent and the partial density matrix part ξ (second part of RHS) is time-dependent. One represents

$$\xi = p/2\begin{pmatrix} 1 & 0 \\ 0 & -1 \end{pmatrix} = -p/2\sigma_0$$

where σ_0 is one of the Pauli spin matrices and $p = (\hbar\gamma B_0)/kT$. Here $\hbar$ is Planck's constant ($=6.18 \times 10^{-27}$ JS), γ the gyromagnetic ratio of a particular nucleus, k the Boltzmann constant ($=1.38 \times 10^{-23}$ J/K), and T the temperature of the ensemble (homonuclear spin-½) of nuclei under consideration.

One should recollect at this stage as follows: When a magnetic field B_0 is applied to an ensemble of nuclei, the spins are only preferably oriented toward the $+z$-axis (positive direction of the applied field). As a matter of fact, there is a distribution of spins around the preferred axis, with a weight p. This weight varies, over the range of angles $0 - 2\pi$, with respect to the z-axis. This variation is caused by the temperature of the ensemble and other distractions, e.g., interactions of the nuclei with the environment in which they are present. The environment includes other nuclei, molecules, fluids, etc. It can be shown that, in general, all possible configurations of a given spin system in the field can be expressed as a linear combination of 2^N basis matrices. Here N is the number of spin-½ nuclei in the system:

Table 3.6 The Spin 1/2 Nuclei System

	$\tilde{U} - \varphi x$	$\tilde{U} - \varphi y$	$\tilde{U} - \varphi z$	$\tilde{U}_H$
σ_x	σ_x	$(\cos\varphi)\sigma_x + (\sin\varphi)\sigma_y$	$(\cos\varphi)\sigma_x - (\sin\varphi)\sigma_y$	$(\cos qt)\sigma_x + (\sin qt)\sigma_y$
σ_y	$(\cos\varphi)\sigma_y(\sin\varphi)\sigma_z$	σ_y	$(\cos\varphi)\sigma_y + (\sin\varphi)\sigma_x$	$(\cos qt)\sigma_y - (\sin qt)\sigma_y$
σ_z	$(\cos\varphi)\sigma_z + (\sin\varphi)\sigma_y$	$(\cos\varphi)\sigma_z - (\sin\varphi)\sigma_x$	σ_z	
	σ_z			

3.2.1.2 A Single-Pulse Experiment [11]

One can show, mathematically, that a positive 90° rotation about the x-axis of a magnetization M, originally positioned toward the z-axis, will carry M away from the z-axis toward the $-y$-axis. Similarly, a negative 90° rotation about the x-axis will rotate M from the z-axis to the y-axis. On equilibrium, the density matrix $\xi_a = \boldsymbol{\xi}(0) = p/2$ changes to $\xi_b = U_{-\varphi x}\xi_a U_{-\varphi x}{}^{-1}$. $U_{-\varphi x}{}^{-1}$ is an inverse matrix of $U_{-\varphi x}$, such that

$$U_{-\varphi x}U_{-\varphi x}^{-1} = 1/\sqrt{2}\begin{pmatrix} 1 & \mathrm{i} \\ \mathrm{i} & 1 \end{pmatrix} 1/\sqrt{2}\begin{pmatrix} 1 & -\mathrm{i} \\ -\mathrm{i} & 1 \end{pmatrix}$$
$$= \begin{pmatrix} 1+1 & -\mathrm{i}+i \\ \mathrm{i}-\mathrm{i} & 1+1 \end{pmatrix} = \begin{pmatrix} 1 & 0 \\ 0 & 1 \end{pmatrix} = 1 \text{ (determinant value)}$$

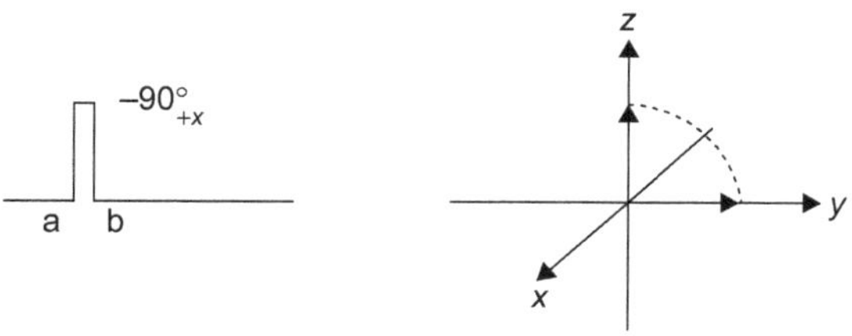

Figure 3.7 A $90°_{-x}$ (or a $-90°_{x}$) pulse rotates the magnetization about the x (or $-x$) axis from a z to a y orientation.

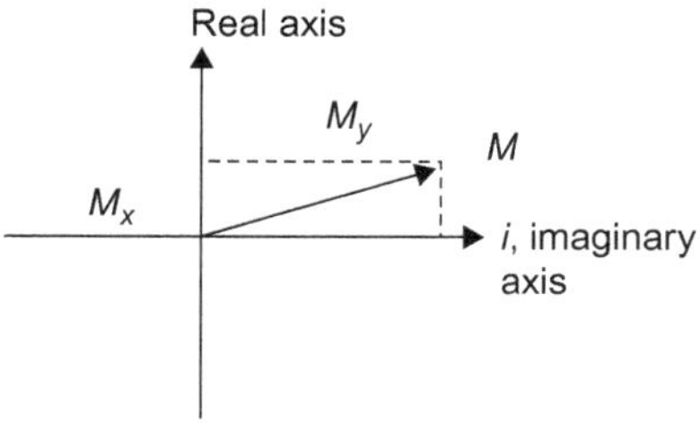

Figure 3.8 The magnetization in the rotating xy plane is represented by a complex number, $M = M_x + \mathrm{i}M_y$.

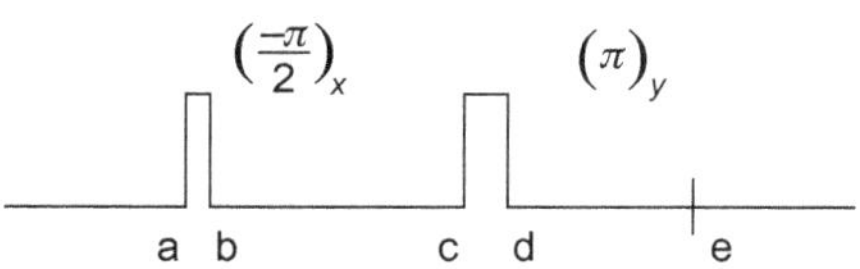

Figure 3.9 A spin-echo pulse sequence for a spin-½ system.

One should note here that $U_{-\phi x}{}^{-1}$ is obtained from $U_{-\phi x}$ by changing rows into columns and columns into rows, and changing i into $-$i. This so-called *transpose matrix* is commonly used in QM. After substituting $U_{-\phi x}$ and $U_{-\phi x}{}^{-1}$, one finds upon matrix multiplication (row by columns, in steps of two, at a time), from the right side,

$$\xi_{\mathrm{b}} = 1/\sqrt{2}\begin{pmatrix} 1 & \mathrm{i} \\ \mathrm{i} & 1 \end{pmatrix} p/2 \begin{pmatrix} 1 & 0 \\ 0 & -1 \end{pmatrix} 1/\sqrt{2}\begin{pmatrix} 1 & -\mathrm{i} \\ -\mathrm{i} & 1 \end{pmatrix}$$
$$= p/4\begin{pmatrix} 1 & \mathrm{i} \\ \mathrm{i} & 1 \end{pmatrix}\begin{pmatrix} 0 & -2\mathrm{i} \\ 2\mathrm{i} & 0 \end{pmatrix} = p/2\begin{pmatrix} 0 & -\mathrm{i} \\ \mathrm{i} & 0 \end{pmatrix} = p/2\sigma_{\mathrm{y}}$$

Thus, the effect of the $-90°_x$ pulse is to equalize the populations of energy levels. It also creates off-diagonal transverse magnetization elements (Figures 3.7–3.9).

3.2.1.3 *Operator Formalism [11]*

One can see that using MM to get the desired result involves multiple steps of multiplications of several matrices. This is a tedious task, but one should remember that only four basic operations are carried out: rotation about the x-, y-, z-axes and the evolution under the influence of the system Hamiltonian. The four fundamental matrices operate only on the four basis combination of the four basis matrices. Any configuration of the system may be expressed as some combination of the four basis matrices. Thus, one can build a multiplication table and represent the various operations by their corresponding operators. Suffice it to say that for a simple 2-spin system, each nucleus with a spin of 1/2, there are four energy levels, and the state functions describing these energy levels are given by $\Psi_1 = \alpha\alpha\rangle$, $\Psi_2 = \alpha\beta\rangle$, $\Psi_3 = \beta\alpha\rangle$, and $\Psi_4 = \beta\beta\rangle$. Given the Hamiltonian for this system, and using the standard rules for the result of an angular-momentum operator operating on the various state functions, one can (in a simple, if somewhat laborious and tedious, fashion) calculate with great precision and accuracy the energy levels (and hence the transition frequencies), and the transition probabilities (the intensities of the various spectra lines).

In the early days of NMR spectroscopy, when only CW spectrometers were available, this theory met most of the needs of the average chemist or NMR spectroscopist. Today, however, with the powerful and sophisticated pulse NMR spectrometers available, one usually needs to know in detail how a system of spins evolves in time. This requires some additional tools. To analyze complex NMR experiments (e.g., 1D and 2D high-resolution experiments, or NMR relaxation-time experiments), it is usually necessary to use density matrix theory. The name of this mathematical formalism makes it sound quite difficult, but in fact, density matrix theory is one of the simplest, most elegant, and most powerful tools available to the modern spectroscopist. In this brief introductory section we give a few simple examples of the power and utility of density matrix theory.

We begin in the usual fashion and describe the system at equilibrium by a state function. A typical state function, describing one of the four energy levels for a 2-spin system, can be written as a linear combination of the basis functions. Consider as follows. A single rotation of φ degrees on the spin-½ system is given as

$$\begin{aligned}\tilde{U} - \varphi x\sigma_z = U\varphi x\sigma_z U\varphi x^{-1} &= p/2\begin{pmatrix} c & \text{is} \\ \text{is} & c \end{pmatrix}\begin{pmatrix} 1 & 0 \\ 0 & -1 \end{pmatrix}\begin{pmatrix} c & -\text{is} \\ -\text{is} & c \end{pmatrix} \\ &= \text{i}p/2\begin{pmatrix} (c^2 - s^2) & -2\text{i}cs \\ 2\text{i}cs & (s^2 - c^2) \end{pmatrix}\end{aligned}$$

where $c = \cos(\varphi/2)$ and $s = \sin(\varphi/2)$. Using the trigonometric identities $\cos^2(\varphi/2) - \sin^2(\varphi/2) = \cos\varphi$ and $\sin\varphi = 2\sin(\varphi/2)\cos(\varphi/2)$, one gets

$$\begin{aligned}\tilde{U} - \varphi x\sigma_z = U\varphi x\sigma_z U\varphi x^{-1} &= \begin{pmatrix} \cos\varphi & -\text{i}\sin\varphi \\ -\text{i}\sin\varphi & \cos\varphi \end{pmatrix} \\ &= (\cos\varphi)\begin{pmatrix} 1 & 0 \\ 0 & -1 \end{pmatrix} + (\sin\varphi)\begin{pmatrix} 0 & -\text{i} \\ \text{i} & 0 \end{pmatrix} \\ &= (\cos\varphi)\sigma_z + \sin\varphi\sigma_y\end{aligned}$$

In the case of time-evolution operation in a typical case on σ_y, one gets

$$\tilde{U}_H\sigma_y = p/2\begin{pmatrix} 0 & -\text{i}\exp\text{i}wt \\ \text{i}\exp - \text{i}wt & 0 \end{pmatrix} = (\cos wt)\sigma_y + (\sin wt)\sigma_x$$

This gives one an expression for the operation of the rotation superoperator on σ_y in the laboratory frame of reference. In the rotating frame of reference, the expression becomes

$$\breve{D}_{\Omega}\tilde{U}_H\sigma_y = \begin{pmatrix} 0 & -\mathrm{iexp}-\mathrm{i}qt \\ \mathrm{iexpi}qt & 0 \end{pmatrix} = (\cos qt)\sigma_y - (\sin qt)\sigma_x$$

One can thus carry out the full spectrum (all possible permutations and combinations) of rotational operations around the x-, y-, and z-directions and the evolution operations on the basis matrices σ_x, σ_y, and σ_z.

3.2.2 Product Operators: 2-Spin Heteronuclear Ensemble [12]

3.2.2.1 Weak Coupling: 2 Spins

In this system, each of the nuclear species will have 4×4 matrices as their basis matrices. Because each species has 2D (2×2) matrices, σ_0, σ_x, σ_y, and σ_z, one can build the 4×4 matrices from these existing 2×2 matrices as follows:

The $\tilde{A}$ operator system

$$\tilde{A}_x = \sigma_x \otimes \sigma_0 = 1/2\begin{pmatrix} 0 & 1 \\ 1 & 0 \end{pmatrix} \otimes \begin{pmatrix} 1 & 0 \\ 0 & 1 \end{pmatrix} = \begin{pmatrix} 0\sigma_0 & 1\sigma_0 \\ 1\sigma_0 & 0\sigma_0 \end{pmatrix}$$

Thus, further expansion gives

$$\tilde{A}_x = 1/2 = \begin{pmatrix} \begin{pmatrix} 0 & 0 \\ 0 & 0 \end{pmatrix} & \begin{pmatrix} 1 & 0 \\ 0 & 1 \end{pmatrix} \\ \begin{pmatrix} 1 & 0 \\ 0 & 1 \end{pmatrix} & \begin{pmatrix} 0 & 0 \\ 0 & 0 \end{pmatrix} \end{pmatrix} = \begin{pmatrix} 0 & 0 & 1 & 0 \\ 0 & 0 & 0 & 1 \\ 1 & 0 & 0 & 0 \\ 0 & 1 & 0 & 0 \end{pmatrix}$$

The symbol $\otimes$ means a cross-product of two matrices. What it means is that each of the elements in the first 2×2 matrix multiplies the second 2×2 matrix throughout, and one retains it as a 2×2 matrix at that element space.

Here,

$$0 = 0 = \begin{pmatrix} 0 & 0 \\ 0 & 0 \end{pmatrix} \text{ is called the } \textit{null matrix} \text{ and } \sigma_0 = \begin{pmatrix} 1 & 0 \\ 0 & 1 \end{pmatrix}$$

One can work out the other three basis matrices of the spin A operators, i.e., $\tilde{A}_0$, $\tilde{A}_y$, $\tilde{A}_z$, etc., as follows:

$$\tilde{A}_0 = 1 = \sigma_0 \otimes \sigma_0 = \begin{pmatrix} \begin{pmatrix} 1 & 0 \\ 0 & 1 \end{pmatrix} & \begin{pmatrix} 0 & 0 \\ 0 & 0 \end{pmatrix} \\ \begin{pmatrix} 0 & 0 \\ 0 & 0 \end{pmatrix} & \begin{pmatrix} 1 & 0 \\ 0 & 1 \end{pmatrix} \end{pmatrix} = \begin{pmatrix} 1 & 0 & 0 & 0 \\ 0 & 1 & 0 & 0 \\ 0 & 0 & 1 & 0 \\ 0 & 0 & 0 & 1 \end{pmatrix}$$

$$\tilde{A}_y = \sigma_y \otimes \sigma_0 = 1/2\begin{pmatrix} 0 & -i \\ i & 0 \end{pmatrix} \otimes \begin{pmatrix} 1 & 0 \\ 0 & 1 \end{pmatrix}$$

$$= 1/2\begin{pmatrix} 0\sigma_0 & -\mathrm{i}\sigma_0 \\ \mathrm{i}\sigma_0 & 0\sigma_0 \end{pmatrix} = \begin{pmatrix} 0\begin{pmatrix} 1 & 0 \\ 0 & 1 \end{pmatrix} & -\mathrm{i}\begin{pmatrix} 1 & 0 \\ 0 & 1 \end{pmatrix} \\ \mathrm{i}\begin{pmatrix} 1 & 0 \\ 0 & 1 \end{pmatrix} & 0\begin{pmatrix} 1 & 0 \\ 0 & 1 \end{pmatrix} \end{pmatrix} = \begin{pmatrix} 0 & 0 & -\mathrm{i} & 0 \\ 0 & 0 & 0 & -\mathrm{i} \\ \mathrm{i} & 0 & 0 & 0 \\ 0 & \mathrm{i} & 0 & 0 \end{pmatrix}$$

$$\tilde{A}_z = 1/2\sigma_z \otimes \sigma_0 = 1/2\begin{pmatrix} 1 & 0 \\ 0 & -1 \end{pmatrix} \otimes \begin{pmatrix} 1 & 0 \\ 0 & 1 \end{pmatrix}$$

$$= 1/2\begin{pmatrix} 1\begin{pmatrix} 1 & 0 \\ 0 & 1 \end{pmatrix} & 0\begin{pmatrix} 1 & 0 \\ 0 & 1 \end{pmatrix} \\ 0\begin{pmatrix} 1 & 0 \\ 0 & 1 \end{pmatrix} & -1\begin{pmatrix} 1 & 0 \\ 0 & 1 \end{pmatrix} \end{pmatrix} = \begin{pmatrix} 1 & 0 & 1 & 0 \\ 0 & 1 & 0 & 1 \\ 1 & 0 & -1 & 0 \\ 0 & 1 & 0 & -1 \end{pmatrix}$$

$$A_0 = 1/2\sigma_0 \otimes \sigma_0 = 1/2\begin{pmatrix} 1 & 0 \\ 0 & 1 \end{pmatrix} \otimes \begin{pmatrix} 1 & 0 \\ 0 & 1 \end{pmatrix}$$

$$= 1/2\begin{pmatrix} 1\begin{pmatrix} 1 & 0 \\ 0 & 1 \end{pmatrix} & 0\begin{pmatrix} 1 & 0 \\ 0 & 1 \end{pmatrix} \\ 0\begin{pmatrix} 1 & 0 \\ 0 & 1 \end{pmatrix} & 1\begin{pmatrix} 1 & 0 \\ 0 & 1 \end{pmatrix} \end{pmatrix} = \begin{pmatrix} 1 & 0 & 1 & 0 \\ 0 & 1 & 0 & 1 \\ 1 & 0 & 1 & 0 \\ 0 & 1 & 0 & 1 \end{pmatrix}$$

Similarly, for the second spin:

$$K_0 = 1/2\sigma_0 \otimes \sigma_0 = \begin{pmatrix} 1 & 0 \\ 0 & 1 \end{pmatrix} \otimes \begin{pmatrix} 1 & 0 \\ 0 & 1 \end{pmatrix}$$

$$= 1/2\begin{pmatrix} 1\begin{pmatrix} 1 & 0 \\ 0 & 1 \end{pmatrix} & 0\begin{pmatrix} 1 & 0 \\ 0 & 1 \end{pmatrix} \\ 0\begin{pmatrix} 1 & 0 \\ 0 & 1 \end{pmatrix} & 1\begin{pmatrix} 1 & 0 \\ 0 & 1 \end{pmatrix} \end{pmatrix} = \begin{pmatrix} 1 & 0 & 1 & 0 \\ 0 & 1 & 0 & 1 \\ 1 & 0 & 1 & 0 \\ 0 & 1 & 0 & 1 \end{pmatrix}$$

$$1 = \sigma_0 \otimes \sigma = 2K_0 = 2A_0$$

Now one can write, for the other components of the second spin:

$$K_x = 1/2\sigma_0 \otimes \sigma_x = 1/2\begin{pmatrix} 1 & 0 \\ 0 & 1 \end{pmatrix} \otimes \begin{pmatrix} 0 & 1 \\ 1 & 0 \end{pmatrix} = 1/2\begin{pmatrix} \sigma_x & 0 \\ 0 & \sigma_x \end{pmatrix}$$

$$= 1/2\begin{pmatrix} 1\begin{pmatrix} 0 & 1 \\ 1 & 0 \end{pmatrix} & 0\begin{pmatrix} 0 & 1 \\ 1 & 0 \end{pmatrix} \\ 0\begin{pmatrix} 0 & 1 \\ 1 & 0 \end{pmatrix} & 1\begin{pmatrix} 0 & 1 \\ 1 & 0 \end{pmatrix} \end{pmatrix} = \begin{pmatrix} 0 & 1 & 0 & 0 \\ 1 & 0 & 0 & 0 \\ 0 & 0 & 0 & 1 \\ 0 & 0 & 1 & 0 \end{pmatrix}$$

$$K_y = 1/2\sigma_0 \otimes \sigma_y$$

$$= \begin{pmatrix} \begin{pmatrix} 0 & -i \\ 0 & 0 \end{pmatrix} & \begin{pmatrix} 0 & 0 \\ 0 & 0 \end{pmatrix} \\ \begin{pmatrix} 0 & 0 \\ 0 & 0 \end{pmatrix} & \begin{pmatrix} 0 & -i \\ i & 0 \end{pmatrix} \end{pmatrix} = \begin{pmatrix} 0 & -i & 0 & 0 \\ i & 0 & 0 & 0 \\ 0 & 0 & 0 & -i \\ 0 & 0 & i & 0 \end{pmatrix}$$

$$K_z = 1/2\sigma_0 \otimes \sigma_z$$

$$= 1/2\begin{pmatrix} \begin{pmatrix} 1 & 0 \\ 0 & -1 \end{pmatrix} & \begin{pmatrix} 0 & 0 \\ 0 & 0 \end{pmatrix} \\ \begin{pmatrix} 0 & 0 \\ 0 & 0 \end{pmatrix} & \begin{pmatrix} 1 & 0 \\ 0 & -1 \end{pmatrix} \end{pmatrix} = \begin{pmatrix} 1 & 0 & 0 & 0 \\ 0 & -1 & 0 & 0 \\ 0 & 0 & 1 & 0 \\ 0 & 0 & 0 & -1 \end{pmatrix}$$

The direct product of the matrices A_x, A_y, A_z, and A_0 with K_x, K_y, K_z, and K_0 will create the following 16 matrices:

Table 3.7 The Direct Product of the Matrices

	$\mathbf{A_x}$	$\mathbf{A_y}$	$\mathbf{A_z}$	$\mathbf{A_0}$
K_x	A_xK_x ZQ/DQ	A_yK_x ZQ/DQ	A_zK_x (AP) K_x	A_0K_x
K_y	A_xK_y ZQ/DQ	A_yK_y ZQ/DQ	A_zK_y (AP) K_y	A_0K_y
K_z	A_xK_z (AP) A_x	A_yK_z (AP) A_y	A_sK_s (AP) A_z, K_z	A_0K_z
K_0	A_xK_0	A_yK_0	A_zK_0	A_0K_0

One can work out all the direct-product combination matrices in the table. Some typical examples follow:

$$\begin{aligned} A_xK_z &= (1/2\sigma_x \otimes \sigma_0)(1/2\sigma_0 \otimes \sigma_z) \\ &= 1/4(\sigma_x \otimes \sigma_z) \\ &= 1/4\left[\begin{pmatrix} 0 & 1 \\ 1 & 0 \end{pmatrix} \otimes (\sigma_0)\right]\left[\begin{pmatrix} 1 & 0 \\ 0 & 1 \end{pmatrix} \otimes \sigma_z\right] \\ &= 1/4\begin{pmatrix} 0 & \sigma_0 \\ \sigma_0 & 0 \end{pmatrix}\begin{pmatrix} \sigma_z & 0 \\ 0 & \sigma_z \end{pmatrix} \\ &= 1/4\begin{pmatrix} 0 & \sigma_0\sigma_z \\ \sigma_0\sigma_z & 0 \end{pmatrix} \end{aligned}$$

$$= \begin{pmatrix} \begin{pmatrix} 0 & 0 \\ 0 & 0 \end{pmatrix} & \begin{pmatrix} 1 & 0 \\ 0 & 1 \end{pmatrix}\begin{pmatrix} 1 & 0 \\ 0 & -1 \end{pmatrix} \\ \begin{pmatrix} 1 & 0 \\ 0 & 1 \end{pmatrix}\begin{pmatrix} 1 & 0 \\ 0 & -1 \end{pmatrix} & \begin{pmatrix} 0 & 0 \\ 0 & 0 \end{pmatrix} \end{pmatrix}$$

$$= \begin{pmatrix} \begin{pmatrix} 0 & 0 \\ 0 & 0 \end{pmatrix} & \begin{pmatrix} 1 & 0 \\ 0 & -1 \end{pmatrix} \\ \begin{pmatrix} 1 & 0 \\ 0 & -1 \end{pmatrix} & \begin{pmatrix} 0 & 0 \\ 0 & 0 \end{pmatrix} \end{pmatrix} = \begin{pmatrix} 0 & 0 & 1 & 0 \\ 0 & 0 & 0 & -1 \\ 1 & 0 & 0 & 0 \\ 0 & -1 & 0 & 0 \end{pmatrix}$$

$$A_xK_x = (1/2\sigma_x) \otimes (1/2\sigma_x) = \begin{pmatrix} 0 & 1 \\ 1 & 0 \end{pmatrix} \otimes \sigma_x = 1/4\begin{pmatrix} 0 & \sigma_x \\ \sigma_x & 0 \end{pmatrix}$$

$$= 1/4\begin{pmatrix} \begin{pmatrix} 0 & 0 \\ 0 & 0 \end{pmatrix} & \begin{pmatrix} 0 & 1 \\ 1 & 0 \end{pmatrix} \\ \begin{pmatrix} 0 & 1 \\ 1 & 0 \end{pmatrix} & \begin{pmatrix} 0 & 0 \\ 0 & 0 \end{pmatrix} \end{pmatrix} = 1/4\begin{pmatrix} 0 & 0 & 0 & 1 \\ 0 & 0 & 1 & 0 \\ 0 & 1 & 0 & 0 \\ 1 & 0 & 0 & 0 \end{pmatrix}$$

Each of the 16 matrices represents a physical configuration that can be measured in the laboratory. Because they are all direct matrix products of single-spin matrices, they are called *product matrices*, and the operators associated with them are called *product operators*. One can see that A_x, A_y, and A_z are the matrix representations of the x, y, and z components of the A-spin magnetization, respectively. Similarly, K_x, K_y, and K_z are the matrix representations of the x, y, and z components of the K-spin magnetization, respectively.

The matrices A_xK_x, A_xK_y, A_yK_x, and A_yK_y represent combination ZQCs and DCQs. The form of these matrices is quite distinctive, in that they have nonzero matrix elements only along the reverse diagonal of the matrix. The antiphase x magnetization has two components, one along the positive x-axis and one along the negative x-axis, and similarly for the antiphase y magnetization. The matrices A_zK_x and A_zK_y represent antiphase x and y, A- and K-spin magnetization, respectively. A_zK_z represents antiphase z magnetization for both spins. This state is sometimes referred to as a J-ordered spin state.

It is useful to remember that the matrices that have nonzero matrix elements only along the diagonal or reverse diagonal of the matrix represent magnetizations that cannot be observed in an NMR experiment. There is no way to observe the longitudinal A and K magnetizations represented by the matrices A_z and K_z. The ZQ and DQ magnetizations represented by A_xK_x, A_xK_y, A_yK_x, and A_yK_y, and the J-ordered state represented by A_zK_z are observable. One may, however, indirectly map out the time evolution of the ZQCs and DQCs. One can show that the equilibrium partial density matrix for the 2-spin-½ system consisting of a proton (A) and a carbon-13 (K) is given by

$$\xi = 4A_z + K_z = 4(1/2)\begin{pmatrix} \begin{pmatrix} 1 & 0 \\ 0 & 1 \end{pmatrix} & \begin{pmatrix} 0 & 0 \\ 0 & 0 \end{pmatrix} \\ \begin{pmatrix} 1 & 0 \\ 0 & 1 \end{pmatrix} & \begin{pmatrix} -1 & 0 \\ 0 & -1 \end{pmatrix} \end{pmatrix} + 1/2$$

$$= 2\begin{pmatrix} 1 & 0 & 0 & 0 \\ 0 & 1 & 0 & 0 \\ 0 & 0 & 1 & 0 \\ 0 & 0 & 0 & -1 \end{pmatrix} + 1/2\begin{pmatrix} 1 & 0 & 0 & 0 \\ 0 & 1 & 0 & 0 \\ 0 & 0 & -1 & 0 \\ 0 & 0 & 0 & -1 \end{pmatrix}$$

$$= \begin{pmatrix} 5/2 & 0 & 0 & 0 \\ 0 & 3/2 & 0 & 0 \\ 0 & 0 & -3/2 & 0 \\ 0 & 0 & 0 & -5/2 \end{pmatrix}\begin{pmatrix} \begin{pmatrix} 1 & 0 \\ 0 & -1 \end{pmatrix} & \begin{pmatrix} 0 & 0 \\ 0 & 0 \end{pmatrix} \\ \begin{pmatrix} 0 & 0 \\ 0 & 0 \end{pmatrix} & \begin{pmatrix} 1 & 0 \\ 0 & -1 \end{pmatrix} \end{pmatrix}$$

One can evaluate the effects of various RF pulses and the time evolution of the 2-spin system in exactly the same way as for a single-spin system. In the 2-spin system, the matrices are just a bit larger. In this case, the two proton lines are given by the coherences ξ_{13} and ξ_{24}. The two carbon lines are given by the SQCs ξ_{12} and ξ_{34}. ξ_{23} is a ZQC and ξ_{14} is a DQC. The ZQC and DQC cannot be observed, but they are important because they play a central role in cross-polarization or selective sensitivity-enhancement experiments (Figure 3.10).

3.2.2.2 Rotation Operators (RF Pulses)

$$U^K \varphi x = \sigma_0 \otimes U\varphi x = \begin{pmatrix} 1 & 0 \\ 0 & 1 \end{pmatrix} \otimes \begin{pmatrix} c & -is \\ -is & c \end{pmatrix} = \begin{pmatrix} c & -is & 0 & 0 \\ -is & c & 0 & 0 \\ 0 & 0 & c & -is \\ 0 & 0 & -is & c \end{pmatrix}$$

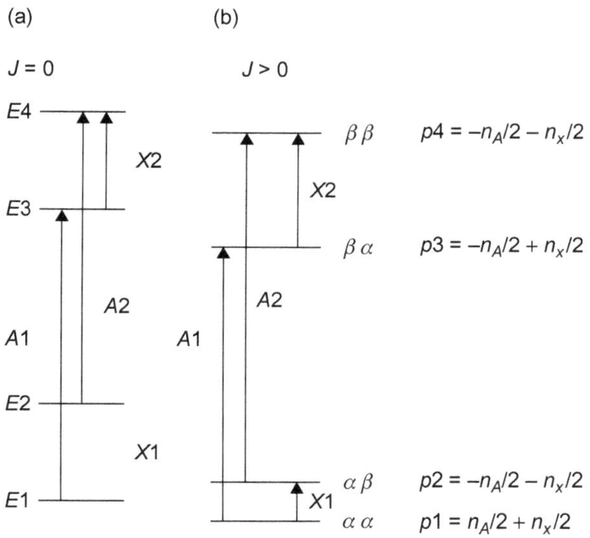

Figure 3.10 The energy-level diagram for a 2-spin system, AX. Both nuclei have a spin of 1/2. The terms pi ($i = 1 - 4$) represent populations of levels i. (a) No spin–spin coupling, $A1 = A2$, $X1 = X2$. (b) Spin–spin coupling > 0.

where, as before, $c = \cos(\varphi/2)$ and $s = \sin(\varphi/2)$. Time evolution of the partial density matrix (due to chemical shift):

$$U^A\omega = U\omega \otimes \sigma_0 = \begin{pmatrix} \exp(i\omega_A t/2) & 0 & 0 & 0 \\ 0 & \exp(-i\omega_A t/2) & 0 & 0 \\ 0 & 0 & \exp(i\omega_A t/2) & 0 \\ 0 & 0 & 0 & \exp(-i\omega_A t/2) \end{pmatrix}$$

$$U^K\omega = \sigma_0 \otimes U\omega = \begin{pmatrix} 1 & 0 \\ 0 & 1 \end{pmatrix} \otimes \begin{pmatrix} \exp(i\omega_K t/2) & 0 \\ 0 & \exp(-i\omega_K t/2) \end{pmatrix}$$

$$= \begin{pmatrix} \exp(i\omega_K t/2) & 0 & 0 & 0 \\ 0 & \exp(-i\omega_K t/2) & 0 & 0 \\ 0 & 0 & \exp(i\omega_K t/2) & 0 \\ 0 & 0 & 0 & \exp(-i\omega_K t/2) \end{pmatrix}$$

These chemical-shift evolution processes are present, at the same time, for the AK 2-spin system.

These can be calculated in a serial fashion. They can also be treated simultaneously by using the Hamiltonian chemical-shift operator. $U_\omega{}^{AK}$ is the simple matrix product (not the direct product $\otimes$).

For systems of two or more unlike spins, an important additional term must be added to the Hamiltonian: the spin-coupling interaction J_{AK}. The Hamiltonian for the AK 2-spin system is

$$U_J = \begin{pmatrix} UH1 & 0 & 1 & 0 \\ 0 & UH2 & 0 & 1 \\ 1 & 0 & UH3 & 0 \\ 0 & 1 & 0 & UH4 \end{pmatrix}$$

where $U_{H1} = it/2(\omega_A + \omega_K - J/2)$; $U_{H2} = it/2(\omega_A - \omega_K + J/2)$; $U_{H3} = -it/2(\omega_A - \omega_K + J/2)$; and $U_{H4} = -it/2(\omega_A + \omega_K - J/2)$. With these mathematical tools, all possible NMR experiments can be described in detail. For example, a $90°_{-x}$ pulse on the proton produces the following effect:

$$\xi(t) = U^A{}_{-90x}\xi(0)U^A{}_{-90x}{}^{-1} = 4A_y + K_z = \begin{pmatrix} 1 & 0 & -4i & 0 \\ 0 & -1 & 0 & -4i \\ -4i & 0 & 1 & 0 \\ 0 & -4i & 0 & -1 \end{pmatrix}$$

One should note that this proton pulse leaves unchanged any parts of the diagonal matrix elements that arise from the carbon spins and generates the off-diagonal elements ξ_{13}, ξ_{24}, ξ_{31}, and ξ_{42}. This reveals that one has created SQCs between energy levels 1 and 3 and between 2 and 4. These coherences give rise to the proton NMR signals. Thus, each element of the density matrix corresponds to a physical process that can be measured. The SQC can be directly observed. The ZQC and DQC can be observed indirectly.

This formalism can easily be extended to a 3-spin system. In that system there will be 64 basis matrices. Each of these 64 matrices will be a direct product of the single-spin matrices. As an example, in the 3-spin system, the x component of the A-spin magnetization in the transverse plane $A_{1x}A_{2x}K_z$ is given by $\sigma_x \otimes \sigma_x \otimes \sigma_z$.

3.2.2.3 The Rotation Operator (RF Pulses): Typical Example

$$U^K \varphi x = \sigma_0 \otimes U \varphi x = \begin{pmatrix} 1 & 0 \\ 0 & 1 \end{pmatrix} \otimes \begin{pmatrix} c & -is \\ -is & c \end{pmatrix} = \begin{pmatrix} c & -is & 0 & 0 \\ -is & c & 0 & 0 \\ 0 & 0 & c & -is \\ 0 & 0 & -is & c \end{pmatrix}$$

where, as before, $c = \cos(\varphi/2)$ and $s = \sin(\varphi/2)$. Time evolution of the partial density matrix (due to chemical shift):

$$U^A \omega = U \omega \otimes \sigma_0 = \begin{pmatrix} \exp(i\omega_A t/2) & 0 \\ 0 & \exp(-i\omega_A t/2) \end{pmatrix} \otimes \begin{pmatrix} 1 & 0 \\ 0 & 1 \end{pmatrix}$$

$$= \begin{pmatrix} \exp(i\omega_A t/2) & 0 & 0 & 0 \\ 0 & \exp(-i\omega_A t/2) & 0 & 0 \\ 0 & 0 & \exp(i\omega_A t/2) & 0 \\ 0 & 0 & 0 & \exp(-i\omega_A t/2) \end{pmatrix}$$

3.2.3 Coherence Transfer Pathways and Multiquantum Correlations

3.2.3.1 Density Matrix Product Operators: 2-Spin (I−S) System [13]

The expectation value of the z component of the magnetic moment $\langle \mu_z \rangle$, and that of the observable z component of the magnetization M_z (magnetic moment per unit volume of the whole sample), can be written in matrix form as follows: First, the $\langle \mu_z \rangle = (\psi | \mu_z |) = \gamma \hbar (\psi | I_z | \psi)$ and the $M_z = N_0 \gamma \hbar (\psi | I_z | \psi)$. Here N_0 is the number of spins per unit volume and γ the gyromagnetic ratio (GMR). By expressing the wave function as a linear combination of basis functions ψ_i (α and β for a single type of n spins),

$$\psi = \sum_{i=1} c_i \psi_i$$

Therefore one gets, on substitution

$$M_z = N_0\gamma\hbar\ [c_1\, c_2 \cdots c_n]$$

$$I_z \begin{bmatrix} c_1 \\ c_2 \\ \cdot \\ \cdot \\ \cdot \\ c_n \end{bmatrix}$$

$$\text{where } I_z = \begin{bmatrix} I_z & I_z & \cdot & I_z \\ I_z & \cdot & \cdot & \cdot \\ \cdot & \cdot & \cdot & \cdot \\ I_z & \cdot & \cdot & I_z \end{bmatrix} I_z = (\psi_i|I_z|\psi_j);\psi_i = [c_1\, c_2 \cdots c_n] \text{ (row matrix) and } \psi_j$$

$$= \begin{bmatrix} c_1 \\ c_2 \\ \cdot \\ \cdot \\ \cdot \\ c_n \end{bmatrix} \text{(column matrix)}$$

Then $I_z^{ij} = (\psi_i|I_z|\psi_j) = \Sigma_n\Sigma_m I_z^{nm} c_m c_n{}^*$ and $M_z = N_0\gamma\ \hbar\Sigma_n\Sigma_m I_z^{nm} c_m c_n{}^*$. Only the elements $c_m c_n{}^*$ vary from system to system in an ensemble. The average $c_m c_n{}^*$ elements make up the density matrix. All changes in a system that occur as a result of precession, relaxation, *J*-coupling, or pulses occur through modification of the density matrix. The time dependence of the state of the system is contained in the density matrix. From this, knowledge of the system can be fully derived.

An alternative method of obtaining the expectation value is to use the trace relation, defined as $\langle M_z \rangle = M_z = N_0\gamma\ \hbar \mathrm{Tr}(Iz\rho)$. Here ρ is the density matrix. The trace Tr = the sum of the diagonal elements of the direct product of the matrices Iz and ρ. The expectation value of any observable can be obtained by interrogating the density matrix with the appropriate spin angular-momentum operator.

In a 2-spin system, the z component matrix of the system is

$$I_z = 1/2 \begin{bmatrix} 1 & 0 & 0 & 0 \\ 0 & 1 & 0 & 0 \\ 0 & 0 & -1 & 0 \\ 0 & 0 & 0 & -1 \end{bmatrix}$$

The transitions that connect states (e.g., which differ by the flip-flop of both the spins, as in $\alpha\beta \rightarrow \beta\alpha$) are the zero-quantum transitions (ZQTs). Those which differ by the flipping of 1 spin are single-quantum transitions (SQTs). The flipping of both spins in the same state (i.e., $\alpha\alpha \rightarrow \beta\beta$ and $\beta\beta \rightarrow \alpha\alpha$) represents DQTs. At equilibrium, the off-diagonal elements in the density matrix (DM) should be zero, and the diagonal elements should be proportional to the equilibrium populations of each of the states. For two protons in thermal equilibrium, the relative populations (probability of occupation) of the state i and j can be written, using Boltzmann statistics, as $P_i/P_j = e[-(E_i - E_j)/kT]$. E_i and E_j are the two energy levels under consideration, k is the Boltzmann constant, and T is the equilibrium temperature of the ensemble. Let $\alpha\beta$ be the zero-state energy.

In this case, $P_1/P_2 = \exp[-(-h\nu_S/kT)] \approx 1 + h\nu_S/kT = 1 + p$. Here h is Planck's constant. The symbol $\hbar = h/2\pi$ is the one normally used in QM equations and ν is the frequency of radiation, corresponding to an energy transition. $P_2/P_2 = P_3/P_3 = \exp(0) = 1$ and $P_4/P_2 = \exp[(h\nu_I/kT)] \approx 1 - (h\nu_I/kT)] = 1 - q$; $p = h\nu_S/kT$ and $q = h\nu_I/kT$. From these results, one gets $P_1 = (1 + p)P_2$; $P_2 = P_2$; $P_3 = P_2$; $P_4 = (1 - P)P_2$ and adding $P_1 + P_2 + P_3 + P_4 = 1 = 4P_2$. So, $P_1 = (1/4)(1 + p)$; $P_2 = P_3 = 1/4$; $P_4 = (1/4)(1 - p)$.

$$\text{The DM at equilibrium} = 1/4\begin{bmatrix}1&0&0&0\\0&1&0&0\\0&0&1&0\\0&0&0&1\end{bmatrix} + p/4\begin{bmatrix}1&0&0&0\\0&0&0&0\\0&0&0&0\\0&0&0&-1\end{bmatrix}$$

The first part is the identity matrix **I**, which is not affected by application of the RF pulses or scalar coupling. So, for practical purposes, density matrix (DW) at equilibrium is

$$\mathrm{DM}(0) = (p/4)\begin{bmatrix}1&0&0&0\\0&0&0&0\\0&0&0&0\\0&0&0&-1\end{bmatrix}$$

$$\equiv (p/4)[1/2\begin{bmatrix}1&0&0&0\\0&1&0&0\\0&0&-1&0\\0&0&0&-1\end{bmatrix} + 1/2\begin{bmatrix}1&0&0&0\\0&-1&0&0\\0&0&1&0\\0&0&0&-1\end{bmatrix}]$$

$$= (p/4hbox)(I_z + S_x)$$

The time dependence of the expectation value is contained in the DM (Table 3.8, Figures 3.11 and 3.12).

Table 3.8 Product Operators in the Cartesian Basis for a 2-Spin System Showing the Matrix Forms Along with Vector Representations

O	Matrix	I Spins	S Spins	O	Matrix	I Spins	S Spins
	$\begin{bmatrix}1&0&0&0\\0&1&0&0\\0&0&1&0\\0&0&0&1\end{bmatrix}$	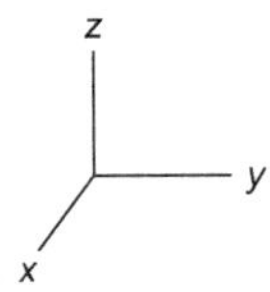	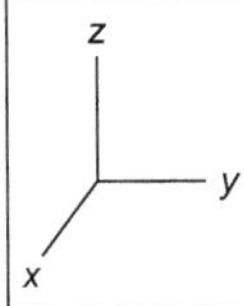	I_yS_z	$1/4\begin{bmatrix}0&0&-1&0\\0&0&0&1\\1&0&0&0\\0&-1&0&1\end{bmatrix}$	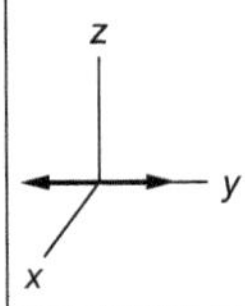	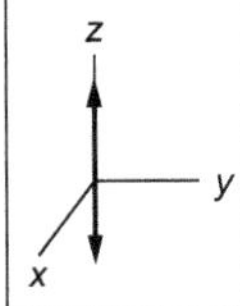
x	$1/2\begin{bmatrix}0&0&1&0\\0&0&0&1\\1&0&0&0\\0&1&0&0\end{bmatrix}$	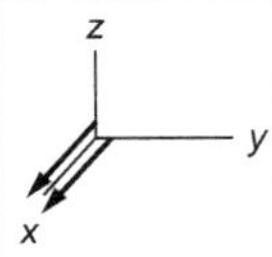	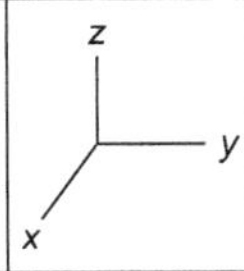	I_yS_z	$1/4\begin{bmatrix}1&0&0&0\\0&-1&0&0\\0&0&-1&0\\0&0&0&1\end{bmatrix}$	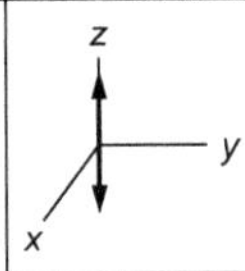	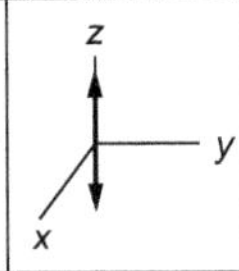
y	$1/2\begin{bmatrix}0&0&-1&0\\0&0&0&-1\\1&0&0&0\\0&1&0&0\end{bmatrix}$	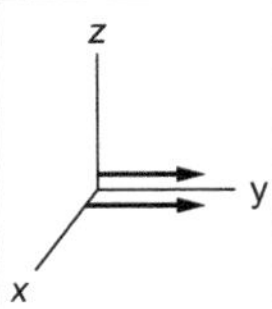	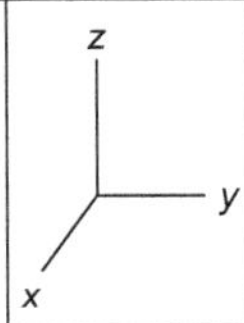	I_zS_x	$1/4\begin{bmatrix}0&1&0&0\\1&0&0&0\\0&0&0&-1\\0&0&-1&0\end{bmatrix}$	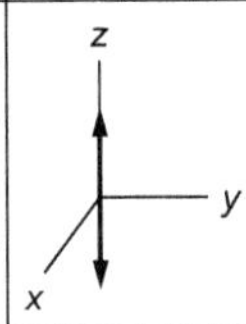	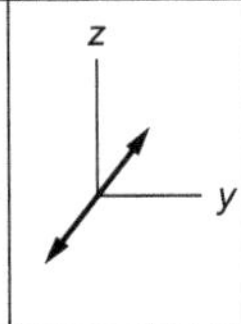
z	$1/2\begin{bmatrix}1&0&0&0\\0&1&0&0\\0&0&-1&0\\0&0&0&-1\end{bmatrix}$	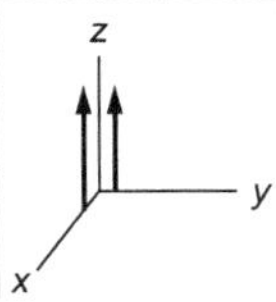	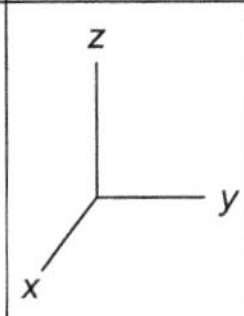	I_zS7	$1/4\begin{bmatrix}0&-1&0&0\\1&0&0&0\\0&0&0&1\\0&0&-1&0\end{bmatrix}$	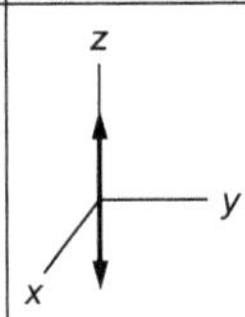	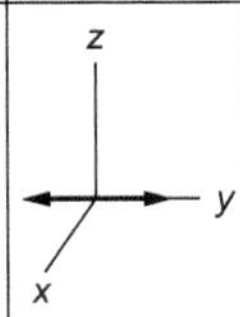
S_x	$1/2\begin{bmatrix}0&1&0&0\\1&0&0&0\\0&0&0&1\\0&0&1&0\end{bmatrix}$	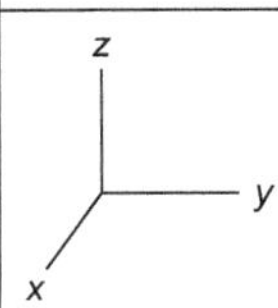	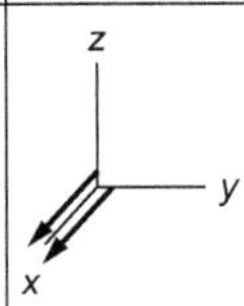	I_xS_x	$1/4\begin{bmatrix}0&0&0&1\\0&0&1&0\\0&1&0&0\\1&0&0&1\end{bmatrix}$	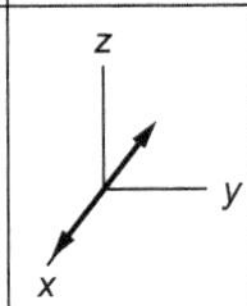	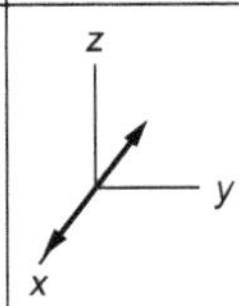
S_y	$1/2\begin{bmatrix}0&-1&0&0\\1&0&0&0\\0&0&0&-1\\0&0&1&0\end{bmatrix}$	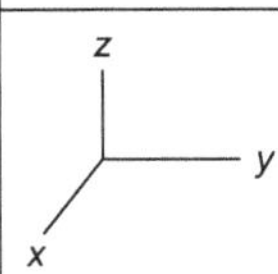	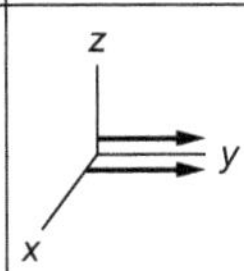	I_yS_y	$1/4\begin{bmatrix}1&0&0&-1\\0&0&1&0\\0&1&0&0\\-1&0&0&0\end{bmatrix}$	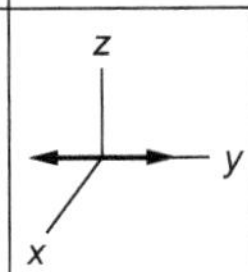	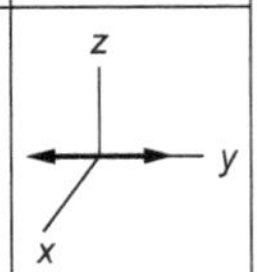
S_z	$1/2\begin{bmatrix}1&0&0&0\\0&-1&0&0\\0&0&1&0\\0&0&0&-1\end{bmatrix}$	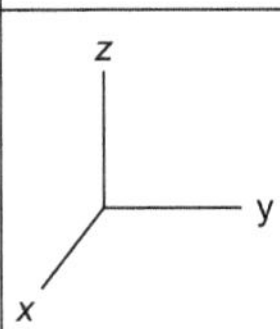	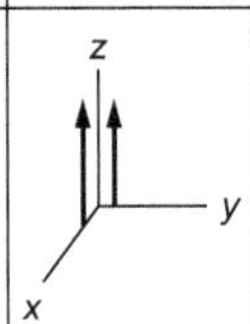	I_xS_y	$1/4\begin{bmatrix}0&0&0&-1\\0&0&1&0\\0&-1&0&0\\1&0&0&0\end{bmatrix}$	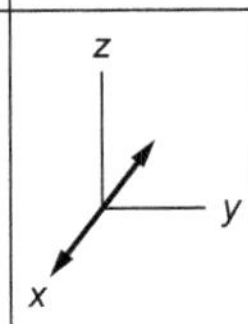	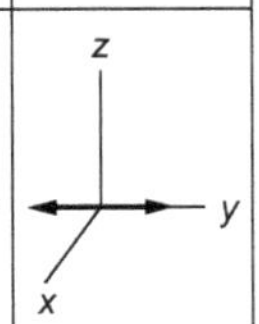
I_xS_z	$1/4\begin{bmatrix}0&0&1&0\\0&0&0&-1\\1&0&0&0\\0&-1&0&0\end{bmatrix}$	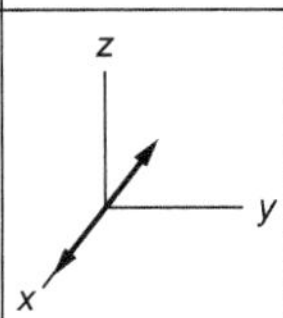	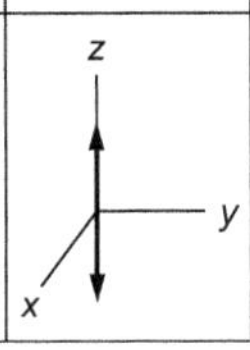	I_yS_x	$1/4\begin{bmatrix}0&0&0&-1\\0&0&-1&0\\0&1&0&0\\1&0&0&0\end{bmatrix}$	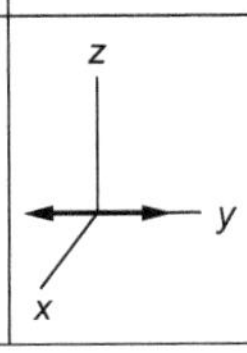	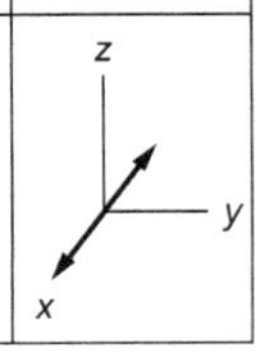

O, operator.

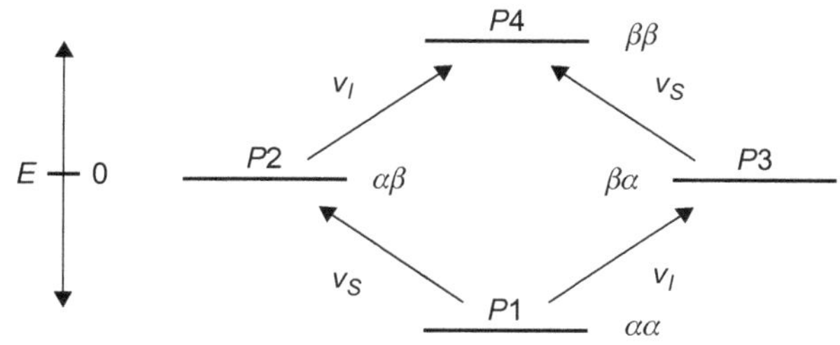

Figure 3.11 Energy-level diagram for a 2-spin (*I*, *S*) system. The first character of the wave function indicates the state of the *I* spin; the second character shows that of the *S* spin. The zero-energy level is defined as the energy of the *p* state. The populations of the four levels are *P*1, *P*2, *P*3, and *P*4.

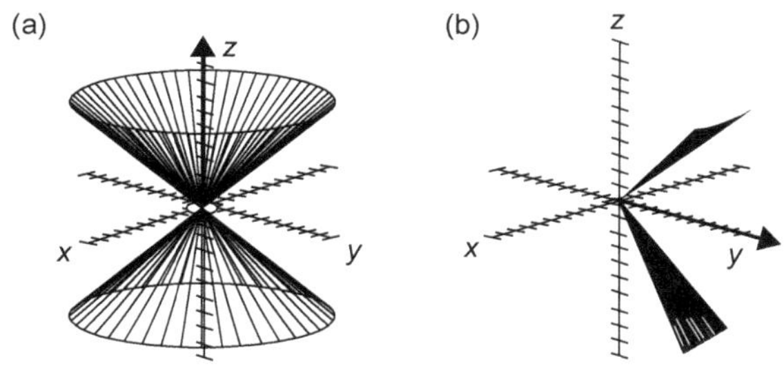

Figure 3.12 Creation of coherence with the application of a 90° pulse to an ensemble of spins at thermal equilibrium. (a) At equilibrium, an excess of spins is aligned with the static magnetic field *B*. They precess about the field with random phase, resulting in a net magnetization along the *z*-axis. With the application of an RF field on resonance along the *x*-axis, the spins precess about the applied field, and the net magnetization is left aligned along the axis. (b) When the RF field is turned off, the spins are again quantized in the *B* field, and there are equal numbers of spins with their *z* components of angular momentum aligned with and against the *B* field. The nonequilibrium net magnetization is the result of the creation of coherence in the phases of the precessing nuclei in the ensemble.

3.2.3.2 Selective Application of a 90° Pulse: Spin Equilibrium Magnetization [13]

A 90° pulse gives $\mathrm{DM}(1) = R^{-1}{}_{I90}, \mathrm{DM}(0)R_{I90}$

$$= (p/4)\,1/\sqrt{2}\begin{bmatrix} 1 & 0 & -i & 0 \\ 0 & 1 & 0 & -i \\ -i & 0 & 1 & 0 \\ 0 & -i & 0 & 1 \end{bmatrix}\begin{bmatrix} 1 & 0 & i & 0 \\ 0 & 0 & 0 & 0 \\ 0 & 0 & 0 & 0 \\ 0 & 0 & 0 & -1 \end{bmatrix} 1/\sqrt{2}\begin{bmatrix} 1 & 0 & i & 0 \\ 0 & 1 & 0 & i \\ i & 0 & 1 & 0 \\ 0 & i & 0 & 1 \end{bmatrix}$$

$$= (p/4)\,1/2\begin{bmatrix} 1 & 0 & -i & 0 \\ 0 & 1 & 0 & -i \\ -i & 0 & 1 & 0 \\ 0 & -i & 0 & 1 \end{bmatrix}\begin{bmatrix} 1 & 0 & 0 & 0 \\ 0 & 0 & 0 & -i \\ i & 0 & 0 & 0 \\ 0 & 0 & 0 & -1 \end{bmatrix}$$

$$= (p/4)\,1/2 \begin{bmatrix} 1 & 0 & i & 0 \\ 0 & -1 & 0 & i \\ -i & 0 & 1 & 0 \\ 0 & -i & 0 & -1 \end{bmatrix}$$

Here we have used the 90° rotation operator for the I spin due to RF radiation, and counterclockwise rotation and precession is used as positive, as a convention.

$$R_{I90} = 1/\sqrt{2} \begin{bmatrix} 1 & 0 & i & 0 \\ 0 & 1 & 0 & i \\ i & 0 & 1 & 0 \\ 0 & i & 0 & i \end{bmatrix} \text{(the rotation operator)}$$

$$R^{-1} = 1/\sqrt{2} \begin{bmatrix} 1 & 0 & -i & 0 \\ 0 & 1 & 0 & -i \\ -i & 0 & i & 0 \\ 0 & -i & 0 & 1 \end{bmatrix} \text{(the inverse rotation operator)}$$

One can verify that

$$R^{-1} R_{I}90 = I$$

$$= 1/\sqrt{2} \begin{bmatrix} 1 & 0 & -i & 0 \\ 0 & 1 & 0 & -i \\ -i & 0 & i & 0 \\ 0 & -i & 0 & 1 \end{bmatrix} 1/\sqrt{2} \begin{bmatrix} 1 & 0 & i & 0 \\ 0 & 1 & 0 & i \\ i & 0 & i & 0 \\ 0 & i & 0 & 1 \end{bmatrix} \begin{bmatrix} 1 & 0 & 0 & 0 \\ 0 & 1 & 0 & 0 \\ 0 & 0 & 1 & 0 \\ 0 & 0 & 0 & 1 \end{bmatrix}$$

$$= \text{(the identity matrix } I \textit{ or } 1)$$

Alternatively, DM(1) can be written as follows:

$$\mathrm{DM}(1) = (p/4)\,1/2 \begin{bmatrix} 1 & 0 & i & 0 \\ 0 & -1 & 0 & i \\ -i & 0 & 1 & 0 \\ 0 & -i & 0 & -1 \end{bmatrix}$$

$$= (p/4)(1/2) \left[\begin{bmatrix} 0 & 0 & i & 0 \\ 0 & 0 & 0 & i \\ -i & 0 & 0 & 0 \\ 0 & -i & 0 & 0 \end{bmatrix} + \begin{bmatrix} 1 & 0 & 0 & 0 \\ 0 & -1 & 0 & 0 \\ 0 & 0 & 1 & 0 \\ 0 & 0 & 0 & -1 \end{bmatrix} \right]$$

$$= (p/4)(1/2)(I_y + S_z)$$

Because S_z, it is seen, does not change on rotation (as it should), I_y is given as

$$I_y = 1/2\begin{bmatrix} 0 & 0 & -\mathrm{i} & 0 \\ 0 & 0 & 0 & -\mathrm{i} \\ \mathrm{i} & 0 & 0 & 0 \\ 0 & \mathrm{i} & 0 & 0 \end{bmatrix} \text{ and } S_z = 1/2\begin{bmatrix} 1 & 0 & 0 & 0 \\ 0 & -1 & 0 & 0 \\ 0 & 0 & 1 & 0 \\ 0 & 0 & 0 & -1 \end{bmatrix} \text{ as before}$$

Application of a 90° pulse rotation, for the S spins, can be performed similarly, as follows:

$$R_{S90} = 1/\sqrt{2}\begin{bmatrix} 1 & \mathrm{i} & 0 & 0 \\ \mathrm{i} & 1 & 0 & 0 \\ 0 & 0 & 1 & \mathrm{i} \\ 0 & 0 & \mathrm{i} & 1 \end{bmatrix} \text{ and } \mathrm{DM}(2) = R^{-1},\ DM(1)R_{S90}$$

$$= (p/4)(1/2)\begin{bmatrix} 0 & \mathrm{i} & \mathrm{i} & 0 \\ -\mathrm{i} & 0 & 0 & \mathrm{i} \\ -\mathrm{i} & 0 & 0 & \mathrm{i} \\ 0 & -\mathrm{i} & -\mathrm{i} & 0 \end{bmatrix}$$

$$= -I_y - S_y = -(p/4)(1/2)\begin{bmatrix} 0 & 0 & -\mathrm{i} & 0 \\ 0 & 0 & 0 & -\mathrm{i} \\ \mathrm{i} & 0 & 0 & 0 \\ 0 & \mathrm{i} & 0 & 0 \end{bmatrix}$$

$$-(p/4)(1/2)\begin{bmatrix} 0 & -\mathrm{i} & 0 & 0 \\ \mathrm{i} & 0 & 0 & 0 \\ 0 & 0 & 0 & -\mathrm{i} \\ 0 & 0 & i & 0 \end{bmatrix}$$

(the net transverse magnetization)

One can work out that

$$S_y = 1/2\begin{bmatrix} 0 & -\mathrm{i} & 0 & 0 \\ \mathrm{i} & 0 & 0 & 0 \\ 0 & 0 & 0 & -\mathrm{i} \\ 0 & 0 & \mathrm{i} & 0 \end{bmatrix}$$

One can see that application of single 90° pulses to the I and S spins (initially aligned along the z-direction by the static magnetic field) results in transverse magnetization (I_y and S_y). These can be measured as an FID relaxation time in a spin-echo experiment. Different nuclei would have different relaxation times (due to different gyromagnetic ratios).

This is the source of MRI when the brain is scanned using RF radiation applied perpendicular to the z-direction. The off-diagonal elements in DM(2) indicate that application of the 90° pulses has induced SQTs, $\mathrm{DM}_{12} \equiv \alpha\alpha \rightarrow \beta\alpha$ and $\mathrm{DM}_{21} \equiv \beta\alpha \rightarrow \alpha\alpha$. The creation of the y magnetization results from transforming a collection of spins precessing about the z-axis, with a random phase at thermal equilibrium, to a collection of spins with coherence in the $x-y$ plane and a resultant net magnetization along the y-axis. The coherent superposition of states is called SQC. It is possible, with further pulses, to change the $DM(2)_{14}$ and $DM(2)_{41}$ elements of the matrix to become nonzero. Therefore, DQCs can also be created. In MRI, one uses multipulse experiments. Thus, the DM increases in complexity of its elements and is not easy to determine. Product-operator formalism offers an easier alternative, as described in Section 3.2.3.3.

3.2.3.3 Product-Operator Formalism: Trace of a Matrix [13]

One can write DM as $\mathrm{DM} = \Sigma_k b_k O_k$, a linear combination of orthogonal matrices/operators (O_k).

The orthogonality is defined using the trace relation, $\mathrm{Tr}(O_i O_j) = 0$. Now we use the complete orthogonal product-operator basis set of the familiar angular-momentum matrices I_x, I_y, I_z, etc. Then, because of the orthogonality, the only component of the expansion that survives in the trace relation is the product operator that corresponds to the angular-momentum operator, the value of which is being sought. As an example, if one is looking for the y component of the magnetization of the I spins, one interrogates the density matrix with the product operator I_y, to obtain $M_y = (N_0\gamma\hbar)\mathrm{Tr}(I_y\mathrm{DM}) = (N_0\gamma\hbar)\mathrm{Tr}[I_y\Sigma_k b_k O_k] = (N_0\gamma\hbar)n_{Iy}b_{Iy}$. Here n_{Iy} is constant, because the operators are not normalized, and b_{Iy} is the expansion coefficient for the I_y term. The expectation value of an operator is the coefficient of the corresponding product operator in the expansion. The trace relation is not used explicitly. Instead, if one requires M_y, for instance, one simply notes the coefficients for the product operator I_y. The initial product-operator expression is usually composed of only one or two terms to represent the thermal equilibrium magnetization.

For a 2-spin system, the initial operator expression would be $I_z + S_z$. Application of pulses, precession, and J-coupling modifies the product operators, annihilates some operators, and creates new operators in the expansion, according to a defined set of rules. At the end of the experiment, only terms in the expansion are physically relevant. I_y is important if the receiver (detector) is set to find an oscillating signal (FID) along the y-axis. The product-operator basis sets can be worked out, including Cartesian (x, y, z) and spherical (r, θ, ψ). The first is useful for visualization of an experiment, and the latter for following coherence transfer during multiple-pulse experiments. It is now clear that a 2-spin-½ system can be described by a 4×4 density matrix. The 16-element matrix requires a basis set of 16 orthogonal matrices. The most straightforward method by which to achieve this is to construct operator matrices.

This is done using the six Cartesian spin angular-momentum operators I_x, I_y, I_z, S_x, S_y, and S_z. The various possible products, taking two at a time, are $\{\mathbf{I_xS_x}, \mathbf{I_xS_y}, \mathbf{I_xS_z}\}$, $\{\mathbf{I_yS_x}, \mathbf{I_yS_y}, \mathbf{I_yS_z}\}$, and $\{\mathbf{I_zS_x}, \mathbf{I_zS_y}, \mathbf{I_zS_z}\}$. Then there is the identity operator **1**, which transforms any operator into itself. It is important to learn how each of these product operators behaves under the action of an arbitrary pulse, precession, and J-coupling. Refer to the typical results set out in Table 3.9 [13].

Application Examples: A 90° pulse along the x-axis $\rightarrow I_z\cos(90°) - I_y\sin(90°) = -\mathbf{I_y}$. Evolution due to pulses, precession, and J-coupling of each of the

Table 3.9 Effect of Application of Pulses, Precession, and *J*-Coupling

Pulse along the *x*-axis through an angle $\theta = \gamma B_1 t$
$I_x \rightarrow I_x$
$I_y \rightarrow I_y\cos\theta + I_z\sin\theta$
$I_z \rightarrow I_z\cos\theta - I_y\sin\theta$
Pulse along the *y*-axis through an angle $\theta = \gamma B_1 t$
$I_x \rightarrow I_x\cos\theta - I_z\sin\theta$
$I_y \rightarrow I_y$
$I_z \rightarrow I_z\cos\theta + I_x\sin\theta$
Precession about the *z*-axis with frequency $= \omega$
$I_x \rightarrow I_x\cos\omega t + I_y\sin\omega t$
$I_y \rightarrow I_y\cos\omega t - I_z\sin\omega t$
$I_z \rightarrow I_z$
J-coupling with $\lambda = 1/2Jt$
$I_x \rightarrow I_x\cos\lambda + 2I_yS_z\sin\lambda$
$I_y \rightarrow I_y\cos\lambda - 2I_xS_z\sin\lambda$
$I_z \rightarrow I_z$

B, the pulse field strength; γ, the gyromagnetic ratio; *J*, the *J*-coupling constant; and *t*, the time of the pulse, precession, and *J*-coupling.

individual operators in a two- (or more) spin product operator (e.g., $\mathbf{I_xS_y}$) is independent. As an example, the effect of applying a 90° pulse along the *y*-axis to **I** spins on the operator $\mathbf{I_xS_z}$ is obtained by substituting the appropriate rule for $\mathbf{I_x}$, and multiplying to get $-\mathbf{I_zS_z}$. The following relations can be worked out: $\mathbf{I_xI_x} = \mathbf{I_yI_y} = \mathbf{I_zI_z} = 1/4$; $\mathbf{I_xI_y} = 1/2\mathbf{I_zI_yI_z} = 1/2\mathbf{I_x}$, $\mathbf{I_zI_x} = 1/2\mathbf{I_y}$ (*x*, *y*, *z* cyclic); $\mathbf{I_yI_x} = -\mathbf{I_xI_y}$, $\mathbf{I_zI_y} = -\mathbf{I_yI_z}$, $\mathbf{I_xI_z} = -\mathbf{I_zI_x}$ (anticommutation). In addition, operators for different spins commute, so that $\mathbf{I_zS_z} = \mathbf{S_zI_z}$ and $\mathbf{S_zI_xS_z} = \mathbf{I_xS_z}^2 = 1/4\mathbf{I_x}$. In a 2-spin system, the initial equilibrium is represented by $\mathbf{I_z} + \mathbf{S_z}$. Application of a 90° pulse only to the *I* spins along the *x*-axis leads to $\mathbf{I_z} + \mathbf{S_z} \rightarrow \mathbf{I_z\cos\theta} - \mathbf{I_y\sin\theta} + \mathbf{S_z} = -\mathbf{I_y} + \mathbf{S_z}$. It is expected, from the classical vector model (CVM), that the two components of the *I* doublets (spin up and spin down) will fan out because of *J*-coupling. One will precess at *J*/2 and the other at $-J/2$ Hz. In product-operator formalism (POF), $-\mathbf{I_y}$ is transformed (Table 3.9 [13]) by *J*-coupling according to $-\mathbf{I_y} \rightarrow -\mathbf{I_y\cos(\pi Jt)} + \mathbf{2I_xS_z\sin(\pi Jt)}$. Thus, the addition of $-\mathbf{I_y}$ and $\mathbf{2I_xS_z}$ leads to the two components of the doublet precessing in opposite directions (Figure 3.13 [13]). This means that $\mathbf{I_xS_z}$ represents antiphase magnetization of the *I* spins. This allows the *J*-coupling to continue for one-quarter of a cycle.

It happens for a period of time $t = 1/2J$, leading to two components being aligned in opposite directions along the *x*-axis. The product simply reduces to $-\mathbf{2I_xS_z}$. The term *antiphase magnetization* (APM) is used to indicate that half of the spins are aligned along the positive *x*-axis, and the other half in the opposite direction. After the first 90° pulse is applied to the *I* spins along the *x*-axis, followed by a delay of $1/2J$, application of a 90° pulse to the *I* spins along the *y*-axis transforms $\mathbf{2I_xS_z}$. If this is followed immediately by the application of a 90° pulse to the *S* spins along the *y*-axis, $\mathbf{-2I_zS_z} \rightarrow -\mathbf{I_z}(\mathbf{S_z\cos\theta} + \mathbf{S_x\sin\theta}) = -2\mathbf{I_zS_x}$. This is the opposite configuration of $2\mathbf{I_xS_z}$. It is clear now that if $2\mathbf{I_xS_z}$ represents the APM of the *I* spins, then $\mathbf{2I_zS_x}$ must represent the APM of the *S* spins. The ability to create this APM of the *S* spins in

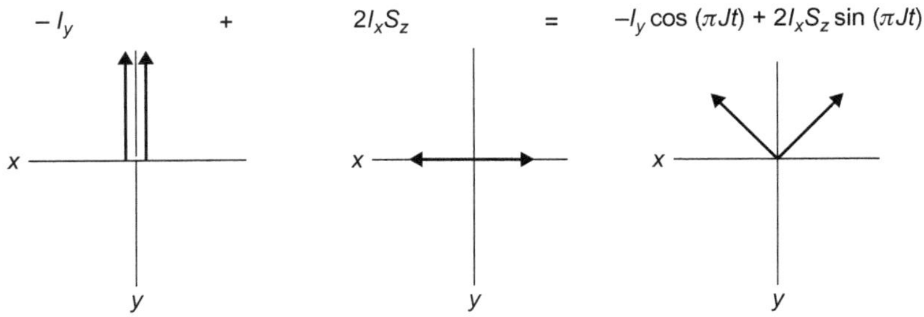

Figure 3.13 The precession of the two components of a doublet is represented by addition of the product operators in the Cartesian basis I_y and $2I_xS_z$.

$\mathbf{2I_zS_x}$, with a 90° pulse, implies that antiphase z magnetization (half of the spins up and half down) of the S spins exists in $\mathbf{2I_zS_z}$, a fact that can be confirmed by viewing the state of the system with a 90° "read" pulse applied to the S spins to get an antiphase doublet (i.e., one line of the doublet up and one down). The nature of $\mathbf{2I_zS_x}$ may also be seen by inspection of the matrix representation of the operator (Table 3.8 [13]). The $\alpha\alpha$ and $\beta\beta$ states are equally populated and exceed the populations of the equally populated $\alpha\beta$ and $\beta\alpha$ states. One line in both the I and S doublets will be positive and one will be negative. Thus, the so-called longitudinal 2-spin-ordered state $\mathbf{2I_zS_z}$ represents selectively inverted I and S spins, and contains antiphase z magnetization for both I and S spins. It can also be concluded that if antiphase z magnetization exists for the S spins in $\mathbf{2I_zS_z}$, then it must be present in $\mathbf{2I_zS_x}$ as well. This is because the former is created from the latter by pulsing only the I spins.

The independent evolution of each of the individual spin operators in product operators is derived from two or more angular-momentum operators. The vector representation of the product operators containing ZQC and DQC cannot be derived from an inspection of the matrix representations. The representations are derived from knowledge that the individual operators evolve independently.

Now one can say that all product operators containing products of two (or more) angular-momentum operators represent antiphase magnetizations for all spins that cannot be observed directly. They are derived from the effects of scalar coupling only. Further, they can be converted into observable magnetization through scalar coupling. All ZQCs and MQCs are represented by one or more such operators (Figure 3.14).

3.2.3.4 Product-Operator Vector Model: MQCs [13]

For purposes of illustration, it is easiest to consider a 2-spin homonuclear-spin system (Figure 3.15 [13]). To start with, apply the first 90° pulse only to spin I_z, along the x-axis. Then allow for precession and J-coupling for time interval t_1. The result is $I_z \rightarrow I_y\cos(\pi Jt_1)\cos(\omega It_1) + 2I_xS_z\sin(\pi Jt_1)\cos(\omega It_1) + I_x\cos(\pi Jt_1)\cos(\omega It_1) + 2I_yS_z \sin(\pi Jt_1)\cos(\omega It_1)$. Application of the second 90° pulse along the x-axis at the end of t_1 gives $-I_z\cos(\pi Jt_1)\cos(\omega It_1) - 2I_xS_y\sin(\pi Jt_1)\cos(\omega It_1) + I_x\cos(\pi Jt_1)\cos(\omega It_1) - 2I_zS_y\sin(\pi Jt_1)\cos(\omega It_1)$. When one has precession and J-coupling over time interval

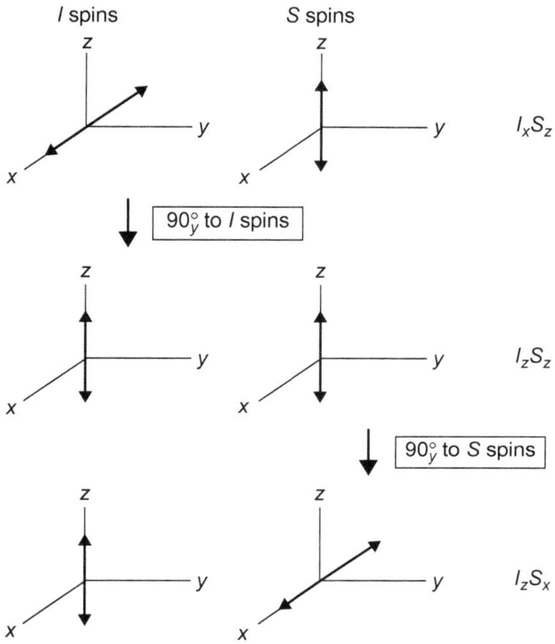

Figure 3.14 The creation of the J-ordered state with antiphase z magnetization for the I and S spins indicates that I, S represents antiphase z magnetization.

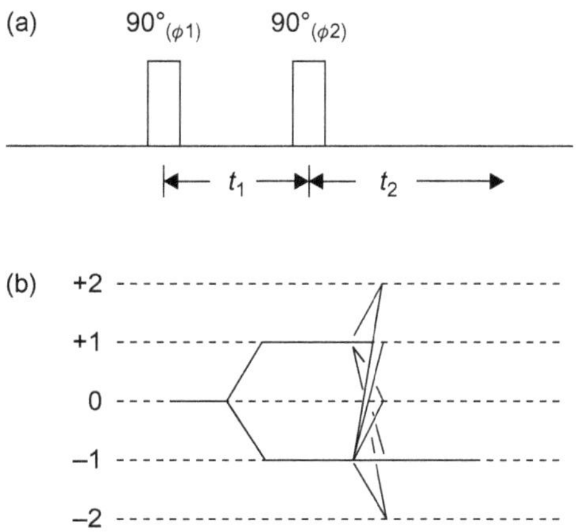

Figure 3.15 Transfer pathways for a 2-spin system indicating coherence orders from -2 to $+2$. (a) Pulse sequence for the 2D NMR COSY experiment. (b) Coherence.

t_2, the result will include the following terms: I_z, I_x, I_y; $2I_xS_x$, $2I_xS_y$, $2I_xS_z$; $2I_yS_x$, $2I_yS_y$, $2I_yS_z$; $2I_yS_z$; I_zS_x, $2I_zS_y$; and S_x, S_y.

Now one can see the presence of ZQC, single spins, and DQC. In COSY, the receiver is turned on after the second pulse. As the detector is along the y-axis, one needs to be concerned only with the I_y and S_y terms.

The detected signal will be $I_y\cos(\pi Jt_1)\cos(\pi Jt_2)\sin(\omega It_1)\sin(\omega It_2) + S_y\sin(\pi Jt_1)\sin(\pi Jt_2)\sin(\omega It_1)\sin(\omega It_2) + S_y\cos(\pi Jt_1)\cos(\pi Jt_2)\sin(\omega St_1)\sin(\omega St_2) + I_y\sin(\pi Jt_1)\sin(\pi Jt_2)\sin(\omega St_1)\sin(\omega It_2)$. This result now includes the effect of applying, in the beginning, a 90° pulse to S spins simultaneously. The first and third terms give peaks at ωI, ωI, and ωS, ωS, after 2D FT and therefore the diagonal peaks in the 2D COSY spectrum. The second and fourth terms give off-diagonal peaks at ωI, ωs, and ωS, ωI, which are the correlation peaks. The cos and sin functions of $(\pi J t_1)$ and $(\pi J t_2)$ result in multiplets, with the diagonal peaks 90° out of phase with respect to the cross-peaks.

DQC: When one looks at Table 3.8, one can recognize the matrices corresponding to the product operators $\mathbf{I_xS_x}$, $\mathbf{I_yS_y}$, $\mathbf{I_xS_y}$, and $\mathbf{I_yS_x}$. They contain ZQC and DQC (nonzero d_{14}, d_{41}, d_{23}, and d_{32} elements), but they cannot be referred to as pure ZQCs and DQCs. Pure DQC is obtained through linear combinations as follows:

$$1/2(2\mathrm{I_xS_x} - 2\mathrm{I_yS_y}) = 1/2\begin{bmatrix} 0 & 0 & 0 & 1 \\ 0 & 0 & 0 & 0 \\ 0 & 0 & 0 & 0 \\ 1 & 0 & 0 & 0 \end{bmatrix}$$

$$1/2(2\mathrm{I_xS_y} + 2\mathrm{I_yS_x}) = \mathrm{i}/2\begin{bmatrix} 0 & 0 & 0 & -1 \\ 0 & 0 & 0 & 0 \\ 0 & 0 & 0 & 0 \\ 1 & 0 & 0 & 0 \end{bmatrix}$$

Similarly, pure ZQC is given by

$$1/2(2\mathrm{I_xS_x} + 2\mathrm{I_yS_y}) = 1/2\begin{bmatrix} 0 & 0 & 0 & 0 \\ 0 & 0 & 1 & 0 \\ 0 & 1 & 0 & 0 \\ 0 & 0 & 0 & 0 \end{bmatrix}$$

$$1/2(2\mathrm{I_yS_x} - 2\mathrm{I_xS_y}) = \mathrm{i}/2\begin{bmatrix} 0 & 0 & 0 & 0 \\ 0 & 0 & -1 & 0 \\ 0 & -1 & 0 & 0 \\ 0 & 0 & 0 & 0 \end{bmatrix}$$

These are not directly observable. They can be obtained only when there are two or more J-coupling spins. They also cannot be created with a single pulse applied to an equilibrium spin system; as expected from a classical energy-level diagram, their creation requires at least two pulses. Application of another 90° pulse converts these to product operators that contain SQCs, which are converted into observable magnetization if they are allowed to precess with J-coupling. At the end of the second pulse in a COSY experiment (before precession and J-coupling), the product-operator expression is as given in Figure 3.13 (pulsing I spins/S spins similarly).

One should note that the diagonal (first and third terms) and the cross-peaks (second and fourth terms) have the same phase, contrary to what was observed with COSY.

3.2.3.5 Phase Cycling: Selection of Coherence Transfer Pathways [13]

Experiments are designed so that one selects a particular coherence transfer pathway (CTP) and eliminates others.

The phases of one or more pulses are systematically incremented through a cycle of N steps. The number N of steps is determined by the selectivity desired. A desired pathway is collected in every experiment by setting the receiver in phase with the pathway. Unwanted pathways are not collected in phase with the receiver in every experiment, and sum to the receiver phase so that it follows the accumulated pulse phase terms. Consider, as an example, selection of the $+1$, $+1$, -3 pathway. It requires setting the receiver in every experiment to follow the accumulated phase in the exponential term, such that $[\text{expi}(\phi)\text{rec}] = [\exp\{-\text{i}(\phi)\}]$. Therefore, (ϕ) rec $= -(1.\phi_1 + 1.\phi_2 - 3.\phi_3)$. ϕ_1, ϕ_2, and ϕ_3 are the phases of the three pulses. The selectivity of the phase cycle that is required for a given pulse is determined by the size of the desired coherence transfer steps (CTSs). Now, using the spherical basis set (SBS), one sees that the transfer steps that one wants to retain with the third pulse, in a double-quantum filtering COSY experiment, are $+1$ and -3. Therefore, one needs to eliminate all transfer steps between $+1$ and -3; i.e., we require a selectivity of 4. It is known that the phase increment, $\Delta\phi$, for the phase cycle is given by $\Delta\phi = 2\pi/N$, with N being the selectivity. Thus, the phase of the third pulse will have to be incremented through the cycle 0 $\pi/2$, π, $3\pi/2$. The simplest phase cycle for the DQF COSY would be as follows: (1) No filter is required for the first pulse because only SQCs can be created. (2) Selection for transfer to DQC is not necessary with the second pulse, if one selects for transfer from DQC. (3) Cycling the third pulse through 0, $\pi/2$, π, $3\pi/2$ allows for elimination of all CTSs between -3 and $+1$. The phase cycle will be as shown below.

One uses the customary notation for pulse phase of 0, 1, 2 and 3 for 0, $\pi/2$, π, $3\pi/2$, or the x-, y-, $-x$-, and $-y$-axes of the rotating frame.

Table 3.10 After Second Pulse along X-Axis, before Pulsing (I Spins), Precession and J-Coupling

$-I_x$	$\cos(\pi J t_1)$	$\cos(\omega I t_1)$
$-2I_xS_y$	$\sin(\pi J t_1)$	$\cos(\omega I t_1)$
$+I_x$	$\cos(\pi J t_1)$	$\sin(\omega I t_1)$
$-2I_zS_y$	$\sin(\pi J t_1)$	$\sin(\omega I t_1)$

Immediate application of a 90° pulse ($t_2 = 0$) along the y-axis (Figure 3.16 [13]), followed by precession and J-coupling, converts the DQC term $\mathbf{2I_xS_y}$ to one that contains an observable signal.

Table 3.11 Immediately ($t_2 = 0$) after Pulsing, along Y-Axis, Precession and J-Coupling

$-2I_zS_y$	$\sin(\pi J t_1)$	$\cos(\omega I t_1)$	$\cos(\pi J t_3)$	$\cos(\omega S t_3)$
$-S_x$	$\sin(\pi J t_1)$	$\cos(\omega I t_1)$	$\sin(\pi J t_3)$	$\cos(\omega S t_3)$
$-2I_zS_x$	$\sin(\pi J t_1)$	$\cos(\omega I t_1)$	$\cos(\pi J t_3)$	$\cos(\omega S t_3)$
$-S_y$	$\sin(\pi J t_1)$	$\cos(\omega I t_1)$	$\sin(\pi J t_3)$	$\cos(\omega S t_3)$

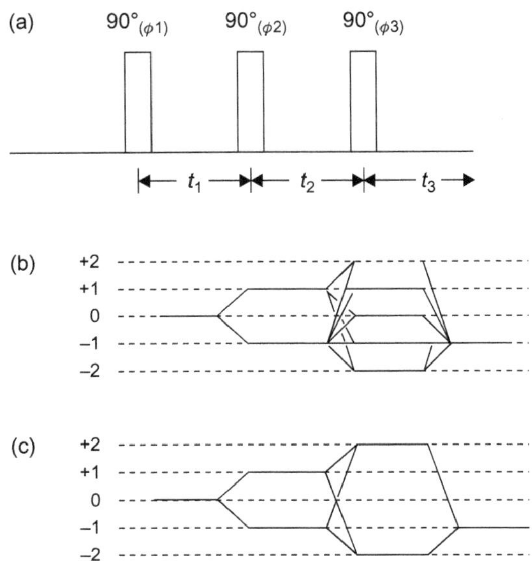

Figure 3.16 (a) Pulse sequence for the 2D NMR DQF COSY experiment. (b) Coherence pathways for a 2-spin system. (c) Pathways selected by the phase cycle described in the text.

It is possible to construct a DQF such that the only signal that arises from the magnetization passing through the DQC survives by summing four separate experiments, each with appropriate setting of the pulse for the third pulse and the receiver. The following shows phase cycling for the third pulse (i.e., the pulse is applied along the x-axis in the first experiment, along the y-axis in the second experiment, and so on).

Table 3.12 The Phase Cycling Experiment

Experiment 1	**Experiment 2**	**Experiment 3**	**Experiment 4**
x	y	$-x$	$-y$
The $-2I_xS_y$ operator becomes (dropping the 2-spin operators)			
$-I_y + I_x$	$-S_x - S_y$	$I_y - I_x$	$S_x + S_y$
and with the receiver along the following axes:			
x	$-y$	$-x$	y
the signal detected is			
FID	FID	FID	FID
and the FIDs accumulate when summed. In contrast, the $-\mathbf{I_z}$ operator is			
$I_y - I_x$	$-I_x - I_y$	$-I_y + I_x$	$I_x + I_y$
and, with a similar receiver phase cycle, the signal detected is			
$I_x + I_y$	$-I_z$	$I_x + I_y$	I_z
and the $-2I_zS_y$ operator becomes			
$-I_x - I_y$	None	$-I_x - I_y$	None

Here "None" indicates no detectable single-operator terms. Using the receiver cycle given above, no detectable signal arises from these contributions. Thus, the only signal accumulation in the four-step cycle is that which has passed through DQC.

The complete signal with appropriate coefficients is as follows:

Table 3.13 The Complete Signal

$2I_x$	$\cos(\omega It_1)$	$\sin(\pi Jt_1)$	$\sin(\omega It_3)$	$\sin(\pi Jt_3)$
$-2I_x$	$\cos(\omega It_1)$	$\sin(\pi Jt_1)$	$\sin(\omega It_3)$	$\sin(\pi Jt_3)$
$2S_x$	$\cos(\omega It_1)$	$\sin(\pi Jt_1)$	$\sin(\omega It_3)$	$\sin(\pi Jt_3)$
$-2S_x$	$\cos(\omega It_1)$	$\sin(\pi Jt_1)$	$\sin(\omega It_3)$	$\sin(\pi Jt_3)$

Table 3.14 The Phase Cycling Results

	Experiment 1	**Experiment 2**	**Experiment 3**	**Experiment 4**
Φ_1	0	0	0	0
Φ_2	0	0	0	0
Φ_3	0	1	2	3
ϕ_{rec}*	0	3	6	9
Or	0	3	2	1
Or	X	$-y$	$-x$	Y
ϕ_{rec}	0	−1	−2	−3
Or	0	3	2	1
ϕ_{signal}	0	1	2	3
	+	−	+	−

Note: Pulse phases, $0 = 0$, $1 = \pi/2$, $2 = \pi$, $3 = 3\pi/2$.
Selection for the −1, +3, −3 pathway (P-type) is accomplished only for a − 3 transfer in the last step. Selection for the other two pathways through the DQC can be shown to require a receiver phase cycle of *; the receiver must be set for the +1, +1, 3 pathway, N-type; **the receiver phase, the signal will sum to zero.

3.2.3.6 Spherical Basis of the Product Operators [13]

It is well known that the spherical tensor basis for the expansion of the density matrix is much better suited for analysis of phase cycling. Its advantage lies in the fact that each product operator is associated with a single coherence level or order. Its disadvantage is that it is unable to provide a vector diagram through which we can visualize the operators. Conversion from the Cartesian to the spherical basis is easily achieved as follows: $I^+ = -(1/\sqrt{2})(I_x + iI_y)$; $I_0 = I_z$; $I^- = -(1/\sqrt{2})(I_x - iI_y)$. The 16 basis operators are $\mathbf{1}$, $\mathbf{I_+}$, $\mathbf{I_0}$, $\mathbf{I_-}$, $\mathbf{S_+}$, $\mathbf{S_0}$, $\mathbf{S_-}$, $\mathbf{I_+S_0}$, $\mathbf{I_-S_0}$, $\mathbf{I_0S_+}$, $\mathbf{I_0S_-}$, $\mathbf{I_0S_0}$, $\mathbf{I_{-+}S_+}$, $\mathbf{I_{-+}S_-}$, $\mathbf{I_{---}S_+}$, $\mathbf{I_{--}S_-}$. The corresponding matrices are presented in Table 3.15 [13].

The pure DQC is represented by $\mathbf{I_+S_+}$ and $\mathbf{I_-S_-}$, whereas pure ZQC is given by $\mathbf{I_+S_-}$ and $\mathbf{I_-S_+}$. SQC is represented by terms such as $\mathbf{I_+}$, $\mathbf{I_-}$, $\mathbf{I_0S_+}$, and $\mathbf{I_0S_-}$. Note that the coherence is given by the sum of the indices. One immediate benefit of this representation is the ability to use a single operator to obtain the phase-sensitive quadrature signal from the density operator, rather than simply $\mathbf{I_x}$ or $\mathbf{I_y}$. If the real part of the quadrature signal is along the y-axis and the imaginary part is along the x-axis, then a pure absorption signal is obtained from $I_y + iI_x = i\sqrt{2}\mathbf{I_-}$. The effect of the pulses, precession, and J-coupling is obtained by substitution, using the rules of the Cartesian operators, and is given in Table 3.16 [13].

Table 3.15 Product Operators in the Spherical Basis for a 2-Spin System, Showing the Coherence Order and Matrix Representations

O	C	Matrix	O	C	Matrix
I	–	$\begin{bmatrix} 1 & 0 & 0 & 0 \\ 0 & 1 & 0 & 0 \\ 0 & 0 & 1 & 0 \\ 0 & 0 & 0 & 1 \end{bmatrix}$	I_0S_+	+1	$1/2\sqrt{2}\begin{bmatrix} 0 & -1 & 0 & 0 \\ 0 & 0 & 0 & 0 \\ 0 & 0 & 0 & 1 \\ 0 & 0 & 0 & 0 \end{bmatrix}$
I_0	–	$1/2\begin{bmatrix} 1 & 0 & 0 & 0 \\ 0 & 1 & 0 & 0 \\ 0 & 0 & -1 & 0 \\ 0 & 0 & 0 & -1 \end{bmatrix}$	I_0S_-	–1	$1/2\sqrt{2}\begin{bmatrix} 0 & 0 & 0 & 0 \\ 1 & 0 & 0 & 0 \\ 0 & 0 & 0 & 0 \\ 0 & 0 & -1 & 0 \end{bmatrix}$
I_+	+1	$1/\sqrt{2}\begin{bmatrix} 0 & 0 & -1 & 0 \\ 0 & 0 & 0 & -1 \\ 0 & 0 & 0 & 0 \\ 0 & 0 & 0 & 0 \end{bmatrix}$	I_+S_0	+1	$1/2\sqrt{2}\begin{bmatrix} 0 & 0 & -1 & 0 \\ 0 & 0 & 0 & 1 \\ 0 & 0 & 0 & 0 \\ 0 & 0 & 0 & 0 \end{bmatrix}$
I_-	–1	$1/\sqrt{2}\begin{bmatrix} 0 & 0 & 0 & 0 \\ 0 & 0 & 0 & 0 \\ 1 & 0 & 0 & 0 \\ 0 & 1 & 0 & 0 \end{bmatrix}$	I_-S_0	–1	$1/2\sqrt{2}\begin{bmatrix} 0 & 0 & 0 & 0 \\ 0 & 0 & 0 & 0 \\ 1 & 0 & 0 & 0 \\ 0 & -1 & 0 & 0 \end{bmatrix}$
S_0	–	$1/2\begin{bmatrix} 1 & 0 & 0 & 0 \\ 0 & -1 & 0 & 0 \\ 0 & 0 & 1 & 0 \\ 0 & 0 & 0 & -1 \end{bmatrix}$	I_+S_+	+2	$1/2\begin{bmatrix} 0 & 0 & 0 & 1 \\ 0 & 0 & 0 & 0 \\ 0 & 0 & 0 & 0 \\ 0 & 0 & 0 & 0 \end{bmatrix}$
S_+	+1	$1/\sqrt{2}\begin{bmatrix} 0 & -1 & 0 & 0 \\ 0 & 0 & 0 & 0 \\ 0 & 0 & 0 & -1 \\ 0 & 0 & 0 & 0 \end{bmatrix}$	I_-S_-	–2	$1/2\begin{bmatrix} 0 & 0 & 0 & 0 \\ 0 & 0 & 0 & 0 \\ 0 & 0 & 0 & 0 \\ 1 & 0 & 0 & 0 \end{bmatrix}$
S_-	–1	$1/\sqrt{2}\begin{bmatrix} 0 & 0 & 0 & 0 \\ 1 & 0 & 0 & 0 \\ 0 & 0 & 0 & 0 \\ 0 & 0 & 1 & 0 \end{bmatrix}$	I_+S_-	0	$1/2\begin{bmatrix} 0 & 0 & 0 & 0 \\ 0 & 0 & -1 & 0 \\ 0 & 0 & 0 & 0 \\ 0 & 0 & 0 & 0 \end{bmatrix}$
I_0 S_0	–	$1/4\begin{bmatrix} 0 & 0 & 0 & 0 \\ 0 & -1 & 0 & 0 \\ 1 & 0 & -1 & 0 \\ 0 & 0 & 0 & 1 \end{bmatrix}$	I_-S_+	0	$1/2\begin{bmatrix} 0 & 0 & 0 & 0 \\ 0 & 0 & 0 & 0 \\ 0 & -1 & 0 & 0 \\ 0 & 0 & 0 & 0 \end{bmatrix}$

O, operator; *C*, coherence.

The following simple manipulations should help to clarify the importance of the spherical basis. Application of a 90° pulse along the *x*-axis to the *I* spins of a 2-spin system at equilibrium converts $\mathbf{I_0}$ as follows (ϕ_1 is the phase of the pulse): $\mathbf{I_0} \rightarrow -0.707\text{i}$ $(\exp-(-\text{i}\phi_1))I_- - 0.707(\exp(-\text{i}\phi_1))I_+ = -0.707\text{i}I_- - 0.707\text{i}I_+$. This has equal

Table 3.16 Effect of Application of Pulses, Precession, and *J*-Coupling on Spherical Product Operators[a]

Pulse along the ϕ axis[b] through an angle $\theta = \gamma B_1 t$
$I_0 \rightarrow I_0\cos\theta - i/\sqrt{2}I_+ \sin\theta\exp(-i\phi) - i/\sqrt{2}I_- \sin\theta\exp(i\phi)$
$I_+ \rightarrow 1/2I_+\cos(\theta+1) - i/\sqrt{2}I_0\sin\theta\exp(i\phi) + 1/2I_+\cos(\theta - 1)\exp(2i\phi)$
$I_- \rightarrow 1/2I_-\cos(\theta+1) - i/\sqrt{2}I_0\sin\theta\exp(-i\phi) + 1/2I_+\cos(\theta-1)\exp(-2i\phi)$
Precession about the *z*-axis with frequency y = ω
$I_+ \rightarrow I_+\exp(-i\omega t)$
$I_- \rightarrow I_-\exp(i\omega t)$
$I_0 \rightarrow I_0$
J-coupling with $\lambda = 1/2Jt$
$I_+ \rightarrow I_+\cos\lambda - i2I_+S_0\sin\lambda$
$I_- \rightarrow I_-\cos\lambda + i2I_-S_0\sin\lambda$
$I_0 \rightarrow I_0$

[a] *B* is the pulse field strength; γ, the gyromagnetic ratio; *J*, the *J*-coupling constant; and *t*, the time of the pulse, precession, and *J*-coupling.
[b] ϕ is the axis about which the pulse field is applied; it is 0 for the *x*-axis, *n*/2 for the *y*-axis, *n* for the −*x*-axis, and 3*n*/2 for the −*y*-axis.

parts I_+ and I_-, indicating that ZQC has been converted into equal amounts of −1 and + 1 coherence. Although the expression is simplified by substituting 0 for ϕ_1, it is usually best not to substitute explicit values for each of the pulse phase factors (to maintain a history of the coherence transfer). Precession under the influence of chemical shift will be ignored for the moment, because it does not affect the coherence level, but instead multiplies the coherence by an additional phase factor. *J*-coupling also does not affect the coherence level, but it does create a new 2-spin product-operator term, necessary for movement to a coherence level other than +1, 0, or −1, with the application of another pulse. Thus, *J*-coupling for time *t* leads to $-0.707i\ (\exp-(-i\phi_1)\ I_-\ -0.707i\exp(-i\phi_1)I_+ \rightarrow -0.707iI_-\cos(\pi Jt)(\exp - (-i\phi_1))\ -0.707iI_+\cos(\pi Jt)\ \exp - i\phi_1) + 1.414iI_-S_0\sin(\pi Jt)\exp(\exp - (-i\phi_1)) - 1.414i\ I_+S_0\sin(\pi Jt)\exp(\exp\ (-i\phi_1))$. All the four terms are SQCs, so the coherence level does not change. Application of another 90° pulse (with phase ϕ_2) to the *I* spins leads to 18 terms (Table 3.17), with coherence levels from −2 to +2, including ZQCs, SQCs, and DQCs.

The advantage of raising the pulse phase factors is now clear. Each term contains the coherence level in the product operator. At the same time, though, the CTP is indicated in the exponent that contains the pulse phases. As an example, the last term has a coherence level of $+2(\mathrm{I}_+\mathbf{S}_+)$, which was arrived at with a +1 coherence transition, with the first pulse indicated by a +1 coefficient for ϕ_2.

In contrast, the 17th term also has a coherence level of +2, but its CTP is −1 for the first pulse and +3 for the second pulse. CTPs are shown below. The receiver is normally set to detect only coherences of level −1 for the spins *I* and *S*, so that the pertinent terms detected in a COSY experiment for the *I* spins are as below. Here, for the sake of completeness, one calculates the effects of precession and *J*-coupling, during the evolution (t_1) and detection (t_2) time periods: $0.354iI_-\cos(\pi Jt_1)\cos(\pi Jt_2)\exp[-i(-\phi_1)]\exp[-i(-\omega It_1 - \omega It_2)] + 0.354iI_-\cos(\pi Jt_1)$

$\cos(Jt_2)\exp[-i(-\phi_1-2\phi_2)]\exp[-i(\omega It_1-\omega It_2)]$. The terms for the S spins are similar. The first term represents CTSs of -1 and 0. This is because the coefficients for ϕ_2 must be 0. It is missing in the exponential expression. The second term represents steps of $+1$ and -2.

In the first term, the sign of precession is the same during t_1 and t_2. This is to give the P-type signal or the coherence transfer echo (CTE). For the second term, precession occurs in opposite directions during t_1 and t_2. This is to give the N-type signal or CTE. To obtain pure 2D absorption peaks without the phase twist, both N- and P-type signals are required. If quadrature detection without pure 2D absorption is acceptable, then either the N- or the P-type peaks, but not both, can be selected with appropriate phase cycling. Upon performing analysis of the DQF COSY in a spherical basis Table 3.18, below, for the 2-spin system one gets 16 terms. In this situation, t_2 is assumed to be brief, so that J-coupling and precession are neglected during this time for simplicity. The four asterisked terms represent the I_-, SQC (in t_3) that existed as DQC between the second and third pulses. The CTPs for these four terms are shown in Figure 3.16 [13], above.

Table 3.17 Correlating Spectroscopy (COSY)

1	$-0.5I_0\cos(\pi Jt)\exp[-i(-\phi_1+\phi_2)]$
2	$-0.5I_0\cos(\pi Jt)\exp[-i(\phi_1-\phi_2)]$
3	$-0.354iI_-\cos(\pi Jt)\exp[-i(-\phi_1)]$
4	$+0.354iI_-\cos(\pi Jt)\exp[-i\phi_1-2\phi_2)]$
5	$-0.354I_+\cos(\pi Jt)\exp[-i(\phi_1)]$
6	$+0.354I_+\cos(\pi Jt)\exp[-i(-\phi_1+2\phi_2)]$
7	$-0.707I_0S_-\sin(\pi Jt)\exp[-i(\phi_1)]$
8	$+0.707I_0S_-\sin(\pi Jt)\exp[-i(\phi_1-2\phi_2)]$
9	$+0.707I_0S_+\sin(\pi Jt)\exp[-i(\phi_1)]$
10	$-0.707I_0S_-\sin(\pi Jt)\exp[-i(-\phi_1+2\phi_2)]$
11	$-0.5I_-S_-\sin(\pi Jt)\exp[-i(\phi_1-\phi_2)]$
12	$-0.5I_-S_-\sin(\pi Jt)\exp[-i(\phi_1-3\phi_2)]$
13	$+0.5I_+S_-\sin(\pi Jt)\exp[-i(-\phi_1+\phi_2)]$
14	$+0.5I_+S_-\sin(\pi Jt)\exp[-i(\phi_1-\phi_2)]$
15	$-0.5I_-S_+\sin(\pi Jt)\exp[-i(-\phi_1+\phi_2)]$
16	$-0.5I_-S_+\sin(\pi Jt)\exp[-i(\phi_1-\phi_2)]$
17	$+0.5I_+S_+\sin(\pi Jt)\exp[-i(-\phi_1+3\phi_2)]$
18	$+0.5I_+S_+\sin(\pi Jt)\exp[-i(\phi_1+\phi_2)]$

The advantage of retaining the pulse phase factors can now be seen. Not only does each term contain the coherence level in the product operator, but the CTP is also indicated in the exponent that contains the pulse phases. For example, entry 18 in the preceding table has a coherence level of $+2(I_+S_+)$, which was arrived at with a $+1$ coherence transition, with the first pulse indicated by the $+1$ coefficient for ϕ_1, and a $+1$ coherence transition with the second pulse indicated by the $+1$ coefficient for ϕ_2. In contrast, entry 17 has a coherence level of $+2$, but its CTP is -1 for the first pulse and $+3$ for the second pulse. CTPs are depicted in Figure 3.15 [13].

Table 3.18 Double Quantum Filter (DQF) Correlating Spectroscopy (COSY)

1	$0.177I_{-}\cos(\pi Jt_1)\cos(\pi Jt_3)\exp[-i(\phi_1-2\phi_2)\exp[-i(\omega It_1-\omega It_3)$
2	$0.354I_{-}\cos(\pi Jt_1)\cos(\pi Jt_3)\exp[-i(\phi_1-\phi_2-\phi_3)\exp[-i(\omega It_1-\omega It_3)]$
3	$0.177I_{-}\cos(\pi Jt_1)\cos(\pi Jt_3)\exp[-i(\phi_1-2\phi_3)\exp[-i(\omega It_1-\omega It_3)]$
4	$-0.177I_{-}\cos(\pi Jt_1)\cos(\pi Jt_3)\exp[-i(-\phi_1)]\exp[-i(-\omega It_1-\omega It_3)]$
5	$0.354I_{-}\cos(\pi Jt_1)\cos(\pi Jt_3)\exp[-i(-\phi+\phi-\phi_3)]\exp[-i(-\omega It_1-\omega It_3)]$
6	$-0.177I_{-}\cos(\pi Jt_1)\cos(\pi Jt_3)\exp[-i(-\phi_1+2\phi_2-2\phi_3)]\exp[-i(-\omega It_1-\omega It_3)]$
7	$-0.177I_{-}\sin(\pi Jt_1)\sin(\pi Jt_3)\exp[-i(\phi_1-2\phi_2)]\exp[-i(\omega It_1-\omega It_3)]$
8	$-0.088I_{-}\sin(\pi Jt_1)\sin(\pi Jt_3)\exp[-i(\phi_1-3\phi_2+\phi_3)]\exp[-i(\omega It_1-\omega It_3)]^*$
9	$-0.177I_{-}\sin(\pi Jt_1)\sin(\pi Jt_3)\exp[-i(\phi_1-\phi_2-\phi_3)]\exp[-i(\omega It_1-\omega It_3)]$
10	$-0.177I_{-}\sin(\pi Jt_1)\sin(\pi Jt_3)\exp[-i(\phi_1-2\phi_2)]\exp[-i(\omega It_1-\omega It_3)]$
11	$-0.088I_{-}\sin(\pi Jt_1)\sin(\pi t_3)\exp[-i(\phi_1+\phi_2-3\phi_3)]\exp[-i(\omega It_1-\omega It_3)]^*$
12	$0.177I_{-}\sin(\pi Jt_1)\sin(\pi Jt_3)\exp[-i(-\phi_1)]\exp[-i(-\omega It_1-\omega It_3)]$
13	$-0.088I_{-}\sin(\pi Jt_1)\sin(\pi Jt_3)\exp[-i(-\phi_1-\phi_2+\phi_3)]\exp[-i(\omega It_1-\omega It_3)]^*$
14	$-0.177I_{-}\sin(\pi Jt_1)\sin(\pi Jt_3)\exp[-i(-\phi_1+\phi_2-\phi_3)]\exp[-i(-\omega It_1-\omega It_3)]$
15	$0.177I_{-}\sin(\pi Jt_1)\sin(\pi Jt_3)\exp[-i(-\phi_1+2\phi_2-2\phi_3)]\exp[-i(-\omega It_1-\omega It_3)]$
16	$-0.088I_{-}\sin(\pi Jt_1)\sin(\pi Jt_3)\exp[-i(\phi_1+3\phi_2-3\phi_3)]\exp[-i(-\omega It_1-\omega It_3)]^*$

Spherical basis in COSY. The four asterisked terms represent the I_{-}, SQCs (in t_3) that existed as DQCs between the second and third pulses (Figure 3.16), above.

3.2.4 *Multiple CTPs [14]*

3.2.4.1 *Multiple CTPs*

Suppose there is a pair of eigenstates, with an arbitrary difference in quantum numbers, $p_{rs} = M_r - M_s$.

The transverse magnetization corresponds to a particular class of coherence associated with a change in quantum number p. The coherences can be described as a coherent superposition of the two eigenstates, as $\Psi_{rs} = a_r|r\rangle a_s|s\rangle$. In terms of a density operator σ, a coherence between the states is expressed by the existence of the nonzero density matrix elements. These elements indicate a transition in the progress between two connected states. Each transition is associated with two coherences, σ_{rs} and σ_{sr}, with coherence orders of opposite sign. Each coherence σ_{rs} conserves its quantum number, p_{rs}, in the course of free precession. RF pulses may induce a transfer between coherences σ_{rs} and σ_{tu}, a process that may change the coherence order. One can write the various terms of the density operator according to the coherence order of p: $\sigma_t = \Sigma_p \sigma_t^P(t)$. For a system of K-spins-½, p extends from $-K$ to K. The characteristic properties of coherence of order p are determined by the transformation, under rotations, about the z-axis: $[\exp\{-i\psi F_z\}\sigma^{\mathrm{P}}\exp\{i\psi F_z\}] = [\sigma^{\mathrm{P}}\exp\{-ip\psi\}]$, where $F_z = \Sigma_{k=1}{}^{N} I_{kz}$. The events are depicted in Figures 3.17 and 3.18.

3.2.5 *MQC Spectroscopy*

3.2.5.1 *Two-Spin* AX *Weak Coupling: Point-Resolved Spin Spectroscopy [15]*

Point-Resolved Spectroscopy sequence (PRESS): A PRESS employs three selective RF pulses in the presence of mutually orthogonal B_0 (static magnetic field)

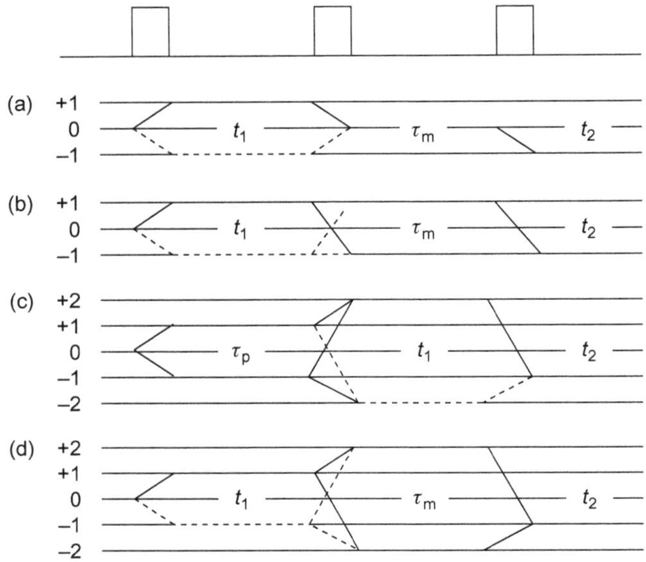

Figure 3.17 Coherence transfer (CT) maps for various 2D experiments involving three pulses. Solid lines indicate pathways that involve a single-order p in the evolution period. For a basic understanding of the experiments, these pathways suffice. If pure-phase lineshapes are not essential (e.g., if composite lineshapes or absolute-value plots are acceptable), it is sufficient to select the pathways shown by solid lines. Mirror-image pathways, with $-p$ in t, are indicated by dashed lines. For pure-phase spectra (i.e., pure 2D absorption lineshapes), both solid and dashed pathways must be retained. Four experimental schemes are shown. (a) 2D exchange spectroscopy; (b) relayed COSY (pathways shown for fixed mixing interval τ_m); (c) double-quantum spectroscopy; and (d) 2D COSY with DQF ($\tau_m = 0$).

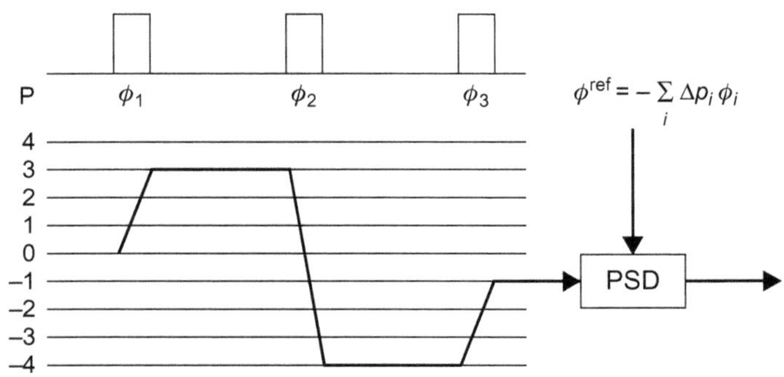

Figure 3.18 The selection of a CTP, characterized in this hypothetical example by the changes in coherence order $\Delta p_1 = +3$, $\Delta p_2 = -7$, and $\Delta p_3 = +3$, can be achieved by cycling the phases of the three CT pulses and by shifting the phase of the reference channel of the phase-sensitive detector (PSD).

gradients. This is to generate a spin-echo of the spins inside the VOI. For the purpose of grasping the theoretical concept, the experiment can be approximated by a spin-echo sequence with only one refocusing pulse (Figure 3.19 [15]).

This approximation is valid if a strongly asymmetric PRESS is used, where the time between excitation and the first refocusing pulse is short compared to $1/J$. In this situation, the coupling effects influence only the second refocusing. A spin-echo experiment begins with a selective 90° RF pulse.

It generates transverse magnetization (Figure 3.19 [15]). During subsequent interval τ, this magnetization is dephased by the gradient $\Omega(x)$. Additionally, it is periodically transformed into in-phase (upper path) and antiphase (lower path) magnetization, because of J-coupling. Thus, at $\tau = 0, 1/J, 2/J, \ldots$, the state of the spin system is described by in-phase magnetization, which is completely equivalent to ordinary transverse magnetization occurring in uncoupled spin systems. Because the resulting signal consists of a doublet whose two peaks are in phase, it is symbolized by parallel bars (upper path). In contrast, at $\tau = 1/2J, 3/2J$, the spin system is in a state of antiphase magnetization, which leads to a signal where the peaks of the doublet are antiphase (Figure 3.19 [15], lower path). Any modification due to J-coupling originates from this kind of quantization, which does not occur in uncoupled systems. At arbitrary echo times, the state of the spin system is described by an admixture of in-phase and antiphase magnetization. The further

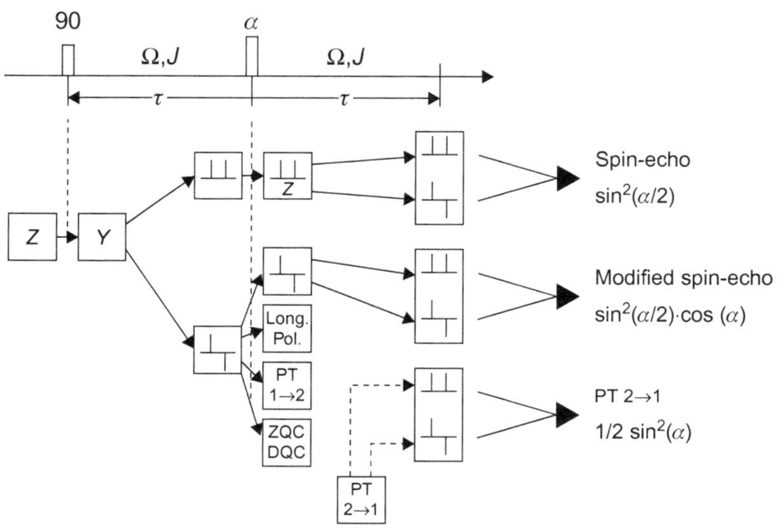

Figure 3.19 Coherence pathways of an AX system in a spin-echo experiment with one refocusing pulse of flip angle α. The symbols Q and J stand for the evolution under chemical shift, i.e., gradients and J-coupling, respectively. The upper pathway shows the evolution of the in-phase part of magnetization, indicated by parallel bars, that finally leads to the well-known spin-echo. Antiphase magnetization, which evolves through the lower path, leads to the formation of an echo with modified flip-angle dependence.

evolution of these kinds of magnetizations differs significantly. The second RF pulse of flip angle α transforms the (dephased) transverse magnetization into z- and transverse magnetization. This is still dephased, but is inverted as compared to the situation prior to the pulse. Thus, applying a further switched gradient $\Omega(x)$, within the second τ interval, leads to a spin-echo at $T_E = 2\tau$. Its dependence on the flip angle α is shown.

One should note that the observed signal does not necessarily consist of pure in-phase magnetization. This is because J-coupling also acts during the second τ period. The effect of the refocusing pulse on antiphase magnetization differs considerably from its effect on in-phase magnetization. The dephased antiphase magnetization is partially refocused, which after application of the gradient $\Omega(x)$ leads to a spin-echo at $T_E = 2\tau$. The dependence of its amplitude on the flip angle is modified by additional factor $\cos\alpha$.

Thus, at $\alpha = 180°$, refocusing is perfect but the signal is inverted. For $90° < \alpha < 180°$, the antiphase magnetization is refocused less efficiently, and the signals are even destructive for $0° < \alpha < 90°$.

At $\alpha = 90°$, antiphase magnetization is not refocused at all. This behavior is valid for any group coupled to a single-spin-½ particle. Refocusing of antiphase magnetization is not, however, the only effect of the second RF pulse. It additionally creates longitudinal polarization, ZQCs, and DQCs. These states do not exist in uncoupled spin systems. They cannot be observed directly. Moreover, the refocusing pulse leads to polarization transfer (PT), which means that the antiphase magnetization of spin species 1 (the signal of which is to be observed) is transformed into antiphase magnetization of spin species 2 (PT1→PT2). The inverse process of transferring polarization from species 2 to species 1 (PT2→PT1) contributes to the observed signal. Its dependence on the flip angle is valid only in *AX* systems. Variations of the echo time lead to modulations of the resulting signal, as the corresponding magnetization evolves with the frequency of species 2 within the first τ interval, and the frequency of species 1 within the second τ interval.

3.2.5.2 Two-Spin AX *Weak Coupling: STEAM [15]*

In a STEAM situation (Figure 3.20 [15]), the evolution up to the second RF pulse is identical to that in Figure 3.19 [15]. In the middle interval t_m after the second RF pulse, strong dephasing (x) gradients δ are switched to dephase transverse magnetization. Z magnetization that is created by an α pulse is unaffected by gradients. The third RF pulse flips it back into the transverse plane so that application of a further gradient $\Omega(x)$ leads to the generation of a stimulated echo at $T_E = 2\tau$. The dependence of its amplitude on the flip angle is shown in Figure 3.20 [15]. One can see that the stimulated echo is maximal at $\alpha = \beta = 90°$.

Its efficiency relative to PRESS is 50%, as one-half of the magnetization is dephased in the t_m period.

As for PRESS, the antiphase part of the magnetization prior to the second RF pulse is transformed into (refocused) antiphase magnetization, longitudinal polarization, ZQCs, and DQCs, apart from polarization transfer (lower path). In analogy

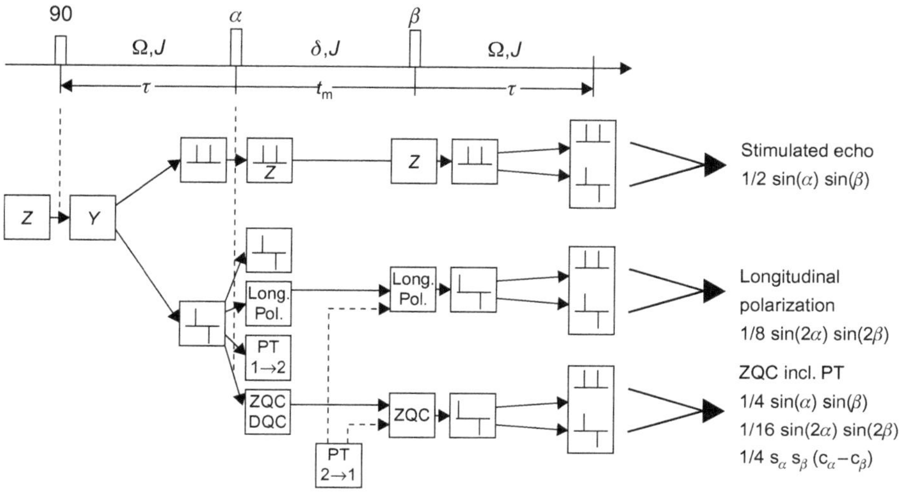

Figure 3.20 Coherence pathways of an *AX* system in a stimulated-echo experiment. The symbols Ω and *J* stand for the evolution under chemical shift, i.e., gradients and *J*-coupling, respectively. Within the t_m interval, strong dephasing gradients *d* are applied. The upper pathway shows the evolution of the in-phase part of magnetization, indicated by parallel bars, that finally leads to the well-known stimulated echo. Antiphase magnetization, which is symbolized by antiparallel bars, evolves through the lower path. For further details, see the text.

to ordinary transverse magnetization, antiphase magnetization is dephased by the gradient $\delta(x)$ and does not contribute to the observed signal. The same holds for DQCs, which dephase twice as fast. It can be shown, however, that neither ZQCs nor longitudinal polarization is affected by gradients, a property they share with *z* magnetization. The third RF pulse transforms them into antiphase magnetization, which refocuses during the last interval τ (lower path) and which finally leads to an echo at $T_E = 2\tau$. Longitudinal polarization consists of a coherent superposition of *z* magnetization of both coupled spin species. The dependence of its amplitude on the flip angle is shown below. The maximum efficiency is 12.5% at $\alpha = \beta = 45°$, whereas no evolution through longitudinal polarization takes place at $\alpha = 90°$ or $\beta = 90°$. Several ZQCs are generated by an RF pulse of arbitrary flip angle α, all of which differ in their flip-angle dependence. In the optimum case of $\alpha = \beta = 90°$, however, the evolution t_m is predominated by one ZQC only, whose efficiency is 25%. It can be shown that this value also sets the upper limit for the efficiency of single ZQCs in arbitrary coupled systems. Because ZQCs evolve at a frequency determined by the difference of chemical shifts of species 1 and 2, amplitude modulations can be observed when t_m is varied. Moreover, polarization is transferred through ZQCs, so that the signal may also be modulated when T_E is varied. Therefore, the acquired signal, which clearly is not "stimulated echo" in its original sense, consists of a rather complicated superposition of signals, all of which depend differently on flip angle, T_E and T_m.

3.2.5.3 Two-Spin AX_2 (Heteronuclear): Weak Coupling [15]

3.2.5.3.1 PRESS Experiment

Here the in-phase magnetization and "ordinary" antiphase magnetization (antiphase with respect to 1 spin only) are created by *J*-coupling. The evolution of these two kinds of magnetization is equivalent to the *AX* case. Additionally, magnetization is created which is antiphase to both *A* spins. It is like the magnetization that determines the state of the spin system at $\tau = 1/2J$, $3/2J$, ... The amplitude of the spin-echo resulting from refocusing by an RF pulse of flip angle α differs from that of uncoupled spin systems by an additional $\cos^2\alpha$. This is valid for any group coupled weakly to 2-spin-½ particles (i.e., to a CH_2 group).

One can say that refocusing is less efficient than for uncoupled spins. No magnetization is refocused at $\alpha = 90°$, as in the *AX* system. However, the sign of the spin-echo is not reversed at $\alpha = 90°$, so nondestructive interference occurs.

3.2.5.3.2 STEAM Experiment

Magnetization antiphase to both *A* spins is partially refocused and partially transferred into longitudinal polarization, ZQCs, DQCs, and triple-quantum coherences (3QCs) by the RF pulse α, apart from polarization transfer. Again, only ZQCs and longitudinal polarization are not dephased during the t_m interval, thus contributing to the observed signal. One can work out that the predominating process (at $\alpha = \beta = 90°$) is a ZQC consisting of a coherent superposition of transverse magnetization of two *A* spins with *z* magnetization of the *X* spin, i.e., a ZQC of the magnetically equivalent *A* spins antiphase to the *X* spin. Thus, the process dominating evolution through the t_m period does not occur and t_m variations do not influence the acquired signal, in complete contrast to the *AX* case. The spectrum of lactate consists of a doublet of the CH_3 protons at 1.3 ppm and a quartet of the CH protons at 4.3 ppm, which is usually too close to the water peak (at 4.7 ppm) to be observed in vivo. These two groups are weakly coupled with a coupling constant of $J = 7.35$ Hz, i.e., $1/J = 135$ ms. As lactate is an AX_3 system, many of the formulae of the *AX* system apply to its *X* spins.

The effect of *J*-coupling on the spectra of glutamate and glutamine proves to be nearly identical, due to their similar chemical structure. The spectrum consists of the resonances of α-CH at 3.8 ppm, β-CH_2 at 2.1 ppm, and γ-CH_2 at 2.1 ppm. The α and β groups are weakly coupled with a coupling constant of approximately 7.3 Hz, whereas the β and γ peaks partially overlap at 2-T magnetic field strength. Hence, they cannot be represented by a first-order coupling network.

The spectrum of GABA consists of three resonances. Of these, the β-CH_2 at 1.9 ppm and γ-CH_2 at 3.0 ppm are weakly coupled, whereas the α and β resonances partially overlap at 85 MHz. Strong modulations occur for $100\,\text{ms} \le T_E \le 170\,\text{ms}$, which are less pronounced in PRESS than in STEAM. The resonances are refocused quite completely at $T_E = 275$ ms.

For the same reasons as pertain to glutamate, spectra acquired with STEAM are hardly influenced by t_m variations. The spectrum of NAA consists of three resonances. The well-known singlet of the CH_3 proton is at 2.0 ppm, where three major lines are observed in the normal brain. The CH_2 and CH protons are weakly coupled with a coupling constant of 7 Hz. This leads to a doublet at 2.85 ppm and a triplet at 4.55 ppm, which is too close to the water peak to be observed in vivo. The spectrum of myoinositol is produced by six protons, all of which are magnetically nonequivalent. They are coupled with different coupling constants.

Additionally, several spatially distinct configurations exist, so the resulting spectrum consists of two partially overlapping multiplets at 3.55 and 4.1 ppm. These are basically observed (Figures 3.21–3.27).

3.2.6 Cartesian Product-Operator Formalism

Cartesian product-operator formalism is based on the resolved spin terms in the Cartesian coordinate frame of reference. The zero-level transformation is $I_x(\theta) \equiv [I_x \cos\theta + I_y \sin\theta] = [I_x c_\theta + I_y s_\theta]$; $I_y(\theta) \equiv I_y \cos\theta - I_x \sin\theta = [I_y c_\theta - I_x s_\theta]$ (Figure 3.28 [16]). Here $c_\theta = \cos\theta$ and $s_\theta = \sin\theta$. Level one transformation is expressed: $I_x(\phi)c_\alpha = (1/2)[I_x(\phi+\alpha) + I_x(\phi-\alpha)]$; $I_x(\phi)s_\alpha = (1/2)[-I_y(\phi+\alpha) + I_y(\phi-\alpha)]$; $I_y(\phi)c_\alpha = (1/2)[I_y(\phi+\alpha) + I_y(\phi-\alpha)]$; $I_y(\phi)s_\alpha = (1/2)[I_x(\phi+\alpha) - I_x(\phi-\alpha)]$ (Figure 3.28 [16]). One can transform these equations a level further, into another rotated coordinated frame of reference, as in Figure 3.29 [16].

Figures 3.28 and 3.29 depict that level one transformation is equivalent to the summation of two vectors, with equal magnitudes and different phases, $(\phi+\alpha)$ and $(\phi-\alpha)$. The equations can be rearranged to emphasize the fact that after an RF pulse, transverse magnetization is split into two new coherence pathways, with effective flip angles 0° and 180°.

3.2.7 Computer Simulations in NMR

3.2.7.1 The Computer Interface

The microscopic imaging produced by NMR involves the use of quantum mechanical operators, represented by square-type matrices. One has to deal with the rotation of matrices, the diagonalization of the matrices, their evolution in time, and so on. Thus, the mathematical language needed in producing a computer interface with the MRI machine is different from those normally used. The computer language C++ has all the features required to create a successful and convenient computer interface.

3.2.7.2 Quantum Mechanical Operators: Density Matrix

The main purpose of the mathematical module, which is the core of learning concern here, is the convenient programming of quantum mechanical computations in operator formalism. This has to be achieved in HS, with as little user involvement or interface as possible regarding the matrix calculations. The HS calculations may be called *operator calculus*, in analogy to the *differential calculus* of NM. For those who have learned QM in their educational program, the use of HS becomes a

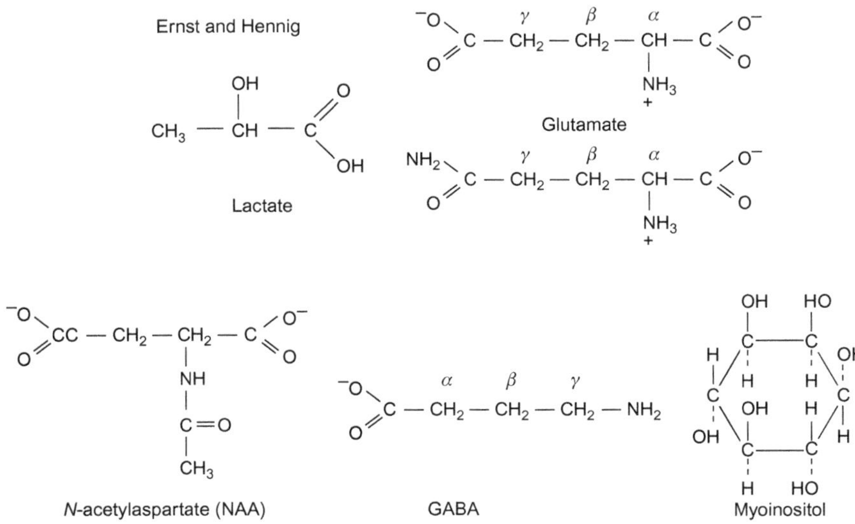

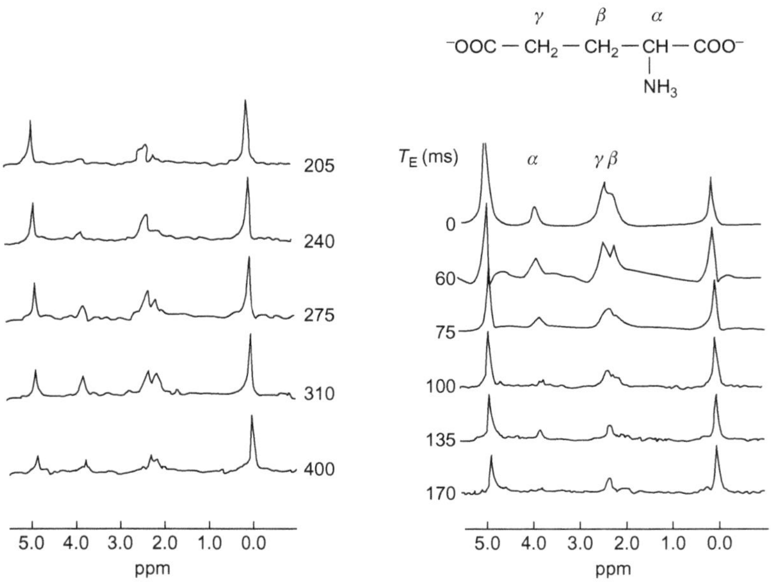

Figure 3.21 ^{1}H spectra of glutamate acquired with a STEAM sequence at various echo times ($f = 100$ ms).

commonplace, just as differential calculus is in NM. But a computer program, such as general approach to magnetic resonance using mathematical analysis (GAMMA), is designed in such a manner as to be user-friendly to a person who is not a specialist in QM. This may be a medical person, who is primarily interested

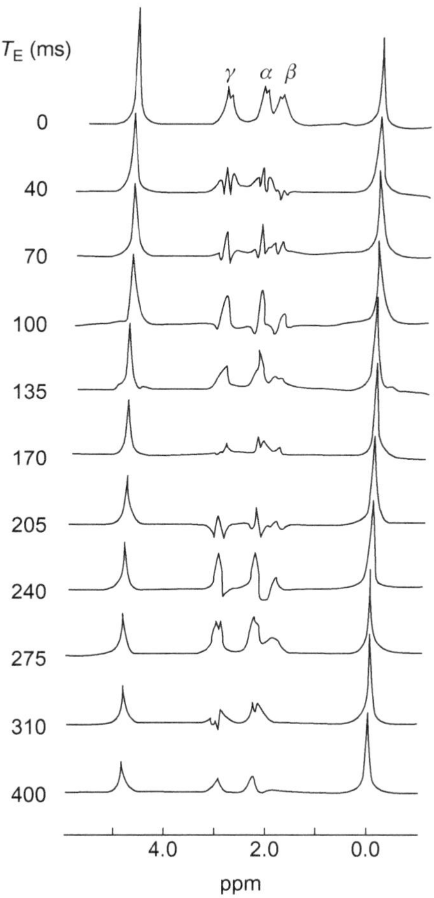

Figure 3.22 ^{1}H spectra of GABA acquired with a PRESS at various echo times.

in using the MRI machine as a diagnostic tool, but at the same time wishes to understand more than just the picture output from the machine (e.g., when trying to work out the contributions of various molecules to metabolic activities, neuron transmission).

3.2.7.3 Quantum Operators and Computations

The expectation value of an observable operator A is obtained by performing a trace operation, $\langle A \rangle = \mathrm{Tr}\{A\sigma\}$. This largely sets the stage for the operation needed within the mathematical module.

The spectroscopic module introduces higher-level concepts, familiar to the MR spectroscopist but probably not to other users. Of importance is the definition of a spin system and its associated Hamiltonian, the evolution of the density operator under the Hamiltonian, and the effect of idealized selective pulses. The definition

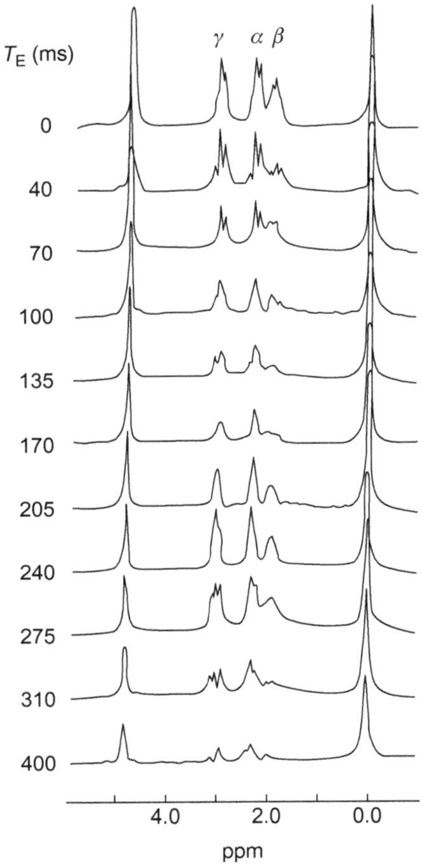

Figure 3.23 ^{1}H spectra of GABA acquired with a STEAM sequence at various echo times ($t = 70$ ms).

of a relaxation-time-dependent operator, for a given relaxation mechanism, also belongs in the spectroscopic module.

3.2.7.4 Computer Modeling: Spin Operators, Matrices, and Other Data [17]

Abstract data types are defined as a set of objects coupled with a set of functions. The objects are handled as entities whose internal structure need not be known in order to execute any associated function.

The object matrix and the "copy" operation, taken together, may form an abstract data type in C++. The user need not know the structure of the matrix to apply the copy function. This is no longer true, however, when the operation "multiply" is included. The matrix multiplication must be excited elementwise.

The user must provide the dimensions of the matrices involved, and often much more of the matrix structure is required. In this case, a matrix forms a data

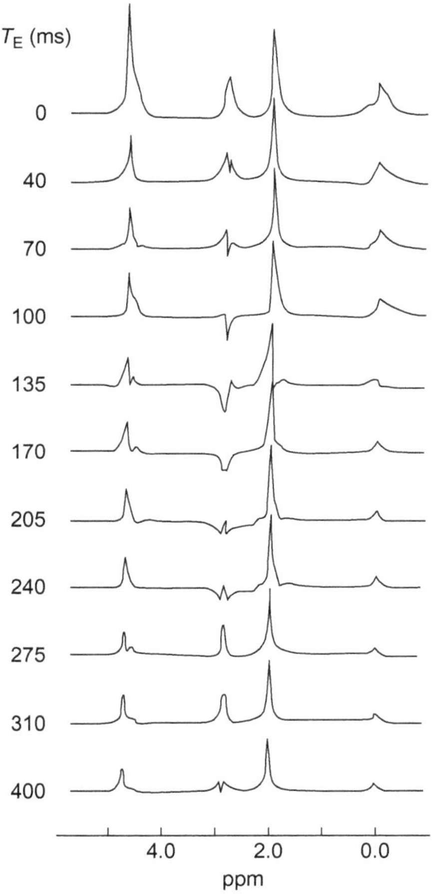

Figure 3.24 ^1H spectra of NAA acquired with a PRESS at various echo times ($t_m = 100\,ms$).

aggregate, and a common manifestation of its aggregate nature is that the user must contend with an elaborate scheme of subroutines that handle matrix manipulations. It is possible to create user-defined abstract data types of an arbitrary kind. This makes it feasible to define a matrix as an abstract data type. It is done with respect to addition, multiplication, and other matrix functions.

A coherent general matrix algebra may then be formulated irrespective of matrix dimensions, particular matrix symmetries, or special matrix storage methods.

A class definition of data type may be made as the class interface and the class implementation.

The class interface should provide all information required by the user, in particular the definition of the objects and the associated functions. It consists of the algorithms that execute the defined functions and the computer code for the representation of the objects. The class implementation, though accessible, is not the concern of the end user. In fact, the system programmer can change the class

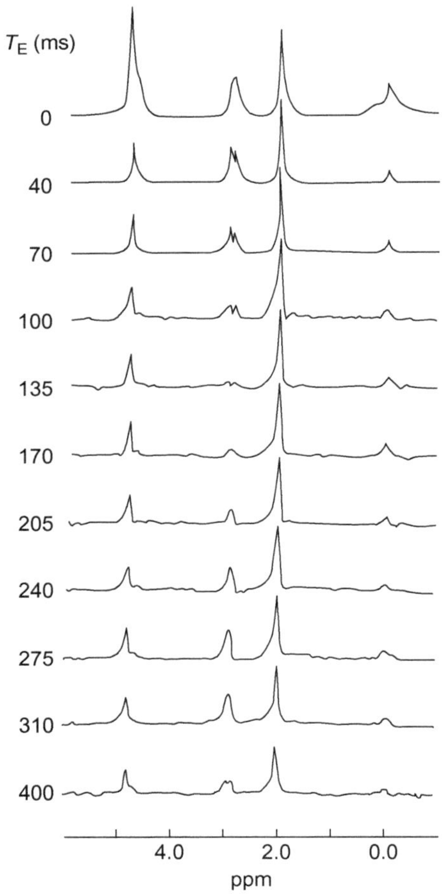

Figure 3.25 ^{1}H spectra of NAA acquired with a STEAM sequence at various echo times ($t_m = 100$ ms).

implementation without having to rewrite the end-user programs. On the basis of existing classes, or base classes, it is also possible to construct derived classes. A derived class inherits the abilities of its base class and provides means for the programmer to extend its functionality and take advantage of previously written code.

For the spectroscopic layer, it is beneficial to define an extended class that contains the isotope types (including the GMRs, the magnetic field strength, and the parameters needed to construct the Hamiltonian, including chemical shifts and coupling constants J). The different layers of spin-system classes are summarized in later subsections. The class implementation will include operations of the form $I_{1z}I_1^+ - I_1^+ I_{1z} = I_1^+$; $I_1^- I_{1z} - I_{1z}I_1^- = I_1^-$; $I_1^+ I_1^- - I_1^- I_1^+ = 2I_{1z}$. For differently ordered operator products, the commutation relations can be applied to obtain a sum of products of ordered spin operators, of which the trace can be computed with the above equations. It is implicitly assumed that all spin operators belong to the same spin system. Operations combining spin operators from different spin systems are not defined.

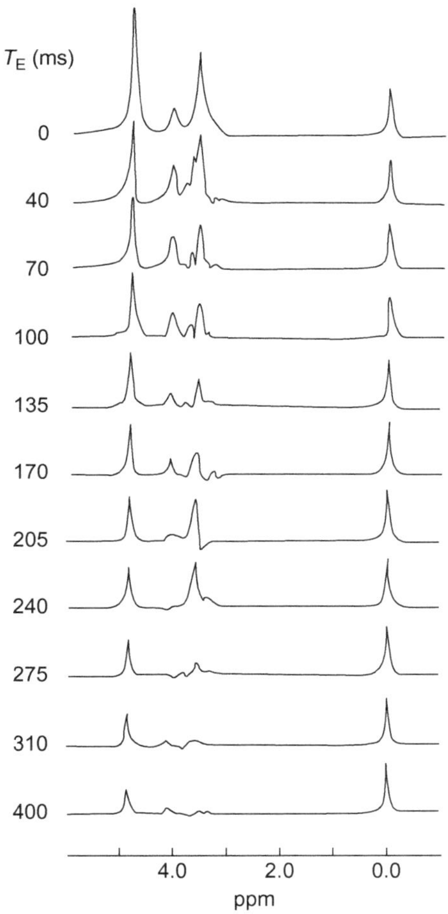

Figure 3.26 ^{1}H spectra of myoinositol acquired with a PRESS at various echo times ($t_m = 100\,\text{ms}$).

The number-crunching part of most simulations is predominantly located within the class matrix, making it of crucial importance to GAMMA's overall performance. Diagonal, block-diagonal, Hermitian, symmetric, and sparse matrices often appear in NMR applications. The class matrix is designed to take full advantage of these special matrix types, and to do so automatically. The user deals with a single data type matrix, but the implementation invokes the appropriate functions for the specific matrix types. Technically, all matrices' data types are derived classes from a basic class. The mathematical writing of expressions with matrices implies much copying of matrices. Therefore, the concept of deferred copying is implemented to optimize storage and computation time. Deferred copying delays the process of copying a matrix until it can no longer be avoided; until then, only a reference is stored. The copying process is necessary when the matrix is to be modified while the original is retained. This is often not necessary if the matrix was copied to a temporary variable that was destroyed without

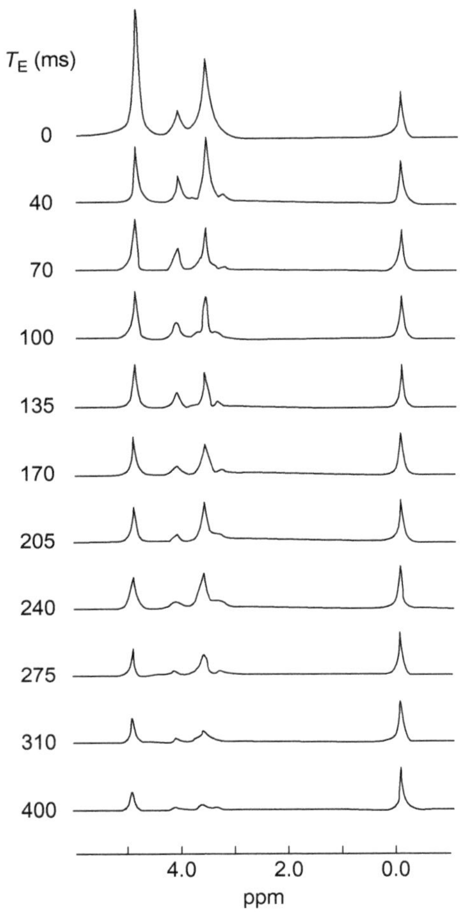

Figure 3.27 ^{1}H spectra of myoinositol acquired with a STEAM sequence at various echo times ($t_m = 100\,\text{ms}$).

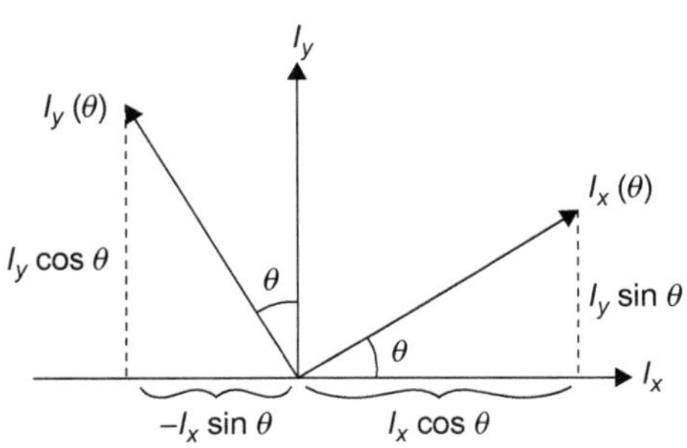

Figure 3.28 Rotation of a vector through an angle θ.

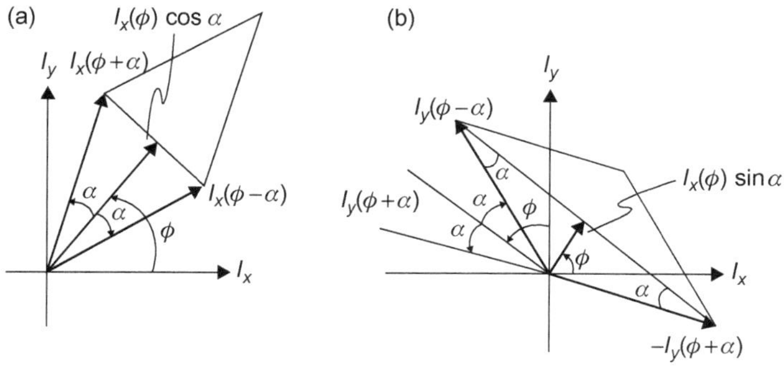

Figure 3.29 Decomposition of a vector into the sum of two vectors with equal magnitudes and different phases. (a) Diagram for $I_x(\phi)\cos\alpha = (1/2)[I_x(\phi+\alpha) + I_x(\phi-\alpha)]$. (b) Diagram for $I_x(\phi)\sin\alpha = (1/2)[-I_y(\phi+\alpha) + I_y(\phi-\alpha)]$.

any modifications. The usual operator representation in a computer is by a matrix in a HS spanned by a complete set of basis functions.

Usually, the Hamiltonian and the initial density operator are set upon the product basis, for convenience. The time evolution of the density operator, in contrast, is most easily computed in the eigenbasis of the Hamiltonian. It can be advantageous to maintain several representations of an operator simultaneously, as repeated conversions in the course of a simulation may then be avoided. In order to treat the relaxation and the chemical exchange in coupled spin systems, a Liouville space superoperator description is indispensable.

3.2.7.5 Computer Simulation GAMMA [17]

Computer simulations are carried out using the Liouville–von Neumann equation. In this case, it is $d/dt[\sigma] = -i2\pi H\sigma - \Gamma\{\sigma - \sigma_{\text{inf}}\}$ and $H\sigma = [H, \sigma] = [H\sigma - \sigma H]$ (anti-commutation). Here the density operator σ is a column vector and the superoperators H and Γ are represented by matrices of the dimension $\Pi_{k=0}^{N-1}[q_k(q_k-1)]^2$. The symbol q_k represents the spin quantum numbers of the N involved spins. Although GAMMA provides the tools with which to simulate spectra in these more general cases, we restrict the discussion here to the laboratory-frame relaxation where $[H, \sigma_{\text{inf}}] = 0$. In this particular instance, one uses two simulated 2D nuclear Overhauser enhancement spectroscopy (NOESY) spectra for a mixing time of 250 ms, a proton resonance frequency of 600 MHz, and correlation times (assuming isotropic tumbling) of 0.1 and 1 ns, respectively. These values are typical for solutions of small and medium-size peptides at room temperatures. For the correlation time of $\tau_c = 0.1$ ns ($\tau_c\omega_0 = 0.377$), small negative NOESY cross-peaks are found in the simulated spectrum.

This is in agreement with expectations. For $\tau_c = 1$ ns ($\tau_c\omega_0 = 3.77$), positive cross-peaks are found, again in agreement with expectations. For the spectra, an additional line-broadening effect caused by a random field, leading to 1.5-Hz line broadening (in the absence of other relaxation mechanisms) was taken into account. It may reflect the influence of additional spins on the 4-spin system. Cross-sections along the ω_1 direction of the 2D spectra, taken at $\omega_2 \approx 3.19$ ppm, were calculated. This is for a correlation time of 0.1 ns. The mentioned sign change of the cross-peaks is seen. A comparable cross-section was calculated using a correlation time of 0.1 ns and a short mixing time of 25 ms. Antiphase cross-peak contributions, caused by transfer through a ZQC that cannot be eliminated by phase cycling, now dominate the cross-peak patterns. For longer mixing times, these contributions are attenuated by relaxation. They may be eliminated by special procedures (Figures 3.30 and 3.31).

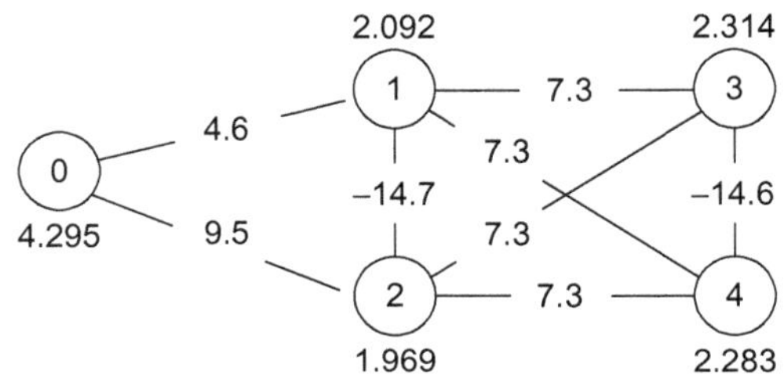

Figure 3.30 Chemical shifts (ppm (k)) and coupling constants (k, l) of glutamic acid in D_2O solution. The spin-system parameters replicate the file used as an input to GAMMA with the following assignment: spin 0 is H_α, spins 1 and 2 are $H_\beta{}^{1,2}$, and spins 3 and 4 are $H_\gamma{}^{1,2}$. A line contains the name of the parameter, the type of the parameter as a number, the value of the parameter, and a description of the parameter.

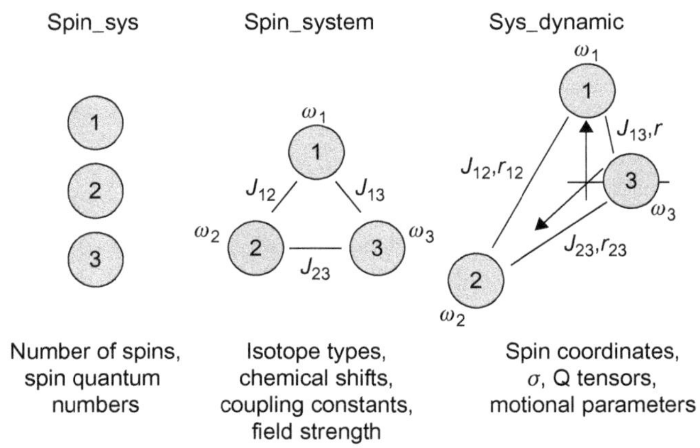

Figure 3.31 The three classes that GAMMA uses to describe spin systems. The lowest level is the class spin_sys, which contains only the properties necessary for the definition of a spin HS. A more detailed spin-system description is given by the class spin_system. It contains all parameters needed for construction of an isotropic liquid state NMR Hamiltonian, and is a derived class of the class spin_sys. The class sys_dynamic contains spatial and dynamic information used in computing spectral densities and relaxation phenomena.

3.2.8 Computer Simulation CTPs: Phase Cycling and Pulsed Field Gradients [18]

The properties of the RF pulse (flip angle, offset effects, inhomogeneity) and transport phenomena (diffusion flow), in conjunction with gradients, cause a weighting of the different computer simulation CTPs.

Pulse sequences are usually created by designing wanted and unwanted CTPs. RF pulses are generally nonselective with respect to the excitation or interconversion of different coherence orders, which necessitates use of CTP filtering techniques. Phase cycling (PC) and sequences of pulsed field gradients (PGFs) are the two most important coherence filtering techniques. PC allows the selection of parallel pathways, but may take a long time, especially in sequences with many pulses, where high coherence orders and good selectivity (in terms of CTPs) are desired. A compromise between CTP selectivity and time requirement can be achieved by combining PGFs and PC, while reducing the number of phase steps as much as possible. The number of possible CTPs in a pulse sequence is of the order $(2p_{\max}+1)^n$, where $p_{\max}$ is the maximum possible coherence order and n is the number of pulses.

One can devise an effective scheme that can be used to simulate the CTP selection process and that involves only simple matrix calculations (Figures 3.32 and 3.33).

3.2.9 Classical Orbital: Single-Particle Spin [19]

3.2.9.1 Quantized Angular Momentum and Energy [19]

During the early years of the twentieth century, Bohr and Thomas established a vector model for the circular motion of negatively charged electrons around the positively charged nucleus. Although the basics of their model have stayed the same, many further developments have been made in the model as scientific knowledge about the atomic structure of matter improved over decades. Though this is called the classical (NM) physics model, it led to the birth of quantum physics (quantum mechanical physics). This chapter's descriptions are cast in terms of an atom, rather than the nucleus, but there is a close analogy between the magnetic properties of an electron and the nucleus. This has been experimentally verified.

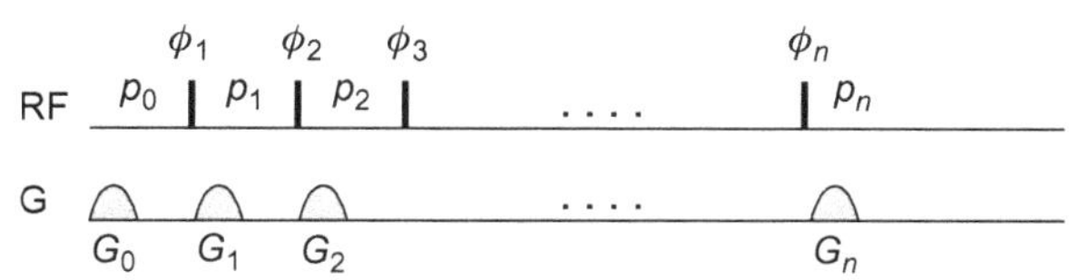

Figure 3.32 Simple scheme for a general pulse sequence: n pulses separate $n+1$ periods of free evolution during which gradient pulses G_i may be switched on with different strengths and directions. The coherence orders p_i (and the effective GMRs γ^*) can change only during the pulses.

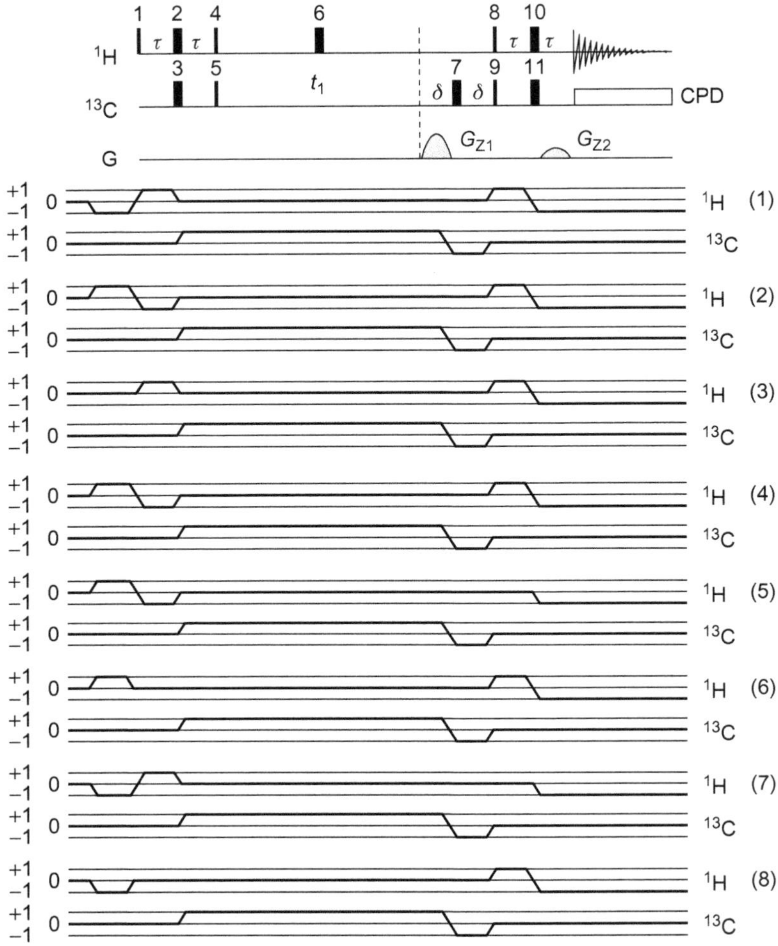

No.	total	ψ	norm	pc	grad	diff	pulse
1	0.4491	90	1.0000	1.0000	1.0000	0.9995	0.4493
2	0.4489	90	0.9995	1.0000	1.0000	0.9995	−0.4491
3	0.0039	270	0.0087	1.0000	1.0000	0.9995	−0.0039
4	0.0039	270	0.0087	1.0000	1.0000	0.9995	0.0039
5	0.0039	270	0.0086	1.0000	1.0000	0.9995	0.0039
6	0.0038	0	0.0085	1.0000	1.0000	0.9995	0.0038
7	0.0038	270	0.0084	1.0000	1.0000	0.9995	−0.0038
8	0.0038	180	0.0084	1.0000	1.0000	0.9995	−0.0038

Note: total = total signal attenuation, ψ = phases of the pathways, norm = normalized to the strongest pathway, pc = attenuation due to phase cycle, grad = attenuation due to the gradients, diff = attenuation due to the diffusion,

Figure 3.33 Complete calculation of coherence pathways: Simulation of a heteronuclear-spin quantum coherence experiment including a density matrix calculation for an *AMX* system. The phase cycle is $\phi(4) = y$; $\phi(5) = (x, -x)$; $\phi(6) = x, x, -x, -x$; $\phi(7) = \phi(9) = 4x, 4(-x)$; $\phi(\text{receiver}) = 2(x, -x), 2(-x, x)$; all other phases are constant at x. For the diffusion weighting, only the delay $\delta + \tau$, which was set to 3.7 ms, is relevant. The pulses were set 5% short, and the evolution delays 1.7, 1.7, 15, 17, 2, 1.7, and 1.7 ms were used. The gradient strengths were 0.199 and 0.05 T/m at 1-ms duration. For the eight strongest pathways, the partial coherence orders of protons and carbons are displayed separately.

Thus, it is very instructive to go through the Thomas–Bohr model of quantized electron orbits.

The model is based on the central attractive positive force on the orbiting electron exerted by the positively charged nucleus at the center (Figure 3.34). The motion is stable in an orbit, and the electron does not emit any radiation while it stays in that orbit. Electrons can change their orbit, but under the constraint that they can change their angular momentum, L, only in quantum jumps of $L = n\hbar$, $n = 1, 2, 3$. Here $\hbar$ is Planck's constant divided by 2π. The change of orbit can happen by absorption of some energy, in which case the electron jumps from a lower orbit (of energy E_i) to a higher orbit (of energy E_f). The electron stays in the higher orbit for a time of a microsecond, and then jumps back down to its original orbit. In this process, energy is reemitted due to the electron jump from the higher orbit (of energy E_f) to the lower orbit (of energy E_i). The frequency of the radiation emitted or absorbed is given by $f = |E_i - E_f|/h$. The electron stays in an orbit because of the balance between the attractive coulomb force toward the center $= 1/(4\pi\varepsilon_0 v^2)Ze^2/r^2$, toward the nucleus, and the centripetal force $= mv^2/r$. Here Z is the atomic number (the number of protons in the nucleus), e is the electron charge, r is the radius of the electron orbit, and ε_0 is the permittivity of the medium in which the atom is present, which in the present case is assumed to be vacuum. One can find v from this balance, in terms of the fundamental parameters of the atom. Further, one uses Bohr's postulate that angular momentum of the electron in orbits can take only quantized values of $L = mvr = n\hbar$. With very little mathematical juggling, one can write as follows: $Ze^2 = 4\pi\varepsilon_0 mv^2 r = (4\pi\varepsilon_0)[n^2h^2/(mr)^2]$. This gives the radius of the orbit, $r = (4\pi\varepsilon_0)(n^2h^2/(mZe^2))$, $n = 1, 2, 3, \ldots$ and the velocity (tangential at any point in the orbit), $v(=\omega/r) = (n\hbar)/(mr) = [1/(4\pi\varepsilon_0)(Ze^2/(n\hbar))]$, $n = 1, 2, 3, \ldots$. Sometimes the result is expressed in terms of angular velocity ω (rad/s), rather than linear velocity (m/s).

The potential energy of the electron in any orbit (bound state) is given as $V = -\int(Ze^2)/(4\pi\varepsilon_0 r^2)\mathrm{d}r = -(Ze^2)/(4\pi\varepsilon_0 r)$ (− sign means attractive potential). Now kinetic energy (KE) $= (1/2mv^2) = (Ze^2)/[(4\pi\varepsilon_0)(2r)]$ (substituting $mv^2 = 1/(4\pi\varepsilon_0 v^2/)$ Ze^2/r, from above). The total energy (TE) is $\mathrm{TE} = -(Ze^2)/[2(4\pi\varepsilon_0 r)] = -K = [\{-(mZ^2e^4)\}/\{(4\pi\varepsilon_0)^2\hbar^2\}](1/n^2)$, $n = 1, 2, 3, \ldots$. Thus, one can infer that quantization of angular momentum leads to quantization of energy. The frequency of radiation emitted for a transition from a final state (n_f) to an initial state (n_i) is

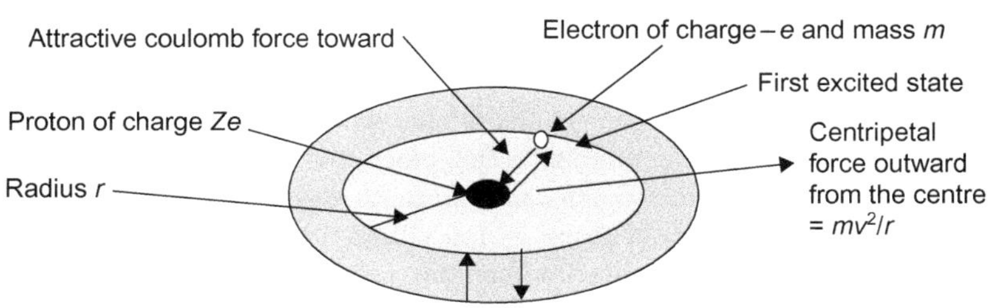

Figure 3.34 Electron orbits in a hydrogen atom.

$f = E_f - E_i/\hbar = [1/(4\pi\varepsilon_0)]^2[(mZ^2e^4)/(4\pi\hbar^3)][1/n_f^2 - 1/n_i^2]$. In terms of the linear wave number k (which is the inverse of the linear wavelength λ of the radiation), the above expression becomes (for $k = 1/\lambda = f/c$, $c =$ the velocity of light) k (wave number) $= [1/(4\pi\varepsilon_0)]^2[(mZ^2e^4)/(4\pi\hbar^3c)][1/n_f^2 - 1/n_i^2] = R_\infty Z^2[1/n_f^2 - 1/n_i^2]$. Here $R_\infty = [1/(4\pi\varepsilon_0)]^2[(me^4)/(4\pi\hbar^3c)]$.

Consider the simplest case, which is the hydrogen (H) atom. The spectroscopy of hydrogen, in a typical choice of orbits as below (spectral lines emitted), will consist of $k = R_\infty Z^2[1/n_f^2 - 1/n_i^2]$, $n_f = 2$ and $n_i > n_f = R_\infty Z^2[1/2^2 - 1/n_i^2]$, $n_i = 3, 4, 5, \ldots$. This series of hydrogen spectral lines (called the *Balmer series*) has been verified experimentally. In fact, the emission (nd absorption) spectral lines for all the elements in the periodic table have been studied and verified; the results are available in the form of standard tables. In the preceding analysis, a tacit assumption was made that the mass of the nucleus is infinite. This means that we assumed the nucleus remains fixed in space. In practice, this is not true: The mass of the nucleus is actually finite. One can correct the analysis by taking into account that the electron and nucleus move about their common center of mass. The mass of the electron should be corrected to $\mu = [mM/(m + M)]$, where m is the mass of the electron and M that of the nucleus. Thus, the quantization of angular momentum, $mvr = n\hbar$, in the above analysis should be replaced by $\mu vr = n\hbar$.

The wave number k then becomes $k = R_M Z^2[1/n_f^2 - 1/n_i^2]$, where $R_M = (M/m + M)R_\infty = (\mu/m)R_\infty$. The quantity R_M is called the *Rydberg constant* of a nucleus of mass M. One can see that as $M/m \to \infty$, $R_M \to R_\infty$.

In general, one can say that the constant R_M is less than R_∞ by a factor $1/(1 + m/M)$. In the case of the lightest nucleus, $M/m = 1836$ and R_M is less than R_∞ by about one part in 2000. By substitution of the fundamental constants, electron charge, mass, etc., one finds that for hydrogen $R_H = 10{,}968{,}100\,\mathrm{m}^{-1}$.

As noted, the constant R_M (or R_∞) is referred to in the scientific literature as the *Rydberg constant*. As a matter of interest, by substituting the fundamental constants into the expression for r, for the case of $n = 1$ (closest orbit), for the H atom, one gets the radius r of the orbit $= 5.3 \times 10^{-11}\,\mathrm{m} = 0.053 \times 10^{-9}\,\mathrm{m} \sim 0.05\,\mathrm{nm}$.

One may interpret the radius of the orbit as a measure of the radius of the hydrogen atom in its normal (ground) state. Similarly, calculating the orbital velocity v using the expression as above, one gets $v = 2.2 \times 10^6\,\mathrm{m/s}$. This is less than 1% of the velocity of light ($\sim 3 \times 10^8$). In the sense of the preceding analysis, this model of the atom is referred to in the literature as the *classical vector model*. These results are very fundamental to an understanding of NMR. The electron charge moving in orbit is equivalent to a circulating current (charge/s) in the orbit. This gives an intrinsic dipole (two faces of the orbit) magnetic moment to the electron. The nucleus (proton) is also a charged particle and analogously has an intrinsic dipole magnetic moment. The magnetic moment is related to the orbital angular momentum. When an external magnetic field (static along the z-direction) and an RF field along x- and y-directions are applied, the nucleus absorbs RF radiation. The angular-momentum vector, and thus the dipole magnetic moment, of an individual nucleus can thus be made to oscillate in resonance with the applied RF field.

The variation of this resonance over space can be converted into electrical signal variations, and thus imaging of the brain can be achieved. Now, as a chosen space (a *voxel*) in the brain is scanned, the nuclei in that region carry with them their interactions with neighboring and distant nuclei. This appears as a modulation of the signals obtained in the receiver coil (sensor). Thus, point-to-point variation in the image carries information about regional variations, which can be chemical, diffusionary, etc., in nature.

A time-and-space map can be obtained from voxel to voxel, and thus an overall picture of the brain in space and time can be constructed. It is in the interpretation of this picture that the knowledge and competence of the individual (a doctor, technician, etc.) is really tested.

3.2.9.2 Magnetic Dipole (Orbital): Magnetic Moment [19]

In an MRI situation, it is the response to the applied RF field of the dipole magnetic moment (DMM) of a nucleus, μ_I, that one is after. But the electron model of orbit was verified long before the nuclear-magnetic moment was discovered. It is assumed that the same model applies to a nucleus. In fact, experimentally, that is what is observed. The electron while orbiting the nucleus (proton) has a time period of $T = 2\pi r/v$.

The electron motion per unit time T (in one period) produces an electrical current of $i = e$ (electron charge)/$T = ev/2\pi r$. It is a basic law of physics that c circulating current (current loop) produces a magnetic field, which is the same at a distance from the loop as that of a magnetic dipole placed at the center of the loop and oriented perpendicular to its plane. For a current i in the loop of area A, the orbital dipole magnetic moment (ODMM) μ_I of the equivalent dipole is given as $\mu_I = \mathrm{i}A = (ev/2\pi r)\pi r^2 = evr/2$.

However, orbital angular momentum is $L = mvr$. Hence, one gets the ratio $\mu_L/L = (e/2m)$. One chooses to write $\mu_b = (e\hbar/2m)$, where $\hbar$ is Planck's constant divided by 2π. This is called the *Bohr magneton*, and is taken as the microscopic unit of the magnetic moment (microscopic because we are dealing with microscopic particles). Consequently, μ_L/L simplifies to $(\mu_b/\hbar)$. The unit $\mu_b = (e\hbar/2m) = 0.927 \times 10^{-23}\,\mathrm{A\,m^2}$ is the smallest (i.e., the most fundamental) magnitude of a magnetic moment. Thus, μ_b is taken as the unit of μ_L. One can rewrite this as $\mu_L = -[L(\mu_b/\hbar)]$. The negative sign is placed to emphasize the fact that the vector μ_L is antiparallel to L. It is interesting to note that the ratio μ_L/L is independent of the radius of the orbit.

It should be pointed out here that, using quantum mechanical principles, one finds that $L = \sqrt{(l(l+1)}\hbar$. Thus, $\mu_L = \mu_b\sqrt{(l(l+1)}$. For the z component of μ_L, one gets, in quantum mechanical language, $\mu_{L_z} = -(\mu_b/\hbar L_z) = -(\mu_b/\hbar m_l)$. Here l is the quantum number in general space and m_l in the z-direction.

When an atom (e.g., hydrogen) is placed in a magnetic field B, the ODMM will experience a torque (rotational force/moment), $\tau = \mu_L \times B$. The symbol $\times$ means the cross-product of the two vectors μ_L and B. In magnitude, $\tau = \mu_L B \sin\theta$. Here θ is the angle between μ_L and B. The direction of the torque will be perpendicular to

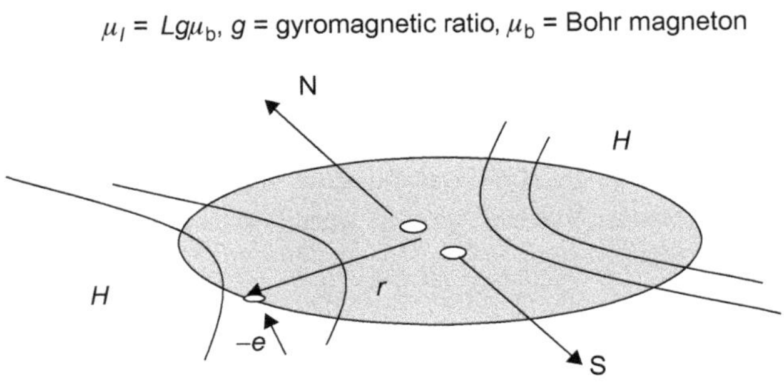

Figure 3.35 The orbital angular momentum L and the orbital magnetic dipole moment μ_L of an electron moving in a closed orbit.

both μ_L and B, in the sense when μ_L is rotated toward B (like a screw). Associated with this torque is a potential energy of orientation $\Delta E = -\mu_L B$. This is the dot product of the two vectors. In magnitude, it would be $\mu_L B \cos\theta$. The energy is a scalar quantity, and θ is the same. If there is no way for a system consisting of a magnetic dipole moment μ_L in a magnetic field B to dissipate energy, the orientational potential energy, ΔE, of the system must remain constant. In these circumstances, μ_L cannot align itself with B. Instead, μ_L will precess around B in such a way that the angle between these two vectors will remain constant. The technical meaning of *precession* is that the tip of the vector of the magnetic moment executes a circular orbit around its stationary axis. Also, the magnitude of both the vectors will remain constant (Figure 3.35).

This precessional motion is a consequence of the fact that the torque acting on the dipole is always perpendicular to the plane of the angular momentum, in complete analogy to the case of a spinning top. The torque on an ODMM gives rise to a changed dL in the angular momentum, during time dt. According to Newton's law, $\mathrm{d}L/\mathrm{d}t = \tau$ (the torque). The changed dL causes L to precess through an angle ($\omega\,\mathrm{d}t$). Here ω is the precessional angular velocity. One can write (refer to Figure 3.36), $\mathrm{d}L = LH \sin\theta\omega\,\mathrm{d}t$ or $L \sin\theta\omega = \mathrm{d}L/\mathrm{d}t = \tau = (\mu_b/\hbar)LB \sin\theta$. Thus, $\omega = (\mu_b/\hbar)B$. This equation also makes it clear that the sense of the precession is in the direction of B. This phenomenon is known as the *Larmor precession*, and ω is called the *Larmor frequency*.

3.2.9.3 Effect of Magnetic Field Gradient [19]

In the presence of a uniform magnetic field in space, there will be no net translational force acting on the ODMM, but there certainly will be torque acting on it. However, if the field is nonuniform (e.g., there is a gradient in one or more than one direction), there will be one or more translational forces acting on the magnetic dipole moment. Figure 3.37 shows the situation in a region in which the field B is converging and the electron is moving with velocity v through a circular orbit.

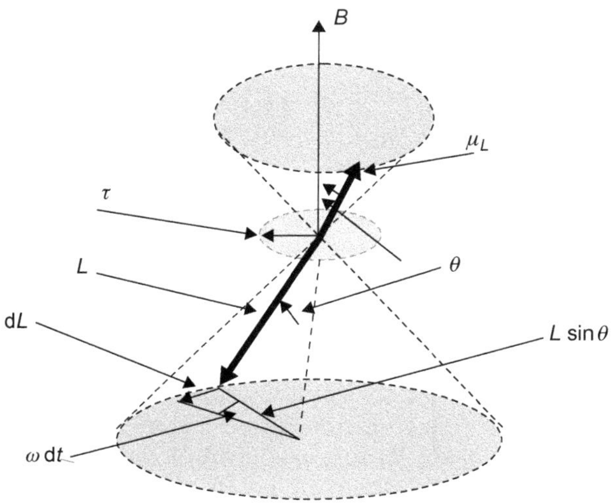

Figure 3.36 A torque = $\mu_L \times B = -(\mu_b/\hbar)(L \times B)$ arises as the atom's magnetic dipole moment μ_L interacts with the applied magnetic field *B*. *Note*: The symbols *B* and *H* are used interchangeably for the magnetic field.

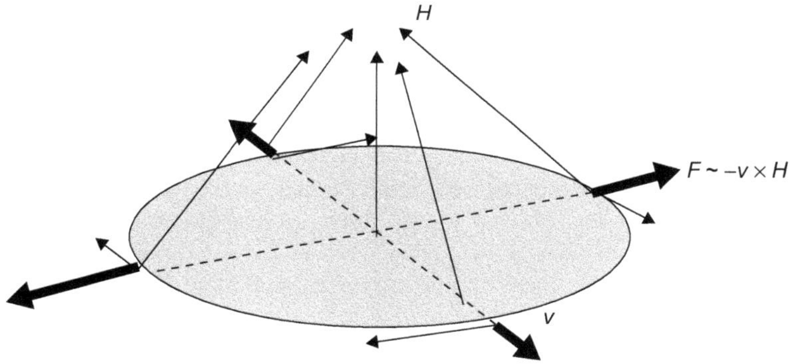

Figure 3.37 In a magnetic field gradient converging in the *z*-direction, the field will exert a translational force *F*, proportional to $v \times B$, in the direction of the convergence of the field, on the electron in the orbit.

It feels a force proportional to $-v \times B$ (the cross-product; they are oriented at 90° to each other here). The force will have a translational component in the direction in which the field becomes more intense. The effect can also be visualized via analogy to the force between a fictitious magnetic dipole in a nonuniform magnetic field and an electric dipole in a nonuniform electric field (Figure 3.38). One can write this converging force as $F_z \sim (\partial B/\partial z)\mu_{Lz}$. Here z is the coordinate axis in the direction of the field strength, and $(\partial B/\partial z)$ is the rate of increase in field per unit change in space. A magnetic dipole in a nonuniform magnetic field experiences a torque, which will cause precession and also create a translational force causing displacement. Now, the orbital magnetic moment vector can have any value from

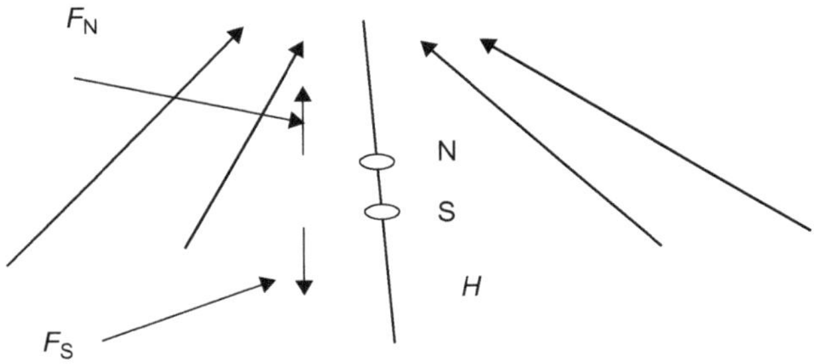

Figure 3.38 The forces F_N and F_S acting on the poles of a fictitious magnetic dipole (equivalent to the circulating electron in an orbit).

$-\mu_l$ (the symbol L has been replaced by l, to signify its connection to quantum number l) to $+\mu_l$. This signifies a continuous spread. According to QM, μ_{lz} can only have quantized values, given as $\mu_{lz} = -g_l \mu_b m_l$, where m_l is a quantum number as $m_1 = -l, -l+1, \ldots, 0, +1, +l-1, +l, l$. Here, a multiplying factor g_l, called the *spectroscopic splitting factor*, has been added in μ_{lz}. Its significance will become clear shortly. It was found experimentally for a beam of atoms (silver) in a gradient field; the beam was split into only two components, $+z$ and $-z$. This supported the quantum prediction, but only qualitatively established the quantization.

In actual practice, the number of possible values of μ_l is equal to the number of possible values of m_l, which is $2l+1$. Because l is an integer, this value $(2l+1)$ is always an odd number. Also, for any value of l, one of the possible values of m_l is 0. The hydrogen atom, which has a single electron, is interesting in this respect, for viewing in detail the intricacies of QM. Suppose that we pass H gas through a magnetic field. We keep relatively low temperatures so that the atoms remain in their ground state. This means that $l=0$. Then there is only one possible value of m_l, i.e., $m_l = 0$. Consequently, there should be no deflection of the beam in the gradiated magnetic field.

However, we find that the beam is still split into two symmetrically deflected components. This deflection in quantification experiments was established to be due to the electron having an intrinsic (built-in) magnetic dipole moment μ_s. This is because it has an intrinsic angular momentum S, called its spin. From a classical point of view, one can think, at least crudely, of the electron producing the external magnetic field of a magnetic dipole because of the current loops associated with its spinning charge.

We also assume that the magnitude of S and the z component S_z of the spin angular momentum are related to two quantum numbers, s and m_s, by quantization relations, which are identical to those for the orbital angular momentum. Thus, $S = \sqrt{s(s+1)}\hbar$; $Sz = m_s\hbar$. One should note that S_x and S_y are not quantized, nor are L_x and L_y. We also assume that the relation between the spin magnetic dipole moment and the spin angular momentum is of the same form as the relation for the orbital case. This means that $\mu_s = -g_s(\mu_b/\hbar)S$; $\mu_{sz} = -g_s m_s(\mu_b)$. Vector g_s is called spin g factor. It was found by experimental observation that the beam of H atoms is split into two symmetrically deflected components. Thus, it is clear that μ_{sz} can assume just two values, which are equal in magnitude but opposite in sign. If we make the final assumption that the possible values of m_s differ by one, ranging from $-s$ to $+s$, as is true of the quantum numbers m_l and l for orbital angular momentum, then one can conclude that the two possible values of m_s are $m_s = -1/2$ and $+1/2$, and that s has the single value, $s = 1/2$. By measuring the splitting of the beam of H atoms, it is possible to evaluate the net force F_z the atoms feel while traversing the magnetic field. The force is $F_z = -(\partial B_z/\partial z)\mu_b g_s m_s$. Because μ_b is known and $\partial B_z/\partial z$ can be measured, the experiments determine the value of the quantity $g_s m_s$. Within their accuracy, it was found that $g_s m_s = \pm 1$. Because we have concluded that $m_s = \pm 1/2$, it makes $g_s = 2$. These calculations are confirmed by many experiments. For instance, in the Zeeman effect, an external magnetic field is applied to a collection of atoms, and measurements are made of the potential energies of orientation in the field of the dipole moments of the atoms. This is done by measuring the splitting of the spectral line emitted when the atoms decay from some higher energy level to their ground-state energy level.

The splitting of the line occurs because the levels themselves are split according to the different values assumed by the orientational potential energy of the atoms. A simple example is the Zeeman effect for H atoms. In their ground state, these atoms have no orbital angular momentum, and therefore no orbital magnetic dipole moments. However, the measurements show that their ground-state energy level is split by the applied magnetic field into two components, symmetrically disposed about the energy of the ground state in the absence of a field. This splitting reflects the two possible values of the orientational energy, $\Delta E = -\mu_s B = -\mu_{sz} B = g_s \mu_b m_s B = \pm g_s \mu_b B/2$, where the z-axis is taken in the direction of the applied field.

The fact that the level is symmetrically split into two components confirms the conclusion that $m_s = \pm 1/2$, and the measured magnitude of the splitting confirms the conclusion that $g_s = 2$. More accurate spectroscopic measurements show that $g_s = 2.00232$. However, in almost all situations, it is quite adequate to say simply that the spin g factor for an electron is twice as large as its orbital g factor. This means that the spin dipole magnetic moment is twice as large, compared to the spin angular momentum, as the orbital magnetic dipole is compared to the orbital angular momentum. In contrast, μ_s and S are antiparallel, just like μ_l and L. This is because the relative orientation of either pair of vectors depends only on the fact that the electron has a negative charge.

3.2.9.4 Spin–Orbit Interaction [19]

The origin of the internal magnetic field experienced by an electron moving in a one-electron atom is easy to understand. Think of the motion of the nucleus from the point of view of the electron(s); i.e., a frame of reference fixed on the electron. The charged nucleus moves around the electron.

The electron is, in effect, located inside a current loop. The nucleus produces a magnetic field. One can note that the charged nucleus, moving with velocity $-v$, constitutes a current j where $j = -Zev$. According to Ampere's law, this produces magnetic field B which, at the position of the electron, is $B = (\mu_0/4\pi)[(j \times r)/r^3] = -Ze$

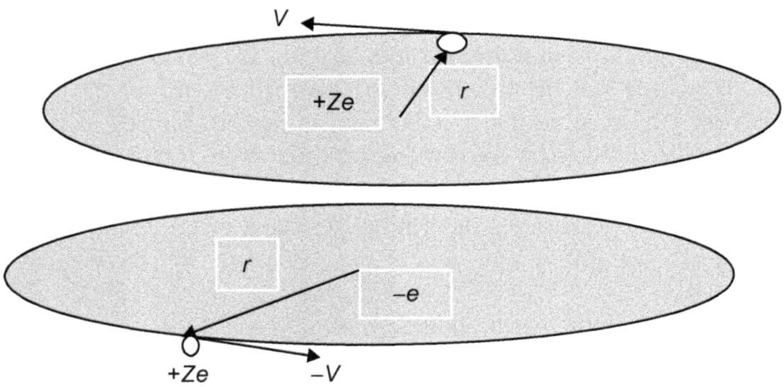

Figure 3.39 An electron moving in a circular orbit. The first sketch (top) is the motion as seen by the nucleus. One can instead think from the point of view of the electron, i.e., the nucleus moves around the electron. The second sketch (bottom) is motion, as if the nucleus was moving around the electron. In the second case, the magnetic field B experienced by the electron is in the direction out of the page at the electron's position.

$(\mu_0/4\pi)[(v \times r)/r^3]$. It is convenient to express this in terms of the electric field E acting on the electron. According to Coulomb's law, $E = (Ze/4\pi\varepsilon_0)(r/r^3)$. From these two equations, one gets $B = -\varepsilon_0\mu_0 v \times E$ or $B = (-1/c^2)v \times E$, since $c = \sqrt{(\varepsilon_0\mu_0)}$. Here ε_0 is the permittivity, and μ_0 the permeability, of the medium in which the charges are placed. The quantity B is the magnetic field strength experienced by the electron when it is moving with velocity v with respect to the nucleus, and therefore through the electric field of strength E which the nucleus exerts on it. The electron and its spin magnetic dipole moment can assume different orientations in the internal magnetic field of the atoms, and its potential energy (PE) is different for each of the orientations. If we evaluate the orientational PE of the magnetic dipole moment in the magnetic field, one gets $= -\mu_s B = (g_s\mu_b/\hbar)$S.B. With this result, one can express the spin−orbit interaction energy (SOIE) as $\Delta E = (g_s\mu_b)$ $(1/2em^2c^2\hbar)(1/r)(\mathrm{d}V(r)/\mathrm{d}r)$S.L. Now $|\Delta E| \sim \mu_s B$. $\mu_s \sim \mu_b \sim 10^{-23}\,\mathrm{A\,m^2}$, so $B \sim 10^{-23}\,\mathrm{J}$ (H atom, $n = 2$, $l = 1$ state)$/10^{-23}\,\mathrm{A\,m^2} \sim 1\,\mathrm{T}$. This is a high field, in the sense of an electromagnetic coil, producing it. Thus, the electrons' spin magnetic dipole moment feels a strong magnetic field, because the electron is moving at a high speed through the strong electric field surrounding the nucleus.

One should note the choice for the $n = 2$, $l = 1$ state of the H atom. There is no SOIE in the $n = 1$ state, because for $n = 1$, the only possible value for l is $l = 0$, which means $L = 0$ (Figure 3.39).

3.2.9.5 The Total Angular Momentum (J = L + S) *[19]*

If there is no spin−orbit interaction (SOI), the orbital and spin angular momenta L and S of an atomic electron would be independent of each other, and so they would independently obey the quantum mechanical angular conservation law. That is, when an atom without SOI is in free space, there would be no torques acting on either L or S, so both these vectors would precess randomly about the z-axis in such a way that their magnitude and z components, L, L_z, S, S_z, would have fixed values. The fixed values are the ones specified by the quantum numbers l, m_l, s, and m_s. However, there is SOI: a strong internal magnetic field is acting on the atomic electron, the orientation of which is determined by L, and produces a torque on its spin dipole magnetic moment, the orientation of which is determined by S. As in the case of Larmor precession, the torque will not change the magnitude of S. Nor will the reaction torque acting on L change its magnitude. The torque does, however, enforce a coupling between L and S that makes the orientation of each dependent on the orientation of the other. As a result, these angular-momentum vectors undergo a different motion than they would if there were no SOI. They precess around their sum instead of both precessing around the z-axis. Because these vectors do not precess around the z-axis, their z components, L_z and S_z, do not have fixed values when there is SOI (Figure 3.40). They must move in such a way that their sum, the total angular momentum J, satisfies the quantum mechanical angular-momentum conservation law. That is, if the sum is in free space so that no external torques act on it, its total angular momentum, $J = L + S$, maintains a fixed magnitude J and a z component J_z, while x and y components J_x and J_y fluctuate about zero. The vectors L and S precess around their sum J, and their components in the direction of J remain fixed so that its magnitude J is fixed. Simultaneously, J precesses around the z-axis, maintaining a fixed component J_z.

It can be shown that the magnitude and z component of the total angular J are specified by two quantum numbers j and m_j according to usual quantization conditions $J = \sqrt{j(j+1)}\hbar$ and $J_z = m_j\hbar$, $m_j = -j, \ -j+1, \ \ldots +j-1, \ +j$. The possible values of the quantum number j can be determined as follows: The z component of J is $J_z = L_z + S_z$, which is $m_j\hbar = m_l\hbar + m_s\hbar$ or $m_j = m_l + m_s$. Because the maximum possible value of m_l is il, and the maximum possible value of m_s is $s = 1/2$, the maximum possible value of m_j is $(m_j)_{\max} = l + 1/2$. But this is also the maximum possible value of j. In common with the other angular quantum numbers, the possible values of j differ by integers. Therefore, the values must be members of the decreasing series, $j = l + 1/2, \ l, \ -1, \ 2, \ l - 3/2, \ l, \ -5/2$; to determine where the series terminates, one may use the vector inequality $|L+S| \geq |L| - |S|$ whose validity can be easily demonstrated by inspecting Figure 3.41. Writing $L+S$ as J, we have from the inequality $|J| \geq |L| - |S|$or $\sqrt{j(j+1)}\hbar \geq |\sqrt{1(l+1)}\hbar - \sqrt{s(s+1)}\hbar|$. From this it can be shown that because $s = 1/2$, generally there are two members of the series that satisfy the inequality. These are $j = l, \ 1/2, \ -1/2$. It is even more apparent that if $l = 0$, there is only one possible value of j, namely

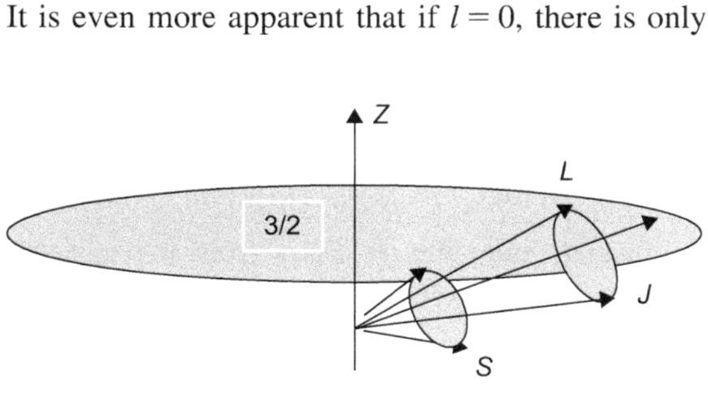

Figure 3.40 The angular-momentum vectors L, S, and J for a typical case of a state with $l = 2$, $j = 5/2$, and $m_j = 3/2$. The vectors L and S precess uniformly about their sum J, as J precesses randomly about the z-axis.

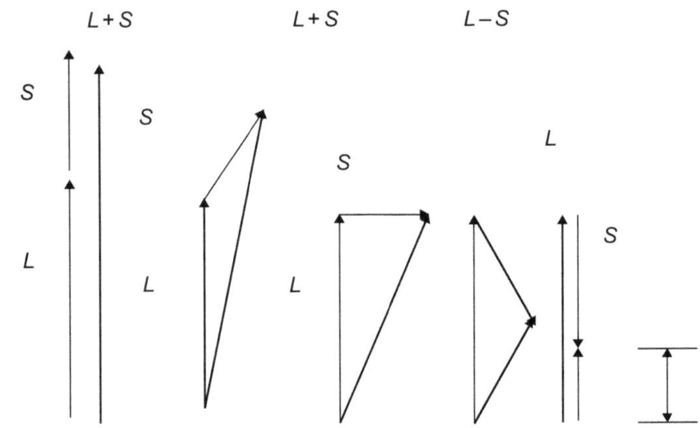

Figure 3.41 The vector diagram shows that for any two vectors L and S, the magnitude $|L+S|$ of their sum is always at least as large as the magnitude of the difference in their magnitudes, $|L-S|$. This is the case for $|L|>|S|$, as shown. The conclusion is the same if $|L|<|S|$.

$j=1/2$, if $l=0$. The content of the equations stating the possible values of the quantum numbers m_j and j can be represented in terms of the order of the rules of the vector addition, by constructing a set of vectors whose lengths are proportional to the values of the quantum numbers l, s, and j.

3.2.9.6 Spin–Orbit Interaction Energy

The SOIE was evaluated before as $\Delta E=(1/2m^2c^2)[(1/r\mathrm{d}V(r)/\mathrm{d}r)]$ $(\mathbf{S}\cdot\mathbf{L})$. One can express this in terms of the quantum numbers l, s, and j. We know that $\mathbf{J}=(\mathbf{L}+\mathbf{S})$. Now $\mathbf{J}^2=(\mathbf{L}+\mathbf{S})\cdot(\mathbf{L}+\mathbf{S})=\mathbf{L}\cdot\mathbf{L}+\mathbf{S}\cdot\mathbf{S}+2\mathbf{S}\cdot\mathbf{L}$ (since $\mathbf{S}\cdot\mathbf{L}=\mathbf{L}\cdot\mathbf{S}$). So $\mathbf{S}\cdot\mathbf{L}=[J^2-L^2-S^2]/2$. In the quantum state associated with quantum numbers l, s, and j, each atom (each term on the right) has a fixed value, and $\mathbf{S}\cdot\mathbf{L}$ has the fixed value $\mathbf{S}\cdot\mathbf{L}=\hbar^2/2[j(j+1)-l(l+1)-s(s+1)]$. Thus, $\Delta E=(\hbar^2/4m^2c^2)[j(j+1)-l(l+1)-s(s+1)][(1/r\mathrm{d}V(r)/\mathrm{d}r)]$. The spin–orbit energy of the state is just the expectation value of this quantity. Thus, $\Delta E=(\hbar^2/4m^2c^2)[j(j+1)-l(l+1)-s(s+1)]$ $[(1/r\mathrm{d}V(r)/\mathrm{d}r)]$.

The expectation value for $[(1/r\mathrm{d}V(r)/\mathrm{d}r)]$ is calculated using the potential function $V(r)$ for the system and the probability density.

3.2.10 Nuclear Spin

3.2.10.1 The Interactions: Nuclear-Orbital Electron Spins (Internal Field)

Atomic nuclei also have a dipole magnetic moment. It arises from the intrinsic dipole magnetic moments of the protons and neutrons in the nuclei, and from the currents circulating in the nuclei due to the motion of the protons. The nuclear dipole moment (NDP) interacts with atomic electrons, which produce magnetic fields of their own. This results in hyperfine splitting of atomic energy levels. The amount of the interaction energy is $\Delta E = C[f(f+1) - i(i+1) - j(j+1)]$. Here j, i, and f are quantum numbers specifying the magnitude of the atom's total electronic angular momentum (multielectron), total nuclear angular momentum (protons and neutrons), and the grand total angular momentum, respectively. The equation is analogous to that describing the atomic SOIE. The constant C is proportional to the magnitude of the nuclear dipole magnetic moment μ. Measurements of ΔE and C show that for all nuclei, μ is on the order of the nuclear magneton. The quantity is $\mu_n = (e\hbar/2M) = 0.505 \times 10^{-26}\,\mathrm{A\,m^2} \approx 10^{-3}\mu_b$. Here M is the proton mass and μ_b is the Bohr magneton. Measurements of the hyperfine splitting also show that the sign of the nuclear magnetic dipole moment (NDM) (giving the relative orientation of the dipole magnetic moment vector and the angular momentum of the nucleus) is positive (parallel) in some cases and negative (antiparallel) in others.

Nuclei for which both A (mass) and Z (atomic number) are even have $\mu_n = 0$. The total nuclear angular momentum quantum number i, usually called the spin, can be obtained simply by counting the number of energy levels of a hyperfine splitting multiplet. If the multiplet is associated with a value of j larger than i, then one can assume $2i + 1$ different energy levels. It is found that i is an integer for nuclei of even A, with $i = 0$, if Z is also even, and that i is a half-integer for nuclei of odd A. The magnitude I of the total nuclear angular momentum is given in terms of i, by the usual relation $I = \sqrt{[i(i+1)\hbar]}$. The total angular momentum of a nucleus arises from the intrinsic spin angular momenta of its protons and neutrons, and also from the orbital momenta due to the motion of these particles within the nucleus. It should be emphasized that in nuclear physics the word *spin* frequently refers to total angular momentum of the nucleus, in contrast to atomic physics, where the word refers to the intrinsic angular momentum only. It would be wiser to use the term *intrinsic spin angular momentum* to avoid possible confusion. One can then use the symbol s when referring to that part of the angular momentum of a single particle that has nothing to do with orbital angular momentum (e.g., the intrinsic spin angular momentum of both protons and electrons is given by $s = 1/2$).

Closely related to the spin of a nucleus is the symmetry character of the eigenfunction for a system containing two or more nuclei of the same species. This is studied by analyzing the spectra of diatomic molecules containing two identical nuclei. It is found that nuclei with integral spin quantum number I (nuclei of even A) are of symmetric type, i.e., they are bosons (uncharged particles, e.g., neutrons), whereas nuclei with half-integer i (nuclei of odd A) are of antisymmetric type, i.e., they are fermions (charged particles, e.g., electrons and protons). Such molecular spectra also provide independent measurements of i that confirm the values obtained from hyperfine splitting. We know that nuclei are composed of protons and neutrons. The neutron is an uncharged particle of nearly the same mass as a proton, and precisely the same intrinsic spin angular momentum and symmetry character ($s = 1/2$, antisymmetric). A nucleus with mass number A and atomic number Z contains A nucleons (a word used for both protons and neutrons), of which Z are protons and $A - Z$ neutrons. This rule obviously leads to a mass and charge rule, as per the periodic table.

It may be useful for some readers to briefly discuss what this means. At the high school level, one learns that the mass of a nucleus is only slightly less than the mass of an atom (the mass of electrons being much smaller than that of nuclei). Thus, the nuclear mass is approximately equal to A times the mass of a hydrogen atom, or approximately equal to A times the mass of a proton (which constitutes the nucleus of a hydrogen atom). The integer A, called the *mass number*, is the one closest to the atomic weight of the atom containing the nucleus in question. One knows that the charge of a nucleus is exactly equal to the atomic number Z of the corresponding atom, times the negative of the charge of the electron, or Z times the charge of the proton. The atomic number gives the location of an atom, in the periodic table of the elements. The table shows 'periodic table of elements' that A is roughly equal to $2Z$, except for the proton, for which $Z = A = 1$.

3.2.10.2 The Nuclear Dipole Magnetic Moment: Externally Applied Magnetic Field [20]

It is known that nuclei possess angular momentum, which is measured by the quantum number I. There is also an associated magnetic moment μ. The latter allows interaction of the nuclei with the outer electron clouds and the environment in which they are present. A nucleus is an assembly of neutrons and protons. Each nucleon (i.e., each neutron or proton) has an effective radius of about 1.4×10^{15} m, and an intrinsic angular momentum corresponding to a quantum number 1/2. Nucleons also posses orbital angular momentum. When combined with their intrinsic spin, this produces a resultant angular momentum, which is by tradition called the *nuclear spin* and is denoted by I.

The I is given by $|I| = \hbar[I(I+1)]^{1/2}$, where I is the appropriate quantum number. The resultant nuclear spin may take quite large values. The nuclear magneton, in analogy with electron magneton, is defined as $\mu_N = (e\hbar)/2M = 5.05 \times 10^{-27}$ J/T (joules per Tesla). Here M is the mass of the proton. The proton does not have a magnetic moment equal to one nuclear magneton, but rather $2.79\mu_N$. The neutron has no net charge, but it has magnetic moment $-1.9\mu_N$ (i.e., the magnetic moment is antiparallel to I). The nuclear-magnetic moment is expressed as $\mu = \gamma I$, γ being the magnetomechanical ratio (also referred to as the gyromagnetic ratio). It varies from nucleus to nucleus. Although the ratio is usually positive in sign, there are nuclei with negative values, e.g., ^{107}Ag has $I = 1/2$, $\mu = -0.133\mu_N$. Owing to spatial quantization, a nucleus with quantum number I possesses $2I + 1$ levels, which in the absence of an external field are degenerate. Application of an external field leads to the occurrence of $2I + 1$ discrete nuclear Zeeman levels, each of which is separated from its neighbors by an energy $\mu\gamma B_0$. The frequency ω_0, also known as the Larmor frequency, is defined as $\omega_0 = \gamma B_0$. This is the fluency with which the nuclear moment, μ, precesses around the direction of the applied static magnetic field B_0. Note that the symbols B and B_0 are often used to mean a static magnetic field. This field is assumed to be directed along the z-axis. For the proton in a magnetic field of 1 T, the energy split $\hbar\gamma B_0$ is 1.76×10^{-7} eV.

This, when expressed in frequency, corresponds to 42.5 MHz. This is in the range of the RF part of the electromagnetic radiation spectrum. The electrostatic potential along, say, the z-axis of an axially symmetric distribution of electric charge $\rho(r)$ can be represented as a power series in r/z ($r \ll z$), leading to the following expression for the potential: [$\phi_P = (1/z)\int \rho \mathrm{d}v + (1/z^2)\int \rho r v \cos\theta \, \mathrm{d}v + (1/z^3) \int \rho r^2 (3\cos\theta^2 - 1)\mathrm{d}v$].

The coefficient of the first term is often called the *monopole contribution*. This is a measure of the net charge density in a neutral charge distribution, and thus is zero. The second term is the *dipole-moment contribution* to the potential, which arises when the positive and the charge distributions do not have a common center of gravity; in a symmetrical charge distribution, this is also zero. The third term arises when the charge distribution has a center of gravity but is not spherically symmetric. Nuclei with spin $I \geq 1$ often have an ellipsoidal form of charge distribution. The nuclear charge distribution adopts an oblate (flattened at the poles) or prolate (elongated at the poles) form, leading to nuclear quadrupole moments of negative and positive sign, respectively. The quadruple moment is Q, where eQ is the coefficient of the term in z^{-3}.

In the presence of an axially symmetric electric field gradient, the quadruple moment causes a precession of the angular momentum and a splitting of the nuclear spin levels, even in the absence of a magnetic field. In the presence of a magnetic field, the Zeeman splittings are distorted. In crystals with cubic symmetry (i.e., in the SS), the electric field gradient at a nucleus is zero, and quadrupole effects are absent, but if the cubic symmetry is disturbed by mechanical strain, by the presence of dislocations, or by alloying, quadrupole effects are again introduced (Figures 3.42–3.44).

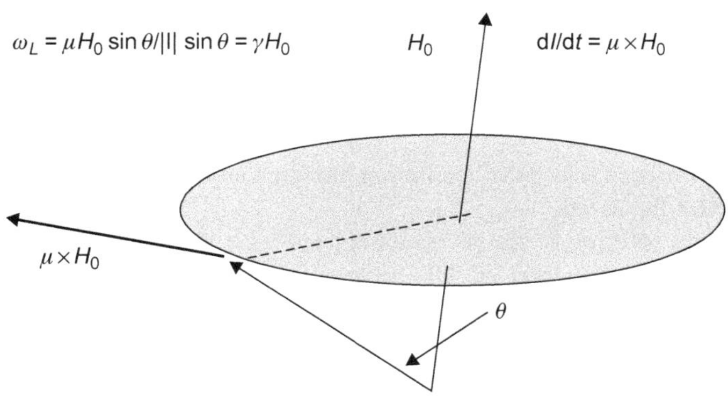

Figure 3.42 The precession of the nuclear spin I in an external magnetic field B_0. The rate change of the angular momentum ($\omega_0 I \sin\theta$) is equated to the torque ($\mu \times B_0$) produced by the magnetic field.

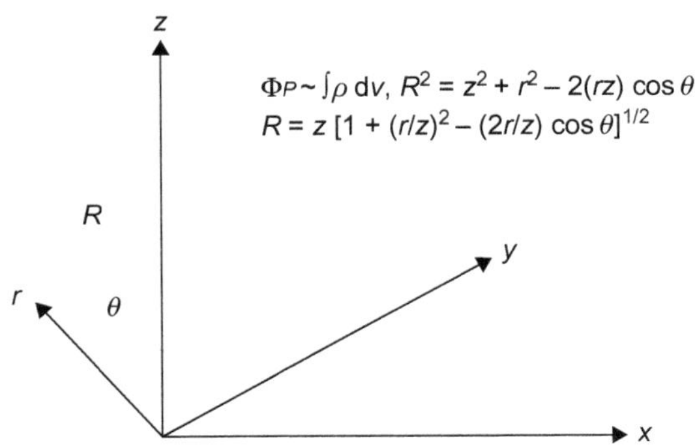

Figure 3.43 The geometry for the charge distribution. One imagines an axially symmetric distribution of charge and needs to calculate the potential at a point P on the z-axis.

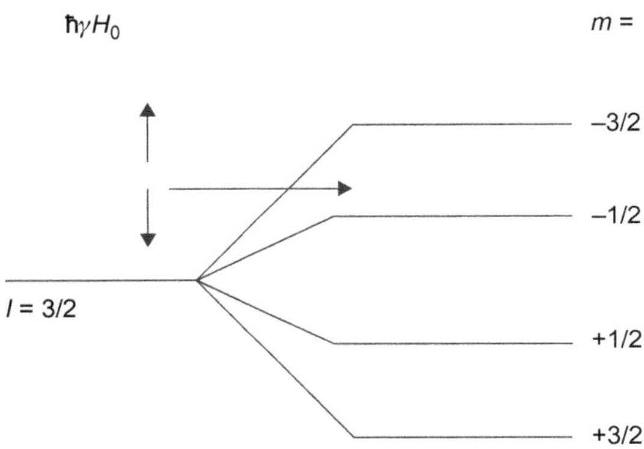

Figure 3.44 The splitting of nuclear levels by an external magnetic field. The symbol H_0 means the same thing as B_0 (used in the text).

3.2.10.3 NMR: The SS

The temperature of a substance is associated with the vibratory motion of the atoms. However, a solid is not always a single homogenous arrangement of atoms. One often needs to divide it into the phonon system, the electron system, the atomic-magnet system, and the nuclear-magnetic system. Each of these systems has its own characteristic energy domain. The phonons have the mean energy on the order of 10^{-2} eV.

Electron in metals may have excitation energies anywhere from 0 eV to as high as one can heat the metal.

The Zeeman splitting of the levels associated with an atomic-magnetic moment is $g\mu BB_0$, which for a field of 1 T corresponds to 1.16×10^{-4} eV, when $g = 2$ (or the equivalent frequency 28.0 GHz). However, the nuclear Zeeman levels have separation on the order of 10^{-7} eV for a proton. These different systems are in thermal equilibrium with one another only if energy exchange between them can occur rapidly—just how rapidly depends on the timescale of the measurement being made. The phonon spectrum extends over a vast range of frequencies, from 0 to 10^{14} Hz.

3.3 Diagnostics of Human Brain Disorders/Tumors: MRI

3.3.1 MRI Diagnostics

3.3.1.1 Human Brain Biochemistry and Brain Tumors [21]

N-acetyl-aspartyl-glutamate (NAAG) is co-localized with NAA in neurons and releases NAA and glutamate when it is cleaved by *N*-acetylated alpha-linked dipeptidase. Glutamate and possibly NAAG are excitatory amino acids, and in physiological concentrations glutamate can be neurotoxic. NAAG may be the form in which the neuron stores glutamate to protect the cell from the excitatory and potentially neurotoxic action of glutamate. The excitatory amino acids may play a role in various neurological disorders. In addition to stroke (compounds that block glutamate decrease infarct size), a potentially neurotoxic role for glutamate has been suggested in status epilepticus, olivopontocerebellar (OPCA) degeneration, Huntington's disease, hypoxemia, degenerative motor systems, and other conditions. The analysis of NAA by proton NMR can improve understanding of other neurological conditions. Levels of NAA may decrease in patients with chronic MS, where there is loss of axons. Studies relating immunocytochemistry and proton NMR could be highly beneficial.

Creatine (Cr), phosphocreatine, and other creatine-related compounds are taken in the diet, and are synthesized within the liver, kidneys, and pancreas. Arginine, glycine, and *S*-adenosylmethionine are precursor molecules for creatine. Humans have approximately 120 g of creatine; creatine phosphate is present in muscle and neurons and approximately 2 g/day are replaced by diet and de novo synthesis. ^{31}P NMR measures only creatine phosphate, but both creatine phosphate and creatine are measured as a single peak with ^{1}H NMR. Creatine phosphate serves as a reserve for high-energy phosphates in the cytosol of muscles and neurons and buffers cellular ATP/ADP reservoirs. Creatine kinase converts creatine to creatine phosphate by utilizing ATP. Tissues where the largest changes in energy metabolism occur, like muscle and brain, have the highest concentrations of creatine kinase.

It is possible that because the creatine peak is relatively stable with various diseases, it might be used as a control value, for comparison with more volatile changes seen with NAA and choline (Cho). Studies correlating in vivo concentrations of Cr with chemical measurements in vivo may help to determine how to use this peak for the study of a disease.

Choline compounds are nutrients present in most foods, and thus are absorbed into the human body through diet. Choline is the precursor for two important molecules, acetylcholine (ACho) and phosphatidylcholine (PtdCho). Synthesis of ACho occurs only within cholinergic neurons, whereas all cells use Cho to synthesize PtdCho, which is a major constituent of the cell membrane. PtdCho can be synthesized de novo within the liver from phosphatidylethanolamine (PtdEth) via PtdEth-*N*-ethyltransferase.

Cho is transported into all cells via a low-affinity transport system; Cho entering the cell through this mechanism is coupled to synthesis of PtdCho. Cho is first phosphorylated by choline phosphotransferase and phosphocholine is subsequently used to synthesize PtdCho via the Kennedy pathway. Phosphocholine is the first molecule in the synthesis of PtdCho. It tends to be high in concentration in the developing brain, but the level subsequently falls with maturation. Several enzymes are responsible for the catabolism of PtdCho. Phospholipase A_1 and A_2 break down PtdCho to lysophosphatidylcholine, a cell surfactant with potential for causing cellular toxicity. For this reason, lysophosphatidylcholine is rapidly converted to a fatty acid and glycerophosphocholine by lysophospholipase. Also, there is an exchange system in which the fatty acid from one lysolecithin is exchanged with a second lysolecithin molecule, yielding PtdCho and glycerophosphocholine. Phospholipase D cleaves Cho from glycerophosphocholine to form phosphatidic acid.

PMRS has been used extensively to find which of the Cho-containing compounds contributes to the 3.2-ppm peak. This is assigned to phosphocholine, but its cortical concentration is in the 0.3-mM range, below the range of PMRS resolution. Glycerophosphocholine is seen to be present in a 0.3-mM concentration, and free Cho has 0.028-mM concentration. Significant changes in Cho-containing compounds are associated with certain disease states. For example, free brain Cho rises with brain ischemia. Cho measurements with PMRS may be helpful to better define the time of infarct, the infarct size, and the predicted outcome in stroke patients. In individuals with Alzheimer's disease (AD), one observes that cortical ACho is decreased. The decline of cortical ACho is due to the loss of cholinergic cells projecting from the basal forebrain, the nucleus basalis of Meynert. These cholinergic cells appear to be important for memory and for many aspects of cognition. On the whole, observations support the contention that PMRS holds great promise for improving differential diagnosis of various brain diseases.

Three main peaks are observed in ^{1}H (proton) NMR spectroscopy of the human brain. Their chemical constitution and operative functions can be broadly summarized as noted earlier in this section (Figures 3.45–3.47).

3.3.1.2 Identification of Cerebral Metabolites [22]

It has been observed that localized ^{1}H NMR spectra of the human brain in vivo are affected by signal overlap, strong spin–spin coupling, and complex J modulation.

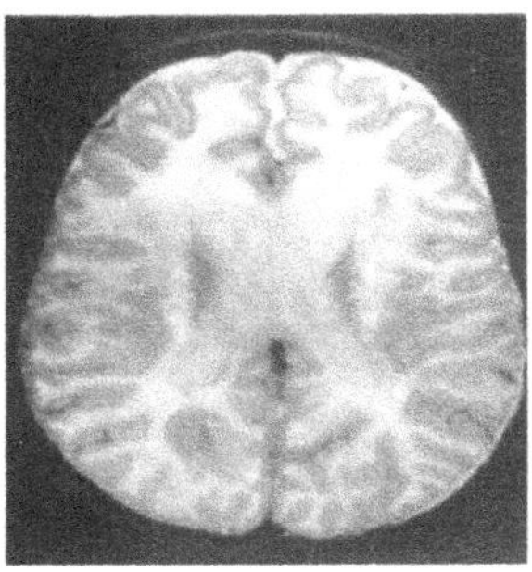

Figure 3.45 A T_2-weighted MRI scan of a 3-year-old girl with Canavan's disease, demonstrating extensive hyperintensity throughout the WM.

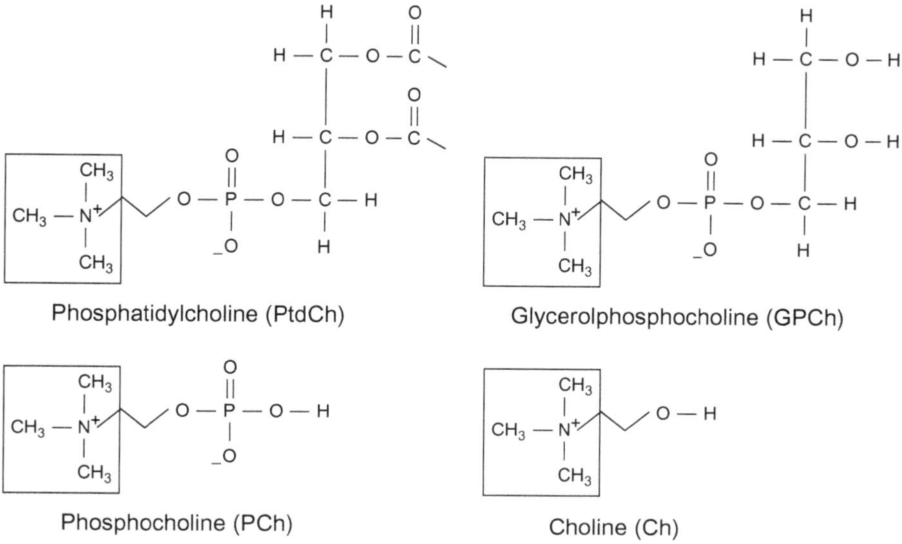

Figure 3.46 This demonstrates the chemical structure of the major choline-containing compounds within the brain: choline, phosphatidylcholine, glycerophosphocholine, and phosphocholine.

One is looking for the assignment of ^{1}H resonances within cerebral resonances under the experimental conditions used for human investigations. Conventional 7 T (Tesla), FID, localized short echo time (TE), and STEAM spectra (TE = 20 ms) of aqueous metabolic solutions can be compared to in vivo brain spectra of humans. In particular, stimulated-echo localization sequences with short echo times reveal a large number of components in spectra of the human brain in vivo.

The respective resonances are linked to cerebral metabolites that are of major relevance for the biochemistry of the functioning brain. NMR spectroscopy not only offers noninvasive strategies for clinical diagnosis based on metabolic information, but also opens the way to completely new area of research, such as human neurochemistry. To fully develop recent methodological achievements into valuable tools for studies of both normal and pathological metabolism, the spectroscopic characterization of a localized region of tissue must be transformed into a metabolic characterization in terms of absolute concentrations. The first step in this direction is an unambiguous identification of metabolite resonances acquired under optimized experimental conditions. These conditions require a reliable localization technique, and further necessitate the use of short echo times, to detect resonances with small T_2 relaxation times and to reduce J modulation effects on strongly spin-coupled resonances. Using NMR technology, it is possible to identify ^{1}H NMR-visible metabolites that are observable in human brain in vivo. It is also possible to make detailed assignments using singlet resonances.

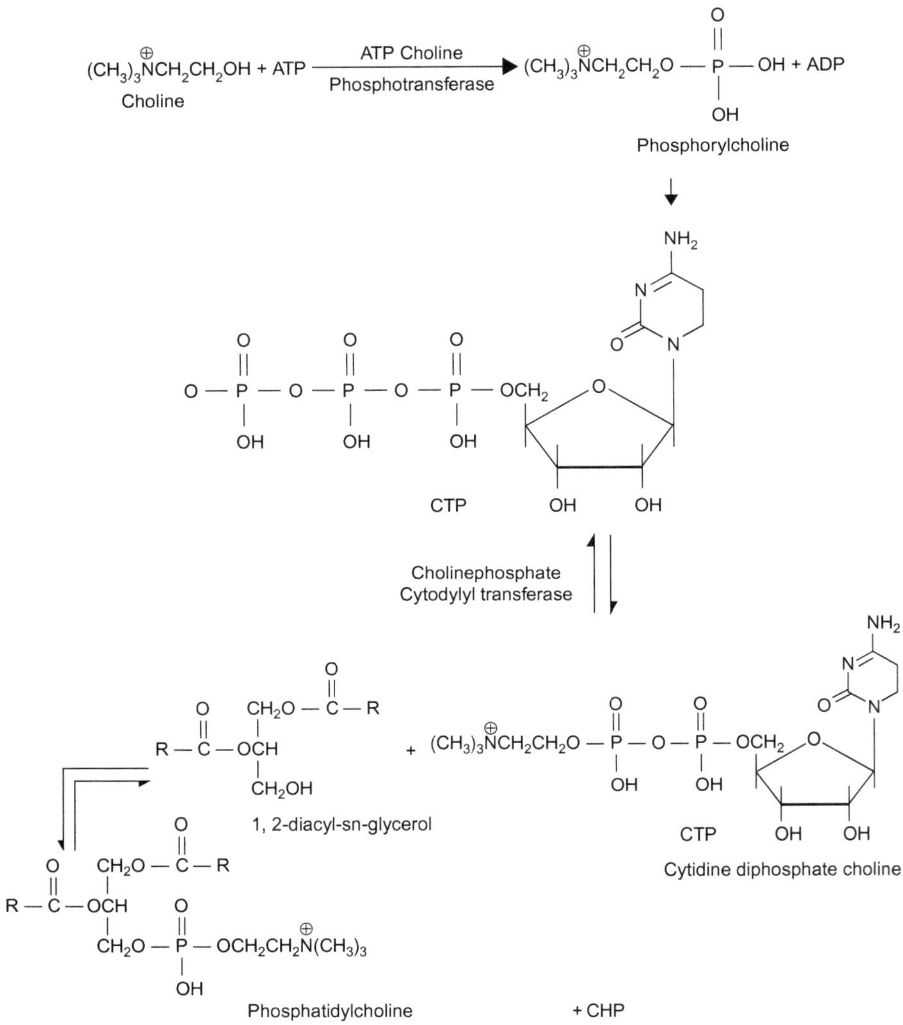

Figure 3.47 The Kennedy pathway, the major route for synthesis of phosphatidylcholine.

In contrast, the chemical-shift assignments for most multiplet structures of spin-coupled resonances are not transferable to lower fields. This is not due to the increased resonance line widths in spectra of living tissue, nor to potential complications from signal overlap, but is caused by the influence of strong coupling at the relatively low-field strengths of 1 and 4 T in use in human NMR spectroscopy. Strong coupling occurs when the chemical-shift difference $\Delta\delta$ between 2 spin-coupled partners approaches the same order of magnitude as the respective spin–spin coupling constant J, i.e., $0.1 \le J/\Delta\delta \le 1$. Because chemical-shift differences in ^{1}H NMR spectra of amino acids are within 0.1–1.0 ppm (i.e., 8–85 MHz at 2 T), as compared to field-independent coupling constants of 7±2 Hz, strongly coupled multiplet resonances are the rule rather than the exception for most spin-coupled ^{1}H NMR resonances. The resulting multiplets exhibit major distortions relative to high-field spectra, including variations in the number of resonance lines as well as in the intensities and chemical-shift values. A further consequence is distortion of the linear relationship between peak area and the number of spins, which must be accounted for when evaluating metabolite concentrations. Good examples of the qualitative spectral alterations that have to be dealt with in low-field, in vivo NMR spectra are shown in Figure 3.48 [22], for myoinositol and taurine, respectively. In both cases, considerable lineshapes are observed in going from a 7-T FID spectrum to a 2-T localized STEAM spectrum at an echo time of TE = 20 ms. The 8-ml VOI was chosen inside a 50-ml beaker of an aqueous metabolite solution, placed in the center of the whole-body magnet.

These conditions correspond to the situation encountered for in vivo ^{1}H studies of human brain.

For myoinositol, the main resonance (at 3.56 ppm) originates from a collapse of the H1, H3 and H4, H6 multiplets at 7 T. The intensity of the distorted H5 at 3.28 ppm is significantly reduced at 2 T, although the less strongly coupled H2 triplet at 4.06 ppm remains. In the taurine spectrum, the distorted N-CH_2 (3.26 ppm) and S-CH_2 (3.43 ppm) triplets at 7 T merge to an apparent quartet at 2 T, with effective chemical shifts of 3.26, 3.32, 3.38, and 3.43 ppm. The spectra in Figures 3.48 (c) and 3.49(c) [22] represent artificially broadened versions of the 2T metabolites in Figures 3.48(b) and 3.49(b) [22], to mimic the in vivo line widths and the influence of limited resolution. Obviously, the enhanced filtering by Gaussian multiplication (3.6-Hz line broadening) results in improved SNRs, while the peak intensities of the sharp resonances in Figures (3.48(b) and 3.49(b)) [22] drop when using the same amplitude scale. The lower parts of Figures 3.48 and 3.49 [22] offer a comparison between the line-broadened metabolite spectra and in vivo ^{1}H NMR brain spectra (8-ml VOI, TE = 20 ms) of a 4-year-old child, Figures 3.48(d) and 3.49(d) [22]. In fact, both inositol and taurine resonances are found to be higher in spectra obtained from children than in those from adults.

The spectrum in Figure 3.48 [22] was realized in the thalamus, whereas Figure 3.49 [22] refers to a location in the parietal WM of one hemisphere. It should be noted that the in vivo resonances of inositol and taurine partially overlap those from other metabolites. This particularly applies to the taurine resonances at 3.26 and 3.32 ppm, which in most cases are covered by the choline-containing signals at 3.22 ppm.

The weak resonances of inositol at 3.65 and 3.77 ppm overlap with the 3.68 and 3.75 ppm resonances of the less strongly coupled α-CH triplet of glutamine. It is further noteworthy that the in vivo resonance line widths are significantly broader than those obtained from metabolite solutions, under identical experimental conditions. Thus, the resolution of the human brain spectra presented here is determined by biological susceptibility differences rather than limited by magnetic field

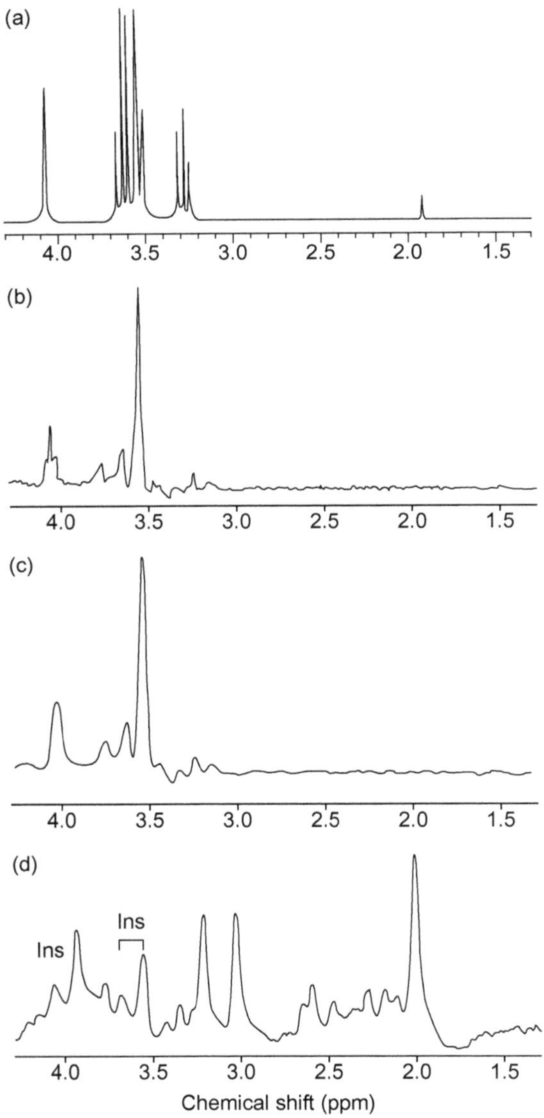

Figure 3.48 ^{1}H NMR spectra of an aqueous solution of myoinositol at (a) 7.0 T and (b, c) 2.0 T in comparison to (d) a 2.0-T ^{1}H NMR spectrum of human brain *in vivo*. (a) Fully relaxed FID spectrum of myoinositol (pH 7.1, 50 mM, 7.0 T, acetate reference); (b) fully relaxed, localized STEAM spectrum of myoinositol (8-ml VOI, TE = 20 ms, pH 7.1, 100 rnM in a 250-ml beaker, 2.0 T); and (c) line-broadened version of spectrum (b) to mimic the line widths obtained in spectra of human brain *in vivo*. (d) ^{1}H NMR spectrum of human brain *in vivo* localized in the thalamus of a 4-year-old patient without pathological findings (8-ml VOI, TE = 20 ms, TR = 1500 ms, 256 scans, 2.0 T).

inhomogeneities of the whole-body magnet. The origin of the in vivo inositol signal is expected to be largely due to myoinositol.

Significant contributions from inositol headgroups of membrane-bound phospholipids, or from inositol phosphate, can be excluded on the basis of their different ^{1}H NMR spectra, e.g., for the chemical shift of the H2 resonance at 4.06 ppm.

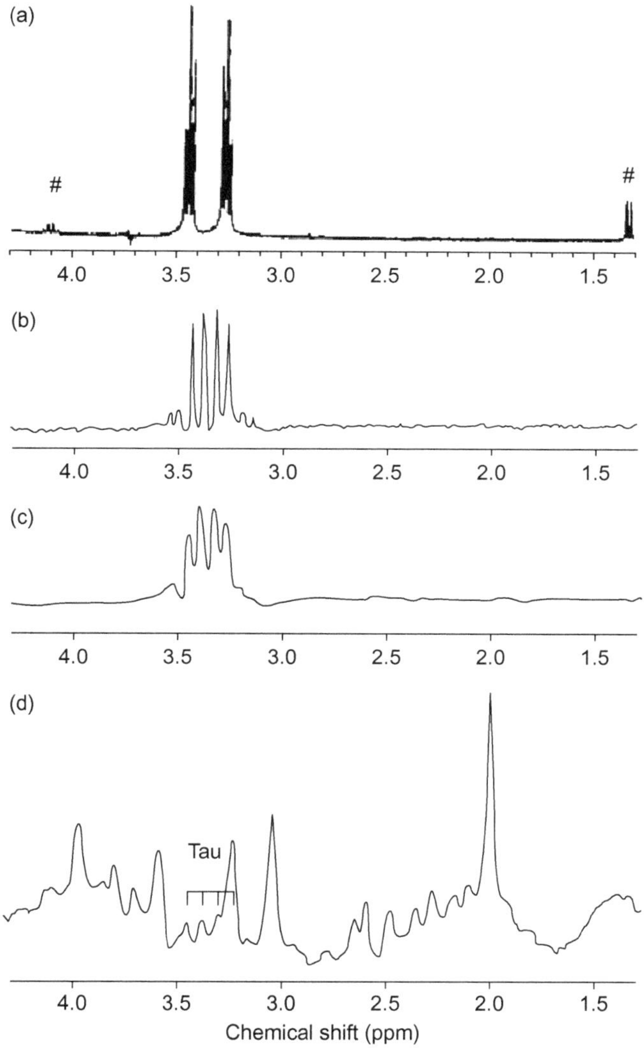

Figure 3.49 ^{1}H NMR spectra of an aqueous solution of taurine at (a) 7.0 T and (b, c) 2.0 T in comparison to (d) a 2.0-T ^{1}H NMR spectrum of human brain *in vivo*. (a) Fully relaxed FID spectrum of taurine (pH 7.2, 50 mM, 7.0 T, *lactate reference); (b) fully relaxed, localized STEAM spectrum of taurine (8-ml VOI, TE = 20 ms, pH 7.2, 50 mM in a 250-ml beaker, 2.0 T); (c) line-broadened version of spectrum (b); and (d) ^{1}H NMR spectrum of human brain *in vivo* localized in the parietal WM of a 4-year-old patient without pathological findings (8-ml VOI, TE = 20 ms, TR = 1500 ms, 256 scans, 2.0 T).

This finding is further supported by the fact that homonuclear off-resonance irradiation did not show a decrease of any ^{1}H NMR-visible metabolite resonances in the human brain, although the water signal was considerably reduced by a transfer of saturated magnetization from macromolecular structures to neighboring water molecules. Thus, the observed inositol resonances are not in close contact with immobilized membrane structures, as would be the case for phospholipid inositols.

One can see from these examples that the identification problems associated with alterations in ^{1}H NMR spectra at low fields versus high fields may be overcome by qualitative assignments through lineshape comparisons of spectra. The spectra obtained are of the metabolite solutions in vitro and of the human brain in vivo, respectively. However, because even minor variations in the field strength may cause spectral changes in the presence of strong spin–spin coupling, the present data are valid only for a field strength of 2 T. Another consequence of the result is that, with the prominent exception of lactate, homonuclear editing techniques with evolution periods on the order of $1/J$ will not be applicable to cerebral metabolites. This is due to their unavoidable T_2 losses and the incomplete refocusing of multiplets due to complex J modulation.

The identification and detailed understanding of the spectral appearance of a particular metabolite are prerequisites for a quantification of the absolute concentration of that metabolite in vivo. Current efforts, using a large number of volunteers, are aimed at establishing the biological reproducibility, regional distribution, age dependence, and physiological variability of brain spectra. In addition, alternative strategies are being tested for consistency of absolute metabolite concentrations. So far, relative concentrations of cerebral metabolites may be reported that are based on a comparison of metabolite spectra from fully relaxed (TR = 6000 ms) cases with ^{1}H NMR spectra of WM in the parietal lobe of young adults. Assuming a total creatine about 10 mm, the relative values are 18.9 mm for NAA, 3.2 mm for choline, 11.1 mm for glutamate, 7.0 mm for inositol, and 7.4 mm for taurine. The concentrations of glutamine and glucose are too low to be detected under normal conditions. Cerebral lactate varies depending on the physiological state (Table 3.20).

3.3.1.3 MRI Diagnostics: High-Grade Astrocytomas [23]

In vivo ^{1}H MRS has great potential to provide new variables and information for brain tumor diagnosis. With echo time (TE) values of 135 or 270 ms, three resonances are typically observed:

Table 3.19 Spectroscopic Peaks Assigned

Resonance	**Assigned to**
3.2 ppm	$N(CH_3)_3$ group of choline/phosphocholine/ glycerophosphocholine (Cho/PCho/GPCho)
3.0 ppm	NCH_3 group of creatine/phosphocreatine (Cr/PCr)
2.0 ppm	*N*-acetyl CH_3 group of *N*-acetylaspartate (NAA)
1.3 ppm	Lactate (Lac) (on the basis of its inversion at a TE of 135 ms)

Table 3.20 Chemical Shifts δ and Multiplicities m of ^1H NMR Resonances of Cerebral Metabolites as Detected in 7.0 T FID Spectra of Model Solutions and in 2.0 T Localized STEAM Spectra (TE = 20 ms) of Human Brain In Vivo

Metabolite	Molecular Group	δ^a, m^b	δ^c
Lactate	CH_3	1.33, d	1.33
Alanine	CH_3	1.48, d	1.48
NAA	CH_3	2.01, s	2.01
	CH_2	2.49/2.67, dd/dd	2.48/2.60/2.66
Glutamate	β-CH_2	2.09, c	2.00/2.05/2.12/2.18
	γ-CH_2	2.36, t	2.28/2.36/2.45
	α-CH	3.75, c	3.68/3.75/3.82
Glutamate	β-CH_2	2.13, c	1.97/2.06/2.16/2.24
	γ-CH_2	2.45, t	2.37/2.45/.255
	α-CH	3.77, c	3.70/3.77/3.85
PCr/Cr	CH_3	3.05, s/3.03, s	3.03
	CH_2	3.96, s/3.93, s	3.94
Cholines	N-$(CH_3)_3$	3.19−3.25	3.22
Taurine	N-CH_2/S-CH_2	3.26, t/3.43, t	3.26/3.32/3.38/3.43
Glucose	H_{2-6}	3.21−3.93, c	3.43/3.80
Myoinositol	H_5	3.28, t	3.28
	H_1,3/$H_{4,6}$	3.52, dd/3.62, t	3.56/3.65/3.77
	H_2	4.06, t	4.06

[a]Chemical shifts are given in ppm and referenced to 2.01 ppm for the CH_3 resonance of NAA.
[b]s, Singlet; d, doublet; t, triplet; dd, two doublets; c, complex.
[c]Major resonances of strongly spin-coupled multiplets are underlined. The spectral resolution corresponds to $k = 0.01$ ppm.

In vivo MRS of brain tumors shows a trend toward a decrease in the NAA/Cho ratio from normal or contralateral brain to low-grade astrocytomas to high-grade astrocytomas. However, most reports indicate that high-grade astrocytomas, notably glioblastoma multiforme or grade 4 astrocytomas (depending on the classification system used), have a resonance intensity at 2.0 ppm—much higher than expected from the extract studies. Methyl and methylene resonances, which may be due to mobile lipids, have been reported in high-grade astrocytomas. Neoplasms with cellular pleomorphism, vascular proliferation, necrosis, and pseudopallisading were called *grade 4 astrocytomas*. The pathological diagnosis is based on the surgical biopsy submitted for routine histopathology. In addition, two 5-μm sections obtained from the center of each sample investigated by MRS and stained with hematoxylon and eosin are evaluated. The amounts of viable and necrotic tumor and normal and GM/WM in the sections are estimated. Other histological features (e.g., vascular proliferation, thrombosed vessels, hemosiderin accumulation, degrees of calcification, surgical artifacts) are detailed. Figure 3.50 [23] shows absolute-intensity 1D spectra and Figure 3.51 shows 2D COSY spectra of a sample of tumor margin, consisting mainly of cortex with very little tumor (Figures 3.50(a) and 3.51(a) [23]) and 100% necrotic tumor tissues (Figures 3.50(b) and 3.51(b) [23]). The spectra of the tumor margin show resonances/cross-peaks from multiple

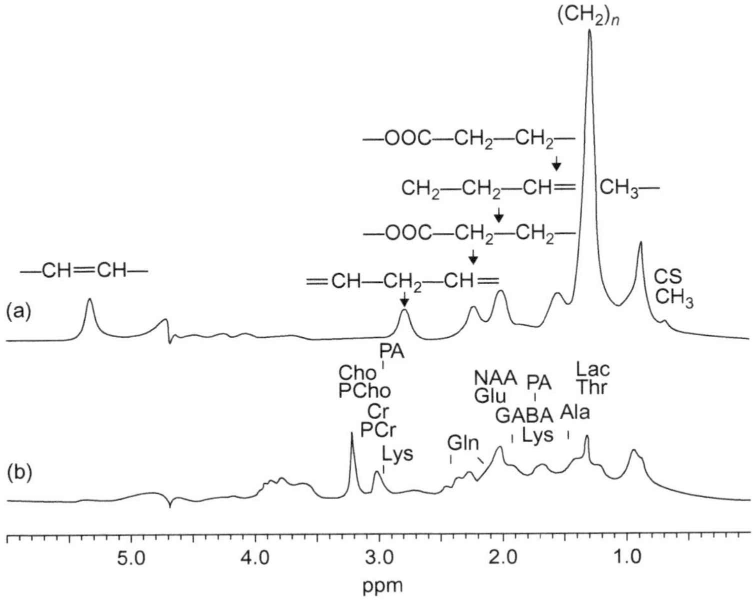

Figure 3.50 Absolute-intensity ^{1}H MR spectra, ex vivo, at 37°C, of (a) tumor margin with mainly cortex and very little tumor tissue (46 mg) and (b) 100% necrotic tumor tissue (35 mg). Abbreviations: Ala, alanine; Cho, choline; Cs, cholesterol; Cr/PCr, creatine/phosphocreatine; GABA, γ-aminobutyrate; Gln, glutamine; Glu, glutamate; Lac, lactate; Lys, lysine; NAA, *N*-acetylaspartate; PA, polyamines; PCho, phosphocholine; Thr, threonine. Assignments do not imply that these are the only substances contributing to a particular peak.

metabolites, assigned in Figures 3.50(a) and 3.51(a) [23]. In contrast, the spectra of the necrotic tumor tissue are dominated by resonance/cross-peaks from fatty acids and cholesterol, assigned in Figures 3.50(b) and 3.51(b) [23]. For samples for which only 1D spectra were acquired, the presence of mobile lipids was inferred from the intensity and lineshape of the resonances at 0.9, 1.3, 2.0, and 2.8 ppm, and the presence of the CH = CH resonances at 5.3 ppm. Figure 3.52 [23] shows spectra from two samples of 100% viable tumor and one sample containing 40% viable and 60% necrotic tumor.

The spectrum in Figure 3.52(a) [23] is quite similar to that of cortex at the tumor margin, Figure 3.50(a) [23]. It represents one of the two viable tumor samples in which no lipids were observed. The spectrum in Figure 3.52(b) [23] is typical of those obtained from samples containing mainly viable tumor.

The resonances of lipids at 5.3 and 1.3 ppm are intense relative to the resonances at 3.2 and 3.0 ppm. The spectrum in Figure 3.52(c) [23] is typical for those obtained from samples with high-intensity lipid resonances and a high percentage of necrosis. Comparison of the intensity of the lipid resonances with the histological evaluation indicates a correlation with the extent of necrosis in the sample.

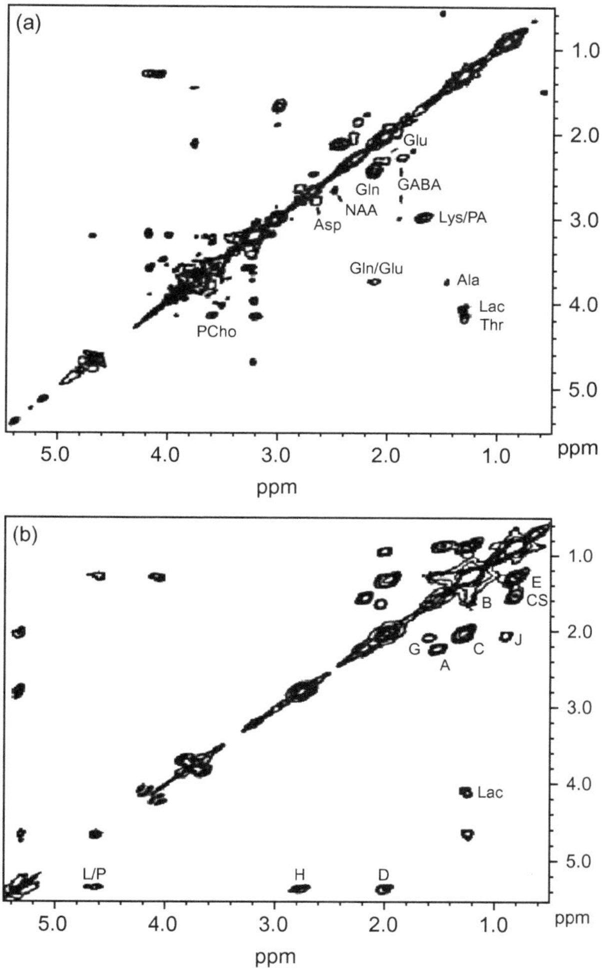

Figure 3.51 2D COSY spectra, ex vivo, at 37°C of (a) tumor margin with mainly cortex and very little tumor tissue (46 mg) and (b) 100% necrotic tumor tissue (35 mg). For abbreviations, see legend to Figure 3.50; assignments of fatty acid cross-peaks *A*–*P* in (b).

The 2D spectra of 16 out of 17 tumor tissue samples show that the broad 2.0 ppm signal in the 1D spectra contains resonances from lipids and glutamine (Gln), but only small, if any, contributions from glutamate (Glu), GABA, and NAA. This is consistent with results from perchloric acid extracts of malignant astrocytomas. One can see that the presence of lipid/fatty acid resonances distinguishes the majority of grade 4 astrocytoma samples from tumor margin samples (i.e., cortex and WM), and that the amount of mobile lipids increases with the extent of necrosis. Thus, necrosis is associated with an accumulation of mobile lipids considerably above the level present in viable tumor.

One cannot say with absolute certainty that activity of lipolytic enzymes after excision of the biopsies is responsible for the mobile lipids observed, though this is unlikely. The accumulation of large amounts of mobile lipids/fatty acids may not be restricted to astrocytomas. Samples from a malignant subependymal giant cell

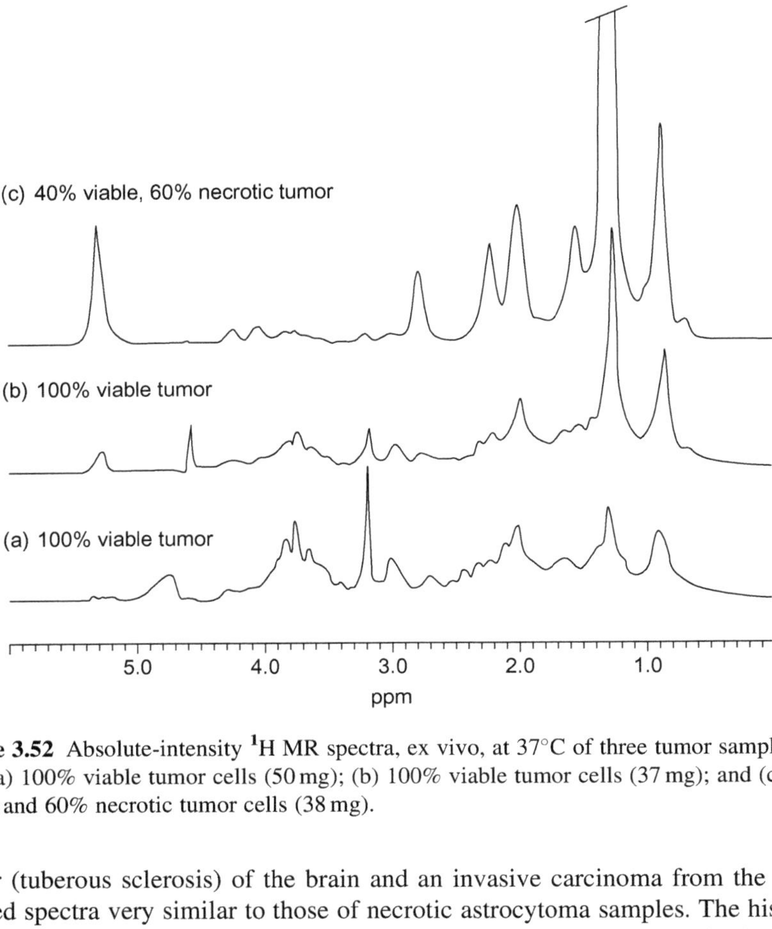

Figure 3.52 Absolute-intensity ^{1}H MR spectra, ex vivo, at 37°C of three tumor samples with (a) 100% viable tumor cells (50 mg); (b) 100% viable tumor cells (37 mg); and (c) 40% viable and 60% necrotic tumor cells (38 mg).

tumor (tuberous sclerosis) of the brain and an invasive carcinoma from the cervix yielded spectra very similar to those of necrotic astrocytoma samples. The histological reports on the NMR samples of the giant cell tumor and the surgical samples of the cervix carcinoma noted extensive necrosis.

Because metastatic carcinomas are frequently necrotic, they may be indistinguishable from grade 4 astrocytoma using in vivo ^{1}H MR spectra. It is also seen that the intensity of lipid resonances is higher in cultures of malignant tumors than in cultures of benign colon tumor cell lines. Thus, it is likely that accumulation of mobile lipids occurs in viable tumor cells but that it is accelerated in association with necrosis. Not only the amount but also the source, type, or environment of mobile lipids in viable and in necrotic cells may be different. This is indicated by the different phase-transition temperature of lipids in viable and necrotic tumor tissue. Thus, although the extraction and subsequent analysis of lipids by a multitude of methods is the easiest path, this method cannot avoid all the problems. The extraction destroys the supermolecular organization of all lipids in the tissue and thus makes it impossible to distinguish between those that were mobile MR-visible. Any alternative approach to better characterization of the lipids (e.g., NMR investigations of isolated membranes, lipid vesicles, electron microscope investigations) will have to preserve the supermolecular organization of the lipids as much as possible.

The data show a clear relation between lipid intensities and the extent of necrosis, but the range of intensities within each of the classes of necrosis defined is largely relative to the differences between the classes. Three factors limit the accuracy with which correlation between necrosis and amounts of mobile lipids can be quantified. The methods currently available for correlation are spectral quantification, histology-based estimates of necrotic and viable tumor, and the history of tumor. The lipid signals may be useful to discriminate high-grade from low-grade gliomas. The results suggest that low amounts of mobile lipids are present in viable brain tumor tissue and also show that tumor necrosis is associated with very high amounts of mobile lipids, which could be detected in vivo. Thus, there is a basis for linking the presence of lipid signals in the in vivo ^{1}H MR spectra to the presence of necrosis in the tumor. Necrosis distinguishes high-grade astrocytomas from low-grade astrocytomas in the majority of histological classification systems and correlates inversely with survival time. Hence, the detection of lipid signals in in vivo spectra could be of immense diagnostic and prognostic importance.

3.3.1.4 Imaging Diagnostics: PMRS [24]

One can develop a computerized prototype system for classifying ^{1}H single-voxel spectra, obtained at an echo time of 135 ms, of the four most common types of brain tumors: meningiomas (MMs), astrocytomas (ASTs), oligodendrogliomas (ODs), and metastases (MEs) and cysts.

Cancers of the brain can be loosely classified into the following three main groups: (1) meningeal tumors, (2) tumors of the glia, and (3) metastatic tumors that arise from cancers elsewhere in the body.

MRI plays an important role in classifying or differentiating lesion types. However, it is only a topographic and morphological technique, and contrast agents show only the blood–brain barrier breakdown that is characteristic of tumors and many other lesions. The ability of MRS to detect metabolites is very useful in daily radiological practice. As an example, finding acetate, lactate, and amino acids in a cystic lesion would almost certainly suggest an abscess, whereas the presence of lactate alone would indicate a cystic lesion in a tumor. Both these lesions have very similar imaging patterns. Different types and grades of tumor can also be characterized by their spectra. The primary potential role for pattern recognition in MRS would, therefore, seem to be concerned with noninvasive diagnosis and grading, particularly because the MR examinations can be repeated without hazard to the patient. To diagnose and grade tumors noninvasively, it will be necessary to have reliable methods for classifying their spectra. Methods for distinguishing between different groups, and for developing rules for classifying spectra of unknown origin, are provided by discriminant analysis. The rules are developed using a training set of data that has previously been classified, e.g., in the case of brain tumor spectra, by histopathological tests.

Each spectrum in the training set is represented by a set of features, such as peak height, peak area, or intensity rules. These features are then combined in such a way as to maximize the separation between and distinction among the classes in the training set. This is achieved by dividing the feature space into as many regions as there are classes. The boundaries between them—the *decision boundaries*—can be used to assign a spectrum whose class is unknown to a particular set. The decision boundaries (on which the rules are based) may be linear or nonlinear. Linear discriminant analysis (LDA) is generally more appropriate when the number of samples is small, and has given good results in a number of studies of MRS data.

LDA is available in most multivariate statistical software packages. The decision rules are formulated using discriminant functions (DFs), which are a linear combination of spectral features. Each spectrum is assigned a score, which is the weighted combination of its values of the discriminating variables. The decision as to whether it comes from one group or another is based on measuring the distance between its particular score and the mean (centroid) scores of different groups, and comparing the probabilities of its membership in each class. This is equivalent to determining which decision region the spectrum belongs to. If c is the number of classes, generally $(c - 1)$ DFs will be used. The DFs can be evaluated using a test set of individuals of unknown class. When the number of total individuals is small, it is normal to use the "leave one out" method to create the test set. That is, one uses all the spectra except one as the training set, and then uses the excluded case as the test set, repeating the process until all the spectra have been tested. LDA may also be used to select an optimal subset of features for discrimination, and most LDA programs provide methods for doing this. However, if the number of original features is large, as will be the case if the values of the spectral data points are used as variables, it is generally preferable to use another feature selection/reduction method before LDA.

To develop a standard pattern-recognition methodology for MRS spectra, one needs to take into account the standardized practices. The first problem that must be considered is the protocol to be used for acquiring the spectrum. It should be standardized, so that the unknown spectrum is comparable to the original database from which the training spectra were obtained (the training set). One can provide a practical program as part of the routine software on a clinical MRI/MRS instrument. At least three different methods are currently in use for ^{1}H MRS in the brain.

Both STEAM and PRES give single-voxel spectra, and CSI gives a matrix of spectra in one, two, or three dimensions (Figure 3.53). Each method has different characteristics that affect the resulting spectra. Furthermore, the choice of the repetition-time (TR) and echo-time values (among other parameters) will also influence the intensity of the signals from individual metabolites, as well as their lineshapes, couplings, etc. A simplest initial solution would be to define, exactly, the protocol used for obtaining the spectra in the training set, and to collect the unknown spectrum in the same way. In the longer term, it would be necessary to develop new training sets for each new MRS protocol. Another possibility is development (and validation) of procedures for correcting unknown spectra that were obtained with slightly different acquisition parameters. Changes in acquisition parameters that introduce simple linear changes in spectra (e.g., a small alteration in TR that induces a change in the degree of saturation of each peak) might be correctable. Indeed, as most metabolites tend to have similar T_1 values, the relative

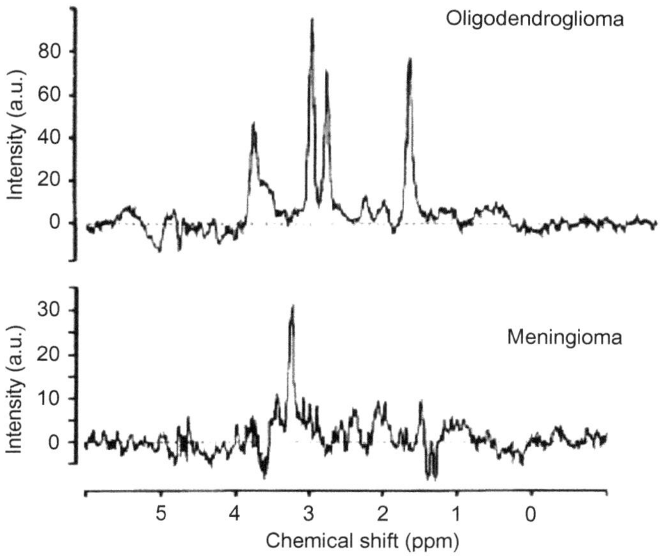

Figure 3.53 Two typical single-voxel spectra from the database, obtained and processed. Spectra are displayed before normalization.

intensities of their signals might be changed very little. A large change in TE, however, might cause complex changes in the peak shape, because of coupling effects that would be hard to predict, particularly when one is, in effect, dealing with mixtures of unknown substances. Differences in eddy current behavior in magnet/gradient sets could also introduce subtle changes in the ^{1}H spectra. Other conditions to consider are choice of voxel position, preprocessing of spectra (to achieve identical acquisition), residual water processing, phasing, frequency referencing, signal processing, normalization of spectra, and so on. The ultimate aim is to develop an automated system for classifying spectra according to tumor type. In summary, one has to create a mean spectrum for each group (set), by averaging the values of each data point from each normalized spectrum.

These mean spectra are very useful for reference and for preliminary identification of differences between classes (Figure 3.54 [24]). As to the number of features to be used in the classifier, a general rule of thumb that works well is to use no more than about $n/3$ features for developing the DFs. Here n is the number of samples in the smallest group involved in the comparison n. Following the steps explained earlier, a simple prototype classification system can be developed. This program takes FID as input, and outputs a discriminant score and classification for each pairwise comparison (Figure 3.54 [24]). The steps in this program are as follows:

1. Process the spectrum using signal processing.
2. Extract the intensity values of the selected data points.

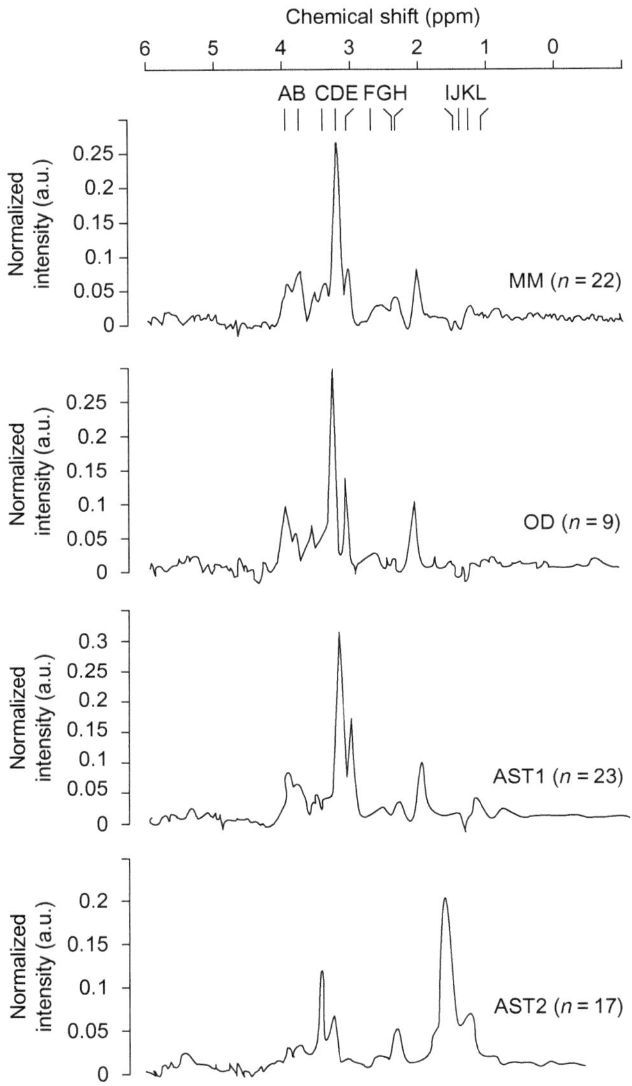

Figure 3.54 Mean spectra for each group used. The results of the pattern-recognition analysis for group descriptions. The capital letters identify the spectral features used for discrimination among classes (see Tables 3.21 and 3.22).

3. Calculate the discriminant scores by multiplying these values by the discriminating coefficients (training set data).
4. Using these scores, assign the spectrum to one of the two classes for each binary comparison.

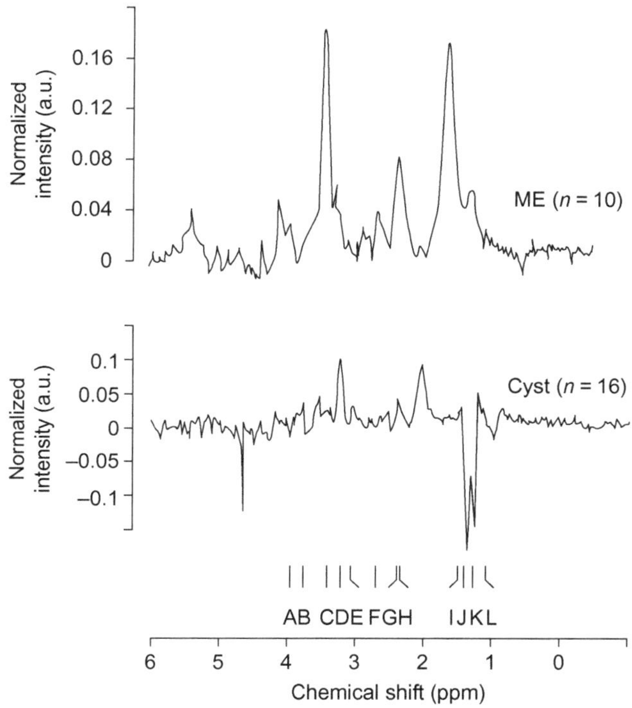

Figure 3.54 continued

These latest additions to the database (which were not included in the initial training set) were classified using the prototype program described above. The results are given in Tables 3.21 and 3.22 respectively [24] (Figure 3.55).

3.3.1.5 Diagnostics of Metastatic Brain Tumors: Lipids and Lactates (PMRS) [25]

It is very helpful to use gadolinium-diethylenetriamine pentaacetic acid-dimeglumine (Gd-DPTA) contrast enhancement in MRI/MRSI. This proves to be a very useful technique in studies of metastatic brain tumors as well. A T_1 (longitudinal FID relaxation time)-weighted MRI is carried out 5 min after the contrast agent Gd-DPTA (0.2 mmol/kg) is administered to the patient. After Gd-DPTA contrast, with inclusion of nonenhancing central regions, mean signal intensities and standard deviations from the largest transverse 6-mm-thickness MRI cross-section of each lesion are calculated. Post-Gd-DPTA MRI contrast (enhancement) is defined as the percentage difference between the mean signals in the largest postcontrast MRI cross-section of the metastasis in contralateral brain tissue. This characterizes the lesion as a whole.

Table 3.21 Key for the Labels Used in Figures. A Tentative Assignment for Possible Major Contributing Compounds Is Given When Available, Based on the Literature or Our Own Analysis of PCA Extract Data of Postsurgical Biopsies

Label	Chemical Shift (ppm)	Assignment
A	3.98	(P)Cr, PE
B	3.75–3.37*	Gln, Glu, Ala, Lys
C	3.41	Tau
D	3.23–3.24*	TMA, Tau, PE, Ino
E	3.05–3.10*	(P)Cr
F	2.70	Aspartate (Asp), FA
G	2.40	Gln, FA
H	2.32–2.36*	Glu, FA
I	1.48	Ala, FA
J	1.35–1.40*	Lac, FA
K	1.21–1.25*	FA
L	1.06	FA

Ala, alanine; Asp, Aspartate; FA, fatty acyl chain; Gln, glutamine; Glu, glutamic acid; Ino, myoinositol; Lac, lactate; Lys, lysine; (P)Cr, (phospho)creatine; PE, phosphoethanolamine; Tau, taurine; TMA, trimethyl-amine-containing compounds (e.g., choline, phosphocholine).
*Different spectral features in this ppm range were used for different class pairs.

Table 3.22 Classification Results for the Independent Test Set

	MM ($n = 2$)	AST ($n = 9$)	OD ($n = 2$)
ME ($n = 2$)	100%-(4/4) B,D,E,F	55%-(6/11)[a] D,E,F,K	100%-(4/4) E,L
OD ($n = 2$)	100%-(4/4) B,E,G,H	70%-(7/10)[b] D,H,I,K	
AST ($n = 9$)	82%-(9/11)[c] B,C,D,E,H,J		

The number of correctly classified samples and the total number of samples involved in each comparison are given between brackets. Capital letters identify the spectral features used for discrimination.
[a]5 AST (1 Al and 4 GL) incorrectly classified.
[b]3 AST (2 Al and 1 GL) incorrectly classified.
[c]2 AST (Al) incorrectly classified.

The central nonenhancing necrosis is reflected in lower means and higher standard deviations in the tumor pixel values. The volume of each metastasis is estimated from the dimensions in postcontrast MRI.

The acquisition of a sagittal localizer is followed by proton density and T_2 (transverse FID relaxation time)-weighted imaging. Double spin-echo PMRS, also referred to as PRESS, was used with TE = 135 ms when, due to J modulation, the doublet of lactate (Lac) is inverted (180° out of phase) relative to the signals of Cho, Cr, and NAA. The 2.56 ms sinc-shaped RF pulses were preceded by 25.6 ms Gaussian-shaped RF pulses for CHESS excitation and subsequent spoiling of the water signal. The second half of the spin-echo was collected using 1024 data

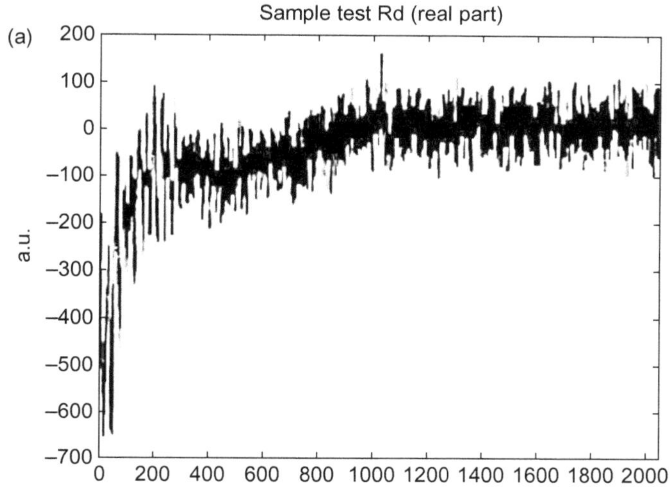

Figure 3.55 Diagram of the steps involved in the classification of an unknown sample (sample test) using the prototype program. The sample to be classified (a) is preprocessed (b) before Fourier transformation using the MRUI software. Automatic phasing and water filtering are achieved by applying the QUALITY and HLSVD algorithms, respectively. Discriminant scores (labeled as "*x*" in the discriminant plots) (c) are calculated for each comparison using the previously obtained DFs. In the case shown here, the sample is consistently classified in the program output (d, e) as a meningioma.

points, a spectral width of 500 Hz, 256 acquisitions, 4 prescans, and TR = 1600 ms (acquisition time 6.57 min). Quantitative analysis of patient spectra was confined to Cho (chemical shift 3.21 ppm), Cr (3.02 ppm), and NAA (2.01 ppm). Lactate (inverted doublet centered at 1.32 ppm with a coupling constant of 7 Hz) and lipid signals (0.9 and 1.3 ppm) were estimated roughly by correcting for contributions of normal brain tissue included in the MRS voxel, without correcting for saturation effects, lesion metabolite signals; peak integrals were normalized to NAA in contracentral occipital voxel. One uses Cho* = Cho − (1/2)NAA, Cr* = Cr − (1/2) NAA. The metabolite ratios measured in the occipital control voxels were similar to those encountered in healthy volunteers (Cho = 38±6 and Cr = 51±13 SD relative to NAA = 100). The brain tissue (and edema) directly surrounding the tumor do not have the same metabolite concentrations as the normal brain tissue. Also, there are differences between different regions of the human brain, as well as between adjacent WM and GM.

Brain metastases of lung cancers, mammary carcinomas, melanomas, and those originating from other tumors have similar Cho and Cr signals. These do not relate directly to intracellular metabolic concentrations, because highly variable amounts of necrosis are included in the tumor voxels. As with post-Gd-DPTA MRI contrast

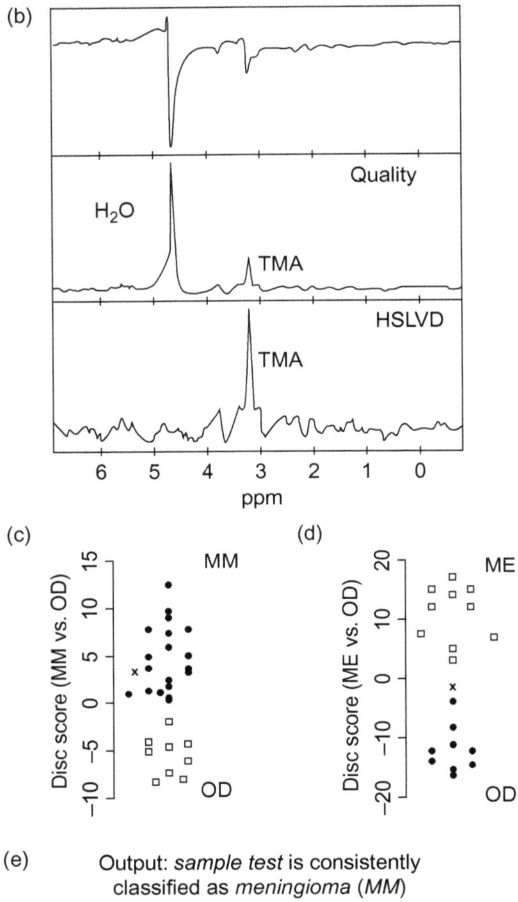

Figure 3.55 continued

enhancement, the values provide an average for the lesions as a whole. The central nonenhancing necrosis is reflected in reduced tumor metabolic signals. Information regarding metabolic tumor heterogeneity was obtained in only a minority of cases. In those cases, additional MRS measurements were performed on tumor subregions. Tumor heterogeneity (i.e., the presence of necrotic tissue) is reflected by the standard deviations in the MRI tumor pixel values, after administration of contrast agent. Metastases without lipid and/or lactate signals in the PMRS are comparatively small in size, with a post-Gd-DPTA MRI contrast of 33±5% and a Cho concentration higher than in contralateral normal brain tissue (Figure 3.56 [25]). Representative MR spectra of another group of patients are shown in Figure 3.57 [25]. In all metastases, combined Gd-DPTA enhancement was found to show significant linear correlation with Cho.

This correlation is improved by correcting for the contributions of normal brain tissue included in the MRS voxel.

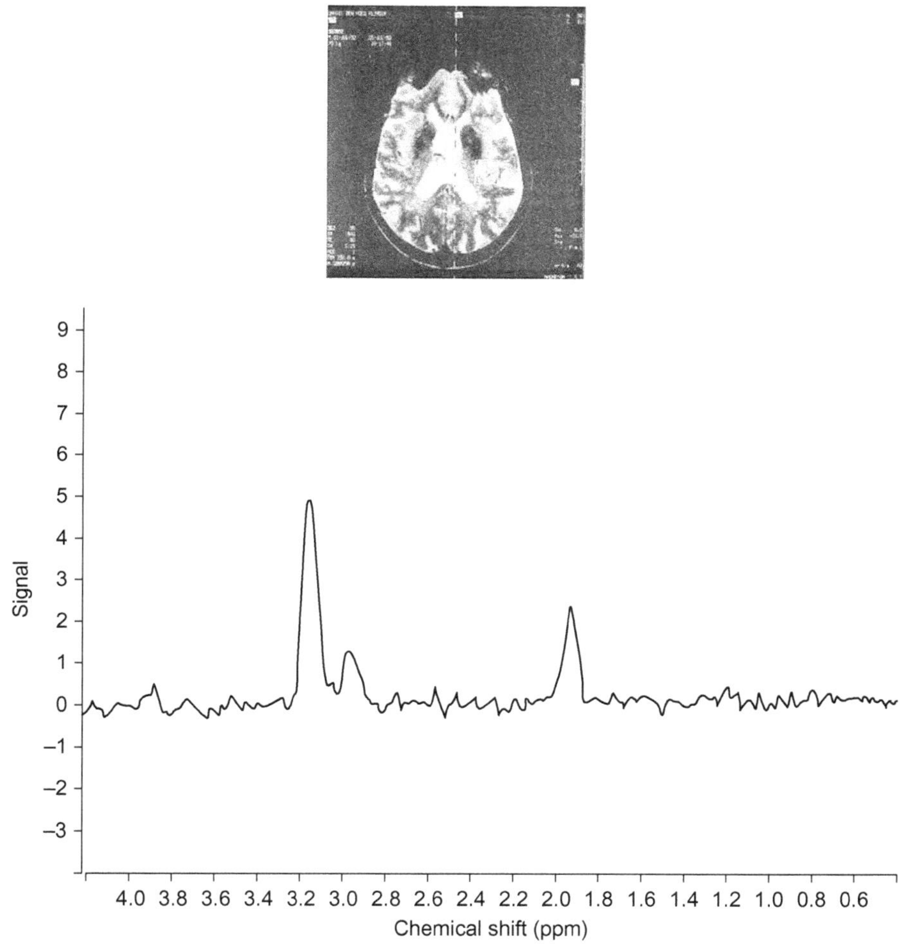

Figure 3.56 ^{1}H MRS spectra of 8-cm^3 voxels containing comparatively small amounts of metastatic brain tumor (lung carcinoma, 3.2 cm^3, above parts 1, 2) and occipital control tissue (below parts 3, 4), shown with the axial MRIs showing the location of the lesion and its appearance on postcontrast T_1-weighted MRI. Spectral assignments—Cho, 3.21; Cr, 3.02 (and 3.91); NAA, 2.01 ppm—relative to the unsuppressed water signal served as secondary chemical-shift reference (4.70 ppm).

3.3.1.6 Brain Tumor Diagnostics: Spin-Echo Point-Resolved Spectroscopy [26]

Spectroscopy promises new insights by adding metabolic information about tumors. One-third of brain neoplasms consist of gliomas. The histological classification is based on the origin of the cells and a grading scheme. Normally, one chooses rapid acquisition with relaxation enhancement (RARE) to obtain the reference image, because this sequence produces an image with excellent T_2

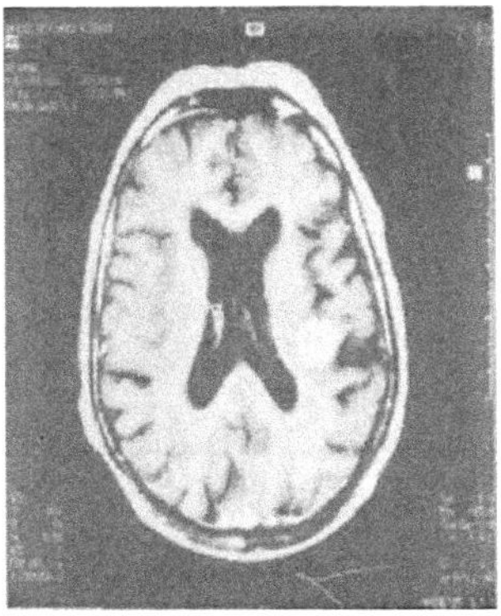

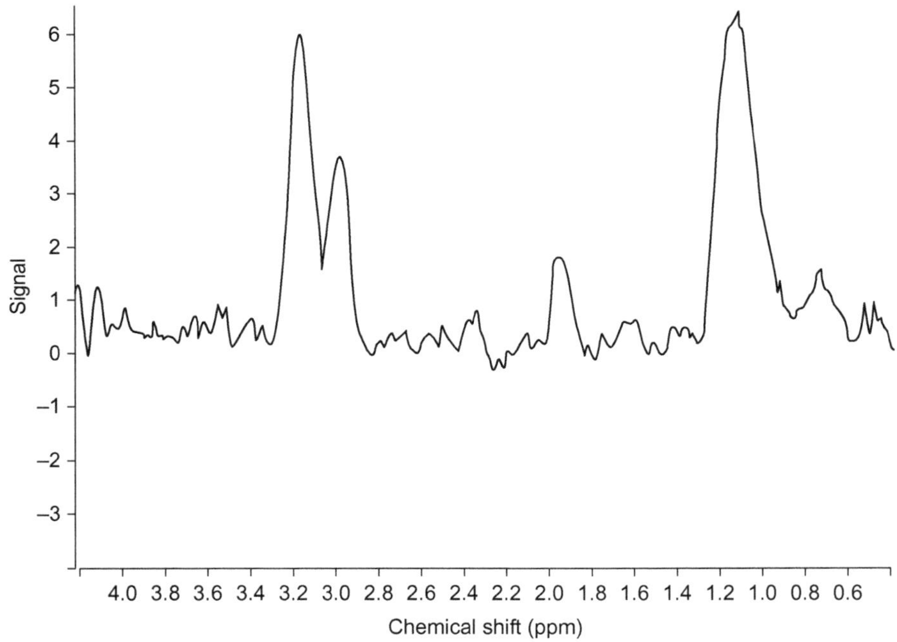

Figure 3.56 continued

contrast within a short acquisition time. Typical models for the examination were the acquisition of spectra from the nominal contralateral side of the brain as a reference and/or the acquisition of spectra from different locations in one tumor to investigate the spatial heterogeneity of the tumor. For optimization of the acquisition process, two spectra from different locations were obtained simultaneously, in an interleaved manner, by means of spectroscopy performed with tilted axes for bilocalized examination (STABLE). In two cases of suspected metastasis with meningiomatosis, positive findings upon cytological

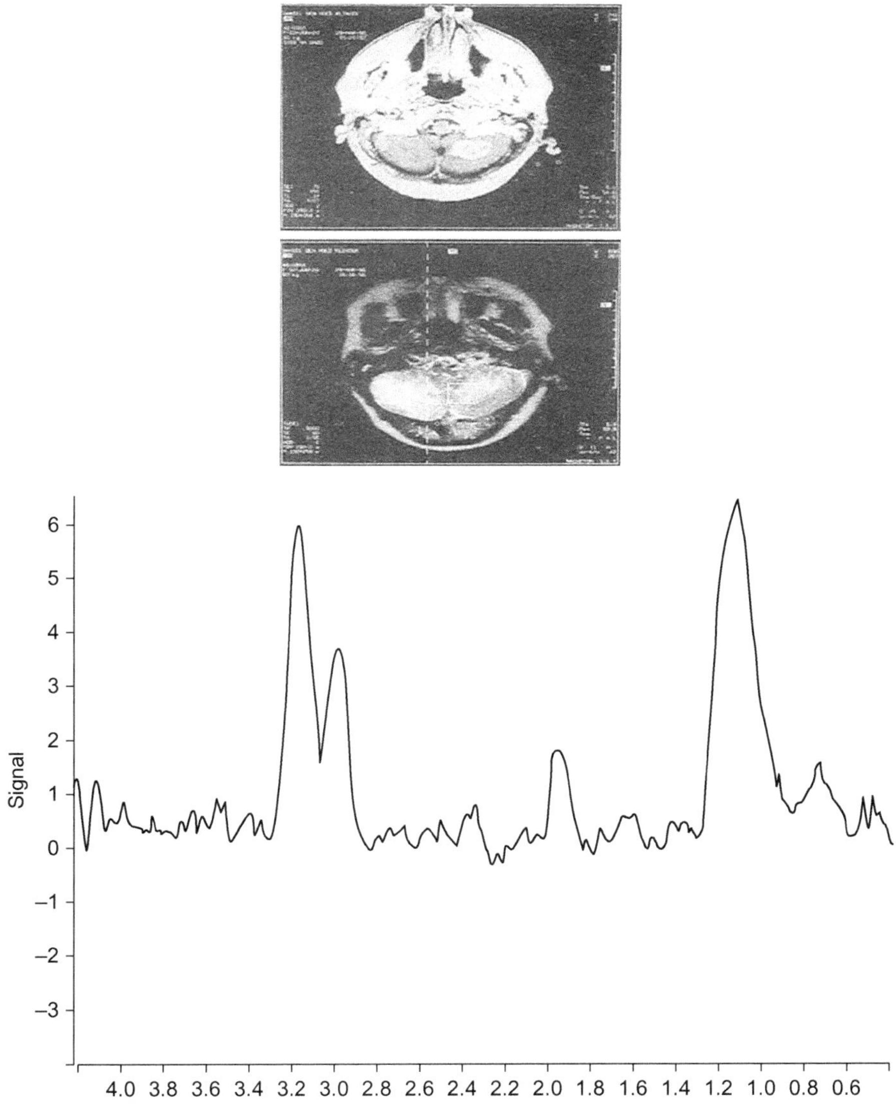

Figure 3.57 ^{1}H MRS spectra of 8-cm^3 voxels centered on a large brain metastasis (mammary carcinoma of 5.0 cm^3, parts 1, 2) and a very large metastasis (mammary carcinoma of 12 cm^3, parts 3, 4), displayed with the axial MRIs showing the location of the lesion and its appearance on postcontrast T_1-weighted MRI. Spectra; assignments: Cho, 3.21; Cr, 3.02 (and 3.91); NAA, 2.01; Lac doublets 1.26 and 1.36 ppm; lipids 0.9 and 1.3 ppm.

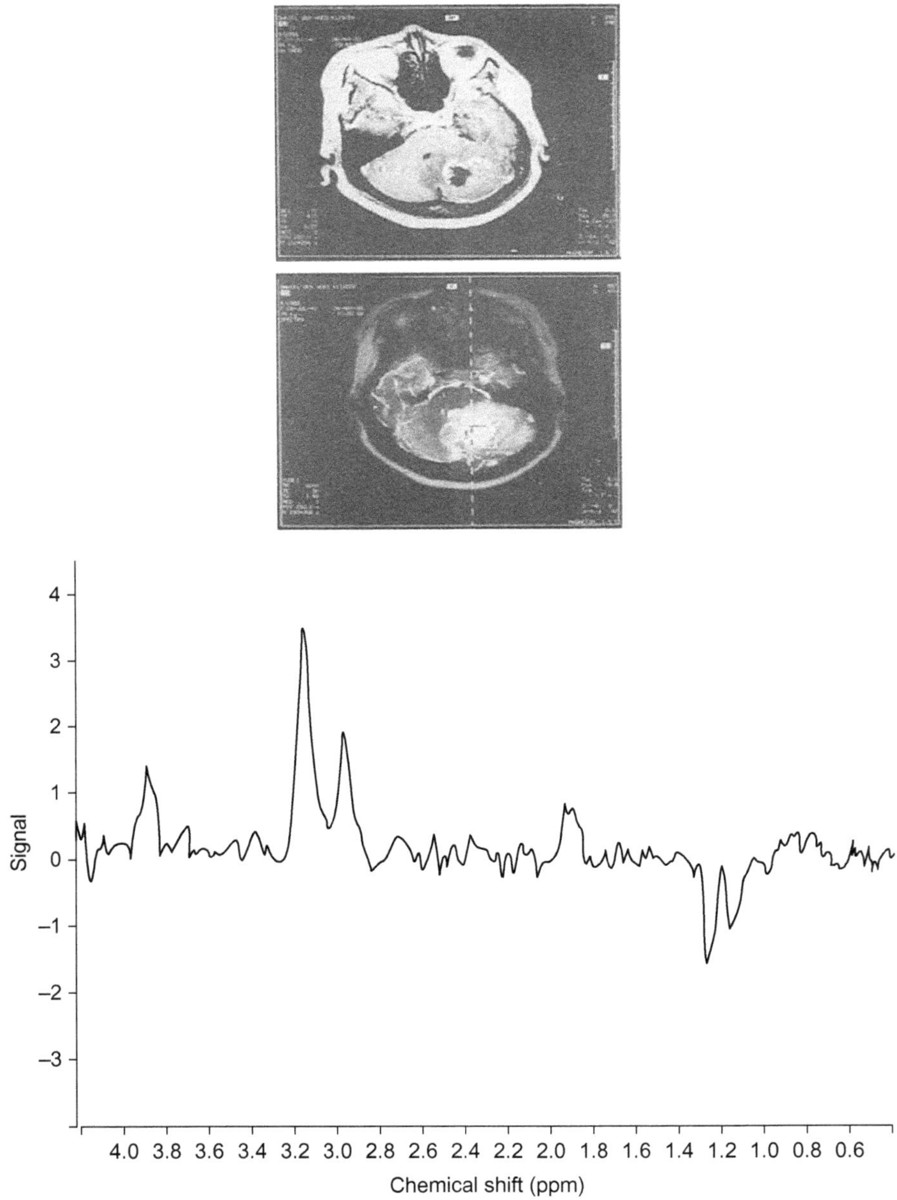

Figure 3.57 continued

examination of CSF provided the diagnosis. In the patients with glioma, a semi-quantitative approach was used, with planimetric integration of the various peaks. The NAA/Cho, phosphocreatine and creatine (Cr/Cho), lactate/Cr, and other detectable metabolite/Cr ratios were calculated. In spectra obtained with

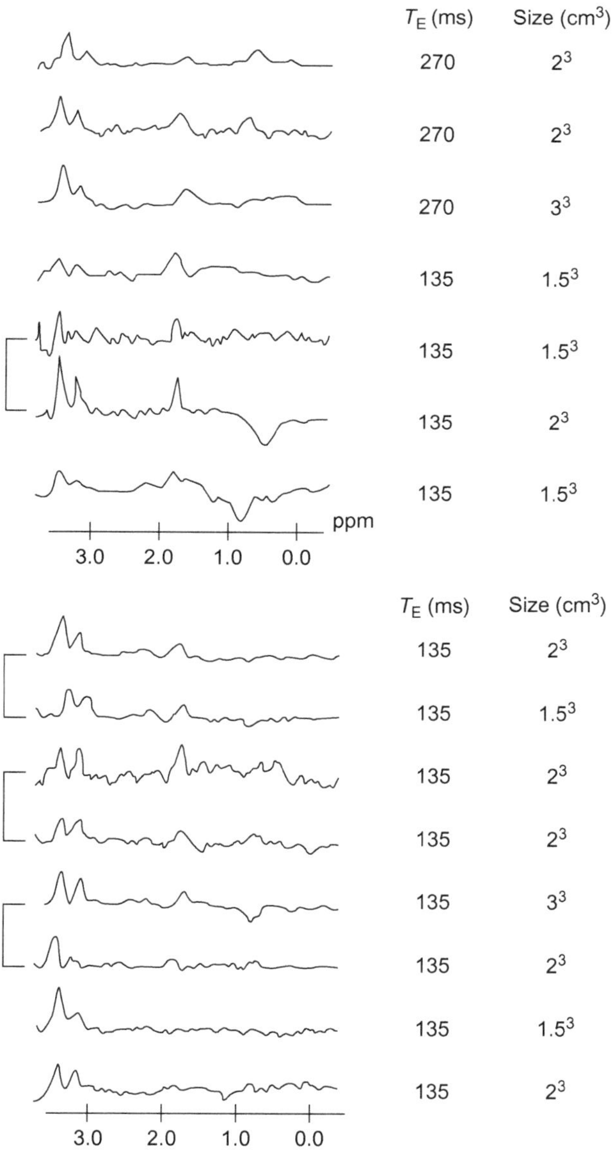

Figure 3.58 (a, b) Proton spectra of low-grade astrocytomas (WHO grade 2). The TEs (T_E) and the voxel size were due to local inhomogeneity (the voxel sizes are indicated to the right of each spectrum). The brackets at left connect spectra from patients with different locations in the same patient. ms = milliseconds.

STABLE techniques, one calculates the percentage of Cho, Cr, and NAA integrals within the tumor, compared with respective integrals of the reference tissue. Figure 3.58 [26] shows spectra from representative patients with low-grade astrocytomas (WHO grade 2, except one infratentorial astrocytoma classified as WHO grade 1). Figure 3.59 [26] is a T_2-weighted image of a grade 2 astrocytoma, with insertion of two different voxel sizes.

For grading of tumors, the classification discriminators are cellularity, presence of mitosis, necrosis, and hemorrhage. More recent neuropathologic examinations include cell kinetic studies that improve predictability of the tumor growth rate (Figures 3.60–3.63). Different tumor cells are not expected to differ in their overall metabolism detected by means of in vivo PMRS. This explains the lack of specificity of spectra for either histological subtypes or even general classification.

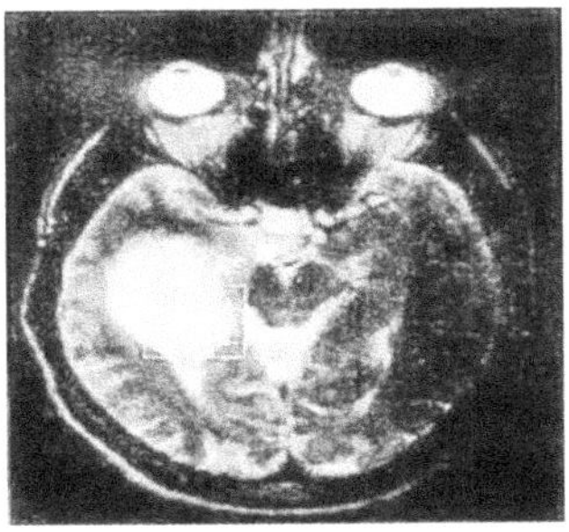

Figure 3.59 Reference T_2-weighted image of a low-grade astrocytoma with inserted concentric voxel sizes ($3\,cm^3 \times 3\,cm^3 \times 3\,cm^3$ versus $2\,cm^3 \times 2\,cm^3 \times 2\,cm^3$). The two spectra from these voxels are shown in the fifth and sixth rows from the top in Figure 3.58(b).

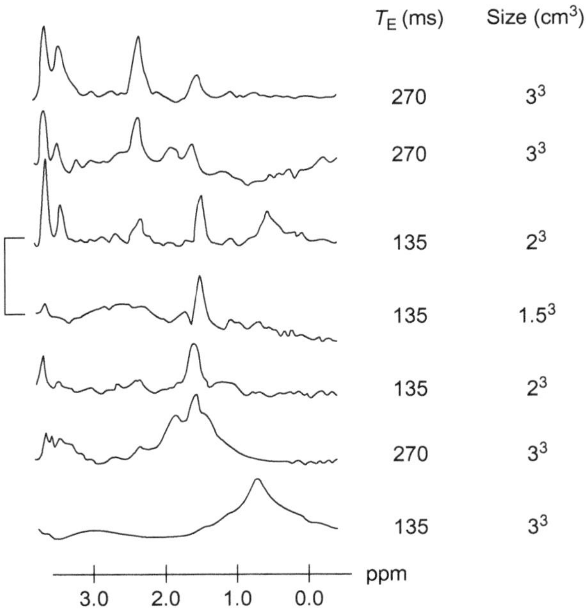

Figure 3.60 Proton spectra of glioblastomas. Spectra from different locations in the same patient are connected by the bracket. T_E (ms) = TE (ms).

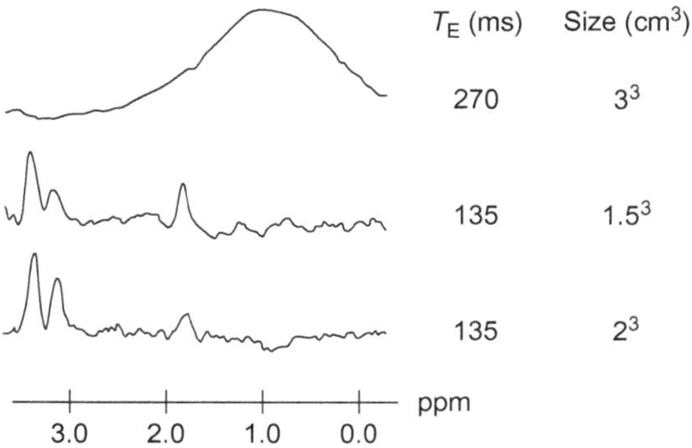

Figure 3.61 Proton spectra of anaplastic astrocytomas (WHO grade 3). T_E (ms) = TE (ms).

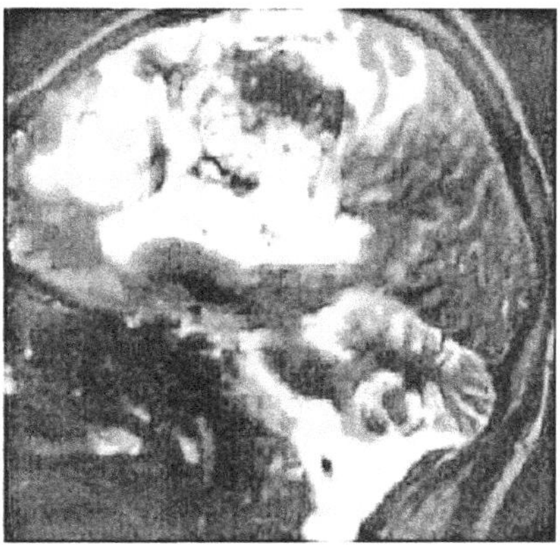

Figure 3.62 Parasagittal reference image of an anaplastic astrocytoma. This MR image was obtained by means of a RARE sequence with the same parameters.

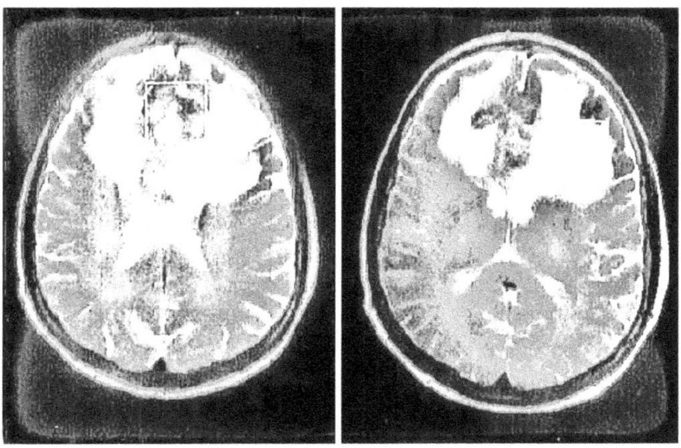

Figure 3.63 Reference MR images (obtained by means of a RARE sequence with the same parameters) of a meningiosarcoma and peritumoral edema with inserted voxels. The spectrum obtained in the image at left corresponds to the bottom spectrum [26]. The spectrum from edematous tissue did show lactate and a slight reduction in the level of NAA; otherwise it was normal.

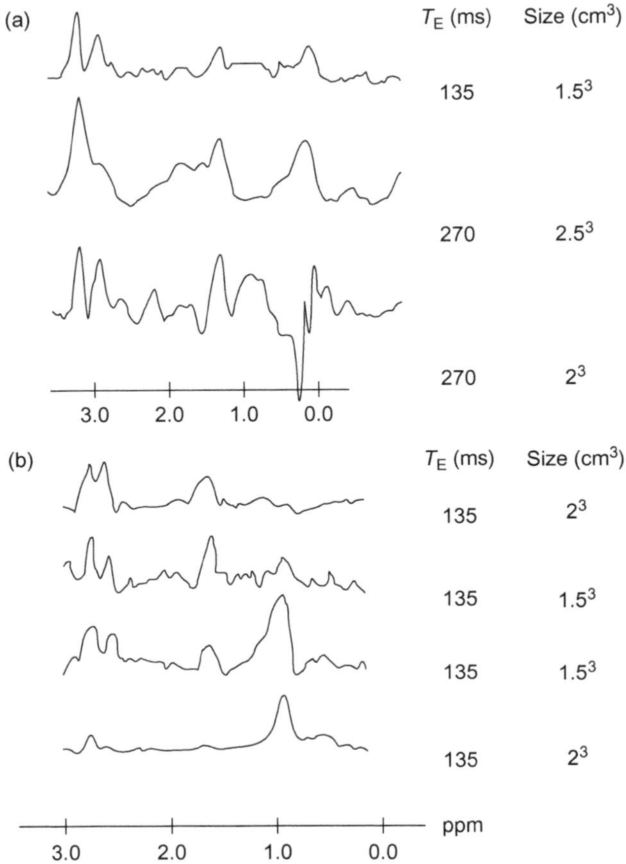

Figure 3.64 Proton spectra of metastases from (a) breast carcinoma and (b) bronchial carcinoma. T_E (ms) = TE (ms).

Examples of spectra from brain metastases are shown in Figure 3.64(a) and (b) [26]. The complete necrosis that occurs in fast-growing tumors of all histological groups, as well as in an abscess, can demonstrate similar spectra, as shown in one particular case in Figure 3.65 [26]. Thus, the differential diagnosis of a solitary mass with ring enhancement on CT scans or MR images must include glioblastoma, metastasis, or abscess, even after spectroscopic evaluation. One of the major drawbacks of in vivo PMRS, as opposed to specimen MRS, is the rather large voxel size, which makes it necessary to look at given spectra as an admixture of spectra of various tissue components.

Cell cultures and biopsy examinations have shown that NAA is more or less solely found in neuronal cells. Hence, NAA appears to be a valid indicator of the presence of functioning neurons. Therefore, any process that destroys neurons

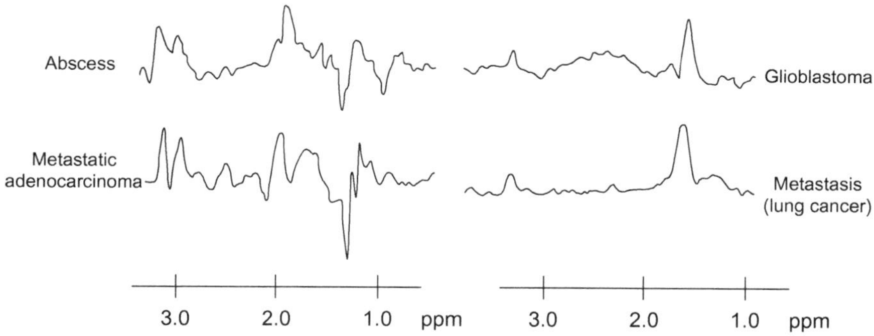

Figure 3.65 Comparison of proton spectrum of an abcess (top left) with that of a metastatic adenocarcinoma (bottom left) and proton spectrum of a glioblastoma (top right) with a metastasis from lung cancer (bottom right).

displays the same reduction of the NAA peak. The variable reduction in NAA can be explained on the basis of the fitting of the voxel to the tumor margins. For tumors only slightly larger than the voxel size, signal contamination from surrounding normal tissue occurs because of imperfect voxel definition. A second possibility for contamination of NAA from normal neurons within a tumor voxel lies in the growth pattern of gliomas. In low-grade gliomas, it is difficult to differentiate tumor margins from edematous neuron-containing parts of the brain that are otherwise normal. Imaging methods alone can only poorly define tumor margins. At present, the major contribution of PMRS seems to be provision of information about histologic grade. It has proven possible to obtain spectra with short readout times that reveal the presence of other substances, such as inositol, glutamine, and glutamate. Although the metabolic pathways for these substances are different in different types of cells, such as neurons glial cells or tumor cells, it is doubtful whether these differences are sufficient to improve the differential diagnosis of brain tumors.

3.3.1.7 PMRS Pattern-Guided Diagnostics: Human Brain Disorders [27]

Localized, in vivo MRS enables investigation of specific metabolites within tissue in situ. The combination of MRI and MRS is fast developing ways to enable routine study or examination of the human brain in a clinical environment. In the diagnosis of intracranial tumors, the major aim is differentiation according to their origin (cerebral or noncerebral) and their grade of malignancy as a prerequisite for planning adequate treatment. ^{31}P MRS can be used to investigate aspects of energy and membrane lipid metabolism of tumors in vivo. Clinically oriented studies have shown the differences in spectra originating from meningiomas, pituitary adenomas, and gliomas.

One major disadvantage of phosphorous in clinically used MR systems is its relatively low intensity, which restricts measurements in brain to tissue volumes no smaller than approximately 30 ml. This is required if one is to use methods for image-guided volume selection. PMRS, in contrast, is based on proton (^{1}H) spectroscopy that allows examination of considerably smaller volumes. The spectra of this (proton) nucleus give information complementary to that obtained with ^{31}P. In whole-body MR systems, PMRS primarily demonstrates signals of choline-containing compounds (Cho), phosphocreatine and creatine (Cr), nitrogen acetylaspartate, and lactate.

To prepare for a spectroscopic measurement, the homogeneity of the magnetic field is optimized over the selected volume, by observing the ^{1}H MR of tissue water measured with the spatially selective $90°-180°-180°$ Hahn double spin-echo (SE) sequence. Typical line widths (FWHM) of 2–4 Hz can be achieved. The same sequence can be used to acquire the spectra for suppressing the water signal; a frequency-selective inversion pulse at the water resonance precedes the volume-selective double SE sequence. It starts at the zero crossing of the water signal. The zero crossing time is adjusted individually for each spectrum. Inversion times are ~350 ms in healthy tissue. In tumors, inversion times are longer, at approximately 600 ms. This water suppression does not affect signals outside the range of the inversion pulses (0.5 ppm). Double SE times (TE) of 136 ms are used. This leads to an inversion of doublets within spin coupling, of about 7.35 Hz (e.g., lactate). Repetition time is 2 s, resulting in a total acquisition time of about 8 min. All spectra are brought to a spectral resolution of 0.977 Hz per point. The chemical shifts (δ) are expressed relative to trimethylsilylpropionate (set to 0 ppm), with NAA ($\delta = 2.02$ ppm) or Cho ($\delta = 3.21$ ppm) as internal standards.

The spectra are evaluated by comparing intensities of the (real) resonance signals. The peak areas are determined by calculating the product of peak height and the width at half maximum, assuming Lorentzian lineshape. The peak heights are determined relative to a flat baseline drawn through the noise. The peak intensities

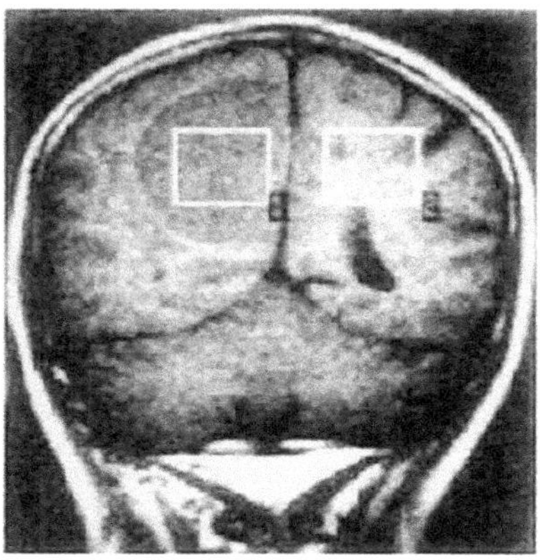

Figure 3.66 Coronal T_1-weighted SE image of a 53-year-old woman with a fibromatous meningioma. The 2.5 × 2.5 × 2.5 cm VOIs selected for spectroscopy are indicated.

are often given in relation to the signal of Cr, as this metabolite has been suggested as an internal concentration standard, assuming a normal concentration of approximately 10 mmol/l in brain tissue. In tumors, concentration of Cr is variable, and sometimes one detects no Cr at all. In the present case, peak intensities are therefore expressed in relation to Cho, even though its intensity varied as well, because Cho is detectable in all evaluated spectra and thus allows the calculation of finite ratios. In healthy brain tissue, the most prominent signals originate from the CH_3 groups of Cho at 3.21 ppm, Cr and phosphocreatine at 3.04 ppm, and NAA at 2.02 ppm, assigned according to their chemical shifts. A small peak of lactate, at 1.3 ppm, is visible in some spectra. Control spectra of nontumorous cerebral tissue were obtained from uninvolved patients. Refer to Figures 3.66–3.81 [27], for a statistical analysis of the data, and an appreciation of the techniques involved in spectroscopic-imaging correlations, in real cases (Tables 3.23 and 3.24).

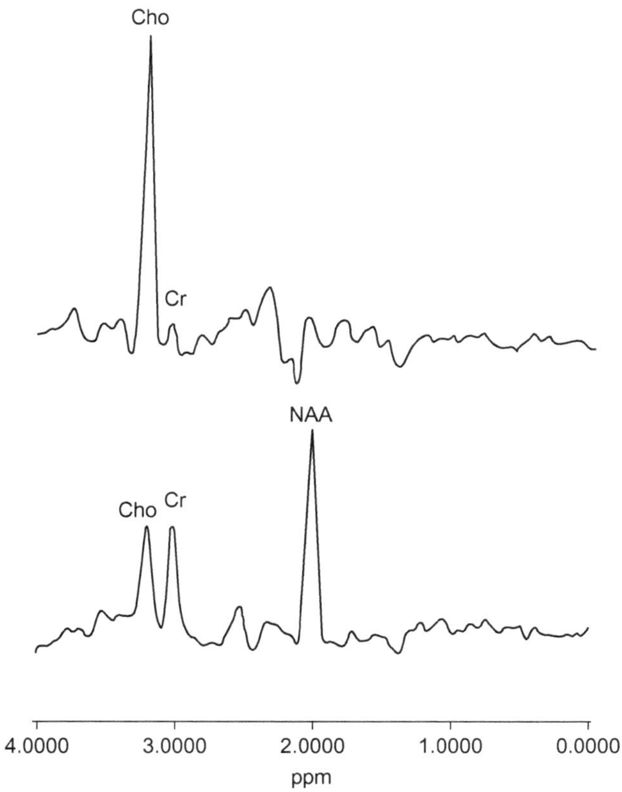

Figure 3.67 Proton MR spectra originate from the tumor (top curve) and from normal (bottom curve) tissue in the contralateral hemisphere. The most striking finding is a drastic increase of the Cho signal beside a strong decrease of the Cr signal and missing NAA.

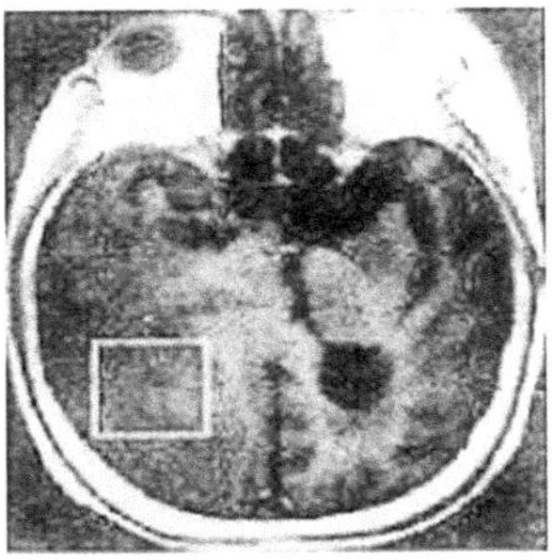

Figure 3.68 Transverse T_1-weighted SE image of an 81-year-old woman with an endotheliomatous meningioma.

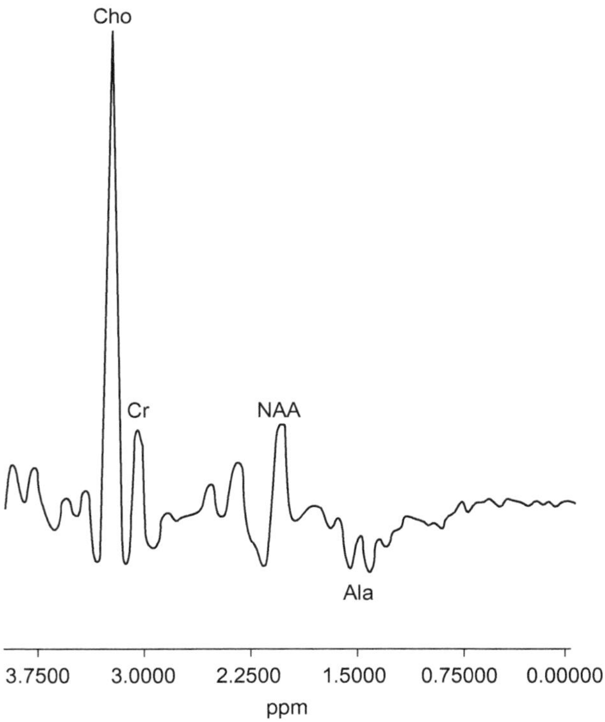

Figure 3.69 Tumor spectrum shows a small inverted doublet of alanine (Ala), increased Cho/Cr ratio, and a small remaining signal of NAA or another *N*-acetyl compound (NA).

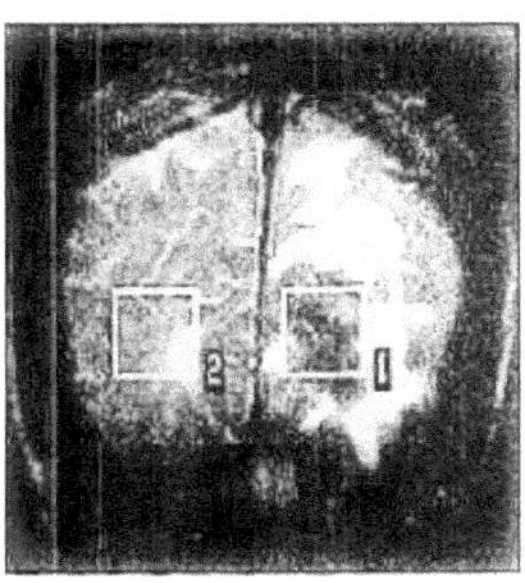

Figure 3.70 Coronal T_2-weighted SE image of a 65-year-old patient with a metastasis of a colon carcinoma.

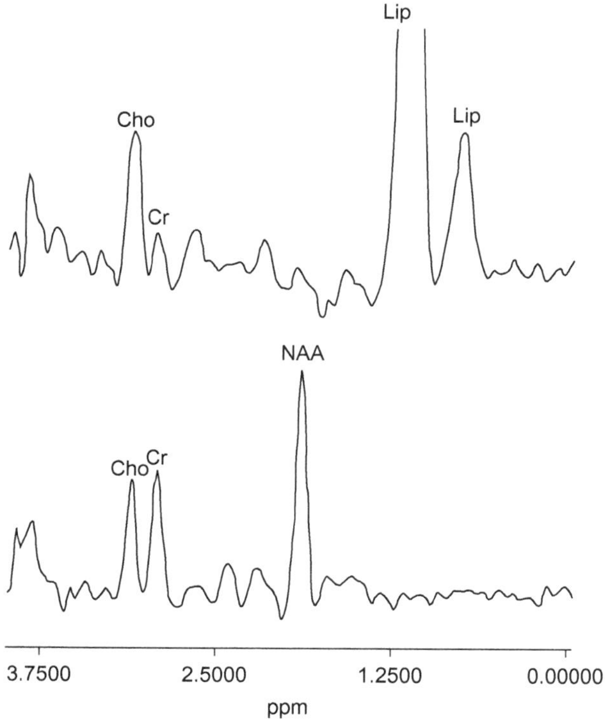

Figure 3.71 Spectra were obtained from volumes (2 cm × 2 cm × 2 cm) indicated on the image. Comparison of normal tissue (bottom curve) and tumor (top curve) demonstrates typical characteristics of noncerebral tumors: small Cr signal and no NAA signal. Typical for metastases are two signals assigned to mobile lipids; here the larger lipid signal at 1.3 ppm is truncated at approximately 40% of its full height. In the tumor, a small inverted signal originating from lactate appears at 1.3 ppm.

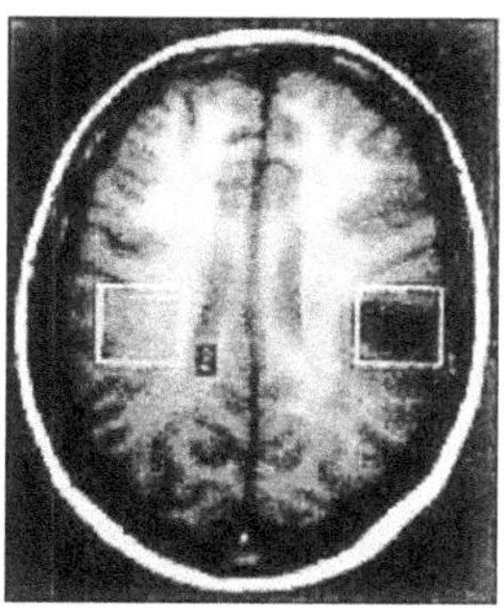

Figure 3.72 Transverse T_1-weighted image of a 31-year-old woman with a grade 2 astrocytoma.

3.3.1.8 Diagnostic of Human Brain Gliomas: PMRS and Positron Emission Tomography [28]

Evaluation of the regional glucose utilization rate (GUR) with fluorine-18 fluorodeoxyglucose (FDG) positron emission tomography (PET) has been shown to be valuable in the diagnosis and staging of brain tumors. Conceivably, technologies in which radionuclides are not used may help in appraising metabolic features and offer additional information for the diagnosis and management of brain tumors.

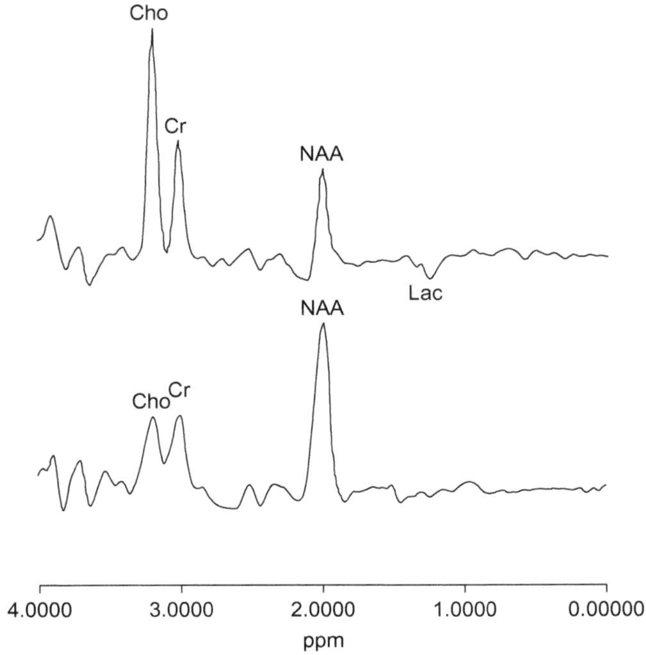

Figure 3.73 Spectra from normal tissue (bottom curve) and from the tumor (top curve) were obtained from the boxed areas in Figure 3.72. Cho is increased and NAA is decreased in the tumor compared with the contralateral hemisphere. Cr has the same intensity in both spectra. Because of less-than-perfectshim, signals in the spectrum of normal tissue are broader, resulting in a decreased signal height. In the tumor, a small inverted signal originating from lactate appears at 1.3 ppm.

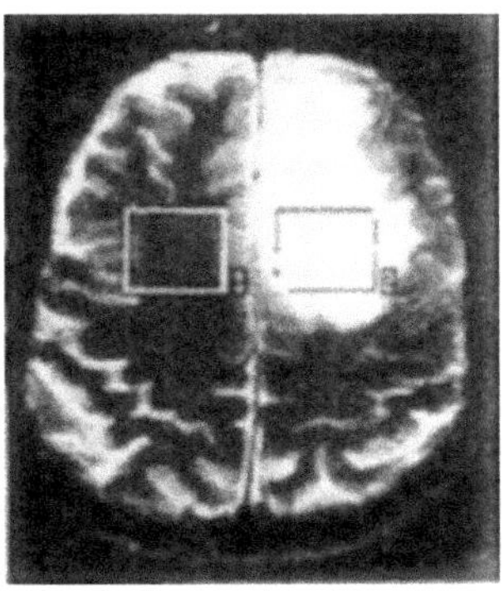

Figure 3.74 Transverse T_2-weighted image of a 54-year-old man with a grade 2 astrocytoma.

^{1}H (proton) MRS provides specific metabolic information about the brain tumors. There may be no systematic correlation between phosphorous magnetic resonance spectroscopy (PMRS) and FDG PET measurements, but comparative FDG PET and ^{31}P MRS studies have been carried out. The technique in ^{1}H MRS is similar to that of

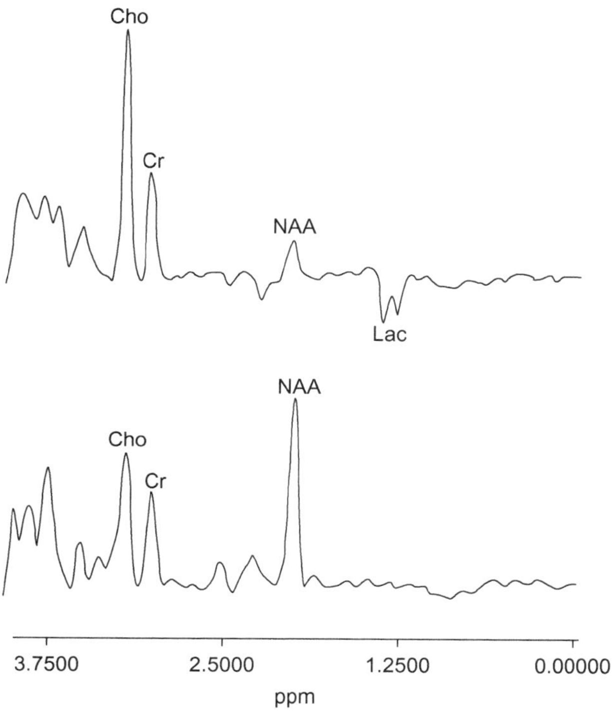

Figure 3.75 Spectra obtained from boxed volumes indicated in Figure 3.74 are of tumor (top) and contralateral normal tissue (bottom). Spectral changes similar to those shown in Figures 3.72 and 3.73 can be observed.

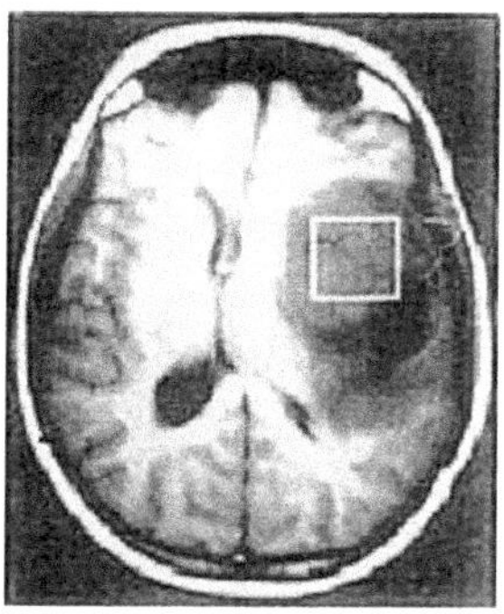

Figure 3.76 T_1-weighted SE image of a 29-year-old woman with a recurrent tumor 16 months after resection of a grade 2 astrocytoma.

the ^{31}P1 phosphorous resonance spectroscopy (PRS) that has been used for brain tumor analysis. In both, NMR signals from constituent nuclei of metabolic molecules are detected in situ. In the case of ^{1}H spectroscopy, it is known that signals from NAA, creatine-containing compounds, and choline-containing compounds can be detected with relative ease, by using MRI devices operating in the 1.5–2.0-T range. Lactate can also be detected if it is sufficiently elevated above its normal concentration in the brain. Figure 3.82 [28] illustrates the chemical structure of the detectable compounds and displays a simulation of the MR spectrum they produce.

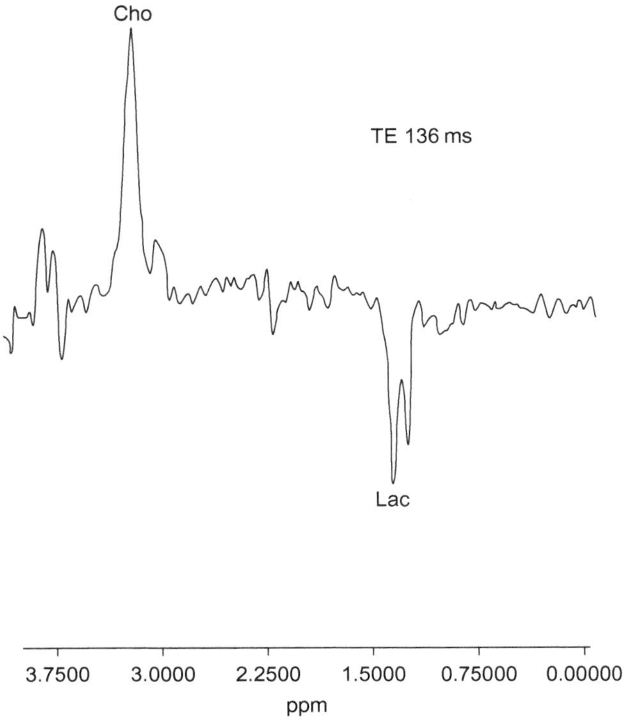

Figure 3.77 The spectrum of a pretreated tumor differs from that of other grade 2 gliomas in that no NAA is observable, Cr is strongly reduced, and there is a large inverted signal of lactate (Lac).

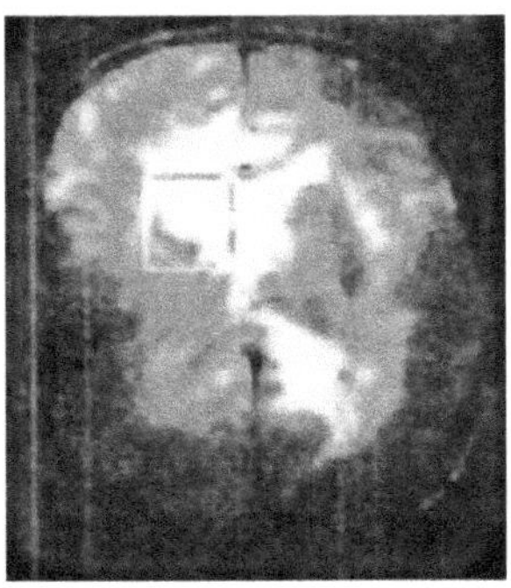

Figure 3.78 Transverse T_2-weighted SE image of a 47-year-old woman with a glioblastoma or a gliomatosis cerebra.

Other chemical constituents of the brain, such as glutamate, glutamine, aspartate, and alanine, may also be detected. Studies of acid extracts of brain tissue reveal that the ease of in vivo detection of these four compounds arises primarily from their high intracellular concentrations, their relatively long T_2 (TRT) values, and the fact that each has an MR signal that is not extensively split into multiple lines. The apparently

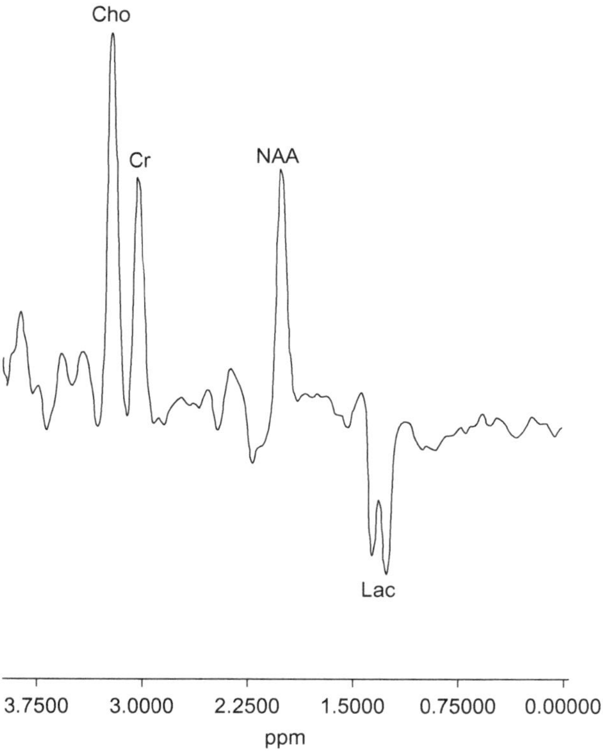

Figure 3.79 The spectrum from a lesion occupying the anterior space is an example of the first group of glioblastoma spectra, which have similar ratios of Cr/Cho and NAA/Cho as spectra of grade 2 astrocytomas. Additionally, this tumor exhibits a strong lactate signal (Lac).

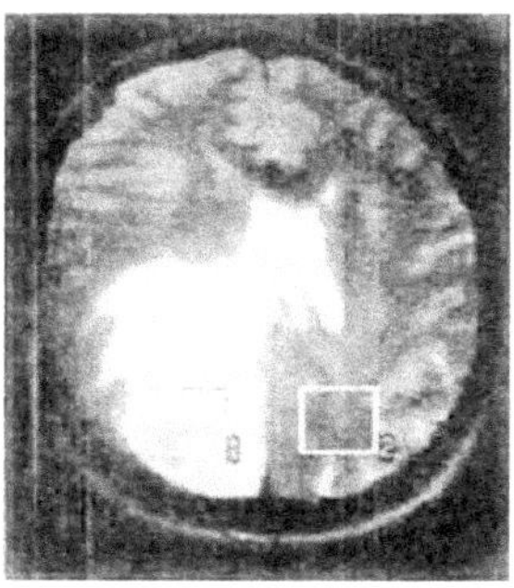

Figure 3.80 Transverse image of a 27-year-old patient with a glioblastoma.

short T_2 of the myelin lipid signal contributes to the ease with which it is detected in the brain, because it prevents the lipid signal from obscuring neighboring signals.

There have been efforts toward achieving a high degree of spatial localization of the signal sources. This has resulted in the introduction of a variety of techniques that are related to the section-selection and phase-encoding procedures commonly used in MRI. Such procedures are being incorporated into clinical imaging protocols. The simplest method, STEAM, collects one spectrum at a time from a single voxel, and is currently in use in many studies. The spectral imaging methods by

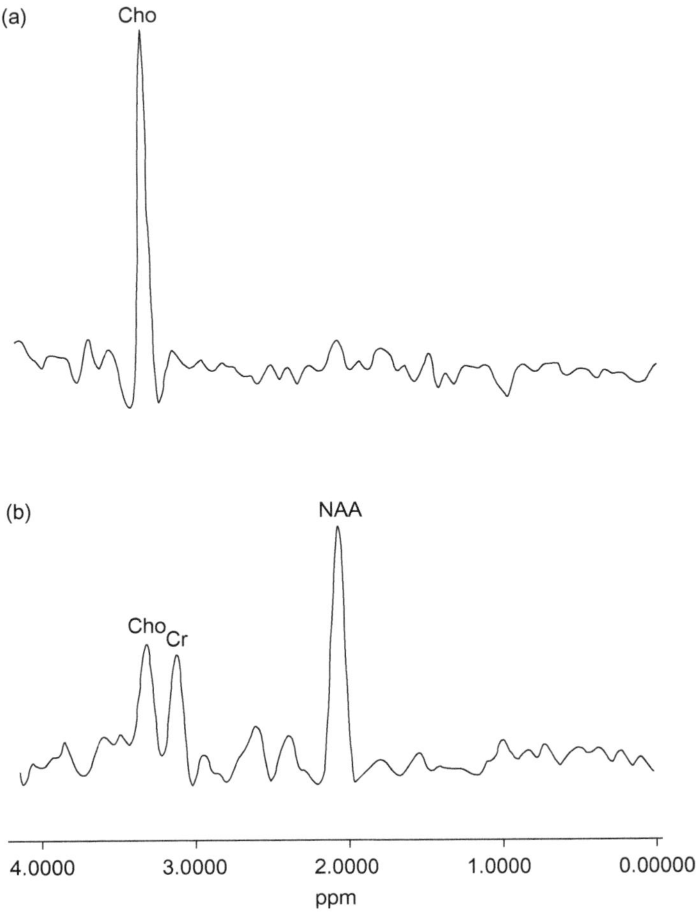

Figure 3.81 Spectra are from the tumor (a) and from the contralateral side (b) (boxed areas in Figure 3.80). This case is an extreme example of the second group of glioblastoma spectra, which show no NAA. The Cho signal of the tumor is doubled. Additionally, Cr is missing from this inhomogeneous, partly necrotic tumor.

which data are accumulated simultaneously from multiple voxels are much more advantageous. One should note that the minimum VOI that can be achieved in NMRS is limited by the SNR produced by the various metabolites. Thus, it is necessary to collect data from large volumes to make up for the weak signals produced by metabolites such as NAA, which has a concentration 4000- to 6000-fold lower than that of water. This requirement has limited localized ^1H MRS of the brain to voxels of $8\,cm^3$ or greater, when whole-head coils are used. Surface-coil detection results in better spatial resolution, provided the VOI is relatively superficial.

Studies have shown that human gliomas may have a GUR that is either increased (hypermetabolic) or decreased (hypometabolic) relative to surrounding

Table 3.23 ^{1}H MRS Signal Intensity Ratios Arranged According to Histological Tumor Types

Diagnosis	*N*	Age	MA	Cr/Cho[a]	NAA/Cho[a]	NAA/Cr	Lactate/Cho	NWA	NWL
Normal brain tissue									
Predominantly WM of healthy volunteers	27	19–32	25	0.90±0.2	2.0± −.05	2.2±0.5	+[b]	0	0
Contralateral sides of tumor of patients	12	27–31	60	0.94±0.25	1.79±0.46	1.95±0.52	+[b]	0	0
Noncerebral tumors									
Meningiomas	7[c]	32–81	51	0.17±0.19	0.06±0.08	0 to 2.4[d]	0	5	0
Metastases	3[e]	58–65	62	0.29±0.08	0.43±0.38	0 to 3.0	0	0	3
Acoustic neurinoma	1	50	...	0.19	0.46	2.38	0	0	0
B-cell lymphoma	1	67	...	0	0.10	...[d]	0.17[f]	0	0
Cerebral tumors									
Gliomas grade 2 (four astrocytomas, two oligodendrogliomas)	6	26–59	34	0.48±0.02	0.41±0.19	0.85±0.04	0.11–	0	0
Recurrent astrocytoma grade 2	1	29	...	0.17	0	0	0.66[g]	0	0
Gliomas grade 3 (two oligodendrogliomas, three oligoastrocytomas)	5	32–58	44	0.38±0.19	0.17±0.06	0.53±0.28	0.77[h]	0	0
Glioblastomas (both groups)	10	27–73	56	0.46±0.33	0.32±0.28	0 to 1.14[d]	0	3	1[#]
Group I[**]	5	47–73	52	0.54±0.12	0.54±0.21	0.97±0.24	0.11–		
Group[**]	5	27–61	57	0.36±0.47	0.10±0.07	0 to 0.56[d]	0.85[i]		

Note: Numbers are mean ± standard deviation.

NWM, normal white matter; A, NWM alanine; NWL, NW Lipid; MA, Median Age.

[a]Significance of tumors versus contralateral sides, $P < 0.005$; of gliomas grade 2 versus meningiomas, $P < 0.01$; of gliomas grade 3 versus meningiomas, $P < 0.1$.

[b]Small lactate signal detectable in some volunteers and patients.

[c]One atypical meningioma with high level of NAA not included in calculation of NAA/Cho (unlike metastases, one single signal at 1.3 ppm; grouping according to level of NAA signal).

[d]Ratio not defined in three meningiomas, one neurinoma, and one glioblastoma without Cr.

[e]One metastasis of a malignant melanoma not evaluated.

[f]$n = 1$.

[g]$n = 3$.

[h]$n = 1$.

[i]$n = 5$.

[#]Unlike metasatses, one single signal at 1.3 ppm.

[**]Grouping according to level of NAA signal

Table 3.24 Spectroscopic Imaging Correlations

Diagnosis	*n*	Age (Year)	Cho_t/Cho_n[a]	Cr_t/Cr_n	NAA_t/NAA_n
Meningiomas	2	53, 66	1/4/3.0	0/0.3	0
Metastases	2	63, 65	1.5/1.9	0.3/0.4	0/0.5
Gliomas grade 2 (three astrocytomas, one oligodendroglioma)	4	31–59	1.6–1.8[b]	0.9–1.4[b]	0.2–0.5[b]
Glioblastomas Group I	2	61, 73	1.0/1.4	0.5/1.0	0.3/0.7
Group II	2	27, 61	1.7/1.9	0/2.3	0.13/0.18

[a]Subscript t = tumor, subscript n = normal.
[b]Range of level times level in normal tissue.

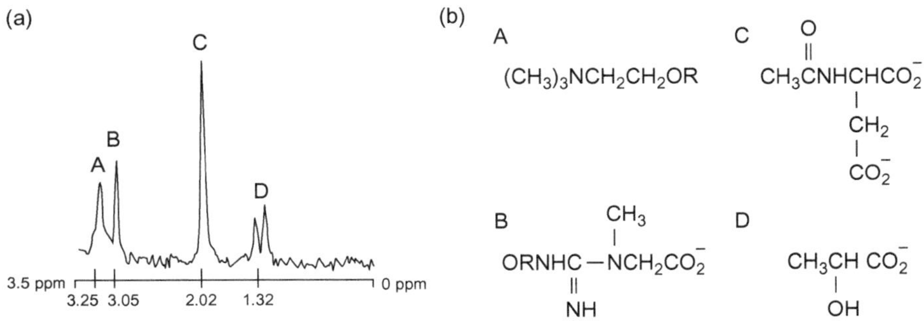

Figure 3.82 (a) Computer-simulated proton spectrum with the detectable metabolites and chemical-shift assignments. (b) Chemical structures of the detectable metabolites. The highlighted *H* in each chemical structure produces the resonances simulated in (a). A = choline, B = creatine, C = NAA, D = lactate.

brain tissues. The hypermetabolism has been attributed to an attenuation of respiratory metabolism and a derangement of the respiratory control of glycolysis, resulting in an increased dependence on the inefficient anaerobic glucose utilization pathway. This suggests that hypermetabolic tumors should exhibit relatively high rates of net lactic acid production. However, studies suggest that high-grade brain tumors, usually hypermetabolic, are not acidic. Multiple-section T_1- and T_2-weighted MR images are obtained after administration of 0.01 mmol/kg of gadopentetate dimeglumine. At least 24 h was allowed for clearance of the paramagnetic contrast agent before ^{1}H MRS was performed, to prevent artifacts associated with magnetic field inhomogeneity. The gradient-echo images proved useful for visualizing the location of the lesions and for identifying magnetic field inhomogeneities due to bone and hemorrhage. The magnetic field homogeneity over the entire head was optimized by maximizing the amplitude of the unlocalized water ^{1}H MR signal. Localization of rectangular volume was achieved with one of the two modified STEAM pulse sequences. These sequences produced three section-selective RF pulses that, together with additional gradient pulses, caused the nuclear spins at the

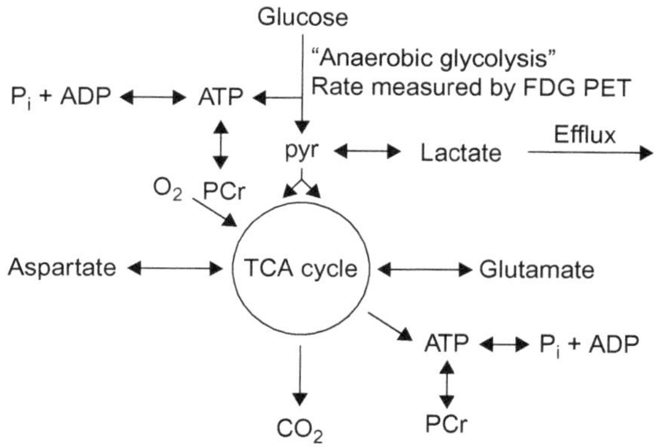

Figure 3.83 Illustration of cerebral glucose metabolism as viewed with MRS and FDG PET. The GUR is measured with FDG PET. The ^{1}H MR spectroscopic signal intensity areas are proportional to the steady-state levels of lactate and other compounds. Adenosine triphosphate (ATP) is produced through an anaerobic pathway (glucose pyruvate [pyr]-1 lactate) and through a respiratory pathway (pyruvate → carbon dioxide via the TCA cycle). In normal aerobic tissues, glucose is oxidized completely to carbon dioxide, and lactate is maintained at a low concentration not easily detected with MRS. Tumors of the CNS often exhibit abnormally high rates for the anaerobic pathway and abnormally low rates for the respiratory pathway. As a result, lactate becomes an end product of glucose utilization. However, the lactate formed must be eliminated either by the slowed respiratory pathway and/or by efflux (passive or active transport). ADP = adenosine diphosphate, PCr = phosphocreatine, P = inorganic diphosphate.

intersection of the three chosen planes to form a simulated echo signal. Water suppression is achieved in this sequence by using highly frequency-selective pulses and the appropriate gradient pulses before the section-selective pulses. PET was performed 30 min after the intravenous injection of 5–10 m Ci of FDG. The FDG PET scans were analyzed both visually and quantitatively. For both the visual and the numerical assessments, the WM of the centrum semiovale was used for the reference GUR, because gliomas arise from the neuroglia that provide the bulk of the WM GUR. A tumoral lesion was regarded as hypometabolic (Figure 3.84(d) [28]) if its GUR was less than that of the contralateral centrum semiovale. A hypermetabolic lesion was characterized by a GUR greater than that of the reference WM (Figures 3.86(d) and 3.87 [28]).

Examples of the correlation between GUR and the lactate signal intensity levels in gliomas are shown in the Figures 3.82–3.87. In each clinical examination, FDG, PET, and MR spectroscopic data are presented. Figures 3.82–3.87 demonstrate that lactate signal intensities can be detected in cases in which the tumoral GUR is lower than that of the surrounding brain. The lactate signal intensity elevation persisted in two separate examinations over a period of 9 months. An opposite finding is illustrated in Figure 3.84, showing a low-grade glioma with no detectable lactate signal intensity.

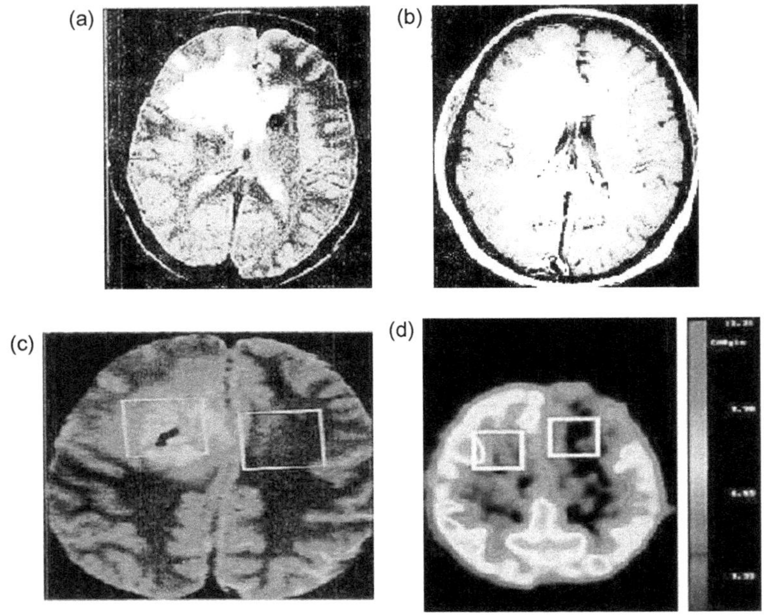

Figure 3.84 MR images and spectra from an untreated low-grade astrocytoma. (a) Axial T_2-weighted image (1.5 T, SE 2000/80) shows tumor in the right frontal lobe, extending across the midline. (b) T_1-weighted image after gadopentetate dimeglumine administration (1.5 T, SE 600/20) demonstrates line-broadening (LB) 2-Hz contrast enhancement of the tumoral mass. (c) Multiplanar gradient-echo image (600/30, 10° flip angle) from the middle section, through the 27-cm^3 MR spectroscopic VOIs (squares) of the tumor and mainly normal contralateral brain. (d) On this and all subsequent FDG PET scans, the patient's right appears on the reader's right, and the approximate LB: 2-Hz locations of the MR spectroscopic VOIs are outlined by squares. Note hypometabolism of the WM and cortex in the right frontal lobe (the peak GUR in the tumor VOI is 0.8 mg/min, and the GUR in the control WM is 2 mg/mm). (e) ^{1}H MR spectra from the lesion (bottom curve) and normal control tissue (top curve). Both the choline (3.25 ppm) and creatine (3.05 ppm) levels appear to be elevated, and the NAA peak (2.02 ppm) is markedly decreased in the lesion spectrum compared with that in the control spectrum. Lactate is clearly present at 1.33 ppm in the lesion. In this and subsequent figures, all lesion and control spectra are scaled for direct comparison. Also listed are the type of pulse sequence used (TR/TE/number of averages), VOI location, and the amount of line broadening used in processing. VOI locations are given in millimeters relative to the magnet reference system. A = anterior, P = posterior, S = superior, I = inferior, L = left, R = right.

3.3.1.9 Diagnostics of Changes in Acute and Subacute Cerebral Infarctions (PMRS) [29]

MRSI allows visualization of metabolic changes in stroke with reasonable spatial resolution. T_2-weighted spin-echo MR images are highly sensitive for the detection of ischemic brain lesions. It has been shown that the enhanced signal intensity in

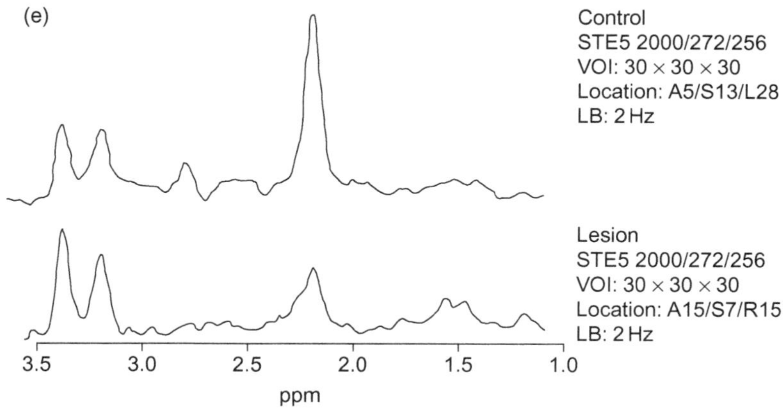

Figure 3.84 continued

such areas is due to a variety of pathologic tissue changes, ranging from colliquative necrosis to edema. Although differences of signal intensity may be detected within larger infarcted regions, information about the specific state of the brain tissue cannot be drawn directly from the image. The method of investigating specific aspects of brain metabolism in infarcts is localized in PRMS. Single-volume studies, as well as studies using 1D spectroscopic imaging and 2D MRSI, show loss of *N*-acetyl compounds like NAA and a reduction in choline and total creatine in acute infarctions. A marked increase in lactate is generally observed in the infarcted area. The NAA signal may indicate the presence of neuronal tissue, and there are signals from compounds containing trimethyl amino residues (often referred to simply as choline). The latter are related to membrane metabolism, and creatine is related to energy metabolism. Lactate accumulates if its precursor, pyruvate, which is produced by glycolysis, is not used as a substrate for oxidative energy production in the Krebs cycle.

One can use PMRS imaging to obtain further information about the spatial distribution of spectroscopically visible metabolites in normal and ischemic human brain tissues. The data can be presented as a set of spectra from individual volume elements or as metabolic maps to display distributions of specific metabolites within the section of interest. Here, one discovers the advantages of spectroscopic imaging in acute and subacute cerebral infarcts. Spectroscopic imaging permits evaluation of the entire extent of a pathologic lesion and comparison with contralateral brain tissue. Of particular interest is the extent to which such spectroscopic changes correspond to "conventional" imaging findings and whether spectroscopic data permit or support allocation of different regions, within large infarcts, to different extents of brain alteration.

Standard MRI is performed before spectroscopic measurements are obtained to define a transverse 2–2.5-cm-thick section comprising the ischemic lesion. The spectroscopic images are obtained by using a combination of localized excitation

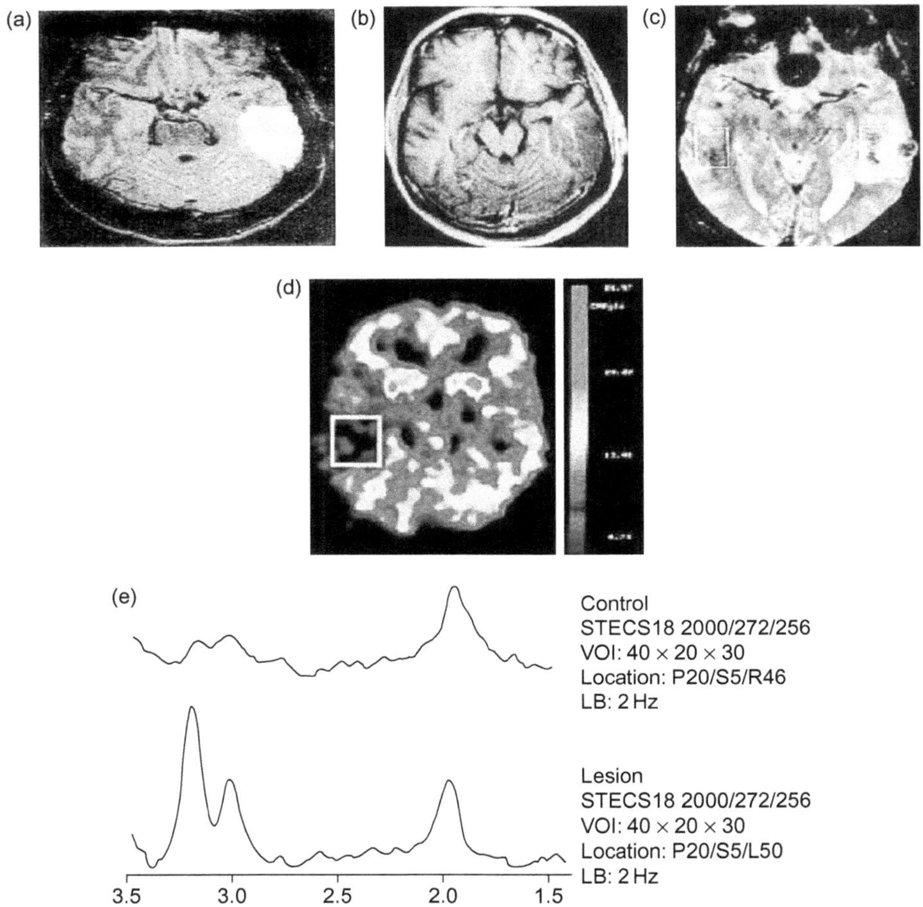

Figure 3.85 Low-grade astrocytoma, previously treated with radiation therapy. (a) Axial T_2-weighted image (0.5 T, SE 2067/80). (b) T_1-weighted image after gadopentetate dimeglumine administration (0.5 T, SE 500/i2) demonstrates a central area of contrast enhancement in the left temporal lobe tumor. (c) Multiplanar gradient-echo image shows the 24-cm^3 tumor and normal contralateral VOIs (squares) used for MRS. (d) FDG PET scan shows hypometabolism in the left temporal lobe (peak GUR in the tumor VOI [square] is 2.7 mg/100 g/mm, and the GUR in the control WM is 3 mg/min). (e) MR spectra in which both choline (3.25 ppm) and creatine (3.05 ppm) signal intensities appear to be elevated and the NAA peak area (2.02 ppm) is decreased relative to the control area. Lactate signal intensities are not apparent. See Figure 3.84 legend for expansion of abbreviations.

and phase-encoded acquisition. Restricting the excitation to an area of interest within the skull prevents contamination of the spectra with strong signals from skull fat, which would mask the lactate resonance. Volume-selective excitation is performed with a double SE sequence, with an echo time of 272 ms. Resonance

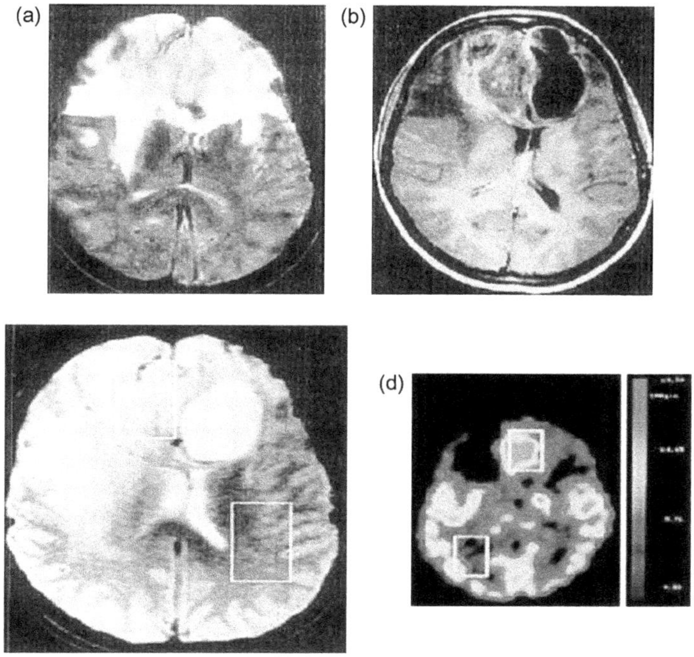

Figure 3.86 Recurrent anaplastic (grade 3) astrocytoma. (a) Axial T_2-weighted image (0.5 T, SE 2067/80). (b) Axial T_1-weighted image (SE 500/12) after gadopentetate dimeglumine administration demonstrates contrast enhancement in the left frontal lobe extending into the opposite hemisphere. A left frontal postsurgical cystic cavity is present. (c) Multiplanar gradient-echo image shows the 40-cm^3 VOIs (squares) of the tumor and normal brain (contralateral occipitoparietal region) used for MRS. (d) FDG PET scan shows a midline hypermetabolic lesion (peak GUR in tumor VOI is 20.6 mg/min, and the GUR in the control WM is 2.6 mg/mm). (e) MR spectra show elevation of the choline (3.25 ppm) signal intensity. In this case, NAA is not detected on the basis of the qualitative scoring criteria used in this study. The broad peak observed between 2.4 and 1.8 ppm represents undetermined metabolites that have a resonance position similar to that of NAA. Strong lactate signal intensities appear at 1.33 ppm. See Figure 3.84 legend for expansion of abbreviations.

signals are acquired from a FOV of 25 mm × 225 mm, applying 32 × 32 phase-encoding steps with a repetition time of 2000 ms, which results in a nominal pixel size of 7 mm × 7 mm. The data can be presented as maps representing the distribution of specific biochemical compounds and as sets of spatially resolved spectra; the latter allows more precise evaluation of specific foci of interest. Signal positions in the spectra are given as ppm on a relative frequency scale, referenced to the position of the trimethylsilylpropionate set at 0 ppm, with NAA ($\delta = 2.02$ ppm) or choline ($\delta = 3.21$ ppm) as internal standards. Signal intensity ratios are determined directly from the spectra selected from the central region of the infarct and for comparison from unaffected brain tissue. All infarcts were seen as

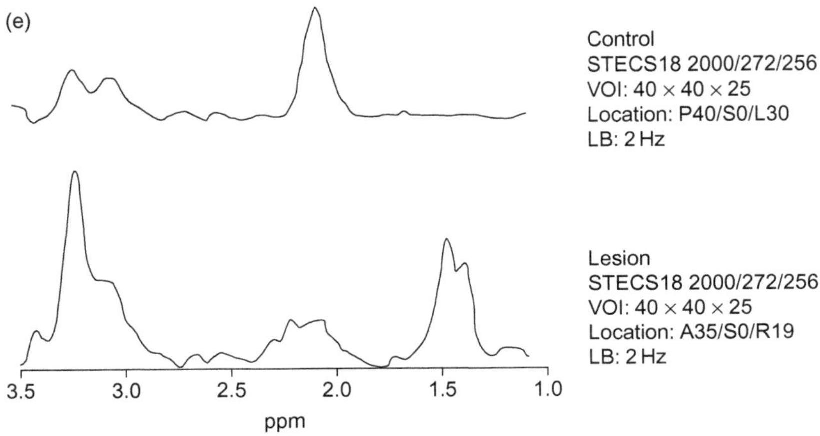

Figure 3.86 continued

high-signal-intensity areas on the T_2-weighted SE image, Figure 3.88(a) [29]. The spatial distributions of NAA, choline, and lactate are displayed as color-coded intensity maps in Figure 3.88(b)–(d) [29]. The NAA map (Figure 3.88(b) [29]) shows a severe decrease in NAA in the freshly infarcted area in the left occipital lobe and the old infarct in the right occipital lobe, but no decrease in the small, fresh infarct in the posterior thalamus. In another patient (Figure 3.89(a) [29]), the anterior part of the infarcted area shows hemorrhagic changes, and is characterized by a nearly complete loss of signal intensity from choline, creatine, and NAA.

The unevaluated area corresponds to the white area on the color-coded metabolite maps (Figure 3.89(b)–(d) [29]). Lactate is seen present in and beyond these areas, with persistent choline. CT scans, obtained the same day as the MR spectroscopic images, showed residual brain tissue in the posterior mesial part of the infarction (Figure 3.89(e) [29]), corresponding to the normal low-signal-intensity areas on the T_2-weighted MR image. Two months later, there was complete loss of this brain tissue (Figure 3.89(f) [29]).

3.3.1.10 Diagnostics of Differentiation of Brain Abscess from Cystic or Necrotic Brain Tumor [30]

MRS has been widely applied in patients with a variety of conditions, including tumors, infarcts, demyelinating diseases, encephalitises, seizure disorders, and others. It can be used to differentiate noninvasively between brain abscess and tumor (Figures 3.90–3.96). Table 3.25 [30] lists the medical details of the patients involved in the clinical study. In one group, the cystic or necrotic masses appeared as a ring-shaped area of high signal intensity or of contrast material enhancement on MR or CT images. One patient had multiple brain abscesses, two of which were examined with MRS.

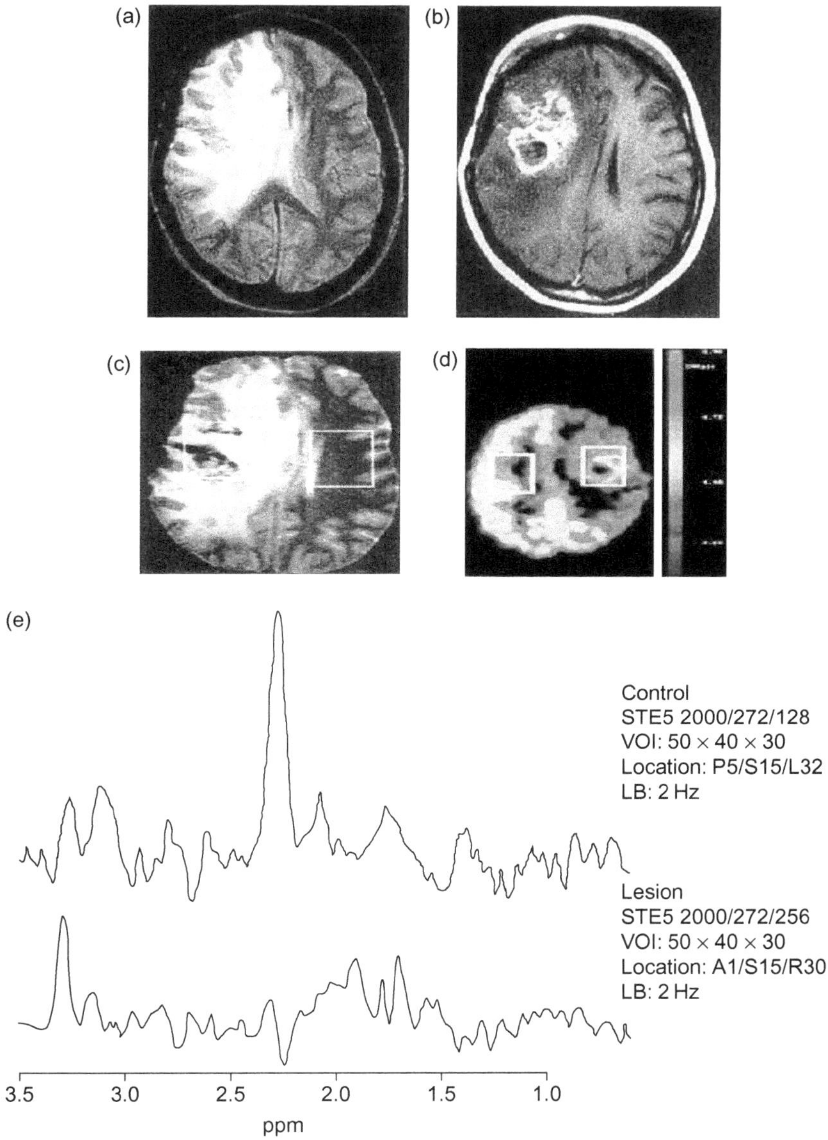

Figure 3.87 Recurrent glioblastoma multiforme. (a) Axial T_2-weighted image (0.5 T, SE 2067/80). (b) T_1-weighted image (0.5 T, SE 450/12) after gadopentetate dimeglumine administration shows an irregular area of contrast enhancement in the right parietal lobe. (c) Multiplanar gradient-echo image shows the 60-cm^3 tumor and control VOIs (squares) used for MRS. The area of hypointensity within the tumor most likely represents an area of necrosis and/or hemorrhage. (d) FDG PET scan shows a right frontoparietal lesion composed of a crescent-shaped hypermetabolic area encircling an ametabolic core (peak GUR in the tumor VOI is 7.5 mg/100 g/min, and the GUR of the control WM is 0.5 mg/100 g/min). (e) MR spectra show a tumor spectrum with elevated choline signal intensity and decreased NAA signal intensity. No lactate signal intensities are apparent. See Figure 3.84 legend for expansion of abbreviations.

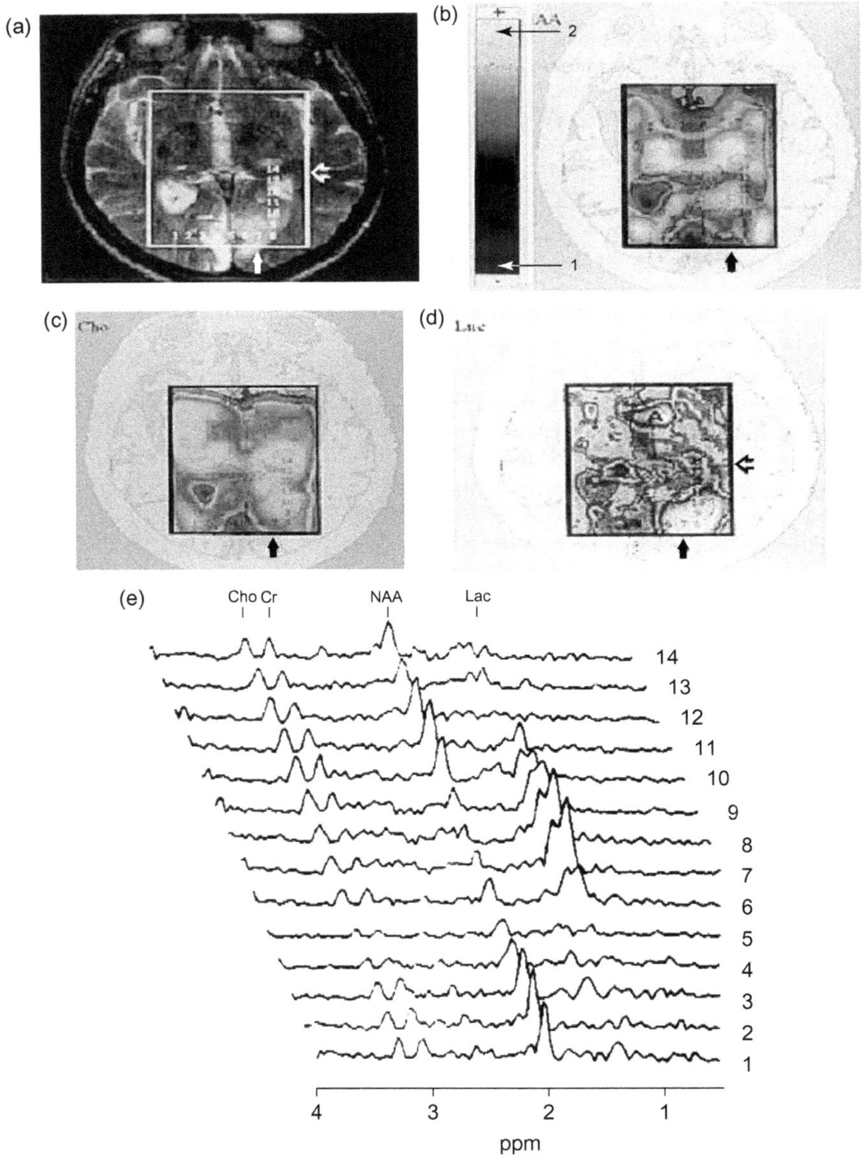

Figure 3.88 (a) T_2-weighted SE MR image (2200/100 [repetition time ms/echo time ms]) shows 2-day-old infarcts in the left occipital lobe (solid arrow) and also in (b–d) and left posterior thalamus (open arrow). A 3-year-old infarct is seen in the right occipital lobe. (b–d) Color-coded signal intensity maps projected onto a contour plan (a) show spatial distributions of (b) NAA, (c) choline (Cho), and (d) lactate (Lac). The color scale is renormalized to minimal (blue) and maximal (yellow) signal intensity for each metabolite map. Signal intensity from fatty bone marrow in the clivus is indicated by *A* (artifact) in (d). (e) Fourteen spectra originating from a path through the lesion. The origins of the numbered spectra correspond to numbers on the MR image and metabolite maps. Cho = choline, Cr = creatine, Lac = lactate.

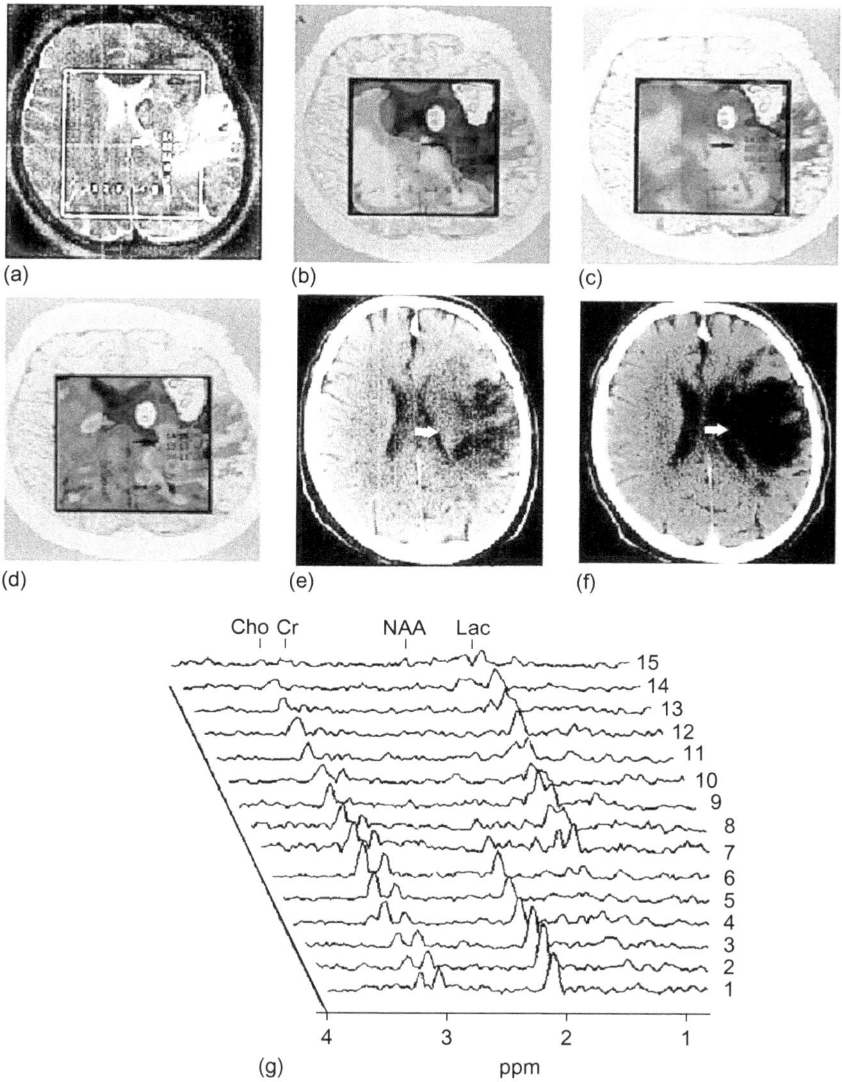

Figure 3.89 (a) T_2-weighted SE MR image (2200/100) shows a 35-day-old infarct with hemorrhage in the frontal left lobe and a 2-day-old infarct in the precentral left lobe. (b–d) Color-coded signal intensity maps projected onto a contour plan (a) show partial distributions of (b) NAA, (c) choline, and (d) lactate. The color scale is renormalized for each metabolite. The hemorrhagic areas that were not evaluated are in white. (e) A CT scan, obtained the same day as the MR spectroscopic examination, shows residual brain tissue in the posterior mesial part of the infarction (arrow), corresponding to the normal areas of low signal intensity on the T_2-weighted MR image (arrow in (a)), as well as persistent choline, increased lactate, and complete loss of NAA (arrow in b–d) on the metabolite maps. (f) CT scan obtained 2 months after (e) shows a complete loss of this brain tissue (arrow). (g) Fifteen spectra originating from a path through the lesion. The origins of the numbered spectra correspond to numbers on the MR image and metabolite maps. Cho = choline, Cr = creatine, Lac = lactate.

Table 3.25 Summary of Clinical Data in 14 Patients

Patient No./Sex/Age (Year)	Chief Complaint	DOS	SOI Fever	ESR	ND	Location	Final Diagnosis
1/M/56	Headache, RH	1 mo	Absent	N	RHCB	M	BADTEI
2/F/41	Fever, ST	1 mo	Present	E	None	LF	BADTStrII
3/M/70	RH	10d	Absent	N	MARH	LF	BADTStrII
4/M/47	Headache	12d	Absent	N	LH	RT	BADTStrII
5/F/34	SST	. . .	Absent	N	None	RT	BADTG-POI
6/M/38	AM	25d	AITP	N	LPRH	BS	BADTG-POI
7/M/25	Fever, chill	20d	Present	N	None	LT	BADTStaEI
8/M/46	Seizure	1 mo	Absent	N	RH	LP	GM
9/F/38	Headache	20d	Present	N	LHPr	RP	ROGM
10/F/60	Headache Fever	20d	Present	N	DLH	RF	GM
11/M/59	Headache	1 mo	Absent	N	None	RT	GM
12/F/29	Headache	Insidious	Absent	N	LW	LC	PA
13/M/29	LLEW	1 mo	Absent	N	LLEWLSP	RP	GM
14/M/45	Headache	6 mo	Absent	N	None	M	MFNSLC

AITP, absent initially and then present; ESR, erythrocyte sedimentation rate; DOS, duration of symptoms; ND, neurological deficit; SOI, sigmatta of infection; N, normal; E, elevated; BADTStrII, brain abscess due to streptococcus intermedius infection; BADTStaEI, brain abscess due to *Staphylococcus epidermidis* infection; DLH, dysarthria left hemiparesthesia; RHCB, right hemiparesis cortical blindness; LH, left hemiparesis; RH, right hemiparesis; LHPr, left Hemiparesthesia; LP, left ptosis; MARH, motor aphasia right hemiparesis; ROGM, recurrence of glioblastoma multiforme; GM, glioblastoma multiforme; LLEW, left lower extremity weakness; LLEWLSP, left lower extremity weakness left sole paresthesia; MFNSLC, metastasis from non-small-cell lung cancer; BADTEI, brain abscess due to *Elentum* infection; ST, sore throat; LW, left weakness; LC, left cerebellum; RP, right parietal; SST, single seizure; AM, attack mentality; LT, left temporal; RT, right temporal; LF, left frontal; M, multiple; BS, brainstem; LP, left parietal.

The spectroscopic studies used a clinical 1.5-T system, with a circularly polarized head coil, for imaging and spectroscopy. Before MRS, T_1-weighted or T_2-weighted MR images in three orthogonal planes were obtained to define 2 cm × 2 cm × 2 cm VOIs. The spectra were obtained using a PRESS with an echo time of 270 ms and a repetition time of 2000 ms; 128 signals were acquired for all patients. In four patients, additional MR spectra were obtained, with a 135 ms TE, to confirm the phase reversal associated with *J*-coupled metabolites such as lactate and amino acids. Before spectroscopic measurement, field homogeneity was optimized for the selected VOIs, by measuring the ^{1}H MR signal of water in tissue with the spatial-selective PRESS. Typical FWHM of 4–8 Hz were achieved in all examinations. The water signal was suppressed by means of a frequency-selective saturation at the water resonance. A sweep width of 1000 Hz was used, with a data size of 1024 points. Only the second half of the echo was acquired. After the zero filling of 4096 points, for all three FID data, an exponential line broadening (center, 0 ms, half time, and 150 ms) was applied before Fourier transformation.

Resonance peaks were assigned to one of the three grades on the basis of the ratio of the integral of the metabolite peak to the integral of the unsuppressed water peak. The ratios of all metabolite peaks show a wide range. The range is arbitrarily divided into three groups. Peaks that are measured in the lower one-third of the range are assigned a grade of "small." Those that are measured in the middle are "moderate." The peaks that are measured in the upper third are scored as "large." The MRS findings are summarized in Table 3.26 [30]. Variable combinations of the resonance were attributed to lactate (1.3 ppm), amino acids (including valine, alanine, and leucine [0.9, 1.5, and 3.6 ppm, respectively]), organic acids such as acetate (1.9 ppm) and succinate (2.4 ppm), and unidentified metabolites (2.2, 2.9, 3.2, 3.4, and 3.8 ppm). The resonance peaks attributed to lactate were mainly moderate and large ($n = 6$). One peak was scored as moderate, and one was scored as small. Two resonances, attributed to lactate and amino acids such as valine or leucine (0.9 ppm, γ-CH_3 group), were detected.

These findings were confirmed with phase inversion at reexamination, with a 135 ms TE. Of four patients in whom the VOIs were contaminated by neighboring solid tissue, spectra in only one patient showed small resonance peaks representative of NAA (2.1 ppm), choline compounds (3.2 ppm), and creatine–phosphocreatine complex (3.0 ppm). One of the unassigned peaks was observed in patient 7, Figure 3.93 [30]. In six of seven patients with brain tumors, spectra show only the peak attributed to lactate. In one patient with metastatic brain tumor from lung cancer, the spectra show two identifiable peaks, one at 1.3 ppm (attributed to lactate) and the broad signal band around 0.9 ppm, presumably attributed to the CH_2 and CH_3 lipid components (0.8–1.2 ppm), to amino acid (0.9 ppm), or to both (Figure 3.96 [30]).

^{1}H MRS patterns in patients with brain abscesses differ markedly from those in patients with cystic or necrotic brain tumors (lactate with acetate, succinate, and/or amino acids versus lactate with or without small lipids). The results indicate that most brain abscesses may be differentiated from cystic or necrotic tumors on the basis of in vivo ^{1}H MRS. Discrimination between amino acids

Table 3.26 MRS Findings in Patients with Brain Abscess or Tumor

Patient No.	Final Diagnosis	Metabolites Detected[a]
1	Brain abscess	Lactate (++), acetate (+++), succinate (++), valine or leucine (++), alanine (+), unknown metabolite A (2.2 ppm), unknown metabolite E (3.8 ppm)
1[b]	Brain abscess	Lactate (+++), acetate (++), succinate (++), valine or leucine (++), alanine (+), unknown metabolite A (2.2 ppm)
2	Brain abscess	Lactate (+++), amino acid or lipid (+++), leucine (+++), unknown metabolite B (2.9 ppm), unknown metabolite C (3.2 ppm), unknown metabolite D (3.4 ppm)
3	Brain abscess	Lactate (+++), acetate (++), amino acid or lipid (+++), alanine (++), leucine (++)
4	Brain abscess	Lactate (+++), valine or leucine (++)
5	Brain abscess	Lactate (+++), acetate (+), amino acid or lipid (++)
6	Brain abscess	Lactate or lipid (+), succinate (+), alanine (+), NAA, creatine-phosphocreatine complex
7	Brain abscess	Lactate (+++), acetate (+), succinate (+), amino acid or lipid (+), unknown metabolite C (3.2 ppm)
8	Glioblastoma multiforme	Lactate (++)
9	Recurrent glioblastoma multiforme	Lactate (+)
10	Glioblastoma multiforme	Lactate (+++)
11	Glioblastoma multiforme	Lactate (+++)
12	Pilocytic astrocytoma	Lactate (+)
13	Glioblastoma multiforme	Lactate (+)
14	Metastasis from non-small-cell lung cancer	Lactate (++), amino acid or lipid (+)

[a]Data in parentheses indicate peak size. +, small peak; ++, moderate peak; +++, large peak.
[b]Two spectra were obtained in patient 1, who had two brain abscesses.

(i.e., valine or leucine at 0.9 ppm) and lipids (0.8−1.2 ppm) is important because lactate and lipid signals may be found in the spectra of both brain tumor and brain abscess. However, valine and leucine seem to be components of a key marker of brain abscess. It is known that a TE of 135 ms causes the resonance peaks of the lactate doublet and amino acid multiplets to invert, on account of *J*-coupling (the interaction of two nuclear spins, resulting from the distortion in the electron clouds of the nuclei, on the same molecule). In contrast, resonances from lipids do not invert, because of the uncoupled spins. Spectra in one case actually

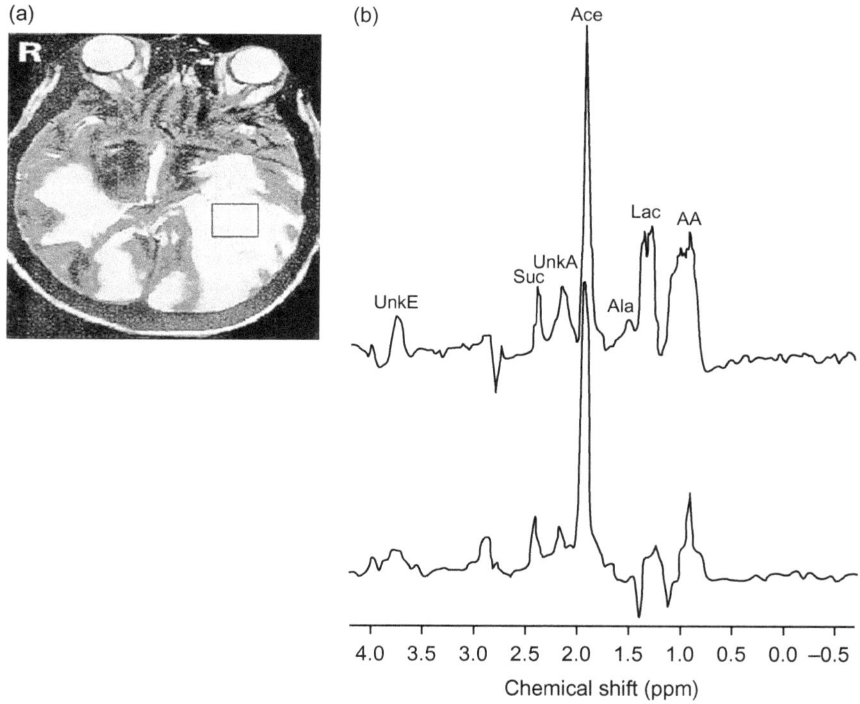

Figure 3.90 Multiple pyogenic brain abscesses in a 56-year-old man. (a) Axial T_2-weighted fast spin-echo MR image (repetition time/TE 3500/13 ms) shows a 4 cm × 3 cm abscess with edema in the left posterior temporal lobe. The $2\,cm^3 \times 2\,cm^3 \times 2\,cm^3$ VOI (box) is confined to the abscess cavity. There were multifocal areas of edema in the right temporal lobe and in both occipital lobes. (b) MR spectra from the abscess cavity were obtained with a TE of 270 (top spectrum) and 135 ms (bottom spectrum). The phase reversal of lactate (Lac) (1.3 ppm) and the amino acid (AA) (valine or leucine, 0.9 ppm) is well depicted. Ace = acetate, Ala = alanine, Suc = succinate, UnkA = unknown metabolite A, UnkE = unknown metabolite E.

showed two resonance peaks attributed to lactate and amino acids (valine or leucine). This finding was confirmed with phase inversion of the peaks, when a 135-ms TE is used. If resonance peaks at around 0.9–1.5 ppm are found in ^{1}H MRS, with TE of 270 ms, additional ^{1}H MRS should be performed at 135-ms TE, to distinguish, on the basis of J-coupling-associated phase inversion, lactate or amino acid signals from lipid signals. The use of both 270- and 135-ms TEs is helpful in the differential diagnosis of brain abscess and tumor, even though the ^{1}H MRS patterns may vary with the evolution of the abscess. One can see that ^{1}H MRS might be a useful tool in the evaluation of patients in the early, encephalitic stage of focal infection (Figures 3.90–3.96).

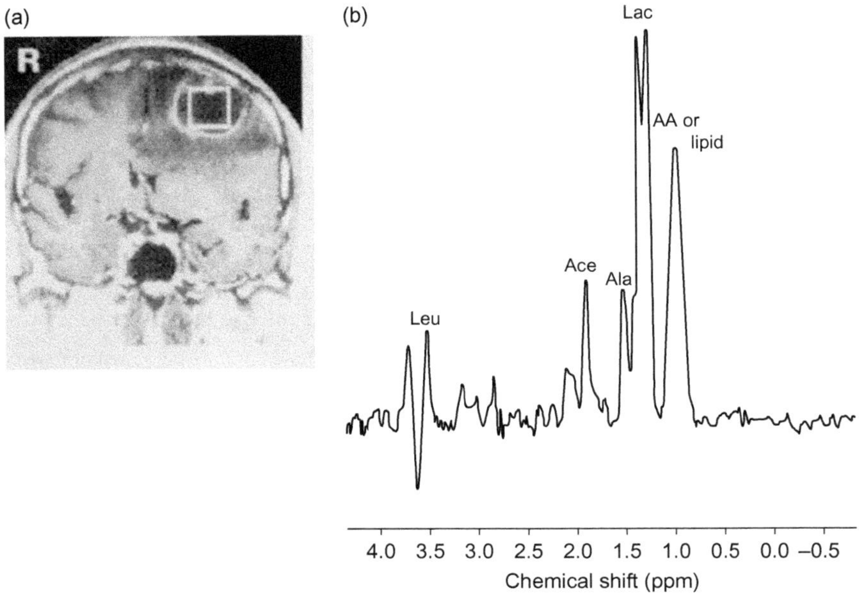

Figure 3.91 Pyogenic brain abscess in a 70-year-old man. (a) Postcontrast coronal T_1-weighted spin-echo MR image (550/14) shows a 4 cm × 4 cm abscess with ring-shaped contrast enhancement and surrounding edema in the left frontal lobe. The 2 cm^3 × 2 cm^3 × 2 cm^3 (box) is confined to the abscess cavity. It was difficult to differentiate abscess from a cystic or necrotic tumor on the basis of clinical and radiologic findings before ^1H MRS. (b) MR spectrum from the abscess cavity was obtained with a TE of 270 ms. The resonance peak at 1.9 ppm (attributed to acetate [Ace]) is well depicted. To our knowledge, this peak has not been reported in a case of brain tumor. AA = amino acid, Ala = alanine, Lac = lactate, Leu = leucine.

3.3.2 Structural and Spectroscopic NMRI

3.3.2.1 Structural Imaging: Prepolarized Magnetic Resonance Imaging [31]

In clinical imaging situations, it is very convenient to have smaller magnetic fields and more mobile imaging systems. This illustration (Figures 3.97–3.102) is an example of such a system. Prepolarized magnetic resonance imaging (PMRI) uses two separate magnetic fields for the two different functions: the polarization (orientation of spins in the z-direction) field, B_p, and the usual readout of the signals, the RF field. The electromagnets do not use superconducting magnets; rather, the magnetic coils are water-cooled. This reduces the sophistication of the system, as well as its bulk and expense. In PMRI, susceptibility artifacts are greatly reduced by acquiring data at low-field strengths, thus allowing diagnostic-quality imaging. The specific absorption of the radiation is also reduced (affording the tests a more non-invasive character).

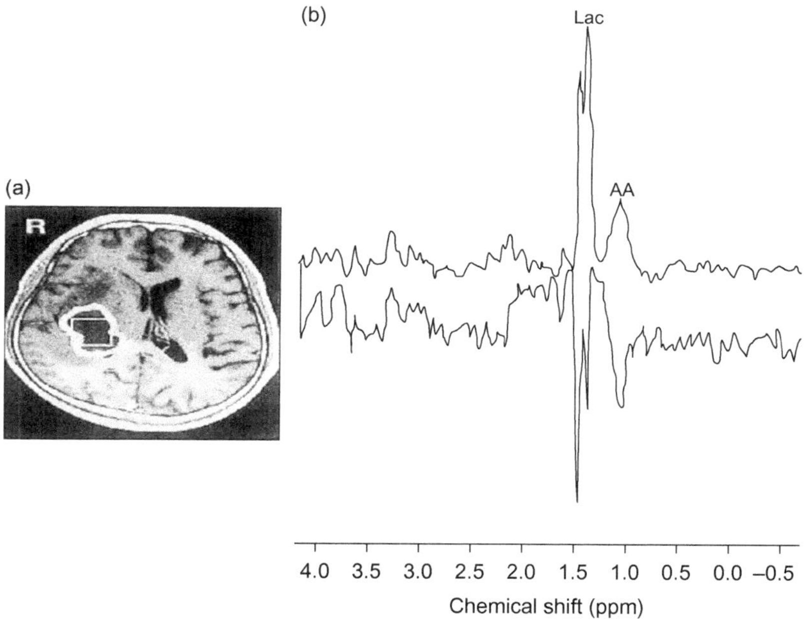

Figure 3.92 Pyogenic brain abscess in a 47-year-old man. (a) Postcontrast axial T_1-weighted spin-echo MR image (600/15) reveals a 3 cm × 2 cm bilobate mass with ring-shaped contrast enhancement and surrounding edema in the area of the right basal ganglia and thalamus. The 2 cm × 2 cm × 2 cm VOI (box) includes part of the abscess wall. (b) MR spectra from the abscess cavity were obtained with a TE of 270 ms (top spectrum) and 135 ms (bottom spectrum). Only two resonance peaks are identified. However, the phase reversal of the amino acid (AA) signal (valine or leucine, 0.9 ppm) is depicted on the MR spectrum obtained with a 135-ms TE, which is indicative of a pyogenic brain abscess. The phase reversal in this case does not follow the classic pattern, because of the presence of some phase distortion.

In PMRI, 3D imaging of volumetric data is acquired within a reasonable time. Technically, this system is called RARE. The sequence is used with a repetition time (TR) that is many times longer than the data acquisition window for each slice.

In conventional MRI system, a single superconducting coil generates a strong (~3-T) static magnetic field B_0, or both longitudinal magnetization and the Larmor precision of spins (the RF field). In PMRI, two separate fields, B_p and B_0 (from two separate magnetic coils), are used. A RARE sequence consists of an RF excitation pulse followed by a series of refocusing pulses to give an echo train, as in Figure 3.99 [31].

In conventional clinical MRI, a commonly prescribed protocol includes T_1 (longitudinal FID)-weighted images for anatomical reference and T_2-weighted images with fat suppression for finding edema, which has a long T_2. Edema is often an indication of pathology, so it is important to achieve simultaneous fat suppression

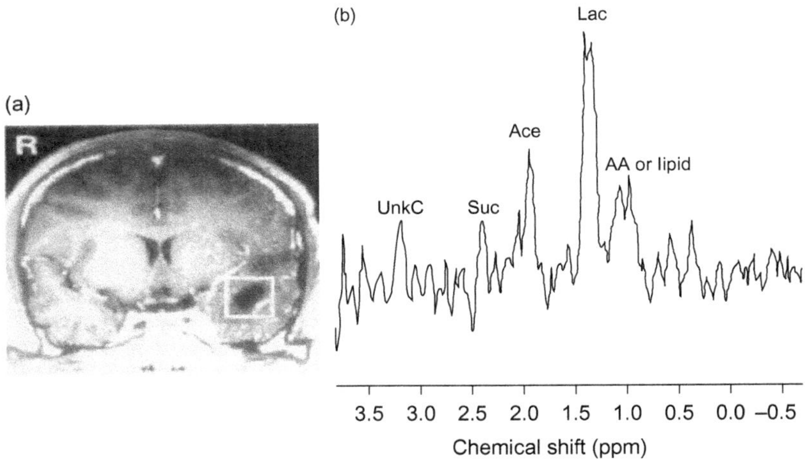

Figure 3.93 (a) Pyogenic brain abscess in a 25-year-old man. Postcontrast coronal T_1-weighted spin-echo MR image (550/14) shows diffuse swelling of the left temporal lobe and a central necrotic portion with low signal intensity. The $2\,cm^3 \times 2\,cm^3 \times 2\,cm^3$ VOI (box) includes the central necrotic portion of the abscess and adjacent, solid tissue. The absence of an identifiable contrast-enhanced wall suggests that the abscess is in the late encephalitis stage. (b) MR spectrum from the abscess cavity and adjacent, solid, inflamed tissue was obtained with a 270-ms TE. AA = amino acid, Ace = acetate, Lac = lactate, Suc = succinate, Link = unknown metabolite C.

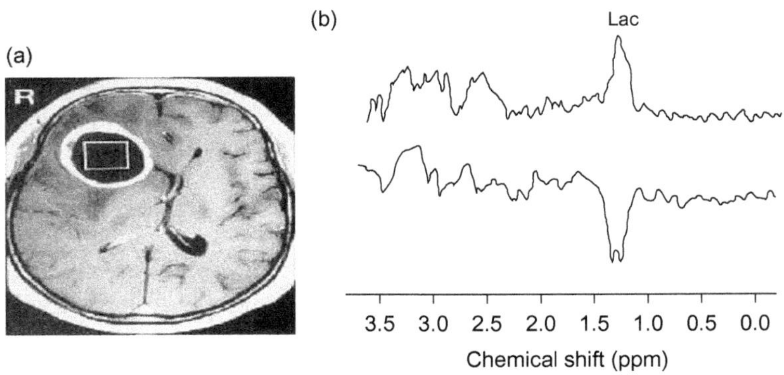

Figure 3.94 Glioblastoma multiforme in a 60-year-old woman. (a) Postcontrast axial T_1-weighted spin-echo MR image (600/15) shows a $5\,cm^2 \times 5\,cm^2$ mass with ring-shaped contrast enhancement and surrounding edema in the right frontal lobe. The $2\,cm^3 \times 2\,cm^3 \times 2\,cm^3$ VOI (box) is within the central necrotic cavity. Before MRS, a high-grade glioma (grade 3 or 4) could not be differentiated from brain abscess on the basis of clinical and MRI findings. (b) MR spectra from the cystic or necrotic portion of the tumor obtained with a TE of 270 ms (top spectrum) and 135 ms (bottom spectrum) show only the peak attributed to lactate (Lac). Phase reversal of the lactate signal is demonstrated on the spectrum obtained with a 135-ms TE. The broad lactate peak appears to have been caused by poor shimming.

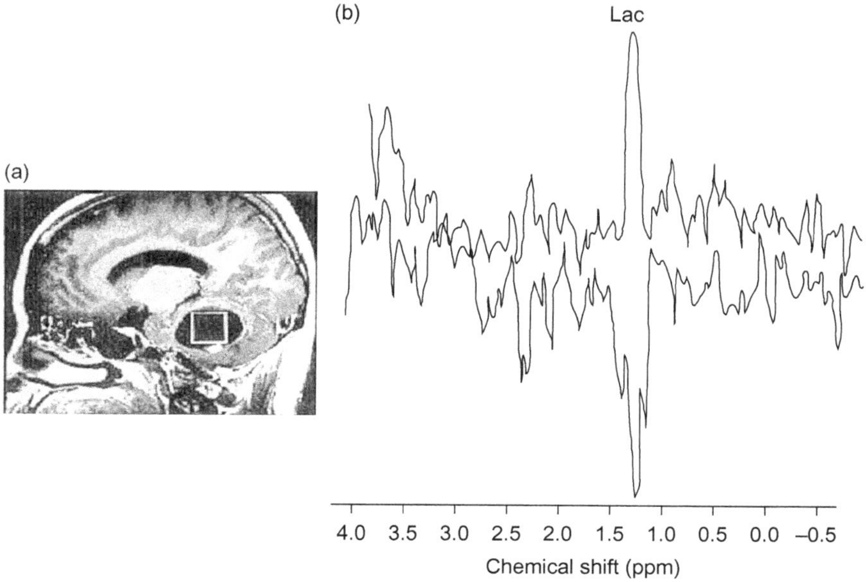

Figure 3.95 Pilocytic astrocytoma in a 29-year-old woman. (a) Postcontrast sagittal T_1-weighted spin-echo MR image (550/14) shows a 4 cm × 5 cm cystic tumor with a small mural nodule in the cerebellar hemisphere. The $2\,cm^3 \times 2\,cm^3 \times 2\,cm^3$ VOI (box) includes only the cystic fluid. (b) MR spectra from the cystic fluid obtained with a TE of 270 ms (top spectrum) and 135 ms (bottom spectrum) show a tall peak attributed to lactate (Lac). Phase reversal of the lactate peak is depicted on the spectrum obtained with a 135-ms TE. Multiple small peaks at various frequencies are present; these peaks may represent noise or unassigned metabolites.

and T_2 weighting to create strong contrast among edema, muscle, and fat. 3D RARE offers the flexibility of implementing both types of image weighting. One can appreciate the ability of PMRI RARE to generate contrast among three phantoms, with relaxation parameters similar to those of human fat, muscle, and edema. In the proton-density-weighted image, all three phantoms were relatively bright, with the variation determined by the proton density of the sample. In the T_1-weighted images, the water was brightest, because it had a much longer T_2. RARE can be used in PMRI much as it is in static-field (conventional) imaging, except without the possibility of using spectroscopic or frequency-selective techniques. Magnetization preparation contrast techniques are well suited to PMRI RARE, because the polarization interval is itself essentially a magnetization preparation. It can be appended to modify the contrast.

Driven equilibrium can also be used to shorten the TR. One can see in vivo PMRI RARE images with clinical resolution in a normal clinical scan time of approximately 7 min. There are two impediments to use of RARE in PMRI, which are seldom significant in conventional static-field MRI. These are concomitant gradient field effects and temporal B_0 variations. Both of these can disrupt

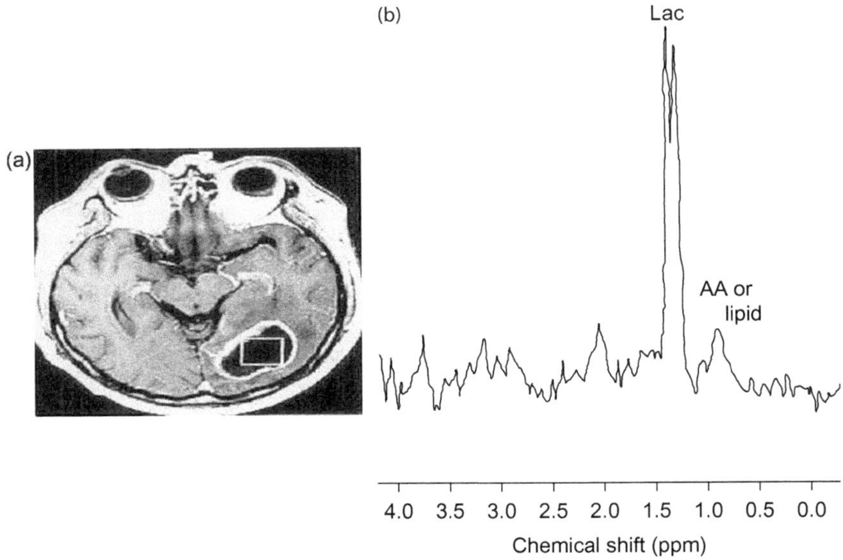

Figure 3.96 Metastasis from non-small-cell lung cancer in a 45-year-old man. (a) Postcontrast axial T_1-weighted spin-echo MR image (600/14) shows a 3 cm × 5 cm cystic mass with ring-shaped contrast enhancement and surrounding edema. This patient had multiple cystic masses with ring-shaped contrast enhancement in other portions of the brain (not shown). The 2 cm × 2 cm × 2 cm VOI (box) is confined to the central necrotic cavity. (b) MR spectrum from the cystic or necrotic portion of the tumor was obtained with a TE of 270 ms. Two peaks, one attributed to lactate (Lac) and another centered at 0.9 ppm, are shown. Note the similarity of this spectral pattern to that of a brain abscess. AA = amino acid.

Table 3.27 Phantom Experiment

Phantom	T_1 at 0.4 T (ms)	T_2 at 52 mT (ms)
Canola oil	196	70
Chicken muscle	356	45
3 mM Cu^{++}	329	168

Table 3.28 Proton Density And T_1, T_2 Experiment

Weighting	t_p (ms)	k-space	t_e (ms)	TE_{eff} (ms)
Proton density	700	Centric	7.6	7.6
T_1	250	Centric	7.6	7.6
T_2	700	Sequential	8.8	71
STIR-T_2	700	Sequential	8.8	71

phase-sensitive Carr–Purcell–Meiboom–Gill (CPMG) echo trains. These are key to implementing RARE, with insensitivity to flip-angle errors and B_0 spatial inhomogeneity (Tables 3.27 and 3.28 and Figures 3.97–3.102).

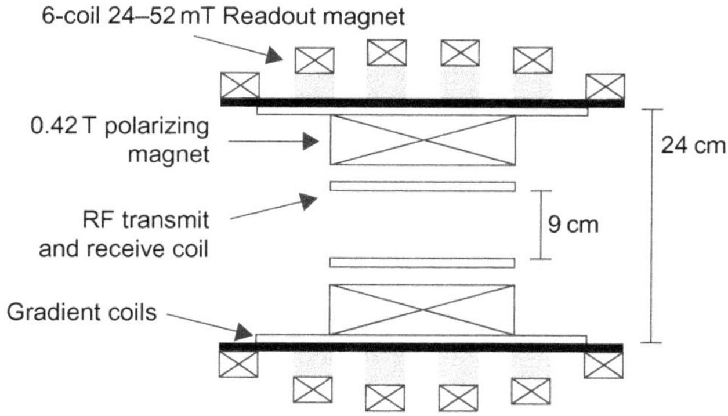

Figure 3.97 Schematic cross-section of the Stanford PMRI magnet system (all magnets wound onto cylindrical surfaces). The readout and polarizing magnets are concentric but independent copper-wire solenoids with parallel field alignment.

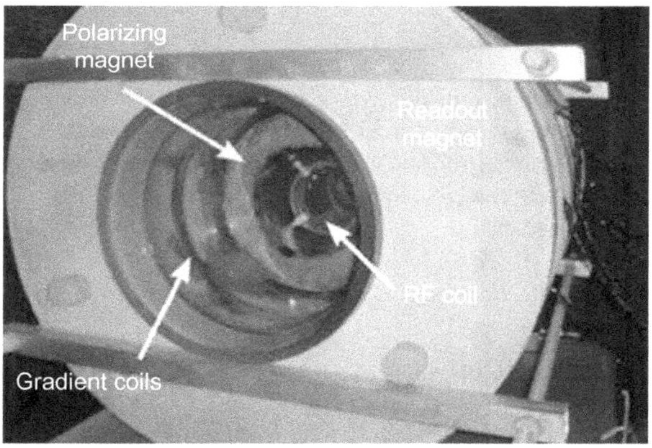

Figure 3.98 Photograph of the Stanford PMRI magnet system. The bore is large enough for imaging *in vivo* human wrists (see Figure 3.97 for dimensions). (Color figure can be viewed in the online issue, which is available at www.interscience.wiley.com.)

3.3.2.2 Spectroscopic Imaging: STEAM [32]

The human brain has many key metabolites, several of which have strongly coupled spins. Their response to STEAM has been calculated using QM. For selective pulses, the RF envelope was divided into 7.5-μs segments and the gradient-induced frequency distribution was typically incremented to give rise to millimeter spatial resolution. Such a resolution enabled the 90–10% roll-off of the 90° pulse to be captured over 30–40 spiral intervals. The temporal evolution of any of the various coherences, or of the ultimate transverse magnetization, emerging from the sequence can easily be worked out from the trace of the product of the density operator, with the corresponding coherence or magnetization operator. One can represent the STEAM sequences pictorially, as shown in Figure 3.103. There are three independent sets of gradients: the slice selection, the echo, and the mixing gradients. One can depict the interpulse evolution of coherence types that are influential in causing the STEAM output variability as shown in Figures 3.104 and 3.105.

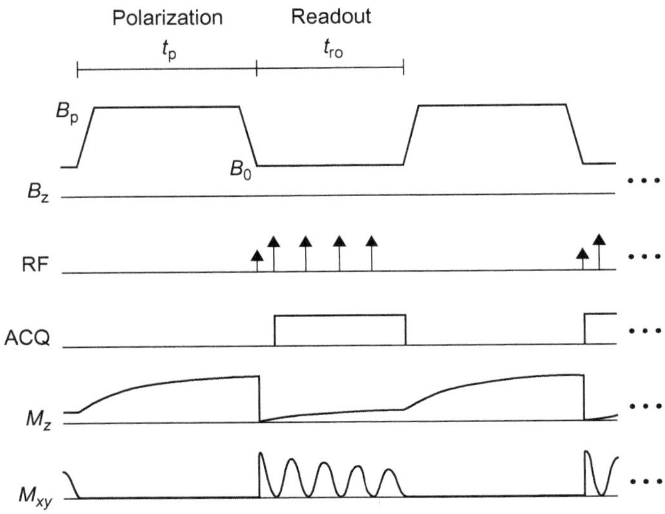

Figure 3.99 Pulse sequence and magnetization diagram for echo trains in PMRI. B_z shows the magnitude of the total main magnetic field; M_z and M_{xy} show the longitudinal and transverse magnetization; ACQ shows the data acquisition window timing; and RF shows the RF pulse timing. M_z growth occurs predominantly during the polarization interval (t_p). To maximize signal and SNR efficiency, we used 90° and 180° flip angles for the excitation pulse and refocusing pulses, respectively. In our current system, $t_p = 50-1000\,\text{ms}$, $t_{ro} = 10-200\,\text{ms}$, $B_p = 0.4\,\text{T}$, and $B_0 = 27-55\,\text{mT}$.

The first of these, arising after the first 90° pulse, is the mixture of in-phase and antiphase coherences that evolves from in-phase transverse magnetization during the first TE/2 period. The evolution is governed primarily by the scalar coupling, but is also influenced by the field-strength-dependent chemical-shift difference. The chemical-shift (and inhomogeneous field) dephasing is ultimately recovered for the spins of the target voxel, through the action of the pair of echo signals. The second influential coherence type is the ZQC group, which is generated by the second 90° pulse and exhibits an oscillatory evolution during the mixing period TM. The magnitude of the antiphase terms (at the end of the first TE/2) that are convertible to MQCs by the second pulse (this includes ZQC) is governed by the choice of TE. In the subsequent TM, the mixing gradient (a filter mechanism designed to destroy all gradient-sensitive magnetization) does not affect either the ZQCs or the longitudinal magnetization. However, it radically dephases transverse magnetization and higher orders of MQC. The evolution of ZQCs during TM determines the magnitude and character of an important set of antiphase coherence (APC) terms, subsequently produced by the third 90° pulse. That set evolves to contribute to the signal during the acquisition period.

It is known that three separate mechanisms related to selective pulses complicate the evolution of the spins that are coupled. These are the finite roll-off of the tip

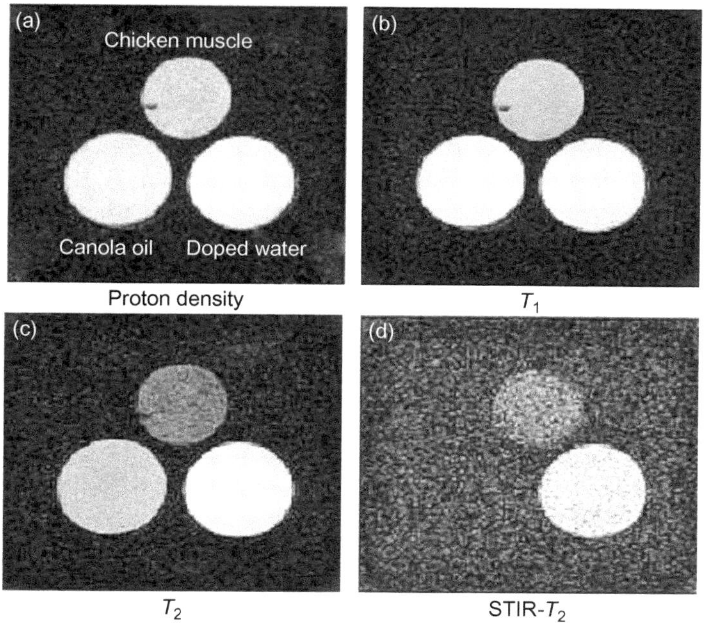

Figure 3.100 Basic contrast demonstration using 3D prepolarized RARE. By varying the polarizing time-and-phase-encoding acquisition order in prepolarized RARE, we achieved (a) classic proton density, (b) T_1, and (c, d) T_2 weighting, with phantoms mimicking human muscle, fat, and edema. We also nulled the fat signal in a T_2-weighted image (d) by adding a stimulated inversion recovery (STIR) preparation following each polarizing pulse. Table 3.27 [31] gives relaxation parameters for the phantoms, and Table 3.28 shows the pulse sequence parameter modifications used for the different types of contrast. Scan parameters: FOV = 8 cm × 8 cm × 8 cm; matrix = 196 × 196 × 16; BW = 48 kHz; $B_p = 0.4$ T; $B_0 = 52$ mT; NEX = (a) 2, (b) 4, (c) 2, (d) 2; scan time = (a) 336 s, (b) 327 s, (c) 344 s, (d) 388 s; inversion recovery = 58 ms for STIR in (d).

angle at the edges of the excitation band; the spatial displacement of the excitation bands of proton species that have different chemical shifts; and the continued evolution of the spins under scalar and chemically shifted Zeeman interactions, during application of the selective pulse. In the application of the first 90° pulse, only longitudinal magnetization is being transformed. For the remaining two pulses, the isolation is not as straightforward in STEAM as it is in PRESS. In PRESS, the selective 180° refocusing pulses can be treated independently of each other. In contrast to PRESS, the second and third selective 90° pulses of STEAM, together with echo and mixing-time gradients, act in concert to generate the stimulated echo.

To demonstrate the difference between a selective-RF-pulse package and a hard-pulse-gradient pulse package, with the same 90° tip angles, it is necessary to reduce TM artificially, to eliminate interpulse evolutions and hence the effects of the pulses. The difference in outcome between these packages is summarized by transformation matrices.

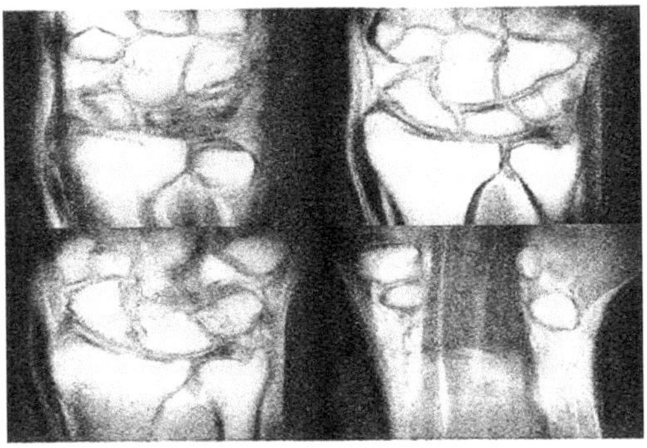

Figure 3.101 Coronal T_1-weighted 3D prepolarized RARE on a healthy volunteer's wrist. Four slices from the 20-slice dataset are shown. The wrist bones and joint are clearly depicted and the images have no artifacts related to field instability or concomitant gradients. Scan parameters: FOV = 8 cm × 8 cm × 6 cm; matrix = 192 × 192 × 20; BW = 48 kHz; PMRI t_p = 350 ms; NEX = 3; scan time = 323 s; TE = 7.6 ms; B_p = 0.4 T; B_0 = 52 mT. Cropped to 180 × 180 for detail.

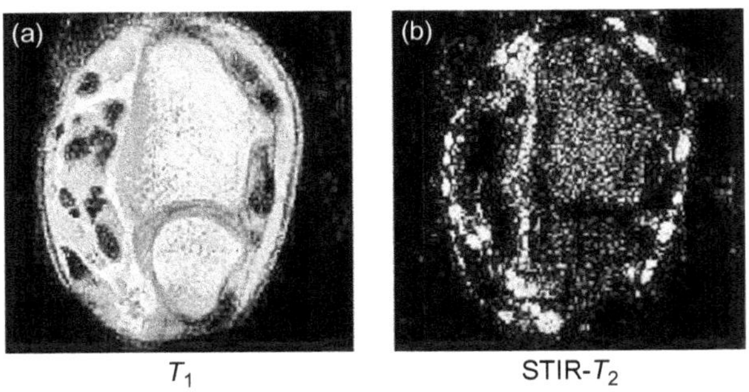

Figure 3.102 (a) Axial T_1-weighted and (b) STIR fat-suppressed T_2-weighted 3D prepolarized RARE images of a healthy volunteer's wrist. Blood vessels appear bright in (b) since they are not suppressed by the STIR. Scan parameters (a)/(b): FOV = 8 cm × 8 cm × 8 cm; matrix = 196 × 196 × 20; TE_{eff} = 7.6 ms/60 ms; BW = 48 kHz; B_p = 0.4 T; B_0 = 52 mT; t_p = 350 ms/750 ms; NEX = 3/2; imaging time = 326 s/420 s. The STIR image used TI = 60 ms cropped to 143 × 143 for detail.

In these matrices, the elements of each row reflect the weighting factors of the individual output coherences into which the RF-gradient package would convert the particular density operator component listed at the input side of the matrix. It should be noted that these matrix elements are only a summary, because they represent the spatially integrated effect of the pulses across a selective band. Comparing Figure 3.105(a) and 3.105(c) [32], it is evident that both hard 90° pulses and practical 90° pulses are comparable in their ability to interchange coherence. In both

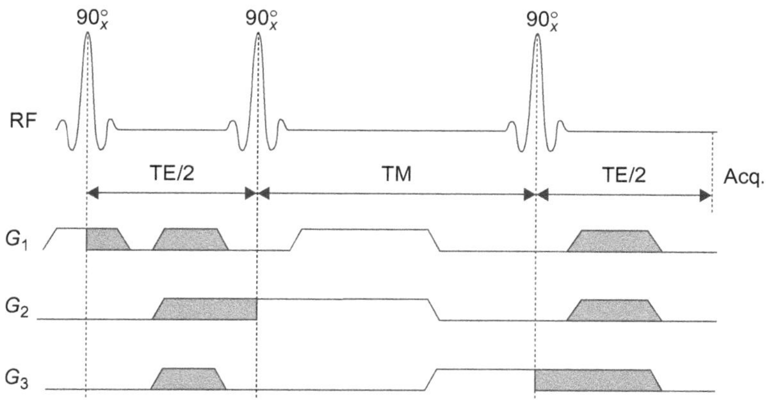

Figure 3.103 An illustration of the STEAM sequence used for single-voxel *in vivo* spectroscopy. The shaded gradients (G_i) applied prior to the midpoint of the second 90° pulse are matched by the shaded gradients following the midpoint of the third 90° pulse, refocusing both the slice and spoiler gradient evolutions. The optimized RF pulses (optsinc) have sinc envelopes optimized using the Shinnar–LaRoux procedure to give a length bandwidth product of 7.35.

cases the generation of off-diagonal elements occurs, primarily because of the coupled-spin evolutions arising from the scalar coupling and chemical-shift interactions of the coupled spins. Nevertheless, there are measurable, though not substantial, differences between the distributions of alternative coherence pathways produced by the two selective pulses, which can clearly be influenced by the design of the selective RF pulses.

The second principal consequence of the selective pulse design is its effect on the voxel size, shape, and location for each individual coherence term, because this also affects the ultimate metabolite yield (and lineshape). The spatial distribution of coherences produced by a slice-selective pulse is not reflected in the transformation matrix of Figure 3.105 [32], because its elements display only the spatially integrated influence of the pulse. Figure 3.104 [32] shows that all except the longitudinal and ZQC terms of the density matrix (DM) are gradient sensitive and dephased in TM by the mixing gradient. Notwithstanding the fact that only in-phase transverse magnetization terms are detectable, the evolution between terms predominates, such that a metabolic yield and lineshape will be governed by the distribution over antiphase, as well as in-phase, transverse coherences at the onset of the signal acquisition period. Stated mathematically, the response of an N-spin, $I = 1/2$, system is determined by the density operator at the start of acquisition. This can be expressed as a weighted sum of the complete set of product operator basis terms. Although each of the $(2 \times 2)^N$ single-quantum (transverse) terms will evolve during an acquisition, it is the evolution that produces the lineshape component characteristic of the originating coherence. In vivo proton spectra and their computer simulations are shown in Figure 3.106 [32].

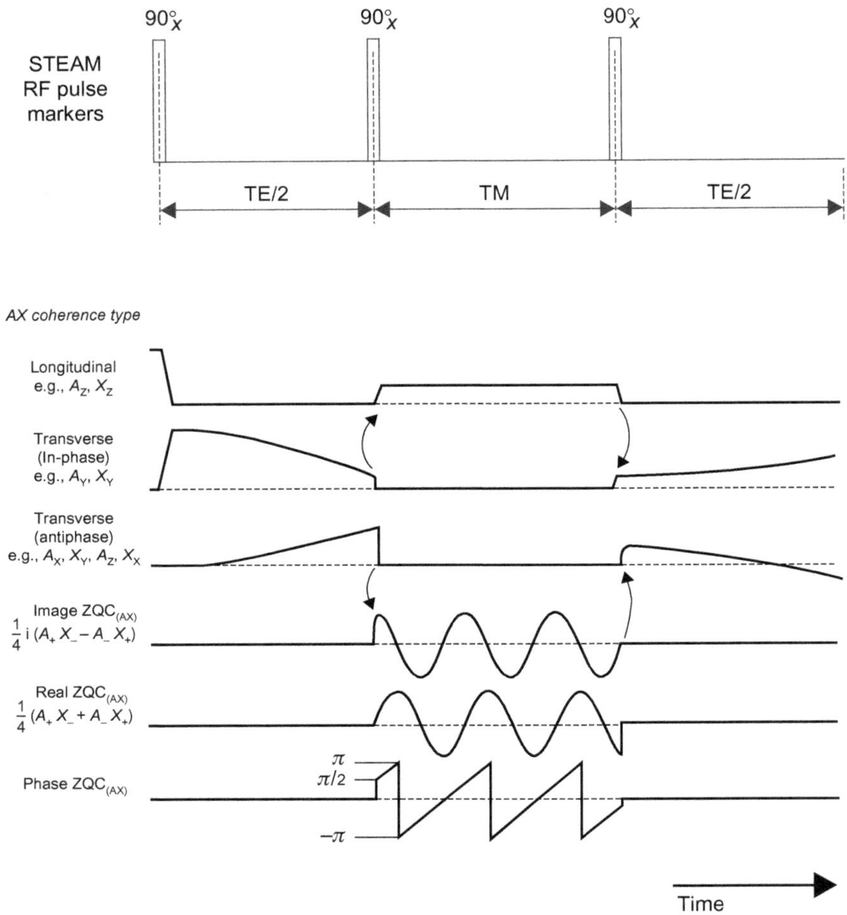

Figure 3.104 A schematic illustration of the coherence evolution process that is key to the coupled-spin response to the STEAM sequence. The spoiler gradients commonly applied in both the echo and mixing periods are omitted for clarity. Following excitation, the transverse magnetization evolves (in the first TE/2 period) into a mixture of in-phase and antiphase coherences. The second 90° pulse transforms the in-phase coherences to longitudinal magnetization and the antiphase coherences to a mixture of ZQCs and higher-order coherences. Any spoiler gradient applied during TM will dephase all coherences except the gradient-insensitive terms, namely, the longitudinal and ZQC terms. During TM, the ZQCs evolve between real and imaginary states, while the longitudinal terms are static. The final 90° pulse tips the longitudinal magnetization back to the transverse plane, but transforms only the imaginary part of the ZQCs back to antiphase coherence, because of the phase sensitivity of this process. This phase sensitivity results in a TM-dependent production of antiphase coherences by the last 90° pulse. During the final TE/2 evolution period, the in-phase and antiphase coherences evolve into the final mixture of transverse coherences that will determine the metabolite lineshape and yield.

(a)

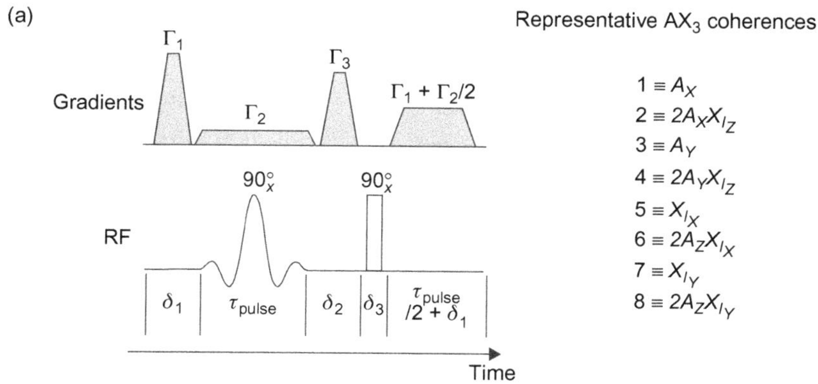

(b) 90° Variant 1 (Hard pulse)

Input \ Output	1	2	3	4	5	6	7	8
1	49.9	−0.5	0	0.8	0	0.7	0	−0.8
2	−0.5	24.3	−0.7	0	−0.7	23.7	0.9	5.5
3	0	0.8	−49.9	0.5	0	−0.8	0	−0.7
4	−0.7	0	0.5	−24.3	0.9	5.5	0.7	−23.7
5	0	−0.7	0	−0.8	49.9	0.5	0	0.8
6	0.7	23.7	0.9	−5.5	0.5	24.3	−0.7	0
7	0	−0.8	0	0.7	0	0.7	−49.9	−0.5
8	0.9	−5.5	−0.7	−23.7	−0.7	0	−0.5	−24.3

(c) 90° Variant 2 (Optimized sinc pulse $\Delta \equiv 2$ kHz)

Input \ Output	1	2	3	4	5	6	7	8
1	49.9	−0.3	0	0.9	0.1	0.9	0	−0.7
2	−0.3	23.7	−0.7	3.6	−0.6	22.2	1.1	8.9
3	0	0.9	−49.9	0.4	0	−0.7	0	−0.8
4	−0.7	3.6	0.4	−23.6	1.1	8.9	0.7	−22.1
5	0.1	−0.8	0	−0.7	49.9	0.4	0	0.9
6	0.7	22.2	1.1	−8.8	0.4	23.7	−0.7	−3.5
7	0	−0.7	0	0.9	0	0.9	−49.9	−0.3
8	1.1	−8.8	−0.6	−22.1	−0.7	−3.5	−0.5	−23.6

Figure 3.105 To illustrate the coherence transformation properties of a realistic selective 90° pulse in comparison with an idealized, hard 90° pulse, each of these pulses is, in turn, packaged with a second, idealized hard 90° pulse, as shown for the soft pulse. The matrices delineating the transformation of eight representative lactate coherences are presented in (b) and (c), for an idealized hard pulse and a realistic soft pulse, respectively, packaged with a hard final pulse. The realistic 90° pulse had a bandwidth of 2 kHz. To ensure a valid comparison between these very different packages, δ_1 and δ_2 were set infinitesimally small and *t* pulse set to 3.675 ms for (c), whereas in (b), to keep the total sequence length identical for both cases, δ_1 and δ_2 were each set to 3.675/2 ms.

The numerical spectra (Figure 3.106(a)) are calculated for the same sequence timings: namely, TM = 20 ms and 34 ms at a TE of 30 ms, as were used in the in vivo acquisition of 3 T. The in vivo spectra arise from a $2\,cm^3 \times 2\,cm^3 \times 2\,cm^3$ volume located in the parietal lobe of the brain of a healthy volunteer, and are shown in Figure 3.106(b) [32], where the stability of the methyl singlets of NAA (2.02 ppm), Cr (3.05 ppm), and Cho (3.24 ppm) contrasts with the significant TM sensitivity of the coupled-spin region of the spectrum (2.1 and 2.8 ppm). In the spectral region between 2 and 3 ppm, several coupled-spin metabolites contribute in vivo protein spectra, in addition to the strongly coupled NAA and glutamate (Glu) illustrated. They include aspartate (Asp), which is similar in behavior to the asparaxyl group of NAA; glutamine (Gln), the spectral characteristics of which are very similar to Glu; and GABA. They also include, albeit at lower concentrations, NAAG, which contains both aspartyl and glutamate groups, and taurine (Tau), which contains an A_2B spin system (Table 3.29).

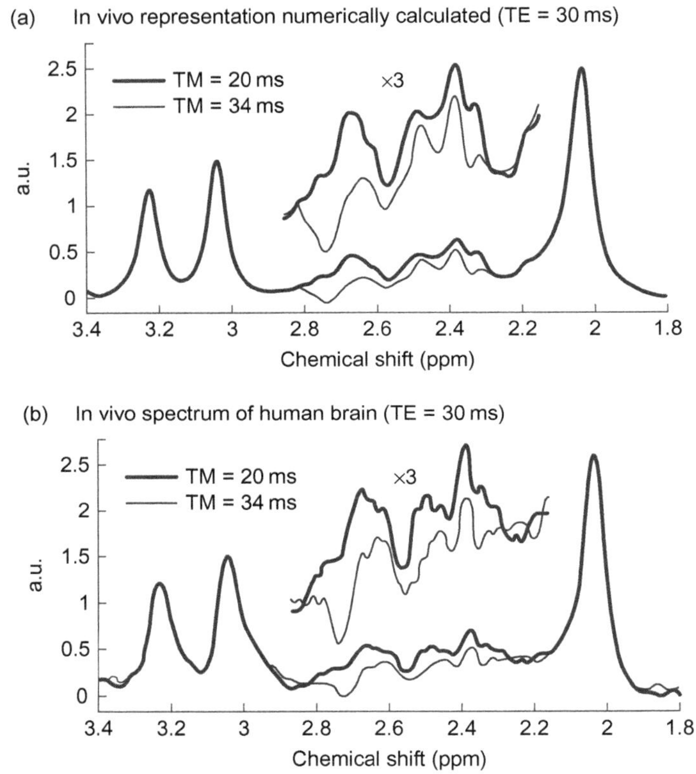

Figure 3.106 A demonstration of the correspondence of the calculated (a) and the *in vivo* (b) spectra from human brain, as well as an *in vivo* demonstration of the sensitivity of coupled-spin response to the STEAM mixing time, TM. In vivo spectra were acquired at 3 T from an 8-cm^3 single cubic voxel located in the parietal lobe in 256 scans with TR = 2.5 ms, and utilizing pulses and timings identical to those used in the numerical simulations. The heavier lines correspond to (TE, TM) timings of (30 ms, 20 ms), whereas the lighter lines correspond to (30 ms, 34 ms). An exponential line broadening of 1.5 Hz was applied to the *in vivo* data and the calculated spectra were broadened to a 7-Hz Lorentzian lineshape. The relative metabolite concentrations utilized for the calculations were (NAA: NAAG: Glu: Gln: Cr: Cho: Asp: GABA: Tau) (8.10: 1.60: 7.45: 2.65: 6.10: 1.39: 1.40: 1.40: 1.00). The individual contributions to a form—Glu, Gln, and NAA (*AB* spins)—are displayed in (c).

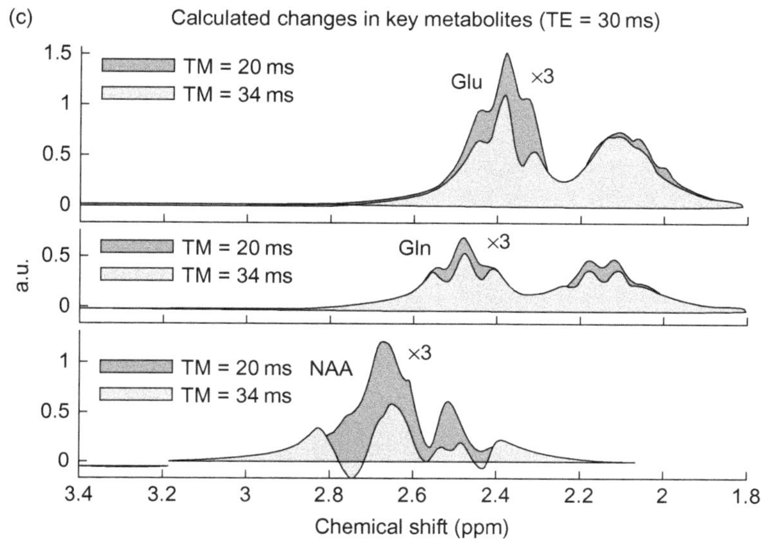

Figure 3.106 continued

3.3.2.3 Spectroscopic Imaging Versus Spatial Imaging/Sensitivity-Encoded Spectroscopic Imaging [33]

Nowadays PRMS has become more or less a customary tool in studies of both the healthy and the pathologic human brain. For the investigation of local metabolic concentrations, the advantage of spatial imaging (SI) over single-voxel spectroscopy (SVS) lies in the spatial resolution obtained by acquiring a whole-grid spectra. These spectra show only a slight reduction in the SNR per unit volume and time, with respect to SVS. Also, displaying the resulting metabolite levels as images allow easy comparison of local metabolism changes with other data, such as anatomical images and functional activation maps. The major disadvantage of present spectroscopic-imaging techniques is the long acquisition time they require to yield a high-resolution image. The long acquisition time of spectroscopic-imaging measurements with two spatial dimensions arises from the large number of samples collected. This refers to $K_x \times K_y \times N_t$ samples in the two spatial dimensions and one time/spectral dimension. The K_x, K_y are the wave (linear) number points in the k ($=1/\gamma$, $\lambda =$ wavelength of the wave, corresponding to a particular spectroscopic excitation in the voxel)-space. In conventional spectroscopic imaging, k-space is sampled by acquiring N_t data points in the time dimension, per excitation. This requires $K_x \times K_y$ excitations for a complete 2D spatial image. The idea is to sample k-space in the most efficient way possible. Acquiring multiple spin-echos results in uneven T_2 weighting of k-space, which adversely affects the point-spread function (PSF) of the technique.

Table 3.29 Chemical Shifts, δ ppm, and Scalar Coupling Constants, J Hz, Used in the Calculations

		Chemical Shifts						
Metabolite	**Spin System**	δ_A	δ_B	δ_M	δ_N	δ_P	δ_Q	δ_X
Creatine	Uncoupled	3.02						
Choline	Uncoupled	3.22						
NAA	Uncoupled	2.023						
NAAG	Uncoupled	2.05						
Lactate	AX_3	4.0908	–	–	–	–	–	1.3125
GABA	$A_2M_2X_2$	3.01	–	1.91	–	–	–	2.31
Aspartate	ABX	2.82	2.69	–	–	–	–	3.90
NAA	ABX	2.52	2.70	–	–	–	–	4.40
NAAG	ABX	2.75	2.55	–	–	–	–	4.62
NAAG	$AMNP_2$	4.15	–	1.91	2.07	2.42	–	–
Glutamate	AMNPQ	3.75	–	2.05	2.14	2.36	2.37	–
Glutamine	AMNPQ	3.76	–	2.13	2.15	2.45	2.47	–

		Scalar Couplings										
Metabolite	**Spin System**	J_{AB}	J_{AM}	J_{AN}	J_{AX}	J_{BX}	J_{MN}	J_{MP}	J_{MQ}	J_{MX}	J_{NP}	J_{NQ}	J_{PQ}
Lactate	AX_3	–	–	–	6.933	–	–	–	–	–	–	–	–
GABA	$A_2M_2X_2$	–	7.3	–	–	–	–	–	–	7.3	–	–	–
Aspartate	ABX	17.6	–	–	4.0	8.2	–	–	–	–	–	–	–
NAA	ABX	15.5	–	–	10.1	3.7	–	–	–	–	–	–	–
NAAG	ABX	−15.9	–	–	4.37	9.5	–	–	–	–	–	–	–
NAAG	$AMNP_2$	–	8.45	4.6	–	–	−14.0	8.3	8.5	–	8.1	7.3	–
Glutamate	$AMNPQ$	–	7.33	4.65	–	–	−14.85	6.43	8.47	–	8.39	6.89	−15.89
Glutamine	$AMNPQ$	–	6.53	5.84	–	–	−14.45	6.33	9.25	–	9.16	6.35	−15.55

Another type of fast spectroscopic-imaging technique uses readout magnetic field gradient spatial encoding. Proton echo planar spectroscopic imaging (PEPSI) saves time by acquiring $K_x \times N_t$ data points, per excitation, using echo planar trajectories. An alternative, intended to achieve higher efficiency, is the acquisition of $K_x \times K_y$ data points per excitation, using trajectories in which the chemical-shift information is encoded by incrementing the evolution period for each excitation. In these methods, the spatial resolution and the scan time can be varied independently, with the scan time determining the spectral resolution. Throughout, these fast spectroscopic-imaging methods require sampling a full set of $K_x \times K_y \times N_t$ data points. Scan time reduction is achieved by using a greater number of points per excitation.

In a different approach, as is described here, a resolution/sensitivity enhancement is achieved with little effect on overall performance. Here the spectroscopic imaging is based on the concept of SENSE, originally described for fast MRI. It permits one to undersample *k*-space by using multiple receiver coils for single acquisition. In SENSE reconstruction, the spatial-encoding information lost through undersampling is recovered through knowledge of the distinct individual coil sensitivities. By applying this concept to each spatial phase-encoding dimension, the scan times required for spectroscopic imaging can be substantially reduced, without compromising spatial or spectral resolution.

SENSE utilizes spatial information related to the coil sensitivities of a receiver array for signal localization, while using the homogenous RF field of the body coil for excitation. This permits scan time reduction because it decreases the sampling density in *k*-space, which is equivalent to reducing the FOV. The sensitivities of the receiver coils provide complementary information, enabling image reconstruction free of the aliasing artifacts that would otherwise result from undersampling *k*-space.

The minimum number of receiver channels required for each reconstruction is given by the reduction factor of *k*-space sampling density. For most efficient scan time reduction, in conventional spectroscopic imaging, the SENSE method can be applied to both spatial-encoding dimensions, x and y. For a SENSE spectroscopic-imaging experiment, the FOV is reduced by a factor R_x in the x-direction and a factor R_y in the y-direction. In this way, only a fraction of *k*-space positions are sampled, as compared with full *k*-space sampling, leading to scan time reduction by the factor $R = R_x \times R_y$. Thus, if the full FOV is to be resolved by $n \times n$ spectra, only $(n/R_x) \times (n/R_y)$ individual signals need to be sampled. In the imaging domain, this sampling scheme corresponds to an $(n/R_x) \times (n/R_y)$ grid with a reduced FOV. The propagation of white noise in SENSE reconstruction leads to the relation SNR = (SNRconv (conventional)/$g\sqrt{R}$), $g \geq 1$. Here $R = R_x \times R_y$ denotes the net reduction factor, and g the local geometry factor. It depends on the shape, size, and placement of the receiver coils. The geometry factor describes the ability of the coil configuration to separate pixels superimposed by aliasing and is optimally equal to 1. In a spectroscopic-imaging experiment, the SNR of the metabolite images closely relates to the SNR of the spectra; thus, only the SNR of the spectra is considered here.

The receiver coil array consists of six elements. Two of these are circular, the other four are rectangular. The placement of these coils with respect to each other and to the object is chosen such that the geometry is close to 1. This corresponds to minimal loss in SNR. In vivo, the two circular elements are placed in the anterior and posterior positions, and the rectangular elements are arranged in pairs to the left and right of the head (Figure 3.107 [33]). The results of in vivo experiments on

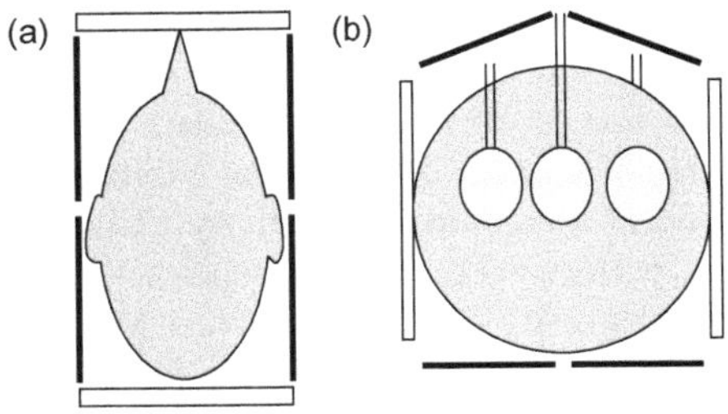

Figure 3.107 Experimental setup of the coil array *in vivo* (a) and in the phantom experiment (b). Solid black lines represent rectangular elements (10 cm × 20 cm), void lines represent circular elements (diameter 20 cm).

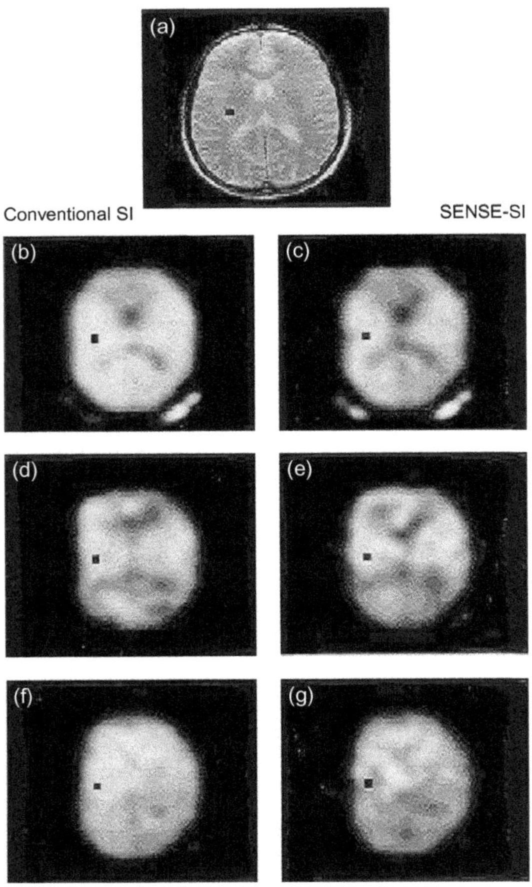

Figure 3.108 In vivo results. Comparison of 32 × 32 metabolite images from the slice shown in (a). The metabolite maps on the left-hand side—(b) NAA, (d) CRE, and (f) CHO—were acquired with conventional spectroscopic imaging in 26 min. The images on the right-hand side—(c) NAA, (e) CRE, and (g) CHO—were acquired with SENSE spectroscopic imaging within 6.5 min. All metabolite images were interpolated to 256 × 256 pixels.

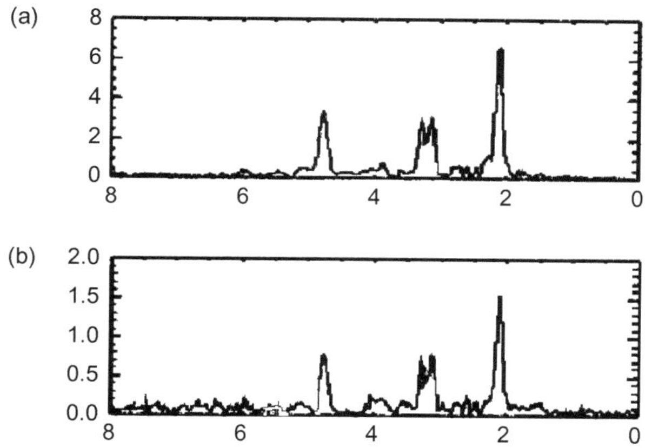

Figure 3.109 ^{1}H spectrum from the black voxel depicted in Figure 3.108, acquired with conventional spectroscopic imaging (a) and with SENSE spectroscopic imaging (b). The only discernible difference is the lower SNR in (b).

healthy volunteers are shown in Figure 3.108 [33]. For one slice of the brain, first a conventional spectroscopic-imaging experiment was performed. This took 26 min, including 4 min for the B_0 map. The following SENSE spectroscopic-imaging measurement lasted 6.5 min, including 1 min for the B_0 map.

Many fast methods exhibit loss in spectral resolution and/or restrictions in the chosen spectral bandwidth or echo time. No parameter restrictions of any kind are necessary to achieve time-saving in SENSE spectroscopic imaging, and both spectral and spatial resolutions are maintained; nor does this method show increased sensitivity to lipid contamination. An important advantage of SENSE spectroscopic imaging is its character of being a multicoil acquisition rather than a specific pulse sequence (Figure 3.109).

3.3.2.4 Spatial-Phase-Encoding (Magnetic Field Gradients): Reconstruction of Spectral Resolution/FT [34]

In MRSI, all spatial (real x, y, z coordinates on the object) information is marked by gradient phase encoding while irradiating a voxel (a selected region on an object). This information marking is then followed by a regular selection of points in phase space, or k-space, by the receiver or the sensor coil during the acquisition phase of the imaging. Those who are not familiar with the meaning of k-space should recollect the interference experiment using a diffraction grating. (This is normally a part of the first-year physics course at the university undergraduate science level—but it could well be part of the high school curriculum.) Once the light irradiates the grating, on the other side of the grating, one sees lines of different colors, separated on a screen. The longer wavelengths are located farther away from the center, following the relation $d \sin\theta \sim n\lambda$, n being the order of the spectrum and λ the wavelength. Quite similar things happen in MRSI. First the atoms and nuclei are excited by the irradiation coil (the induction coil). When the particles undergo transitions to their ground state, the emitted radiation is detected by the receiver coil. These emitted radiations will have different relaxation times, depending upon the nuclei and their location in the voxel selected. This weighting of signals is then reconstructed in k-space, producing spatial as well as spectroscopic imaging. The k-space is basically a wavelength space, because $k = 1/\lambda$, the linear wave number, or $k = 2\pi/\lambda$, the radial wave number (rad/m). This analogy should be of some help to those readers who do not have an adequate physics background.

Conventionally, the extent of the sampled k-space distribution describes a rectangular or cubic region for two or three spatial dimensions, respectively. Spectral information is sampled directly and represents an additional data dimension. The MRSI reconstruction is obtained by Fourier transformation of the resultant 3D or 4D arrays. For 3D MRSI (i.e., three spatial dimensions), the total number of acquisitions can become very large, even when a small number of phase-encoding measurements are used in each dimension. Excessively long acquisition times may then be required. One also sees overall image distortion (Gibbs ringing, GR) and signal contamination (Figure 3.110). One should remember that the phase-encoding measurements also perform the function of signal averaging. However, the effect of averaging over all k-space acquisitions may not be optimal, due to signal losses inherent in the gradient phase-encoding method, and may result in a degradation of spatial resolution. It is known that weighted acquisitions can provide improved sensitivity. It may therefore be possible to reduce the total number of phase-encoding steps and increase signal averaging, to provide improved sensitivity while maintaining the same total acquisition time. A reduced k-space acquisition method would allow weighted averaging to be used more effectively. Clinical studies carried out at typical available field strengths use a single acquisition for each phase-encoded measurement. The problem thus boils down to finding methods that either

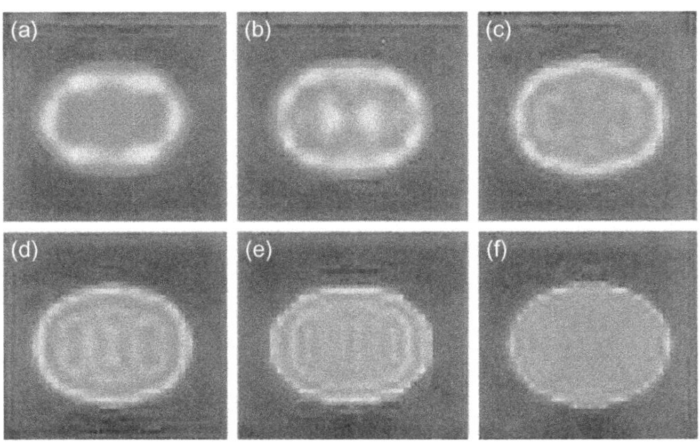

Figure 3.110 Computer simulation of the MRSI acquisition showing GR effects for different numbers of k-space points. Data are shown for a single slice through a 3D MRSI dataset with a prolate spheroidal object, and for cubic k-space dimensions of 9, 13, 17, 21, 25, and 31 points for images (a) through (f), respectively.

allow a reduction in the total number of phase-encoding measurements required or yield greater spatial resolution with the same number of encoding measurements.

To examine spatial resolution, one uses the spatial response function (SRF). This is the relative signal obtained from spins distributed over the imaging FOV, contributing to the resultant signal at a given voxel location. A computer simulation of the 4D MRSI time data, $S = S(t, l, m, n)$, is carried out for a k-space distribution described by $K = K(l, m, n)$. Here, $K = 0$ is where no k-space point is acquired; otherwise, $K = 1$. K and S are taken to be symmetrical functions, i.e., the index 1 runs from $-L/2$ to $L/2$ (extent of the object), and $S(t)$ describes a spin-echo acquisition. The object definition, $O = O(f, x, y, z)$, describes the distribution of a single spectral resonance. This is for either a solid object described as a prolate spheroid, or a solid object filling approximately 1/2 of the total FOV. The MRSI time data is given by $S(t, l, m, n) = \int x \int y \int z OK\{\exp - [i(\omega_n t + \varphi_{l,m,n}) dx\, dy\, dz]\}$; $\varphi_{l,m,n} = \gamma \int_{t=0}^{t_{\text{phenc}}}(\{l \times 2G_x/L\} + \{my2G_y/M\} + \{nz2G_z/L\})dt$. Here $\varphi_{l,m,n}$ is the phase-encoding term; γ is the GMR; G_x, G_y, and G_z are the maximum encoding gradient strengths; L, M, and N are the number of k-space measurements x, y, z; and t_{phenc} is the duration of the phase-encoding gradient application. The object is assumed to contain a homogenous distribution that is described with a data array that has 4 times greater resolution in each dimension than the normal voxel size. Data reconstruction is carried out using standard frequency FT (FFT), with zero filling used to interpolate acquired data to two points in each dimension. Below is shown the nature of the distortions (GR) that occur with 3D spectroscopic imaging. A simple slice through the center of the prolate spheroid object is shown, generated by computer simulation, with different numbers of k-space sample points and a cubic distribution. It can be seen that a large portion of the image is significantly affected by the GR intensity variations. A comparison of the spectral quality obtained is shown in Figure 3.111 [34] for the ^{31}P phantom study. This figure shows the spectroscopic images obtained using cubic k-space with $10 \times 10 \times 10$ encodings and spherical k-space acquisition with a diameter of 12 points. The total number of acquisitions is 1000 and 922. Metabolite images are shown for a single slice from the spectral region, corresponding to P_i resonance only. An examination of the spectra and the spectroscopic images show that there is no significant difference in spatial or spectroscopic quality between the two k-space distributions. This is to be expected, as approximately the same number of acquisitions were used for each method.

An even distribution of image intensity over the outer P_i volume of the phantom is obtained in both cases. An alternative method of reducing acquisition time is by using reduced k-space acquisition, which is commonly done in MRI: the encoding dimension is only partially sampled, which in the limit means that only that one-half of the data need to be acquired. This approach is suitable only if complete spin-echo (i.e., symmetrical) data are available in the other dimensions. Because the acquired data are asymmetric, the image data will have a significant dispersive component following FFT. The magnitude of the image data is therefore no longer suitable for viewing, and the real part of the image data is required. This requires accurate phase correction of the data. Repetition of the simulations, but using a reduced number of encoding measurements in one dimension, with index 1 running

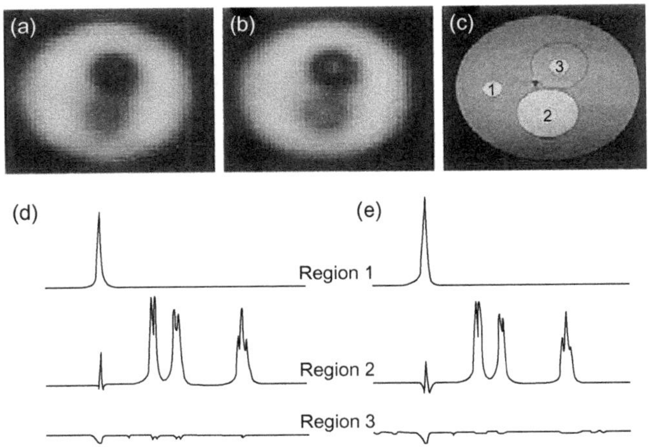

Figure 3.111 Phantom study comparing spectra obtained using cubic and spherical encoding. (a) Spectroscopic image of the P_1 region obtained using cubic k-space with $10 \times 10 \times 10$ encodings. (b) Corresponding image obtained using spherical k-space with a diameter of 12 points. (c) MRI at the same section, showing the individual compartments in the phantom. The numbers indicate the regions from which spectra were obtained, where (1) is in the outer region that contained P_1 only, (2) is in the section containing ATP and P_1, and (3) is the zero-signal region. The corresponding spectra from the cubic and spherical encoding acquisitions are shown in (d) and (e), respectively. The real parts of these spectra are shown.

from 0 to $L/2$, produced identical results: i.e., the real part of the data was observed. The major effect of reduced k-space acquisition is a reduction in SNR of the resultant data. An additional concern is the necessity to display real images in the presence of the spatially dependent phase shifts caused by the eddy current distortions.

With MRSI, because all spatial information is acquired via phase encoding, this effect is much reduced, and one observes no phase variation in ^{1}H MRS studies. Shown below are the results of an experimental ^{1}H MRS acquisition of an axial section through a rat brain, for full and 1/2 k-space acquisition. The MRSI method obtains complete spectral information. Only the NAA distribution is shown, though comparison of metabolic images created for creatine and choline show similar results. Figure 3.112(a) [34] shows a single slice from a 3D ^{1}H MRS, obtained using $16 \times 16 \times 16$ phase-encoding measurements and spin-echo acquisition. In Figure 3.112(b), the acquisition matrix is reduced to $16 \times 16 \times 9$ and the real part of the NAA images from the slice is displayed. The images appear very similar, except for the reduced SNR of Figure 3.112(b) [34]. Figure 3.112(c) [34] shows the T_2-weighted MRI for better visualization of the lower-resolution NAA images.

The use of 1/2 k-space in more than one dimension results in a spatially dependent phase shift of the data, causing severe image distortions. This applies to any

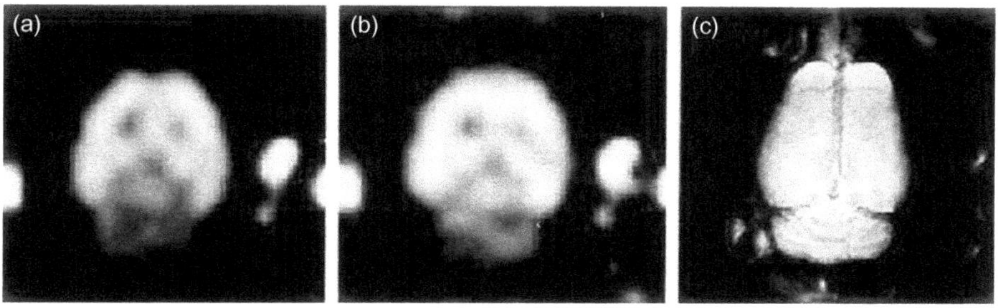

Figure 3.112 NAA image obtained from an ^1H MRSI acquisition of rat brain, obtained with (a) complete *k*-space distribution, and (b) 1/2 *k*-space in one dimension. (c) The MRI from the same section.

two dimensions, thus also making 1/2 *k*-space acquisition unsuitable for FID or half-echo acquisition.

3.3.2.5 Computer Simulation: Magnetic Field Gradient Spatial Imaging and Spectral Metabolite Characteristics [35]

The QM formulations, e.g., MM algebra, though very logical and straightforward, run into volume (size) difficulties as the number of species of spins in a system becomes larger than two. Unfortunately, a real system such as the human brain has several different species of nuclei and compounds. Thus, one naturally turns to use of a computer to simulate theoretical models, and to obtain experimental data to compare with the models, to work out spectral characteristics. One often uses a STEAM sequence experimentally to obtain spectroscopic imaging of the human brain. In a computer model, one must consider spatially dependent gradient-induced phases for spin coherences. The approach includes a search for optimal values for multiple-sequence parameters for observation of a variety of spin systems, significant reduction in simulation times, and the like.

One can take advantage of insights gained from product-operator formalism, regarding coherence transfer, to generate a computationally efficient simulation scheme for simulating gradients, and therefore for simulating STEAM. The effects of the gradients in the initial and final TE periods shown in Figure 3.113 [35] are to change the phase terms of coherence order p, by an amount $-p\gamma(Gr)t$. Here γ is the GMR of the nucleus being considered, G is the gradient vector, r is the position vector of the spin system, and t is the duration of the gradient pulse (Figure 3.114).

3.3.2.6 Spectroscopic Imaging: A Prior GAMMA Computer Simulation [36]

A standard approach for analyzing signal contributions in MR spectroscopic data is to select a signal model based on the known characteristics of the system under investigation and then to determine the model parameters that best describe the data. It is a significant advantage to have an automated procedure for determining these parameters, and a number of suitable techniques have been described for

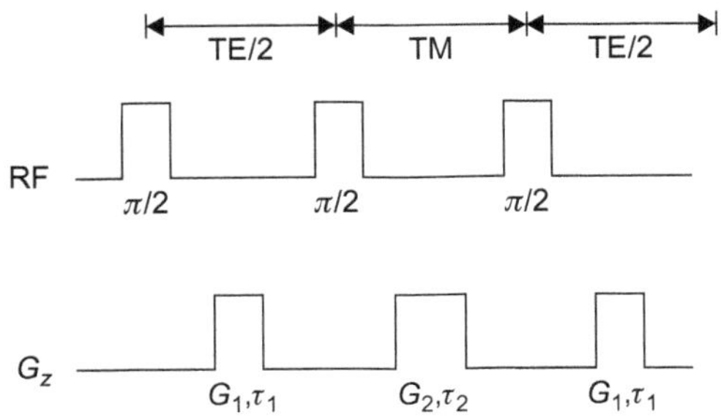

Figure 3.113 Schematic of the STEAM pulse sequence. This shows RF pulses with the same phase, and gradient pulses in the z-direction, as was used experimentally.

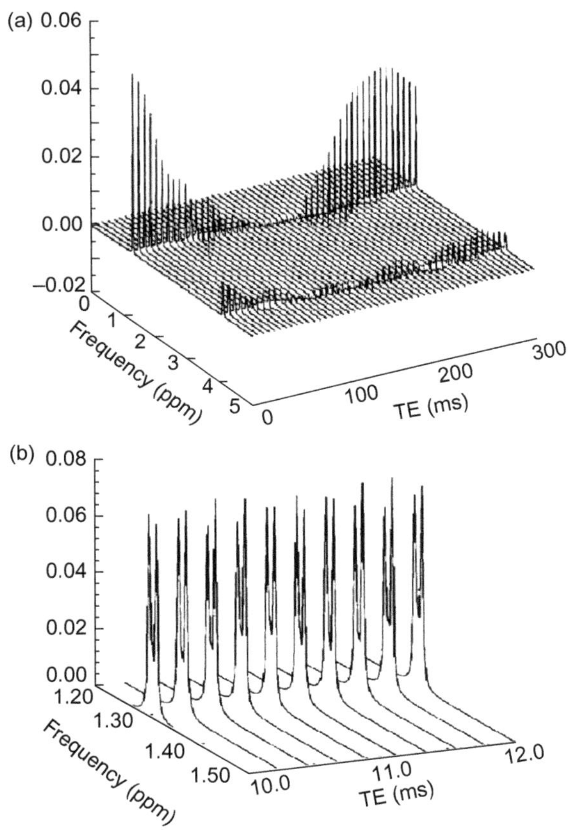

Figure 3.114 (a) Stack plot of GAMMA simulations for the STEAM sequence of lactate as a function of echo time, and a fixed measurement-time value of 12 ms. The plot illustrates the variation in amplitude of the doublet and quartet resonance groups due to the combined effects of chemical shift and coupling evolution. (b) Stack plot of the evolution of the doublet resonance group for TE values between 10 and 12 ms, and a fixed TM value of 12 ms. The plot illustrates the phase variation of the doublet that occurs at a frequency of 1.7 KHz.

automated analysis of in vivo MRS data. Methods that use least-squares optimization and include known spectral characteristics to minimize parameter selection have been shown to offer improved accuracy and robustness over other nonparametric methods. In the absence of efficient global optimization techniques, the accuracy and speed of parameter optimization is significantly improved by the use of accurate a priori information, good initial parameter estimates, and optimization constraints. For in vivo MRS, many observable tissue metabolites have been identified and their spectra characterized, providing a strong foundation for spectral modeling methods. However, the available information may be incomplete when spin-echo acquisition sequences are used, as is frequently the case for in vivo localization techniques.

For these sequences, the phase of resonances from spin-coupled compounds will be altered due to the dynamics of coupled-spin systems. The subsequent overlap between neighboring resonances, as well as superposition with resonances from other compounds, may result in considerable alteration of the expected spectral patterns. One can address this difficulty by obtaining an in vitro sample spectrum for each metabolite to be analyzed, using the exact pulse sequence as for the subsequent MRS measurement. The resultant suite of spectra is then used as model functions for a parametric analysis routine.

This method requires regeneration of model functions for each pulse sequence and field strength used, which necessitates careful sample preparation and measurement procedures. In this discussion, we address the need to conveniently and accurately generate the a priori information used in a parameter optimization procedure for analysis of data obtained using any acquisition method. The method utilizes a powerful NMR spectral simulation package called general approach to magnetic resonance by mathematical analysis (GAMMA) (code and examples are available at http://www.gamma.magnet.fsu.edu/), to generate a priori spectral information based upon knowledge of the molecular interactions underlying the MRS data acquisition. Known chemical shifts, scalar spin couplings, and acquisition sequences are used to create a spectral database of resonance frequencies, phases, and relative amplitudes. One particular use of this information is to provide initial estimates and model parameters for automated parametric spectral analysis of in vivo ^{1}H spectra.

GAMMA is an object-oriented library written in C++ that facilitates software development for simulation of NMR experiments. It takes full advantage of object-oriented design concepts and requires only basic programming expertise on the part of the user. The entities and operations common to NMR experiments are provided as programming objects that extend the C++ language, allowing matrices, operators, commutators, spin systems, and RF pulses to be simply described and applied in program procedures. GAMMA is a broad and flexible tool for NMR analysis. It is known that for a particular field strength and pulse sequence, the "a priori knowledge (API)" can be generated for a parametric spectral model that would include the frequencies, phases, and relative amplitudes on all resonances. This is for all compounds to be analyzed. To gain this information, it is first necessary to provide the chemical shift and the spin–spin coupling constants for each compound, the Hamiltonian of the spin interactions, the nuclei and isotopes type (^{1}H, ^{13}C, ^{31}P, etc.), the pulse sequence type and timings, and the spectrometer frequency. The API can then be used for analysis of in vivo spectra from brain tissue. The coupling constants and chemical-shift values can be obtained from the literature. Alternatively, one can perform one's own high-resolution NMR studies of metabolites in solution.

The parameters returned from the spectral simulation are stored in a database, and this information is made available to a parametric spectral analysis procedure for any new experiment. The spectral data returned by the GAMMA simulation program include all possible transitions between every nucleus, many of which are degenerate or have insignificant amplitude. The program combines the degenerate ones and eliminates the ones with negligible amplitudes. This information is transferred back to the database. This is done by having the spectral simulation program return the spectral data with a finite line broadening of 1 Hz. A minimal amplitude threshold and a peak-picking process are then put in place, so as to return only the significant resonances. The phase of each resonance is determined by evaluating the complex data value at the location of the maximum, of the magnitude value, of the peak. The accuracy of this method depends on the sampling resolution used. The spectral simulation data are generated with a finite sampling resolution, using 64 K data points. This results in an accuracy in the phase measurement of better than 0.4°. Finally, the database resonance amplitudes are scaled according to the number of protons contributing to the signal, i.e., the returned values are divided by the total signal integral and multiplied by the number of protons in that compound. This normalization simplifies subsequent scaling of integral values, obtained from a parametric fit procedure, to obtain relative concentrations. One is then ready for a new database entry. The parameters returned from the spectral simulation are stored in a standard database format. The input parameters for the simulation are stored in two sets of tables. One set has information on the pulse sequence, including the timings and the RF pulse information. The other holds the chemical shifts and the coupling constants of all the compounds. The results from the spectral simulation are stored in tables containing multiple records, with one record for

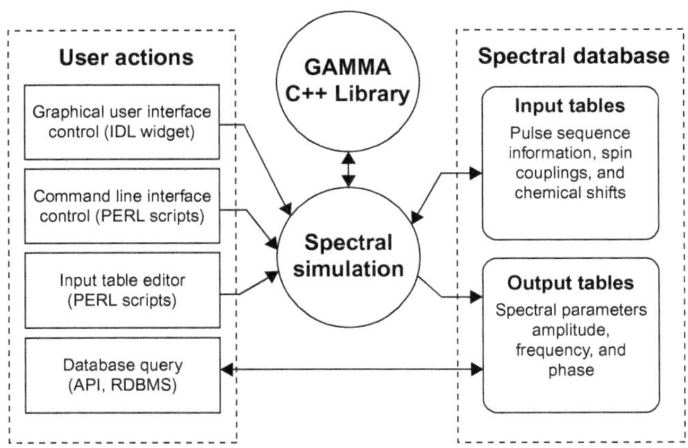

Figure 3.115 Schematic of the relationship between user interface modules, GAMMA simulation, and the relational database. Interface modules automatically generate C++ code that calls routines from the GAMMA library, and results are stored as ASCII records in a relational database. Users can edit the values in the input databases that contain acquisition sequence descriptors and spin-coupling and chemical-shift information.

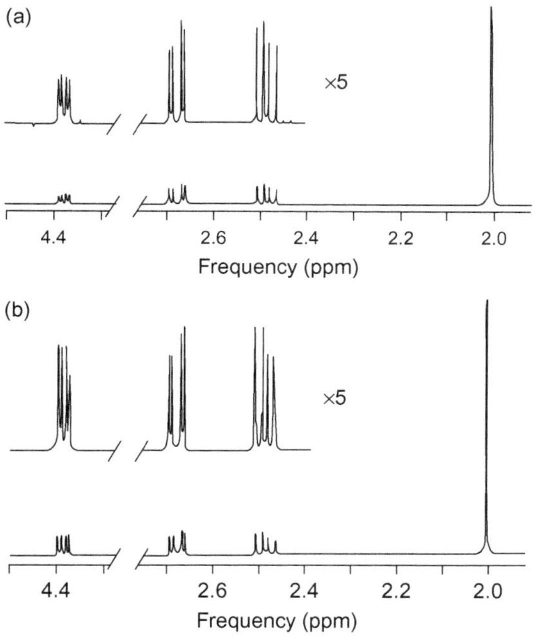

Figure 3.116 Comparison of (a) experimental acquisition and (b) spectral simulation for NAA at 600 MHz for a pulse-acquire sequence.

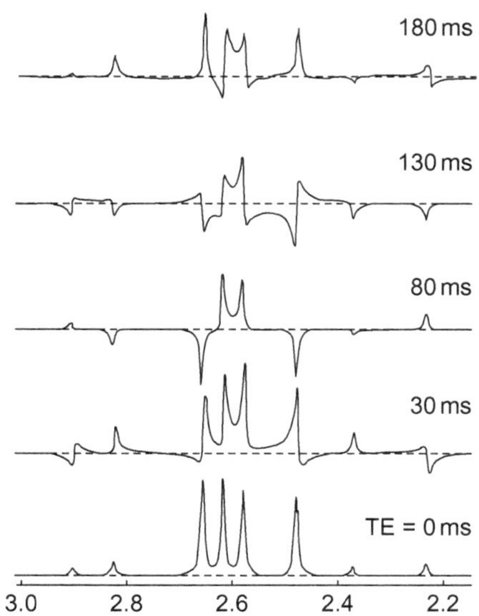

Figure 3.117 Example of the simulation output for a spin-echo acquisition. Resonances of the CH_2 multiplet group of NAA, centered at 2.6 ppm, are shown for different TE values at 64 MHz. A 1.0-Hz line broadening has been applied.

each combination of nucleus, pulse sequence type, acquisition timing, and field strength. All records are keyed, by name attribute, to the input information used in their creation. Figures 3.115–3.117 [36] provide a rough sketch that indicates the connections between the user interface, the simulation program, and the relational database.

3.3.2.7 Density Matrix: AX_3 Spin System [37]

One can work back and forth between experiment and a computer simulation in NMR. Such an approach involves both spatial imaging and spectroscopic imaging. One can keep improving the input parameters of the simulation (frequency of resonance, chemical shift, *J*-coupling, etc.) until there is perfect agreement between the computer model and the observed results. At that point, one has worked out the spectral characteristics of the metabolites one is looking for. The spectroscopic image can be analyzed on a voxel-by-voxel basis, thus revealing more and more localized (direct) spin-coupling characteristics of, say, a large molecule-like structure in each voxel. As the voxel size is gradually increased, one looks for more and more global characteristics about the phenomena that are happening in the brain as a whole. An alternative approach is also available. With that approach, one first develops an analytic mathematical expression for the intensity profile, using density-operator formalism. This is then plotted alongside the observed results to get the desired characteristics results, e.g., chemical shift, *J*-coupling, and so forth.

This approach is easy when a fairly small number of spin species (say, three or four) are coupled. However, as the number of spins increases, so does the volume of the analytic mathematical expressions. This problem can be alleviated, to a large extent, by using the Mathematica computer package. This program allows long symbolic mathematical expressions to be easily derived, and can be kept in storage in the computer, thus avoiding the hassle of printing out large volumes. The density operators involved in a spin AX_3 (doublet) system, in the STEAM sequence using a spherical coordinate system, are briefly expressed:

Table 3.30 AX_3 Spin system

Density Operator Spin *X*	Density Operator Spin *A*
After the first 90° pulse, the initial density operator is $\sigma_x(0+) = -(\mathrm{i}/\sqrt{2})(X_+ + X_-)$. The starting operator is $\sigma(0) = X_z + A_z$. Here *X* represents the doublet terms, *A* is the quartet term. $X_u = \Sigma_i X_{iu}$ with $u = x, y, z$, and $i = 1, 2, 3$. Here attention is confined to the doublet terms.	After the first 90° pulse, the initial density operator is $\sigma_A(0+) = -(\mathrm{i}/\sqrt{2})(A_+ + A_-)$. The ultimate doublet signal will result first from terms that reflect the evolution of the initial doublet term in the density operator, and second from magnetization transferred between quartet and doublet.
The evolution during TE/2, and under gradient G_1, results in decoherence, and the result is $\sigma_x(\mathrm{TE}/2-) = [-(\mathrm{i}/\sqrt{2})(X_+\exp - \mathrm{i}\theta + X_-\ \exp\mathrm{i}\theta)\cos(\pi \mathrm{JTE}/2)] - [\{\sqrt{2}(X_+\exp - \mathrm{i}\theta - X_-\exp\mathrm{i}\theta)\}\{A_z\sin(\pi\mathrm{JTE}/2)\}]$. There is dephasing of coherence of order *p* by $-p\gamma(G_1 r)t_1$, where *r* is the position vector and *G* is the GMR. $\theta = [\omega_x \mathrm{TE}/2 + \gamma(G_1 r)t_1]$ represents the dephasing due to the doublet chemical shift ω_X relative to the TF and to the gradient G_1.	Following the first 90° pulse, the magnetization transfer between quartet and doublet takes place. The terms that cannot transfer transverse magnetization to the doublet are negligible. The equilibrium operator for the quartet A_z evolves under *J*-couplings, chemical shift, and gradient field, giving $G_1\ \sigma_A(\mathrm{TE}/2-) = [\sqrt{2}(A_-\exp\mathrm{i}\varphi - A_+\exp - \mathrm{i}\theta)X_z][\cos^2(\pi\mathrm{JTE}/2)\sin(\pi\mathrm{JTE}/2)]$. $\varphi = [\omega_A \mathrm{TE}/2 + \gamma(G_1 r)t_1]$.

Table 3.30 (continued)

Density Operator Spin X	**Density Operator Spin A**
In anticipation of the refocusing effects of the gradient in the second TE/2 period, one can confine attention in the TM (measurement) period to those terms which are insensitive to G_2. So, immediately after the second $90°_x$ pulse, one gets $\sigma_x(\text{TE/2}+)$ (Figure 3.118).	Following the second 90° pulse, the terms in the equation above, which are insensitive to gradients during TM, evolve as $\sigma_A(\text{TE/2+TM}-)$. The terms that are not refocused in the acquisition are neglected.
In the TM period, one gets $\sigma_x(\text{TE/2+TM}-)$.	In the the TM period, one gets $\sigma_A(\text{TE+TM}-)$
The final $90°x$: $\sigma_x(\text{TE}+\text{TM})$	The final $90°x$: $\sigma_A(\text{TE}+\text{TM})$
The final result is $\sigma_x\ (\text{TE}+\text{TM}+)=(\text{i}/4\sqrt{2})[\{X_+ + X_-\}\{2\cos^2(\pi\text{JTE}/2)-\sin^2(\pi\text{JTE}/2)\cos^2(\pi\text{JTM})\cos(\delta\omega\text{TM})\}]$; $\delta\omega=\omega_X-\omega_A$ is the chemical-shift difference between the doublet and the quartet. The antiphase 2-spin coherence terms that give rise to no observable magnetization are omitted.	In the quadrature detection, 90° receiver phase shift (aligned along the y-axis), the net doublet signal is observed as $S(t)=(1/2)[\cos^2(\pi\text{JTE}/2)-(1/2)\sin^2(\pi/2)\{1-\cos^2(\pi\text{JTE}/2)\cos(\delta\omega\text{TE}/2)\}\cos^2(\pi\text{JTM})\cos(\delta\omega\text{TM})]$.

Using the STEAM sequence, the doublet density operator (σ) in the AX_3 spin system is shown in the preceding table. The equation for $S(t)$ (Table 3.30) and Figure 3.119 emphasize, in addition to the TM dependence of the ZQC component, the significant effects that TE has on that TM dependence (Figure 3.118).

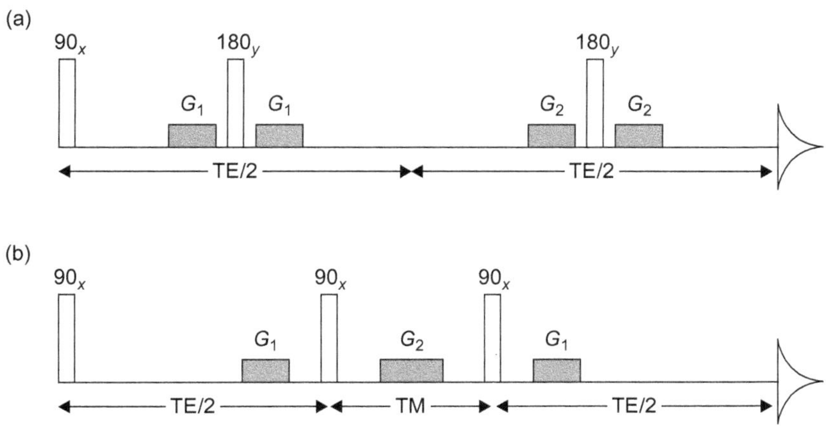

Figure 3.118 The pulse sequences used for experimental and theoretical analysis. (a) An approximation of the PRESS. (b) An approximation of the STEAM sequence. G_1 and G_2 are orthogonal field gradients.

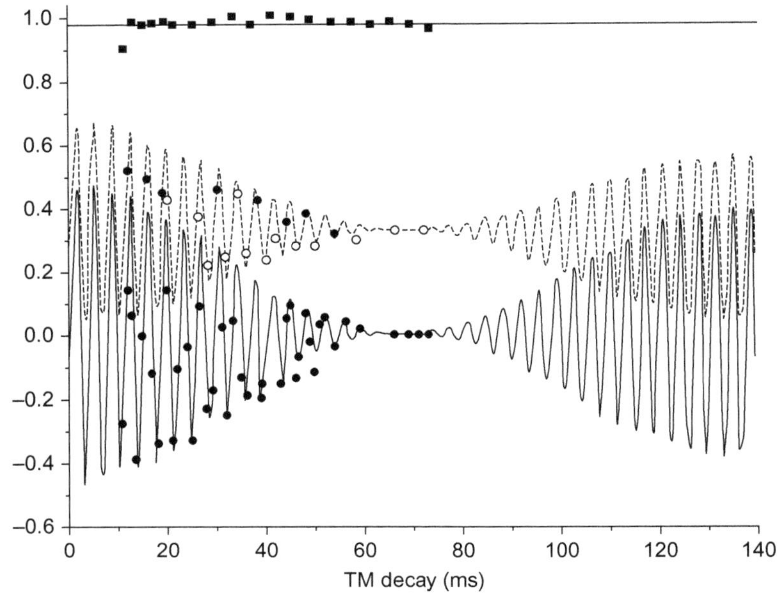

Figure 3.119 An illustration of that TM periodicities of the lactate doublet intensity when observed by STEAM at the half-echo times TE/2 = 8 ms. Dark square −40 ms and dark circle − 70 ms. The lines represent the theoretical curves, and the points are experimental data. $J = 7.3$ Hz was assumed.

3.3.2.8 Human Brain Metabolite Quantification: PRESS Spins Weak–Strong Coupling [38]

The response of coupled spins to a localization MRS pulse sequence is quite variable. First, it gives rise to misleading comparisons with the response of uncoupled spins; second, it is a potential source of error in comparisons between different sites using different pulse designs. Advanced numerical methods are now available to treat systems with five or more spins that are strongly coupled. Five and more spin species are widely spread in the human brain, including the amino acids, glutamate (Glu), and glutamine (Gln). The coupled-spin response to the PRESS in Figure 3.120 [38] is substantially more variable than that of the uncoupled spins. In the first instance, there is dependence on the length (bandwidth) and on the spectral roll-off of the RF pulses. One can demonstrate the effects for the strongly coupled 5-spin *AMNPQ* system of Glu. The consequences of providing alternative coherence pathways and signal loss mechanisms through the slice-selection procedures do not seem to have been fully clarified, even in the weak coupling limit. The AX_3 spin system of lactate (weak coupling) is used to characterize the intrapulse effects. It is appropriate to mention here that in most in vivo studies in the past, the action of the pulse sequence has not been an issue. This is because these studies have confined their attention to the uncoupled singlet resonances of NAA at 2.02 ppm, of creatine

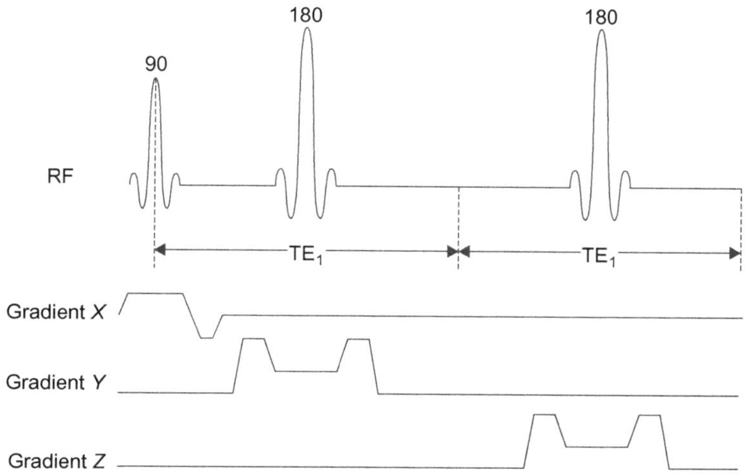

Figure 3.120 A schematic diagram of a PRESS. The excitation 90° pulse is a 2000-Hz bandwidth sinc-gaussian of length 3.5 ms. The refocusing pulses are optimized sinc pulses of 1200-Hz bandwidth and a length of 3.5 ms, interspersed between 2-ms, 10-mT/m spoiler gradients.

at 3.05 ppm, and of choline at 3.2 ppm, for which the spin dynamics are straightforward.

It is known that most metabolites contain only coupled spins, and if metabolic quantification is to go beyond uncoupled singlets, the modification of the lineshapes and the loss in integrated signal intensity of the coupled-spin multiplets must be quantitatively understood. To demonstrate the action of different slice-selective 180° designs, truncated transformation matrices (containing only eight representative coherences of the AX_3 system) are illustrated for five progressive variants of an $180°_x$ in Figure 3.121 [38]. The pulses illustrated in Figure 3.121 (a)−(c) were chosen to demonstrate the three intrapulse effects.

The first pulse variant, an ideal sharp rectangular pulse with an infinitely wide bandwidth and applied without a concurrent magnetic field gradient, eliminates the edge distribution of tip (spin) angle, the excitation band shift, and evolutions occurring during a selective pulse. This variant is equivalent to the hard pulses assumed in product operator mathematics and its transformation matrix. Figure 3.121(a) [38] reflects its ideal nature, as 100% of each input coherence term is recovered following the pulse.

A sign change in the transformation matrix denotes inversion of the input. A second variant, designed to prevent any spatial shift of the excitation band, as well as any intrapulse chemical shift or scalar-coupling evolutions, can still be made to illustrate the tip-angle effects. This is done with time-scaling, by 10^{-3},

$1 \equiv A_x$
$2 \equiv 2A_xX_{1z}$
$3 \equiv A_Y$
$4 \equiv 2A_yX_{1z}$
$5 \equiv X_{1x}$
$6 \equiv 2A_zX_{1x}$
$7 \equiv X_{1y}$
$8 \equiv 2A_zX_{1y}$

(a) Variant 1 *(hard pulses)*

Input \ Output	1	2	3	4	5	6	7	8
1	100	0	0	0	0	0	0	0
2	0	−100	0	0	0	0	0	0
3	0	0	−100	0	0	0	0	0
4	0	0	0	100	0	0	0	0
5	0	0	0	0	100	0	0	0
6	0	0	0	0	0	−100	0	0
7	0	0	0	0	0	0	−100	0
8	0	0	0	0	0	0	0	100

(b) Variant 2 (Optimized sinc pulse Δ = 1.2 kHz)

Input \ Output	1	2	3	4	5	6	7	8
1	100	0	0	0	0	0	0	0
2	0	−76	0	0	0	24	0	0
3	0	0	−100	0	0	0	0	0
4	0	0	0	76	0	0	0	−24
5	0	0	0	0	100	0	0	0
6	0	24	0	0	0	−76	0	0
7	0	0	0	0	0	0	−100	0
8	0	0	0	−24	0	0	0	76

(c) Variant 3 (Optimized sinc pulse Δ = 1.2 kHz)

Input \ Output	1	2	3	4	5	6	7	8
1	98	0	0	5	0	−2	0	1
2	0	−38	−5	3	2	1	−1	6
3	0	5	−98	0	0	1	0	2
4	−5	3	0	38	−1	6	−2	−1
5	0	2	0	1	98	0	0	5
6	−2	1	−1	−6	0	−38	−5	−3
7	0	1	0	−2	0	5	−98	0
8	−1	−6	2	−1	−5	−3	0	38

(d) Variant 4 (Sinc-Hamming pulse Δ = 1.2 kHz)

Input \ Output	1	2	3	4	5	6	7	8
1	99	0	0	5	0	−2	0	2
2	0	−33	−5	−4	2	−19	−2	5
3	0	5	−99	0	0	2	0	2
4	−5	−4	0	33	−2	5	−2	19
5	0	2	0	2	100	0	0	5
6	−2	−19	−2	−5	0	−33	−5	4
7	0	2	0	−2	0	5	−100	0
8	−2	−5	2	19	−5	4	0	33

(e) Variant 5 (Gaussian pulse Δ = 1.2 kHz)

Input \ Output	1	2	3	4	5	6	7	8
1	100	−1	0	3	0	−3	0	0
2	0	−25	−3	−3	3	−18	1	−39
3	0	3	−100	1	0	0	0	3
4	−3	−3	0	25	1	−39	−3	18
5	0	3	0	0	100	1	0	3
6	−3	−18	1	39	0	−25	−3	3
7	0	0	0	−3	0	3	−100	−1
8	1	39	3	18	−3	3	0	25

Figure 3.121 Transformation matrices of a representative set of lactate coherences for five variants of a 180° refocusing pulse; specifically, an ideal hard refocusing pulse (a), an artificial wideband-selective pulse (b), and three pulses with realistic bandwidths of 1.2 kHz, namely, an optimized sinc (c), a sinc-Hamming pulse (d), and a Gaussian pulse (e). The highlighted elements represent the nonzero contributions and illustrate that the off-diagonal terms increase in both number and magnitude with worsening of pulse quality from (a) to (e).

an acceptable selective pulse envelope to an artificially short scheme. The form of this variant is a three-lobe-optimized sinc pulse, of length 3.5 μs, applied in conjunction with a 1.7-T/m gradient pulse to refocus a slice of approximately 1.66-cm thick. Its transformation matrix, shown in Figure 3.121(b) [38], demonstrates that each antiphase coherence input (the even input elements) is transformed to a mixture of antiphase terms following the slice-selective 180° pulse. Several important brain metabolites are strongly coupled at in vivo field strengths (1.5–4 T), including, Glu, Gln, aspartate, NAA, NAAG, taurine, and myoinositol. Although the resonance of the simplest of these—namely, the *AB* part of the *ABX* spin system of Asp and NAAG—has been explored, using both product-operator formalism and the full numerical solution, the treatment of larger, strongly coupled spin systems is still progressing. As an illustration, one should look at the response of the strongly coupled *AMNPQ* system of Glu to PRESS, where Glu was defined by the parameters $\omega_A = 3.74$ ppm, $\omega_M = 2.042$, $\omega_N = 2.121$ ppm, $\omega_Q = 2.351$, $J_{AM} = 7.33$ Hz, $J_{AN} = 4.65$ Hz, $J_{MP} = 6.41$ Hz, $J_{MQ} = 8.4$ Hz, $J_{NP} = 8.48$ Hz, $J_{NQ} = 6.88$ Hz, and $J_{PQ} = 15.93$ Hz. Focusing on the 2.0–2.45-ppm range of the proton spectrum, one can calculate the *MN* (2.08 ppm) and *PQ* (2.34) multiplets of the 5-spin system. The irregular positive and negative excursions of the *PQ* and *MN* multiplets, coupled to their complex evolution with echo time, can make a simple area measurement a somewhat misleading measure of the spectral yield.

Although peak height is not entirely reliable for quantification either, the principal peak of the Glu *PQ* multiplet, centered at 2.34 ppm, is a stable feature of this multiplet evolution, and so it is used to reflect the variability of Glu yield from PRESS. In contrast to the weak coupling limit, coherence transfer takes place even when the ideal hard 180° pulses are employed. During the strongly coupled evolution, the coherence terms proliferate, and substantial polarization transfer takes place between *M*, *N*, *P*, and *Q* spins.

This is demonstrated in Figure 3.122 [38]. Bearing in mind that in practice the initial state comprises the *x* magnetization of the *A*, *M*, *N*, *P*, and *Q* spins, Figure 3.122(a) [38] assumes an initial state exclusively of P_x spin magnetization and emphasizes the proliferation by providing a snapshot (at a single TE = 30 ms) of the amplitudes of all the 160 single-quantum terms that were tracked for Glu. Figure 3.122(b) [38], in contrast, illustrates how several representative in-phase and antiphase coherences have evolved at TE_1, in a hard-pulse PRESS (equivalent to a spin-echo experiment) where $0 < TE_1 < 100$ ms.

Some of the terms illustrated have no *P*-spin operator involved, emphasizing the polarization transfer.

The ultimate influence of the slice-selection pulse design is therefore to mitigate the reduction in the intensity of the *PQ* multiplet by minimizing the redistribution, Figure 3.122(b) [38]. Using the Cr 3.05 ppm line as the intensity standard, the experimental and calculated results are compared for both the *MNPQ* multiplet shape and the peak amplitude of the 2.34 ppm across 45° cuts through (TE_1-TE_2) space that maintained TE_1-TE_2 (symmetric PRESS) and also maintained constant TE, Figure 3.123 [38].

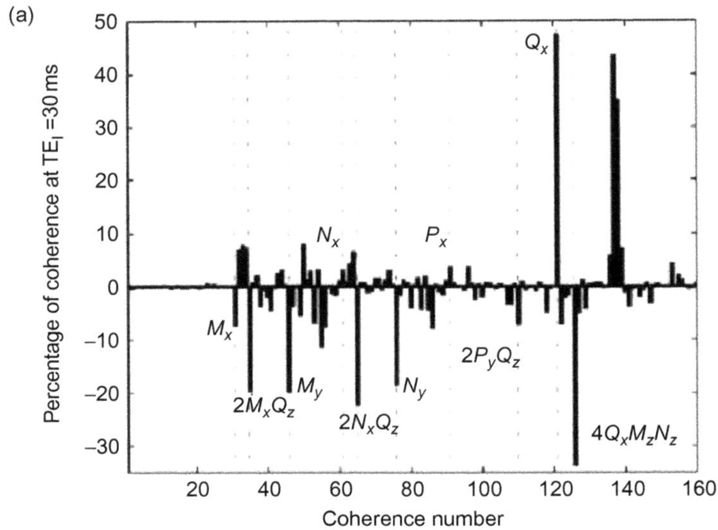

Figure 3.122 (a) Snapshot at TE_1 of 30 ms of the distribution across 160 single-quantum terms of the *AMNPQ* spin system of Glu. (b) The variation of several of these terms at TE_1 as TE_1 is varied from 0 to 100 ms, showing the evolution of Glu *P*-spin magnetization into the coherences of the *M*, *N*, and *Q* spins.

3.3.2.9 3D Imaging of Human Brain In Vivo: 1D Hadamard Spectroscopic Imaging and 2D CSI and Assigning Spectroscopy of Metabolites [39]

Imaging of the human brain requires suppression of the subcutaneous lipid and bone marrow signals, to remove contamination of the much smaller brain metabolite signals. This is done most effectively by outer-volume suppression (OVS) combined with selective excitation of the VOI.

Finer localization within that volume is achieved with 2D CSI (Figure 3.124 [39]). OVS favors a VOI within an axial or paraxial slice, so that the fat at the rim can easily be saturated by spatially selective RF pulses. For these pulses to contain the skull's curvature, the slice must be thin (a few centimeters), ruling out the use of CSI in that direction. To perform localization, a one-voxel-thick slice is excited and 2D CSI performed in its plane. The 3D coverage is achieved by sequentially interleaving *N* (usually four) single slices. To keep acquisition time small, one acquires all *N* slices in the VOI with a hybrid of CSI and Hadamard spectroscopic imaging (HSI). Because both methods excite the spins in the entire VOI, the repetition time can be made equal to the acquisition time T_{acq}, to produce an *N*-fold increase in the number of acquisitions per given time, and a $\sqrt{N}$ gain in the SNR, compared with slice-interleaving. In addition, HSI, which isolates well with four partitions, is suited to the short axis of the VOI. CSI is widely used nowadays to obtain 2D arrays of localized ^1H (proton) spectra in the human brain. Because CSI-obtained spectra can be voxel-shifted in postprocessing, precise time-consuming

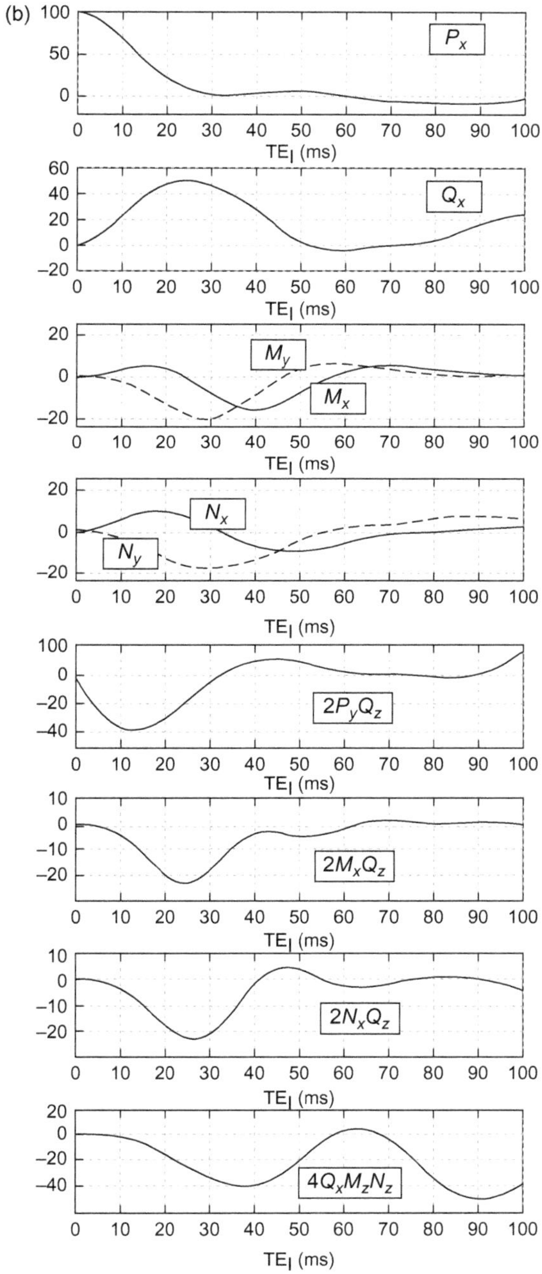

Figure 3.122 continued

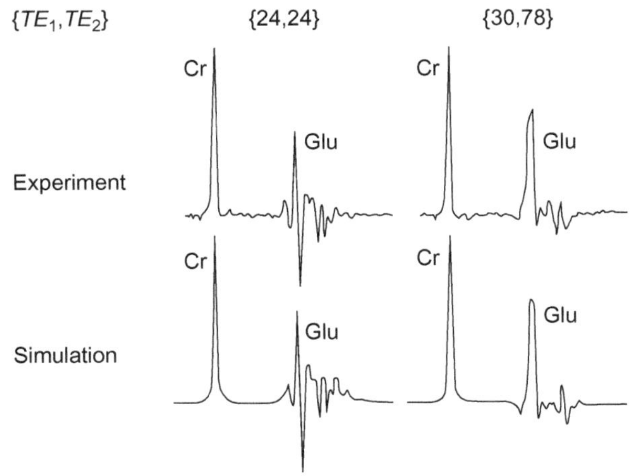

Figure 3.123 Two illustrations of the experimental and calculated Glu response to a localizing PRESS, incorporating the optimized sinc refocusing pulses. The experimental solution spectra were also broadened to 5 Hz through postprocessing with an exponential multiplication in the time domain. For both timing examples, the measured *PQ* multiplet peak value is within 2% of the numerically predicted value.

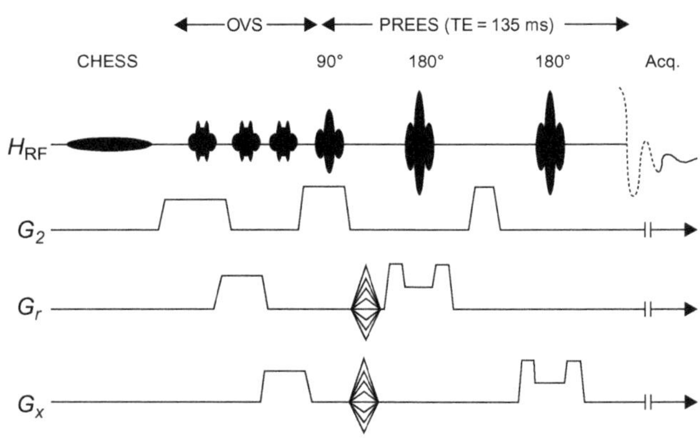

Figure 3.124 The hybrid HSI-CSI localization sequence used in the 3D ^{1}H MRS experiment. A 25.6-ms, 60-Hz bandwidth CHESS was followed by three 5.12-ms-long, dual-lobe, time-shifted $90 + C_i$, $C_i = 17°$, 10°, and 5°, OVS sinc pulses. The VOI was selectively excited by PRESS. Its 5.12 ms 90± ("±"denotes 0/180° phase alternation) pulse and a fourth-order HSI encoded along *Z* under a 3-mT/M gradient. The PRESS 180° pulses were 5.12 ms long under 1 mT/M. 16 × 16 2D CSI was performed during the echo time by phase-encoding gradient pulses along *x* and *y*.

placement of the VOI over the pathology during examination is unnecessary. However, CSI suffers from two disadvantages. First, unless selective excitation and/or OVS are used, the FOV must be larger than the object, to prevent aliasing extraneous signals. Second, spectral contamination from outside the voxel occurs as a result of the point spread function (PSF). This intrinsic artifact, generally known as *voxel bleed*, is caused by restricted k-space sampling, and worsens for similar ratios of FOV/number of phase-encodes. In transverse HSI, the spin's relative phases, $\varphi = 0$ or $180°$, denoted " + " or " − ," correspond to the ± 1s of the jth row of the Hadamard matrix of order N, $N = 2^n$, $n = 1, 2, 3, \ldots$. This is accomplished by applying a shaped RF pulse in the presence of a gradient.

The experiment is repeated N times, each for a different row in the matrix. In an in vivo human brain situation, the hybrid was performed, using a 7.5 cm × {QJ} 7.5 cm × 6 cm PRESS box. The PRESS box was excited, within the brain, with the OVS shown in Figure 3.125 [39]. The FOV was portioned into 16 × 16 2D CSI and fourth-order HSI to yield 1.5 cm × 1.5 cm × 1.5 cm voxels. At TR = 1.6 s, optimal for metabolite T_1s of 1−1.4 s and 90° nutation angle, the measurement required 27 min. The spectra from the HSI slices (a)−(d) at the coronal projection in Figure 3.125 [39] are superimposed on the axial images in Figure 3.126 [39]. It shows a good correlation between spectral characteristics and the underlying anatomy. There are larger signals in GM slices ((a) and (b), center column), and smaller signals where the voxels involve ventricle slices (Figure 3.126(c) and (d)). None of the 100 spectra display any extraneous fat contamination up field from the NAA signal at 2.0 ppm.

This qualifies the hybrid for studies of abnormal lipid, alanine, and lactate resonances associated with tumors and strokes. The SNR is sufficient for analysis of peak areas of NAA at 2.0 ppm, choline at 3.0 ppm, and creatine/phosphocreatine

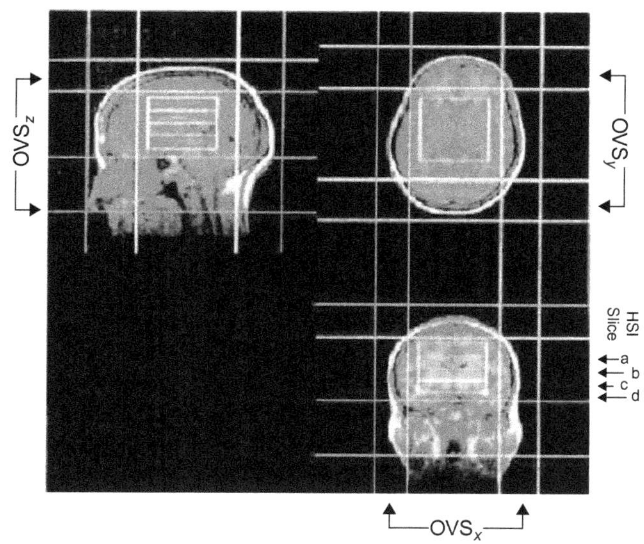

Figure 3.125 The graphic "front-end" of the image-guided hybrid MRS software. The PRESS box is interactively placed on any image with the correct perspectives automatically maintained on the others. The HSI direction is then chosen and OVS bands individually placed in each direction. The software then generates all the shaped pulses in Figure 3.124.

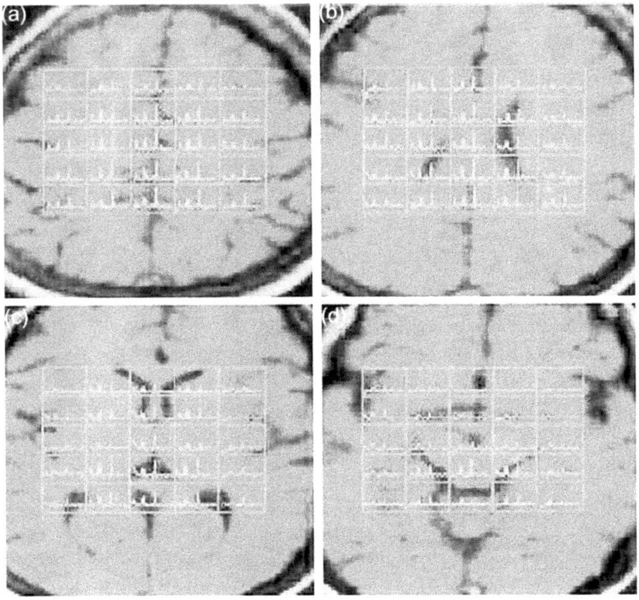

Figure 3.126 The full 3D data from a hybrid experiment. The location of the HSI slices (a–d) is shown in Figure 3.125. The PRESS box was 7.5 × 7.5 × 6 and the voxel size 1.5 cm × 1.5 cm × 1.5 cm. The spectra from each HSI partition are superimposed on the axial image and the grid makes the normal voxel boundaries. The horizontal (0.3–4.2 ppm region) and vertical scales are common. The spectrum is marked with an asterisk in (b) is expanded in Figure 3.127.

(Cr/PCr) at 2.0 ppm, ratios of which are commonly used to distinguish between normal brain and pathologic brain. Comparison of the spectra shows a slight asymmetry between the left and right sides in Figure 3.126(b) and (c) [39]. This occurred in part because the PRESS box was not precisely centered, as seen from the position of the brain midline in the center column of voxels; and in part because the patient's head was slightly tilted, as seen from comparing the ventricles in Figure 3.126(b). The deterioration of the spectra in the anterior rows of Figure 3.126(d) [39] reflects susceptibility anisotropy from air–issue interface, in the sinuses below, as seen in the sagittal position of the box in Figure 3.125 [39]. To demonstrate the performance of a 27-min hybrid, a spectrum of a single 3.4-ml voxel is expanded in Figure 3.127 [39]. In addition to the main signals, other less intense signals are visible. There is PCr/Cr at 3.96 ppm, glutamate/glutamine at 3.75 and 2.38, inositol at 4.57, and NAA at 2.53. The intensities of the peaks are different in the single spectrum and the sum, due to metabolite-level heterogeneity over the VOI.

To assess the SAR associated with the hybrid, one assumes that most of the RF power of sinc is in its central lobe. We approximated this lobe with a rectangular pulse of duration equal to the inverse bandwidth yields that, at TR = 1.6 s each

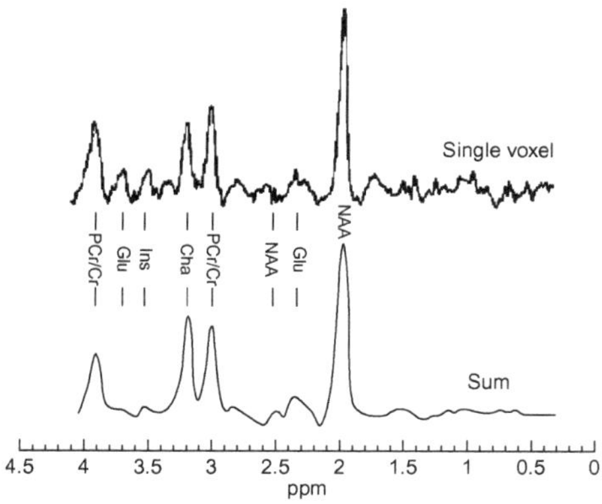

Figure 3.127 Expanded ^{1}H spectrum from the voxel with the asterisk in Figure 3.126 (b) (thick line, top) and an average sum of the spectra from all 100 voxels (thin line, bottom).

OVS, deposited approximately 0.3 W power. The PRESS pulses contributed approximately 1.2 W each. Assuming 3 kg for an average head, one gets approximately 1.5 W/kg. This is under the FDA's (UK) restriction of 3.2 W/kg. It is also less than the international standard (IEC) of 3 W/kg.

3.3.2.10 Human Brain Metabolic Imaging (Maps): Combined Volume and Spectroscopic (Proton) Imaging [40]

Most volume-selection techniques are based on spin-echo sequences. Volume-selection techniques include stimulated-echo (STEAM) volume selection and localization by means of a modification of this sequence with the addition of a section-selective refocusing pulse at the end of the sequence and replacement of the first excitation pulse by a binomial water-suppression pulse. The combination of spectroscopic imaging and volume selection has made it possible to obtain multiple ^{1}H (proton) MR spectra in a single experimental session. With the use of phase-gradient-encoding techniques in one dimension and volume selection based on refocused stimulated echoes, a series of ^{1}H MR spectra could be obtained from contiguous sections from one side of the head to the other. From proton spectroscopic imaging, it is possible to show that there are differences among ^{1}H MR spectra from different parts of the tumor region. In particular, it can be demonstrated that in some tumors, the choline level is increased in one region of the tumor and decreased in another. Also, the lactate level may be elevated only in certain region of the tumor and not elsewhere. An implementation of Carr–Purcell (CP) spin-echo volume-selection techniques has further improved the sensitivity of localized ^{1}H MRS. This is because signal loss of a factor of two in the stimulated-echo sequence is prevented. With this better sensitivity, it is now possible to record spectroscopic-imaging datasets by means of phase encoding in two orthogonal directions.

The phase encoding still has to be combined with spatially selective excitation in three dimensions. The first dimension selected is the one orthogonal to the phase-encoding directions. This is to accomplish section selection. The selective excitation in the other two directions restricts the sensitive volume to a rectangular ROI inside the brain. This is to cut off the very intense lipid signals from bone marrow and subcutaneous fat. Fourier transformation in the two spatial directions, and along the chemical-shift frequency axis, results in spectroscopic images. They may be represented as a set of spatially resolved spectra or as chemical-shift images, representing the density distribution of a biochemical compound. Such metabolite maps are constructed to visualize the spatial distribution of choline, creatine, NAA, and lactate with a nominal resolution of 7 mm and a section width of 2.5 cm. This extension of ^{1}H MRSI, from one to two spatial dimensions, allows comparison of ^{1}H metabolic maps. Thus, one can use conventional 2D FT MR images for the purpose of correcting variations in metabolic concentrations, with heterogeneities visible on the images. In particular, one can address questions concerning the origin of variations in choline concentrations across the tumor. For measurements, the standard ^{1}H, 30-cm head coil is used for MRI spectroscopy.

^{1}H MR precedes spectroscopy to locate the intracranial tumors. To avoid time-consuming measurements, T_1-weighted spin-echo sequences with a repetition time of 300 ms and echo time of 15 ms were used. On the basis of these scout images, a rectangular VOI was defined to exclude lipid signals from the skull. Typical dimensions used in the spectroscopic measurements were 80 mm $\times$ 80 mm $\times$ 25 mm (anteroposterior, left to right, craniocaudal). For the ^{1}H MRSI measurements, water suppression and volume selection were combined with gradient phase encoding in two dimensions. A large VOI was selected (typically 80 mm $\times$ 80 mm $\times$ 25 mm), and the FOV for the spectroscopic imaging was chosen to be greater than that of the selected volume (typically 120 mm $\times$ 120 mm $\times$ 120 mm). The ^{1}H MR spectroscopic images were obtained at 2000/272. Sampling was carried out immediately after the last 180° pulse, so as to acquire not only the second half of the echo, but also part of the first half. A 2D ^{1}H MR spectroscopic image is obtained over the selected volume with 16 $\times$ 16 phase-encoding steps and a typical FOV of 120 mm^3 $\times$ 120 mm^3 $\times$ 120 mm^3, which leads to a nominal voxel size of 7 mm^3 $\times$ 7 mm^3 $\times$ 25 mm^3. The single-volume data (Figure 3.128 [40]) were processed by applying Lorentz–Gauss windowing in the time domain (exponential narrowing of 4 Hz and Gaussian broadening of 5 Hz) for noise reduction and spectral resolution enhancement, followed by zero filling to 4096 data points. A linear phase correction was applied in the frequency domain to obtain absorption-mode spectra from these half-echo measurements. Selected frequency-domain spectra, shown in Figure 3.128, were obtained from the spectroscopic dataset by processing the ^{1}H MR spectroscopic datasets by means of windowing, preprocessing, and zero filling. To obtain the metabolic maps in Figure 3.129 [40], spectroscopic datasets were obtained. No zero filling was applied in the time domain. Instead, zero filling to 64 $\times$ 64 was applied. After appropriate windowing and zero filling were achieved, 3D Fourier transformation was performed, and the magnitude spectra were calculated. Integration boundaries were selected on the basis of visual inspection of frequency-domain spectra. Because of volume-selective shimming, the peak positions did not vary by more than 10 Hz across the image. Thus, possible cross-contamination of the different metabolite maps was minimized.

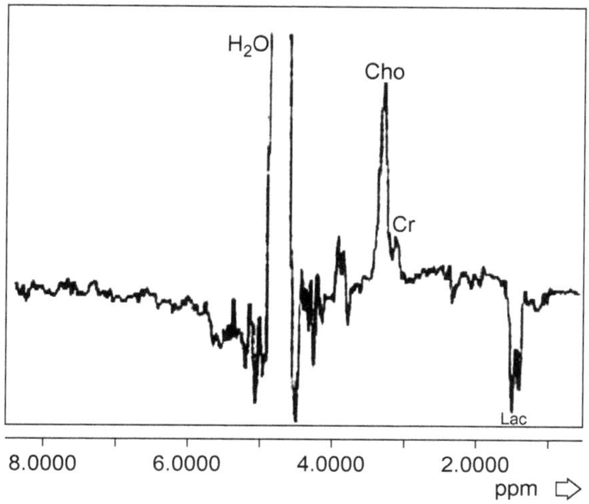

Figure 3.128 Single-volume ^{1}H MR spectrum (2000/136) of patient 2, who had a grade 2 astrocytoma. The spectrum originates from a 25 mm × 25 mm × 25mm volume in the solid part of the tumor (Figure 3.129(a) [40]). The spectrum shows high lactate concentration (inverted because a TE of 136 ms was used) and high choline concentration, but no NAA. The inversion of the lactate signal at 3 ppm, at a TE of 136 ms, is indicative of lactate. This single-volume spectrum is similar to the spectroscopic spectrum selected from the same region (Figure 3.129(f) [40], spectrum C). Cho = choline, Cr = creatine, H_2O = water, Lac = lactate.

The spectra are not aligned, to account for residual magnetic field inhomogeneity. Metabolite maps are reconstructed for choline, creatine, NAA, and lactate by calculating the peak integrals for all spectra in the 3D dataset. These integrals were displayed on a linear gray scale to obtain metabolite maps of 64 × 54 × 64 voxels. Conventional MR images obtained separately assist in the interpretation of these chemical-shift images.

3.3.2.11 Multiple- (Double-)Quantum Filtering ($A_2M_2X_2$ Spin System): GABA [41]

It is well known that GABA is one of the inhibitor neurotransmitters in the human brain. It plays an important part in several brain disorders. There is a comparatively low concentration of GABA in the brain, which makes it very difficult to detect. It is also masked by the multiplet resonances in the brain spectrum. Echo differential spectroscopy using PMRS is one technique currently in use in the field. Multiple-quantum filtering provides a good alternative for editing the A_2GABA triplet. This editing effectively allows one to suppress the large creatine singlet resonance.

This suppression is required because the Cr resonance obscures the A_2GABA triplet (3.0 ppm). However, due to the large length of the focused filter,

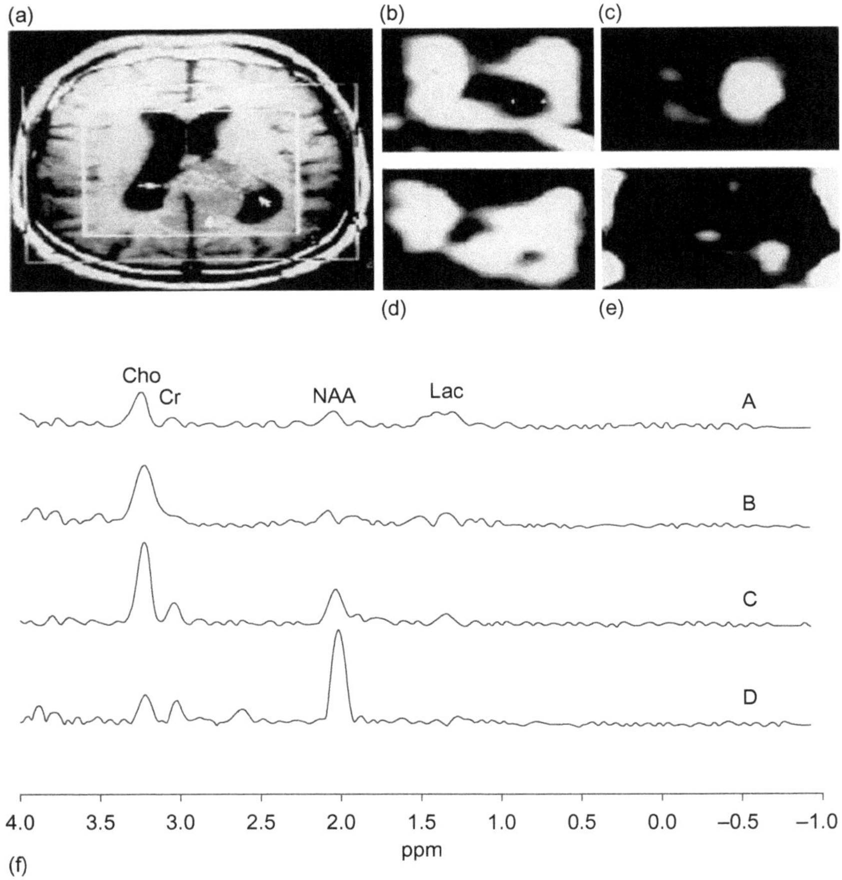

Figure 3.129 (a) ^{1}H MR scout image of patient 1, who had a grade 2/3 oligodendroglioma extending from the midline through the CC. Arrows = extent of the tumor. The inner box (1) represents the selected volume used in the corresponding volume-selective spectroscopic measurement, and the outer box (2) indicates the FOV of the spectroscopic measurement. Metabolite maps of NAA (b), choline (c), creatine (d), and lactate (e) were reconstructed from the spectroscopic dataset measured in patient 1. The maps were originally scaled to correspond in size to the outer box (2) in (a) (original scaling has not been retained). Choline concentration is elevated in the tumor, while NAA concentration is decreased in the tumor region and ventricles. However, the NAA level is reduced by a factor of two in the anterior part of the tumor and by a factor of four in the posterior part. Lactate is observed in the cornu posterius of the left lateral ventricle. (The high signal intensity along the edges of the lactate map is due to lipid signals from the skull that coresonate with lactate; on the metabolite map, the lipid signal is off scale, to bring out the relatively low lactate signal intensity.) A–D in (b) are the locations from which the selected spectra in (f) were obtained. (f) Selected ^{1}H MR spectra from the spectroscopic dataset obtained in patient 1. The spectrum from the region of the cornu posterius of the left lateral ventricle (a) shows low levels of choline (Cho), creatine (Cr), and NAA, but an elevated lactate (Lac) (doublet signal at 1.3 ppm). The tumor spectra (b from the posterior part of the tumor, c from the anterior part) show a high choline level compared with that of a control spectrum from the right hemisphere (d). The NAA level is higher in the anterior part of the tumor than in the posterior part.

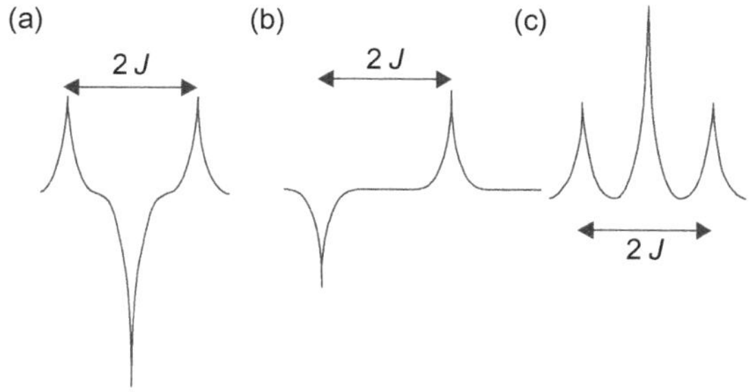

Figure 3.130 The real FT of possible magnetization signals that can arise from the A2 triplet of GABA, depending on the start of the acquisition. At the start of acquisition, the density operator terms were assumed to be the following: (a) 3-spin antiphase coherence (e.g., $A_yM_{1z}M2_z$), (b) 2-spin antiphase coherence (e.g., A_xM_{1z}), and (c) in-phase magnetization (e.g., A_y). In all cases, the separation between the outer two peaks is 2 J or 14.6 Hz for GABA.

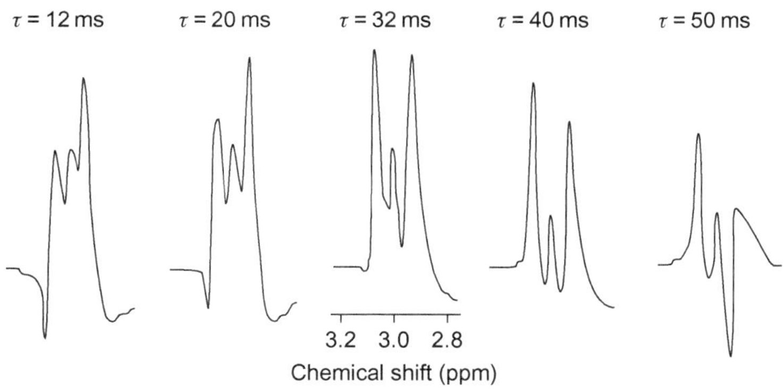

Figure 3.131 An illustration of the time dependence of the shape of the GABA 3.0 ppm multiplet obtained at 2.35 T using the DQ filter of Figure 3.133(a). The changing shape arises from the evolving combinations of in-phase and antiphase magnetization that appear at the start of acquisition and thenceforth evolve during the acquisition period.

approximately $1/J = 137$ ms, it leads to significant transverse relaxation losses. One can reduce the duration of double-quantum filtering to approximately $1/4J$, and thereby bring transverse magnetization losses to a minimum. The sequence shown in Figure 3.130 [41] also reduces intrinsic noise in the filter. It is important that the first two 90° pulses excite both the 3.0 and the 1.9 GABA multiplets, so that this transfer can take place. Because the t_1 chemical-shift factor is now global, the spectrum may still be phased to yield the maximum, even when $\cos(2\omega t_1) \neq \pm 1$ (Figure 3.131).

Table 3.31 The Percentage A_2 Triplet Signal Intensity from GABA and the Creatine Suppression Factors for Various Editing Sequences

Pulse Sequence	**Theoretical Intensity**[a] $t_1 = 0$	**Theoretical Intensity**[b] $t_1 = 9$ **ms**	**Experimental Intensity**[c] $t_1 = 9$ **ms**	**Creatine Suppression Factor**[c]
(a) DQC	25	14	14	250
(a) ZQC	50	50	50	1
(b) TQC with $\theta = 90°$	12.5	11	10	1600
(c) TQC with selective read pulse on *M* spins	25	21	18	1600
(d) TQC plus SQC with $G_2 = 2G_1$	100	65	59	2
(e) TQC plus SQC with $G_2 = 4G_1$	25	16	12	1000
(f) Correlated *z* order	100	100	99	2

[a]The intensities are expressed as a percentage of that obtainable in a simple spin-echo experiment.
[b]The values in this column take account of the evolution under weak scalar coupling during t_1.
[c]The range of experimental variations from three repeated measurements is less than ±5% and the creatine suppression factors have been rounded to two significant integers.

3.3.2.12 Coherence Pathway Quantum Filtering: Quantum Filtering/GABA Single-Quantum Editing [42]

The MQFs are divided into two groups: even-order (zero and double) and odd-order (single and triple) filters. The pulse sequences that give rise to these filters are illustrated in Table 3.31 [42]. The evaluation criteria are as follows: First, there is preservation of the magnitude of the output GABA signal for each editing strategy. Second, there is the efficiency with which each pulse sequence suppresses unwanted neighboring signals. The product operators A_{ju}, M_{ju}, and X_{ju} are the spin operators for the *A*, *M*, and *X* spins, respectively, in the *AMX* spin system. Here $j = \alpha, \beta$ denotes a particular spin of each pair, and where $u = x, y, z$ denotes the rotating frame of reference. The effects of both longitudinal and transverse relaxation are neglected. Irrespective of the order of the filter, each of the pulse sequences begins with a $(90° - \tau/2 - 180° - \tau/2)$ subsequence, where $\tau = 1/2J$, and J is the scalar-coupling constant ($J_{MX} = J_{AM} = J$). The best compromise between the suppression of Cr and the preservation of GABA is found to be the 3QC filter, with selective read excitation of the 1.9-ppm GABA quintet (Figure 3.132).

3.4 Summary and Conclusion: Future Directions in MRI and Imaging Diagnostics

3.4.1 Future Directions in Imaging

3.4.1.1 Chemical-Shift Imaging and Single-Voxel Imaging [43]

A few important points in CSI are highly demanding. Some of these are a short echo time, absolute quantification of metabolite concentrations in units of mM, and

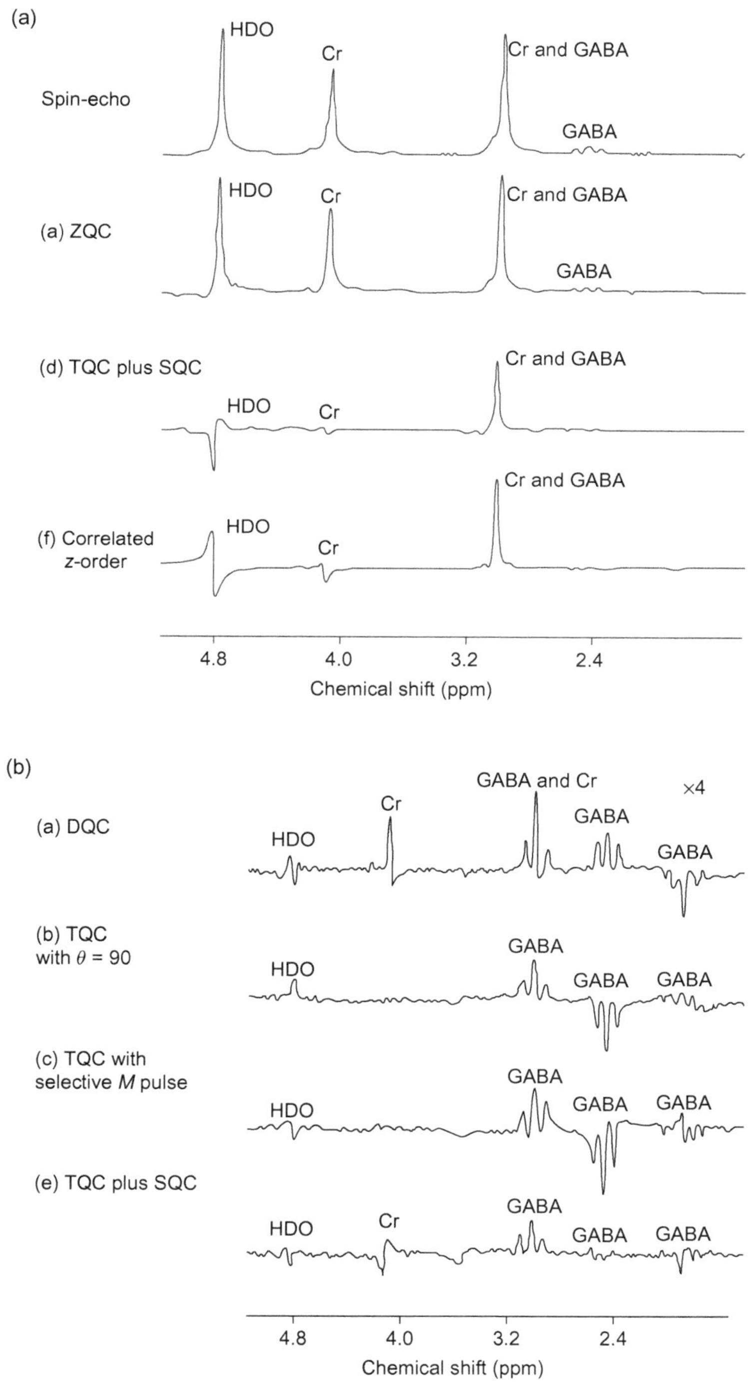

Figure 3.132 The 100-MHz proton spectra obtained from a sample of 400 mM creatine and 70 mM GABA in D_2O using the editing pulse sequences listed to the left. (a) The spectra corresponding to unsuccessful editing of the 3.0-ppm triplet. All spectra were acquired with one scan. (b) The spectra corresponding to successful editing of the 3.0-ppm triplet. All spectra were obtained with 32 signal-averaging scans but no phase cycling. The creatine suppression ratios are listed in Table 3.31.

an informative resolution (say, a nominal size of about 0.5 cm^3) such that chemically distinct regions can begin to be correlated with specific brain structures and/or different parts of a lesion, among others. These points seem to have been widely attended to in SVS. Proton (^{1}H) CSI requires short relaxation times (TE < 50 ms). Short-TE ^{1}H CSI can resolve additional metabolite signals, such as myoinositol (MI), glutamate (Glu), glutamine (Gln), and glucose (Glc), that are not detectable in long-echo-time ^{1}H spectroscopy (TE > 100 ms). This results from *J*-coupling modulations. MI, Glu, Gln, and Glc all play key roles in the metabolic pathways of the brain. Myoinositol is thought to be a glial marker, and is involved in intracellular signal pathways. Glutamate is the major excitatory neurotransmitter, and Gln is involved in neurotransmitter synthesis. Glucose plays a control role in energy metabolism of the brain.

The ability to detect these metabolites and visualize their distribution provides a more comprehensive neurochemical profile, which should ultimately translate into greater sensitivity to changes in tissue biochemistry related to disease. Characterization of the macromolecule baseline is a significant challenge for the quantification of short-echo-time spectra, both in normal brain and pathological conditions. One can use the macromolecule subtraction method to include short-TE ^{1}H CSI. Here, localization by adiabatic selective refocusing (LASER) uses six AFP pulses to achieve voxel selection with sharp excitation profiles. Supplementary crushers dephase magnetization outside the VOI and make OVS unnecessary. When the time between AFP pulses is kept short, LASER also has the added benefit of reducing *J*-coupling modulation. One can optimize and validate a short-echo-time ^{1}H CSI for FOV reduction, and incorporate subtraction of macromolecule signals for precise quantitative measurement of metabolite signals.

It is possible to make a direct comparison between the metabolite measurements from a single-voxel spectrum localized by LASER and a spectrum acquired from the same tissue region with a multi-voxel 2D phase-encoding acquisition (with matched SNR 0). In this setup, the slice and FOV are localized using LASER. It is expected that the CSI sequence would achieve equivalent macromolecule signal elimination, but that water suppression and line widths would be somewhat variable throughout the chemical-shift image, due to small variations in B_0. It is also expected that metabolic measurements from LASER SVS and the matched LASER CSI voxel would be in agreement within the range of variability exhibited by each method, and that LASER CSI would therefore represent a viable and attractive avenue for future ^{1}H NMR spectroscopy.

In CSI acquisition, the center of *k*-space is acquired first and all averages for a single *k*-space element were consecutively acquired. Acquisition of full spectra and macromolecule spectra was interleaved. No averaging was used in the acquisition of water-unsuppressed data. Single-voxel spectra (TE = 46 ms) were acquired from a volume equal to the effective voxel size of the CSI imaging protocol (1.5 $cm^3 \times$ 1.5 $cm^3 \times$ 1.5 cm^3). The single voxel selected was located in the most homogenous region of the WM, in the same plane. Single acquisitions of full spectra (TR = 2150 ms averages, 128 averages) and macromolecule spectra (TR = 4250, 128 averages) were alternated, for a total acquisition time of 14 min.

The unsuppressed water spectrum (used to remove lineshape distortions and act as an internal standard for concentration reference) was averaged 8 times. After spatial reconstruction of the full, macromolecule, and water spectra, a phase offset was applied to the *k*-space CSI data that corresponded to the required shift in image space, to retrospectively align with the location of the single-voxel spectrum. Following voxel alignment, CSI spectra were processed identically to single-voxel spectra, on a voxel-by-voxel basis. The macromolecule contribution to the full spectrum was removed by subtracting the macromolecule spectrum from the full spectrum. Residual water signal was removed by subtracting resonances between 4.1 and 5.1 ppm (center water frequency 4.7 ppm). The levels were normalized to the level of unsuppressed water within each voxel and then corrected for T_1 and T_2 relaxation effects, using values from the literature. T_1 and T_2 differences between water and the various metabolite resonances were approximated, using literature values for NAA, Cr, Cho, and water, and estimated for all other measurable metabolites by averaging the relaxation times for NAA, Cr, and Cho. The fractional water content was assumed to be 71% in WM. For quantification purposes, the water reference *k*-space data for CSI were scaled by a filter function that matched the variable averaging scheme employed during acquisition of the water-suppressed data. This was to ensure that the water reference CSI data had the same PSF as the water-suppressed data.

Table 3.32 [43] compares various spectral parameters between SVS and CSI acquisitions. The average metabolite concentration measured by each method ($N = 8$) is listed in Table 3.33 [43]. One of the advantages of LASER is the sharp excitation profiles that can be achieved by adiabatic slice selection. LASER acquisitions also benefit from the use of high-bandwidth RF pulses, approximately 4.28 KHz. Despite careful alignment of the CSI nominal voxel with the single-voxel acquisition, direct comparison is difficult, due to differences in the PSF. Variability in metabolic measurements between methods is probably also a

Table 3.32 Spectral Characteristics

	SVS ($N = 8$) Mean ± SD	CSI ($N = 8$) Mean ± SD	*P*-Value[c]
Water suppression (*I* of suppressed water/*I* of NAA_{CH3})	1.1±0.7	2.2±1.1	0.04
Macromolecule subtraction (SD in macromolecule region[a]/SD of baseline noise[b])	1.08±0.09	1.04±0.04	0.20
SNR (*I* of NAA_{CH3} /SD of baseline noise[b])	10.5±2.5	13.9±4.5	0.13
Line width of unsuppressed water peak (FWHM in Hz)	7.1±0.6	8.7±2.7	0.14
Line width of NAA_{CH33} (FWHM in Hz)	6.6±0.9	8.3±4.1	0.31

SD, standard deviation; I, signal intensity; FWHM, full width at half maximum; SVS, single voxel spectroscopy.
[a]SD of macromolecule region was calculated between 0.75 and 1.8 ppm.
[b]SD of baseline noise was calculated between 7.6 and 9.9 ppm.
[c]*P*-value from a two-tailed paired student's *t*-test.

Table 3.33 Metabolite Levels Measured in WM

Metabolites	Single-Voxel Level[a] ± SD	CSI, Level[a] ± SD	Mean of Absolute Differences[b]	*P*-Value[c]
NAA	12.4±1.6	12.6±1.8	1.7	0.9
Glutamate	6.4±1.7	7.8±2.2	1.3	0.06
Glucose	4.7±1.5	5.1±1.8	1.7	0.57
Myoinositol	7.7±1.8	8.1±2.3	1.8	0.62
Creatine	9.6±1.2	9.8±2.1	1.8	0.91
Total choline	3.2±0.5	3.1±0.5	0.6	0.57

SD, standard deviation.
[a]Levels reported in mM incorporating T_1 and T_2 relaxation times taken from literature for NAA, Cr, and GPC/PC, and approximated for all other metabolites.
[b]Mean absolute differences were calculated by taking the absolute difference between each SVS/CSI pair and then averaging these absolute differences for the eight datasets.
[c]*P*-value from a two-tailed paired student's *t*-test.

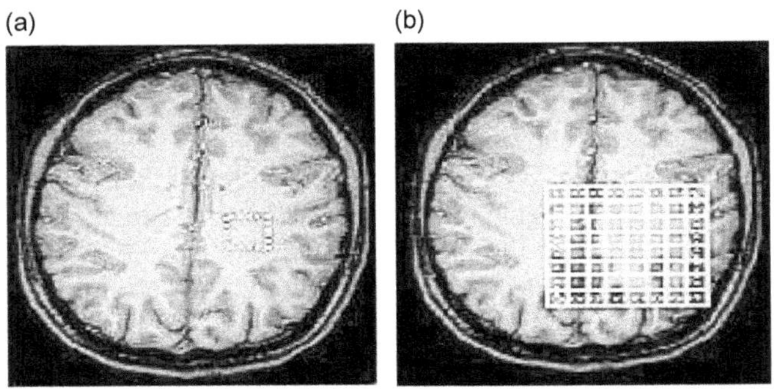

Figure 3.133 The position of (a) a 1.5 cm^3 × 0.5 cm^3 × 1.0 cm^3 single LASER voxel (dotted line) in WM, and (b) a 5.4 cm^3 × 5.4 cm^3 × 1.0 cm^3 LASER imaging volume (dotted line) with an 8 × 8 grid (FOV = 6.0 cm^2 × 6.0 cm^2) outlining the nominal voxels corresponding to the chemical-shift image acquisition (solid lines) in the same subject. Axial images (FOV = 22 cm^2 × 22 cm^2, 256 × 256 acquisition matrix, slice thickness = 2.5 mm) acquired with an inversion-prepared (T_1 = 500 ms), segmented turbo FLASH sequence were used to plan the location of spectroscopy acquisitions.

reflection of spectral quality. Apart from motion artifacts, off-resonance effects are one of the biggest challenges associated with CSI. Spectral resolution, water suppression, signal localization, and crusher gradients are all highly dependent on a homogenous B_0. One of the most distinctive features of the LASER CSI protocol is the acquisition and subsequent subtraction of macromolecule resonances. Although this is a highly effective method for removing macromolecule signals from short-TE spectra, another benefit to acquiring a macromolecule is that it may provide additional diagnostic information. Some pathologic conditions are characterized by an abnormal concentration of mobile lipids and/or pathologically altered macromolecules. For example, high-grade brain tumors, which typically exhibit necrotic

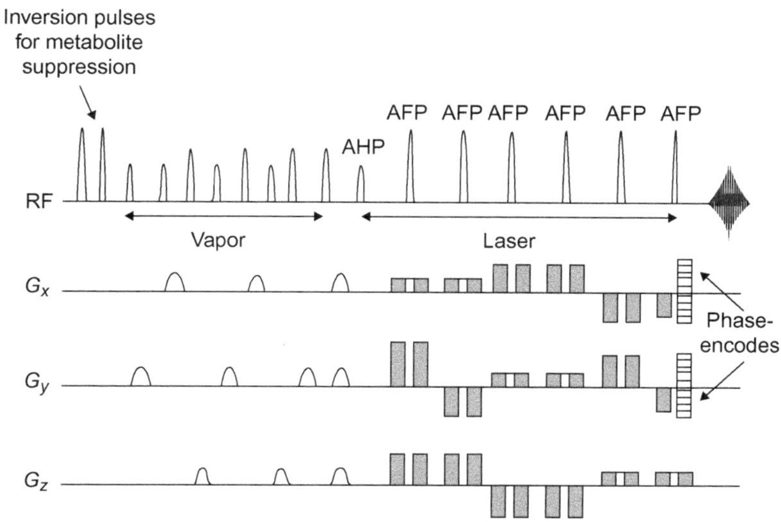

Figure 3.134 A diagram of the LASER CSI pulse sequence. The pulse sequence commences with two 180° pulses that are executed only for the metabolite-suppressed acquisitions. The inversion pulses are followed by a water-suppression scheme called variable power and optimized relaxation (VAPOR) delays. Following VAPOR, the LASER portion of the pulse sequence commences with an AHP pulse that excites all magnetization in an unselective manner. Three pairs of AFP pulses, in combination with *x*, *y*, and *z* gradients, refocus magnetization from three orthogonal planes such that only magnetization from the intersection of these three orthogonal planes will be refocused in the final echo. Crusher gradients (gray) bracket each AFP pulse and serve to eliminate unwanted stimulated echoes and spin-echos. Incremental phase-encode gradients in *x* and *y* bracket the final AFP pulse (superimposed on the crusher gradients that also bracket this RF pulse).

cores, are associated with increased lipids; increased lipid concentrations have also been reported in stroke and MS (Figures 3.133 and 3.134).

3.4.1.2 Parallel Imaging: Coil Sensitivity [44]

This parallel imaging (PI) technique is based upon extraction of spatial information from an array of multiple surface coils, to speed up the acquisition of an image. It has been demonstrated that by using some simplifying assumptions, all different PI techniques can be treated as different solutions to the same set of linear imaging equations. The various paths taken toward solving this inverse (reconstruction of image) are what primarily distinguishes the various PI methods from one another. One of the most essential elements of any PI method is the information describing the coil sensitivity distribution throughout the sample. The way in which this information is obtained is one of the intriguing features of PI methods.

Generalized autocalibrated partially parallel acquisition (GRAPPA) is the most widely used autocalibration (among the coils) technique, and can be used as a basis for understanding other methods. As with all autocalibrating methods, a few extra lines are acquired temporally close to the image acquisition. The required reconstruction parameters are then determined directly in *k*-space by fitting one or several lines to other lines in this calibration dataset. By fitting data to data, a pure coil sensitivity map is not needed, only the few lines of extra data. No body coil image and intensity thresholds are needed, so one generates normal-appearing images, even in the background. In addition, aliasing in the reconstructed image is not a problem, thereby allowing slightly folded images to be acquired without any problems in the reconstruction. The primary difference between the various *k*-space-based methods is the way in which the reference data are used to determine the reconstruction parameters.

Although most early *k*-space methods, such as autocalibrated simultaneous acquisition of spatial harmonics (AUTO-SMASH) and variable-density AUTO-SMASH, used relatively simple fitting models and relatively little reference data, most modern methods, including GRAPPA, use expanded fitting along with more advanced methods for acquiring and using reference data. Nevertheless, there are image domain techniques, such as modified sensitivity encoding (mSENSE), which essentially use a modified SENSE formulation for the reconstructed image, based on an autocalibrated acquisition. GRAPPA is a general example of the *k*-space methods. In many cases, the same sequence as the imaging sequence is used for acquisition of the coil sensitivity information. In general, this is done through the acquisition of several additional lines in the middle of *k*-space. One advantage of using the same imaging sequence for sensitivity estimation and imaging is that the additionally acquired lines can be included in the reconstruction for improving both artifact levels and SNR. In some applications, especially the single-shot acquisitions, use of the same sequence for sensitivity mapping and imaging may not be particularly beneficial. In some cases, it can be faster to acquire the sensitivity information using a fast imaging block directly, before the imaging experiment. Another case in which a prescan might be advantageous is situations where motion, and in particular flow, might occur during acquisition of the calibration information. This could easily occur during the acquisition of slower, spin-echo-type sequences.

An example of these phenomena is shown in Figure 3.135 [44]. In this example, the same underlying spin-echo acquisition is reconstructed with coil sensitivity information from either the spin acquisition itself or a separate fast low-angle shot (FLASH) acquisition. Although both reconstructions are of relatively high quality, the reconstruction using the spin-echo data has a somewhat lower SNR, compared with the reconstruction from the FLASH data. Upon closer inspection, this is most likely due to flow artifacts in the spin-echo data. These flow artifacts artificially increase the size of the head, thereby increasing the total width of the object that has to unfold in the reconstruction. Additionally, the coil sensitivity content of these flow artifacts is from the original location of the flowing blood, not the final location in the image. These two effects combine to significantly reduce SNR

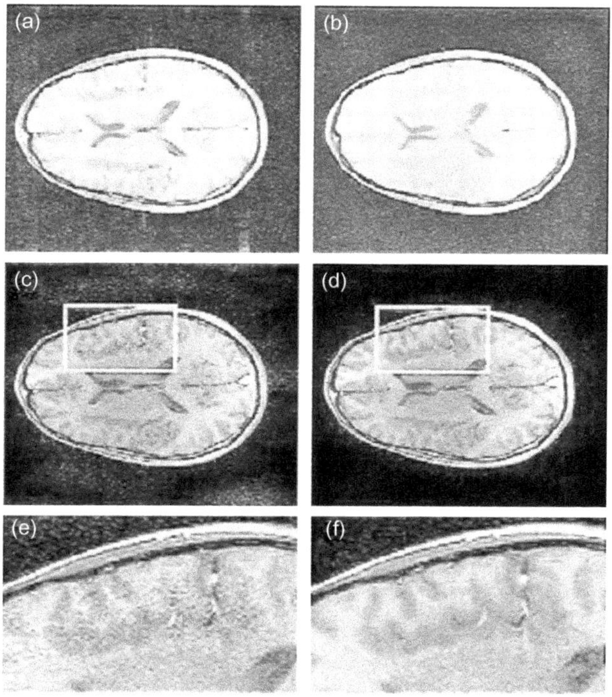

Figure 3.135 (a) Spin-echo acquisition with substantial flow artifacts. (b) FLASH acquisition of the same slice without any substantial flow artifacts. Both (a) and (b) are shown with a brightened contrast to highlight the flow artifacts in (a). (c) Simulated fourfold accelerated GRAPPA reconstruction of the spin-echo data using intrinsic reference lines. (d) Simulated fourfold accelerated GRAPPA reconstruction of the spin-echo data using reference lines from the FLASH acquisition. (e and f) Details of (c) and (d). Note the reduced SNR in (c) and (e), presumably due to the flow artifacts in the spin-echo reference lines. Acquisition parameters: Siemens 1.5-T Quantum Symphony, eight-channel head array. Spin-echo: TR = 500 ms, TE = 12 ms. FLASH: TR = 100 ms, TE = 4.76 ms, flip angle = 40°. Both sequences: slice thickness = 5 mm, FOV = 230 mm^2 × 230 mm^2, matrix = 256 × 256.

performance, as more data points have been unaliased in the reconstruction process, and more of these points have similar coil sensitivity information.

Another case in which *k*-space methods are advantageous is the use of PI in combination with EPI. Here the primary goal is the reduction of distortion or chemical-shift artifacts. Methods such as temporal SENSE (TENSE) and temporal GRAPPA (TGRAPPA) are particularly well suited for dynamic imaging during free breathing, where the coils may potentially move significantly. This is demonstrated in Figure 3.136 [44]. This figure shows the benefits of real-time coil sensitivity calibration for real-time cardiac imaging. The right column shows the results using the TGRAPPA reconstruction for this dataset, when the reconstruction

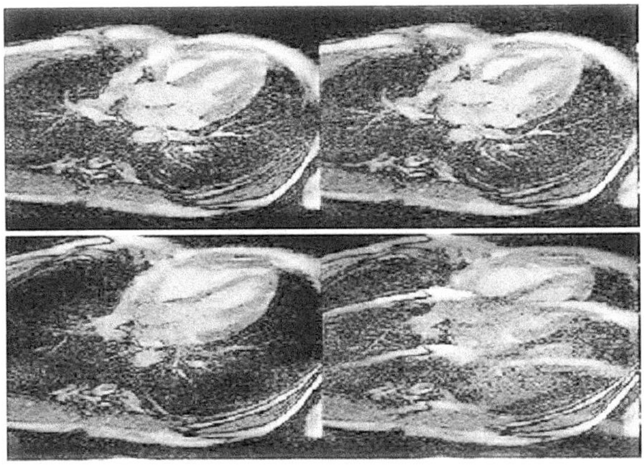

Figure 3.136 Real-time fourfold accelerated free-breathing TGRAPPA acquisitions at end of expiration (top) and inspiration (bottom), using either a real-time update of the coil sensitivity information (left), or only the information from a prescan (right), which was obtained during expiration. Large artifacts occur whenever the coil positions change dramatically. Note the different positions of the chest wall in the top row compared with the bottom row. Acquisition parameters: Siemens 1.5-T Avanto, 12-channel body array, TR = 2.7 ms, TE = 1.34 ms, flip angle = 55°, slice thickness = 8 mm, FOV = 260 mm^2 × 260 mm^2, matrix = 256 × 256.

weights were only determined from the first eight frames of the dataset. As can be seen, significant artifacts are visible when the coils move substantially. However, using real-time coil sensitivity calibration, these errors disappear, and high-quality reconstructions are seen throughout.

3.4.1.3 PI: Functional Imaging [45]

SENSE MRI, with an acceleration factor R, leads to an image SNR reduction of $g/\sqrt{R}$, with g being the coil geometry factor. Activation levels in BOLD experiments are not directly derived from the signal amplitude, but follow from the difference in signal between rest and active states. Subsequent division by the level of temporal instabilities of the signal determines the statistical significance of the signal change. The level of temporal instabilities, typically on the order of 1%, is commonly expressed as the relative temporal standard deviation σ_t. Both the intrinsic image noise σ_i and the physiological (the standard deviation of physiological fluctuations) noise σ_{ph} contribute to σ_t. If image SNR ($\sim\sigma_i^{-1}$) exceeds temporal stability, it can be sacrificed through use of PI to achieve artifact reduction, without affecting the statistical power of MRI. If σ_i and σ_{ph} (physiological) are totally independent, then $(\boldsymbol{\sigma}_{\mathbf{t,\ PI}}/\boldsymbol{\sigma}_{\mathbf{t,\ noPI}}) = \sqrt{[\mathbf{1} + (\mathbf{g}^2\mathbf{R} - \mathbf{1})(\boldsymbol{\sigma}_{\mathbf{i,\ noPI}}/\boldsymbol{\sigma}_{\mathbf{t,\ noPI}})^2]}$. This means that the penalty for SENSE use in functional MRI (fMRI) might not be as severe as one might think, based on a decrease in image SNR.

In an experiment where temporal signal stability was completely dominated by σ_i (and thus by image SNR), the measured fMRI action did indeed suffer the full $g/\sqrt{R}$ penalty. However, if physiological noise is the dominant noise source, application of SENSE will not affect the sensitivity of the fMRI experiment at all, see Figure 3.137 [45]. Geometric distortions are a result of the use of a long readout train in the presence of the field inhomogeneities in the human head. Such inhomogeneities tend to be most severe in brain regions close to bone–tissue interfaces; air cavities, such as the nasal cavity; and the ears. Owing to such inhomogeneities, spins in some areas of the brain are off-resonance, causing the phase of their signal to change during data acquisition. This phase change is incremented during the acquisition of a train of echoes, as is used in EPI, leading to linear phase gradient over *k*-space, in the phase-encode direction for the spins, which in turn leads to a shift of the signal in the reconstructed image. The effect scales linearly

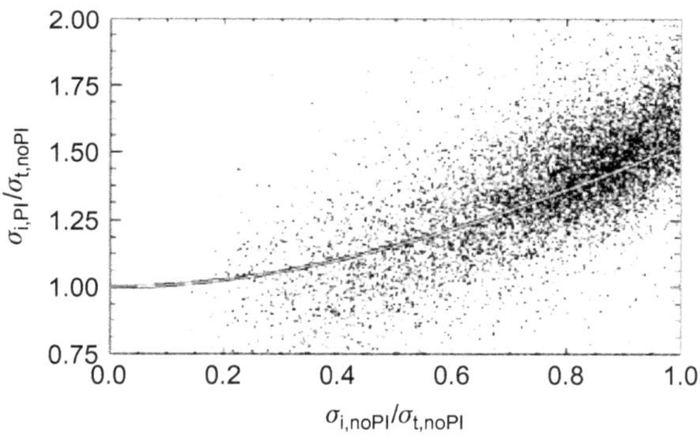

Figure 3.137 Plot of the effect of SENSE use on the temporal stability of fMRI data. This penalty for SENSE use is computed for individual pixels (.) in the superior part of the brain of a volunteer. The solid line shows the dependence predicted by a model, in which the average *g* factor of the experimental data was used for *g*. In turn, experimental data points were corrected if the *g* factor in that voxel differed from the average *g* factor. In a voxel in which the intrinsic (thermal) noise is the dominant noise source (right side of the plot), the penalty to the fMRI experiment is the full factor $g\sqrt{2}$ predicted, whereas fMRI sensitivity is completely unaffected by SENSE in voxels that are dominated by temporal instabilities (left side of the plot). Experimental data were acquired using a four-element dome coil; voxel size was $3.4\,mm^3 \times 3.4\,mm^3 \times 4\,mm^3$ (reconstructed matrix size 64 times). Twelve slices were acquired, using both standard single-shot gradient-echo EPI and rate-2 SENSE EPI, on a 1.5 TGE Signa LX with 2000-ms repetition time and 40-ms echo time. The EPI readout duration was 24.1 ms for standard EPI and 12.4 ms for SENSE EPI, respectively.

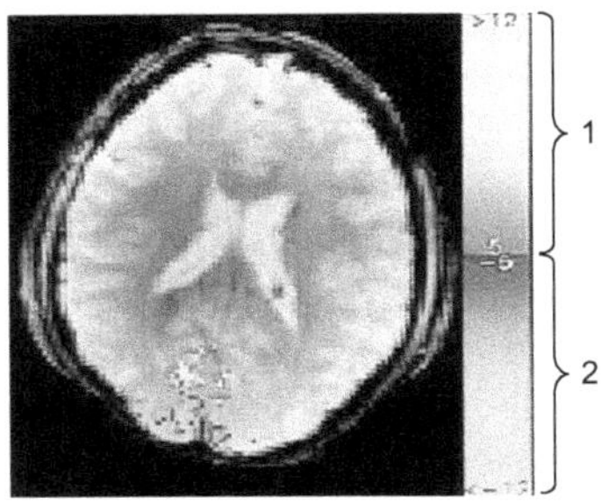

Figure 3.138 Results of a high-resolution single-shot rate-2 SENSE EPI perfusion-based fMRI experiment. Data were acquired at 3.0-T field strength with a 16-channel head coil, connected to an in-house-built 16-channel receiver. Scans were performed using a 3 s TR, 25 ms TE, 128 × 96 matrix size (128 × 48 acquired), 220 mm^2 × 165 mm^2 FOV, and 2-mm slice thickness, resulting in a nominal spatial resolution of 1.7 mm^3 × 1.7 mm^3 × 2.0 mm^3. The length of the EPI readout train was 30 ms. Perfusion labeling was done with the SSPL method, using two adiabatic foci pulses, creating a selective inversion at 1500 ms prior to image excitation and nonselective inversion at 300 ms prior to the excitation pulse. The color overlay indicates *t*-score, derived from a 5-min visual stimulus paradigm that stimulates alternating peripheral and foveal vision in 30-s-long blocks. The positive scale (1) marks the significance of peripheral activation; negative *t*-values (2) mark foveal activation.

with the field strength, thus limiting the applicability of single-shot EPI at high-field strength (~1 T). When using rate-*R* PI, the length of the readout can also be shortened by a factor *R*. This is the result of the extent of *k*-space being traversed *R* times faster, leading to a factor *R* decrease in the steepness of the aforementioned phase gradient. The reduced TE is an advantage for perfusion-based fMRI methods, like flow-sensitive alternating inversion recovery (FAIR) and single-shot perfusion labeling (SSPL). If the TE is kept short, the amount of signal and thus SNR can be increased, as well as its stability. At the same time, the contribution of BOLD signal changes to the experiment can be minimized.

An illustration of a single-shot rate-2 SENSE EPI-based perfusion experiment is shown in Figure 3.138 [45]. A potential benefit of higher-field strength for fMRI, in addition to increased contrast-to-noise ratio (CNR), is an improved specificity to parenchymal activation, partly because of the improved suppression of signal that originates from larger vessels due to the short T_2 of deoxygenated blood. The BOLD signal strength has been found to increase, at a higher rate than linearly, with field strength. Both the increased NMR signal strength at higher field and the increased CNR in BOLD-based imaging would allow fMRI at reduced voxel volumes, with sensitivity similar to that of conventional experiments at 1.5 T. An illustration of high-resolution single-shot gradient-echo SENSE EPI is shown in Figure 3.139 [45].

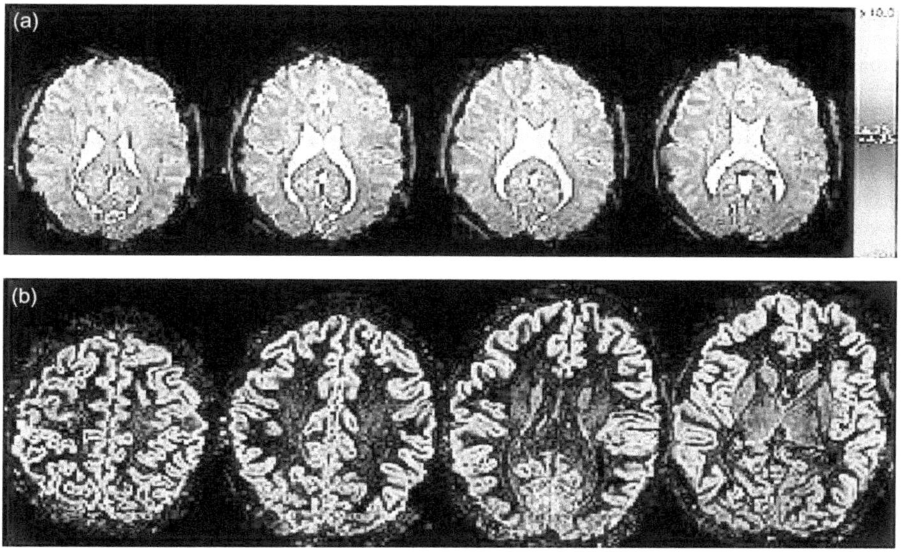

Figure 3.139 (a) High-resolution SENSE EPI BOLD fMRI data acquired at 3.0-T field strength with a 16-channel head coil, connected to an in-house-built 16-channel receiver. Single-shot gradient-echo fMRI experiment with a nominal voxel size of $1.1\ mm^3 \times 1.1\ mm^3 \times 1.5\ mm^3$ ($211\ mm^2 \times 158\ mm^2$ FOV with 1.5-mm slice thickness and $192 \times$ acquisition matrix). Fourteen slices were acquired with 2-s TR and 48-ms TE using the 5-min visual stimulus paradigm described in the caption of Figure 3.138. (b) Functional MRI data acquired with the same hardware, superimposed on a GM-only image that was acquired with a double-inversion version of the same EPI sequence. Parameters for the GM-only scan were: 6-s TR; 48-ms TE; rate-2 SENSE EPI; $192 \times 144 \times 96$ reconstructed matrix size; $1.15\ mm^3 \times 1.15\ mm^3 \times 1.15\ mm^3$ nominal voxel size ($220\ mm^3 \times 165\ mm^3 \times 110\ mm^3$ FOV); 621-s total acquisition time (including SENSE reference). Parameters for the fMRI scan were similar to those in (a) except for a $1.15\ mm^3 \times 1.15\ mm^3 \times 1.8\ mm^3$ and 45 ms TE.

3.4.1.4 *Accelerated Parallel MRI: Selective RF Excitation [46]*

Conventional MRI and parallel MRI similarities and differences can be briefly compared as follows:

Table 3.34 Comparison of Conventional and Parallel MRI

Conventional MRI	**Parallel MRI**
Selective RF (SRF) excitation (SRE) techniques have been used as a measure for spatial encoding in MRI, such as slice and line selection in Fourier-encoded MRI (FE-MRI). These techniques are also able to excite a wide variety of spatial profiles, allowing for alternative approaches called non-Fourier-encoded MRI (NFE-MRI).	PI techniques, such as SMASH, SENSE (sensitivity encoding), SPACE-RIP (sensitivity profiles from an array of coils for encoding and reconstruction in parallel), and GRAPPA, have shown the ability to increase acquisition efficiency in MRI, and have shown great utility in reducing acquisition time. These

(Continued)

Table 3.34 (continued)

Conventional MRI	Parallel MRI
These encoding approaches hold interesting potential for providing diverse benefits, including increased information throughput, interview motion compensation, elongated effective tissue relaxation times, 3D imaging with only a small number of slices, and variable resolution images. Spatial RF encoding (i.e., encoding by spatially varying the applied magnetic field), to produce arbitrary profiles of transverse magnetization, can be guided by predetermined mathematical basis sets, such as wavelets or Hadamard, or can be varied in real time to optimally encode or follow dynamic changes that are deemed to be occurring in the FOV. Dynamic adaptive encoding can offer benefits by exploiting prior information about the imaged sample. This is possible because selective RF encoding enables acquisition of the MR image using a potentially more efficient representation of image data, i.e., one that compacts the information content of the particular image in the acquired signal space. Despite its apparent potential benefits in MRI, a number of practical limitations have prevented the use of more complicated selective excitation patterns from reaching the clinical realm. These limitations include, but are not limited to, a considerable lengthening of the RF pulse when high-definition 2D or 3D profiles are sought, and added complexity throughout the MRI system, from acquisition, encoding, and reconstruction to RF excitation design.	techniques reconstruct MR images from subsampled k-space data acquired in parallel using multiple RF receiver coils, and exploit the coil sensitivity profiles as complementary encoding functions. However, owing to the intrinsic nature of parallel MRI, the ability of these methods to suppress aliasing artifacts decreases as the acceleration rate increases, if noise amplification is to be constrained within diagnostic use. Parallel MRI and SRE both provide an encoding mechanism beyond the Fourier methods normally employed in MRI. One can use a combination of these mechanisms as a new technique. An SRF excitation in combination with a parallel MR acquisition technique can be used to improve the numerical conditioning of the image reconstruction problem, enabling more robust reconstructions at high-acceleration factors. PMRI is based on the notion of using multiple RF receiver coils, where each coli provides independent information about the imaged sample. The image is reconstructed through a weighted combination of the (possibly subsampled) data acquired from each coil.

When a 2D SRF pulse is used, the MR signal received in each coil can be written as $G_l(S_e{}^m, G_y{}^q, t) = \int\int \rho(x, y) W_l(x, y) S_e{}^m(x, y) \exp[i\gamma(G_x xt + G_y{}^q y\tau)]dx\,dy$. Here $S_e{}^m(x, y)$ = the spatial excitation profile in the sample, during the mth excitation; $W_l(x, y)$ = the 2D sensitivity profile of the lth coil; $\rho(x, y)$ = the sample spin density distribution weighted by the specific imaging factors, such as relaxation and hardware characteristics; G_x = the readout gradient amplitude in the x-direction; x and y = the coordinates within the spatial FOV; τ = the pulse width of the qth phase-encoding gradient $G_y{}^q$ applied after the excitation; and γ = the GMR. In cases where SRF is not used, $S_e{}^m(x, y) = 1$ for all pixels in the image. In contrast to phase-encoded MRI, which utilizes magnetic field manipulations to extract a phase-encoded signal from the excited tissue volume, NFE-MRI relies entirely on the spatially selective excitation to encode the MR sample. Good examples are phase offset multiplanar (POMP) imaging and the Hadamard slice selection. Because the spatially selective excitations perform the encoding, they must necessarily be varied proportionally to the number of locations being encoded along y, so that $S_e{}^k = S_e{}^l$ only when $k \neq l$ when averaging over multiple excitations is desired (Figure 3.140).

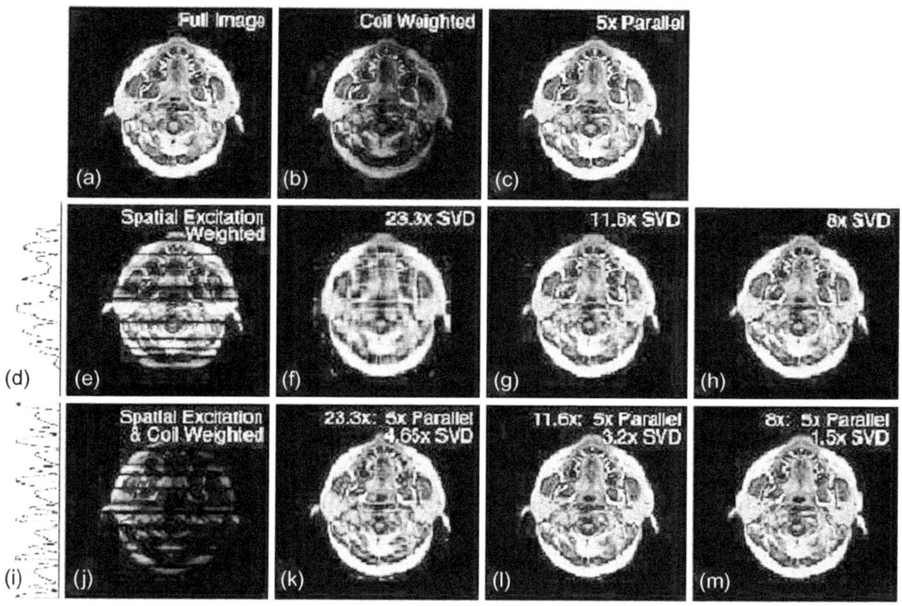

Figure 3.140 Top row: (a) Axial head image used in simulations. (b) One of eight simulated coil sensitivity weighted images. (c) Fivefold parallel accelerated SPACE-RIP reconstruction. Middle row: (d) One encoding function, ${}^S5_e(y)$, used to non-Fourier-encode the MR image. (e) Sample after excitation using that encoding function and unit uniform receiver sensitivity as in (a). (f–h) Images reconstructed from simulated non-Fourier acquisition using 11, 22, and 32 single value decomposition (SVD) encoding functions, respectively. Bottom row: (i) One encoding function ${}^S3_e(y)$ used to non-Fourier-encode the FOV when using a fivefold parallel acceleration factor. (j) Sample after excitation using that encoding function as seen through the same coil sensitivity used in (b). (k–m) Images reconstructed from 11, 22, and 32 SVD encoding functions, in conjunction with a fivefold parallel reconstruction. For the same acceleration rate, image quality is significantly improved when the acceleration is separated between the two spatial-encoding approaches (bottom row).

3.4.1.5 Parallel MRI: Encoding and Reconstruction [47]

In a standard Fourier MRI, the spatial resolution within an image plane (or 3D volume) relies exclusively on gradient fields, which impose plane-wave modulations on the transverse magnetization $M(r)$. Processing is done in the main field B_0 (static magnetic field along the z-direction) and the superimposed gradient fields. The magnetization generates RF electromagnetic fields (EMFs). These are detected with one or multiple receiver coils. Neglecting constant factors, a data sample taken with a homogenously sensitive coil is given by $d_k = \int M(r)\exp(\mathrm{i}k_k r)\mathrm{d}r$. Here i denotes the imaginary unit ($\sqrt{-1}$) and k_k is the wave vector, describing the k-space plane-wave encoding (Figure 3.141 [47]). The gradient fields directly manipulate

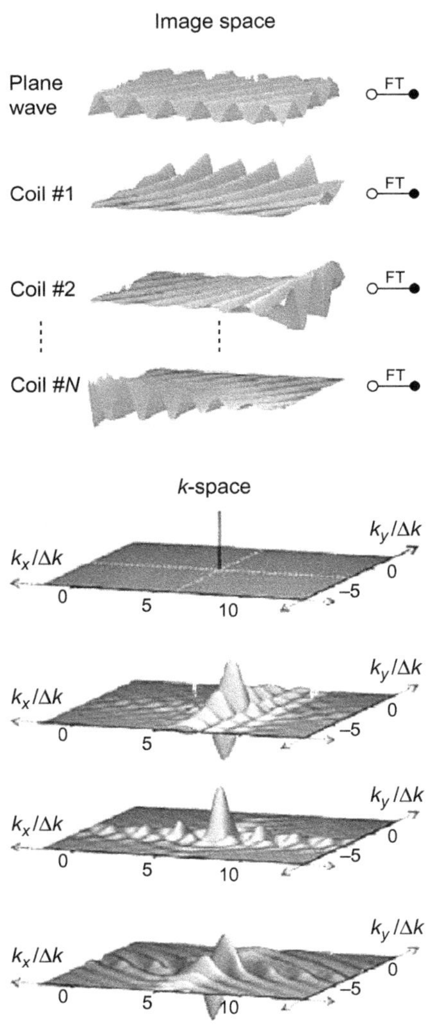

Figure 3.141 Encoding functions. Top row: Example of a plane-wave-encoding function as created by gradient fields in standard Fourier MRI ($k_x = 6 \times 2\pi/\text{FOV}$; $k_y = -3 \times 2\pi/\text{FOV}$). Its k-space representation is a single Dirac peak at (k_x, k_y). Other rows: Hybrid encoding functions resulting from the same plane wave multiplied with different coil sensitivities. In k-space, each hybrid encoding function has a distinct shape, which is equal to the FT of the respective coil sensitivity shifted by (k_x, k_y).

the magnetization to be depicted. Thus, each plane-wave encoding corresponds to a certain magnetization state. As a consequence, only one such encoding can be performed at a time.

This results in long scan durations. The signal from the γth coil is $d_{\gamma,k} = \int M(r) S\gamma(r)\exp(ik_k r)dr$. $S\gamma(r)$ is the coil's complex-valued, spatially varying sensitivity. With full Fourier encoding and standard Fourier reconstruction, each coil yields an individual image weighted by the coil sensitivity. In a single-coil reconstruction, $S\gamma(r)$ is effectively treated as a modulation of $M(r)$, causing a modulation of the resulting image. In parallel MRI, $S\gamma(r)$ is considered a modulation of the plane-wave-encoding function, leading to the hybrid encoding basis $\mathrm{enc}_{\gamma,k} = S\gamma(r)\exp(ik_k r)$. The encoding equation then becomes $d_{\gamma,k} = \int M(r)\mathrm{enc}_{\gamma,k}(r)dr$ (Figure 3.141 [47]). The hybrid encoding functions are no longer plane waves, but plane waves multiplied by coil sensitivity. In their equivalent k-space representations, they are no longer Dirac (single sharp) peaks, but instead have distinct shapes and a significant extent. Each encoding no longer yields a genuine k-space sample, but rather a weighted integral of data from a certain k-space neighborhood. The encoding operation is no longer a sampling due to the F of $M(r)$. The above integral represents a scalar product that can be interpreted as the projection of $M(r)$ onto $\mathrm{enc}_{\gamma,k}(r)$. The result is that image reconstruction can no longer be accomplished by FT only, but amounts to recovering $M(r)$ from a set of more general projections. The gradient mechanism encodes the position of the magnetization vector in the frequency and phase of its precession. It relies on the QM of the spins. The frequency and phase modulation then translate to the EMFs that the magnetization generates. This means that the gradient encoding stores image information only in the temporal degrees of freedom of the RF fields detected. It is a natural consequence of the behind-the-scenes spectroscopy involved.

The EMFs generated by the input gradient signals also have spatial degrees of freedom. Even a small portion of magnetization generates a characteristic spatial distribution of EMFs. The EMFs' spatial degrees of freedom store a significant amount of image information. In conventional MRI, with a single coil, this information is lost when the spatially varying fields are collapsed into a single voltage. In contrast, in MRI with multiple receiver coils at different positions, at least some of the inherent spatial variation is preserved. In this way, the image information encoded in the spatial EMFs' degrees of freedom is partly recovered. Thus, gradient encoding uses only the temporal degrees of freedom of EMFs of image formation. Only one such degree of freedom can be read at any single point in time. The spatial degrees of freedom of the electrodynamics are, per se, the carriers of substantial image information in an MRI experiment. Parallel detection is a way of tapping inherently parallel information channels. This yields data of distinct information content simultaneously.

In principle, this offers an infinite number of spatial degrees of freedom. However, the amount of image information that can be extracted from them is limited, primarily because the field components that exhibit the strongest spatial variation decay rapidly, according to the distance from their source. Therefore, the ability to detect them outside the object is greatly reduced. The space and time degrees of

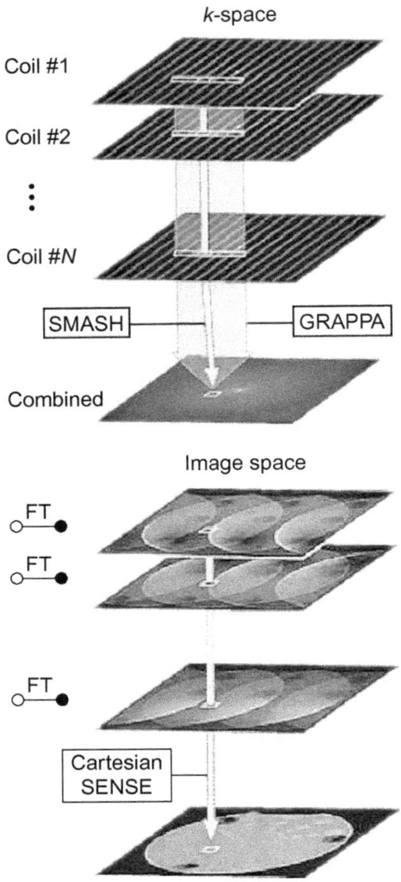

Figure 3.142 Image reconstruction from Cartesian data. The SMASH and GRAPPA techniques operate in *k*-space. For reconstructing one *k*-space value of the target image, SMASH retrieves one *k*-space sample per coil, whereas GRAPPA involves neighboring data along the phase-encoding direction. Cartesian SENSE operates in the domain, calculating each pixel from the corresponding set of pixels in the aliased single-coil images.

freedom do not interfere and can thus be freely combined. The contrast mechanisms most commonly used in MRI, the gradient encoding, are based on the QM of spins. As a consequence, the additional encoding via the electrodynamics pathway does not perturb the image. The illustrations shown in Figures 3.142–3.144 represent how the various techniques of PI can be implemented, including PI with augmented radius in *k*-space (PARS), SPACE-RIP, PI with localized sensitivity (PILS), GRAPPA, etc.

3.4.1.6 Combination of PI and High Fields (7 T) [48]

High-field conditions (3–7 T) complement PI in more than one way. In comparison with gradient-encoding schemes, spatial encoding is carried out by means of distinct coil sensitivities. It is unique in the sense that it does not interfere with the nuclear spin magnetization that it encodes. There are some advantages in PI, but they come at some cost, e.g., the reduced SNR. In fact, PI and high fields act complementary to each other. It is known that phased-array coils increase SNR. This

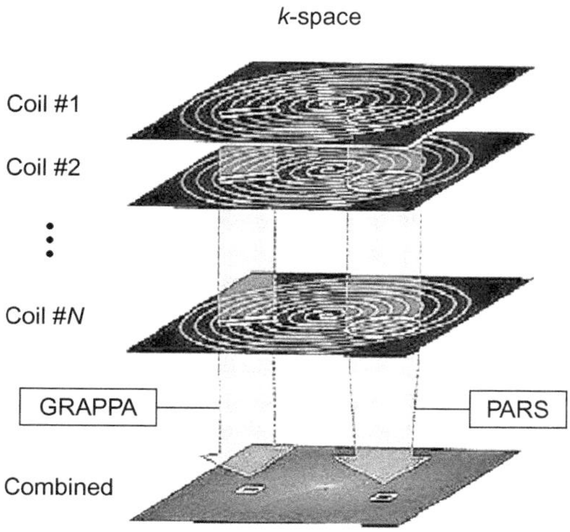

Figure 3.143 Image reconstruction from non-Cartesian data. With non-Cartesian data, the GRAPPA approach can still be used in subregions with regular sampling structure. Another *k*-space approach, PARS, involves all single-coil data within a certain *k*-space radius for constructing a target *k*-space value.

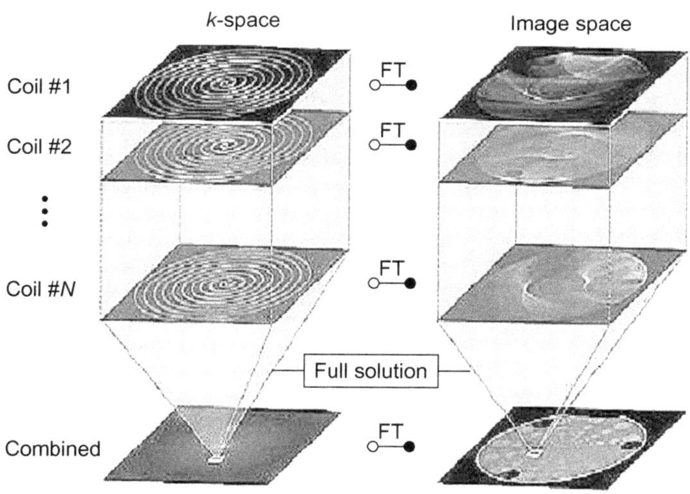

Figure 3.144 Full solution of the general reconstruction problem. With nonlattice sampling, an exact reconstruction will generally involve all raw data in the reconstruction of each target *k*-space value. Likewise, in the image domain view, all pixels in single-coil images contribute to each pixel in the reconstructed image.

can be used in combination with multiple coils having individual localized characteristics. In PI, the parallel nature of array reception is additionally utilized for supplementary spatial encoding. The SNR benefits, but the encoding power of coil arrays depends crucially on the individual coil's reception characteristics. The latter depends on signal frequency and B_0 (the longitudinal static field). The coil sensitivities s_c correspond to the transverse components of the magnetic fields, i.e., $s_c(r) = \mu(r)[H_x^c(r) - iH_y^c(r)]$, where c counts the array elements, r denotes position in 3D space, μ is the magnetic permeability of the scanned object, and B_0 is the static magnetic field applied along the z-direction. The noise characteristics can be described by the noise covariance matrix ψ, which is related to the electric fields, i.e., $\psi_{c,c*} = \int \sigma(r)E^c(r)\sigma(r)E^c(r)^* d^3r$. Here σ denotes the electrical conductivity of the scanned object and * indicates complex conjugation.

In the common case of Cartesian k-space sampling, the PI reconstruction can be performed by image domain unfolding. This is essentially done by inverting a matrix of sensitivity values for each set of pixels that alias in single-coil reconstruction. Suppose that ρ counts the number of pixels in any such set and r_0 denotes their positions. The sensitivity matrix for the set would then be given by $S_{c,\rho} = S_c(r)$. Similar to the mathematical description of phased-array imaging with full k-space density, the SNR of each of the reconstructed pixel values can now be written as $\mathrm{SNR}_\rho^{\mathrm{PI}} = B_0^2/[\{\sqrt{R}\{S^H\psi^{-1}S\}^{-1}\}_{\rho,\rho}]$. Here the reduction factor R indicates the factor by which k-space was undersampled and the superscript H denotes the complex conjugate transpose.

The above mathematical relations illustrate the close relationship between the SNR yield and the electrodynamics of the receiver coil array. In practice, the SNR yield in PI is often stated relative to the SNR that would be achieved with full-density k-space sampling, i.e., $\mathrm{SNR}_\rho^{\mathrm{PI}} = (\mathrm{SNR}_\rho \mathrm{FULL})/(\sqrt{R}g_\rho)$ (Table 3.35). Here g is the geometry factor. This relation emphasizes that the characteristic SNR loss in PI may be viewed as due to two independent mechanisms. The square root of R reflects reduced overall data acquisition and hence reduced intrinsic signal averaging. Conversely, the g factor describes noise amplification related to the conditioning of the unfolding operation. Hence, it reflects the suitability of the coil configuration for the specific PI task, which is characterized by the size, shape, and dielectric properties of the object; the imaged slice or volume; and the undersampling scheme. In a broad sense, the g factor may as well be regarded as a measure of distinctness among the set of coil sensitivities.

Table 3.35 Characteristics of High-Field and Parallel Imaging

	High Field	Parallel Imaging	Combination
SNR	Increased	Decreased	+
Scan efficiency time, resolution, coverage	×××	Increased	×××
ΔB_0 artifacts blurring, ghosting, distortions	Increased	Decreased	+
Motion artifacts	×××	Decreased	×××
RF energy absorption	Increased	Decreased	+
Acoustic gradient noise	Increased	Decreased	+
Intrinsic PPI performance	Increased	×××	+

"×××" indicates that there is no immediate effect. The two concepts are highly complementary and form various synergies when combined, as indicated by the "+" symbols. In this context, parallel imaging can be regarded as a converter that puts the SNR benefits of high field to work in lieu of alternative advantages.

Table 3.36 RF Wavelength (λ_{RF}) and Skin Depth (δ_{RF}) at the ^{1}H Larmor Frequency as a Function of B_0, Calculated for Average Material Properties of In Vivo Brain; the Material Properties Are Frequency Dependent

B_0 (T)	λ_{RF} (m)	δ_{RF} (m)
1.5	0.44	0.14
3.0	0.27	0.10
4.7	0.19	0.084
7.0	0.13	0.073
9.4	0.10	0.065
11.5	0.086	0.061

Highly distinct coil sensitivities yield low (i.e., favorable) g factors; these, however, can never drop below the optimal value of 1. In proton MRI, in high B_0, the near-field approximation (NFA) is often violated.

One can write the significance of the size of the object in relation to the wave number as follows: $k = [\pi(L/\lambda) + i(L/\delta)]$. Here δ denotes the RF skin depth. For the NFA to be valid, the real part of the dimensionless wave number should be valid well below 2π, and thus L/λ is <1. Typical values for human head imaging are shown in Table 3.36 [48]. At higher fields or with larger objects, the RF electrodynamics becomes considerably more complex, because of the onset of wave propagation and diffraction within the object and the increasing influence of the object characteristics. This regime is referred to as the *wave regime*. Wave behavior within the object, including propagation, reflection, and interference, generally translates into more structured and hence more distinct sensitivities. Therefore, it may be expected that wave-regime conditions permit more effective spatial encoding by coil sensitivities, which would be reflected in lower g factors and hence higher SNR.

A promising solution to some of these challenges is introduced in the form of transceiver transmission-line arrays. This design concept is specifically suited for high-field applications, because each transmission-line element is intrinsically shielded and the array as a whole exhibits conventional broadband decoupling characteristics. Further, if the transceiver coil array permits independent phase and amplitude control, it also enables RF shimming, i.e., the optimization of transmission-field uniformity. Even more promising are the prospects of operating each transmitting channel with an individual RF waveform. This approach permits the acceleration of multidimensional SRE (transmit SENSE) and mixes a promising means of addressing B_1 (transverse-field) inhomogeneity and SAR.

Using transceiver transmission-line arrays and the extended sensitivity calibration scheme, the feasibility of SENSE at 7 T has been demonstrated. From a practical point of view, PI with ultrahigh fields (>7 T) is especially appealing for EPI and spiral k-space acquisitions, which are commonly used for functional and diffusion-weighted MRI. In addition, increasing field strength changes the electrodynamics of the resonance signal in favor of PI performance. Higher resonance frequency results in reduced RF wavelength, giving rise to enhanced RF

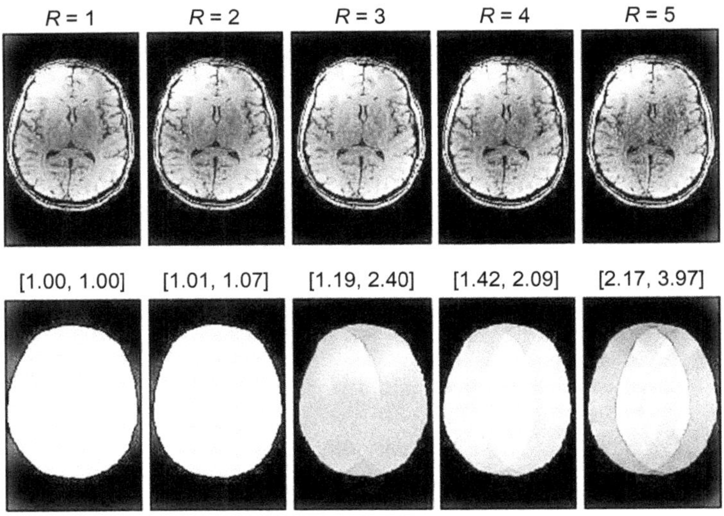

Figure 3.145 Feasibility of parallel SENSE imaging at 7 T using a 16-element transceiver coil array. Top row: T_1-weighted SENSE images with reduction factors $R = 1, 2, 3, 4$, and 5. Bottom row: Corresponding g factor maps. The numbers in brackets indicate the mean (left) and the maximum (right) g factor.

propagation and interference effects. These translate into more distinct coil sensitivities and hence into more effective sensitivity encoding (Figures 3.145).

3.4.1.7 In Vivo MRI Detection of Axon Firing: Human Optic Nerve [49]

It is known that in MEG, spontaneous and evoked magnetic fields, measured at a distance of approximately 2–4 cm, are close to 10^{-12} T. If the source of these magnetic fields is modeled as a current dipole, the magnetic field is expected to be much larger near the current source, because they are inversely proportional to the square of the distance ($\sim 1/d^2$). If the source were a quadrupole moment, the MEG result expected would be even weaker ($\sim 1/d^3$). In MRI, the measurement is based on modulation of the dipole spin moments that directly give rise to the signal. Recent in vivo studies have investigated motor paradigms, visuomotor tasks, spontaneous alpha waves, strobe simulation of the optic nerve, electrical activity in epilepsy seizures, and a sophisticated visual stimulation paradigm to acquire both BOLD and neuronal detection measurements in a single scan. The experiment involving the optic nerve under photopic illumination used synthesized data bursts with durations of tens of milliseconds. Upon excitation by an AP, the conductance of the sodium ion channels increases dramatically. The Na^+ ions flow through the axon membrane for about 1 ms. The motion is perpendicular to the long axis of the ions, at the nodes of Ranvier, which are at 1–2-μm intervals in the myelination of the axon, found approximately every 1–2 mm.

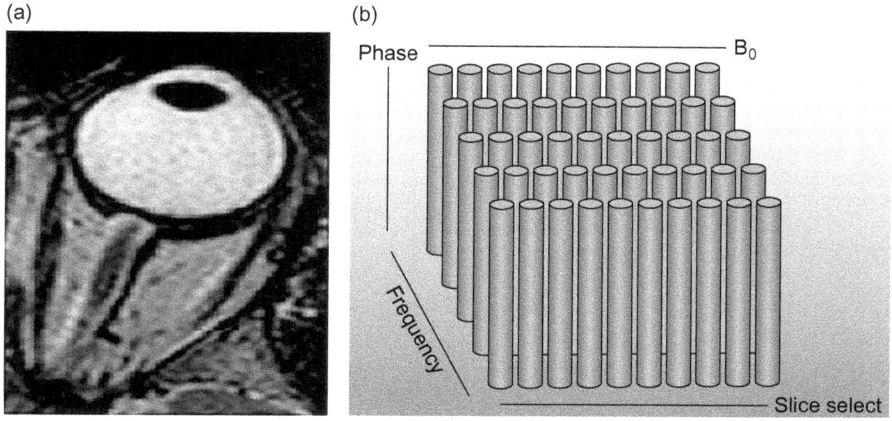

Figure 3.146 (a) High-resolution fast spin-echo image of the optic nerve and orbit acquired using a six-channel SENSE head coil showing the parallel axons within the dural sheath to illustrate the highly symmetrical anatomy. The optic nerve actually forms part of the CNS and is continuous with the brain. Sequence parameters were TR = 3000 ms, TE = 120 ms, in-plane resolution = 0.45 mm, SLT = 1 mm, NEX = 1, ETL = 15. (b) Schematic diagram showing the location of the optic nerve axons relative to B_0 and the image-encoding axes. Experiments were also performed with the phase and frequency directions swapped to assess motion effects. The optic nerve contains 9.26 × parallel axons with an average diameter of 0.72 A packed into an extended cylinder of diameter approximately 3000 A.

Current then flows passively down the axon to the next node, at a rate of between 6 and 15 mm/ms, depending on the axonal diameter. K^+ ions then flow out, after a short delay, resulting in a slight negative undershoot of the net trans-membrane potential (Figures 3.146 and 3.147).

3.4.1.8 Ultrahigh Magnetic Field (≥7 T) MRI: Finite Difference Time Domain Model [50]

Simple physics tells us that as the Larmor frequency increases due to the increased magnetic field, the length of the conductors (struts) used in head-sized volume coils becomes a significant fraction of the operating wavelength. In this case, the coil struts develop self-resonance, which may degrade coil homogeneity because the currents are no longer uniform along each strut. Thus, the transmission-line properties of the coil struts become significant. It invalidates lower-field-strength assumptions, where these properties could safely be neglected. In contrast to lumped-element designs, distributed-circuit resonators utilize and enhance the transmission-line properties of conductors. This is done by using the intrinsic reactance of transmission-line elements. Also, a significant practical advantage of distributed-circuit coils is seen in ultrahigh-field MRI (UHF MRI). The operation is relatively easy, involving only the tuning process of the coil when loaded (when the object is enclosed). In the limit of long TR of the RF pulse sequences applied to the object, $\gg T_1$ (longitudinal

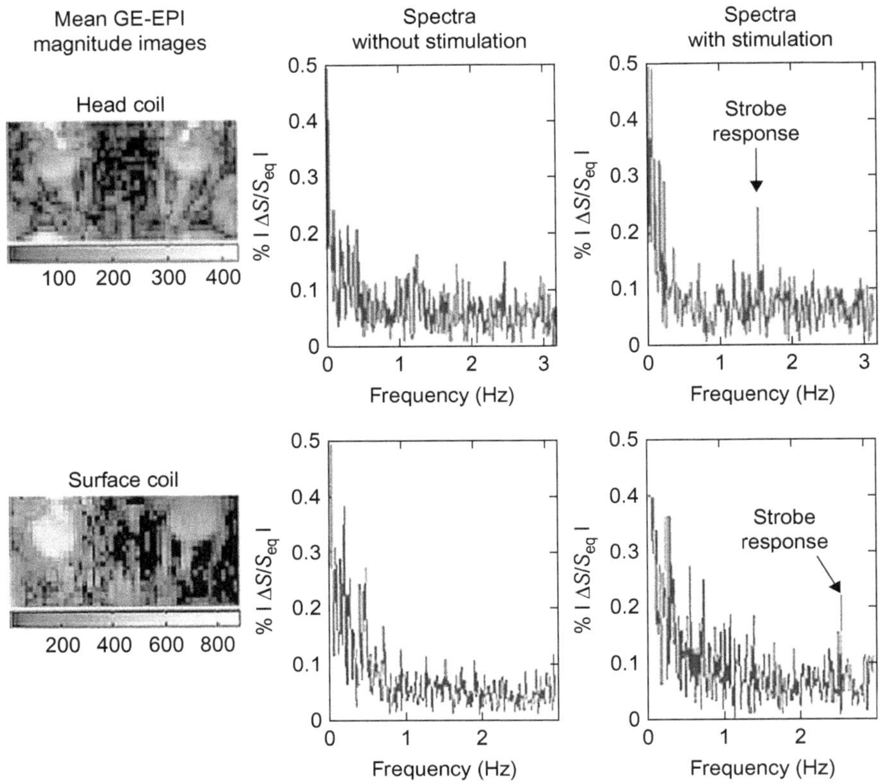

Figure 3.147 Typical measured frequency spectra for ROIs selected in the optic nerve (2×2 voxel ROI, voxel size $1.88\,mm^3 \times 1.88\,mm^3 \times 5\,mm^3$), using either a head coil or a surface coil with 1.5- and 2.5-Hz strobe stimulation, respectively, compared with the control experiment without stimulation. A student's *t*-test ($P<0.05$, degree of freedom = 32) showed that spectra with and without stimulation were significantly different.

relaxation time, along static field B_0), the image intensity S, produced in a voxel located at r, for a gradient-recalled echo sequence can be mathematically expressed as follows: $S(r) = M_0 B_1^-(r) \sin\theta\,(r) \exp(TE/T_2^*)$. Here M_0 is the equilibrium magnetization of the sample. The field $B_1^-(r)$ is the reception sensitivity of the coil and $\theta(r)$ is the effective local flip angle. The local flip angle is proportional to an effective B_1^+ field. It excites the spins in UHF MRI. A voxel is selected such that its flip angle can be determined, and thus serves as the reference. A flip angle of 90° is determined in the reference voxel by maximizing the signal in a voxel-selective stimulated-echo sequence. The flip angle in this voxel is called the *nominal flip angle*. Based on the transmission power supplied to the coil for maximum STE signal, other nominal flip angles can then be specified. One can determine the B_1^+ and B_1^- fields from the images by writing the intensity in the parameterized form as follows: $Si(r) = a_0(r) \sin a_1[(r)]\theta_{nom,i}$. Here Si is the signal in the ith image

($i = 1-24$) from the voxel located at r. $\theta_{nom,i}$ is the nominal flip angle, used for the ith image, and a_0 and a_1 are fitting parameters. One should note that a_0 contains both the proton density and $B_1^-(r)$ field dependence, as a function of position r. The parameter a_1, in contrast, is equivalent to the ratio of the effective flip angle at position r divided by the nominal flip angle. Thus, the parameter maps of a_1 and a_0 describe the B_1^+ and B_1^- fields in relative units. In theoretical simulation, signal intensities $Si(r)$ are fitted as a function of $\theta_{nom,i}$. For in vivo studies, modes 0, 1, 2 of a 16-strut transverse electromagnetic (TEM) coil were tuned to 340 MHz for 8-T field imaging of a healthy human subject.

The first mode of the coil (the mode that occurs at the lowest frequency) is labeled mode 0; the next mode (the second lowest frequency) is mode 1; and so on. The illustrations shown present useful information on progress in the UHF MRI area. Figure 3.148 [50] displays the finite difference time domain (FDTD)-calculated polarization vectors, which are presented across an entire central voxel slice, at a single snapshot in time, for the modes 0–5 of the coil, in a local (close-up) area. The intensities of the polarization vectors are represented by the length of the arrows. The directions of the polarization vectors are represented by the tips of the arrows. The B_1^+ and B_1^- field distributions differ in many regions of the human head and neck mesh (Figure 3.149). Figure 3.150 [50] shows the sagittal low-flip-angle images (top row) of an in vivo human subject and the corresponding simulated results using the human head and neck mesh (bottom row), obtained using the FDTD model for modes 0 and 2. The swirling polarization vectors (Figure 3.148 [50]) at the periphery of modes 2, and above, explain the star-shaped inhomogeneities observed. As with the spherical phantom case, the human head images displayed in Figure 3.150 [50] (both experimentally and numerically obtained) show that modes 0–2 exhibit strong signals near the drive port. Hence, mode 1 is the clear choice for imaging of central brain structures near the ventricles. The remaining modes (0, 2–5) demonstrate significant image intensity only along the

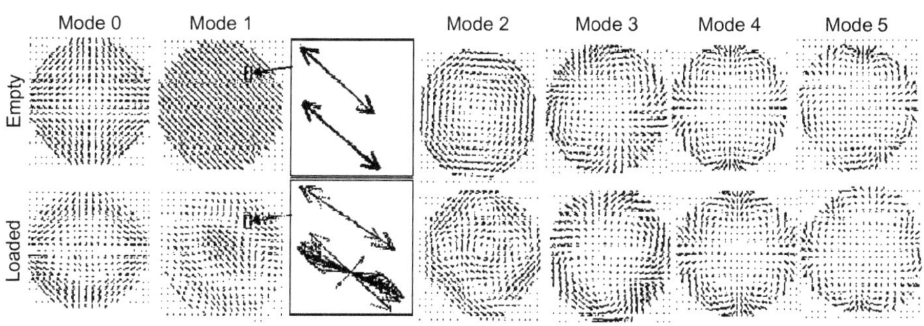

Figure 3.148 Axial slices of FDTD-calculated vector field plots (polarization vectors) presented in a whole slice (at a single snapshot in time) for the first six modes of the coil (modes 0–5) and locally inside a 32 mm^2 area (at many time the snapshots form a complete period, i.e., 2π) for mode 1 (the standard mode of operation of the coil). The FDTD calculations were done in 3D.

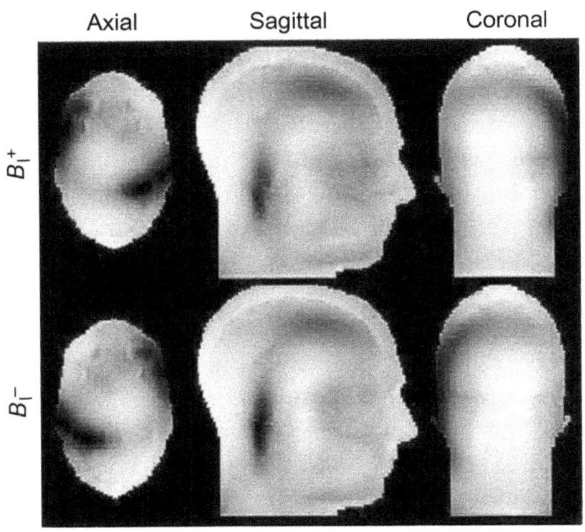

Figure 3.149 Mode 1 distributions of the B_1^+ and B_1^- fields. The results are obtained utilizing the FDTD model for the anatomically detailed human head and neck mesh loaded in a 16-strut TEM resonator operating under linear excitation (back of the head) and tuned to 340 MHz.

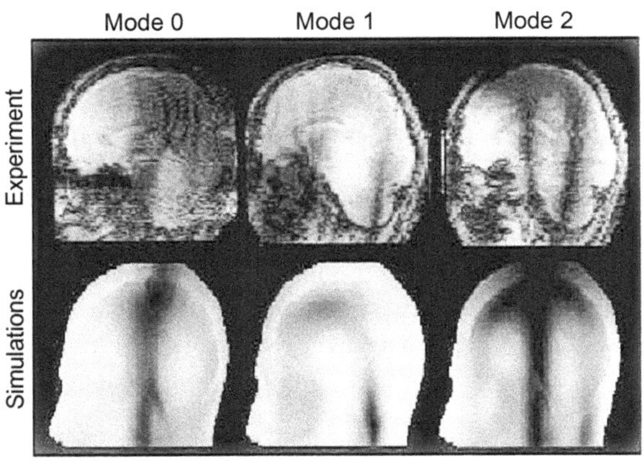

Figure 3.150 Sagittal low-flip-angle images obtained at 8 T (top rows) and their corresponding simulated results obtained at 340 MHz using the FDTD model (bottom rows) for modes 0–2. The excitation was done linearly with the drive port positioned in the back of the head for mode 1 and in the front of the head for modes 0 and 2.

periphery. Mode 0 shows a larger usable volume (i.e., a smaller central signal void). Furthermore, mode 0 is free of the alternating bright/dark patterns of modes 2–5, which are caused by the swirling field lines predicted. Thus, mode 0 should be useful in overall coverage of the cortex or superficial structures.

3.4.1.9 Proton Diffusion MRI: Microscopic Magnetic Field Inhomogeneities/Microvascular Structure [51]

In this technique, no extrinsic precession is induced by the pulse sequence, thus making it prone to any kind of phase imperfections that disturb the perfectly balanced acquisition scheme. One can study the effect of nonuniform spin-phase evolution in balanced steady-state free precession (bSSFP), as generated by eddy currents; that is, their flow or periodic movements. It is seen that even tiny imperfections, such as small but changing spin-phase accruals, may have a large effect on the stability of the steady state. It is thus possible that the proton's random walk through microscopic susceptibility-related magnetic field inhomogeneities keeps the magnetization from achieving a stable steady-state situation, and thereby induces signal loss.

In contrast to simple gradient and spin-echo relaxivity measurements, where microscopic susceptibility variations can be measured as an enhanced transverse relaxation, overall bSSFP signal intensity is drastically reduced. This reduction depends first on external parameters, such as the amount of susceptibility variation $\Delta\chi$, i.e., local concentration of contrast agent. Second, it depends upon the sequence-related parameters, such as flip angle α and repetition time TR. It also depends on the internal parameters, such as the size of the microsphere (radius R) or relaxation times T_1 and T_2. In contrast to spin-echo (SE), gradient-echo (GE), or CPMG measurements, where microscopic relaxation may be ignored for susceptibility analysis, T_1, T_2, and even flip angles play a fundamental role in the determination of steady-state properties. From this point, it is clear that the usual set of three perturbation-related parameters (which are not independent)—$\Delta\chi$, R (radius of contrast agent spheres), and TR—must be complemented by steady-state-related parameters α, T_1, and T_2. Figure 3.151 [51] depicts a hypothetical spin (proton) moving through local field in homogeneities ΔB_z, in the main magnetic field B_0, generated by spherical microparticles of radius R. For the magnetic field perturbation induced by a single sphere, one can write: $\Delta B_z(r, \theta) = (4\pi/3)\Delta\chi(R/r)^3(3\cos^2\theta - 1)\, B_0$. Here r and θ are spherical coordinates between the proton's and the perturber's specific locations, and $\Delta\chi$ is the susceptibility difference between the sphere and the surrounding medium. The proton's stochastic motion along the trajectory $x(t)$ leads to nonconstant spin phase, $\varphi = \gamma\int_0^\tau \Delta B_z[x(t)]\mathrm{d}t$, with processing diffusion time τ, that shows up as an overall dephasing of an initially "phased" spin packet. Traditionally this feature is captured, in an SE- (or GE-) type experiment, as an enhancement in transverse relaxivity of the measured signal time course.

Two extreme cases can be distinguished. In the so-called diffusion or motional narrowing regime (DNR), with small R where dephasing is weak, a higher diffusion constant D leads to reduced relaxivity.

In the static dephasing regime (SDR), with large R, where the diffusion length $l_D = \sqrt{(2D\tau)}$ is much shorter than the perturber radius, diffusion is unimportant, because dephasing takes place in a constant-gradient magnetic field. For the proton's random walk, one can write the spin-phase increment within the nth diffusion time interval as approximately $\Delta_{\varphi n}(\varphi)$. The correlation $G(\tau)$ of $\Delta\varphi$, as a function of diffusion time τ, is shown in Figure 3.151.

The uncorrelated spin-phase variations within any TR lead to a dephasing process analogous to a T_2-like decay. For repetition $T \ll \tau_c$ (the characteristic $G(\tau)$ decay time), changes in $\Delta_{\varphi n}(\varphi)$ are smooth compared with the timescale of the RF pulse train. Thus, bSSFP is not perturbed and comparable signal amplitudes of the perturbed (S) and unperturbed (S_0) steady state can be expected. In the transition range τ_c~TR, however, TR must have a significant impact on the apparent steady-state signal, because relative reactions of protons experience smooth phase changes, compared with protons experiencing randomized phase increments drastically with TR.

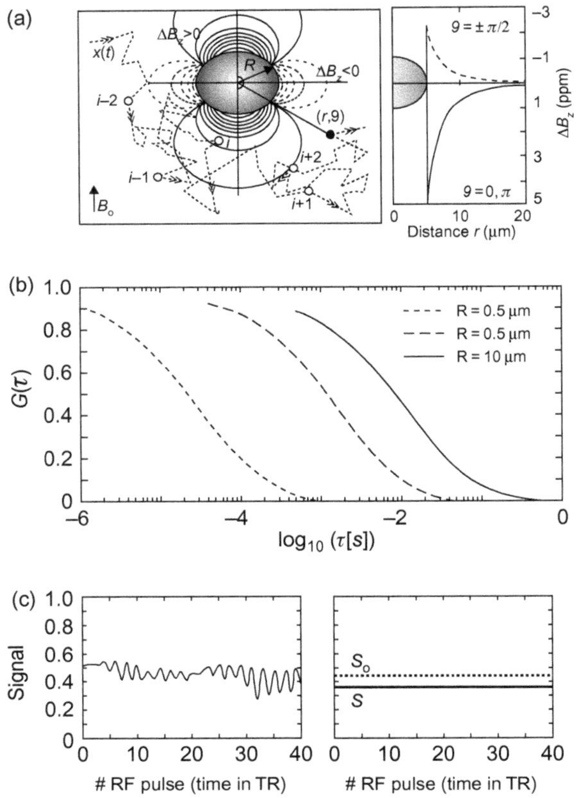

Figure 3.151 (a, left) Sketch of a proton's random walk through local magnetic field inhomogeneities ΔB_z generated by impermeable spherical perturbers of radius R. The open circles on the trajectory $x(t)$ represent the time points of RF excitation in the bSSFP protocol. (a, right) Main magnetic field perturbation as generated by an R. (b) Correlation of spin dephasings $G(t)$ with diffusion time. The gray-shaded region roughly indicates the accessible time range in a bSSFP experiment. The parameters were $B_0 = 1.5\,\text{T}$, $z = 2\%$, $\Delta sx = 0.6 \times 10 \times^{-6}$(cgs), $D = 2\,\mu\text{m}^2/\text{ms}$. (c, left) Individual (single spin) bSSFP signal perturbation in the long time limit ($t \to \infty$) generated by uncorrelated random Gaussian spin dephasings (mean zero, 5° standard deviation) within each TR. (c, right) Unperturbed bSSFP steady-state signal S_0 versus "averaged" bSSFP signal S, using 10,000 spins. The simulation parameters are $T_1 = 590\,\text{ms}$, $T_2 = 490\,\text{ms}$, $\text{TR} = 3\,\text{ms}$, $a = 708$.

3.4.2 Diagnostics

3.4.2.1 Diagnosis of Neurodegenerative Diseases: MRI and MRS [52]

The term *neurodegenerative diseases* refers to a large, clinically and pathologically heterogeneous group that encompasses all neurological disorders leading to dysfunction and finally death of subsets of neurons in specific functional anatomical systems. The most common neurodegenerative diseases of the brain are

Alzheimer's disease, Parkinson's disease (PD), dementia with Lewy bodies, Huntington's disease, and amyotrophic lateral sclerosis (ALS). The definition of neurodegenerative disease used herein is fairly broad; it also allows for the inclusion of a disorder that usually is not considered to be a typical neurodegenerative disease: vascular dementia (VD). VD, particularly the subcortical variant, can clinically mimic more typical neurodegenerative diseases, aggravate them, or even coexist with typical neurodegenerative diseases (e.g., in the form of mixed dementia with AD). Although some of the drugs developed for the treatment of typical neurodegenerative diseases can also yield beneficial effects in VD or cerebrovascular disease, the coexistence of vascular lesions might reduce treatment efficacy in other cases. Owing to these characteristics and interactions with typical neurodegenerative diseases, inclusion of VD in this group seems to be justified.

Increasing age is the single most consistent risk factor for the development of neurodegenerative diseases; their incidence and socioeconomic impact are expected to grow with increasing life expectancy in developed nations. Owing to intensive research over the past few years, some of the basic pathological mechanisms leading to neurodegeneration are now slowly being revealed; e.g., many of these diseases share the phenomenon of protein aggregation (e.g., amyloidal plaques in AD, Lewy bodies in PD, polyglutamine aggregates in Huntington's disease). However, it is still not clear if those aggregates cause neurodegeneration, are an incidental epiphenomenon, or may even be involved in a protective mechanism. Nonetheless, several drugs that promise to modify the neurodegenerative processes, or at least effectively alleviate their consequences, are now in development.

After the mechanism of action and safety of a new compound have been established, through testing in cell cultures and animal systems, its efficacy in humans has to be assessed in clinical trials. This clinical stage of drug development usually has three phases. In Phase I, safety and pharmacokinetics of the drug in humans are established. In Phase II, the efficacy of the treatment is established in small patient samples and information necessary for the planning of Phase III (e.g., determination of appropriate dosages, outcome measures, size of study population, and duration of trial) is obtained. Phase III studies are performed for regulatory approval and must always include longitudinal placebo–treatment comparisons. The best method for establishing the efficacy of a neuroprotective treatment would be to determine directly the number and function of neurons surviving because of the treatment. However, this is impossible in patients; instead, surrogate outcome markers, which are supposed to reliably reflect the number of surviving neurons in a clinically meaningful way, are used. Currently, clinical measures of disease severity (e.g., degree of cognitive impairment and disability in AD, or muscle strength and forced vital capacity in ALS) are most often used for this purpose. However, although clinical measures unquestionably reflect a very important aspect of disease progression, i.e., impairment of function, they also suffer from limitations. Probably the most important limitation is that clinical outcome measures usually do not allow for a distinction between disease-modifying drug effects and purely symptomatic drug effects, i.e., functional improvement but unchanged disease progress, or progress only in complicated trial designs.

Moreover, clinical symptoms become manifest only when the amount of neuron loss/dysfunction has reached a certain threshold (usually around 50–70%), i.e., relatively late in the disease process as a whole. Therefore, clinical outcome measures are not suited to detect disease-modifying actions of a drug in the preclinical stage, i.e., in the phase when an effective treatment would have the most impact. Furthermore, it is also possible that even if a treatment has an immediate effect on the disease process, its effects on the clinical outcome measure become apparent only after a delay, and hence the drug might be wrongly dismissed as ineffective.

Another limitation of clinical outcome measures is their poor test–retest reliability, which is mostly due to the fact that clinical outcome measures are influenced not only by the disease process but also by a number of other factors that are difficult to control: the patient's motivation, presence of other illnesses or adverse effects of other drugs, events in the patient's life, and learning effects in neuropsychological tests, to name a few. Furthermore, many of the clinical scales (e.g., the unified PD rating scale) are based on subjective, semiquantitative assessments of functions, which suffer from substantial between-rater and between-site variability. All those sources of variance result in considerable between- and within-subject variances that diminish the statistical power to detect a treatment effect; therefore, large study populations and long observation periods are needed when clinical outcome measures are used.

Development of a list of characteristics of an ideal outcome marker is a consequence of the shortcomings of the existing clinical outcome markers; there is an increasing need to complement existing techniques with objective and quantifiable outcome markers. Ideally, such an outcome measure should meet the following criteria:

1. The relationship between the outcome marker and the desired clinical outcome (e.g., prevention of cognitive impairment in AD) should be clearly established.
2. The outcome measure should be objective and have high test–retest reliability, to allow for assessment of treatment efficacy in a single patient as well as assessing group effects.
3. The outcome measure should be representative of the stage of the neurodegenerative process at which the drug is supposed to have its maximum effect. For example, if the drug effect is maximum during the preclinical stage, the outcome measure should reflect the disease process in the preclinical stage.
4. The outcome marker should be representative of the supposed mechanism of action of the drug (e.g., a measure of amyloidal burden if the drug is supposed to prevent amyloidal accumulation).
5. Assessment should be noninvasive and well tolerated.
6. Assessment should be inexpensive and not restricted to specialized centers.

Neuroimaging methods, particularly MRI, fulfill at least some of those criteria (notably criterion 6). Their ability to replace clinical outcome measures for the assessment of putative neuroprotective properties of a new drug is increasingly being investigated. Currently, volumetric MRI (VMRI) measurements are capable of monitoring uses in neurodegenerative diseases. The relationship between neuron loss and volume loss/atrophy is now part of a more-or-less routine technique. Still, neuron loss and thus atrophic changes are not necessarily specific to pathological neurodegenerative processes; they are also a feature of normal aging, about <1% per year (py). In contrast, atrophy rates in neurodegenerative diseases are significantly higher, up to 2–3% py. They affect different structures than do normal aging changes (e.g., increased atrophy rate of limbic and temporal lobe structures in AD).

Diffusion-weighted imaging (DWI) and diffusion tensor imaging (DTI) allow one to study the random motion (*diffusion*) of water in brain tissue. If unconstrained, the random motion of water is equally probable along any direction, and thus isotropic. However, in tissues, the random motion of water is hindered by the physical boundaries of the 3D tissue microstructure. Therefore, it occurs preferentially perpendicular to the boundaries and becomes anisotropic. These properties make DWI ideally suited to detect the effects of acute ischemia. DWI/DTI can be used efficiently to detect neurological diseases affecting the integrity of highly structured tissues, such as WM (e.g., as in MS). In MR perfusion studies, endogenous water molecules are "magnetically" tagged in the arteries providing blood flow to the brain. These targeted water molecules then diffuse across the blood–brain barrier into the brain and alter the local magnetization state of the brain tissue in proportion to the inflow of saturated protons. Extensive neuropathological studies have shown the earliest pathological manifestations of AD, i.e., the neuron loss and accumulation of neurofibrillary tangles and β-amyloid plaques found in the peripheral/enchorial cortex and hippocampus. VMRI is able to detect the volume loss in the hippocampus and enchorial cortex, not only in subjects suffering from probable AD, but also in those with mild cognitive impairment. Figures 3.152–3.154 are included as practical illustrations.

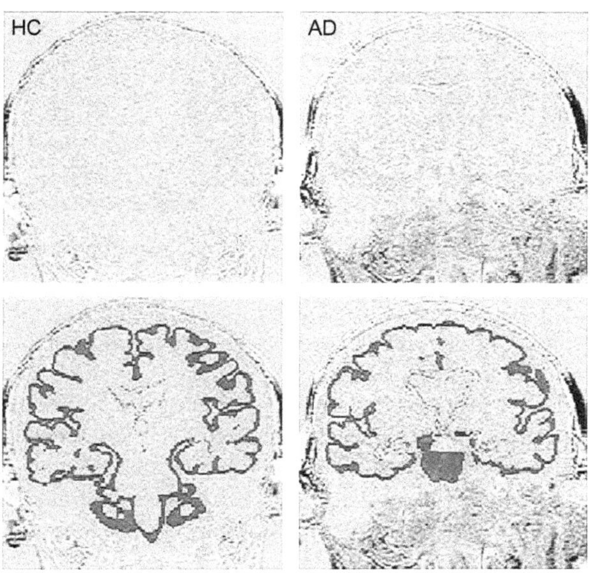

Figure 3.152 Change in cortical and ventricular volume over a 2-year interval in a healthy elderly control (HC) (left) and an AD patient (right). Higher intensity on subtraction image (top) shows greater degree of tissue loss over 2-year interval. Blue shading (bottom) shows the region within which cortical brain structure index (BSI) is measured; yellow shading shows the region within which ventricular BSI is measured.

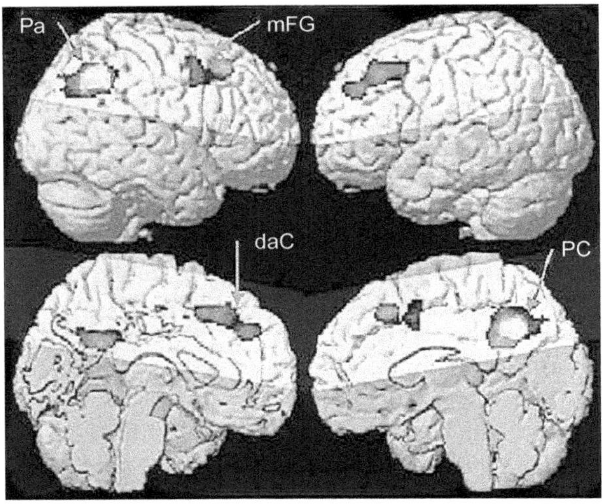

Figure 3.153 Group mean effect of hypoperfusion in AD versus cognitive perfusion for normal elderly, measured by amyotrophic lateral sclerosis (ASL) MRI at cluster level $P<0.05$. Pa, parietal lobe (angular gyrus); PC, posterior cingulate; mFG, medial frontal gyrus; daC, dorsal anterior cingulate area.

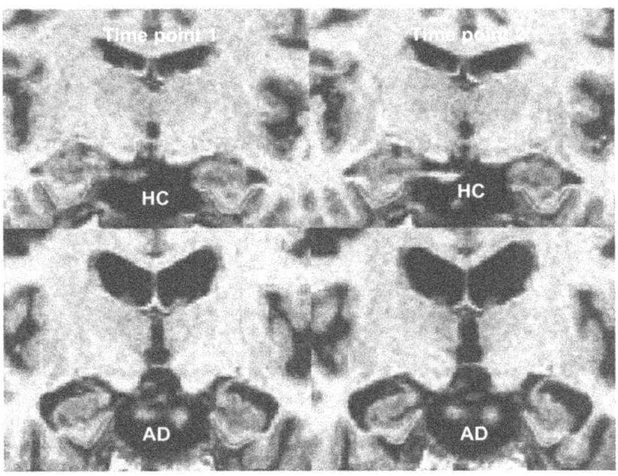

Figure 3.154 Measurement of the entorhinal cortex over 2.2 years in a cognitively normal subject (HC) and an AD patient.

3.4.2.2 Fiber Diffusion Tensor Imaging: Diagnostics of the Human Brain [53]

The brain has three major functional pathways: the motor, sensory, and cerebellar connections.

Within the overall sensory pathway, the primary sensory, auditory, and trigeminal pathways lie in the brainstem. The ventral portion of the brainstem contains fibers, including corticopontine and corticospinal tracts. The superior cerebellar peduncle decussates at the level of the midbrain. The midportion of the brainstem houses transverse pontine fibers, which are the major efferent pathway to the cerebellum. At the medulla, there are olivocerebellar fibers, decussating at the central portion. The medial lemniscus is seen in the midportion of the brainstem. Smaller fibers in the segmental portion of the brainstem run craniocaudally. These include central segmental tracts and medial longitudinal fascicles. DTI can identify the major fibers in different color-coded mapping, although the size and direction of the fibers may limit the visualization of color-mapped images.

In one experiment, the MRI protocol for healthy volunteers was comprised of 3D magnetization-prepared rapid acquisition gradient echo (3D MPRAGE) and DTI. Four patients with clinically diagnosed multisystem atrophy (MSA) were also scanned. Figure 3.155 [53] shows a midline sagittal color-map image, with different eight axial planes: two in the midbrain, three in the pons, and three in the medulla levels. Figure 3.156 [53] shows coronal planes. The slice shows the coronal planes passing through the corticospinal tracts. Slice (b) shows the coronal plane passing through the medial lemniscuses. In the case of MSA (Figure 3.157 [53]), all four patients demonstrated severe cerebellar atrophy. The pons is also atrophic in two cases, though within normal limits in two cases. In the color-coded DTI, the transverse pontine fibers are smaller than normal in all cases.

The brain functions of the neurons are defined by connectivity of the WM. The connectivity of human in vivo WM anatomy is relatively unknown. Even when one looks at myelin-stained pathological specimens, one finds that most of the fibers are mixed together and cannot be separated from one another. The directionality is difficult to assess from the myelin stain, and the information concerning the origin and termination of the fibers is ambiguous. Now, with DTI, one is able to get a tensor map of the fiber orientation based on the water-molecule diffusion in the direction of the myelinated fibers. Using DTI, the fibers in the brainstem, which include cerebellar connections and motor and sensory pathways, can be assessed in terms of their

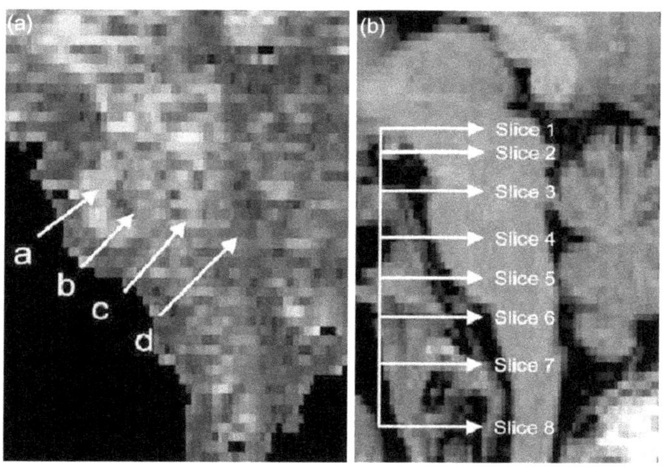

Figure 3.155 A DTI of the sagittal midline section. Arrow (a) shows transverse pontine fibers (red color); arrow (b) shows corticospinal tracts (blue color); arrow (c) shows transverse pontine fibers (red color); arrow (d) shows the medial lemniscus, which is located just behind the transverse pontine fibers (blue color). Structural MRI (MPRAGE) of the sagittal midline section. The levels of the horizontal sections are indicated on the right.

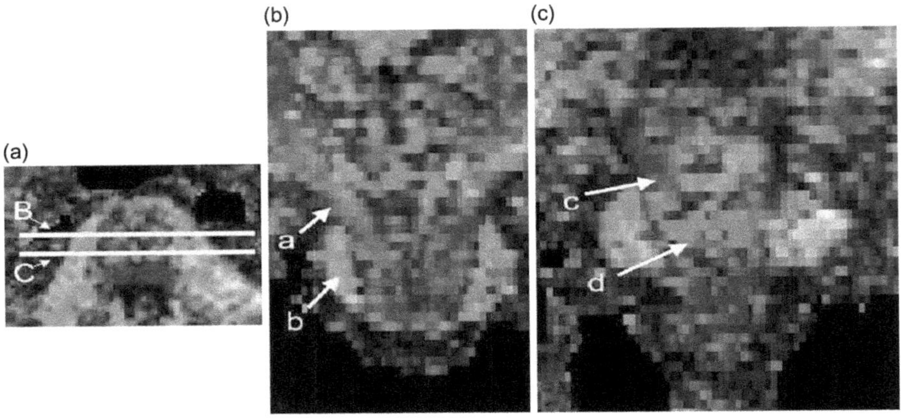

Figure 3.156 (a) Axial DTI image at the mid-pons level. The ventral slice (arrow b) and dorsal slice (arrow c) indicate the positions of the coronal sections. (b) Arrow a indicates blue color area corresponding to the right corticospinal tract; arrow b indicates green color area corresponding to the right middle cerebellar peduncle. (c) Arrow c shows right medial lemniscus (blue color); arrow d shows transverse pontine fibers (red color).

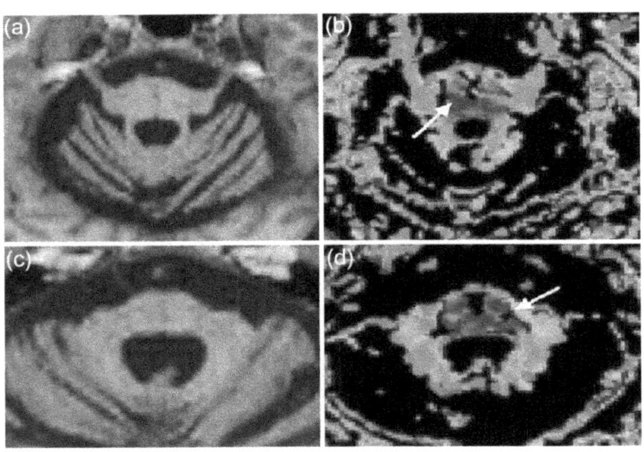

Figure 3.157 (a–d) MSA case. (a) Structural MRI and (b) DTI, both at superior cerebellar peduncle level. (c) Structural MRI and (d) DTI, both at middle cerebellar peduncle level. There is moderate atrophy of the pons and middle cerebellar peduncles. The transverse pontine fibers are thin and the red color and is diminutive.

directionality and integrity. Many applications using DTI have been proposed for evaluation of brainstem structures, including stroke, demyelinating disease, and degenerative disease. WM signal regions, which appear normal with conventional MRI, may be revealed as abnormal with DTI. Hereditary and sporadic ataxias constitute a wide spectrum of neurodegenerative diseases, including spinocerebellar ataxias (SCAs), olivopronto cerebellar atrophy (OPCA), and Friedreich's ataxia with different clinical characteristics. Human pathological specimens and studies with transgenic mice have shown significant loss and disorganization of Purkinje cells in SCA type 1. OPCA patients have significant basal pontine volume loss.

3.4.2.3 Brain Glutamate and Glutamine: Peak Separation Spectroscopic Filtering [54]

Spectral editing is one of the techniques that can be used to find the exact location, spatially and spectroscopically, of the closely located peaks of metabolites in the human brain. This information regarding a region in the brain in a diseased person, as compared to that of a normal person, can be very useful in the process of diagnostics. One uses the spectral difference between the C4 proton resonances of the two systems, Glu and Gln. They are close to 0.1 ppm, approximately 12 Hz at 3 T. Spatially selective RF pulses with a bandwidth of about 12 Hz are employed in this task. Two long (>80 ms) triple-resonance-selective 180° (T180°) RF pulses, implemented within a PRESS, are used to generate the C4 target multiplets, in separate scans. The target signals are well resolved from neighboring resonances of NAA and GSH, at optimized TEs. One needs to combine theoretical quantum mechanical simulation (density matrix), experimental phantom simulation, and in vivo human brain spectroscopy to achieve the aim.

Figures 3.158–3.160 [54] succinctly illustrate the steps involved in the process. The difference between the C4 proton resonances, which is approximately 0.1 ppm,

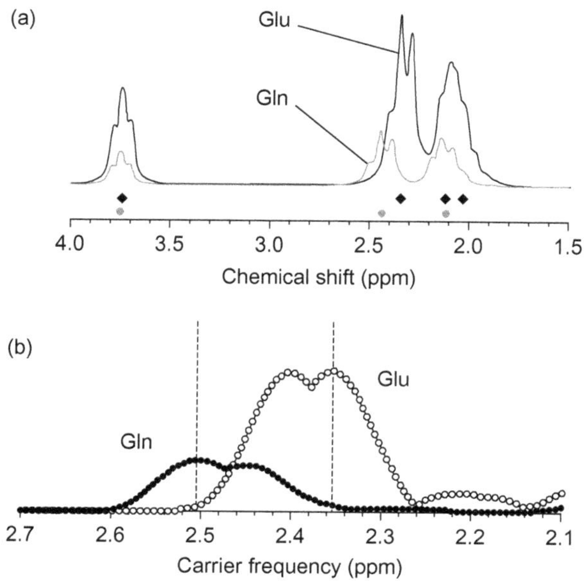

Figure 3.158 (a) Calculated spectra of Glu and Gln following a single-pulse excitation, together with their resonances. (b) Overview of the spectrally selective refocusing method. The height of the calculated C4 proton resonance multiplets of Glu (circles) and Gln (squares) is plotted versus the carrier of the Gaussian 180° RF pulse (86-ms long; truncated at 20%) applied between the 180° pulses of PRESS. TE = 132-ms separation of Glu and Gln signals is achieved at carriers of 2.35 and 2.51 ppm, indicated by vertical dashed lines. A Glu-to-Gln concentration ratio of 3 is assumed; spectral broadening of 5 Hz was used.

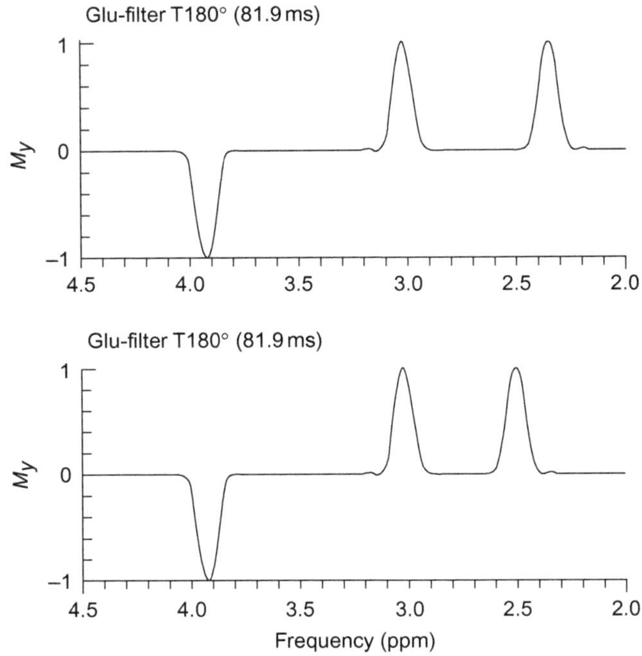

Figure 3.159 Refocusing profiles of two triple-resonance-selective 180° (T180°) RF pulses with a Gaussian waveform, truncated at 20%. The duration of the pulses was 81.9 and 90.4 ms for the Glu and Gln filters, respectively. The former induced refocusing at 2.35, 3.02, and 3.92 ppm, whereas the latter was used for refocusing at 2.51, 3.02, and 3.92 ppm. The 3.92-ppm resonance was refocused 180° out of phase. Bloch simulation for the action of the T180° pulses was made on magnetization M_y, with two-step phase cycling (0 and $\pi/2$) to remove the outer-band components.

is not large enough to discriminate the multiplets reliably. It is possible to use the spectral difference (~13 Hz at 3 T) to filter one or the other, by means of spectrally selective refocusing. Figure 3.158(b) [54] presents the height of the C4 proton resonance multiplets of Glu and Gln as a function of the carrier of an 86-ms-long Gaussian 180° pulse (truncated at 20%) applied between the 180° pulses of PRESS. The full width at half magnitude (FWHM) of the refocusing band of the Gaussian pulse is 12.0 Hz. The overall dependency pattern is similar between Glu and Gln, exhibiting maxima at approximately 2.35 and 2.51 ppm, respectively. Filtering of one or the other can be achieved by selective refocusing at these resonances, where the signal differences between the metabolites are maximal.

Two filters are designed to measure Glu and Gln selectively, in separate scans. The Glu filter utilizes an 81.9-ms-long triple-resonance-selective 180° pulse (T180°). The successive RF phase variations are designed for refocusing at 2.35,

3.02, and 3.92 ppm, Figure 3.159 [54]. The refocusing at 2.35 ppm is for generation of the Glu C4 target at 2.35 ppm. The creatine (Cr) singlets at 3.02 and 3.92 ppm are acquired simultaneously and used for reference in phase correlation. The refocusing of the Cr 3.92 ppm occurs in the direction opposite to that of the other two resonances in the rotating frame, leading to the Cr singlet being 180° out of phase. It is not feasible to obtain refocusing of the three resonances all in phase using the pulse design discussed here. The Gln filter includes a 90.4-ms-long T180°. It is designed for refocusing at 2.51, 3.02, and 3.92 ppm, Figure 3.159. The refocusing at 2.51 ppm generates a Gln C4 multiplet at 2.39 ppm, and the Cr singlets are also generated accordingly.

Density matrix (QM) formalism is employed to calculate the responses of the spin systems, to the filtering sequences. The Hamiltonian of the system includes Zeeman, chemical-shift, and scalar-coupling terms, as well as the gradient and RF pulse waveforms and phase variations. The carrier of the slice-selection RF pulses is set at 3 ppm to ensure that the spectral localization of the resonances of the spin systems include Glu, Cln, Cr, the aspartate moiety of NAA, and the Glu moiety of GSH in the simulation. This would all be within the sample dimension, especially with small bandwidth (1 kHz). The TEs of the Glu and Gln filters are optimized to maximize signal intensity and minimize contamination from neighboring resonances. A primary contaminant in measuring Glu and Gln in the brain is the Glu moiety of NAA. The aspartate (Asp) moiety of GSH is also important. In Glu filtering, the contributions from GABA, succinate, and pyruvate resonances to the Glu target multiplet are negligible, because of their low refocusing ratios and low relative concentrations in vivo. Moreover, the contributions from these resonances are completely eliminated by Glu filtering, because of their greater separation from the target refocusing resonances of the 90.4-ms T180°. Assuming identical T_1 s and T_2 s between Glu, Gln, and Cr, the concentration ratios of Glu and Gln with respect to Cr in prefrontal cortex are estimated, and are shown in Table 3.37. Assuming a Cr concentration of 8 mM, the Glu and Gln concentrations are estimated as 9.7 and 3.0, respectively. The fractions of GM, WM, and CSF within the prefrontal voxel are shown in Table 3.37. The FWHM of the refocusing band of the Gaussian pulse is 12.0 Hz (Table 3.38).

Table 3.37 Results of Spectral Fitting for the Glu-Filtered and Gln-Filtered Spectra from the Prefrontal Cortex of the Seven Healthy Subjects Are Shown Together with the Fractions of GM, WM, and CSF within the Voxel*

Glu Filtering	**Glu Filtering**	**Gln Filtering**	**Gln Filtering**	**Gln Filtering**	**GM (%)**	**WM (%)**	**CSF (%)**
[Glu]/[Cr]	[NAA]/[Cr]	[Gln]/[Cr]	[NAA]/[Cr]	[GSH]/[Cr]			
1.21±0.06	1.44±0.23	0.37±0.09	1.48±0.26	0.42±0.12	63±3	23±2	14±3

*The data are mean±SD.

Table 3.38 Literature Values and Current Results for Average Metabolite Ratios and Intersubject Coefficients of Variation (CV-SD/Mean) for Glu and Gln Measurements in Predominantly GM of Frontal or Prefrontal Cortex in Healthy Adults

Method	Field (T)	Volume (cm^3)	Location	CV (Glu) %	CV (Gln) %	[Gln]/ [Glu]	[Glu]/ [Cr]	[Gln]/ [Cr]
PRESS	1.5	4.0	Frontal	11.0	21.4	0.45	1.43	0.64
STEAM	1.5	8.0	Prefrontal	31.7	75.6	0.41	1.42	0.59
PRESS	1.5	8.0	Frontal	17.7			1.13	
STEAM	2.0	10.0	Frontal	11.8	31.8	0.52	1.33	0.69
Average maps						0.48	1.33	0.64
STEAM	2.9	10.0	Frontal	10.0	24.1	0.41	1.19	0.49
PRESS	3.0	20.0	Frontal	10.0	30.8	0.21	1.21	0.26
CSSF	3.0	16.0	Prefrontal	6.5			1.24	
STEAM	4.0	8.0	Frontal	7.3	22.7	0.17	1.23	0.22
STEAM	4.0	8.0	Prefrontal	9.1			1.21	
STEAM	4.0	1.5	Prefrontal	5.0	59.8	0.26	1.37	0.36
Average maps						0.27	1.24	0.33
SSR	3.0	22.5	Prefrontal	5.0	21.6	0.31	1.21	0.37

SSR, spectrally selective refocusing; CSSF, chemical-shift selective filtering.

3.4.2.4 SENSE MRI of Gliomas: Unaliasing Lipid Contamination [55]

The tumor is usually seen as a hypointense area in T_1-weighted images. After administration of a contrast (Gd-DTPA), the contrast leaks into the areas where the blood–brain barrier is disrupted. This results in hyperintense regions due to the shortening of T_1 (longitudinal relaxation time). The T_2 (TRT)-weighted fluid-attenuated inversion recovery (FLAIR) images show hyperintensity in areas of tumor, edema, and necrosis, due to increased water content. It is difficult to asses the true margin of the tumor border, such as in gliomas, using MRI. This is in part because gliomas exhibit an infiltrative growth pattern in adjacent normal brain tissue. Thus, they do not possess gross definable margins.

It is known that lactate and lipids play an important role in distinguishing between high and low gradable lesions in defining necrosis. Although lipid may indicate cellular membrane breakdown due to necrosis, it is also present in adipose tissue of the scalp. If care is not taken in the data acquisition, it may be aliased into the selected volume even though it is not actually present in the tumor. One can use SENSE as a postprocessing technique to reduce contamination from aliasing lipid resonances that originate from in-slice subcutaneous lipid for 3D MRSI of the brain. This is designed to be effective for data acquired at 3 T, using an eight-channel RF coil, with PRESS localization, and with CHESS pulses for water suppression. Also, very selective suppression (VSS) pulses were

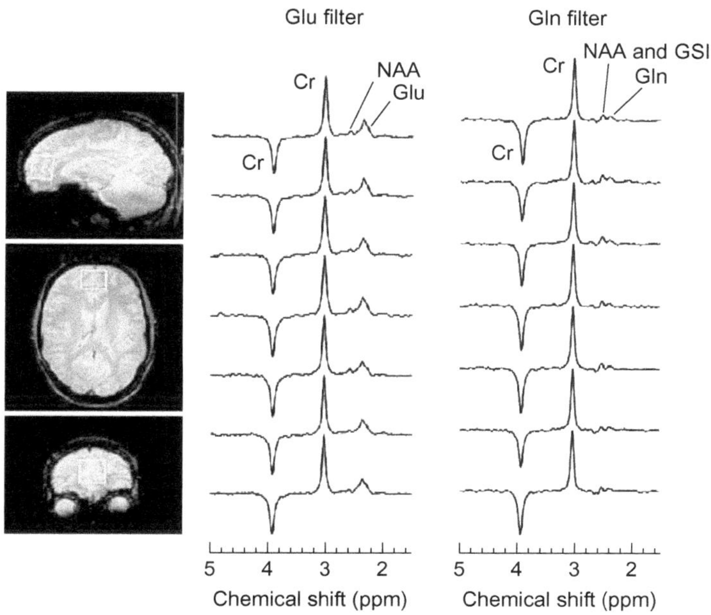

Figure 3.160 A stack of human prefrontal *in vivo* spectra obtained from seven healthy subjects, presented together with a typical voxel positioning. Each spectrum is scaled with respect to its Cr 3.02-ppm peak height. The spectra were all filtered with a 2-Hz exponential and 4-Hz Gaussian function, leading to FWHM of the Cr singlet typically at 9 Hz and mean SNR ratios of the Glu and Gln multiplets of 45 and 15, respectively. The experimental parameters were TR = 2.4 s, TE = 128 ms (Glu) and 158 ms (Gln), and NT = 64 (Glu) and 128 (Gln).

used for improved outer-volume signal suppression. The PRESS pulse had a bandwidth of 2400 Hz for the 90° pulses and 933 Hz for the 180° pulses. The 90° pulse was played along the anterior–posterior (AP) direction and the two 180° pulses were played along the right–left (RL) and superior–inferior (SL) directions. The center frequency was set to be 220 Hz up from the water frequency, so that the selected volumes for metabolites of interest would be close to the desired ROI. The chemical-shift artifact effect on the excitation profiles of different metabolites was estimated from the following relation: $\Delta x = [(\omega - \omega_0)/(BW_{pulse})](X_{box})$. Here Δx is the spatial shift of the excitation for a metabolite with resonance frequency ω, from the center frequency, ω_0, excitation profile. BW_{pulse} is the bandwidth of the PRESS RF excitation in the given direction, and X_{box} = the width of the total excited spectral region. The spectra from the eight coil elements were processed in parallel. Multichannel reconstructed spectra were combined using the coil sensitivity weighted sum (referred to as the *aliased result*) and the lipid aliasing routine. Two sets of spectra were generated for each combined dataset. One was created by phasing on the Cho peak and the other by phasing on the lipid peak, to get accurate estimates of lipid in aliased voxels without affecting the other metabolite levels. Spectral peak heights were determined from these individually phased spectral sets. Lipid-contaminated voxels were determined by selecting the voxels with a lipid peak height higher than the mean Cho height for the patient's data, and higher than one-fourth of the mean Cho height for the phantom data within the selected PRESS box. Lipid heights were normalized with the mean Cho height within the PRESS box, for both the aliased and unaliased datasets.

Figures 3.161 and 3.162 [55] present illustrations of the problems faced and the results observed. Figure 3.161 shows an example of the spectral lipid aliasing problem. The lipid peaks seen on the right edge of the PRESS box were thought to originate from the scalp on the left side of the head, as a result of the chemical-shift artifact. This assumption was supported by calculation of the lipid PRESS box location. Using an 80-mm spectral-box selection along the RL direction with a 1.2 overpress factor, 933-Hz bandwidth of the RF pulse, 220-Hz shift of lipid from the center frequency, and the directionality information of the chemical-shift artifact, the lipid excitation profile could then be estimated to be shifted from the center frequency excitation profile by 22.6 mm toward the left-hand side. This would result in the excitation of the relatively large lipid signals at the indicated subcutaneous fat region, and their aliasing to the other end of the PRESS box due to the small spectral FOV. Although VSS pulses were used for effective suppression of the relatively low metabolite signals outside the PRESS box, they might not have been able to fully suppress the high lipid signal in the scalp region, and could have resulted in aliasing of the spectral signals seen in this example.

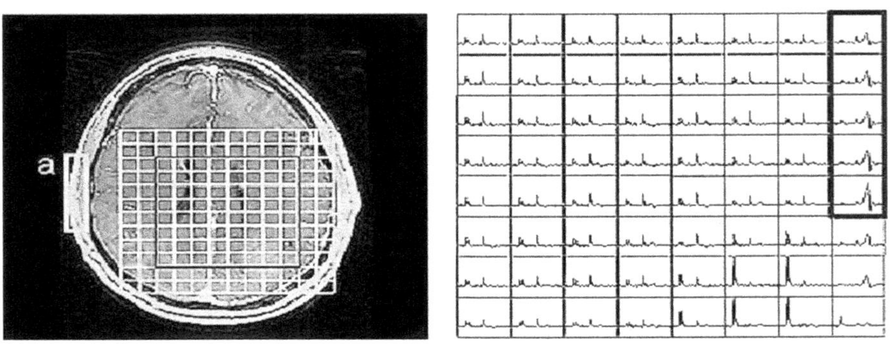

Figure 3.161 Example of the lipid aliasing problem in MR spectra. Left: T_1-weighted spoiler gradient (SPGR) image with the spectral FOV (white grid), and the PRESS-selected region (black box within the grid). Right: MR spectra from the PRESS-selected region. Spectral voxels marked with the black box have aliasing lipid resonances coming from the scalp region one FOV away within the white box marked with "a" on the image.

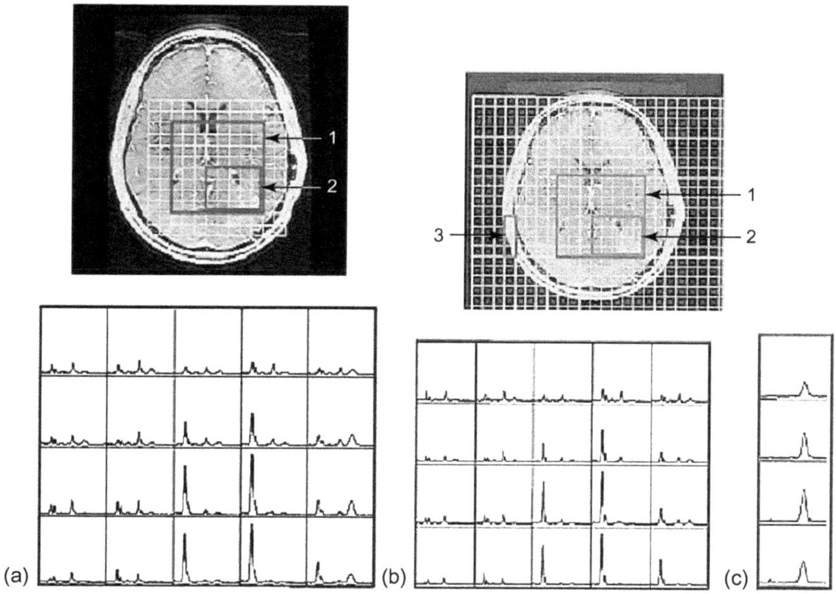

Figure 3.162 Example of data from a glioma patient. Top left: T_1-weighted spoiler gradient (SPGR) image of a slice with the spectral FOV (white grid), a PRESS-selected region (1), and a region with lipid-contaminated voxels (2). Top right: Same image slice with the extended spectral FOV through the unaliasing algorithm (white grid) and the correctly located lipid region (3). Also shown are spectra of the purple box before (a) and after (b) lipid unaliasing, and the lipid spectra unaliased into the red box that is one FOV away (c).

Elements of the eight-channel RF coil were overlapped to minimize the mutual inductance among neighboring elements, and low-impedance preamplifiers were utilized to minimize the interaction with the next-nearest and distant neighbors. Inductive decoupling solenoids were also present, to further reduce coupling between an element and the next-nearest neighbor. It was observed that the noise receiver matrix had high intensities, ranging from 0.54–1 for the phantom to 0.72–1 for sample glioma patient data, in the diagonal entries, whereas the off-diagonal entries showed much smaller values, ranging from 0.08–0.34 for the phantom and 0.004–0.27 for the glioma patient data. Figure 3.162 shows an example of the results for a glioma patient. The original spectral FOV is shown on the top left, and the extended FOV is shown on the top-right image with the white grids. The blue box represents the PRESS-excited volume, and the purple box marks the lipid-contaminated areas. Figure 3.162(a) and (b) shows spectra before and after the unaliasing procedure, respectively. The small red box in the scalp region on the top-right image marks the lipid unfolded areas, and Figure 3.162(c) shows the spectra unfolded into this region. The unaliasing algorithm significantly reduced the normalized lipid heights in individual voxels by a median of $40 \pm 17\%$

for this patient. The maximum and minimum reductions of normalized lipid heights at a given voxel were 76% and 10%, respectively.

3.4.2.5 Activated (Visual Stimulation) Human Visual Cortex Metabolism (7 T): Single-Voxel PMRS [56]

The detection of functional changes of metabolic concentrations can help us to understand the mechanism that sustains the brain at work. Optimization of the sensitivity and accuracy of the methodology for functional MRS studies is important, not only because metabolites are present in low concentrations, but also because concentration changes are expected to be relatively small. The brain metabolite concentrations are likely to be homeostatically controlled in physiological conditions. In this context, ultrahigh (magnetic)-field systems can help in obtaining robust and accurate time courses of metabolites, due to the increased spectral dispersion and SNR. The sensitivity of single-voxel ^{1}H MRS at 7 T enables functional applications, for the purpose of establishing a threshold limit of certainty.

A high number of metabolite concentrations are quantified, during a visual stimulation paradigm. In vivo ^{1}H spectra are acquired using ultrashort-echo-time STEAM, with TE = 6 ms, TM = 32 ms, and TR = 5 s, optimized for applications in humans at UHFs (Table 3.39 and Figure 3.163).

3.4.2.6 Diagnostics of Brain Atrophy in MS: MRI [57]

MRI has already been established as a key investigatory tool for diagnosing MS, and is increasingly used to monitor disease progression and provide outcome measures in therapeutic trials. However, conventional MRI measures of lesions, representing inflammation and demyelination, have been found to correlate poorly with progression of disability. Postmortem and quantitative MR studies have confirmed that widespread abnormalities can occur in normal-appearing brain tissue, and the role of axonal damage and neurodegeneration in the pathogenesis of MS has been highlighted. Neuroaxonal degeneration is irreversible and is likely a relevant mechanism of permanent disability. Measurement of tissue loss (atrophy) has been suggested as a marker of this pathology, and may be a better marker of disease progression than conventional lesion measures. A number of MRI-based techniques used to measure CNS atrophy in vivo have shown progressive brain atrophy above that seen in normal aging in subjects with clinically isolated syndromes (CIS) and MS. Such atrophy has been associated with disability progression. As a result, atrophy measures have been adopted as secondary outcomes in a number of trials of disease-modifying drugs. Here we expand on and update the pathology behind axonal damage, methods developed to measure brain atrophy, the findings of clinical studies that monitored brain atrophy, and the role of atrophy measures in MS research and clinical applications.

Important studies have documented atrophy of the spinal cord and optic nerves in MS, so now one focuses on the brain atrophy. Microscopic examinations of postmortem brain tissue samples from MS patients have provided direct evidence of

Table 3.39 Quantification of Metabolite Concentrations by the LC Model

	Concentration (μmol/g)	CRLB (μmol/g)	CRLB (%)	SD, Voxel OUT (%)	SD, Voxel IN (%)	Simulated BOLD Effect (%)
Asc	1.2	0.21	18	11	6	3.1
Asp	1.0	0.26	28	15	16	3.8
Cr	5.0	0.26	5	3	2	0.8
PCr	3.4	0.24	7	4	3	1.4
GABA	1.0	0.13	14	4	9	1.7
Glc	1.1	0.28	25	13	17	0.3
Gln	2.9	0.16	6	4	3	0.1
Glu	11.0	0.22	2	1	1	1.1
GSH	1.0	0.10	10	8	7	0.9
Mi	6.7	0.20	3	1	2	1.0
Lac	0.8	0.10	13	8	19	0.1
NAA	10.8	0.14	1	1	1	1.2
NAAG	1.4	0.11	8	3	5	− 0.4
PE	1.2	0.19	16	12	7	− 1.4
Scyllo	0.4	0.05	12	9	4	2.4
Tau	1.9	0.18	10	4	4	3.4
GPC + PCho	1.3	0.05	4	2	1	1.4

Values were calculated from all nine spectra obtained during each functional study and by averaging the data from the two subjects. Voxel OUT or IN means outside (control conditions) or inside the visual cortex. The column relative to simulated BOLD effect indicates the estimated artifactual concentration changes derived by the LC model at 0.4 Hz line narrowing.

neuroaxonal damage. Axonal density has been shown to vary between 0% and 100% of normal within chronic and acute WM lesions. In addition, significant decreases in axonal number and density have been observed in normal-appearing WM (NAWM), and axonal transection and loss underlies GM lesions found postmortem. Axonal spheroids, transections, and abnormal constrictions and dilatations have all been observed in lesions and NAWM using immunostaining. In addition, levels of NAA and amino acid found within mature neurons can be assessed in vivo using PMRS, to provide indirect assessment of neuronal integrity. Decreases in NAA levels have been observed within lesions, NAWM, and normal-appearing GM (NAGM) in MS subjects, which implies that neuroaxonal damage is a consistent feature of MS. Notably, these observations have also been made in subjects within the first year of symptom onset, in subjects presenting with CIS, and in early relapsing remitting MS (RRMS). This supports the hypothesis that neuroaxonal damage occurs in the earliest stages of MS, although normal NAA in NAWM has also been noted in CIS. The mechanisms of neuroaxonal degeneration appear to be complicated. Evidence of neuroaxonal degeneration within lesions supports the idea that this pathology may be a consequence of acute focal inflammation. The numbers of $CD8^+$ T-cells increase in inflammatory lesions, and myelin breakdown leaves axons vulnerable to direct attack from toxic inflammatory

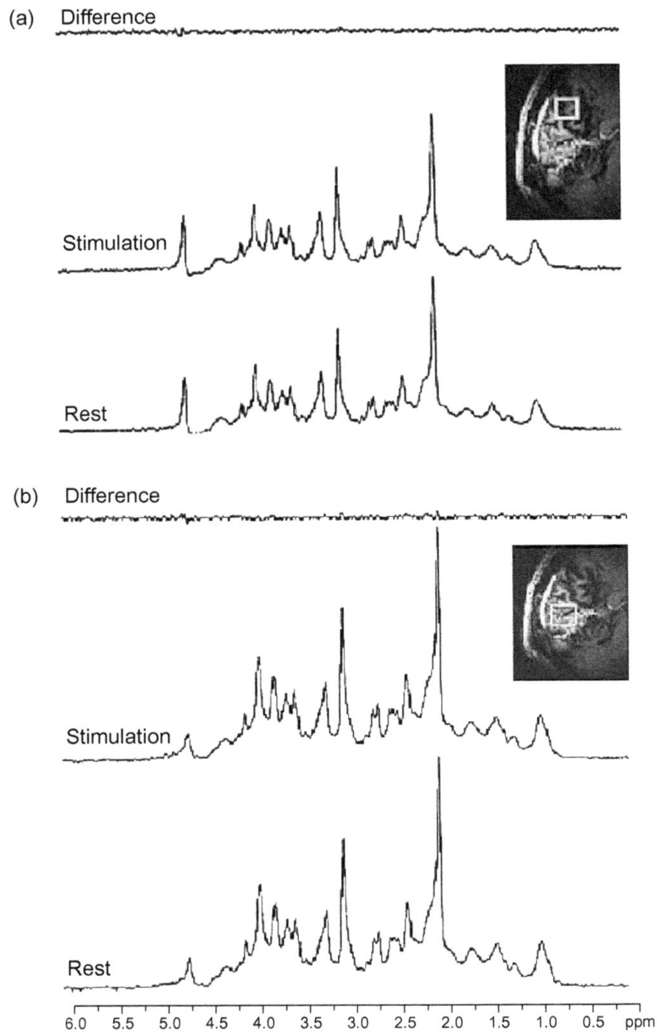

Figure 3.163 Spectra obtained at rest and during stimulation in a single subject, when the VOI ($20\,mm^3 \times 20\,mm^3 \times 20\,mm^3$) was located outside (a) and inside (b) the visual cortex. The insets depict functional maps with the localization of the VOIs (GE EPI: TE = 22 ms, TR = 2.4 s, spatial resolution $2.5\,mm^3 \times 2.5\,mm^3 \times 2.5\,mm^3$; activated pixels correspond to $cc > 0.3$, cluster size ≥ 6 contiguous pixels). Differences between spectra obtained in the two conditions of rest and stimulation are also shown. Spectroscopic parameters: STEAM, TE = 6 ms, TM = 32 ms, TR = 5 s. Processing: frequency and phase corrections of individual scans, summation of 32 scans, residual eddy currents correction, Gaussian multiplication ($\sigma = 0.0865$ s), FFT and zero-order phase correction. No further postprocessing, such as water signal removal or baseline correction, was performed. Spectra in (a) and (b) have the same vertical scale. The small, narrow peaks at 2.0 and 3.0 ppm, visible in the difference spectrum when the VOI was coded inside the visual cortex, were ascribed to the BOLD effect.

mediators such as proteolytic enzymes, cytokines, oxidative products, and free radicals.

A positive correlation of $CD8^+$ T-cells with the extent of axon damage has been shown. In particular, nitric oxide has been identified as contributing to neuroaxonal degeneration and irreversible structural changes within lesions, particularly during high axonal firing, by effecting various changes (e.g., altering mitochondrial DNA and energy metabolism). In addition to direct effects of acute inflammation, secondary effects may result from myelin loss. There is evidence that myelin-forming cells provide trophic support to axons, the loss of which leads to degeneration. Similarly, degeneration has been shown to occur despite intact surrounding myelin. Lastly, correlations observed between focal lesions and NAWM changes are consistent with Wallerian degeneration, wherein axonal transaction within lesions causes distant neuroaxonal damage, possibly due to the loss of pre- and postsynaptic signals. There may also be other causes of axonal loss, less well understood but quantitatively important.

Unlike inflammation, neuroaxonal degeneration is irreversible, and its measurement could therefore provide a particularly relevant marker of MS pathology in terms of clinical disability. Metabolite measures of neuroaxonal damage, such as ^{1}H-MRS-derived estimates of NAA, are subject to reproducibility problems and sampling bias. Brain atrophy measurements may provide a more sensitive marker of diffuse neuroaxonal losses across the whole brain. MRI-derived brain atrophy measurements are widely used in AD, the prototypic neurodegenerative disease. Despite the potential value of brain atrophy as a marker of neuroaxonal loss, other pathological and physiological factors can alter brain volume and should also be taken into account (Table 3.40). A good brain atrophy measurement technique would demonstrate many of the features listed in Table 3.41. Although measurement accuracy and precision are desirable, they are less important in practical terms than sensitivity to disease-related change and measurement reproducibility. Techniques that can be automated and are robust to differences in acquisition are particularly relevant for large multicenter studies. The optimum MRI acquisition for atrophy quantification may to some extent depend on the measurement technique used and the subjects studied. Although atrophy measurements have been successfully applied to 2D acquisitions, 3D (volumetric) acquisitions require less dependence on slice positioning and slice

Table 3.40 Pathological and Physiological Variables with Potential to Affect Brain-Volume Measurements in Multiple Sclerosis

Factors Causing Brain-Volume Increases	Factors Causing Brain-Volume Decreases
Edema	Axon loss
Inflammation	Resolution of inflammation and edema
Gliosis (tissue bulk)	Gliosis (retraction scarring)
Remyelination	Demyelination, dehydration, anit-inflammatory agents, normal aging

Table 3.41 Desirable Qualities of a Brain Atrophy Measure

Quality	Comment
Sensitive to brain atrophy	Allows subtle pathological changes to be detected.
Reproducible	Avoids measurement errors that could lead to erroneous results.
Accurate	Detects actual tissue loss. Accuracy is difficult to verify, however, and small errors are insignificant if consistent between subjects and over time.
Precise	Repeated measurements of the same volume will be of the same value to within a small percentage of the volume.
Automated	Fast to implement, thereby reducing operator time and costs. Operator-dependent errors are minimized.
Robust image quality	Results are more reliable and comparable between subjects and imaging sites.

selection, and may yield more accurate results from automated techniques. Many techniques rely on brain segmentation, which depends on high-contrast boundaries between the brain and CSF. Standard high-resolution 3D acquisitions (acquired in less than 10 min) reduce partial volume effects and provide good CSF/brain and GM/WM contrast, which allow visualization and measurement of small regional structures. In addition, 3D acquisitions allow accurate reslicing of data for registration-based techniques.

Brain volume measurements based on a single scan are difficult to interpret because of wide normal variability. Normalization to intracranial volume should be performed because small volume changes tend to be masked by the biological interindividual variability in absolute brain volumes. Additionally, the progression and rate of atrophy can be assessed only indirectly from cross-sectional measurements, and one must assume that atrophy progresses linearly and at a similar rate among individuals. Longitudinal measurements allow one to monitor disease progression more precisely and identify the extent of true interindividual differences. Linear or 2D measures of brain and ventricular spaces are the simplest atrophy measurements because they can be performed quickly and applied to both 2D and 3D images. Measurement of brain width on axial MRI has demonstrated atrophy rates of -0.64% year^{-1} in RRMS subjects. Two-dimensional measures of ventricular enlargement (a result of tissue loss) have been applied more widely and permit indirect assessment of brain atrophy (Table 3.42). A small loss of brain tissue can result in relatively large increases in CSF compartments. One study that measured third-ventricle and lateral-ventricle width on axial MR images demonstrated annual increases of 4.5% and 5.5%, respectively, in MS subjects, and third-ventricle width has been shown to correlate with third-ventricle volume. Coefficients of variation (COVs) for brain, third-ventricle, and lateral-ventricle width measurements of 1%, 7%, and 4%, respectively, have been shown.

Midsagittal measures of the corpus callosum (CC) area have also been obtained in a number of MS studies. The CC is comprised of axonal tracts connecting the

Table 3.42 Studies of Ventricular Enlargement in Relapsing/Remitting MS

Subjects	Image Type	Enlargement	Findings
42	T15 mm axial	Yes	Median 3.9% $year^{-1}$ volume increase (interquartile range 0.8 to 8.5)
7	3D FSPGR	Yes	Median 4.04% volume increase over 18 months (range 0.68 to 15.16)
15	T13 mm axial	Yes	Median 0.8% $month^{-1}$ volume increase (range −0.5 to 7.6)
6	3D FSPGR	Yes	Median 2.1 ml $year^{-1}$ volume increase (interquartile range 0.7 to 3.7), significantly greater than controls

FSPGR, fast spin gradient echo.

left and right brain hemispheres, and has long been recognized as being particularly affected in MS. Cross-sectional analyses have shown an approximately 22% smaller CC area in MS subjects compared to controls. Significant decreases in the CC area have also been observed longitudinally, with a finding of 4.9% $year^{-1}$ in one study. However, these techniques are limited by methodological variability and the dependence of these 2D measures on slice positioning and selection, which are often based on subjective criteria. Furthermore, slice selection may be more difficult to perform when the head position and orientation within the MR scanner vary among patients and over time. Techniques for measuring cortical thickness and estimating GM atrophy were recently developed using reconstruction of the cortical surface from volumetric MR. The GM/WM and brain/CSF boundaries were found using a deformable surface algorithm, and the thickness of the cortex was computed with a high level of sensitivity. A significant reduction in cortical thickness was shown in RRMS and secondary progressive MS (SPMS) patients relative to controls; however, this method has yet to be applied longitudinally in MS.

Brain segmentation by MRI allows the quantification of global, regional, or GM/WM volume. Brain segmentation methods include manual outlining, semiautomated or fully automated thresholding (driven by the difference in brain/CSF signal intensity), and surface model-based techniques. In comparison with algorithms that perform semiautomated or fully automated segmentation, manual outlining of the brain is time-consuming, subjective, and less reproducible.

3.4.2.7 MRS Diagnostics of Mesial Temporal Lobe Epilepsy and Neocortical Epilepsy [58]

Reductions in the metabolite ratios NAA/Cho or NAA/Cr are commonly interpreted as being linked to neuronal loss or increased gliosis. Studies of cases of mesial temporal lobe epilepsy (mTLE) have found accompanied unilateral hippocampal sclerosis, diagnosed by MRI. Neocortical epilepsy is diagnosed as having discrete lesions on the necrotex and/or ictal onset zones exclusively on the neocortical areas.

MRI diagnosis of hippocampal sclerosis is based on the presence of either definite unilateral atrophy or high T_2 (TRT) signal intensity of the hippocampus, or both. For control subjects, the MRI (1.5 T) procedure involves conventional spin-echo T_1 (longitudinal relaxation time)-weighted sagittal and fast spin-echo T_2-weighted axial and oblique coronal sequences. For patients, a T_2-weighted fast spin-echo sequence, with 3-mm-thick sections and a reformatted T_1-weighted spoiler gradient-echo sequence, with 1.5-mm-thick sections, are obtained in the oblique coronal plane for evaluation of the temporal lobe and the hippocampus. The angle of the oblique coronal imaging is perpendicular to the long axis of the hippocampus. In addition, FLAIR images with 3-mm-thick sections are obtained in the oblique coronal plane in some patients. Single-voxel ^{1}H MRS is carried out immediately after acquiring MR images. Two spectra are obtained for each control subject and patient, from the VOI. The two VOIs are placed bilaterally in the mesial temporal area, including the hippocampus. The VOIs included most of the hippocampal head and parts of its body, the amygdala, the parahippocampal gyrus, and the CSF in the temporal horn. The VOIs are placed so as to avoid potential magnetic susceptibility artifacts from the skull base and the sphenoid sinus, and contamination by fat in the skull base. In all subjects, the spectra are obtained using PRESS, with an echo time of 136 ms, a repetition time of 1.500 ms, and 128 scans. Before spectroscopic measurements, field homogeneity is optimized over the selected VOI by observing the ^{1}H MR signal of tissue water, with a spatially selective PRESS. The water signal is suppressed by a frequency-selective saturation pulse at the water resonance. Figure 3.164 [58] is an illustration of the changes in ratio of the metabolic peaks under examination.

The mean NAA/Cho and NAA/Cr ratios of the right and left sides of the control subjects are compared. These work as thresholds for determining abnormal NAA/Cho and NAA/Cr ratios. The asymmetric index (AI) = (SR − SL)/(SR + SL) is used for patients with bilaterally abnormal ratios of MRS results. Here SR and SL represent the NAA/Cho or NAA/Cr ratios in the right and left hippocampus, respectively.

The NAA/Cho and NAA/Cr are significantly lower in the ipsilateral hippocampus of mTLE patients. In neocortical epilepsy, the NAA/Cho of the hippocampus is also seen to be lower in the side ipsilateral to the epileptogenic hemisphere. NAA/Cr is not significantly different between the ipsilateral and contralateral hippocampus in neocortical epilepsy. NAA/Cho is seen to be abnormally lower in the ipsilateral hippocampus of mLTE patients.

3.4.2.8 Diagnostics of Human Brain Disorders: T_2 *(Transverse Exponential) Decay: Tissue Water Proton Signals [59]*

The total MR signal from the brain includes contributions from protons in water as well as nonacqueous protons in molecules such as lipids, proteins, and nucleic acids. The signal of water in tissue has a T_2 longer than 10 ms. The nonacqueous

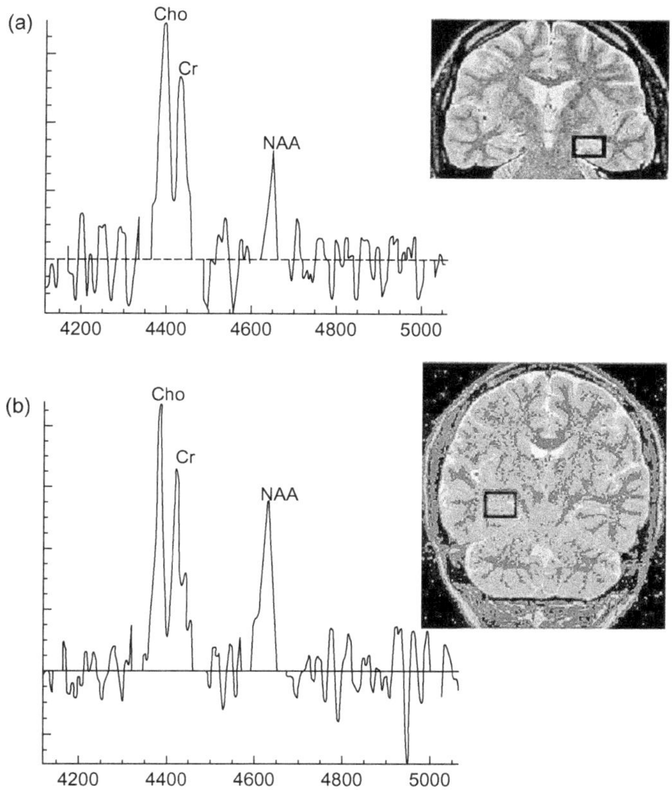

Figure 3.164 (a) Abnormally low NAA/Cho and NAA/Cr ratios in the left hippocampus in the patient with left hippocampal sclerosis. NAA/Cho and NAA/Cr of the left hippocampus are 0.45 and 0.58, respectively. The patient's MRI showed left hippocampal sclerosis. (b) Abnormally low NAA/Cho and NAA/Cr ratios in the right hippocampus in the patient with right frontal lobe epilepsy (FLE). NAA/Cho and NAA/Cr of the right hippocampus are 0.61 and 0.81, respectively. The patient's MRI showed bilaterally normal hippocampus.

proton T_2, in contrast, is approximately 100 μs. This is due to large unaveraged dipole couplings between adjacent protons. Thus, it is easier to measure the MR signal from water in the brain, with no contamination from fast-decaying nonacqueous tissue signals. The key issues for accurate in vivo T_2 decay measurement are maintaining perfect 180° pulses, in the presence of inhomogeneous B_1 (transverse RF magnetic field) and B_0 (static longitudinal magnetic field); and elimination of all contributions from stimulated echoes accruing from signals excited outside the selected slice.

For in vivo human brain studies, the echo spacing should be 10 ms or less, and the echo train should exceed 1 s. This is to measure the shortest T_2 components and be sensitive to T_2~400-ms timescale.

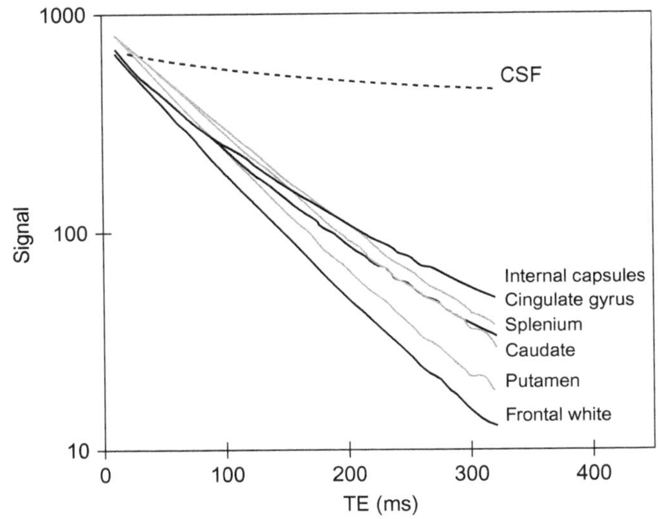

Figure 3.165 T_2 decay curves for various WM (black line) and GM (gray line) structures, as well as CSF (dashed line).

The VOIs contain more than one distinguishable water reservoir, each with its own T_2 exponential decay content. It is observed that the brain T_2 decay curves are best characterized not by a fixed number of discrete T_2 components, but rather by a continuous T_2 distribution. There are T_2 components in normal human brain. The T_2 distribution provides two specific measures related to the CNS. Clinical MR images of MS brain and spinal cord can easily visualize focal areas of the disease, known as *lesions*. Lesions can appear and disappear, though they typically increase in number and volume as the disease progresses (Figures 3.165 and 3.166).

3.4.2.9 Regional Metabolic Distribution: CSI CADSIL Family [60]

Altered regional distribution of brain metabolite concentrations, associated with metabolic pathways of a given neuronal function and structure, can be measured. These measurements provide sensitive indicators of early neurological involvement, and are useful in evaluating the progress of the disease.

Cerebral autosomal dominant arteriopathy with subcortical infarcts and leukoencephalopathy (CADSIL) is an inherited disease of the small vessels of the cerebrum. It is clinically characterized by transient ischemic attacks, recurrent strokes with onset during mid-adulthood, migraine with aura, and progressive cognitive impairment leading to dementia and mood disturbances. In addition to genetic linkage analysis, MRI plays an important role in the diagnosis of CADSIL, by evidencing a constant pattern of lesion distribution in the brain regions of symptomatic patients. MRI studies typically reveal diffuse WM areas with high signal intensity and lacunar infarcts in the centrum semiovale,

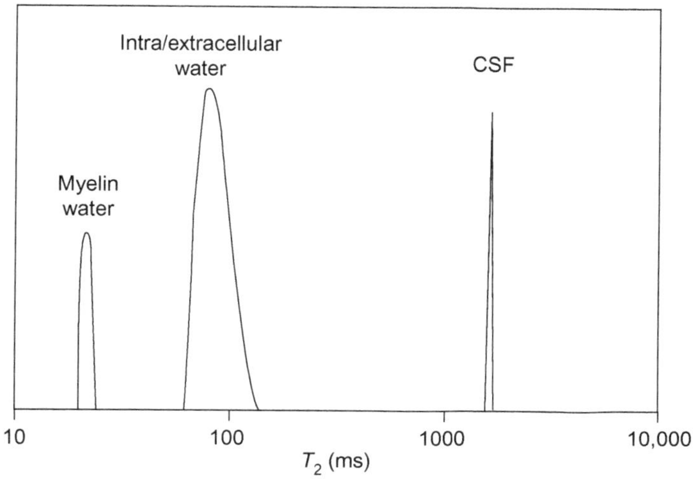

Figure 3.166 An example of a T_2 distribution of human WM *in vivo*, with labeled regions corresponding to compartments of WM. χ^2 was kept within 2–2.5% of its nominal value.

Table 3.43 Regional Metabolite Distribution

Metabolite Ratio	Centrum Semiovale	Mean±SD	CV	*P*-Value
A. Mean metabolite ratios in the patient group				
NAA/Cr	Anterior	1.73±0.15	8%	.0425
	Posterior	1.93±0.05	3%	
NAA/Cho	Anterior	1.63±0.28	17%	.0195
	Posterior	2.01±0.27	13%	
Cho/Cr	Anterior	1.14±0.11	10%	.0046
	Posterior	0.98±0.12	12%	
B. Mean metabolite ratios in the normal controls				
NAA/Cr	Anterior	1.86±0.08	4%	.1902
	Posterior	1.92±0.07	4%	
NAA/Cho	Anterior	1.79±0.24	13%	.0633
	Posterior	1.88±0.22	12%	
Cho/Cr	Anterior	1.18±0.15	13%	0.554
	Posterior	1.04±0.18	17%	

Data are expressed as mean ± SD. CV is the coefficient of variation (metabolite standard deviation divided by the mean metabolite ratio 100%). The *P*-values were obtained by using paired two-tailed *t*-test at the 95% significance level.

thalamus, basal ganglia, and pons, which may be readily detected in presymptomatic CADSIL. Specifically, T_2-weighted and FLAIR MRI disclose early signal intensity changes in the subcortical and periventricular WM in asymptomatic CADSIL subjects. This is particularly true in frontal and temporal lobes with no cortical lesions. Subcortical white-matter hyperintensities (WMHs) on T_2-weighted images are the radiologic hallmark of CADSIL (Table 3.43 and Figures 3.167).

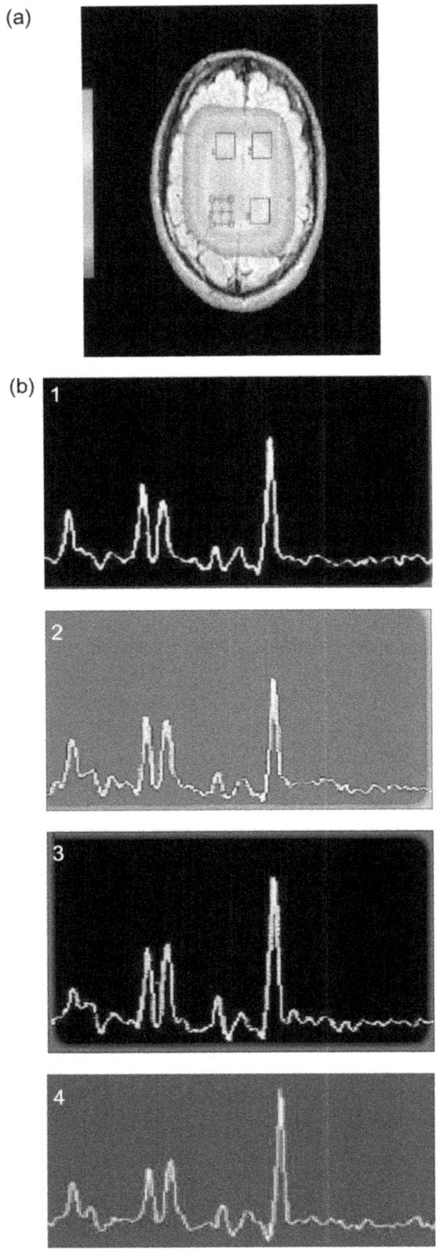

Figure 3.167 (a) A chemical-shift image of a given patient at the level of the centrum semiovale showing the anatomical location of the PRESS selection box. (b) The representative spectra from the four ROIs shown in (a).

3.4.2.10 Metabolics of the Brain: In Vivo ^{13}C NMR Spectroscopy [61]

The ^{13}C NMR spectroscopy is a highly chemical-specific tool. It can distinguish ^{13}C isotope incorporation, not only into different molecules, but also into specific carbon positions within the same molecule (^{13}C isotopomers). It allows one to follow the fate of a ^{13}C label through multiple metabolic pathways. A ^{13}C metabolic study involves (1) the choice of ^{13}C-labeled substrate, (2) the detection of the label incorporation into brain metabolites during infusion of ^{13}C-labeled substrate, and (3) quantitation of ^{13}C spectra to obtain ^{13}C turnover curves. Refer to Figures 3.168 and 3.169 [61] for illustrations of the techniques involved. Because glucose is the main fuel for the brain, ^{13}C-labeled glucose has been the preferred substrate for metabolic studies in the brain. One detects ^{13}C-label incorporation into brain metabolites even during infusion of the ^{13}C-labeled substrates. This is followed by quantification of the resulting NMR spectra. Because most tricarboxylic acid (TCA)-cycle intermediates are not concentrated enough to be detected by NMR *in vivo*, one relies on detection of the ^{13}C label in the larger pools of the brain amino acids, such as glutamate, glutamine, aspartate, and (if sensitivity is sufficient) GABA.

A fundamental decision in the design of in vivo ^{13}C experiments is the choice of a method for detection of the ^{13}C label. Detection methods can be broadly classified into two groups: direct detection at ^{13}C frequency and indirect detection at ^{1}H (proton) frequency. The choice of detection method is important because it determines what experimental data will be obtainable for metabolic modeling. This places constraints on the complexity of the metabolic model used. The choice of detection method is closely related to the choice of the metabolic model to be used. This is because the detection method determines what experimental data can be obtained from NMR spectra, such as which carbon positions can be measured in which amino acid. The main advantage of direct ^{13}C detection lies in the very high chemical specificity of ^{13}C spectroscopy, due to the broad chemical-shift range (~200 ppm). This also allows resolved and simultaneous detection of C4, C3, and C2 resonances of glutamate and glutamine in humans.

This, in turn, allows additional data to be used in the metabolic modeling, making the finite procedure more reliable. Another advantage of direct ^{13}C detection lies in the additional information gained from isotopomers. Molecules labeled at several carbon positions simultaneously appear in ^{13}C spectra as multiplets, whereas molecules labeled at only one carbon position appear as singlets. However, direct detection suffers from relatively low sensitivity, due to the low GMR of ^{13}C. The indirect $^{1}H[^{13}C]$ detection (i.e., detection of a ^{13}C label through the attached protons) offers better sensitivity, due to the fourfold higher GMR of ^{1}H compared to ^{13}C.

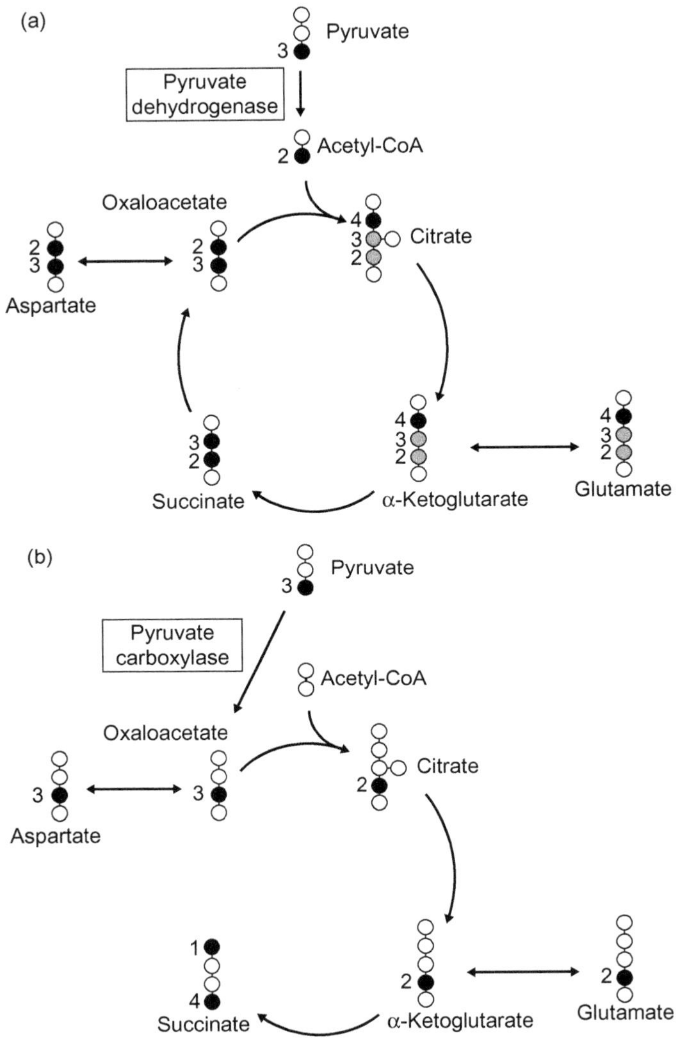

Figure 3.168 ^{13}C flow from [1-^{13}C]glucose into brain metabolites. (a) ^{13}C label flowing through pyruvate dehydrogenase labels the C4 of glutamate in the first turn of the TCA cycle and the C3 and C2 of glutamate in the second turn. (b) ^{13}C label flowing through pyruvate carboxylase labels the C2 of glutamate, but not the C3, leading to differential labeling of the C2 and C3 positions of glutamate.

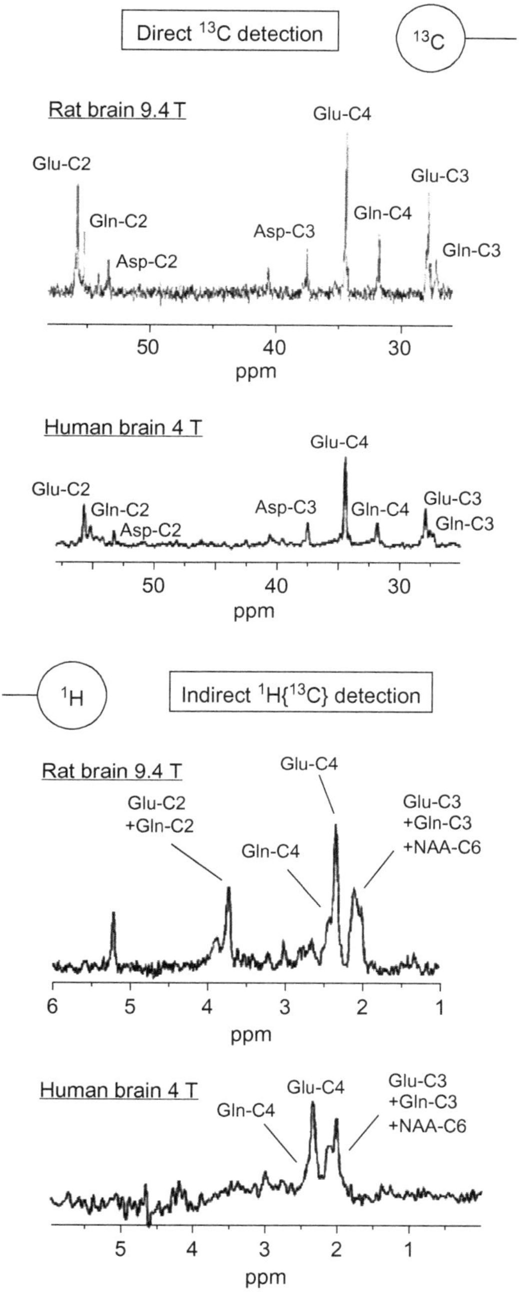

Figure 3.169 *In vivo* localized experimental spectra obtained after ^{13}C-labeled glucose infusion in the rat brain at 9.4 T (top row) and in the human brain at 4 T (bottom row), using direct detection (left column) and indirect detection (right column). Direct-detection spectra were obtained using polarization transfer; indirect-detection spectra were obtained using ^{13}C LASER. Direct detection allowed resolved detection of C4, C3, and C2 resonances of glutamate and glutamine and C3, C2 carbon positions of aspartate. Indirect detection gave better sensitivity (the VOI was smaller), but spectral overlap prevented resolved detection of each individual carbon position. Note the high NAA C6 peak in the human ^{13}C LASER spectrum (bottom right), due to the fact that this spectrum was acquired after a very long infusion time.

3.4.3 The Outlook: Development of New and Existing Concepts and Techniques—Quantification of Metabolite Concentrations/ Diagnosis of Human Brain Disorders

3.4.3.1 Knowledge of Absolute Concentrations: Water Standard Reference [62]

Absolute quantification of metabolite concentrations in the human brain has been the subject of research for many years, and it still has a long way to go. This section describes the various techniques achieved, along with their advantages and drawbacks, so that readers can gain some firsthand knowledge of the mathematics involved. It is hoped that the enormity of the symbols used will not scare readers.

The following is a simplified mathematical quantification relation for the metabolites in the human brain: $S_t^{\mathrm{M}}/K^{\mathrm{M}} = \rho_b c_b^{M}(S_b^{\mathrm{W}}/\beta_b) + \rho_{\mathrm{csf}} c_{\mathrm{csf}}^{\mathrm{M}} S_{\mathrm{csf}}^{\mathrm{W}} + \rho_v c_v^{\mathrm{M}} \varepsilon(S_b^{\mathrm{W}}/\beta_b) f_{v,b}^{\mathrm{M}}$. For a better understanding of the detailed physical and conceptual meaning of these terms, refer to Ref. [62]. A brief explanation is as follows: The superscript M refers to metabolite and W refers to the water standard. The quantities S_t^{M}, S_b^{W}, and $S_{\mathrm{csf}}^{\mathrm{W}}$ are measured quantities; ρ_b, ρ_{csf}, and ρ_v are constant tissue parameters; and c_v^{M} can be monitored in blood samples. The value of $f_{v,b}^{\mathrm{M}}$ is set to unity, although the visibility of metabolite in blood has to be assumed. An estimation of β_b has to be used, unless S_v^{W} (a hypothetical water signal) has been determined experimentally and the preceding equation has been used to calculate it.

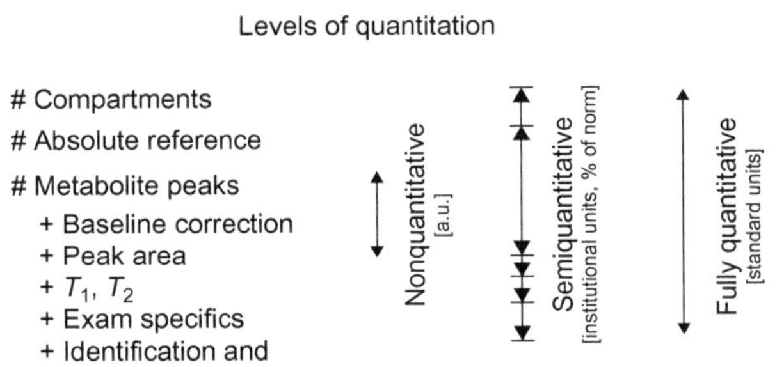

Figure 3.170 The quantitation of NMR spectra can be done at several levels. If it is restricted to the determination of peak area ratios, the evaluation is called *nonquantitative*. If all corrections that will be dealt with in the text are included, one commonly speaks of *absolute* quantitation, and results can be expressed in standard biochemical units (e.g., mmol/kg or mM). In between, there are many possible compromises that will lead to a comparison on an absolute scale between patient data and a normal range, but where results can be expressed only in internal or institutional units, not standard units. Such semiquantitative examinations are often sufficient in applications of screening for a particular differential diagnosis, where the main question may be whether all peaks or a particular one are at normal levels. Also, if spectral abnormalities are to be classified as patterns, semiquantitative approaches are adequate. Only if the abnormalities found are to be interpreted biochemically, or if the MR results are to be used in kinetic or thermodynamic equations, is full absolute quantitation mandatory.

A brief explanation of the meaning of some of the physical terms used in the preceding equation is as follows: $S_t^{\mathrm{M}} = \Sigma_i S_i^{\mathrm{M}}$ (compartments i) $= f_p^{\mathrm{M}} f_E^{\mathrm{M}} f_H f_T V \Sigma_i c_i^{\mathrm{M}} \xi_i \rho_i f_{v,i}^{\mathrm{M}}$ (metabolite signal) $S_t^{\mathrm{W}} = \Sigma_i S_i^{\mathrm{W}}$ (compartments i) $= f_p^{\mathrm{W}} f_E^{\mathrm{W}} f_H f_T V \Sigma_i c_i^{\mathrm{W}} \xi_i \rho_i f_{v,i}$ (water signal); $\beta_i = f_{v,i}^{\mathrm{W}} (\rho_i/\rho_0)(c_i^{\mathrm{W}}/c_0^{\mathrm{W}})$ (content of NMR-visible water compartment i). The index 0 refers to pure water. The total in vivo water signal is $S_t^{\mathrm{W}} = f_p^{\mathrm{W}} f_E^{\mathrm{W}} f_H f_T^{\mathrm{V}} c_0^{\mathrm{W}} \rho_0 V \Sigma_i \xi_i f_{v,i}^{\mathrm{W}}$ (volume of localized ROI); $S_{\mathrm{hyp}}^{\mathrm{W}} = f_p^{\mathrm{W}} f_E^{\mathrm{W}} f_H f_T^{\mathrm{V}} c_0^{\mathrm{W}} \rho_0 V$; $s_i^{\mathrm{M}} = S_{\mathrm{i}}^{\mathrm{M}}/S_{\mathrm{hyp}}^{\mathrm{W}} = (f_p^{\mathrm{M}}/f_p^{\mathrm{W}})(f_E^{\mathrm{M}}/f_E^{\mathrm{W}}) \Sigma_i (c_i^{\mathrm{M}}/C_0^{\mathrm{W}})(\rho_i/\rho_0) \xi_i f_{v,i}^{\mathrm{M}} k^{\mathrm{M}} = (f_p^{\mathrm{M}}/f_p^{\mathrm{W}})(f_E^{\mathrm{M}}/f_E^{\mathrm{W}})(1/\rho_0 C_0^{\mathrm{W}})$. k^{M} is not constant if f_E^{M} and f_E^{W} change independently. This may happen due to patient motion in between the scans, or if any acquisition parameter is readjusted for the reference measurement. Radiation damping or effects of the local dipolar field may also cause the water referencing to be inaccurate. Refer to Figure 3.170 [62] for an illustration of steps involved in metabolite quantification analysis.

Table 3.45 is an illustration of the four-component model of the brain. Brief descriptive meanings of some other terms are included for reader convenience.

Table 3.44 The Meaning of Terms

Term	Meaning
c_i^{M}	Concentration of metabolite M in compartment i (mmol $\mathrm{kg_{ww}}^{-1}$)
ξ_i	Volume fraction of compartment i
ρ_i	Density of compartment i with attenuation factors f relating to
f_p^{M}	particulars of peak P of metabolite M (number of contributing protons, chemical-shift-dependent sensitivity due to water-suppression pulses, off-resonance effects of slice RF pulses and data-processing/fit algorithm baseline correction, postacquisition water filtering, etc.)
f_E^{M}	Particulars of examination (B_1 adjustment [i.e., actual flip angles α_1, α_2, α_3], center frequency adjustment, coil loading characteristics, reciprocity mismatch, signal amplification)
f_v^{M}	Compartment-related NMR visibility (flow, physiologic motion, saturation or magnetization transforms)
f_T	Temperature (equilibrium magnetization, metabolite- and compartment-specific effects for systems in chemical exchange)
f_H	Time-independent hardware (B_0, RF pulse shapes, outer-volume saturation pulses)

Table 3.45 Characteristics of a Four-Compartment Model of Brain, Particularly Relevant for the Quantitation of Glucose or Phenylalanine

Compartment	Abbreviation	C_i^{M}	β_i	ξ_i	$f_{v,i}^{\mathrm{W}}$	S_i^{W}	$T_{2,i}^{\mathrm{W}}$
Brain tissue	b	tbd	tbd or a	tbd	1	$\neq 0$	80
CSF	csf	tbd or a	1	tbd	1	$\neq 0$	1000
Vascular	v	k or a	a[a]	a[b]	1	$\neq 0$	250
Myelin water	m	a[c]	1	a[d]	0	0	<15

Tbd, to be determined by MRS; a, known based on assumption; k, known based on non-MR measurements.
[a]80%;
[b]2–4% of brain tissue volume;
[c]assumed to be zero;
[d]7% of WM volume.

3.4.3.2 In Vivo Human Brain Metabolite Concentrations: LC Model Spectra [63]

In this technique, complete model spectra, rather than individual resonances, are used, in order to incorporate maximum prior information from the analysis. Because the amount of peak overlap depends on variable factors like shimming, peak areas as a measure of metabolite concentration can be inconsistent as well as erroneous. The LC model method analyzes an in vivo spectrum as a linear combination of model in vitro spectra from individual metabolite solutions. By using a model of complete spectra, rather than individual peaks, the complexities of the metabolite spectra become an advantage. This is because two metabolites with nearly identical spectra in one frequency region can still be resolved if they have different signals in other part of the spectrum. By using the same sequence for the library of model in vitro spectra as that used for the in vivo spectra, complications such as multiplet structures due to coupling are automatically accounted for. This way full prior knowledge is also incorporated.

More difficult are the unpredictable forms of the lineshape and baseline. The basic form of the lineshape can vary greatly, depending upon residual eddy currents and field inhomogeneities. Similarly, particularly with the more informative, short-echo-time spectra, macromolecule and lipid resonances (as well as noncausal signal from partially suppressed water) produce a baseline that is also difficult to parameterize. The LC model uses a nearly model-free, constrained regularization method. It attempts to achieve the best possible compromise between the two situations, by finding the smoothest lineshape and baseline consistent with the data. The aim of this regularization method is to allow into the lineshape or baseline only the complexity demanded by the data (on the basis of the statistical tests). Thus, the model adapts itself to the data; no subjective, fixed parameterized model is required. This permits flexibility but at the same time prevents concentration errors due to over- or underparameterization. To account for the extra line broadening in the in vivo spectrum being analyzed, the in vitro model spectra are convoluted with an effectively model-free lineshape function, which is determined by constrained regularization. In addition, each model metabolite spectrum is allowed its own small shift (to account for possible referencing errors) and its own extra broadening (to account for the shorter T_2 in vivo than in vitro). In addition to the lineshape and baseline functions, parameters for the zero-order and first-order phase corrections and the overall (possibly large) referencing shift are included in the model. All the data in the analysis range (typically 4.0–1.0 ppm) are simultaneously used in a constrained least-squares analysis to obtain (approximately maximum-likelihood) estimates of all the model parameters (metabolite concentrations, phases, referencing shift, lineshape, baseline, etc.) and the uncertainties in the concentrations (Cramer–Rao lower bounds).

The analysis is automatic (noninteractive); no subjective input, phasing, referencing, or starting estimates are required. There is one version of the LC model for all field strengths and sequences; only the appropriate model spectra must be selected. The fully relaxed model metabolite spectra are normally acquired once at each laboratory for each sequence (e.g., STEAM/PRESS, TE, TM). However, manufacturers have compiled LC model libraries of model spectra for some of their standard sequences.

These model spectra seem to be very portable among sites. This improves the objective comparison of results, as each site will obtain exactly the same LC model results with the same in vivo data. The LC model was originally developed and tested with 4000 human brain and rat brain spectra at 2.0 and 2.35 T. However, since then the LC model has been used elsewhere at fields ranging from 1.5 to 9.4 T with a wide range of sequences. During glucose (Glc) infusion experiments, the LC model Glc concentration in GM followed the smooth rise and fall of the plasma Glc concentration (with a slight delay), although Glc is considered very difficult to quantify. Similarly, during dynamic studies of ischemia and functional MRS, the time courses of Glc and lactate (Lac) could be clearly followed. In all of these studies, the concentrations of the other metabolites remained constant with time to within experimental error, as would be expected. The flexibility of the regularization method allows the baseline to automatically account for large unexpected distortions, such as residual water signal, susceptibility artifacts, and macromolecular or lipid signals. Even in spectra with no water suppression, the strong climb toward the water resonance is handled by the baseline. Similarly, the regularized lineshape can automatically account for unexpected lineshapes, e.g., extreme eddy current distortions. However, this flexibility is also important for lineshapes that appear reasonable; for example, excellent shimming can result in nearly Lorentzian shapes, whereas less-perfect shimming can lead to very different lineshapes. Figure 3.171 illustrates the large amount of information available with PMRS and excellent shimming.

Intra assay reproducibility was tested with a series of 32 three-minute acquisitions (each with 1/16 the scans of Figure 3.171 [63]). The COV for the LC model concentrations was still below 5% for glutamate (Glu), myoinositol, NAA, taurine, phosphocreatine, and creatine (1.8% for total Cr + PCr). It was below 12% for cholines (total glycerophosphocholine plus phosphocholine; GPC-PC), glutamine (Gln), GSH, lactate, phosphoethanolamine, and Glc. It was between 27% and 57% for alanine (Ala), aspartate (Asp), GABA, scylloinositol, and NAAG, indicating that these metabolites are too weakly represented to be reliably estimated (although in other cases Ala, scylloinositol, and NAAG are strong enough to be reliably estimated). The excellent reproducibility for Tau, Glu, NAA, and MI is not surprising, as some of their detailed characteristic patterns in the in vitro model spectra can even be recognized by visual inspection of the in vivo data (Figure 3.172).

3.4.3.3 Quantitative Metabolite Maps: Human Brain GM/WM Ratio (LC Model) [64]

For most metabolites, a positive correlation between metabolite concentration and GM content is seen. This correlation is useful for producing maps of metabolites, such as glutamate/glutamine (Glx). They exhibit overlapping of *J*-coupled peaks.

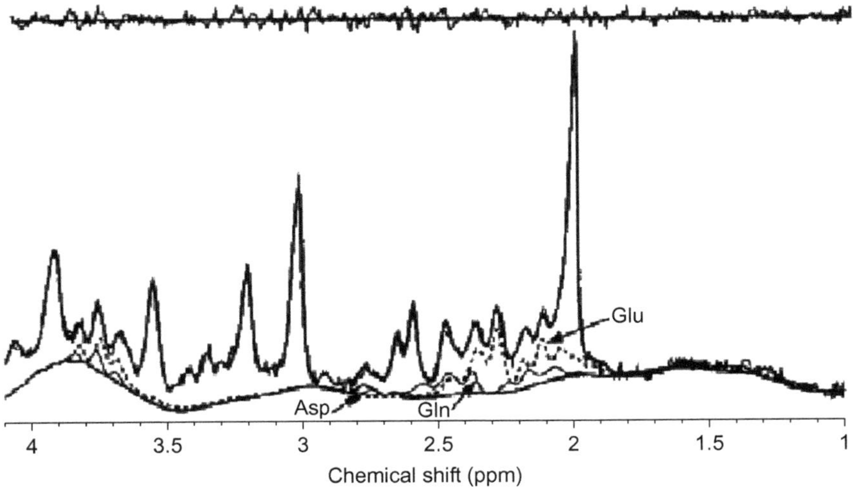

Figure 3.171 LC model analysis (thick curve) of a high-quality (line width about 0.04 ppm) *in vivo* spectrum (thin curve) at 2.0 T. The differences (residues) between these data and the fit are plotted at the top. The contribution of the baseline (plotted at the bottom) has been included in the plotted contributions of the Glu, Gln, and Asp model spectra to the analysis. Glu and Gln can be reliably resolved.

The requirements for quantification of Glx introduce a number of constraints into MRSI image acquisition. Here, short echo times must be used to minimize the signal decay caused by both T_2 and coupling effects. This enhances the quantification of all metabolites because it reduces the need for application of large correction factors. It is especially beneficial if a long repetition time is also used. When short TEs are used, peak fitting must be used in preference to integration to find peak areas, due to the presence of overlapping peaks and broad baseline components. Therefore, a relatively coarse MRSI matrix is helpful to produce a sufficient SNR in each voxel for the peak-fitting analysis. One consequence of the coarse MRSI is that individual voxels of interest are more likely to be at or near the edge of slice profiles of the PRESS-localized volume.

This leads to errors. It may also be confused with chemical-shift artifacts. The chemical shift produces a different excitation profile for each metabolite. It may be better to use a more flexible technique in solving various problems, so as to create more reliable metabolite-concentration maps.

One way of normalizing metabolite images is to use measurements of the excitation profile in vivo, including correction for any chemical-shift artifacts. Any method that attempts to quantify metabolites with coarse resolution needs to estimate the voxel composition of GM, WM, and CSF.

This information can be used both to correct for the diluting effects of CSF, and to study the variation of metabolite content between WM and GM (Table 3.46).

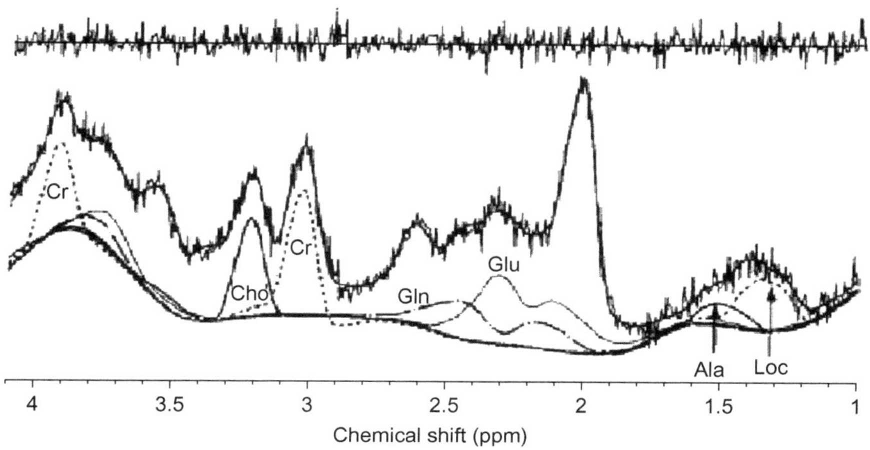

Figure 3.172 LC model analysis of a low-resolution spectrum (line width about 0.11 ppm). Conventions are as in Figure 3.171. The contribution of the basis spectra of Cr plus PCr is labeled Cr. Note that the wide space (compared to Figure 3.171) between the fit and, for example, the Glu curve is entirely filled by other overlapping model metabolites, because the baseline has already been added to the Gly curve. Despite this strong overlap and poorly defined peaks, Cramer–Rao lower bounds of 10% or less for NAA + NAAG, Cr + PCr, and total cholines are obtained.

Table 3.46 Metabolite Concentrations [mM 6 SD] in Pure WM and Pure GM

Metabolite	WM	GM	GM/WM
NAA + NAAG	7.7±0.6 (8%)	8.8±0.3 (3%)	1.2±0.1 (8%)
Glu + Gln	5.9±0.8 (14%)	14±2 (14%)	2.4±0.5 (21%)
Creatine	3.8±0.4 (11%)	6.4±0.6 (9%)	1.7±0.3 (18%)
Myoinositol	3.1±0.4 (13%)	4.9±0.5 (10%)	1.6±0.3 (19%)
Choline	1.2±0.2 (17%)	1.1±0.3 (27%)	0.9±0.2 (22%)
(NAA + NAAG)/Cr	2.0±0.2 (10%)	1.4±0.1 (7%)	0.69±0.08 (12%)

Extrapolated from linear fits of concentration versus GM content in each volunteer ($n = 513$). The coefficient of variation (SD/mean of the group) is shown in parentheses.

The LC model technique fits spectra as a linear combination of model (phantom/physiological concentrations of metabolites in solution) spectra, with very high SNR and narrow line width, known as the *basis set*. MRSI is performed on a 15-mm-thick axial slab, superior to the ventricles, that included WM of the centrum semiovale and GM along the interhemispheric fissure. The nominal voxel size is 2.04 ml (24 × 24 phase-encode steps over a 28-cm FOV).

In FT data processing, peak registration is performed on the single peak of NAA to account for frequency shifts across the FOV. After this minimal processing, a computer program is run to fit the peaks in a spectral analysis window

1.0–3.85 ppm, using the LC model. The size and location of the excitation region is read from the raw MRSI file header, and only spectra within it are extracted for processing. After completing peak fitting, another computer program is used to read the postscript output and produce datasets of the metabolite-concentration ratio to Cr, and the standard deviation figures. Inversion-recovery-prepared, fast spoiled gradient-echo images are acquired (124 mm × 1.5 mm thick), with the FOV matching the MRSI FOV (24–30 cm) and a matrix of 256 × 128, TE/T_1/TR = 4.2/450/16 ms, flip angle 20°. The PRESS excitation profile images are reduced to the resolution of the metabolite maps (24 × 24) by taking the mean pixel intensity from each region corresponding to an MRSI voxel. This effectively gives an estimate of the fraction of maximal PRESS excitation experienced at each voxel position. The quantitative metabolite maps are divided by these excitation maps, to produce normalized metabolite images. Figures 3.173 and 3.174 present illustrations representative of metabolite-concentration results.

3.4.3.4 Effective TRT: NAA and Creatine [65]

One can express the observed (apparent) TRT T_2^* and the intrinsic relaxation time T_2 as $1/T_2^* = (1/T_2) + (1/T_{2\mathrm{Diff}}) + (1/T_{2\mathrm{Exch}})$. It is known that the T_2^* is controlled by interactions of various kinds. Some of these (to mention just a few) are as follows: First is the homonuclear dipole–dipole (D–D) interaction between protons. This is strongly dependent on rotational correlation time τ_c and the hyperfine (contact) interaction of protons with a paramagnetic (permanent magnetic moment) molecule. Second is the cross-relaxation mechanism, which can be significant in dipole-coupled systems. The terms $T_{2\mathrm{Diff}}$ and $T_{2\mathrm{Exch}}$ in the preceding equation are

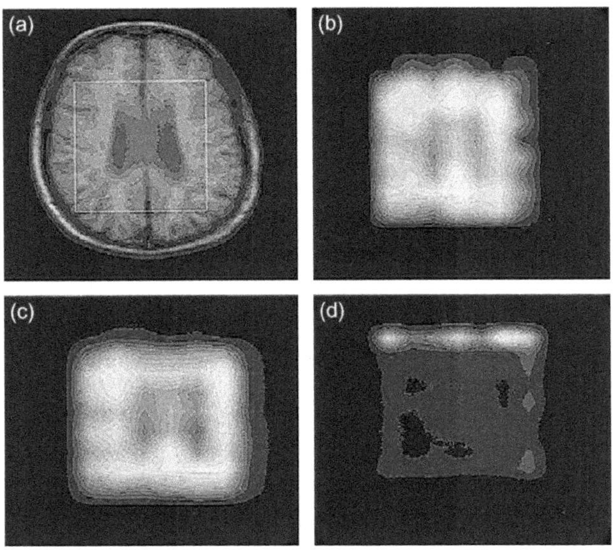

Figure 3.173 Quantitative metabolite maps without correction for CSF contamination or excitation profiles, with bilinear interpolation for display. (a) Spin-echo scout image showing PRESS-selected ROI. (b) Glutamate + glutamine (Glx). (c) *N*acetyl-aspartate + *N*-acetyl-aspartyl-glutamate (NAc). (d) Estimate of percentage error in quantification of NAc.

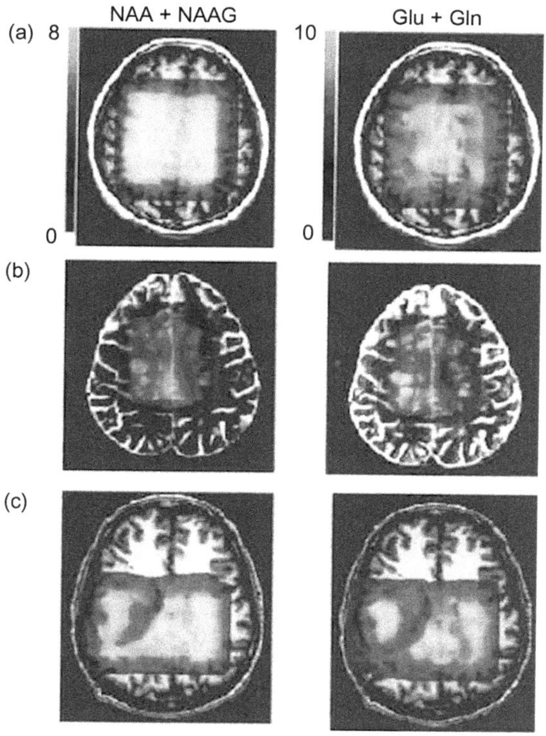

Figure 3.174 Overlays of metabolite concentration (mM) with a quantitative color scale on anatomic scout images. Left side is NAc; right side is Glx. (a) Normal control. (b) MS patient. (c) Patient with a malformation of cortical development and epilepsy.

due to diffusion and exchange of spins between regions with different magnetic field strengths (e.g., magnetic field inhomogeneities). These interactions define a dynamic dephasing regime (DDR). Here the net magnetization in NMR is reduced by diffusion and exchange between regions with different magnetic field strengths. It causes the phases of the different spin packets to average out. The situation in an opposite environment (no DDR) is called the SDR. The NMR signal loss in SDR can be refocused by a SENSE sequence.

It is known that changes induced by neuronal activity can result in modification of the concentration of deoxyhemoglobin, which plays a major role, as the intravascular contrast agent, in fMRI. The contrast in fMR images is related to the difference between the magnetic properties of oxygenated and deoxygenated hemoglobin (diamagnetic and paramagnetic).

It may be interesting for the reader to recollect that in a diamagnetic substance (e.g., oxygen), magnetism arises because of the application of an external magnetic field. It is basically due to the screening (reaction by orbital electrons in an atom) created by the substance in response to the external field. Therefore, the substance is said to have a negative susceptibility (change in magnetization per unit applied field). In a paramagnetic substance, the susceptibility is positive (part of the permanent electronic structure of the molecule).

The variation in $\mathbf{T_2}$* is the source of contrast in fMRI. It is known that local magnetic susceptibility variations can occur in different regions of the brain, such as GM and WM. These variations can be due to the differences in the concentrations, and compartmentalization, of iron and deoxyhemoglobin. One of the factors that alters the $\mathbf{T_2}$* value of water is the presence of ferritin, a protein that stores the largest fraction of iron in the brain. One should remember that changes in physiology and brain activity can lead to changes in the rotational correlation time τ_c and the ADC. The rotational correlation time of a cerebral metabolite strongly depends on the metabolite's environment, such as when the metabolite is interacting with slowly tumbling macromolecules. The changes in metabolite concentrations in the human brain can be studied by using the changes produced in $\mathbf{T_2}$* by the applied fields. Figures 3.175 and 3.176 [65] illustrate how these kind of studies can be performed. The CP adiabatic pulse scheme has been used in this kind of study. It employs LASER technique, done in comparison with PRESS.

3.4.3.5 Multiple-Quantum Filtering: AMNPQ *(5-Spin) Metabolite System in Brain/Numerical Solution of Density Matrix Approach [66]*

Although observation of the uncoupled methyl singlets of choline (Cho), creatine (Cr), and NAA has become a significant impetus to the use of in vivo proton

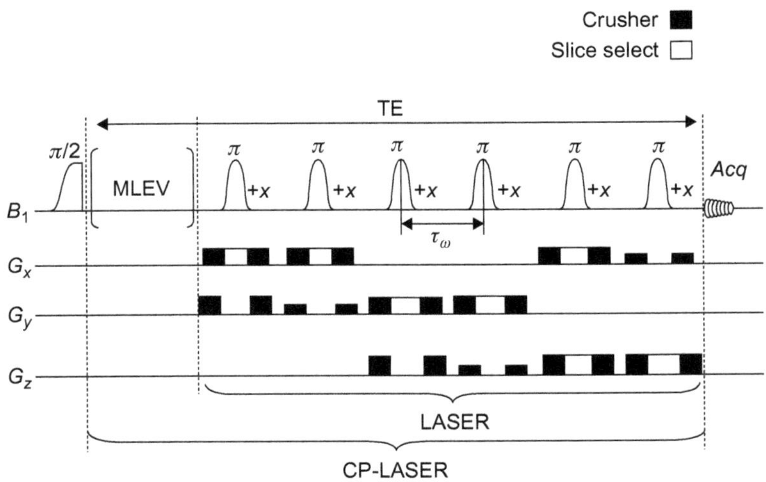

Figure 3.175 CP LASER and LASER sequences based on adiabatic RF pulses. In CP LASER, a CP train of _ pulses was inserted prior to the LASER localization. These additional pulses had RF phases prescribed according to the MLEV scheme, namely: MLEV-4:_X_X − X − X; MLEV-8:_X _X − X X − X_X_ X − X; MLEV-16:_X _X − X − X − X _X _X − X − X − X _X _X _X − X − X _X. No pulsed B_0 gradients were applied during the MLEV portion of CP LASER.

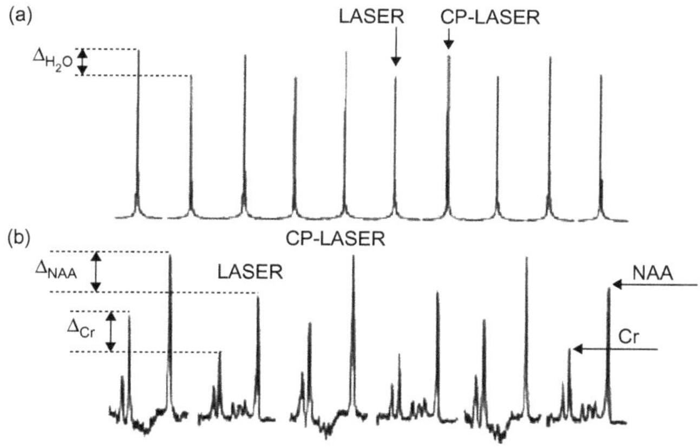

Figure 3.176 (a) Repeated localized ^{1}H spectra (12-cm^3 localized volume) of the water resonance (4.7 ppm, NEX _ 2), obtained with interleaved-LASER (22 AFP pulses, $\tau_{cp} = 6.3$ ms) and LASER (six AFP pulses, $\tau_{cp} = 23.1$ ms) at 7 T using the same TE (138.6 ms). (b) Expanded views of water-suppressed spectra of metabolites: NAA (2 ppm) and Cr (3.03 ppm); NEX = 8; other conditions are the same as in part (a). ΔH_2O, ΔNAA, and ΔCr indicate the difference in the NMR spectroscopic imaging of H_2O, NAA, and Cr, respectively, detected with interleaved CP LASER and LASER sequences at 7 T.

spectroscopy for detecting biochemical abnormalities in the human brain, further substantial progress requires the ability to observe, and to quantify at will, several metabolites with coupled spins, most notably the amino acids. To quantify these metabolites, one must be able to deal with their having similar chemical structures (and, therefore, similar chemical-shift distributions); and with scalar couplings between their spins (and, therefore, multiplets that can overlap when several similar chemical structures are present together). This occurs, for example, with spin systems of 5 and 6 spins that are not always weakly coupled, and sometimes with signal intensities relative to background resonances and noise that approach unity. Methods for dealing with these issues seem to fall into four broad categories: spectral modeling, spectral editing, 2D spectroscopy, and increased field strength (Table 3.47).

One can present a spectral editing procedure, based on MQC filtering, that is designed to accommodate the more challenging of the amino acid-specific editing problems, including that of glutamate (Glu) and glutamine (Gln) (collectively referred to as Glx). The procedure uses numerical methods in the design and optimization of the MQC filter sequence. Although in the past researchers have used the analytical approach of product-operator calculations to evaluate filter performance (e.g., for weakly coupled systems such as lactate and GABA, as well as in strongly coupled systems with a very limited number of spins, such as citrate and the aspartate group), the need to accommodate both strong coupling and an increasing number of spins necessitated the combination of

Table 3.47 Chemical Shifts, δ ppm, and Scalar Couplings, J Hz, for the AMNPQ Spin Systems of Glu and Gln

Chemical Shift	δ_A	δ_M	δ_N	δ_P	δ_Q
Metabolite					
Glu (27)	3.75	2.13	2.05	2.36	2.34
Glu (22)	3.76	2.12	2.04	2.35	2.34
Glu (26)	3.76	2.10	2.06	2.36	2.36
Glu (28)	3.743	2.120	2.042	2.349	2.334
Gln (27)	3.77	2.14	2.13	2.46	2.44
Gln (26)	3.78	2.14	2.14	2.46	2.46
Gln (28)	3.757	2.135	2.115	2.456	2.434

***J*-coupling**	J_{AM}	J_{AN}	J_{MN}	J_{MP}	J_{MQ}	J_{NP}	J_{NQ}	J_{PQ}
Metabolite								
Glu (27)	4.67	7.35	−14.7	6.63	8.86	8.83	6.16	−16.22
Glu (22)	4.5	7.2	−15.2	6.0	9.5	8.0	6.8	−15.7
Glu (26)	4.9	7.3	–	–	–	–	–	–
Glu (28)	4.65	7.33	−14.85	6.43	8.47	8.39	6.89	−15.89
Gln (27)	5.8	6.5	−14.0	6.0	9.8	9.5	5.8	−16.4
Gln (26)	–	–	–	3.6	7.6	7.9	7.9	–
Gln (28)	5.84	6.53	−14.45	6.33	9.25	9.16	6.35	−15.55

analytical understanding with numerical optimization to reach a potential solution to this problem.

The sequence design procedure concurrently incorporates, into a composite sequence, the functions of spatial localization and coherence manipulation; optimizes sequence timings in relation to the overall spin system; and eliminates the solvent dipolar magnetization signals from water in the brain. No weak coupling approximations were made. In the filter design procedure, both Glu and Gln were treated as *AMNPQ* spin systems, reflecting their strong scalar couplings at 3 T. The proton designated *A* is that bonded to the C2 carbon of Glx, and the *MN* and *PQ* designations refer to the methylene protons that are bonded, respectively, to carbons C4 and C3. Strong coupling calculations were also included for the background signals from the *AB* multiplet of the *ABX* spin system of the aspartate groups of aspartic acid (Asp) and NAA. Moreover, when a full calculation of the GABA background was evaluated, quite minor but nevertheless quantifiable deviations from the weak coupling limit were also found. These deviations will be more pronounced at 1.5 T. The efficacy of the resulting design procedure is demonstrated at 3 T, first by the response of aqueous solution phantoms to the Glu filter, and second by the response of the human brain. A diagram illustrating the MQC filter sequence is shown in Figure 3.177.

An issue in the design of a composite sequence intended to target a single metabolite, and incorporate spatial localization and coherence manipulation

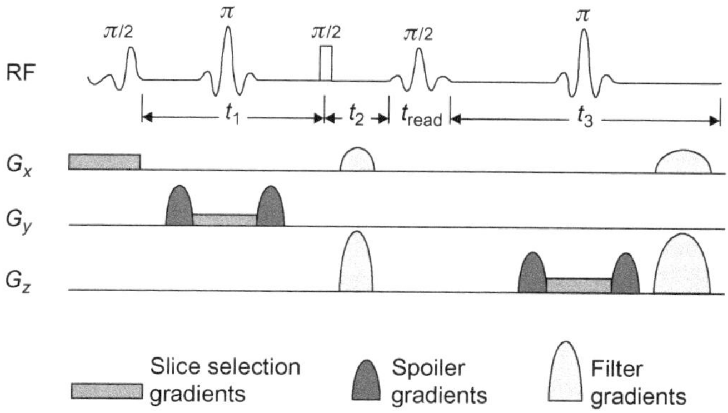

Figure 3.177 A schematic diagram of the DQF pulse sequence localized for a single voxel. The $\pi/2$ excitation pulse was a 4-ms BURP pulse; the π refocusing pulses were 4-ms sinc pulses, modified to fine-tune the slice profile; the $\pi/2$ double-quantum generating pulse was a 250-μs rectangular pulse; and the final $\pi/2$ read pulse was a 5-ms two-lobe sinc Gaussian. The spoiler gradients were 2-ms long and 20-mT/m maximal amplitude, whereas the filter gradients, oriented at the "magic angle," were 2 and 4 ms in length, also with strength of 20 mT/m. The slice-selection gradients were 0.1 mT/m.

concurrently, is design of the pulses and optimization of the timings between them. These issues affect primarily the metabolite specificity, but to a lesser extent also affect the spatial resolution.

Incorporation of spatial selectivity into one or more of the pulses of an in vivo editing acquisition sequence was originally explored several years ago, but has only recently begun to show more promise. Both the conflicting demands of localization and coherence manipulation, and the consequences of strong scalar couplings, have become better understood. Between the earliest and the most recent reports of editing sequences, we find that, in addition to hard pulses, Gaussian, sinc, or even binomial 90° and 180° pulses have been employed in MQC filters, often in conjunction with gradients, to fulfill both the filtering and localizing roles. However, the presence of a chemical-shift distribution and of the scalar coupling interaction complicates the response of the nuclear spin system to soft pulses (particularly if the scalar coupling is strong), and renders soft pulses specifically designed for the imaging of uncoupled water protons less appropriate for localized spectral editing. For example, for an initial spatially selective excitation pulse, chemical-shift dephasing as well as gradient dephasing must be removed if one is to begin the preparation period (t_1 in Figure 3.177) with spins that are all in phase.

This can be done using self-refocusing pulses; several workers have chosen to relax this condition and use sinc-like excitation pulses, together with the incorporation of a phase correction later in the sequence. The self-refocusing methodology was adopted here because of the need to control the postpulse phase distribution of terms in the density operator. It took account of evolutions under gradient, chemical shift, and the slice-positioning frequency offsets. Soft 90° pulses later in the sequence can, if sufficient care is not taken, also give rise to unwanted distributions of coherence at the end of the pulses. The coherence distributions can be quite different from those that would have been produced by the very short, rectangular pulses of the same tip angles that are usually assumed in product-operator analyses. At least three of the five pulses normally used in a single-voxel DQC filter must be made spatially selective. The most appropriate, because of their null role in transferring coherences, are the 90° excitation pulse and the two 180° pulses used to refocus chemical shift and field inhomogeneity dispersions. The remaining 90° pulses—namely, the producer of MQCs (second 90° pulse) and the read pulse (third go pulse)—can then be designed to satisfy filtering considerations alone. The 90° excitation pulse, which is required to produce in-phase transverse magnetization for a range of multiplet resonance frequencies, was provided by a band-selective uniform response pulse phase (BURP) self-refocusing pulse.

If, within the excitation slice, the spins are not all in phase at the beginning of the preparation period, t_0, the transfer of coherences later in the sequence will be compromised. Because 180° sinc pulses are self-refocusing, the refocusing of chemical-shift dephasing and field inhomogeneities in the time periods t_1 and t_3 was carried out using spatially selective sinc pulses modified to optimize their frequency profile. To minimize the mixing of coherences during the second 90° pulse (and, hence, modification of the filtered lineshape or yield), a rectangular (250 ps) wideband (full width at half height = 37.5 ppm) pulse was used. The importance of a short pulse for producing the desired combination of MQCs at the beginning of the period t_2 is accentuated with increasing strength of B, because of the increase in chemical-shift difference between the *MN* and *PQ* spins of Glx, and to a lesser extent because of the strong coupling regime of the scalar coupling. For the read pulse, a spectrally selective sinc-Gaussian pulse was used; its spectral profile excluded the water resonance, but was uniform over the metabolite resonances; its tip angle and ability to promote the mixing of coherences during the pulse were used to advantage, to tailor the filtered lineshape (Figure 3.178).

3.4.3.6 Quantification of Human Brain Metabolites Using PMRSI [67]

Quantification of metabolite concentrations from spectroscopic images has been performed by a variety of approaches, including the use of internal references such as tissue water, external references, and phantom replacement. The internal-reference method, using tissue water, can be difficult, especially in pathological states in which the content and relaxation characteristics of water may depend on the variable severity of edema. The presence of multiple interfaces (e.g., bone, air) inherent in external references can create a susceptibility problem. Furthermore, unless an appropriately doped reference is used, the reference signal (or its solvent) can easily obscure resonances of interest. The phantom replacement approach is probably the most reliable, but it requires a separate study and at high field (Table 3.48).

The competing effects of skin depth and dielectric resonance make this method difficult. In this section we report on use of the acquisition of a ventricular CSF

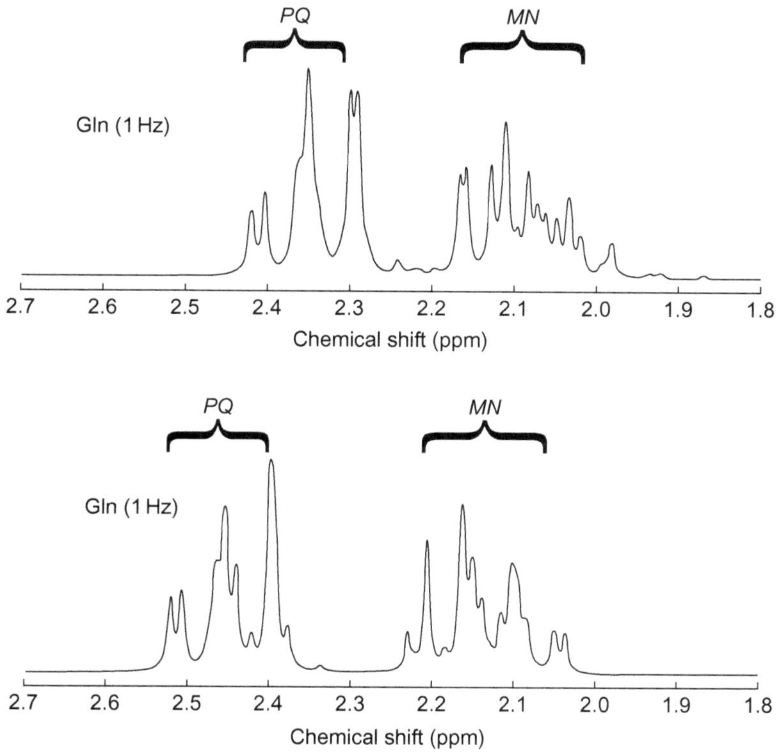

Figure 3.178 Calculated multiplet spectra for the *MNPQ* spins of both Glu and Gln at a field strength of 3 T. The chemical shift and scalar couplings used in the calculations were obtained in this laboratory and are listed in Table 3.47. The line width was set at 1 Hz to provide resolution of the major multiplet peaks and to demonstrate the lack of overlap between Glu and Gln in the 2.3 pm region of the spectrum.

Table 3.48 Metabolite Concentrations (mM)

WM	WM	WM	GM	GM	GM
NAA	Creatine	Choline	NAA	Creatine	Choline
I.[a] 11.1±0.9	6.2±0.8	1.8±0.2	9.4±0.9	8.8±1.6	1.6±0.5
II.[b] 14.6±2.8	6.0±1.2	1.9±0.4			
III.[c] 11.3±1.7	6.7±1.4	2.0±0.3	10.3±2.0	8.5±1.6	1.7±0.8

[a]Present data.
[b]Previous data
[c]Using centrum semiovale WM and frontoparietal GM data.

spectrum to provide a reference and image-based B_1 mapping to measure the concentrations of creatine (Cr), choline (Ch), and NAA in the human brain at 4.1 T. CSF can be described as a dialysate of serum, with the water content of ventricular CSF ranging from 54.6 to 55.1 M. Only under conditions of major hemorrhage or tumor lysis might it deviate significantly. Furthermore, by virtue of the dilute nature of CSF, the relaxation values for CSF are not expected to change significantly between volunteers or patients, which could occur with tissue water measurements. As has been discussed by others, B_1 mapping is critical for accurate quantification. Such B_1 measurements across the spectroscopic-imaging slice are necessary to allow calculation of a spatially dependent correction map that depends on the acquisition sequences for the spectroscopic image as well as the CSF reference.

In the experiment reported here, the correction map, in addition to total brain content normalization, as determined by T_1-based image segmentation and accounting for PSF effects, was applied to the integrated spectroscopic data and CSF spectra. Concentration values were then obtained after final corrections for relaxation effects. A doubly oblique gradient scout imaging was used to localize the slab of interest, taken through the lateral ventricles. The same slice was then used for the spectroscopic study, image segmentation, and B_1 (transverse field) mapping. The acquisition was a single 50-ms spin-echo localization, achieved through 10-mm slice selection and two dimensions (32 × 32) of rectangular phase encoding. T_1 values were calculated and individual pixels were categorized as GM, WM, CSF (negligible contribution) or nonbrain tissue, according to T_1 measured values.

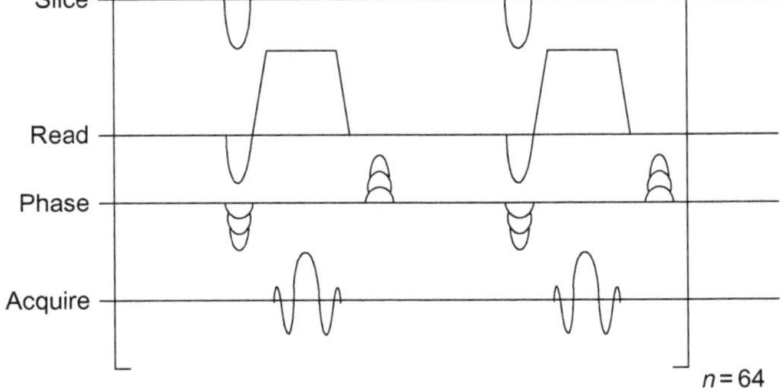

Figure 3.179 Pulse sequence for B_1 mapping. The interval between the two 60° excitations was kept short to minimize T_1 relaxation effects.

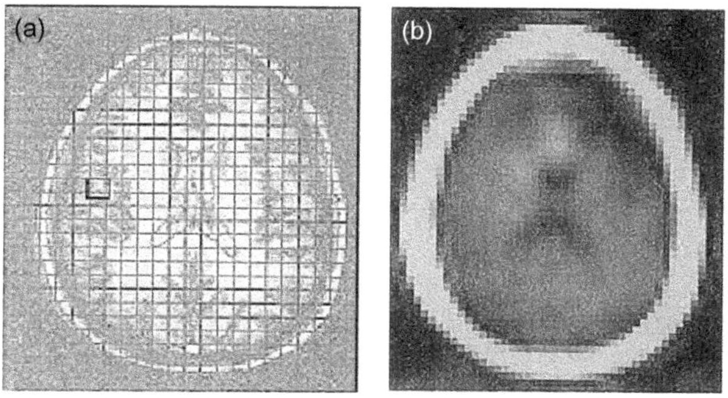

(c)

0	0	0	0	0	0	0	0	0	0	0	0	0	0	0	0	0	0	0	0	0	0	0
0	0	0	0	0	0	0	0	0	0	25	24	0	0	0	0	0	0	0	0	0	0	0
0	0	0	0	0	0	0	14	35	48	54	33	35	51	0	0	0	0	0	0	0	0	0
0	0	0	0	0	0	26	48	49	49	48	40	45	48	46	39	0	0	0	0	0	0	0
0	0	0	0	0	36	46	46	45	44	44	44	45	47	43	38	27	21	0	0	0	0	0
0	0	0	0	10	32	43	47	49	47	46	46	47	49	47	43	43	13	10	0	0	0	0
0	0	0	0	34	32	45	48	50	49	49	50	50	51	49	46	44	32	39	17	0	0	0
0	0	0	11	48	35	48	50	50	51	52	55	55	53	50	48	46	44	32	40	0	0	0
0	0	0	37	46	46	48	50	50	51	52	57	56	53	51	48	47	43	44	44	0	0	0
0	0	28	48	37	47	47	50	53	54	56	55	55	55	54	50	48	45	34	44	0	0	0
0	0	25	45	48	46	48	51	54	55	59	55	58	59	54	53	50	47	45	48	0	0	0
0	0	48	45	35	44	48	51	58	58	59	57	58	59	57	55	50	47	49	49	23	0	0
0	0	45	46	45	42	47	53	56	60	61	62	62	61	58	58	53	47	47	40	24	18	0
0	21	47	34	41	43	49	52	55	59	62	60	62	59	59	57	52	49	48	43	41	14	0
0	23	43	22	43	43	50	52	65	60	63	61	63	60	57	55	52	50	49	36	42	20	0
0	18	35	49	43	45	51	55	57	61	62	62	61	59	58	55	54	49	47	56	42	41	0
0	15	31	41	39	43	50	65	56	58	60	58	59	58	56	55	53	50	48	25	47	30	0
0	19	30	29	41	46	50	52	50	58	61	60	60	58	53	54	52	51	48	53	49	34	0
0	9	30	33	41	46	51	52	55	59	60	60	58	57	51	53	50	47	46	47	35	23	0
0	0	35	24	38	47	49	51	55	57	59	59	57	54	53	51	49	45	45	52	42	23	0
0	0	27	19	19	44	47	51	54	55	56	57	56	52	52	50	50	47	43	12	44	12	0
0	0	11	39	0	35	47	50	54	53	56	55	52	51	51	48	48	46	42	31	32	12	0
0	0	0	11	29	12	45	50	51	51	52	52	49	50	50	46	44	46	30	39	11	0	0
0	0	0	0	9	33	10	31	49	52	53	51	49	49	48	45	35	32	42	10	0	0	0
0	0	0	0	0	11	39	24	11	35	47	19	40	44	54	21	28	44	12	0	0	0	0
0	0	0	0	0	0	0	26	13	23	26	27	25	22	33	48	25	0	0	0	0	0	0
0	0	0	0	0	0	0	0	0	0	0	25	12	0	0	0	0	0	0	0	0	0	0

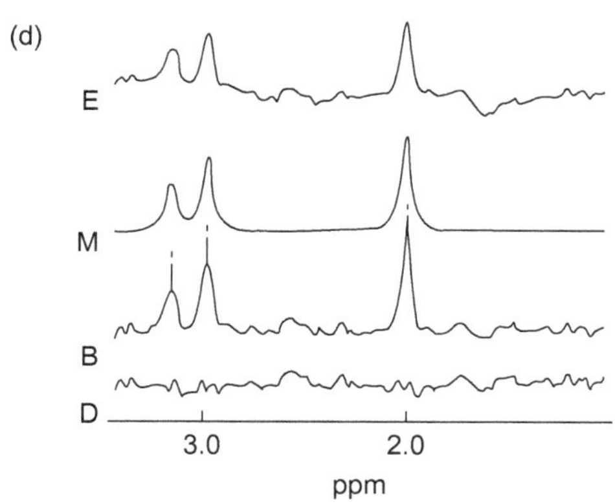

Figure 3.180 (a) The scout. (b) NAA: magnitude image. (c) B_1 map. (d) A typical spectrum showing baseline correction and curve fitting. E refers to the experimental spectrum, B is the spectrum after baseline correction, M is the curve-fitted spectrum with noise removed, and D is the difference spectrum between B and M.

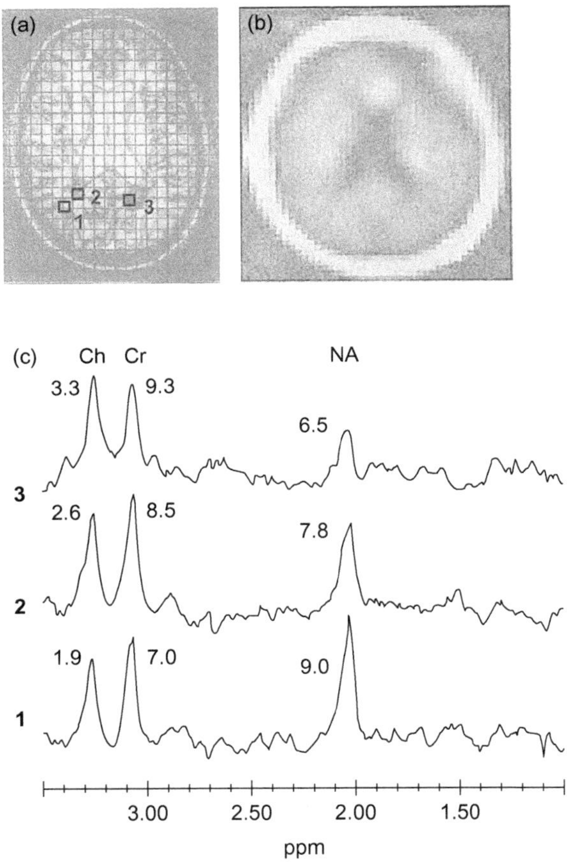

Figure 3.181 (a) The scout. (b) NAA magnitude image. (c) Three spectra arising from the positions labeled 1 through 3, with quantitated values from a patient with SPMS.

The signal at a point i, j can be written as $S_1 i,\ j = M_{i,j,0}\ \sin\theta_{i,j}$. Here $\theta_{i,j}$ is the applied pulse angle and $M_{i,j,0}$ is the equilibrium longitudinal magnetization. Without any T_1 relaxation, the second acquisition has a signal intensity $S_2 i,\ j = M_{i,j,0}\sin\theta_{i,j}\cos\theta_{i,j}$. The ratio is $S_2 i,\ j/S_1 i,\ j = \cos\theta_{i,j}$. Thus, $B_1(i,\ j)$~arc $\cos(S_2 i,\ j/S_1 i,\ j)$, and is described in degrees. The reference signal was acquired using PRESS. The illustrations in Figures 3.179–3.181 are included to enable readers to visualize the concepts and techniques of the future as still connected to past achievements.

3.4.3.7 Prior Spectral Information: Computer Simulation of Metabolites [68]

One can use shorter echo-time values to gain the benefit of higher SNR as well as improved detection of metabolites such as myoinositol and glutamate. These are otherwise subject to excessive signal loss due to *J*-coupled phase modulation and

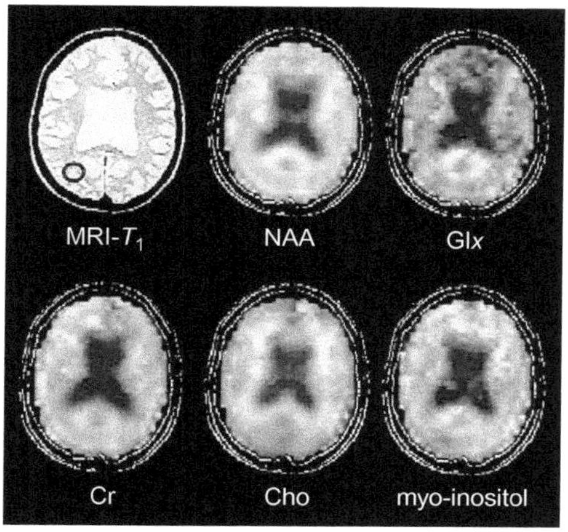

Figure 3.182 T_1-weighted localizer MRI and metabolite maps for NAA, creatine, choline, myoinositol, and Glx from an automated fit of ventricular, 25-ms TE, 1.5-T, spin-echo MRSI data.

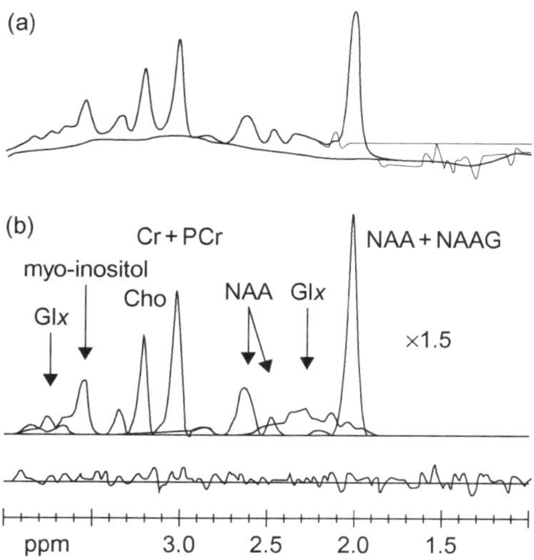

Figure 3.183 Spectra from voxel indicated in the localizer MRI in Figure 3.182. (a) Spectral data (thin line) overlaid with the metabolite + baseline fit (heavy line) and baseline-only fit (heavy line). (b) Individual metabolite fits (scaled by 1.5). At the bottom is the residual spectrum.

fast T_2 relaxation rates. Abnormal levels of these metabolites are seen in several neurological diseases, including AD, Creutzfeld–Jacob disease, and ALS.

Nevertheless, short-TE ^{1}H MRS poses some challenges. Some of these are water suppression and removal of intense signal contribution from subcutaneous lipids. Also, one frequently needs to correct for gradient eddy current distortions. Difficulties of spectral analysis include metabolite signal complexity and large uncharacterized baseline signal contributions. Signals from macromolecules, residual water, and subcutaneous lipids, which generally appear as broad baseline contributions, are much more prominent at short TEs. Figures 3.182–3.185 are some illustrations of this technique.

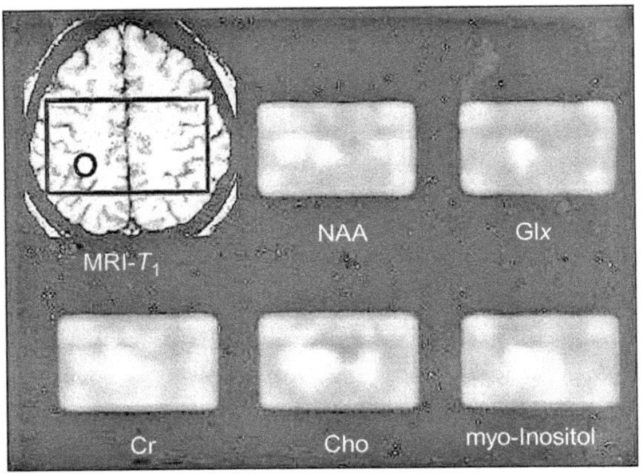

Figure 3.184 T_1-weighted localizer MRI and metabolite maps for NAA, creatine, choline, myoinositol, and Glx from an automated fit of supraventricular, 25-ms TE, 1.5-T, PRESS spectroscopic-imaging data.

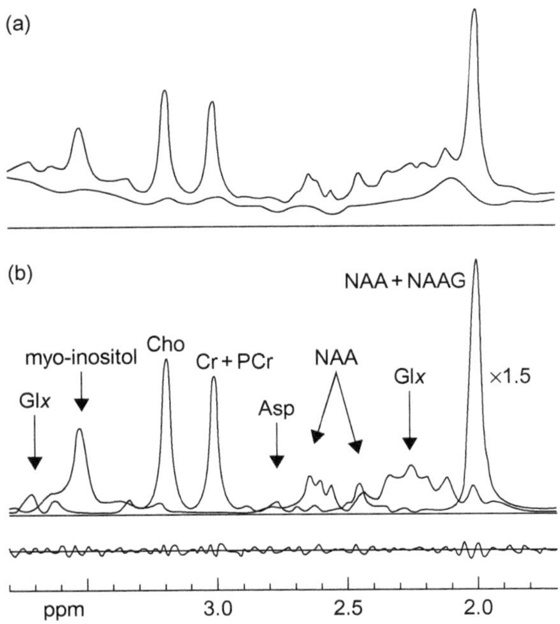

Figure 3.185 Spectra from voxel indicated in the localizer MRI in Figure 3.184. (a) Spectral data (thin line) overlaid with the metabolite + baseline fit (heavy line) and baseline-only fit (heavy line). (b) Individual metabolite fits (scaled by 1.5). At the bottom is the residual spectrum.

References

[1] J. Harolds, M.M. Graham, A. Maurer, M.J. Guiberteau, T.R. Miller, K. Vydareny, Acad. Radiol. 13 (2006) 1405–1409.

[2] C.H. Cagnon, D.D. Cody, M.F. McNitt-Gray, J.A. Seibert, P.F. Judy, D.R. Aberie, Acad. Radiol. 13 (2006) 1431.

[3] S.K. Mukherji, V. Chong, Atlas of Head and Neck Imaging: The Extracranial Head and Neck, New York, USA, 2004.

[4] P.J. Ell, S.S. Gambhir (Eds.), Nuclear Medicine in Clinical Diagnosis and Treatment, Churchill Livingstone, New York, USA, 2004.
[5] E. Bullitt, D. Zeng, G. Gerig, S. Aylward, S. Joshi, J.K. Smith, et al., Acad. Radiol. 12 (2005) 1232–1240.
[6] M. Bergui, J. Zhong, G.B. Bradac, S. Sales, Neuroradiology 43 (2001) 824–829.
[7] W. Moller-Hartmnn, S. Herminghaus, T. Krings, G. Marquardt, H. Lanfermann, U. Pilatus, et al., Neuroradiology 44 (2002) 371–384.
[8] C.-J. Juan, C.-Y. Chen, Y.-J. Liu, H.-W. Chung, S.-C. Chin, C.-J. Hsueh, et al., Neuroradiology 47 (2005) 401–405.
[9] K. Sarac, A. Akinci, A. Alkan, M. Aslam, T. Baysal, C. Ozscan, Neuroradiology 47 (2005) 562–565.
[10] A. Goswami, Quantum Mechanics, Wm. C. Brown Publishers, Dubuque, IA, USA, 2001, ISBN 0-697-11811-8
[11] T.C. Farar, Concepts Magn. Reson. 2 (1990) 1–12.
[12] T.C. Farar, Concepts Magn. Reson. 2 (1990) 55–61.
[13] J. Shriver, Concepts Magn. Reson. 4 (1992) 1–33.
[14] G. Bodenhausen, H. Kogler, R.R. Ernst, J. Magn. Reson. 58 (1984) 370–388.
[15] T.H. Ernst, J. Henning, Magn. Reson. Med. 21 (1991) 82–96.
[16] P.B. Kingsley, J. Magn. Reson. A107 (1994) 14–23.
[17] S.A. Smith, T.O. Levante, B.H. Meier, E.R. Ernst, J. Magn. Reson. A106 (1994) 75–105.
[18] A. Jerschow, N. Muller, J. Magn. Reson. 134 (1998) 17–29.
[19] R. Eisberg, R. Resnick, Quantum Physics of Atoms, Molecules, Solids, Nuclei and Particles, John Wiley & Sons, New York, USA, 1974, ISBN 0-471-23464-8
[20] H.P. Myers, Introducory Solid State Physics, Taylor and Francis, New York, USA, 1990.
[21] B.L. Miller, NMR Biomed. 4 (1991) 47–52.
[22] T. Michaelis, K.-D. Merbeldt, W. Hanice, M.L. Gyngell, H. Brauhn, J. Frahm, NMR Biomed. 4 (1991) 90–98.
[23] A. Kuessel, G.R. Sutherland, W. Halliday, Ian C.P. Smith, NMR Biomed. 7 (1994) 149–155.
[24] A.R. Tate, J.R. Griffiths, I. Martinez-Perez, A. Moreno, I. Barba, M.E. Cabanas, et al., NMR Biomed. 11 (1998) 177–191.
[25] P.E. Sijens, P.C. Levndag, C.J. Vecht, P. van Dijk, M. Oudkerk, NMR Biomed. 9 (1996) 65–71.
[26] D. Ott, J. Hennig, T. Ernst, Radiology 186 (1993) 745–752.
[27] H. Kugel, W. Heindel, R.-I. Ernestus, Radiology 183 (1992) 701–709.
[28] J.R. Alger, J.A. Frank, A. Bizzi, M.J. Fulham, B.X. Desouza, M.O. Duhaney, et al., Radiology 177 (1990) 633–641.
[29] H. Lanfermann, H. Kugel, W. Heindel, K. Herhelz, Wolf-Dieter, K. Lackner, Radiology 196 (1995) 203–210.
[30] S.H. Kim, K.-H. Chang, C. Song, M.H. Han, H.C. Kim, H.S. Kang, et al., Radiology 204 (1997) 239–245.
[31] N.I. Matter, G.C. Scott, R.D. Venook, S.E. Ungersma, T. Grafendorfer, A. Macovski, et al., Magn. Reson. Med. 56 (2006) 1085–1095.
[32] R.B. Thompson, P.S. Allen, Magn. Reson. Med. 45 (2001) 955–965.
[33] U. Dydak, M. Weigner, K.P. Pruessmann, D. Meier, P. Boesiger, Magn. Reson. Med. 46 (2001) 713–722.

[34] A.A. Maudsley, G.B. Matson, J.W. Hugg, M.W. Weiner, Magn. Reson. Med. 31 (1994) 645–651.
[35] K. Young, G.B. Matson, V. Govindaraju, A. Maudsley, J. Mag. Reson. 140 (1999) 146–152.
[36] K. Young, V. Govindaraju, B.J. Soher, A.A. Maudsley, Magn. Reson. Med. 40 (1998) 812–815.
[37] A.H. Wilman, P.S. Allen, J. Magn. Reson. B101 (1993) 102–105.
[38] R.B. Thompson, P.S. Allen, Magn. Reson. Med. 41 (1999) 1162–1169.
[39] O. Gonen, F. Arias-Mendoza, G. Goelman, Magn. Reson. Med. 37 (1997) 644–650.
[40] P.R. Luyten, A.J. Marien, W. Heindel, P.H.J. van Grewen, K. Herholz, J.A. den Holander, et al., Radiology 176 (1990) 791–799.
[41] A.H. Wilman, P.S. Allen, J. Magn. Reson. B109 (1995) 169–174.
[42] A.H. Wilman, P.S. Allen, J. Magn. Reson. B101 (1993) 165–171.
[43] J.A. McNab, R. Bartha, NMR Biomed. 19 (2006) 999–1009.
[44] M.A. Griswold, F. Breuer, M. Blaimer, S. Kannengiesser, R.M. Heidemann, M. Mueller, et al., NMR Biomed. 19 (2006) 316–324.
[45] J.A. de Zwart, P. van Gelderen, X. Golay, V.N. Ikonomidou, J.H. Duyn, NMR Biomed. 19 (2006) 342–351.
[46] W.E. Kyriakos, W.S. Hoge, D. Mitsouras, NMR Biomed. 19 (2006) 379–392.
[47] K.P. Pruessmann, NMR Biomed. 19 (2006) 288–299.
[48] F. Wiesinger, P.-F. Van de Moortele, G. Adriany, N. De zanche, NMR Biomed. 19 (2006) 368–378.
[49] L.S. Chow, G.G. Cook, E. Whitby, M.N.J. Paley, Neuroimage 30 (2006) 835–846.
[50] T.S. Ibrahim, C. Mitchell, R. Abrhm, P. Schmalbrock, NMR Biomed. 20 (2007) 58–68.
[51] O. Bieri, K. Scheffler, NMR Biomed. 20 (2007) 1–10.
[52] S.G. Mueller, N. Schuff, M.W. Weiner, NMR Biomed. 19 (2006) 655–668.
[53] N. Salamon, N. Sicotte, J. Alger, D. Shattuck, S. Perlman, U. Sinha, et al., Neuroradiology 47 (2005) 895–902.
[54] C. Choi, N.J. Coupland, P.P. Bhardwaj, N. Malykhin, D. Gheorghin, P.S.A. Allen, Magn. Reson. Med. 55 (2006) 997–1005.
[55] E. Ozturk-Isik, J.C. Crane, S. Cha, S.M. Chang, M.S. Berger, S.J. Nelson, Magn. Reson. Med. 55 (2006) 1164–1169.
[56] S. Mangia, I. Tkac, R. Gruetter, P.-F. Van De Moortele, F. Giove, B. Maraviglia, et al., Magn. Reson. Im. 24 (2006) 341–348.
[57] V.M. Anderson, N.C. Fox, D.H. Miller, J. Magn. Reson. Im. 23 (2006) 605–618.
[58] S.K. Lee, D.W. Kim, K.K. Kim, C.K. Chung, I.C. Song, K.H. Chang, Neuroradiology 47 (2005) 916–923.
[59] A. MacKay, C. Laule, I. Vavasour, T. Bjarnason, S. Kolind, B. Madler, Magn. Reson. Im. 24 (2006) 515–525.
[60] M.A. Macri, C. Colonnnese, G. Garreffa, F. Fattapposta, R. Restuccia, F. Bianco, et al., Magn. Reson. Im. 24 (2006) 443–447.
[61] P.-G. Henry, G. Adriany, D. Deelchand, R. Gruetter, M. Marjanska, G. Oz, et al., Mag. Reson. Im. 24 (2006) 527–539.
[62] R. Kreis, J. Prog. NMR Spect. 31 (1997) 155–195.
[63] S.W. Provencher, NMR Biomed. 14 (2001) 260–264.
[64] M.A. McLean, F.G. Woermann, G.J. Barker, J.S. Duncan, Magn. Reson. Med. 44 (2000) 401–411.

[65] S. Michaeli, M. Garwood, X.-H. Zhu, L. DelaBaree, P. Andersen, G. Adriany, et al., Magn. Reson. Med. 47 (2002) 629–633.
[66] R.B. Thompson, P.S. Allen, Magn. Reson. Med. 39 (1998) 762–771.
[67] J.W. Pan, D.B. Twieg, H.P. Hetherington, Magn. Reson. Med. 40 (1998) 363–369.
[68] B.J. Soher, P. Vermathen, N. Schuff, D. Wiedermann, D.J. Meyerhoff, M.W. Weiner, et al., Magn. Reson. Im. 18 (2000) 1159–1165.

www.ingramcontent.com/pod-product-compliance
Ingram Content Group UK Ltd.
Pitfield, Milton Keynes, MK11 3LW, UK
UKHW020128290926
15471UKWH00003B/354